Lecture Notes in Artificial Intelligence 8397

Subseries of Lecture Notes in Computer Science

Ngoc Thanh Nguyen   Boonwat Attachoo
Bogdan Trawiński   Kulwadee Somboonviwat (Eds.)

Intelligent Information and Database Systems

6th Asian Conference, ACIIDS 2014
Bangkok, Thailand, April 7-9, 2014
Proceedings, Part I

Volume Editors

Ngoc Thanh Nguyen
Bogdan Trawiński
Wrocław University of Technology, Poland
E-mail: ngoc-thanh.nguyen@pwr.edu.pl
E-mail: bogdan.trawinski@pwr.wroc.pl

Boonwat Attachoo
Kulwadee Somboonviwat
King Mongkut's Institute of Technology Ladkrabang
Bangkok, Thailand
E-mail: {boonwat, kskulwad}@kmitl.ac.th

ISSN 0302-9743 e-ISSN 1611-3349
ISBN 978-3-319-05475-9 e-ISBN 978-3-319-05476-6
DOI 10.1007/978-3-319-05476-6
Springer Cham Heidelberg New York Dordrecht London

Library of Congress Control Number: 2014933552

LNCS Sublibrary: SL 7 – Artificial Intelligence

Typesetting: Camera-ready by author, data conversion by Scientific Publishing Services, Chennai, India

Printed on acid-free paper

Springer is part of Springer Science+Business Media (www.springer.com)

Preface

ACIIDS 2014 was the sixth event in the series of international scientific conferences for research and applications in the field of intelligent information and database systems. The aim of ACIIDS 2014 was to provide an internationally respected forum for scientific research in the technologies and applications of intelligent information and database systems. ACIIDS 2014 was co-organized by King Mongkut's Institute of Technology Ladkrabang (Thailand) and Wrocław University of Technology (Poland) in co-operation with the IEEE SMC Technical Committee on Computational Collective Intelligence, Hue University (Vietnam), University of Information Technology UIT-HCM (Vietnam), and Quang Binh University (Vietnam) and took place in Bangkok (Thailand) during April 7-9, 2014. The first two events, ACIIDS 2009 and ACIIDS 2010, took place in Dong Hoi City and Hue City in Vietnam, respectively. The third event, ACIIDS 2011, took place in Daegu (Korea), while the fourth event, ACIIDS 2012, took place in Kaohsiung (Taiwan). The fifth event, ACIIDS 2013, was held in Kuala Lumpur in Malaysia.

We received almost 300 papers from over 30 countries from around the world. Each paper was peer reviewed by at least two members of the international Program Committee and International Reviewer Board. Only 124 papers with the highest quality were selected for oral presentation and publication in the two volumes of ACIIDS 2014 proceedings.

The papers included in the proceedings cover the following topics: natural language and text processing, intelligent information retrieval, Semantic Web, social networks and recommendation systems, intelligent database systems, technologies for intelligent information systems, decision support systems, computer vision techniques, machine learning and data mining, multiple model approach to machine learning, computational intelligence, engineering knowledge and semantic systems, innovations in intelligent computation and applications, modelling and optimization techniques in information systems, database systems and industrial systems, innovation via collective intelligences and globalization in business management, intelligent supply chains as well as human motion: acquisition, processing, analysis, synthesis, and visualization for massive datasets.

Accepted and presented papers highlight the new trends and challenges of intelligent information and database systems. The presenters showed how new research could lead to new and innovative applications. We hope you will find these results useful and inspiring for your future research.

We would like to express our sincere thanks to the honorary chairs, Prof. Tawil Paungma (Former President of King Mongkut's Institute of Technology Ladkrabang, Thailand), Prof. Tadeusz Więckowski (Rector of Wrocław University of Technology, Poland), and Prof. Andrzej Kasprzak, Vice-Rector of Wrocław University of Technology, Poland, for their support.

Our special thanks go to the program chairs and the members of the international Program Committee for their valuable efforts in the review process, which helped us to guarantee the highest quality of the selected papers for the conference. We cordially thank the organizers and chairs of special sessions who essentially contributed to the success of the conference.

We would also like to express our thanks to the keynote speakers (Prof. Hoai An Le Thi, Prof. Klaus-Robert Müller, Prof. Leszek Rutkowski, Prof. Vilas Wuwongse) for their interesting and informative talks of world-class standard.

We cordially thank our main sponsors, King Mongkut's Institute of Technology Ladkrabang (Thailand), Wrocław University of Technology (Poland), IEEE SMC Technical Committee on Computational Collective Intelligence, Hue University (Vietnam), University of Information Technology UIT-HCM (Vietnam), and Quang Binh University (Vietnam). Our special thanks are also due to Springer for publishing the proceedings, and to the other sponsors for their kind support.

We wish to thank the members of the Organizing Committee for their very substantial work and the members of the local Organizing Committee for their excellent work.

We cordially thank all the authors for their valuable contributions and all the other participants of this conference. The conference would not have been possible without them.

Thanks are also due to many experts who contributed to making the event a success.

April 2014

Ngoc Thanh Nguyen
Boonwat Attachoo
Bogdan Trawiński
Kulwadee Somboonviwat

Conference Organization

Honorary Chairs

Tawil Paungma	Former President of King Mongkut's Institute of Technology Ladkrabang, Thailand
Tadeusz Więckowski	Rector of Wrocław University of Technology, Poland
Andrzej Kasprzak	Vice-Rector of Wrocław University of Technology, Poland

General Chairs

Ngoc Thanh Nguyen	Wrocław University of Technology, Poland
Suphamit Chittayasothorn	King Mongkut's Institute of Technology Ladkrabang, Thailand

Program Chairs

Bogdan Trawiński	Wrocław University of Technology, Poland
Kulwadee Somboonviwat	King Mongkut's Institute of Technology Ladkrabang, Thailand
Tzung-Pei Hong	National University of Kaohsiung, Taiwan
Hamido Fujita	Iwate Prefectural University, Japan

Organizing Chairs

Boonwat Attachoo	King Mongkut's Institute of Technology Ladkrabang, Thailand
Adrianna Kozierkiewicz-Hetmańska	Wrocław University of Technology, Poland

Special Session Chairs

Janusz Sobecki	Wrocław University of Technology, Poland
Veera Boonjing	King Mongkut's Institute of Technology Ladkrabang, Thailand

Publicity Chairs

Kridsada Budsara	King Mongkut's Institute of Technology Ladkrabang, Thailand
Zbigniew Telec	Wrocław University of Technology, Poland

Conference Webmaster

Natthapong Jungteerapanich	King Mongkut's Institute of Technology Ladkrabang, Thailand

Local Organizing Committee

Visit Hirunkitti	King Mongkut's Institute of Technology Ladkrabang, Thailand
Natthapong Jungteerapanich	King Mongkut's Institute of Technology Ladkrabang, Thailand
Sutheera Puntheeranurak	King Mongkut's Institute of Technology Ladkrabang, Thailand
Pitak Thumwarin	King Mongkut's Institute of Technology Ladkrabang, Thailand
Somsak Walairacht	King Mongkut's Institute of Technology Ladkrabang, Thailand
Bernadetta Maleszka	Wrocław University of Technology, Poland
Marcin Maleszka	Wrocław University of Technology, Poland
Marcin Pietranik	Wrocław University of Technology, Poland

Steering Committee

Ngoc Thanh Nguyen (Chair)	Wrocław University of Technology, Poland
Longbing Cao	University of Technology Sydney, Australia
Tu Bao Ho	Japan Advanced Institute of Science and Technology, Japan
Tzung-Pei Hong	National University of Kaohsiung, Taiwan
Lakhmi C. Jain	University of South Australia, Australia
Geun-Sik Jo	Inha University, South Korea
Jason J. Jung	Yeungnam University, South Korea
Hoai An Le Thi	University Paul Verlaine - Metz, France
Toyoaki Nishida	Kyoto University, Japan
Leszek Rutkowski	Częstochowa University of Technology, Poland
Suphamit Chittayasothorn	King Mongkut's Institute of Technology Ladkrabang, Thailand
Ali Selamat	Universiti Teknologi Malaysia, Malyasia

Keynote Speakers

Hoai An Le Thi	University of Lorraine, France
Klaus-Robert Müller	Technische Universität Berlin, Germany
Leszek Rutkowski	Częstochowa University of Technology, Poland
Vilas Wuwongse	Asian Institute of Technology, Thailand

Special Sessions Organizers

1.*Multiple Model Approach to Machine Learning (MMAML 2014)*

Tomasz Kajdanowicz	Wrocław University of Technology, Poland
Edwin Lughofer	Johannes Kepler University Linz, Austria
Bogdan Trawiński	Wrocław University of Technology, Poland

2. *Computational Intelligence (CI 2014)*

Piotr Jędrzejowicz	Gdynia Maritime University, Poland
Urszula Boryczka	University of Silesia, Poland
Ireneusz Czarnowski	Gdynia Maritime University, Poland

3. *Engineering Knowledge and Semantic Systems (IWEKSS 2014)*

Jason J. Jung	Yeungnam University, South Korea
Dariusz Król	Bournemouth University, UK

4. *Innovations in Intelligent Computation and Applications (IICA 2014)*

Shyi-Ming Chen	National Taiwan University of Science and Technology, Taiwan
Shou-Hsiung Cheng	Cheinkuo Technology University, Taiwan

5. *Modelling and Optimization Techniques in Information Systems, Database Systems and Industrial Systems (MOT-ACIIDS 2014)*

Hoai An Le Thi	University of Lorraine, France
Tao Pham Dinh	National Institute for Applied Sciences - Rouen, France

6. *Innovation via Collective Intelligences and Globalization in Business Management (ICIGBM 2014)*

Yuh-Shy Chuang	Chien Hsin University, Taiwan
Chao-Fu Hong	Aletheia University, Taiwan
Pen-Choug Sun	Aletheia University, Taiwan

7. *Intelligent Supply Chains (ISC 2014)*

Arkadiusz Kawa	Poznań University of Economics, Poland
Milena Ratajczak-Mrozek	Poznań University of Economics, Poland
Konrad Fuks	Poznań University of Economics, Poland

8. *Human Motion: Acquisition, Processing, Analysis, Synthesis and Visualization for Massive Datasets (HMMD 2014)*

Konrad Wojciechowski	Polish-Japanese Institute of Information Technology, Poland
Marek Kulbacki	Polish-Japanese Institute of Information Technology, Poland
Jakub Segen	Gest3D, USA

International Program Committee

Ajith Abraham	Machine Intelligence Research Labs, USA
Muhammad Abulaish	King Saud University, Saudi Arabia
El-Houssaine Aghezzaf	Ghent University, Belgium
Jesús Alcalá-Fdez	University of Granada, Spain
Haider M. AlSabbagh	Basra University, Iraq
Troels Andreasen	Roskilde University, Denmark
Toni Anwar	Universiti Teknologi Malaysia, Malaysia
Giuliano Armano	University of Cagliari, Italy
Zeyar Aung	Masdar Institute of Science and Technology, United Arab Emirates
Ahmad-Taher Azar	Benha University, Egypt
Costin Bădică	University of Craiova, Romania
Emili Balaguer-Ballester	Bournemouth University, UK
Amar Balla	Ecole Superieure d'Informatique, France
Zbigniew Banaszak	Warsaw University of Technology, Poland
Dariusz Barbucha	Gdynia Maritime University, Poland
Ramazan Bayindir	Gazi University, Turkey
Maumita Bhattacharya	Charles Sturt University, Australia
Mária Bieliková	Slovak University of Technology in Bratislava, Slovakia
Jacek Błażewicz	Poznań University of Technology, Poland
Veera Boonjing	King Mongkut's Institute of Technology Ladkrabang, Thailand
Mariusz Boryczka	University of Silesia, Poland
Urszula Boryczka	University of Silesia, Poland
Abdelhamid Bouchachia	Bournemouth University, UK
Stephane Bressan	National University of Singapore, Singapore
Peter Brida	University of Žilina, Slovakia

Peter Erdi	Kalamazoo College, USA
Victor Felea	Alexandru Ioan Cuza University of Iasi, Romania
Thomas Fober	University of Marburg, Germany
Dariusz Frejlichowski	West Pomeranian University of Technology, Poland
Mohamed Gaber	University of Portsmouth, UK
Patrick Gallinari	Pierre and Marie Curie University, France
Dariusz Gąsior	Wrocław University of Technology, Poland
Andrey Gavrilov	Novosibirsk State Technical University, Russia
Deajan Gjorgjevikj	Ss. Cyril and Methodius University in Skopje, Macedonia
Daniela Godoy	UNICEN University, Argentina
Fernando Gomide	University of Campinas, Brazil
Vladimir I. Gorodetsky	St.Petersburg Institute for Informatics and Automation, Russia
Manuel Graña	University of Basque Country, Spain
Janis Grundspenkis	Riga Technical University, Latvia
Adam Grzech	Wrocław University of Technology, Poland
Sajjad Haider	Institute of Business Administration, Pakistan
Slimane Hammoudi	ESEO Institute of Science and Technology, France
Jin-Kao Hao	University of Angers, France
Habibollah Haron	Universiti Teknologi Malaysia, Malaysia
Tutut Herawan	Universiti Malaysia Pahang, Malaysia
Francisco Herrera	University of Granada, Spain
Huu Hanh Hoang	Hue University, Vietnam
Tzung-Pei Hong	National University of Kaohsiung, Taiwan
Wei-Chiang Hong	Oriental Institute of Technology, Taiwan
Timo Honkela	Helsinki University of Technology, Finland
Mong-Fong Horng	National Kaohsiung University of Applied Sciences, Taiwan
Dosam Hwang	Yeungnam University, South Korea
Natthakan Iam-On	Mae Fah Luang University, Thailand
Dmitry I. Ignatov	National Research University Higher School of Economics, Moscow, Russia
Lazaros Iliadis	Democritus University of Thrace, Greece
Khalid Jebari	LCS Rabat, Morocco
Joanna Jędrzejowicz	University of Gdańsk, Poland
Piotr Jędrzejowicz	Gdynia Maritime University, Poland
Janusz Jeżewski	Institute of Medical Technology and Equipment ITAM, Poland
Gordan Ježić	University of Zagreb, Croatia
Geun-Sik Jo	Inha University, South Korea
Kang-Hyun Jo	University of Ulsan, South Korea

Jason Jung	Yeungnam University, South Korea
Janusz Kacprzyk	Systems Research Institute of Polish Academy of Science, Poland
Tomasz Kajdanowicz	Wrocław University of Technology, Poland
Radosław Katarzyniak	Wrocław University of Technology, Poland
Tsungfei Khang	University of Malaya, Malaysia
Vladimir F. Khoroshevsky	Dorodnicyn Computing Centre of Russian Academy of Sciences, Russia
Muhammad Khurram Khan	King Saud University, Saudi Arabia
Pan-Koo Kim	Chosun University, South Korea
Yong Seog Kim	Utah State University, USA
Frank Klawonn	Ostfalia University of Applied Sciences, Germany
Joanna Kołodziej	University of Bielsko-Biała, Poland
Marek Kopel	Wrocław University of Technology, Poland
Józef Korbicz	University of Zielona Góra, Poland
Leszek Koszałka	Wrocław University of Technology, Poland
Adrianna Kozierkiewicz-Hetmańska	Wrocław University of Technology, Poland
Worapoj Kreesuradej	King Mongkut's Institute of Technology Ladkrabang, Thailand
Ondřej Krejcar	University of Hradec Králové, Czech Republic
Dalia Kriksciuniene	Vilnius University, Lithuania
Dariusz Król	Bournemouth University, UK
Marzena Kryszkiewicz	Warsaw University of Technology, Poland
Adam Krzyzak	Concordia University, Canada
Kazuhiro Kuwabara	Ritsumeikan University, Japan
Sergei O. Kuznetsov	National Research University Higher School of Economics, Moscow, Russia
Halina Kwaśnicka	Wrocław University of Technology, Poland
Pattarachai Lalitrojwong	King Mongkut's Institute of Technology Ladkrabang, Thailand
Helge Langseth	Norwegian University of Science and Technology, Norway
Henrik Legind Larsen	Aalborg University, Denmark
Mark Last	Ben-Gurion University of the Negev, Israel
Annabel Latham	Manchester Metropolitan University, UK
Nguyen-Thinh Le	Clausthal University of Technology, Germany
Hoai An Le Thi	University of Lorraine, France
Kun Chang Lee	Sungkyunkwan University, South Korea
Philippe Lenca	Telecom Bretagne, France
Thitiporn Lertrusdachakul	Thai-Nichi Institute of Technology, Thailand
Lin Li	Wuhan University of Technology, China
Horst Lichter	RWTH Aachen University, Germany

Christoph Quix	RWTH Aachen University, Germany
Preesan Rakwatin	Geo-Informatics and Space Technology Development Agency, Thailand
Ewa Ratajczak-Ropel	Gdynia Maritime University, Poland
Chotirat Ann Ratanamahatana	Chulalongkorn University, Thailand
Rajesh Reghunadhan	Central University of Bihar, India
Przemysław Różewski	West Pomeranian University of Technology, Poland
Miti Ruchanurucks	Kasetsart University, Thailand
Leszek Rutkowski	Częstochowa University of Technology, Poland
Henryk Rybiński	Warsaw University of Technology, Poland
Alexander Ryjov	Lomonosov Moscow State University, Russia
Virgilijus Sakalauskas	Vilnius University, Lithuania
Sakriani Sakti	Nara Institute of Science and Technology, Japan
Daniel Sánchez	University of Granada, Spain
Jürgen Schmidhuber	Swiss AI Lab IDSIA, Switzerland
Björn Schuller	Technical University of Munich, Germany
Ali Selamat	Universiti Teknologi Malaysia, Malaysia
S.M.N. Arosha Senanayake	University of Brunei Darussalam, Brunei
Alexei Sharpanskykh	Delft University of Technology, The Netherlands
Seema Shedole	M S Ramaiah Institute of Technology, India
Quan Z. Sheng	University of Adelaide, USA
Andrzej Siemiński	Wrocław University of Technology, Poland
Dragan Simić	University of Novi Sad, Serbia
Gia Sirbiladze	Iv. Javakhishvili Tbilisi State University, Georgia
Andrzej Skowron	University of Warsaw, Poland
Janusz Sobecki	Wrocław University of Technology, Poland
Kulwadee Somboonviwat	King Mongkut's Institute of Technology Ladkrabang, Thailand
Jerzy Stefanowski	Poznań Univeristy of Technology, Poland
Serge Stinckwich	UMI UMMISCO, France
Stanimir Stoyanov	University of Plovdiv Paisii Hilendarski, Bulgaria
Nidapan Sureerattanan	Thai-Nichi Institute of Technology, Thailand
Dejvuth Suwimonteerabuth	IBM Solutions Delivery, Thailand
Shinji Suzuki	University of Tokyo, Japan
Jerzy Świątek	Wrocław University of Technology, Poland
Edward Szczerbicki	University of Newcastle, Australia
Julian Szymański	Gdańsk University of Technology, Poland
Ryszard Tadeusiewicz	AGH University of Science and Technology, Poland
Yasufumi Takama	Tokyo Metropolitan University, Japan
Kay Chen Tan	National University of Singapore, Singapore

Pham Dinh Tao	INSA-Rouen, France
Zbigniew Telec	Wrocław University of Technology, Poland
Thanaruk Theeramunkong	Thammasat University, Thailand
Krzysztof Tokarz	Silesian University of Technology, Poland
Behcet Ugur Toreyin	Cankaya University, Turkey
Bogdan Trawiński	Wrocław University of Technology, Poland
Krzysztof Trawiński	European Centre for Soft Computing, Spain
Konstantin Tretyakov	University of Tartu, Estonia
Iwan Tri Riyadi Yanto	Universiti Tun Hussein Onn Malaysia, Malaysia
Hong-Linh Truong	Vienna University of Technology, Austria
George A. Tsihrintzis	University of Piraeus, Greece
Alexey Tsymbal	Siemens AG, Germany
Rainer Unland	University of Duisburg-Essen, Germany
Olgierd Unold	Wrocław University of Technology, Poland
Arlette van Wissen	VU University Amsterdam, The Netherlands
Pandian Vasant	Universiti Teknologi Petronas, Malaysia
Emil Vassev	University of Limerick, Ireland
Jørgen Villadsen	Technical University of Denmark, Denmark
Maria Virvou	University of Piraeus, Greece
Gottfried Vossen	University of Münster, Germany
Wahidin Wahab	University of Indonesia, Indonesia
Kitsana Waiyamai	Kasetsart University, Thailand
Ali Wali	University of Kairouan, Tunisia
Botao Wang	Northeastern University, China
Yitong Wang	Fudan University, China
Yongkun Wang	Rakuten, Inc., Japan
Ukrit Watchareeruetai	King Mongkut's Institute of Technology Ladkrabang, Thailand
Izabela Wierzbowska	Gdynia Maritime University, Poland
Nuwee Wiwatwattana	Srinakharinwirot University, Thailand
Wayne Wobcke	University of New South Wales, Australia
Marek Wojciechowski	Poznań University of Technology, Poland
Dong-Min Woo	Myongji University, South Korea
Michał Woźniak	Wrocław University of Technology, Poland
Marian Wysocki	Rzeszow University of Technology, Poland
Guandong Xu	University of Technology, Sydney, Australia
Xin-She Yang	University of Cambridge, UK
Zhenglu Yang	University of Tokyo, Japan
Keem Siah Yap	Universiti Tenaga Nasional, Malaysia
Lean Yu	Chinese Academy of Sciences, AMSS, China
Sławomir Zadrożny	Systems Research Institute of Polish Academy of Science, Poland
Drago Žagar	University of Osijek, Croatia
Danuta Zakrzewska	Łódź University of Technology, Poland

Faisal Zaman	Kyushu Institute of Technology, Japan
Constantin-Bala Zamfirescu	Lucian Blaga University of Sibiu, Romania
Arkady Zaslavsky	CSIRO, Australia
Aleksander Zgrzywa	Wrocław University of Technology, Poland
Jianwei Zhang	Tsukuba University of Technology, Japan
Rui Zhang	Wuhan University of Technology, China
Zhongwei Zhang	University of Southern Queensland, Australia
Cui Zhihua	Complex System and Computational Intelligence Laboratory, China
Zhi-Hua Zhou	Nanjing University, China
Xingquan Zhu	University of Technology, Sydney, Australia

Program Committees of Special Sessions

Special Session on Multiple Model Approach to Machine Learning (MMAML 2014)

Jesús Alcalá-Fdez	University of Granada, Spain
Emili Balaguer-Ballester	Bournemouth University, UK
Abdelhamid Bouchachia	Bournemouth University, UK
Piotr Bródka	Wrocław University of Technology, Poland
Robert Burduk	Wrocław University of Technology, Poland
Oscar Castillo	Tijuana Institute of Technology, Mexico
Rung-Ching Chen	Chaoyang University of Technology, Taiwan
Suphamit Chittayasothorn	King Mongkut's Institute of Technology Ladkrabang, Thailand
José Alfredo F. Costa	Federal University (UFRN), Brazil
Bogusław Cyganek	AGH University of Science and Technology, Poland
Ireneusz Czarnowski	Gdynia Maritime University, Poland
Patrick Gallinari	Pierre et Marie Curie University, France
Fernando Gomide	State University of Campinas, Brazil
Francisco Herrera	University of Granada, Spain
Tzung-Pei Hong	National University of Kaohsiung, Taiwan
Tomasz Kajdanowicz	Wrocław University of Technology, Poland
Yong Seog Kim	Utah State University, USA
Mark Last	Ben-Gurion University of the Negev, Israel
Kun Chang Lee	Sungkyunkwan University, South Korea
Heitor S. Lopes	Federal University of Technology Paraná, Brazil
Edwin Lughofer	Johannes Kepler University Linz, Austria
Mustafa Mat Deris	Universiti Tun Hussein Onn Malaysia, Malaysia
Dragan Simić	University of Novi Sad, Serbia
Jerzy Stefanowski	Poznań University of Technology, Poland
Zbigniew Telec	Wrocław University of Technology, Poland

Bogdan Trawiński	Wrocław University of Technology,Poland
Olgierd Unold	Wrocław University of Technology, Poland
Pandian Vasant	University Technology Petronas, Malaysia
Michał Woźniak	Wrocław University of Technology, Poland
Faisal Zaman	Kyushu Institute of Technology, Japan
Zhongwei Zhang	University of Southern Queensland, Australia
Zhi-Hua Zhou	Nanjing University, China

Computational Intelligence (CI 2014)

Dariusz Barbucha	Gdynia Maritime University, Poland
Mariusz Boryczka	University of Silesia, Poland
Urszula Boryczka	University of Silesia, Poland
Longbing Cao	University of Technology Sydney, Australia
Bogusław Cyganek	AGH University of Science and Technology, Poland
Ireneusz Czarnowski	Gdynia Maritime University, Poland
Piotr Jędrzejowicz	Gdynia Maritime University, Poland
Tianrui Li	Southwest Jiaotong University, China
Alfonso Mateos Caballero	Universidad Politécnica de Madrid, Spain
Mikhail Moshkov	King Abdullah University of Science and Technology, Saudi Arabia
Agnieszka Nowak-Brzezińska	University of Silesia, Poland
Ewa Ratajczak-Ropel	Gdynia Maritime University, Poland
Rafał Różycki	Poznań University of Technology, Poland
Wiesław Sieńko	Gdynia Maritime University, Poland
Adam Słowik	Koszalin University of Technology, Poland
Rafał Skinderowicz	University of Silesia, Poland
Alicja Wakulicz-Deja	University of Silesia, Poland
Beata Zielosko	University of Silesia, Poland

Engineering Knowledge and Semantic Systems (IWEKSS 2014)

Gonzalo A. Aranda-Corral	Universidad de Sevilla, Spain
David Camacho	Autonomous University of Madrid, Spain
Fred Freitas	Universidade Federal de Pernambuco, Brazil
Daniela Godoy	Unicen University, Argentina
Tutut Herawan	University of Malaya, Malaysia
Adam Jatowt	Kyoto University, Japan
Jason J. Jung	Yeungnam University, Korea
Krzysztof Juszczyszyn	Wrocław University of Technology, Poland
Dariusz Król	Bournemouth University, UK
Monika Lanzenberger	Vienna University of Technology, Austria
Jinjiu Li	UTS, Australia

Innovations in Intelligent Computation and Applications (IICA 2014)

An-Zen Shih	Jinwen University of Science and Technology, Taiwan
Albert B. Jeng	Jinwen University of Science and Technology, Taiwan
Victor R. L. Shen	National Taipei University, New Taipei City, Taiwan
Jeng-Shyang Pan	National Kaohsiung University of Applications, Kaohsiung, Taiwan
Mong-Fong Horng	National Kaohsiung University of Applications, Kaohsiung, Taiwan
Huey-Ming Lee	Chinese Culture University, Taipei, Taiwan
Ying-Tung Hsiao	National Taipei University of Education, Taipei, Taiwan
Shou-Hsiung Cheng	Chienkuo Technology University, Changhua, Taiwan
Chun-Ming Tsai	Taipei Municipal University of Education, Taipei, Taiwan
Cheng-Yi Wang	National Taiwan University of Science and Technology, Taiwan
Shyi-Ming Chen	National Taiwan University of Science and Technology, Taiwan
Heng Li Yang	National Chenchi University, Taiwan
Jium-Ming Lin	Chung Hua University, Taiwan
Chun-Ming Tsai	University of Taipei, Taiwan
Yung-Fa Huang	Chaoyang University of Technology, Taiwan
Ho-Lung Hung	Chienkuo Technology University, Changhua, Taiwan
Chih-Hung Wu	National Taichung University of Education, Taiwan
Jyh-Horng Wen	Tunghai University, Taiwan
Jui-Chung Hung	University of Taipei, Taiwan

Modelling and Optimization Techniques in Information Systems, Database Systems and Industrial Systems (MOT-ACIIDS 2014)

Le Thi Hoai An	University of Lorraine, France
Pham Dinh Tao	INSA-Rouen, France
Pham Duc Truong	University of Cardiff, UK
Raymond Bisdorff	Université du Luxembourg, Luxembourg

Jin-Kao Hao	University of Angers, France
Joaquim Judice	University of Coimbra, Portugal
Yann Guermeur	LORIA, France
Boudjeloud Lydia	University of Lorraine, France
Conan-Guez Brieu	University of Lorraine, France
Gely Alain	University of Lorraine, France
Le Hoai Minh	University of Lorraine, France
Do Thanh Nghi	University of Can Tho, Vietnam
Alexandre Blansché	University of Lorraine, France
Nguyen Duc Manh	ENSTA Bretagne, France
Ta Anh Son	Hanoi University of Science and Technology, Vietnam
Tran Duc Quynh	Hanoi University of Agriculture, Vietnam

Innovation via Collective Intelligences and Globalization in Business Management (ICIGBM 2014)

Tzu-Fu Chiu	Aletheia University, Taiwan
Yi-Chih Lee	Chien Hsin University, Taiwan
Jian-Wei Lin	Chien Hsin University, Taiwan
Kuo-Sui Lin	Aletheia University, Taiwan
Tzu-En Lu	Chien Hsin University, Taiwan
Chia-Ling Hsu	TamKang University, Taiwan
Fang-Cheng Hsu	Aletheia University, Taiwan
Rahat Iqbal	Coventry University, UK
Irene Su	TamKang University, Taiwan
Ai-Ling Wang	TamKang University, Taiwan
Henry Wang	Institude of Software Chinese Academy of Sciences, China
Hung-Ming Wu	Aletheia University, Taiwan
Wei-Li Wu	Chien Hsin University, Taiwan
Feng-Sueng Yang	TamKang University, Taiwan

Intelligent Supply Chains (ISC 2014)

Areti Manataki	The University of Edinburgh, UK
Zbigniew Pasek	University of Windsor, Canada
Arkadiusz Kawa	Poznań University of Economics, Poland
Marcin Hajdul	Institute of Logistics and Warehousing, Poland
Paweł Pawlewski	Poznań University of Technology, Poland
Paulina Golińska	Poznań University of Technology, Poland

Human Motion: Acquisition, Processing, Analysis, Synthesis and Visualization for Massive Datasets (HMMD 2014)

Aldona Drabik	Polish-Japanese Institute of Information Technology, Poland
André Gagalowicz	Inria, France
Ryszard Klempous	Wroclaw University of Technology, Poland
Ryszard Kozera	Warsaw University of Life Science, Poland
Marek Kulbacki	Polish-Japanese Institute of Information Technology, Poland
Aleksander Nawrat	Silesian University of Technology, Poland
Lyle Noaks	The University of Western Australia, Australia
Jerzy Paweł Nowacki	Polish-Japanese Institute of Information Technology, Poland
Andrzej Polański	Polish-Japanese Institute of Information Technology, Poland
Andrzej Przybyszewski	University of Massechusetts, USA
Eric Petajan	LiveClips, USA
Jerzy Rozenbilt	University of Arizona, Tucson, USA
Jakub Segen	Gest3D, USA
Aleksander Sieroń	Medical University of Silesia, Poland
Konrad Wojciechowski	Polish-Japanese Institute of Information Technology, Poland

Table of Contents – Part I

Natural Language and Text Processing

Intelligent Information Retrieval

Semantic Web, Social Networks and Recommendation Systems

Intelligent Database Systems

Intelligent Information Systems

Decision Support Systems

Computer Vision Techniques

Table of Contents – Part II

Machine Learning and Data Mining

Multiple Model Approach to Machine Learning (MMAML 2014)

Computational Intelligence (CI 2014)

Engineering Knowledge and Semantic Systems (IWEKSS 2014)

Innovations in Intelligent Computation and Applications (IICA 2014)

Modelling and Optimization Techniques in Information Systems, Database Systems and Industrial Systems (MOT 2014)

Innovation via Collective Intelligences and Globalization in Business Management (ICIGBM 2014)

Intelligent Supply Chains (ISC 2014)

Human Motion: Acquisition, Processing, Analysis, Synthesis and Visualization for Massive Datasets (HMMD 2014)

A Meta-model Guided Expression Engine

Dominic Girardi[1], Josef Küng[2], and Michael Giretzlehner[1]

[1] RISC Software GmbH - Research Unit Medical Informatics, Hagenberg
{firstname.lastname}@risc.uni-linz.ac.at
[2] Institute for Application Oriented Knowledge Processing, JKU Linz
jkueng@faw.uni-linz.ac.at

Abstract. Data acquisition and handling is known to be one of the most severe technical barriers in (bio-)medical research. In order to counter this problem, we created a generic data acquisition and managing system which can be set up for the given domain of application without the need for programming- or database-skills. The user definitions of the domain data structures are stored into an abstract meta data-model and allow the automatic creation of data-input and -managing interfaces. In order to enable the user to define complex search queries on the data or derive new data out of already existing, a meta-model-guided expression engine was developed. Grammatical and structural meta-data are interwoven in order to provide support in expression generation to the domain expert.

1 Introduction

Scientific research in common and medical research in particular usually base on a large amount of observed data. In case of medical research, this data is sensible and highly structured data of heterogeneous data types (nominal, ordinal, interval, ratio). Although every modern hospital is equipped with a hospital information system (HIS), the data stored in these systems can hardly be used for scientific research, because of its mainly administrative, logistic and economical nature [12]. Data mining on HIS data has already been performed, but the yielded results were less scientifically applicable than for management purposes [18], [19]. Although the drawback of HIS been data real graveyards has already been recognized and first approaches have been made, there are still remarkable challenges to cope with (e.g. privacy, data-integration, cross-institutional standardization, etc.) until scientific research can be performed directly from HIS [16]. This is why data for medical research is (semi-)manually collected and stored in medical disease registries. A medical registry is a systematic collection of a clearly defined set of health and demographic data for patients with specific health characteristics, held in a central database for a predefined purpose [3].

These characteristics in combination with high amount of collected data require a professional data management system, that supports the researches in data acquisition, data organization and data analysis. The selection, setup and maintenance of such a system has already been identified as a major obstacle in biomedical research [7]. In 2007 a survey among biomedical researchers [2]

N.T. Nguyen et al. (Eds.): ACIIDS 2014, Part I, LNAI 8397, pp. 1–10, 2014.

pointed out that data handling had become a major barrier in some research projects. Furthermore, biomedical researchers are often hardy able to cope with the complexity of their own data. The fact that many researchers use general-purpose office applications, which do not provide any support in data handling, worsens the situation. Our experiences also showed that especially medical researchers try to avoid the dependency of the software vendor they are in, when they use a custom made software system.

In order to encounter these drawbacks, we developed an absolutely generic, web-based data acquisition and management system, that can be parametrized to store and manage data of almost arbitrary structure. The system is based upon a generic meta data-model that is able to store data and the corresponding structural information. We focused on the scientific researcher as main user and administrator of the system, who is usually no IT expert. To prepare the generic system for the actual application, the domain-specific data model (DM) must be defined, in order to store the structural information into the meta data-model. Therefore, the user defines what data entities exist in his domain, what kind of attributes they have, and in what kind of relationship they are. This doesn't require any programming, and can be done by using intuitive wizards or web forms. Based on this structural meta information, the rest of the application (data management tool, web input forms, search forms, overview tables, data import and export interfaces, etc.) are created automatically at run-time. This configuration can be changed at any time of the research project. So, researchers are independent from their software vendors and can adapt their system to their needs at any time. Due to the user interface generation at run-time, all changes are propagated immediately. Furthermore, by using the stored meta information about the DM, the system is able support the user in data management and analysis.

One particular aspect, where the user requires very intensive support, is the generation of expressions on the data. Whenever data is stored in a structured way - eg. databases, ontologies, XML - an expression language (SQL for databases, SPARQL for OWL ontologies, xQuery for XML) is provided to define complex search queries and data constrains. Usually the usage of these expression languages is reserved for IT experts who are familiar with the grammar of the language and the internal data structures of the application. In order to provide some of the functionality to the end user as well, predefined expressions are encapsulated in user dialogs or search wizards and then parametrized by the user. This approach, however, only provides a fractional amount of functionality of the expression language to the end user. In this paper we present an approach that allows the user of the system to define expressions on their data structure themselves, and thus use the complete range of functionality of the expression language. This is enabled due to the intensive support by the system, which combines grammar meta information of the expression language and structural meta information from the generic meta data structure.

The paper is organized as follows. In Section 2 we provide an overview over related research projects and publications. In Section 3 our generic data acquisition and management infrastructure is described. Section 4 contains the description of the expression engine. Our conclusions can be found in Section 5.

2 Related Research

The idea of using meta models to automatically create parts (data access layer, user interfaces) of data intensive software systems is wide established method in model driven engineering (MDE) [6]. But the MDE approach in general or concrete realizations like the meta model-based approach for automatic user interface generation by da Cruz et al. [5] - to give an example - are used by software engineers to create source code at development time. Cruz et al. use extended UML meta models to describe the domain data model and use cases and derive source code from it - source code that needs to be compiled. That the main difference between our approach and MDE: Our system derives the structure of the user interface from the meta model at run-time. There is no source code generation. Changes to the domain data model have immediate effect to the user interface, without any compiling or even restart of the application. From this perspective our system is related to the Margitte system by Renggli et al. [17]. While the Margitte system is a general purpose framework based on a self-describing meta-model, our system is based upon a meta-entity-relationship model (see Section 3.1) - stored in relational database - and clearly focused and specialized on scientific data-acquisition and -processing considering the biomedical research as a main user and administrator. Asides from the automatically created web interface it offers a corresponding software tool to handle, pre-process (using the expression engine, described in this paper) and subsequently analyze the collected data.

From a functional point of view, our system is closely related to other electronic data capture systems for medical purposes such as Catalyst Web Tools, OpenClinica and REDCap [11]. These systems are very complex and offer additional features for study management and planning. But still, they are limited to clinical data acquisition and not as generic as our system and only support flat data structures, while our systems is able to cope is arbitrary complex relations among the data elements. Furthermore, our system supports the user in navigating through the data structures and offers a meta-model supported expression engine to aggregate and query data within these structures.

3 Meta-model Based Data Acquisition

3.1 The Meta-model

The meta-model we use is based upon Peter Chen's Entity Relationship (ER) Model [4]. It is basically a data model that is able to store ER models and the corresponding data. The Object Management Group OMG [1] defines four levels (M_0 - M_3) of meta modeling. Each model at level M_i conforms to a model at level M_{i+1}. Level M_0 contains the real world or user data. Each object at M_0

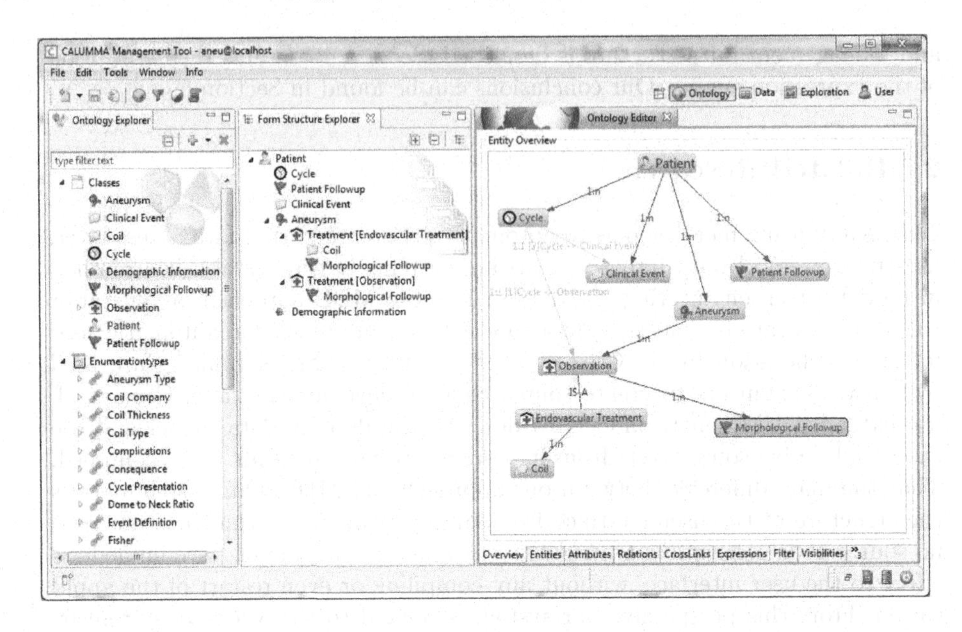

Fig. 1. Domain-specific data structure shown in the CALUMMA Management Tool

is an instance of a model defined in M_1, which is called the model layer. Each model at level M_1 can be seen as an instance of a meta-model at level M_2 - the meta-model layer. Level M_3 contains meta-meta models. A meta-ER model is an ER model describing an ER model. Since ER models are part of the model layer M_1 of the OMG four-level meta-model stack, meta-ER models belong to level M_2. Our M_2 model is able to store the M_1 domain dependent data model as well as the corresponding M_0 data. For more detailed information on the meta-model we use, the reader is referred to our paper [9].

3.2 The CALUMMA System

The CALUMMA (**C**onfigurable **A**daptable **L**ightweihgt and **U**niversal **M**eta **M**odel **A**rchitecture) system is a data acquisition and management infrastructure that is based upon the described meta-model. The three main components are an automatically (at run-time) generated web interface for data acquisition, a Java-based Management Tool for data structure definition, data handling and analysis, and an meta-model based ETL plug-in for the ETL suite Kettle by Pentaho [15] for electronic data import. For a more detailed description of the infra structure the reader is again referred to paper [9] and [10].

In order to prepare the system for the use in an actual domain of application, the user defines the current data structure using either the wizards of the management tool or the administrator back-end of the web interface. These definitions can also be changed during ongoing data acquisition, whereas the system prevents operations that may cause data corruption or loss. Based on these definitions the web interface, the data interface of the Management Tool and the import interface of the ETL plug-in are created at run-time.

3.3 Example

The functionality of the expression engine will be explained by the example of the aneurysm registry, which was the first medical research project, that is based upon the CALUMMA system. The aneurysm registry [14] was established in 2008 by the Institute for Radiology, Landesnervenklinik Wagner Jauregg in Linz, Austria. Its aim is to properly gather and store all relevant information of a patient with cerebral aneurysms. Cerebral aneurysms are balloon-like dilations of brain-located blood vessels [13]. As opposed to hospital information systems, the aneurysm database contains solely medically relevant data in a structured way (no free text or semi-structured information). This allows further automatic processing and analysis of the data, which is used for medical research and internal quality benchmarking.

The domain data structure is organized in a single rooted acyclic entity graph (see Figure 1). The top-most entity, in case of the aneurysm data set, is the entity *Patient*. It contains the basic demographic key data (sex, age) and it has four relations to sub-entities. The first kind of sub-entity is *Cycle*. It encapsulates all relevant information about a single hospital stay of the Patient such as admission date and discharge date, and several medical scores at admission and discharge. In the sub-entity *Clinical Event* all complications are recorded. Each complication is linked to a hospital stay by a cross link relation. The next sub-entity is *Aneurysm*. It contains all information about the aneurysm's morphology and pathology. This entity is the container for all *Treatment* entities where all treatments (observations, surgical or endovascular treatments) are recorded. Each treatment is linked to a cycle by a cross link relation as well. The last sub-entity of Patient is the *Follow up*, which represents follow-up results over an interval of several years.

4 Meta-model Guided Expression Engine

Experiences of the last years showed the need of the user to create arbitrary expression on his data. Three main fields of applications have been discovered: Complex search queries, derivation of new features out of existing values for data analysis, and the definitions of plausibility rules for assuring the data quality. *An expression in the field of computer science is defined as any piece of program code in a high-level language which, when (if) its execution terminates, returns a value. In most programming languages, expressions consist of constants, variables, operators, functions* [8]. This definition already gives a hint, why the usage of expressions in data handling is usually reserved for IT experts. At first, an expression is a piece of programming code. Most of our users are not able to do programming, because they are (bio-)medical researchers. Secondly, the code must be valid in terms of the grammar of a high-level language in order to execute; so the user must be familiar with the grammar (syntax) of the language.

Thirdly the definitions implies that the expression language includes an amount of operators and functions. In order to use the expression language the user should be familiar with at least some of those operators. Additionally, when the expressions are executed on a data structure, they must be correct in terms of this structure as well; concerning compatibility of variables (from the data structure) and operators (part of the grammar).

To enable the user to cope with this complexity, intensive support from the software system is needed. The developed expression engine uses grammatical and structural information to allow the user to build expression.

4.1 Expressions and Grammar

Like any other expression language, the CALUMMA expression language is based upon a grammar. The range of functions is based on well known query languages like SQL and contains numeric functions, aggregations, date and Boolean functions. The following excerpt helps to explain the main concepts of the grammar on the example of the numeric expression grammar branch.

Listing 1.1. Excerpt of the expression grammar

```
Expression = NumericExpression | BooleanExpression | DateExpression |
    StringExpression | EnumerationExpression;
NumericExpression = NumericRootElement NumericTerm;
NumericTerm = [SingleRecordTransOperator] (Constant | ArithmOp | DateOp |
    AggregationOp | Variable);
AggregationOp = CountOp | AggCalcOp;
AggCalcOp = ( Min | Max | Avg | Sum) MultiRecordTransOperator {
    MultiRecordTransOperator} | NumericExpression;
CountOp = Count | MultiRecordTransOperator {MultiRecordTransOperator};
ArithmOp = Add | Sub | Mul | Div | Mod | Abs | Neg;
Sub = NumericExpression - NumericExpression;
...
DateOp = DayDiff | YearDiff;
YearDiff = DateExpression - DateExpression;

DateExpression = DateRootElement DateTerm;
...
```

For every expression a return type has to be defined. The possible types are those that are available with the CALUMMA system: Number, Boolean, Date, String, Enumeration. So the root of the grammar tree is the non-terminal symbol *Expression* which can be either one of the return-type specific specialization (*NumericExpression*, *BooleanExpression*, etc.). A NumericTerm is defined as an optional *SingleRecordTransOperator* followed by either one of the numeric termtypes *Constant*, *ArithmOp*, etc. While a *NumericTerm* is a well known concept the non-terminal symbol *SingleRecordTransOperator* requires further explanation.

The CALUMMA expression engine helps to build expressions based on the user-defined domain-specific data model (DM). So for every expression, besides a name and the return type, a scope has to be defined. A scope is basically an class of the DM. In the case of the aneurysm database the scope for an expression could be *Patient*, or *Aneurysm*, or any other class of the DM (see Figure 1 and Section 3.3). The binding of the expression to a DM class is necessary for DM-based support. Given the scope of a numeric expression is *Aneurysm*, then valid variables are all numeric attributes of the DM class *Aneurysm* (which are then proposed to the user). If the user wants to combine data from different classes, he can use operators to traverse (traversal operators) the relations between the classes (see connections in Figure 1). It is always possible to follow an upstream connection, which yields exactly one record by following a 1:n relation in the :1 direction or follow a 1:1 connection (*SingleRecordTransOperator*). If a user wants to follow a connection in a :n direction (downstream), he is forced to define an aggregation (see non-terminal symbol *AggregationOp*). A traversal operator re-defines the scope of its whole expression sub-tree. So if e.g. a user wants to calculate the age of the patient at the time an aneurysm was discovered at the scope of the aneurysm (Aneurysm.Date found - Patient.Date of Birth), he needs to use the *YearDiff* operator. For the first *DateExpression* a *SingleRecordTransOperator* upstream to the class *Patient* is needed. Now every expression that comes after this operator works in the scope *Patient*. Now the *Variable* operator can be used to read the birth date of the patient. For the second *DateExpression*, the scope *Aneurysm* (derived from the scope of *YearDiff*) is fine and the *Variable* operator is used to read the discovery date. The *YearDiff* operator now calculates the difference in years out of the two dates yielded. If the scope of the *YearDiff* operator was set to *Patient*, an aggregation (min, max, average) must be chosen, in order to be able to use the downstream operator to *Aneurysm*.

4.2 Expression Editor

To spare the user these theoretical aspects of the expression grammar, an expression editor is needed, that takes all grammatical and structural information in consideration and supports the user. Unlike other expression editors, it is not a text-based editor, but a tree-based approach. When a user creates a new expression, he is asked to define the name of the expression, the return type, and the scope (in case the scope can't be determined out of the context). Based on this definitions a corresponding root element is created (*NumericExpression* for return type numeric, etc.).

The root element is displayed in a tree viewer. With context menu on this root element, the user can now add further expression operators to this node. The context menu works for any node of an expression tree. It is generated at runtime and shows only expression operators that are grammatically correct successors of the currently selected operator and are correct concerning the data structure. There are two advantages in this editing mode. Firstly, the user is not able to create incorrect expressions, because he is not given the chance to. The only way to produce an incorrect expression is to define an incomplete one. Secondly, there is no need to know the complete range of functionality, because every correct possibility is presented.

Listing 1.2. Annotated successors of the *Yeardiff* expression

```
@ExpressionSubElement(name="Minued␣Date", type=DateExpression.class)
protected Expression minuendDate;

@ExpressionSubElement(name="Substrahend␣Date",  type=DateExpression.class)
protected Expression substrahendDate;
```

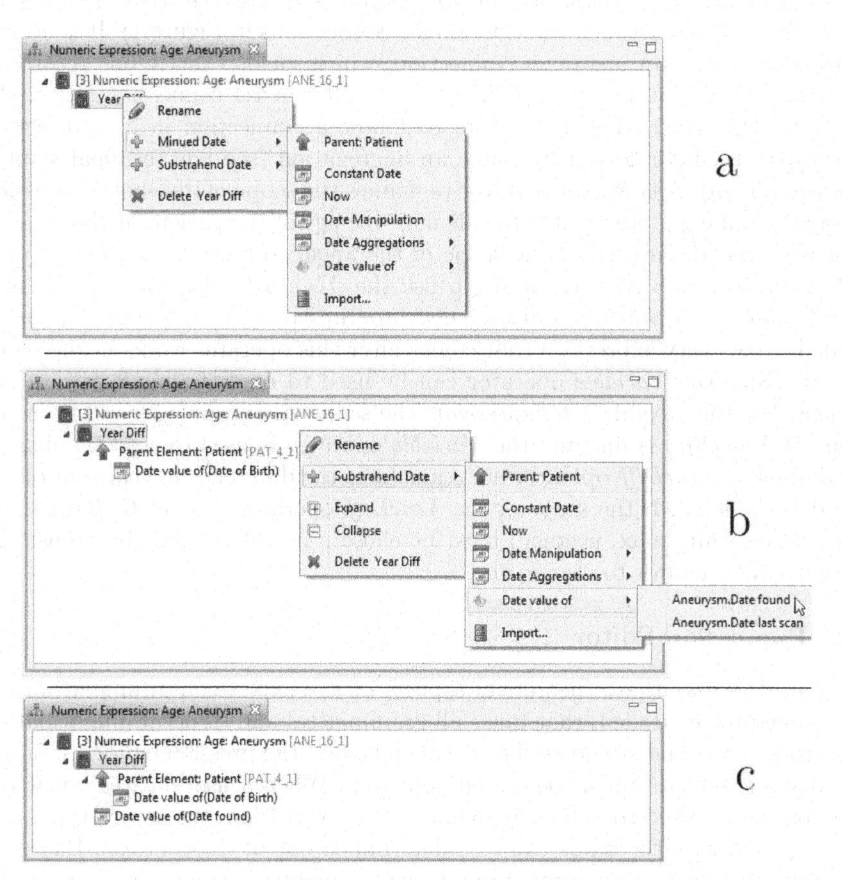

Fig. 2. Expression Editor

The menu generation at runtime is based on Java Annotations. Every expression operator is represented by a Java class derived from the abstract root class Expression. According to the scope, the list of grammatically valid options is further restricted if they are incompatible to the DM. E.g. if there are no relations in the current scope that can be walked in the :1 direction, the menu item for the *SingleRecordTransOperator* is not shown, although it would be grammatically correct. Listing 1.2 shows successor definition of the expression *YearDiff*, used in the example above. *YearDiff* requires two successors (Minuend Date and Subtrahend Date), both of the type *DateExpression*. Out of this definition the context menu, which is shown in Figure 2, is created automatically. The method, that creates the menu based on this definition can be overwritten in every subclass to allow specific behavior.

Figure 2 shows the expression editor during the definition of the already mentioned calculation of the age using the *YearDiff* operator. The root element, the *YearDiff* operator have already been created (a). The root entity is colored red, because the expression is still incomplete. The text "ANE_16_1" in square brackets and the word "Aneurysm" indicate the scope of the root element. In section b, the first successor, Minuend Date, has already been created. The scope after the *SingleRecordTransOperator* "Parent Element" is Patient. The context menu in b shows the options for the success Subtrahend Date. Beside an operator that yields a constant date or the current date, date aggregation (first, last) and date manipulation (add day, add month) operators can be used. Since there are date attributes in the current scope (according to the DM), the usage of these attributes is offered as well. In order to complete this expression as described above, the option Aneurysm.Date found should be chosen. The section c shows the complete definition of the expression.

5 Conclusion

We could show that the usage of abstract meta data-models allows the implementation of generic data acquisition and management systems. Due to the focus on the domain-expert as user, our system can be set up and maintained completely autarchically, without the need of programming or database skills. The additional level of abstraction, requires an expression language on the same level, in order to allow the user to search, aggregate and combine the data on a complex level.

The tree-based input method turned out be the most user-friendly approach. It allows the clear rendering of the logical structure of an expression and offers well known and expected features like drag and drop and expand and collapse. The most important advantage over text-based input methods, is the expression definition by context menu. The context menu allows to absolutely restrict the user to the language grammar (which he doesn't even have to know) and allows the structured display of all currently available options. Furthermore, structural information can also be integrated into the generation of the menu, which enables the user to connect data across his structures.

We would like to thank the Landesnervenklinik Wagner-Jauregg of the GESPAG and the federal state of Upper Austria for their financial support. Special thanks go to Dr. Johannes Trenkler and Dr. Raimund Kleiser for their effort in creating and deploying the aneurysm registry presented in this paper.

References

1. Meta object facility (mof) specification OMG-Document ad/97-08-14 (September 1997)

2. Anderson, N.R., Lee, E.S., Brockenbrough, J.S., Minie, M.E., Fuller, S., Brinkley, J., Tarczy-Hornoch, P.: Issues in biomedical research data management and analysis: Needs and barriers. Journal of the American Medical Informatics Association 14(4), 478–488 (2007)
3. Arts, D.G.T., de Keizer, N.F., Scheffer, G.J.: Defining and improving data quality in medical registries: A literature review, case study, and generic framework. Journal of the American Medical Informatics Association 9(6), 600–611 (2002)
4. Chen, P.P.S.: The entity relationship model - toward a unified view of data. ACM Transactions on Database Systems 1(1), 9–36 (1976)
5. da Cruz, A.M.R., Faria, J.P.: A metamodel-based approach for automatic user interface generation. In: Petriu, D.C., Rouquette, N., Haugen, Ø. (eds.) MODELS 2010, Part I. LNCS, vol. 6394, pp. 256–270. Springer, Heidelberg (2010)
6. Frankel, D.: Model driven architecture: applying MDA to enterprise computing. Wiley, New York (2003)
7. Franklin, J.D., Guidry, A., Brinkley, J.F.: A partnership approach for electronic data capture in small-scale clinical trials. Journal of Biomedical Informatics 44(suppl.1), S103–S108 (2011); AMIA Joint Summits on Translational Science 2011
8. Free On-Line Dictionary of Computing: Free on-line dictionary of computing - expression (October 2013), `http://foldoc.org/expression`
9. Girardi, D., Arthofer, K., Giretzlehner, M.: An ontology-based data acquisition infrastructure. In: Proceedings of 4th International Conference on Knowledge Engineering and Ontology Development, Barcelona, pp. 155–160 (October 2012)
10. Girardi, D., Dirnberger, J., Trenkler, J.: A meta model-based web framework for domain independent data acquisition. In: ICCGI 2013, The Eighth International Multi-Conference on Computing in the Global Information Technology, pp. 133–138 (2013)
11. Harris, P.A., Taylor, R., Thielke, R., Payne, J., Gonzalez, N., Conde, J.G.: Research electronic data capture (redcap)a metadata-driven methodology and workflow process for providing translational research informatics support. Journal of Biomedical Informatics 42(2), 377–381 (2009)
12. Leiner, F., Gaus, W., Haux, R., Knaup-Gregori, P.: Medical Data Management - A Practical Guide. Springer (2003)
13. NIH: Cerebral aneurysm information page (April 2010)
14. OEGNR (May 2013), `www.aneurysmen.at`
15. Pentaho (2013), `http://kettle.pentaho.com/`
16. Prokosch, H.U., Ganslandt, T.: Perspectives for medical informatics. Methods Inf. Med. 48(1), 38–44 (2009)
17. Renggli, L., Ducasse, S., Kuhn, A.: Magritte a meta-driven approach to empower developers and end users. In: Engels, G., Opdyke, B., Schmidt, D.C., Weil, F. (eds.) MODELS 2007. LNCS, vol. 4735, pp. 106–120. Springer, Heidelberg (2007)
18. Tsumoto, S., Hirano, S.: Data mining in hospital information system for hospital management. In: ICME International Conference on Complex Medical Engineering, CME 2009, Tempe, AZ, pp. 1–5 (April 2009)
19. Tsumoto, S., Hirano, S., Tsumoto, Y.: Information reuse in hospital information systems: A data mining approach. In: 2011 IEEE International Conference on Information Reuse and Integration (IRI), pp. 172–176 (August 2011)

Text Clustering Using Novel Hybrid Algorithm

Divya D. Dev[1] and Merlin Jebaruby[2]

[1] Anna University, Computer Science and Engineering, Chennai-25, India
divyaddev@gmail.com
[2] Vel Tech Hi Tech, Electrical and Communication Engineering, Chennai – 90, India
merlinjebaruby@gmail.com

Abstract. Feature clustering has evolved to be a powerful method for clustering text documents. In this paper we propose a hybrid similarity based clustering algorithm for feature clustering. Documents are represented by keywords. These words are grouped into clusters, based on efficient similarity computations. Documents with related words are grouped into clusters. The clusters are characterised by similarity equations, graph based similarity measures and Gaussian parameters. As words are been given into the system, clusters would be generated automatically. The hybrid mechanism works with membership algorithms to identify documents that match with one another and can be grouped into clusters. The method works to find the real distribution of words in the text documents. Experimental results do show that the proposed method is much better when compared against several other clustering methods. The distinguished clusters are identified by a unique group of top keywords, obtained from the documents of a cluster.

1 Introduction

One of the main segments in text mining would be text clustering or document clustering. This is a process by which documents of similar topics or themes can be grouped together. The clusters can be used to improve the reliability, availability and dimensionality of text mining applications. Effective clustering will make the process of information retrieval and document summarization a lot easier. Document clustering revolves around three major problems. The very first one is on how to identify the similarity between documents. The second issue is on how to decide on the final number of document clusters and the third problem deals with the formation of precise clusters. The concept of feature clustering originates from early methods, which would convert the representation of high dimensional data to lower dimensional data sets. Real time applications have made use of linear algorithms due to its efficient and precise nature. The computational complexity is improved by various algorithms as mentioned in [1][2][3]. Feature clustering is one such algorithm that allows documents with pair wise semantic relatedness to be grouped together. Each document will be identified by a minimal number of features or words; hence the overall dimensionality could be reduced by drastic amounts. The motivation of clustering documents with keywords is done due to two aspects. The keywords can be used to reduce the overall

N.T. Nguyen et al. (Eds.): ACIIDS 2014, Part I, LNAI 8397, pp. 11–20, 2014.

dimensionality of a document set. The traditional methods work with huge bags of words that will increase the complexity of text clustering. Thus, as keywords are been extracted, the documents can be indexed with the top collection of words. This will change the representation of documents and result in sparse text documents. The second phase of clustering makes use of the appropriate keywords. This will increase the comprehensibility of clusters. Additionally, the frequent keyword term set will provide documents with contextual and conceptual meanings.

This paper is organised as follows: Section 2 gives an overview of the existing frequent feature term sets based clustering, this includes DC, Word variance based selection, CFWS, FIHC and Fuzzy clustering. Section 3 proposes a novel method called MMMC, a series of Graph Based Similarity measures and efficient Gaussian Computations to construct relevant documents and form optimized clusters. The proposed method normalizes by merging and splitting clusters in an effective manner. Section 4 showcases the experimental evaluation of Hybrid Clustering. Section 5 concludes the paper.

2 Related Work

2.1 Document Frequency

Document frequency is the count of the number of documents in which a term could occur. It is regarded as an easy measure with straightforward criterion by which large datasets can be grouped together at linear computation complexity. The method is simple and is ideal for effective feature selection [4].

2.2 Word Variance Based Selection

Word variance could be used to differentiate the words of a dataset. The algorithm keeps together words with higher variances [4]. The unique variance of terms is then used to sort the words. The cluster size will be equivalent to the number of documents in it. The variance of word w, in document x, and occurrence x(w) in a dataset with N document sets is defined as:

$$\sigma^2 = \frac{1}{N}\sum_x x^2(w) - \left(\frac{1}{N}\sum_x x(w)\right)^2 \tag{1}$$

2.3 Clustering Frequent Word Sequence

Clustering frequent word sequence is a method of clustering proposed in [5]. The technique uses frequent word sequence and K-mismatch for text clustering. The word sequence method takes into consideration the order in which the words are placed. Though K-mismatch is used to form clusters, the presence of transitivity makes certain that documents appear in more than one cluster. When K-mismatch runs extensively, the clusters with become more ambiguous, as a result all documents will be grouped into one cluster. This is called as trivial clustering.

2.4 Frequent Itemset Based Hierarchical Clustering

Frequent itemset based hierarchical clustering was proposed in [6]. The method describes two kinds of frequent items, namely the cluster frequent item and global frequent item. The hierarchical method works through four phases to produce effective document clusters: the frequent itemsets are found, the initial phase of clustering is performed, trees are constructed and finally pruning is done on the final clusters. FIHC does not deal with pair wise similarity. It clusters using the classic method of clustering. FIHC constructs a similarity matrix from which document pairs with the largest number of similarities will be set as zero and are grouped together.

2.5 Fuzzy Self Clustering Feature Clustering

Fuzzy self-clustering feature clustering algorithm is renowned as an incremental approach. It reduces the dimensionality of documents and groups features that are similar to one another. The clusters are identified by statistical mean and deviation. Words that don't fall into the existent clusters will be placed in newly created clusters.

The proposed method is an extension of my previous work which deals with keyword extraction and is proposed in [8]. The optimal output from D4 Keyword Extractor [8] is passed onto the second phase of Hybrid clustering. Most of the existing "feature clustering" algorithms have few common issues, which affects its net output. First, the users have to mention a value for the desired number of clusters. This is a burden on users. The count has to be indicated by a trial and error method, which has to be repeated manually until an appropriate output is generated. Secondly, the underlying variation in clusters is missed by many algorithms. Variance is an important factor that will calculate the similarity between clusters. Appropriate calculation of variances will result in better data distribution. Thirdly, clusters tend to have the same degree of features. Sometimes, the output will be better if the distribution is uneven and certain clusters are made with a larger number of text documents. The proposed Hybrid algorithm deals with the above problems.

3 Proposed Method

3.1 Strategy 1

The main objective of the Hybrid Algorithm is to produce comprehensive document clusters. The keywords are subject to similarity measures by which the relevance of key terms in a given document will be identified. Initially, a Maximal Must and Minimal Cannot (MMMC) algorithm is proposed in conjuncture with the key terms that will contribute to the document's actual meaning. Thus, MMMC would improve the accuracy of the feature itemsets.

Working Process of Strategy 1

MMMC evaluates the accuracy of the extracted keywords. The strategy identifies the similarity between individual keywords of each document, and would eliminate unpromising keywords based on this estimation. MMMC works as a worst case similarity estimation. It guarantees to produce optimal keywords for a given document. The filtering strategy is quite similar to the top-k selection of Fagin's No Random Access (NRA) algorithm [9]. The keywords, t of a document, d are subject to the equations MM and MC. Terms which satisfy the condition, where maximum scores MM is more than the minimal scores MC will be kept in the collection. Terms that fail to satisfy the relationship of similarity will be removed. In this method of MMMC, the actual score of MM is the similarity between a keyword $w_2(t,d)$ and the document's top most keyword which will act as the document's centroid. MM(t) is defined as:

$$MM(t) = \left(\log \left(\frac{1}{\max(W(t,d)-1} \right) - w_2(t,d) \right)^2 - w_2(t,d) \quad (2)$$

Where W(t , d) is a real value that represents the weight of the highest priority keyword. The MC score is a similarity measure that finds a numerical relationship with the least priority keyword and other terms key terms. MC(t) is defined as:

$$MC(t) = w_2(t,d) - \sqrt{2 * w_2(t,d) * \log \left(\frac{1}{1-\min(W(t,d))} \right)} \quad (3)$$

The similarity estimation depends on the numerical weight of terms in a document and not all keywords need to be included in the document summary. The computation works in accordance with the following constraints: a) each term will not be included in the summary of a document if it has a MC that is more than its MM, otherwise the term will be a part of the final keyword set b) the number of keywords present in a document must be more than the median count of keywords. If the top keyword does not give an apt length of keywords, it has to be removed and the second top keyword has to be placed as the next prime keyword.

3.2 Strategy 2

The keywords from MMMC will be passed onto a Graph Based Similarity algorithm [10], which will group the keywords to form preliminary clusters. The algorithm works with three formulations that will identify the dependencies between words, through which the centrality and resulting scores of the words in a document can be found. The algorithm covers over three major tasks: a) the top keywords of the document will form the graph's vertices b) graph dependencies will be constructed with the similar words c) the graph edges are assigned labels with promising scores. The dependencies of words in a document can be represented as a graph. Given a documents D= $\{d_1, d_2, d_3, \ldots d_n\}$ with a sequence of keywords W = $\{w_1, w_2, w_3, \ldots w_n\}$. Similar words of different documents can be connected with admissible labels L= $\{l_1, l_2, l_3, \ldots l_n\}$. We define a label for the Graph G = (V,E) when the keyword weight w_1, w_2 satisfies the three measures of similarity equations [5][6][7]. Note, the graph does not need to be fully connected, as the edges will be labelled only if the graph based similarity measures are satisfied. The information of clustering is drawn from the entire graph. The similarity of keywords in a document is defined as:

$$E_1 = -\log \frac{|w_1 - w_2|}{avg\ (w_1, w_2)} \tag{6}$$

$$E_2 = \frac{low\ (w_1, w_2)}{avg\ (w_1, w_2)} \tag{7}$$

$$E_3 = \frac{\log(\ low\ (w_1, w_2))}{\log(\ w_1) + \log(\ w_2)} \tag{8}$$

$$|E_3\text{-}E_2| <= |E_1\text{-}E_2| <= |E_3\text{-}E_1| \tag{9}$$

The word similarity metrics are derived from Word Net-based implementation. Equation (6) is determined by the similarity measure proposed by Leacock & Chohorow [11], where $|w_1\text{-}w_2|$ gives the difference in weight of the two keywords. Equation (7) is formed with the concept proposed by Lesk [12]. The similarity function identifies the overlapping nature of words. The lowest weight is taken to characterise the two different keywords. Equation (8) is a Wu and Palmer [13] similarity metric which uses the depth of two keywords and the depth of the least common subsume.

3.3 Strategy 3

As a result of the Graph Based Similarity Measure, we are given with a document set D of "n" different documents $d_1,d_2,d_3...d_n$ in "p" different clusters C of $c_1,c_2,c_3...c_p$. Each document will have a unique set of keyword to describe it. Using which we can construct an accurate word pattern x_i for each word w_i, with an occurrence of O_{pi}, is quite similar to what is defined in equation [12].

$$w_{ji}=P(c_1|w_i),\ P(c_2|w_i)......................,\ P(c_p|w_i) = \tag{10}$$

$$P(c_n|w_i) = \frac{\sum_{n=1}^{p} O_{ni} * \delta_{pk}}{\sum_{n=1}^{p} O_{ni}} \tag{11}$$

It is with these word patterns that the session of optimization would work on. The Hybrid clustering algorithm uses Gaussian parameters for optimization. Once the words, w are grouped into clusters in accordance with its word patterns, each cluster can be characterised by a one dimensional Gaussian Function. Gaussian functions are acknowledged as superior functions in terms of its performance. Thus, let C be a cluster with j word patterns x_j. Let $x_j = <x_1,x_2,x_3....x_j>$, and the standard deviation $\sigma = <\sigma_1, \sigma_2,... \sigma_p>$ of each cluster is defined as:

$$x_j = \frac{\sum_{n=1}^{p} w_{ji}}{|C|} \tag{12}$$

$$\sigma_i = \sqrt{\frac{\sum_{j=1}^{p}(w_{ij} - x_{ij})^2}{|C|}} \tag{13}$$

For every 1<j<p, where |C| represents the size of each cluster, i.e. the total number of word patterns in a given cluster C. Optimization through Gaussian parameters makes use of Fuzzy Similarity. Thus for every cluster with word pattern $x_j = \langle x_1, x_2, \ldots x_j \rangle$ and standard deviation $\sigma_j = \langle \sigma_1, \sigma \ldots \sigma_j \rangle$ a membership function is being represented as:

$$\mu_c(w) = \prod_{i=1}^{p} \exp\left[-\left(\frac{w_i - x_j}{\sigma_i}\right)^2\right] \tag{14}$$

Any word pattern that is similar to its mean value will be a part of the cluster. Thus, words with a membership function output equivalent to one ($\mu_c(w) \approx 1$)will be a part of the cluster. If a key term has a word pattern that is quite deviant ($\mu_c(w) \approx 0$) from the cluster's membership function, will hardly be a part of the cluster.

Preliminaries of Strategy 3

Two different cases could occur with the word patterns. Firstly, the word pattern x_i is not similar and it does not fit into the memberships function. Thus, the cluster G_i has to be broken and a new cluster is formed, k=i+1. G_k will have the word pattern x_i, while the word pattern x_i will be removed from G_i. At this point, G_k will have only one word pattern alias word in it. On further iterations, more word patterns could have a membership function which relates to the deviation of x_i. These word patterns can be included in the cluster G_k if and only if the word patterns belonged to documents, which were at least weakly connected with documents of the words in G_k during the Graph Based Similarity Measure. If there is another cluster G_i that produces similar word patterns and membership functions as of G_k, these clusters can be grouped together to form single cluster. Now we will have an optimized number of clusters. These clusters can be stored for future reference. With new training patterns, the algorithm can be run and existing clusters will be modified or new clusters can be created.

4 Experimental Results

In this section, we present the experimental results to show the effectiveness of our hybrid clustering algorithm. Two well known data sets were used to prove our text clustering method: Reuters 21578 and Brown Corpus. Reuters-21578 was obtained from http:kdd.ics.uci.edu/databases/reuters21578/reuters21578. The document collection had 135 different categories. Nevertheless, only 50 different categories were used during the experimentation. Brown Corpus is another data set which contains 500 different samples of text documents. The text documents are distributed around 15 genres. Every word in the document is labelled with part of speech tags. For testing the hybrid algorithm, the complete Brown Corpus document set was used.

4.1 Evaluation Methods

DC, IG, FIHC, CFWS and our Hybrid Clustering algorithm were run on Reuters 21578. The novel method is not compared against traditional methods like k-means and bisecting k-means because the previous methods mentioned in section 2 have been well studied and they are more efficient than the conventional methods. To compare the effectiveness of each method, performance measures in terms of micro-averaged precision (MicroP), micro-averaged recall (MicroR) and micro-averaged F-measure (MicroF1) are used. In the micro averaged formulas, TP_i with respect to each cluster c_i, is the number of documents that are correctly classified to c_i, TN_i is the number of incorrect non-c_i test documents that are classified into the non-c_i clusters. FP_i is the number of non-c_i that is incorrectly classified into c_i. FN_i is the number of false c_i test documents that are classified to non-c_i.

$$MicroP = \frac{\sum_{i=1}^{p} TP_i}{\sum_{i=1}^{p} (TP_i + FP_i)} \qquad MicroR = \frac{\sum_{i=1}^{p} TP_i}{\sum_{i=1}^{p} (TP_i + FN_i)} \qquad MicroF = \frac{2 * MicroP * MicroR}{MicroP + MicroR}$$

4.2 Evaluation Results

Table [1] and Table [2] shows that on Reuters 21578 and Brown corpus, respectively, the novel hybrid algorithm significantly outperforms the other methods used for document clustering.

Table 1. Output for Reuters 21578

No of documents		20	50	80	120	240	500
Microaveraged Precision	FIHC	79.87	79.96	82.82	83.58	83.75	85.38
	WV	91.58	**92.00**	68.82	69.46	76.25	83.74
	DC	81.91	78.35	79.68	79.74	81.42	83.08
	CFWS	85.18	82.96	82.34	82.68	**87.68**	84.73
	Hybrid	**92.34**	91.34	**88.65**	**90.32**	87.28	**87.65**
Microaveraged Recall	FIHC	51.72	53.27	54.93	58.02	61.75	**64.81**
	WV	5.80	11.82	25.72	26.32	36.66	48.16
	DC	49.45	53.05	55.92	56.50	61.32	63.49
	CFWS	**55.33**	61.91	63.41	**63.76**	**65.33**	**66.11**
	Hybrid	54.67	**62.55**	**70.97**	56.67	65.05	60.71
Microaveraged FMeasure	FIHC	62.78	63.94	66.05	68.49	71.09	73.69
	WV	10.91	20.95	37.45	38.17	49.51	61.15
	DC	61.67	63.26	65.72	66.14	69.95	71.98
	CFWS	67.08	70.91	71.65	**72.00**	**74.87**	**74.27**
	Hybrid	**68.68**	**74.25**	**78.85**	69.64	74.54	71.73

Table 2. Output for Brown Corpus

No of documents		20	50	80	120	240	500
Microaveraged Precision	FIHC	88.00	90.97	**91.69**	91.90	92.51	**93.28**
	WV	**91.91**	90.77	89.78	89.07	90.29	89.57
	DC	52.69	56.15	50.12	50.73	48.79	50.54
	CFWS	70.32	78.75	74.45	70.85	62.54	67.89
	Hybrid	82.34	**93.29**	90.40	**92.09**	**93.45**	91.45
Microaveraged Recall	FIHC	62.88	70.80	74.38	73.76	77.54	77.91
	WV	17.75	27.11	30.22	34.16	43.72	52.96
	DC	65.21	74.67	74.96	74.27	**79.84**	**80.61**
	CFWS	70.25	**77.71**	78.32	77.76	78.17	78.11
	Hybrid	**84.32**	77.21	**87.72**	**83.43**	74.55	80.01
Microaveraged FMeasure	FIHC	73.34	79.62	82.13	81.83	**84.36**	84.90
	WV	29.75	41.75	45.21	49.38	58.91	66.56
	DC	58.28	64.09	60.07	60.28	60.56	62.12
	CFWS	70.28	78.23	76.34	74.14	69.49	72.64
	Hybrid	**83.31**	**84.49**	**89.04**	**87.55**	82.94	**85.35**

FIHC outperformed many other methods other than the novel hybrid algorithm. Study showed that FIHC showed favourable amount of performance because it makes use of matching frequent item sets to identify the relationship between clusters and the documents. Nevertheless, FIHC makes use of limited variation. This is because of the method used to form the initial clusters. The very first clusters formed are more skewed i.e. the number of documents present in each cluster varies by considerable amounts, where some clusters have more documents than the others. The difference in terms of documents present in big clusters and the smaller ones is considerably huge. As a result, newer documents introduced into the algorithm are more likely to end in the largest cluster. This is a drawback avoided by the novel hybrid algorithm, where the cluster size does not affect the positioning of documents into clusters. DC and Word Variance Based Clustering failed in most cases because clusters are not formed with the relevance of words in the document but it is based with the number of times a term occurred within the distribution. Thus the clusters are not characterised naturally. To be more precise, the method does not characterise clusters based on the meaningful content of the document. CFWS is another method prone to produce skewed clusters. This is because it uses K-mismatch, which produced very big overlapping between clusters. Thus, the overlapping coefficient in large clusters will always be bigger than what is present in smaller clusters.

5 Conclusion

This paper deals with a hybrid clustering technique that makes use of an incremental approach. The technique reduces document dimensionality and promotes text clustering in a simplified manner. Documents with similar features are grouped together in to a single cluster. Functions based on Gaussian parameters are used to group similar content. This includes an equation to evaluate the word pattern, standard deviation and membership function. If a word does not fall within a given cluster, a new cluster is created for that word. The word pattern and standard deviation of terms are modified automatically, as a new word is positioned into a cluster. The hybrid algorithm works without the help of manual intervention. The desired number of clusters is generated automatically. Furthermore, every cluster is represented by a weighted combination of words. The algorithm uses membership functions to match documents closely. As mentioned previously, the user does not have to mention the number of clusters or the number of documents in each cluster. Thus, errors caused by trial are not present in the hybrid algorithm. Experiments on two different real time data sets proved the effectiveness of our algorithm. On the overall view, the Hybrid Algorithm has a better Microaveraged FMeasure than the other methods. Even as the number of words in each document increased, the hybrid algorithm maintained a better performance. Thus, from the statistical values it is evident that the hybrid algorithm runs better that the existing feature extraction methods.

References

1. Yan, J., Zhang, B., Liu, N., Yan, S., Cheng, Q., Fan, W., Yang, Q., Xi, W., Chen, Z.: Effective And Efficient Dimentionality Reduction For Large Scale And Streaming Data Preprocessing. IEEE Trans. Knowledge and Data Eng. 18(3), 320–333 (2006)
2. Hiraoka, K., Hidai, K., Hamahira, M., Mizoguchi, H., Mishima, T., Yoshizawa, S.: Successive Learning of Linear Discriminat Analysis: Sanger-Type Algorithm. In: Proceedings of IEEE CS Int'l Conf. Pattern Recognition, pp. 2664–2667 (2000)
3. Weng, J., Chang, Y., Hwang, W.S.: Candid Covariance-Free Incremental Principal Component Analysis. IEEE Trans. Pattern Analysis and Machine Intelligence 25(8), 1034–1040 (2003)
4. Yang, Y., Pederson, J.O.: A comparative study on feature selection in text categorization. In: Proc. of the Fourth International Conference on Machine Learning, pp. 412–420 (2007)
5. Li, Y.J., Chungm, S.M., Holt, J.D.: Text document clustering based on frequent word meaning sequences. Data & Knowledge Engineering 64, 381–404 (2008)
6. Fung, B., Wang, K., Ester, M.: Hierarchial Document Clustering Using Frequent Itemsets. In: Proc. of 3rd SIAM International Conference on Data Mining (2003)
7. Khalessizadeh, S.M., Zaefarian, R., Nasseri, S.H., Ardil, E.: Genetic Mining Using Genetic Algorithm For Topic Based On Concept Distribution. World Academy of Science, Engineering and Technology 13 (2006)
8. Rose, J.D., Dev, D.D., Robin, C.R.R.: An Improved Genetic Based Keyword Extraction Technique. In: Terrazas, G., Otero, F.E.B., Masegosa, A.D. (eds.) NICSO 2013. SCI, vol. 512, pp. 153–166. Springer, Heidelberg (2014)
9. Fagin, R., Lotem, A., Naor, M.: Optimal Aggregation Algorithms for Middleware. In: Proc. of the 20th ACM SIGACT-SIGMOD-SIGART Symposium on Principles of Database Systems, pp. 102–113 (2001)
10. Sinha, R., Mihang, R.: Unsupervised Graph Based Word Sense Disambiguation Using Measure Of Word Semantic Similarity. In: IEEE International Conference on Semantic Computing, pp. 363–369 (2007)

11. Leacock, C., Chodorow, M.: Combining Local Context And Wordnet Sense Similatity For Word Sence Identification In Wordnet, An Electronic Lexical Database. The MIT Press (1998)
12. Wu, Z., Palmer, M.: Verb Semantics and Lexical Selection. In: Proc. of the 32nd Annual Meeting of the Association For Computational Linguistics, Las Cruces Mexico (1994)
13. Jiang, J.Y., Liou, R.J., Lee, S.J.: A Fuzzy Self Constructing Feature Clustering Algorithm For Text Classification. IEEE Transactions on Knowledge and Data Engineering 23(3), 335–348 (2011)

Combination of Multi-view Multi-source Language Classifiers for Cross-Lingual Sentiment Classification

Mohammad Sadegh Hajmohammadi, Roliana Ibrahim*,
Ali Selamat, and Alireza Yousefpour

Faculty of Computing, Universiti Teknologi Malaysia (UTM),
81310 Skudai, Johor, Malaysia
{shmohammad2,yalireza3}@live.utm.my, {roliana,aselamat}@utm.my

Abstract. Cross-lingual sentiment classification aims to conduct sentiment classification in a target language using labeled sentiment data in a source language. Most existing research works rely on machine translation to directly project information from one language to another. But cross-lingual classifiers always cannot learn all characteristics of target language data by using only translated data from one language. In this paper, we propose a new learning model that uses labeled sentiment data from more than one language to compensate some of the limitations of resource translation. In this model, we first create different views of sentiment data via machine translation, then train individual classifiers in every view and finally combine the classifiers for final decision. We have applied this model to the sentiment classification datasets in three different languages using different combination methods. The results show that the combination methods improve the performances obtained separately by each individual classifier.

Keywords: Cross-lingual, Sentiment classification, classifier combination, multi-view, multi-language.

1 Introduction

Together with the very rapid increasing of the internet access in the world, the volume of user generated contents have also been increased on the web. Due to the high quantity of user-generated contents, the task of summarizing their information into a useful format is a very hard and challenging problem. This challenge motivates the natural language processing (NLP) communities to design and develop computational methods to analyze these text documents.

Opinion mining or sentiment analysis is one of the most interesting fields in this area that analyzes people's opinions, attitudes and sentiments towards entities such as products, individuals, events, etc. [1]. Text document sentiment classification is the task of determining the sentiment polarity (e.g. positive or negative) of a given text document [2] and has received considerable attention due to its many useful application in product reviews classification [3] and opinion summarization [4].

* Corresponding Author.

N.T. Nguyen et al. (Eds.): ACIIDS 2014, Part I, LNAI 8397, pp. 21–30, 2014.

Up until now, different methods have been used in sentiment classification. These methods can be categorized into two main groups, namely; lexicon based and corpus based. The lexicon based methods classify text documents based on the polarity of words and phrases contained in the text [5, 6]. This group of methods needs a sentiment lexicon to distinguish between the positive and negative terms. In contrast, corpus based methods train a sentiment classifier based on labelled corpus using machine learning classification algorithms [7, 8]. The performance of these methods intensively depends on the quantity and the quality of labelled corpus as the training set.

Sentiment lexicons and annotated sentiment corpora are the most important resources for the sentiment classification. However, since most recent research studies in sentiment classification are in the English language, there are not enough labelled corpus and sentiment lexicons in other languages [9, 10]. Further, manually construction of reliable sentiment resources is a very hard and time-consuming task. Therefore, the challenge is how to utilize labelled sentiment resources in one language (i.e. English) for sentiment classification in another language and leads to an interesting research area called cross-lingual sentiment classification (CLSC). The most direct solution of this problem is the use of machine translation systems to directly project the information of data from one language into the other language[9-15]. The most existing research works develop a sentiment classifier based on the translated labelled data from the source language and use this classifier to determine the sentiment polarity of test data in the target language [12, 13]. Machine translation can be employed in the opposite direction by translating the test documents from the target language into the source language [9, 14, 15]. In this situation, the sentiment classifier is trained based on the original labelled data in the source language and then applied to the translated test data. A few number of research works used both direction of translation to create two different views of the training and the test data to compensate some of the translation limitations [10, 16]. But because the training set and the test set are from two different languages with different writing styles and from different cultures, these methods cannot reach the performance of monolingual sentiment classification methods in which the training and test samples are from the same language. The performance also can be influenced by the low quality of translation because machine translation is still far from satisfactory and therefore the translated text documents cannot cover all the vocabularies in the original text documents. Different term distribution between the original and the translated text documents is another important factor that can reduce the performance of CLSC. It means that a term may be frequently used in one language to express the opinion while the translation of that term is rarely used in the other language.

In this paper, we propose a new cross-lingual sentiment classification model that use more than one language (for example two languages) as the source languages to compensate some of the aforementioned limitations of resource translation in CLSC. We use labelled corpus from two different languages as the training data and also use both translation directions to create three different view of data, one in the target language and two in the source languages. Accordingly, three different classifiers are trained based on these three views and finally the predictions of these classifiers are combined using ensemble method. The proposed model was applied to the book review datasets in three different languages and experiments showed that using multiple source language in multiple views obtains better performance in comparison with the methods that use only one language as the source language.

The reminder of this paper is organized as follows: The next section presents related works on cross-lingual sentiment classification. Section 3 describes the proposed model. The experimental setup is explained in Section 4, while results and discussion are given in Section 5. Finally, Section 6 concludes this paper.

2 Related Work

Cross-lingual sentiment classification has been extensively studied in recent years. These research studies are based on the use of annotated data in the source language (always English) to compensate for the lack of labelled data in the target language. Most approaches focus on resource adaptation from one language to another language with few sentiment resources. For example Mihalcea et al. [17] generate subjectivity analysis resources into a new language from English sentiment resources by using a bilingual dictionary. In other works [13, 18], automatic machine translation engines were used to translate the English resources for subjectivity analysis. Banea et al. [18], showed that automatic machine translation is a viable alternative for the construction of resources for subjectivity analysis in a new language. Wan [19] used unsupervised sentiment polarity classification in Chinese product reviews. He translated Chinese reviews into different English reviews using a variety of machine translation engines and then performed sentiment analysis for both Chinese and English reviews using a lexicon-based technique. Finally, he used ensemble methods to combine the results of analysis. Another approach is that of cross-lingual classification, that is translating the features extracted from labelled documents [20, 21]. The features, selected by a feature selection algorithm, are translated into different languages. Subsequently, based on those translated features; a new model is trained for every language. This approach only needs a bilingual dictionary to translate the selected features. It can, however, suffer from the inaccuracies of dictionary translation, in that, words may have different meanings in different contexts. In another work, Wan [10] used the co-training method to overcome the problem of cross-lingual sentiment classification. In this paper, he exploited a bilingual co-training approach to leverage annotated English resources to sentiment classification in Chinese reviews. In this work, firstly, machine translation services were used to translate English labelled documents (training documents) into Chinese and similarly, Chinese unlabeled documents into English. The author used two different views (English and Chinese) in order to exploit the co-training approach into the classification problem. In an early work, Martin-Validivia et al. [9] proposed a meta classifier system that integrated the corpus based and lexicon based methods in order to create a sentiment classifier in Spanish language. In the first place, they used Spanish corpus along with its translated version in English and create two individual models based on these two corpora and applied machine learning to train the models. Next, they integrated SentiWordNet into the translated data to generate a new lexicon based model. Lastly, they combined these systems by Meta classifiers. To the best of our knowledge, using multiple source languages in multiple views has not yet been investigated in the field of cross-lingual sentiment classification.

3 Proposed Model

In this section we present a new cross-lingual model for sentiment classification that uses multiple languages as source language in multiple views. In this model, after the construction of different views in the target and the source languages, a classifier is trained based on the labeled data in every view and is applied to the test data in corresponding view and finally, the prediction results of each individual classifiers are combined to form the final results.

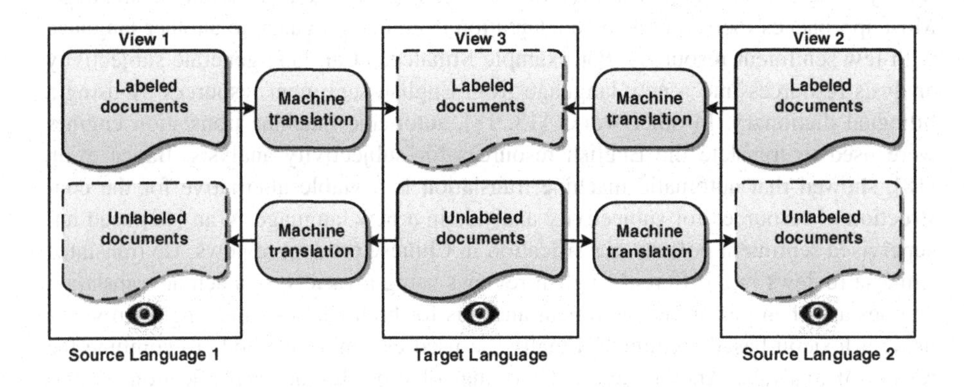

Fig. 1. Creation of multiple views of documents in the case of two source languages, using machine translation

3.1 Multi-view Data Creation

To create multiple views of labeled and unlabeled documents in the source and target languages, we perform machine translation in two different directions. At the first step, unlabeled document (test data) are translated from the target language into the source languages. Next, labeled documents (training data) are translated from the source languages into the target language. Fig. 1 diagrammatically shows the process of multi-view data creation for the situation that two source languages are used. As we can see in this figure, when two different languages are used as source language, we have training and test documents (labeled and unlabeled) in three different views. It means that each document is presented based on three different feature sets, one in the target language and two in the source languages. Therefore, classification process can be performed based on three different feature sets from three different languages.

3.2 Classification Combination in Different Views

Multiple classifiers combination is a well-known learning strategy when a set of classifiers is trained for a same classification. Combination of classifiers is the most reasonable solution when more than one single training set exist or different presentations of the training set are available [22]. Combining multiple classifiers can be advantageous since different classifiers would induce complimentary information for the classification.

In our proposed method, first, the training documents in every view are used to train a member classifier. Then, the trained classifiers are applied to the test set, represented based on the feature sets of corresponding views to determine the prediction label of each sample. After that, a combination rule is used to integrate the output predictions of the member classifiers to make the final classification decision. Several combination algorithms can be used to integrate the results of member classifiers. Definitely, we tried three groups of the most widely used methods, namely: majority voting, fixed rules and stacking, which are explained in the following subsections.

Majority Voting. Majority voting is the simplest method used for classifier combination. In this method, the final predicted class is selected by polling all the classifiers to see which class is the most popular. Whichever class that receives most votes is selected. Majority voting is always successful when the classifiers' output are binary.

Fixed Rules. Individual classifiers always provide not only the predicted label, but also one kind of confidence measure, such as posterior probability. The fixed rule combiner is used to combine these probabilities. Suppose that $p_c(w_j|x)$ is the posterior probability in predicting class w_j for instance x provided by member classifier c.

- Product rule integrates individual classifiers by multiplying the posterior possibilities and use the result for final output based on (1).where m is the number of the individual classifiers.

$$f_{Prod}(x) = \arg\max_{w_j} \prod_{c=1}^{m} p_c(w_j|x) \tag{1}$$

- Sum rule integrates individual classifiers by summing the posterior possibilities and use the results for final output based on (2).

$$f_{sum}(x) = \arg\max_{w_j} \sum_{c=1}^{m} p_c(w_j|x) \tag{2}$$

Stacking. The stacking combiner adds a new classifier (called combiner classifier) that uses the outputs of the member classifiers as input feature vector and learns the best method to integrate these classifiers. In this paper, we used the prediction confidence of every member classifier to form the input feature vector for the combiner classifier. At the first, the member classifiers are trained based on training set and then applied to a development set for prediction task. After that, the prediction confidences are used to form the input feature vector for the combiner classifier. Finally, the combiner classifier is trained based on these new training data to learn the best combination rule. We used four different machine learning algorithms to learn the combination rule: Support Vector Machine (SVM), Naïve Bayes (NB), Artificial Neural Network (ANN) and Linear Least Square (LLS).

4 Experiment

In this section, we evaluate our proposed approach in cross-lingual sentiment classification on three different languages in the book review.

4.1 Experimental Datasets

To create an evaluation dataset, we selected 2000 book reviews (1000 positive and 1000 negative) from Prettenhofer and Stein dataset [23] in three different languages: English, French and German. By combining these three languages, we obtained three different dataset for evaluation. In each dataset, one language is considered as target language and two other languages are the source languages. Documents in each language are translated into two other languages using Google translate service (http://translate.google.com/) to create different views of data. Table 1 shows the characteristics of evaluation datasets. In the pre-processing step, all English reviews are converted into lowercase. Special symbols and other unnecessary characters are eliminated from every review document.

Table 1. Different datasets used for evaluation

Data Set	Languages		
	Source 1	**Source 2**	**Target**
EF-G	English	French	Germany
EG-F	English	Germany	French
FG-E	French	Germany	English

To reduce computational complexity, we performed feature selection using the information gain (IG) technique. We selected 5000 high score unigrams and bi-grams as final features. Every document is represented by a feature vector. Each entry of a feature vector contains a feature weight. We used term presence as feature weights because this method has been confirmed as the most effective feature weighting method in sentiment classification [7, 24].

4.2 Experimental Setup

In all experiments, we used the support vector machine classifier (SVM) as the member classifiers in every view. SVM^{light} (http://svmlight.joachims.org/) is used as the SVM classifier in the experiments with all parameters set to their default values.

However, SVM^{light} does not directly output the posterior probabilities of predicted labels. Therefore we use a strategy that introduced in [25] to compute these probabilities. Labeled documents in every view are randomly divided into training and development sets with the portion of 80% and 20% respectively. The development set is used to train the classifier combiner in stacking methods. For the classifier combiner in stacking methods we used the original MATLAB implementation of machine learning algorithms. For the ANN we used one hidden layer with 20 neurons. Other algorithms were used with default parameters.

5 Results and Discussion

We conducted several experiments with different combination methods in three different languages. In this section, we compare the accuracy of each combination technique with other techniques and also with the accuracy of the member classifiers as base classifiers. The main goal of the classifier combination is to correct the errors of the member classifiers. In our experiment, we can approve the improvement achieved by using this approach since all the combination methods improve the final classification results in compare with individual classifiers.

Table 2. Accuracy of the member classifiers and the combining methods

		Data Sets	EF-G	EG-F	FG-E	Average
Member classifiers		View 1	77.57%	77.10%	74.99%	
		View 2	76.41%	74.85%	74.99%	
		View 3	78.62%	78.40%	80.64%	
Combination Methods		Majority Voting	80.07%	80.45%	80.54%	80.35%
	Fixed Rule	Product	80.82%	80.35%	81.19%	80.79%
		sum	80.47%	80.25%	80.44%	80.39%
	Stacking	SVM	**81.27%**	80.45%	81.29%	81.00%
		NB	80.87%	80.15%	**81.39%**	80.80%
		ANN	**81.27%**	**80.65%**	81.19%	**81.04%**
		LLS	80.97%	79.70%	80.89%	80.52%

As we can see in Table 2, all the classifiers combination methods outperform the member classifiers. This means that the information of multiple source languages can cover each other in predicting the sentiment labels of the target language documents. This table also shows that the accuracy of target language view (View 3) is grater then two other views. Because in this view, features are extracted from the training data that translated from two source languages and therefore cover more vocabularies from target language documents so documents in target language are presented much better with this feature set. Figure 1 also shows the comparison results in graphical format. In this figure, we can also see that stacking methods shows better performance for classifiers combination in comparison with other combination. This is due to the fact that stacking method tries to learn the best combination rule through machine learning.

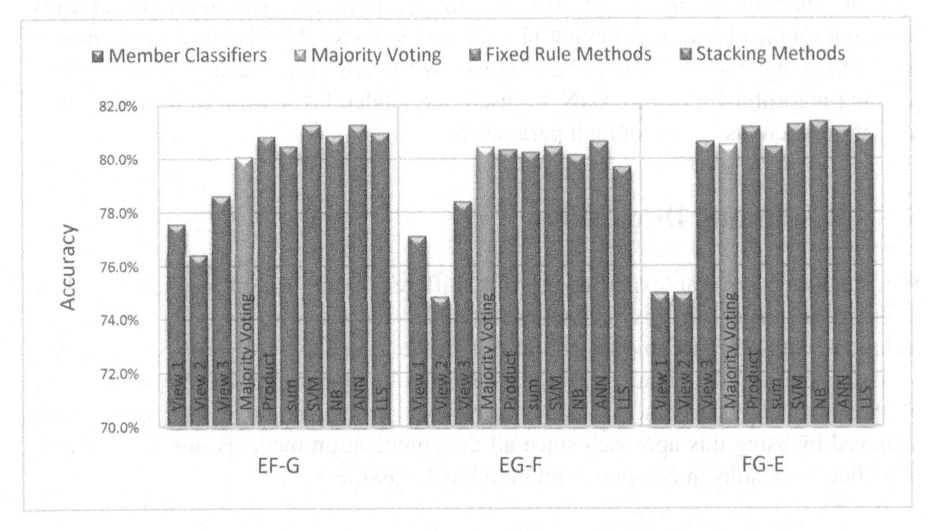

Fig. 2. Comparison results obtained by different combination methods

6 Conclusion and Further Work

In this paper, we have proposed a classifier combination model for using multiple source languages in different views to improve the performance of cross-lingual sentiment classification. In the proposed model, automatic machine translation was used to project the information of target language documents into the source languages and also to translate the training data from the source language into the target language. This bi-direction translation creates different views of classification data and sentiment classification can be performed in every view. Finally, the member classifiers were integrated using different aggregation methods such as fixed rules, majority voting and stacking algorithm. We applied this model to cross-lingual sentiment classification datasets in three different languages and we have shown that the combination methods improve the performances obtained separately by every single classifier. These results shows that using the information of multiple source languages can cover more characteristics of target language in sentiment classification and therefore can improve the performance of sentiment classification in the target language.

In addition, we would like to exploit unlabeled data from target language in this multi-view framework and use semi-supervised multi-view learning algorithms to improve the performance of cross-lingual sentiment classification.

Acknowledgement. This work is supported by the Ministry of Higher Education (MOHE) and Research Management Centre (RMC) at the Universiti Teknologi Malaysia (UTM) under Research University Grant Scheme (Vote No. Q.J130000.2628.07J52)

References

1. Liu, B.: Sentiment Analysis and Opinion Mining. Morgan & Claypool Publishers (2012)
2. Hajmohammadi, M.S., Ibrahim, R., Ali Othman, Z.: Opinion Mining and Sentiment Analysis: A Survey. International Journal of Computers & Technology 2(3), 171–178 (2012)
3. Zhou, S., Chen, Q., Wang, X.: Active deep learning method for semi-supervised sentiment classification. Neurocomputing 120, 536–546 (2013)
4. Ku, L.W., Liang, Y.T., Chen, H.H.: Opinion extraction, summarization and tracking in news and blog corpora. In: Proceedings of AAAI-2006 Spring Symposium on Computational Approaches to Analyzing Weblogs (2006)
5. Taboada, M., Brooke, J., Tofiloski, M., Voll, K., Stede, M.: Lexicon-based methods for sentiment analysis. Comput. Linguist. 37(2), 267–307 (2011)
6. Turney, P.D.: Thumbs up or thumbs down?: Semantic orientation applied to unsupervised classification of reviews. In: Proceedings of the 40th Annual Meeting on Association for Computational Linguistics, pp. 417–424. Association for Computational Linguistics, Philadelphia (2002)
7. Pang, B., Lee, L., Vaithyanathan, S.: Thumbs up?: Sentiment classification using machine learning techniques. In: Proceedings of the ACL 2002 Conference on Empirical Methods in Natural Language Processing, vol. 10, pp. 79–86. Association for Computational Linguistics (2002)
8. Moraes, R., Valiati, J.F., Gavião Neto, W.P.: Document-level sentiment classification: An empirical comparison between SVM and ANN. Expert Systems with Applications 40(2), 621–633 (2013)
9. Martín-Valdivia, M.-T., Martínez-Cámara, E., Perea-Ortega, J.-M., Ureña-López, L.A.: Sentiment polarity detection in Spanish reviews combining supervised and unsupervised approaches. Expert Systems with Applications 40(10), 3934–3942 (2013)
10. Wan, X.: Bilingual co-training for sentiment classification of Chinese product reviews. Comput. Linguist. 37(3), 587–616 (2011)
11. Wan, X.: Co-training for cross-lingual sentiment classification. In: Proceedings of the Joint Conference of the 47th Annual Meeting of the ACL and the 4th International Joint Conference on Natural Language Processing of the AFNLP, pp. 235–243. Association for Computational Linguistics, Suntec (2009)
12. Balahur, A., Turchi, M.: Comparative experiments using supervised learning and machine translation for multilingual sentiment analysis. Computer Speech & Language 28(1), 56–75 (2014)
13. Banea, C., Mihalcea, R., Wiebe, J.: Multilingual subjectivity: are more languages better? In: Proceedings of the 23rd International Conference on Computational Linguistics, pp. 28–36. Association for Computational Linguistics, Beijing (2010)
14. Prettenhofer, P., Stein, B.: Cross-Lingual Adaptation Using Structural Correspondence Learning. ACM Trans. Intell. Syst. Technol. 3(1), 1–22 (2011)
15. Hajmohammadi, M.S., Ibrahim, R., Selamat, A.: Density Based Active Self-training for Cross-Lingual Sentiment Classification. In: Jeong, H.Y., Yen, N.Y., Park, J.J. (eds.) Advanced in Computer Science and its Applications. LNEE, vol. 279, pp. 1053–1059. Springer, Heidelberg (2014)
16. Pan, J., Xue, G.-R., Yu, Y., Wang, Y.: Cross-Lingual Sentiment Classification via Bi-view Non-negative Matrix Tri-Factorization. In: Huang, J.Z., Cao, L., Srivastava, J. (eds.) PAKDD 2011, Part I. LNCS (LNAI), vol. 6634, pp. 289–300. Springer, Heidelberg (2011)

17. Mihalcea, R., Banea, C., Wiebe, J.: Learning multilingual subjective language via cross-lingual projections. In: Proceedings of the 45th Annual Meeting of the Association of Computational Linguistics, pp. 976–983 (2007)
18. Banea, C., Mihalcea, R., Wiebe, J., Hassan, S.: Multilingual subjectivity analysis using machine translation. In: Proceedings of the Conference on Empirical Methods in Natural Language Processing, pp. 127–135. Association for Computational Linguistics, Honolulu (2008)
19. Wan, X.: Using bilingual knowledge and ensemble techniques for unsupervised Chinese sentiment analysis. In: Proceedings of the Conference on Empirical Methods in Natural Language Processing, pp. 553–561. Association for Computational Linguistics, Honolulu (2008)
20. Moh, T.-S., Zhang, Z.: Cross-lingual text classification with model translation and document translation. In: Proceedings of the 50th Annual Southeast Regional Conference, pp. 71–76. ACM, Tuscaloosa (2012)
21. Shi, L., Mihalcea, R., Tian, M.: Cross language text classification by model translation and semi-supervised learning. In: Proceedings of the 2010 Conference on Empirical Methods in Natural Language Processing, pp. 1057–1067. Cambridge, Massachusetts (2010)
22. Jain, A.K., Duin, R.P.W., Jianchang, M.: Statistical pattern recognition: A review. IEEE Transactions on Pattern Analysis and Machine Intelligence 22(1), 4–37 (2000)
23. Prettenhofer, P., Stein, B.: Cross-language text classification using structural correspondence learning. In: Proceedings of the 48th Annual Meeting of the Association for Computational Linguistics, pp. 1118–1127. Association for Computational Linguistics, Uppsala (2010)
24. Xia, R., Zong, C., Li, S.: Ensemble of feature sets and classification algorithms for sentiment classification. Information Sciences 181(6), 1138–1152 (2011)
25. Brefeld, U., Scheffer, T.: Co-EM support vector learning. In: Proceedings of the Twenty-First International Conference on Machine Learning, p. 16. ACM, Banff (2004)

Learning to Simplify Children Stories with Limited Data

Tu Thanh Vu[1], Giang Binh Tran[2], and Son Bao Pham[1]

[1] University of Engineering and Technology, Vietnam National University, Hanoi
{tuvt,sonpb}@vnu.edu.vn
[2] L3S Research Center, Germany
gtran@l3s.de

Abstract. In this paper, we examine children stories and propose a text simplification system to automatically generate simpler versions of the stories and, therefore, make them easier to understand for children, especially ones with difficulty in reading comprehension. Our system learns simplifications from limited data built from a small repository of short English stories for children and can perform important simplification operations, namely splitting, dropping, reordering, and substitution. Our experiment shows that our system outperforms other systems in a variety of automatic measures as well as human judgements with regard to simplicity, grammaticality, and semantic similarity.

Keywords: text simplification, readability, comparable corpora.

1 Introduction

The Internet contains a tremendous amount of documents written in very different readability levels. That fact causes difficulties in understanding the content for readers. For example, children may understand simple fairy-tales or articles about nature and animals, but may not understand technical reports which are more intended for engineers or scientists. Unfortunately, traditional information retrieval systems have not paid much attention to the comprehension ability of users, and very likely deliver some texts that are beyond their readability levels.

A lot of research in psycholinguistics showed that there is a large number of people struggling to understand general text documents, such as children, hearing/deaf people, aphasic readers or second language learners (e.g., [12]). Hence, it is desirable to adapt and simplify written texts to make them more comprehensible to these individuals. The aim of our work is to provide appropriate reading materials for children, however, people with poor reading comprehension skills can also benefit from such simplified texts.

Lots of efforts put into understanding poor comprehenders found that they fail to master high-level cognitive text processing skills, particularly, (s1) coherent use of cohesive markers such as connectives ('because', 'before', 'after') that signal relations in text, (s2) inference-making from different or distant parts of a text, integrating them coherently (e.g., [8,1]), (s3) detection of inconsistencies in texts ([4,9]). Such reasoning skills are very likely to be causally implicated in the development of deep text comprehension - "integration and inference making are

N.T. Nguyen et al. (Eds.): ACIIDS 2014, Part I, LNAI 8397, pp. 31–41, 2014.

crucial for good text comprehension" ([2]). Hence, a good text simplifier should make text documents clear on these text processing skills.

One of the conspicuous challenges for text simplification is the lack of data available. Current trends are to leverage a large repository of simplified language called Simple English Wikipedia (SEW), a simpler version of Main English Wikipedia (MEW). We can generate a large parallel corpus by pairing sentences from MEW with corresponding sentences from SEW (e.g., [13]), and then explore data-driven methods to learn simplification models. However, there is no guarantee that the generated corpus will satisfy (s1), (s2), and (s3). It seems to be more suitable for sentence simplification purpose rather than document simplification purpose. Recently, there is such a corpus of high quality made available by TERENCE[1] consortium which contains a set of English stories[2], of which each is simplified (by experts) into several versions with different reading difficulty levels that reflects (s1), (s2) and (s3). The corpus appears to be a valuable resource for learning simplifications which can mimic simplifying procedures, mirroring high-level text processing skills. Nonetheless, its small size makes it challenging for previous methods in learning efficient simplification models. The work presented in this paper investigates the use of the afore-mentioned corpus to learn a Bayesian probability model for simplifying children documents (or stories).

Summary of Contribution:

- A Bayesian-based framework learnt from limited data for simplifying children stories.
- A method that leverages shallow parsing rather than full parsing in defining the syntax of sentences to improve the performance of simplification model.
- Strategies for rule generalization and manual revision of simplification rules to maximize the benefits of data, especially for small-scale dataset.
- We do quantitative and qualitative evaluations of the proposed methods in comparison with the-state-of-the-art methods.

2 Related Work

Most earlier work in text simplification utilized hand-crafted rules to split long and complicated sentences into several simpler ones ([3]). Other work focuses on lexical simplifications and substitutes low-frequency words by more common synonyms derived from WordNet, or paraphrases found in a predefined vocabulary list ([5]) or their dictionary definitions ([6]). The task can be treated as a monolingual machine translation task with the complex sentence as the source and the simple one as the target and make use of large-scale parallel corpora of paired articles from SEW and MEW for training models ([16,14]). Wikipedia revision history is also an useful resource for learning simplifications ([15]).

[13] investigate the use of an aligned MEW-SEW corpus and SEW edit histories to learn a model based on Quasi-synchronous grammar then use an integer linear programming model to select the most appropriate simplification. Our

[1] `www.terenceproject.eu` - An European project supporting poor comprehenders and their educators).

[2] `http://www.terenceproject.eu/repository/booken/booken.html`

work is similar to [13] in learning rewrite rules which involve sentence splitting, syntactic and lexical simplification. However, rather than exploiting Wikipedia, we examine children documents i.e., stories created and rewritten into different complexity levels by experts. For each type of rules, we suggest a general form suitable for text analysis processes. More concretely, we use shallow parsing rather than full parsing in defining the syntax of sentences for less error-prone. Furthermore, we propose some strategies for rule generalization and manually revise all of rewrite rules to maximize the benefits of the small-scale dataset. Our approach also differs from [13] in the sense that instead of using integer linear programming, we rely on a Bayesian probability model to identify the most appropriate simplification from the space of possible ones.

3 Simplification Dataset

We construct a simplification dataset from two story levels: Level 1 and Level 2 published in TERENCE story repository. Stories at Level 2, which are inherited from all simplifications that reflects high-level text processing skills of human (coherence and cohesion levels), are then simplified on lexicon and grammar to create corresponding stories at Level 1. We therefore consider stories at Level 2 as the original texts, and corresponding stories at Level 1 as the simplified texts. To generate simplification dataset, we first pair each story at Level 2 to its simpler one and then represent them as list of continuous sentences, of which each is pre-processed by tokenization/POS-tagging with the GATE[3] toolkit and the Stanford Parser package ([7]). At the sentence level, we modify METEOR[4] to build up an automatic sentence alignment module based on exact, stem, synonym, and paraphrase matches between words and phrases in sentences. To achieve the best sentence alignment on our dataset, we manually revise all automatically generated alignments to recognize bad ones. Finally, we obtain a list of 1-1 pairs in which each sentence at Level 2 is aligned with only one sentence at Level 1, and 1-n pairs in which each sentence at Level 2 is aligned with several sentences at Level 1. Table 1 shows some statistics for our dataset.

Table 1. Statistics for the simplification dataset. #Sen.Pairs: the number of sentence pairs; #(1-n).Sen.Pairs: the number of 1 to n sentence alignments; Avg.Sen.Len, Avg.Token.Len: the average sentence length and average token length, respectively

#Sen.Pairs	#(1-n).Sen.Pairs	Avg.Sen.Len		Avg.Token.Len	
		Level 2	Level 1	Level 2	Level 1
1050	46	17.17	16.63	3.94	3.80

[3] `http://gate.ac.uk/`

[4] An automatic system for machine translation evaluation (available at `http://www.cs.cmu.edu/~alavie/METEOR/`).

4 Sentence Simplification Model

Our model operates on individual sentences. The simplification process first considers whether or not to split a given original sentence into several shorter ones, then may drop or reorder its components, or substitute difficult words with their simpler synonyms or paraphrases. We rely on a Bayesian probability model to identify the most appropriate simplification for the input sentence. Our model architecture is illustrated in Figure 1.

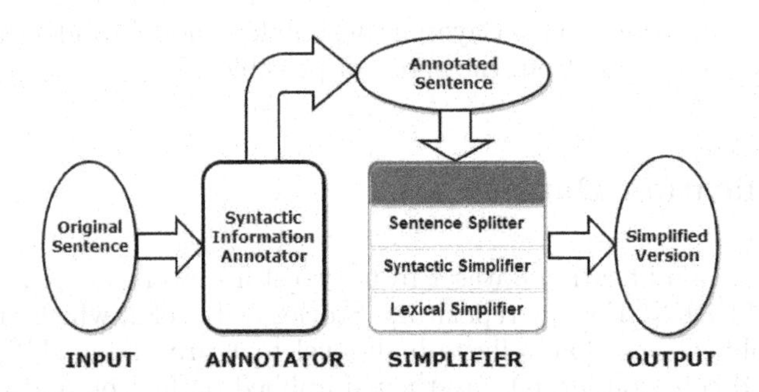

Fig. 1. Sentence simplification model architecture

4.1 Rule Generation

We annotate sentences with syntactic information in the form of phrase structure trees. For each parse tree pairs $T1 \rightarrow T2$, corresponding to an aligned sentence pair $S1 \rightarrow S2$ from the training dataset, we further align nodes in $T1$ with corresponding nodes in $T2$ as follows: first, we construct a list of leaf node alignments from the results of METEOR toolkit as used in preparing simplification data. Next, we recursively align the parent of aligned nodes in $T1$ with corresponding ones in $T2$, as showed in Algorithm 1. After that, rewrite rules are

Algorithm 1: Parent node alignment.

```
ParentNode node1, node2;
for each (node1 in T1, node2 in T2) if (!isAligned(node1, node2))
   if (node1.numChildren() == 1 && node2.numChildren() == 1)
      if(numChildNodeAlignment(node1, node2) == 1)
         align(node1, node2);
   else
      if(numChildNodeAlignment(node1, node2) > 1)
         align(node1, node2);
```

generated from aligned nodes by considering tree-to-tree (lexical and structural) transformation patterns. In what follows we detail how we extract rewrite rules.

(a) Sentence Splitting. Sentence splitting rules are created from 1-n aligned parse tree pairs ($T \rightarrow T1.T2...Tn$), corresponding to 1-n aligned sentence pairs ($S \rightarrow S1.S2...Sn$) in which each sentence at Level 2 is aligned with two or more consecutive sentences at Level 1. These rules allow long and syntactically complicated sentences to be split into several shorter ones. In our work, we use shallow parsing also called "chunking" to describe the syntax in sentences. Concretely, constituents or word groups such as noun and verb phrases are identified, but their internal structures are ignored. This should help the performance of the system become more robust and less error-prone. To illustrate, we use a running example for a given sentence pair:

"Lewis installed big electric fans, while Charles installed big air conditioners." → "Lewis installed big electric fans. In the meanwhile, Charles installed big air conditioners."

After aligning corresponding nodes in the aligned parse tree pair and chunking, we obtain (subscripts show aligned nodes):

[NP Lewis]$_{[1]}$ [VP installed]$_{[2]}$ [NP big electric fans]$_{[3]}$, [SBAR while] [NP Charles]$_{[4]}$ [VP installed]$_{[5]}$ [NP big air conditioners]$_{[6]}$. → [NP Lewis]$_{[1]}$ [VP installed]$_{[2]}$ [NP big electric fans]$_{[3]}$. [PP In] [NP the meanwhile] , [NP Charles]$_{[4]}$ [VP installed]$_{[5]}$ [NP big air conditioners]$_{[6]}$.

A sentence splitting rule can be generated automatically as follows:

NP$_{[1]}$ VP$_{[2]}$ NP$_{[3]}$, while/SBAR NP$_{[4]}$ VP$_{[5]}$ NP$_{[6]}$.

→ NP$_{[1]}$ VP$_{[2]}$ NP$_{[3]}$. In the meanwhile, NP$_{[4]}$ VP$_{[5]}$ NP$_{[6]}$.

(b) Syntactic Simplification. We can drop or reorder components in an original sentence to make it more concise or easier to comprehend than the original. To learn rules for these operations, we consider all aligned sub-tree pairs, where the syntactic structures of the original and simplified sub-trees do not match. We also use shallow parsing technique to render more general resulting rules. For example,

[VBD painted] [NP the car]$_{[1]}$ [JJ white]$_{[2]}$
→ [VBD made] [NP the car]$_{[1]}$ [IN with] [DT the] [JJ white]$_{[2]}$ [NN paint].

A syntactic rule is established automatically as follows:

painted/VBD NP$_{[1]}$ JJ$_{[2]}$ → made NP$_{[1]}$ with the JJ$_{[2]}$ paint.

(c) Lexical Simplification. Lexical simplification rules are learned from leaf node alignments as described above, and also from aligned sub-tree pairs in which the original and simplified sub-trees have the same syntactic structure and only one lexical difference exists. We only extract the words and corresponding POS tags involved rather than take the syntactic context into consideration. For example, "trophy/NN → cup", "turquoise/NN → deep blue", "stunned/VBD → shocked".

Rule Manual Revision. There are still a number of bad rules generated in the process of rules automatic generation. The underlying cause may be due to wrong alignments in the previous steps. In the task of text simplification, we must create simplifications that reduce the reading difficulty of the input text, while maintaining grammaticality and preserving its meaning. To satisfy these constraints, the rewrite rules need to be as good as possible and should be meticulously evaluated before application. We therefore revise all the automatically generated rules to recognize bad ones. Bad rules are then manually adjusted to acquire good versions or removed directly from the training data if necessary.

Rule Generalization. We examine some strategies for generalizing rewrite rules to enrich the training data. Our strategies are mainly based on word form, word similarity, context similarity and grammatical roles. For examples, from a lexical rule "began → started" , we can further obtain other ones (based on word form), in particular, "begin → start", "begun → started", "begins → starts", "beginning →

starting". Similarly, for a syntactic rule "shrunk PRP → made PRP smaller" extracted from an alignment that is "shrunk her → made her smaller" , we can place PRP (personal pronoun) with NP (noun phrase) as well as make the use of different word forms to produce a series of reliable rules as follows:

{"shrink PRP → make PRP smaller", "shrunk PRP → made PRP smaller", "shrinks PRP → makes PRP smaller", "shrinking PRP → making PRP smaller"},

{"shrink NP → make NP smaller", "shrank NP → made NP smaller", "shrunk NP → made NP smaller", "shrinks NP → makes NP smaller", "shrinking NP → making NP smaller"}.

4.2 Rule Identification and Application

Where there is more than one possible simplification, we identify the best sequence of simplifications. In our work, rewrite rules are all written in a general form, LHS → RHS. In each rule, the LHS describes syntactic information of a matching object while the RHS presents its simplified version. Therefore, in a space of all possible rules, we appreciate rules in which the LHS describes more details about syntactic information of objects than in other possible rules. We further construct a function to assess the level of detail of each rule. Besides, we also apply a Bayesian probability model to identify the most likely rules among different possibilities. Specifically, the probability of a rule LHS → RHS from the training data is estimated as the product

$$\mathcal{P}(\mathbf{LHS}).\mathcal{P}(\mathbf{LHS} \rightarrow \mathbf{RHS}|\mathbf{LHS})$$

Note that $\mathcal{P}(\text{LHS}) = \frac{t_{LHS}}{n}$, where n is the total number of rules and t_{LHS} is the number of times LHS appears in the training data. The conditional probability $\mathcal{P}(LHS \rightarrow RHS|LHS)$ is the probability of some event LHS → RHS, given the occurrence of some other event LHS.

5 Experiments

5.1 Experiment Setup

32 English stories from the golden data were randomly partitioned into a training set (25 stories) and a test set (7 stories). We extracted 24 splitting rules, 298 syntactic rules and 464 lexical rules from the training set. The test set contains 230/238 original/simplified sentences excluded from training.

The performance of our system was compared to one of the state-of-the-art systems, namely [13]'s system[5] (**Woodsend**) which learns simplications from Wikipedia (revision histories and MEW-SEW aligned corpus), and a simple baseline (**SpencerK**) that merely relies on lexical substitutions provided by the SEW editor "SpencerK" (Spencer Kelly)[6]. We also included simplifications created by experts as a gold standard (**Experts**).

---

[5] `http://homepages.inf.ed.ac.uk/kwoodsen/demos/simplify.html`

[6] We are grateful to Spencer Kelly for providing us with his list of simple words and simplications.

5.2 Evaluation

We evaluated the system's output in two ways, using automatic evaluation measures and human judgements as follows:

Automatic Evaluation Measures. In line with previous work on text simplification (e.g., [16] [13]), we also used a variety of measures ranging from basic statistics such as the average length of tokens, the average number of tokens in one sentence, the total number of simplified sentences, to more complex measures such as OOV%, FKGL, BLEU, TERp.

Human Judgements. We compared sentences generated by experts (the gold standard) and the three automated systems on our test set of 230 sentences. In total, our material consists of 920 (230 × 4) sentences. We conducted two experiments. First, we randomly selected 80 (20 × 4) original-simplified sentences, corresponding to 20 original sentences from our materials, to manually analyse their quality. Second, we recruited 15 participants to judge the quality of sentences from our material, without awareness of which method was used to generate them. Each participant was asked to rate (in five-point scale) whether each simplified sentence is simpler than the original, is good at grammar, and preserves the main content of the original, respectively. All of participants are college students who are proficient in English and committed to bringing us objective and serious assessments.

5.3 Results and Discussion

Basic Statistics. The first four columns in Table 2 show the results of basic statistics. All simplified sentences get better scores than the input sentences in terms of the average length of tokens (*tokLen*) which may roughly reflect the lexical difficulty. This indicates that at the lexical level, long and difficult words were partly substituted by simpler and shorter ones. The results of this score also show that our system performed lexical simplification better than Woodsend and SpencerK, however it was still not as good as Experts.

Table 2. System performance using basic statistics and automatic evaluation measures

	tokLen	senLen	senLen#S	#sen	OOV%	FKGL	FKGL#S	BLEU	TERp
input	4.01	16.47	13.82	230	59.95	6.78	5.82	100.00	0.000
Experts	**3.86**	15.98	13.92	238	59.57	6.11	**5.33**	**71.86**	**0.207**
OurSystem	3.91	14.92	**13.80**	260	**58.78**	5.87	5.39	74.31	0.090
Woodsend	3.94	**11.74**	13.82	384	62.20	**5.20**	5.81	86.64	0.075
SpencerK	4.00	16.49	13.84	230	59.53	6.77	5.79	96.20	0.015

#sen gives the total number of simplified sentences. We performed the splitting operation a little more than Experts (22 sentences, nearly 10% of the number of the original sentences). Woodsend is much different from the others with 96 simplified sentences. This indicates that our system generates closer output to that of Experts than that of Woodsend. In fact, Experts only performed the splitting operation on 8 sentences in all of the input ones. This small number

indicates that on the story dataset we examine, sentence splitting does not really play a more important role than the other operations.

We now turn to the results of the average number of tokens in one sentence (*senLen*) which roughly reflects the syntactic complexity. All systems except SpencerK usually rendered shorter sentences corresponding to the original ones. This suggests that complex sentences were partly simplified into ones with simpler syntactic structures. SpencerK has the highest *senLen*, showing it performs mostly lexical substitutions. The splitting operation has a significant impact on the results of *senLen* above as this operation reduces the average length, and therefore reduces the syntactic complexity of sentences. To better understand this impact, we measured the average length of simplified sentences that are not generated through the splitting operation (*senLen#S*). The result indicates that our model produced shorter sentences than the others. However, there is no significant difference between all systems in terms of *senLen#S*. Basically, in the absence of the splitting operation, the number of tokens in one sentence is nearly identical for the output of the systems.

As Experts did not produce shorter sentences than the other systems, we learn out that in sentence simplification problem, simplified sentences that are easier to comprehend are not necessarily shorter than the original ones. This notion is also perceived in some previous work (e.g., [13]) or in the SW guidelines[7]

Readability and Translation Assessment. The results of the automatic measures are also displayed in Table 2. OOV is the percentage of words that are not mentioned in the Basic English 850 Words list[8], where lower scores mark sentences that are simpler to read. The results of OOV% show that our system used the most basic English words as it scores best, with 58.58% even lower than Experts. SpencerK also has a good score by dint of lexical substitutions. However, this do not assert that Experts is not good because the core elements here are whether basic words used are reasonable or not and whether their nuanced meanings are preserved. In some cases, lexical substitutions make distortions on the original meaning of sentences.

In terms of the Flesch-Kincaid grade level score[9] (FKGL) which estimates readability, higher scores indicate more complex sentences. Generally, this score is under the major impact of the splitting operation since the fact that it significantly reduces when the total number of sentences increases. However, with regard to the FKGL score for simplified sentences that are not generated through the splitting operation (FKGL#S), Experts scores best with 5.33 (lowest), closely followed by our system (5.39). Again, our system is the closest to Experts and altogether outperforms Woodsend (5.81) and SpencerK (5.79). In the absence of splitting operation, Woodsend has the highest FKGL score compared to the other systems. This roughly reflects that this system uses a considerable number of splitting operations. Besides, the original sentences have the highest reading level.

[7] `http://simple.wikipedia.org/wiki/Main_Page/Introduction`

[8] `http://simple.wikipedia.org/wiki/Wikipedia:Basic_English_alphabetical_wordlist`

[9] $FKGL = 0.39(\frac{total\ words}{total\ sentences}) + 11.8(\frac{total\ syllables}{total\ words}) - 15.59.$

Both BLEU ([10]) and TERp ([11]) are commonly used for automatic machine translation evaluation. In our experiments, BLEU scores the simplified sentences by counting n-gram matches with the original and thus lower BLEU scores are better. TERp measures the number of edit operations (insertion, deletion, substitution, and shift) needed to transform simplified sentences into the original and thus higher TERp scores are better. In both these scores, Experts is the best, followed by our system, Woodsend, and SpencerK. Experts and our system are significantly different from the others. These results also show how close to the gold standard our system is, and once again confirm that our system is more flexible than Woodsend, and SpencerK in using words.

Quality Analysis. The results of this experiment indicate that our system is the closest to the gold standard. Woodsend performed many splitting operations, however most of them made the content of the original sentences become more disjointed and less coherent in fact. Table 3 shows examples of simplifications produced by the systems (we ignore the output of SpencerK that is generally similar to the original sentence). The results also verify that our system can create simplifications that significantly reduce the reading difficulty of the input text while maintaining grammaticality and preserving its meaning.

Table 3. Examples of simplifications produced by the systems

input #1	Effy took the microphone and began to sing.
Experts	Effy took the microphone and began to sing.
OurSystem	Effy took the microphone and started to sing.
Woodsend	Effy took the microphone. It started to sing.
input #2	So she had spent her whole camping holiday practising her singing.
Experts	So she had spent her whole camping holiday practising her singing.
OurSystem	So she had spent her holiday practising her singing.
Woodsend	So she had spent her whole camping holiday practising her singing.
input #3	She smiled because she was surprised at the question, and said it was the Mayor's decision.
Experts	She smiled because she was surprised at the question , and said it was because of the Mayor's decision.
OurSystem	She smiled . In this way, she was surprised at the question, and said it was the Mayor's decision.
Woodsend	She smiled because she was surprised at the question. He said it. It was the Mayor 's decision.

Table 4. Average human ratings for the output of the systems

Systems	Simplicity	Grammaticality	Semantic Similarity
Experts	3.78	4.91	4.88
OurSystem	2.99	4.37	4.02
Woodsend	2.74	4.05	3.83
SpencerK	1.52	4.85	4.82

Average Human Ratings. Table 4 presents the average human ratings for the output of the four systems. With regard to simplicity score, Experts scores highest with 3.78, followed by our system (2.99) and Woodsend (2.74). SpencerK scores lowest with 1.52 as this baseline solely focuses on lexical substitutions.

With regard to grammaticality and semantic similarity, again Experts is rated highest. SpencerK is also rated highly since the fact that it does not change the syntactic structure as well as the main content of the original sentences. Our system continues to score higher than Woodsend.

6 Conclusions

This paper investigates English document simplification task on the domain of children stories. Differently to previous approaches that rely on large-scale parallel corpora derived from MEW and SEW articles, we presented a model which extracts and generalizes simplification rules from a small corpus of 32 stories written by experts and then applies a Bayesian model to select the optimal set of rules to make a text simpler. The corpus is small but reflects different reading comprehension processing skills, making our model promising in generating texts that are suitable for children. Our model is able to perform important simplification operations, namely splitting, dropping, reordering, and substitution. The evaluation shows that our model is close to the gold standard created by experts, and can achieve better results than baseline and state-of-the-art systems in a variety of automatic measures as well as human judgements.

Acknowledgments. We would like to thank The Vietnam National Foundation for Science and Technoloogy Development (NAFOSTED) for financial support.

References

1. Cain, K., Oakhill, J.V., Elbro, C.: The ability to learn new word meanings from context by school-age children with and without language comprehension difficulties. Journal of Child Language 30, 681–694 (2003)
2. Cain, K.: Making sense of text: Skills that support text comprehension and its development. Perspectives on Language and Literacy 35, 11–14 (2009)
3. Carroll, J., Minnen, G., Pearce, D., Canning, Y., Devlin, S., Tait, J.: Simplifying text for language-impaired readers. In: Proceedings of EACL, pp. 269–270 (1999)
4. Ehrlich, M.F., Remond, M., Tardieu, H.: Processing of anaphoric devices in young skilled and less skilled comprehenders: Differences in metacognitive monitoring. Reading 11, 29–63 (1999)
5. Inui, K., Fujita, A., Takahashi, T., Iida, R., Iwakura, T.: Text simplification for reading assistance: A project note. In: Proceedings of IWP, pp. 9–16 (2003)
6. Kaji, N., Kawahara, D., Kurohash, S., Sato, S.: Verb paraphrase based on case frame alignment. In: Proceedings of ACL, pp. 215–222 (2002)
7. Klein, D., Manning, C.D.: Accurate unlexicalized parsing. In: Proceedings of ACL, pp. 423–430 (2003)
8. Oakhill, J., Cain, K.: Inference making and its relation to comprehension failure. Reading and Writing An Interdisciplinary Journal 11, 489–503 (1999)
9. Oakhill, J., Hartt, J., Samols, D.: Levels of comprehension monitoring and working memory in good and poor comprehenders. Reading and Writing 18, 657–686 (2005)
10. Papineni, K., Roukos, S., Ward, T., Zhu, W.J.: Bleu: A method for automatic evaluation of machine translation. In: Proceedings of ACL, pp. 311–318 (2002)
11. Snover, M.G., Madnani, N., Dorr, B., Schwartz, R.: Ter-plus: Paraphrase, semantic, and alignment enhancements to translation edit rate (2010)

12. Wauters, L.N., Bon, W.H.J., Tellings, A.E.J.M.: Reading comprehension of dutch deaf children. Reading and Writing 19, 49–76 (2006)
13. Woodsend, K., Lapata, M.: Learning to simplify sentences with quasi-synchronous grammar and integer programming. In: Proceedings of EMNLP, pp. 409–420 (2011)
14. Wubben, S., van den Bosch, A., Krahmer, E.: Sentence simplification by monolingual machine translation. In: Proceedings of ACL, pp. 1015–1024 (2012)
15. Yatskar, M., Pang, B., Danescu-Niculescu-Mizil, C., Lee, L.: For the sake of simplicity: Unsupervised extraction of lexical simplifications from wikipedia. In: Proceedings of NAACL-HLT, pp. 365–368 (2010)
16. Zhu, Z., Bernhard, D., Gurevych, I.: A monolingual tree-based translation model for sentence simplification. In: Proceedings of COLING, pp. 1353–1361 (2010)

Clustering Based Topic Events Detection on Text Stream

Chunshan Li[1,2], Yunming Ye[1,2], Xiaofeng Zhang[1,2],
Dianhui Chu[3], Shengchun Deng[3], and Xiaofei Xu[3]

[1] Harbin Institute of Technology, Shenzhen Graduate School, China
[2] Shenzhen Key Laboratory of Internet Information Collaboration, China
[3] Department of Computer Science, Harbin Institute of Technology
lichunshan.hit@gmail.com,
{yeyunming,chudianhui,dengshengchun,xuxiaofei}@hit.edu.cn,
zhangxiaofeng@hitsz.edu.cn

Abstract. Detecting and tracking events from the text stream data is critical to social network society and thus attracts more and more research efforts. However, there exist two major limitations in the existing topic detection and tracking models, i.e. noise words and multiple sub-events. In this paper, a novel event detection and tracking algorithm, topic event detection and tracking (TEDT), was proposed to tackle these limitations by clustering the co-occurrent features of the underlying topics in the text stream data and then the evolution of events was analyzed for the event tracking purpose. The evaluation was performed on two real datasets with the promising results demonstrating that (1) the proposed TEDT algorithm is superior to the state-of-the-art topic model with respect to event detection; (2) the proposed TEDT algorithm can successfully track the event changes.

Keywords: social media, event detection, temporal analysis, topic model.

1 Introduction

With the social networking society blossomed into pervasive aspect of human beings, there exist a huge volume of text stream, such as blog posts and tweets. However, the emergence of the important events generally was quickly overwhelmed by the huge amount of non-important events due to the imbalance volume of posts. Therefore detecting these imbalanced but important events has become a hot research issue. Traditional event detection techniques generally involve two major approaches: topic detection and tracking (TDT) [1,2,3] and temporal text mining [4,5].

To detect and track events, text stream data is generally defined as a sequence of chronologically ordered documents, and topic event is a set of co-occurrent words within a short period of time. For example, the 2008 presidential campaign in the USA was reported in several news stories, which could be seen as the text stream data. Naturally, these documents were divided into several subgroups based on their timestamp. Then events detection and tracking algorithm groups words in each time window, e.g. words "obama, speech, iowa", appeared in t, and words "obama, hillari, support, vote" appeared in $t+1$. Subsequently, event words in different time windows could be grouped together into several specific events, i.e. (1) "Barack Obama speak at a campaign rally", (2)"Hillary Clinton leads the list of most admired women with 20%, according to the

N.T. Nguyen et al. (Eds.): ACIIDS 2014, Part I, LNAI 8397, pp. 42–52, 2014.

vote". In the literature, different methods have been proposed to detect events. The topic model [6] was adopted to detect topic by generating multiple topics from text corpus at the same time and each topic, consisting of several group of keywords, can be treated as one event. However, there exist two major drawbacks if we adopt topic model to detect events. First, topic model constructs an event by choosing top k topic keywords manually and thus brings the noisy words. Second, a topic generally includes more than one event. For example, topic model discovers an event containing words like "attack, soldier, militari, luzon, provinc, troop, southern, kill, area, libyan". In this event, it is easily to extract two sub-events: "Libyan people were killed" and "soldier were attacked at island in the southern Luzon".

In this paper, we investigated the problem of detecting and tracking topic event by extending topic model, and proposed topic event detection and tracking(TEDT) algorithm. The main difference between the proposed TEDT algorithm and topic model is how an event is generated. The topic model assumes that an event can be represented by a group of words having similar semantic topics. The proposed TEDT algorithm assumes that the co-occurrence relationship and the semantic topic of words work together to generate an event. In particular, for the TEDT algorithm, the topic features are clustered to generate probabilistic events using a revised topic based approach. The remaining of the paper is organized as follows. Section 2 reviews the related works. The TEDT algorithm is proposed in Section 3. Experiments and evaluation results are given in Section 4. Section 5 concludes the paper.

2 Related Work

The event detection generally involves with several research domains, such as topic detection and tracking [7,8,9], text clustering [10,11,12,13,14] and temporal data analysis. The *topic detection and tracking based approach* is originally proposed to discover the topic hidden in the stream of news stories [7]. Then, Loulwah et al. employed KL-divergence to measure difference among topic over timeline and a threshold was used to detect the occurrence of events [15]. Mei et al. proposed mixture topic model to extract themes (hidden events) [8]. However, these topic-based approaches pay less attention to co-occurrence relation of corpus words and thus either make the generated topics contain noisy words or include fewer sub-topics.

The *clustering based approach* is developed for text mining in its own right and is now adopted for event detection and tracking. To cope with the huge amount of data, authors in [12] treated the words and their co-occurrence as a network and the graph partition algorithm was adopted to discover the densely connected subgraphs as the events. Lin et al. proposed a combined model for event detection from text and community features [10]. In [14], Twitter tags and spatio-temporal features worked together to discover events. Yao et al. [11] proposed to detect the burst of single word and their approach could be extended for the detection of multiple words but at much higher cost. Fund et al. [13] designed a parameter free word clustering approach, called HB-Event, to detect event, which clustered words by co-occurrence relation among words in corpus.

3 Topic Event Detection and Tracking

Although these existing approaches did not resolve aforementioned limitations, they motivate us to extend current topic model by filtering noisy features and unimportant sub-events. In this section, we first formulate the problem of topic event detection, then propose the detailed process to detect topic events. Finally, we discussed how to tract the change of detected events by topics of the events.

3.1 Problem Formulation

Given the text stream data $D = \{d_1, d_2, \ldots, d_i, \ldots\}$, where d_i is a document containing a unique timestamp t_i. All words in d_i are extracted out to form a vocabulary W, where $W = \{w_1 w_2 \ldots\}$. In D, if $i < j$, then $t_i \leq t_j$. The text stream D can be split into L non-overlapping time windows. The topic events detection is trying to find a set of topic events within each time slot, and a topic event in l_i consists of a set of topic features with the highest co-occurrence probability of extracted words. The tracking of topic events is to identify the related events in all timeslots. The Figure 1 explains the proposed word clustering based approach. We first proposed the probabilistic clustering approach to discover the topic events in text stream. Then, we utilized topics of events to track the change of events.

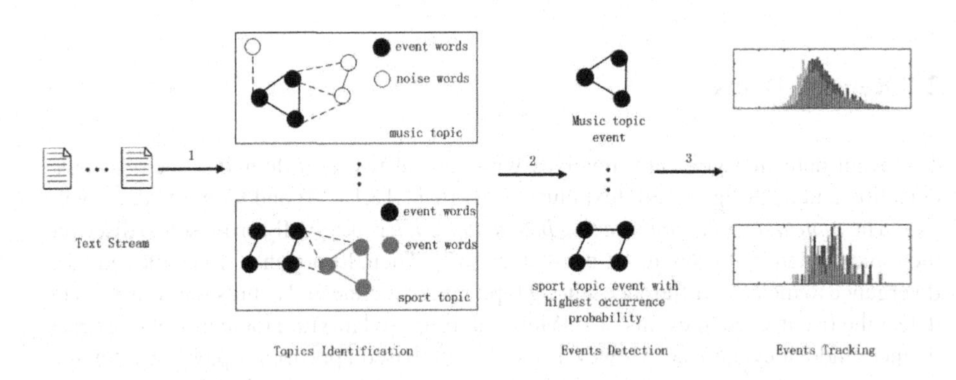

Fig. 1. The Overview of Topic Event Detection and Tracking

3.2 Topic Events Detection

After running online topic model, several sets of topic features were generated in each timeslot. A feature set of topic i can be defined as a word sequence $W_i = \{w_0, \ldots, w_m, w_n, \ldots\}$, in which, if $m < n$, then $probability(w_m|T_i) > probability(w_n|T_i)$. In traditional event detection method, the top k words in W_i are considered as a topic event. To filter the noisy features, these features in W_i will be clustered to form more accurate topic events. Details about the clustering procedure are described as follows.

Let the feature set of topic i be $W_i^l = \{w_0, w_1, ..., w_n\}$, l represents the current time, n is the number of words. A topic event E_i^l in l consists of several topic features and can be defined as $E_i^l = \{\ldots, w_i, w_j, \ldots\}$. E^l represents the set of detected events in l. For example, suppose $W_i^l = \{moive, music, titanic\}$, $E_i = \{moive, titanic\}$ implies that the words in event $\{moive, titanic\}$ have higher co-occurrence probability than $\{music\}$, $\{moive\}$ and $\{titanic\}$ in topic i. In order to find the optimal feature group in W_i^l, we formalize the problem of determining the cluster of topic features under a probabilistic framework, which is to discover occurrent event E_i^l with the highest probability. The objective function is defined as: (see Table 1 for the adopted notations).

$$\begin{aligned} &\max P(E_i|D,T_i) = \max\{P(\mathrm{w}_0,\mathrm{w}_1,...,\mathrm{w}_k|D,T_i)\} \\ &= \max\{\textstyle\prod_{m=1}^{M}\prod_{n}^{|E_i|} f_{m,n}p(\mathrm{w}_{m,n}|z_{m,n,i})p(z_{m,n,i}|d_m)p(d_m)\} \end{aligned} \quad (1)$$

where $z_{m,n,i}$ represents the topic of word $w_{m,n}$ and $f_{m,n}$ is the frequency of $w_{m,n}$.

Table 1. Notations used in the TEDT algorithm

Parameter	description	Parameter	description
T_i	topic i	W_i^l	feature set of topic i in timeslot l
E_i	event i	D	the text corpus
d_m	a document in D	$w_{m,n}$	the word in document d_m for event n
$z_{m,n,i}$	the topic of word $w_{m,n}$		

Taking the logarithm, then the maximization of Eq. 1 is equivalent to maximize:

$$\begin{aligned} &\max\{\log P(E_i|D,T_i)\} \\ &= \max\{\log(\prod_{m=1}^{M}\prod_{n}^{|E_i|} f_{m,n}p(w_{m,n}|z_{m,n,i})p(z_{m,n,i}|d_m)p(d_m))\} \\ &= \max\{\sum_{m}^{M}\sum_{n}^{|E_i|}\log(f_{m,n}p(w_{m,n}|z_{m,n,i})p(z_{m,n,i}|d_m)p(d_m))\} \\ &= \max\{\sum_{m}^{M}\sum_{n}^{|E_i|}(\log f_{m,n} + \log p(w_{m,n}|z_{m,n,i}) + \log p(z_{m,n,i}|d_m) \\ &+ \log p(d_m))\} \end{aligned} \quad (2)$$

Eq. 2 defines the priority of events. In Eq. 2, the document-topic distribution $p(z_{m,n,k}|d_m)$ and topic-word distribution $p(bw_{m,n}|z_{m,n,k})$ will be computed by online topic model [16], which is guaranteed to converge and is able to find comparably precise topics as the original topic model does. Through the preliminary experiments, the number of the topic in our experiments is set to be 20 to get the optimal results.

The proposed TEDT is illustrated in Algorithm 1. The algorithm returns a list of events $\{E_1, \ldots, E_k\}$, where k is the number of topic. The following example will show how the TEDT works with event detection. Suppose $W_i = \{moive, music, titanic\}$, then $p1 = probability(moive|D, T_i)$, $p2 = probability(music|D, T_i)$, $p3 = probability(moive, music|D, T_i)$ is calculated according to the Eq. 2. If $p1 > p3$ and $p2 > p3$, then $\{moive\}$ and $\{music\}$ are considered as two independent events.

If $p3 = probability(moive, titanic|D, T_i)$, which means term "titanic" is considered, then we have $p3 > p1$, $p3 > probability(music, titanic|D, T_i)$ and $p3 > probability(titanic|D, T_i)$. Obviously, the event $\{moive, titanic\}$ has the highest co-occurrence probability in topic i, which is the target event to be extracted out.

Algorithm 1. TEDT algorithm

```
Require: a set of topic feature TW_i, the text corpus D,
   a document-topic distribution matrix θ_{M×T},
   a topic-word distribution matrix β_{T×W}
Ensure: topic event E_i
1: for each w_j ∈ W  do
2:    p1 = log(p(E_i|D, T_i))
3:    p2 = log(p(w_j|D, T_i))
4:    E_i.add(w_j)
5:    p3 = log(p(E_i|D))
6:    p = max{p1, p2, p3}
7:    if p1 == p then
8:       E_i.remove(w_j)
9:    else if p2 == p then
10:        E_i.clear()
11:        E_i.add(w_j)
12:    end if
13: end for
```

In section 3.2, k topic events were already extracted out which verifies the assumed existence of one to one mapping relationship between topic events and topic. In [16], Blei et al. demonstrated that topics in the online topic model is consistent over the timeline, i.e. topic i in l contains correlated content to topic i in $l + 1$. Therefore, correlated topic events can be tracked by the consistency of topics.

4 Empirical Study

To evaluate the performance of the proposed TEDT, the online topic model and the Hot-Bursty-Event detection (HBE) approach [13], are implemented for performance comparison. In online topic model, the top 10 words in topic feature set are considered as an event. The HBE algorithm only considers the co-occurrence relation between topic features to generate topic event. Two real data sets, named *reuters* and *blog*, are used in the experiments. The *reuters* data set contains all news stories collected from 1987-2-26 to 1987-10-20. The general preprocessing steps are performed on *reuters*, such as stopword removal and stemming. Words whose term frequency is less than 3 were filtered out. After all these steps, the data set consists of 19,065 documents and 19,644 distinct words. The *blog* data set contains blog posts collected from 2008-1-1 to 2008-3-5. There are 2 categories in the *blog* which are "technologies" and "politics". After similar pre-processing steps, the date set used in the experiments is composed of 31,831 documents and 48,747 distinct words.

4.1 Performance Evaluation for Event Cohesiveness

Measuring the quality of detected events is a difficult task. Many works employed manual judgement to determine the performance of event detection approach [11]. However, such judgement is extremely expensive and time-consuming, thus is infeasible in reality. Yao et al. [12] proposed a criterion called "bursty cohesiveness", which estimates the performance of approach by quantifying the co-occurrent event words. Similar to Yao's approach, a new metric, called event cohesiveness, is designed to evaluate the quality of detected events. The event cohesiveness assumes event words are generally co-occurrent in the same document and their document frequency is also high enough. By removing these highly co-occurrent words, the event frequency drops quickly. If the co-occurrence of words in an event is higher, then the event is more cohesive. The event cohesiveness is therefore defined as:

$$conhesive(E_i) = \sum_{k}^{|E_i|} (1 - \frac{\text{independence}(bw_k)}{freqd(bw_k)}) \quad (3)$$

where E_i represent a specific event, w_k is an event word in E_i. independence(w_k) indicates the frequency of w_k after removing the co-occurrent words, and $freq(w_k)$ is the frequency of w_k. Subsequently, the cohesiveness of detected event within a timeslot can be estimated as

$$average\_Conhes(E) = \sum_{E_i \in E} \frac{conhesive(E_i)}{|E|}. \quad (4)$$

4.2 Experiments on Event Cohesiveness

In this section, the event cohesiveness is used to evaluate the quality of detected event topics. The cohesiveness of both online topic model and HBE algorithm [13] will be calculated for the comparison. The experimental results are plotted in Figure 2. As can be seen, the value of the TEDT is always higher than that of the HBE and online topic model, which indicates that the TEDT is more cohesive. In general, the TEDT is more sensitive to filter the noisy features and non-important events in a topic. For example, the TEDT achieves a more bigger cohesiveness value on 1987-3-20, and event "the attack in Luzon" is detected with its event words "attack, armi, luzon, davao, rebel", whereas the online topic model detects two events "the attack in island" and "soldier was killed in Libyan". The reason why the TEDT can achieve a much better performance is that the online topic model only utilized semantic topic to find events, whereas the HBE which only consider the co-occurrence of words detected "attack, island, province".

4.3 Experiments on Visualization

To better understand the performance of the algorithms in detecting topic events, the detected topic events are visualized for the performance comparison with node represents event word, edge represents the co-occurrence between words, the size of node represents the word's frequency, and the width of the lines represents the strength of

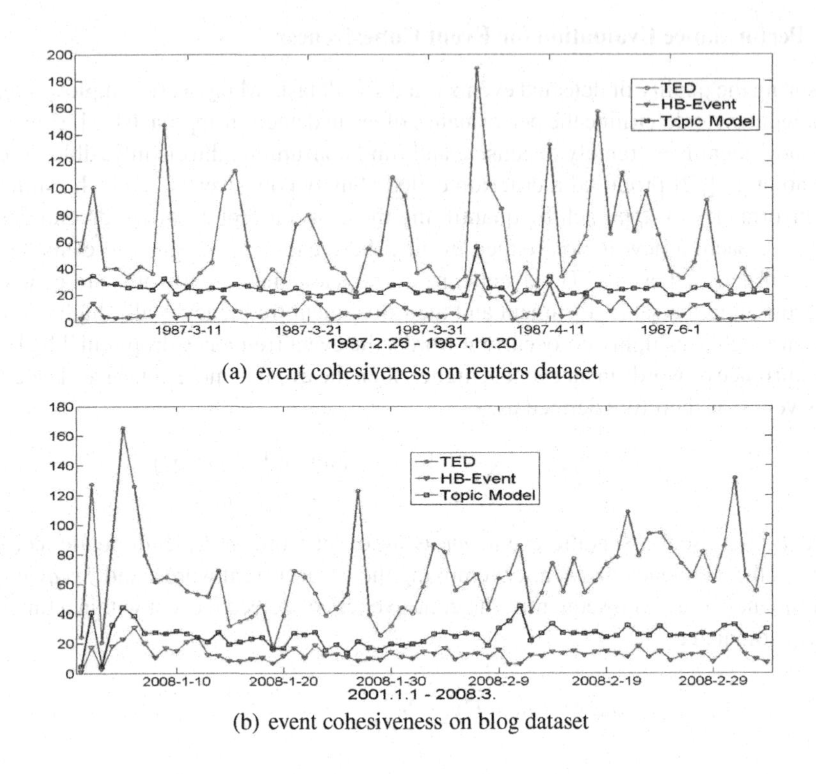

(a) event cohesiveness on reuters dataset

(b) event cohesiveness on blog dataset

Fig. 2. The event cohesiveness of two real datasets

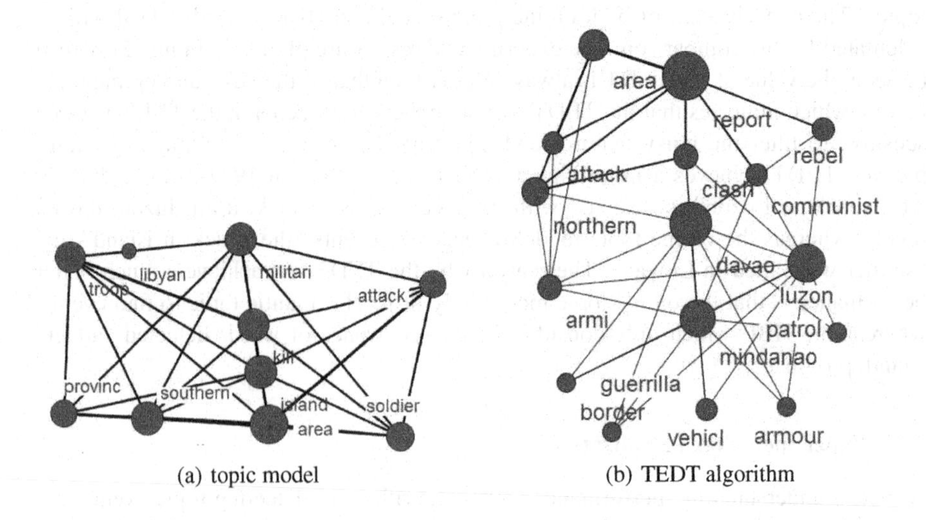

(a) topic model

(b) TEDT algorithm

Fig. 3. Event releasing fight between Philippine government and rebels

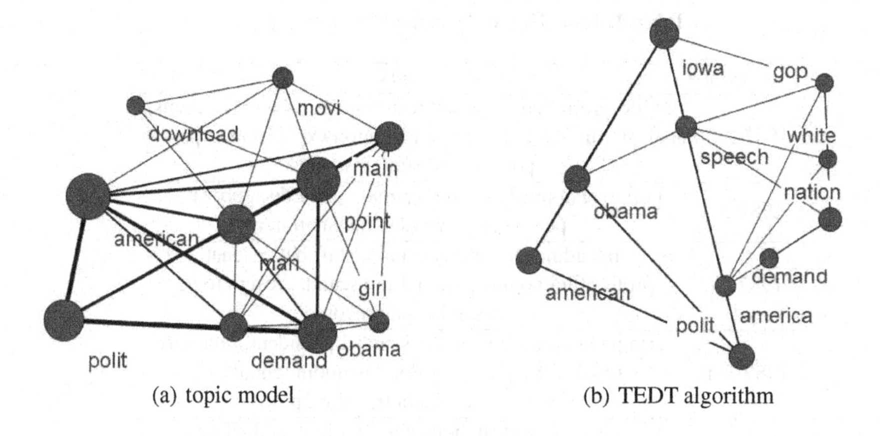

Fig. 4. Event releasing fight between Philippine government and rebels

co-occurrent words. Figure 3 reports the topic event "fight between Philippine government and rebels in the southern Philippine island at 1987". The proposed TEDT detects location of event, "luzon, davao", as well as key words of event "rebel, attack, armi", whereas the online topic model finds two events in one topic, which are "the attack in island" and "soldier was killed in Libyan". The result shows that the TEDT can achieve more accurate event, whereas the online topic model may uncover multiple events in one topic.

Figure 4 reports the ability of the proposed TEDT algorithm to filter out the noisy features of topic. In Figure 4(a), online topic model detects several noisy words, such as "movi, download" and few event words, "obama, american". At the same time, the proposed TEDT algorithm finds "speech, american, iowa, obama", which are closely related to target event.

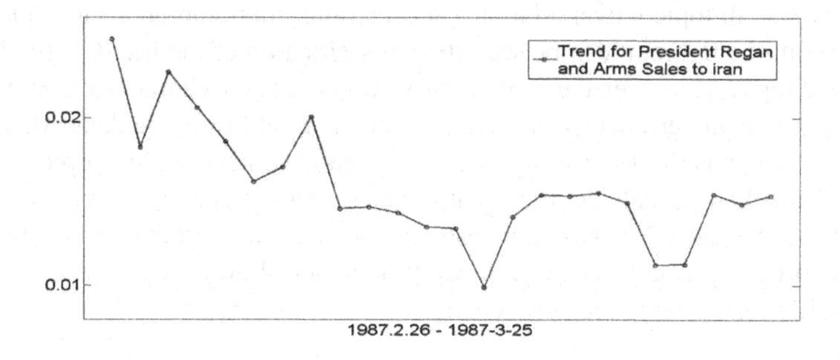

Fig. 5. Hotness for "Iran-Contra and President Regan"

Table 2. Topic Events Tracking On a given Topic

timeslot	events
1987-2-26	report,profit,control,republican,secretari,regan,iran,hous, analyst,arm, white,protect,washington,expect,shultz,person, polit,record,strategist,week
1987-3-1	poll,reagan,presid,conduct,person,three,disapprov,rate, pct,major,newsweek,magazin,iran
1987-3-2	presid,hous,white,gate,chief,staff,baker,senat, profit,offici,nomin,regan,nation,scandal,resign,tower, commiss,sale,record
1987-3-16	arm,presid,hous,iran,walsh,scandal,presidenti,secretari, rebel,white,probe,million,testimoni,tent,affair, chief,senat,wit,contra,georg,profit
1987-3-17	reagan,rebel,arm,hous,contra,iran,presid,white, report,novemb,aid,divers,profit
1987-3-18	reagan,presid,rebel,aid,hous,iran,contra,senat, militari,report,congress
1987-3-19	reagan,presid,aid,white,hous,contra,affair,georg, report,scandal,constitut,profit,told,john
1987-3-23	iran,presid,secretari,missil,radio,report,weinberg,tehran, ship,threat,militari,hawk,iranian
1987-3-24	iran,missil,reagan,strait,hormuz,silkworm,kuwaiti,chief, white,gulf,ship,report,warship,escort,defenc,radio, hous

4.4 Tracking on Topic Events

In this experiment, we will show the ability of the TEDT to track the event changes. Given a topic k, the TEDT first detects all topic events whose topic is k. We only show the results when the topic is "Iran-Contra and President Regan" and the event changes are tracked over all time slots. Part of the tracked events are recorded in Table 4.4.

As can be seen, all event changes related to "Iran-Contra and president Regan" have been detected in the period of ("1987-2-26"–"1987-3-24"). To measure the hotness of an event, the rate of topic k over all topics at each time slot is captured and is plotted in Figure 5. From the figure, it can be seen the tense situation of the Iran-Contra denoted by the rate of topic. Moreover, the evolutionary events can be well tracked. For example, the event "Reagan government sales arms to Iran" is found from the data extracted on "1987-2-26" which is the $1st$ row in the table. From the corresponding topic events on "1987-3-2", it is known that "the President Regan give a report about arms sales". After that on 1987-3-13, the White House investigate the arms sales affairs. In fact, the event changes report every details of the 'Iran-Contra" which shows the effectiveness of the proposed TEDT in tracking the event changes.

5 Conclusion

Detecting events in text stream is a hot research issue in social network society, the conventional approaches only group words by the co-occurrence, which have two major limitations: noisy words and multiple sub-events. In this paper, a novel event detection and tracking algorithm, called topic event detection and tracking (TEDT), was proposed to tackle these limitations. The TEDT extended the topic model and combined the co-occurrence and the topics of words simultaneously. The empirical experiments were performed on two real data sets and the results demonstrated that: (1) the TEDT is superior to the state-of-the-art topic model for event discovery; and (2) the TEDT can track the event changes. For the future work, we are investigating in detecting new events if one topic could appear in several events.

Acknowledgement. This work is supported in part by NSFC under Grant no.61073195, no.61073051, National Commonweal Technology R&D Program of AQSIQ China under Grant No.201310087, Shenzhen Strategic Emerging Industries Program under Grants No.JCYJ20130329142551746, Shenzhen Science and Technology Program under Grant No.CXY201107010163A and No.CXY201107010206A, National Key Technology R&D Program No.2012BAH10F03, 2013BAH17F00 and Science and Technology Development of Shandong Province No.2010GZX20126, 2010GGX10116.

References

1. He, T., Qu, G., Li, S., Tu, X., Zhang, Y., Ren, H.: Semi-automatic hot event detection. In: Li, X., Zaïane, O.R., Li, Z.-h. (eds.) ADMA 2006. LNCS (LNAI), vol. 4093, pp. 1008–1016. Springer, Heidelberg (2006)
2. Wang, C., Zhang, M., Ma, S., Ru, L.: Automatic online news issue construction in web environment. In: Proceedings of the 17th International Conference on World Wide Web, pp. 457–466 (2008)
3. Wang, Y., Xi, Y.H., Wang, L.: Mining the hottest topics on chinese webpage based on the improved k-means partitioning. In: 2009 International Conference on Proceedings of Machine Learning and Cybernetics, vol. 1, pp. 255–260 (2009)
4. Wang, X., Zhai, C., Hu, X., Sproat, R.: Mining correlated bursty topic patterns from coordinated text streams. In: Proceedings of the 13th ACM SIGKDD International Conference on Knowledge Discovery and Data Mining, pp. 784–793 (2007)
5. Hurst, M.F.: Temporal text mining. In: Proceedings of AAAI Spring Symposium: Computational Approaches to Analyzing Weblogs, pp. 73–77 (2006)
6. Eda, T., Yoshikawa, M., Uchiyama, T., Uchiyama, T.: The effectiveness of latent semantic analysis for building up a bottom-up taxonomy from folksonomy tags. In: Proceedings of World Wide Web, pp. 421–440 (2009)
7. Salton, G., Buckley, C.: Term-weighting approaches in automatic text retrieval. Information Processing & Management 24(5), 513–523 (1988)
8. Mei, Q., Zhai, C.: Discovering evolutionary theme patterns from text: an exploration of temporal text mining. In: Proceedings of the Eleventh ACM SIGKDD International Conference on Knowledge Discovery in Data Mining, pp. 198–207 (2005)
9. Osborne, M., Petrovic, S., McCreadie, R., Macdonald, C., Ounis, I.: Bieber no more: First story detection using twitter and wikipedia. In: Proceedings of the SIGIR Workshop on Time-aware Information Access (2012)
10. Lin, C.X., Zhao, B., Mei, Q., Han, J.: Pet: A statistical model for popular events tracking in social communities. In: Proceedings of the 16th ACM SIGKDD International Conference on Knowledge Discovery and Data Mining, pp. 929–938 (2010)

11. Yao, J., Cui, B., Huang, Y., Jin, X.: Temporal and social context based burst detection from folksonomies. In: Proceedings of AAAI (2010)
12. Yao, J., Cui, B., Huang, Y., Zhou, Y.: Bursty event detection from collaborative tags. Proceedings of World Wide Web 15(2), 171–195 (2012)
13. Fung, G.P.C., Yu, J.X., Yu, P.S., Lu, H.: Parameter free bursty events detection in text streams. In: Proceedings of the 31st International Conference on Very Large Data Bases, VLDB Endowment, pp. 181–192 (2005)
14. Singh, V.K., Gao, M., Jain, R.: Social pixels: Genesis and evaluation. In: Proceedings of the International Conference on Multimedia, pp. 481–490 (2010)
15. AlSumait, L., Barbará, D., Domeniconi, C.: On-line lda: Adaptive topic models for mining text streams with applications to topic detection and tracking. In: Proceedings of Eighth IEEE International Conference on Data Mining, pp. 3–12 (2008)
16. Hoffman, M., Bach, F.R., Blei, D.M.: Online learning for latent dirichlet allocation. In: Proceedings of Advances in Neural Information Processing Systems, pp. 856–864 (2010)

Nonterminal Complexity of Weakly Conditional Grammars

Sherzod Turaev[1], Mohd Izzuddin Mohd Tamrin[2], and Norsaremah Salleh[1]

[1] Department of Computer Science
Kulliyyah of Information and Communication Technology
International Islamic University Malaysia
53100 Kuala Lumpur, Malaysia
{sherzod,norsaremah}@iium.edu.my
[2] Department of Information Systems
Kulliyyah of Information and Communication Technology
International Islamic University Malaysia
53100 Kuala Lumpur, Malaysia
izzuddin@iium.edu.my

Abstract. A *weakly conditional grammar* is specified as a pair $K = (G, G')$ where G is a context-free grammar, and G' is a regular grammar such that a production rule of G is only applicable to the sentential form if it belongs to the language generated by G'. The nonterminal complexity $\text{Var}(K)$ of the grammar K is defined as the sum of the numbers of nonterminals of G and G'. This paper studies the nonterminal complexity of weakly conditional grammars, and it proves that every recursively enumerable language can be generated by a weakly conditional grammar with no more than *ten* nonterminals. Moreover, it shows that the number of nonterminals in such grammars without erasing rules leads to an infinite hierarchy of families of languages generated by weakly conditional grammars.

1 Introduction

Conditional grammars (see [1]) are context-free grammars with regulated restriction, in which a regular language is associated to each production, and a production is only applicable to the sentential form if it belongs to the language associated to the given production. Conditional grammars generate all context-sensitive languages if erasing rules are not allowed, and all recursively enumerable languages if erasing rules are allowed (see [1, 2]). Different variants of conditional grammars such as semi-conditional, simple semi-conditional grammars, generalized forbidding grammars, etc. have been studied in several papers (for instance, see [3–7]). Another variant of conditional grammars, called *weakly conditional grammars*, was introduced by Král in [8]. In these grammars a certain regular language is associated to all productions of the underlying context-free grammar. Weakly conditional grammars also generate all recursively-enumerable languages (see [9]) whereas if only λ-free context-free grammars are used, then all context-sensitive languages are generated (see [8]).

N.T. Nguyen et al. (Eds.): ACIIDS 2014, Part I, LNAI 8397, pp. 53–62, 2014.

Since "economical" representation of formal languages has been always important, it is interesting to investigate their grammars from the point of view of descriptional complexity measures such as the number of nonterminals and the number of production rules. In this paper we investigate the nonterminal complexity of weakly conditional grammars with and without erasing rules which is not investigated up to now, though there are a number of papers devoted to the study of the descriptional complexity of other variants of conditional grammars (see [5, 7, 10–12]). We specify a weakly conditional grammar as a pair $K = (G, G')$ where G is a context-free grammar, and G' is a regular grammar such that a production rule of G is only applicable to the sentential form if it belongs to the language generated by G'. We define the nonterminal complexity $\text{Var}(K)$ of the grammar K as the sum of the numbers of nonterminals of G and G'. By this approach, we do not only take the number of nonterminals of G, but also add the number of nonterminals of G' , i.e., we also measure the complexity of the control device (the same idea was also used in [13–15]).

We show that every recursively enumerable language is generated by a weakly conditional grammar with at most *ten* nonterminals. Moreover, we prove that the number of nonterminals in weakly conditional grammars without erasing rules leads to an infinite hierarchy of families of languages generated by weakly conditional grammars.

2 Definitions

We assume that the reader is familiar with formal language theory (see [16, 17]). Let T^* denote the set of all words over an alphabet T. The empty word is denoted by λ. The cardinality of a set X is denoted by $|X|$.

A *context-free grammar* is specified as a quadruple $G = (N, T, P, S)$ where N and T are two disjoint alphabets of *nonterminals* and *terminals*, respectively, $S \in N$ is the start symbol, and $P \subseteq N \times (N \cup T)^*$ is a finite set of (production) rules. A rule (A, x) is written as $A \to x$.

A word $x \in (N \cup T)^+$ *directly derives* $y \in (N \cup T)^*$, written as $x \Rightarrow y$, if and only if there is a rule $A \to \alpha \in P$ such that $x = x_1 A x_2$ and $y = x_1 \alpha x_2$. The reflexive and transitive closure of $\Rightarrow$ is denoted by $\Rightarrow^*$.

A word $w \in (N \cup T)^*$ such that $S \Rightarrow^* w$ is called a *sentential form*. The *language* generated by G is defined by $L(G) = \{w \in T^* : S \Rightarrow^* w\}$.

A grammar is called *regular* if all its rules are of the form $A \to wB$ or $A \to w$ with $A, B \in N$ and $w \in T^*$.

By $\text{Var}(G)$ we denote the number of the nonterminals of a grammar G, i.e., $\text{Var}(G) = |N|$.

A *conditional grammar* is a quintuple $K = (N, T, P, S, M)$ where $G = (N, T, P, S)$ is a context-free grammar, and

$$M = \{\gamma(r) : r \in P \text{ and } \gamma(r) \text{ is a regular set}\}.$$

For $x, y \in (N \cup T)^*$, $x \Rightarrow y$ iff $x = x_1 A x_2$, $y = x_1 \alpha x_2$ and there is a rule $r : A \to \alpha \in P$ such that $x \in \gamma(r)$. A transitive and reflexive closure of $\Rightarrow$ is denoted by $\Rightarrow^*$. The language of K is defined as $L(K) = \{w \in T^* : S \Rightarrow^* w\}$.

A conditional grammar $K = (N, T, P, S, M)$ is called *weak* (i.e., a *weakly conditional grammar*) if $\gamma(r_1) = \gamma(r_2)$ for any two rules $r_1, r_2 \in P$.

The family of all weakly conditional grammars (without erasing rules) is denoted by $w\mathcal{K}$ ($w\mathcal{K} - \lambda$).

Since $M = L(G')$ for some regular grammar $G' = (N', T', P', S')$, the weakly conditional grammar K can be given as a pair $K = (G, G')$. We define the nonterminal complexity of a weakly conditional grammar K as

$$\mathrm{Var}(K) = \mathrm{Var}(G) + \mathrm{Var}(G').$$

For a weakly conditional language L and $\mathcal{G} \in \{w\mathcal{K}, w\mathcal{K} - \lambda\}$, we set

$$\begin{aligned}\mathrm{Var}_{\mathcal{G}}(L) = \min\{\mathrm{Var}(K) : K = (G, G'), G \text{ is a context-free grammar,} \\ G' \text{ is a regular grammar and } L(K) = L\}.\end{aligned}$$

By $\mathcal{L}_k(\mathcal{G})$, $\mathcal{G} \in \{w\mathcal{K}, w\mathcal{K} - \lambda\}$, we denote the families of all languages L generated by weakly conditional grammars such that $\mathrm{Var}_{\mathcal{G}}(L) \leq k$, $k \geq 2$.

By the definition, the following lemma is obvious.

Lemma 1. *For* $\mathcal{G} \in \{w\mathcal{K}, w\mathcal{K} - \lambda\}$,

$$\mathcal{L}_2(\mathcal{G}) \subseteq \mathcal{L}_3(\mathcal{G}) \subseteq \cdots \subseteq \mathcal{L}_n(\mathcal{G}) \subseteq \cdots. \quad (1)$$

3 A Bound for Recursively Enumerable Languages

In this section we prove that ten nonterminals is sufficient for weakly conditional grammars to generate all recursively enumerable languages.

In [18], it was proven that every recursively enumerable language is generated by a grammar

$$G = (\{S, A, B\}, T, P \cup \{ABBBA \to \lambda\}, S),$$

where P consists of context-free productions of the forms

(1) $S \to uSa$, where $u \in \{AB, ABB\}^*$ and $a \in T$;
(2) $S \to uSv$, where $u \in \{AB, ABB\}^*$ and $v \in \{BA, BBA\}^*$;
(3) $S \to \lambda$.

With respect to the rules above, the derivation of a word $w \in T^*$ can be divided into the following phases:

(a) $S \Rightarrow^* w'Sw$, where $w' \in \{AB, ABB\}^*$ and $w \in T^*$, by rules of the form $S \to uSa$, where $u \in \{AB, ABB\}^*$ and $a \in T$;

(b) $w'Sw \Rightarrow^* w_1 w_2 w$, where $w_1 \in \{AB, ABB\}^*$ and $w_2 \in \{BA, BBA\}^*$, by rules of the form $S \rightarrow uSv$ and $S \rightarrow \lambda$, where $u \in \{AB, ABB\}^*$ and $v \in \{BA, BBA\}^*$;
(c) $w_1 w_2 w \Rightarrow^* w$ by $ABBBA \rightarrow \lambda$.

Theorem 1. *Every recursively enumerable language can be generated by a weakly conditional grammar with no more than ten nonterminals.*

Proof. Let $L \subseteq T^*$ be a recursively enumerable language generated by the grammar

$$G = (\{S, A, B\}, T, P \cup \{ABBBA \rightarrow \lambda\}, S)$$

defined as above. We construct a weakly conditional grammar

$$K = (N, T, P \cup R, S, M)$$

where

$$\begin{aligned} N =& \{S, A, B, A', B', \$, \#\}, \\ R =& \{A \rightarrow A'\$, B \rightarrow B'\$, \$ \rightarrow \#\} \cup \{X \rightarrow \lambda : X \in \{A', B', \#\}\}, \\ M =& (\{S, A, B\} \cup T)^* \cup \{A, B\}^* X \{A, B\}^* T^* \end{aligned}$$

with

$$\begin{aligned} X = \{& A'\$BBBA, A'\$B'\$BBA, A'\$B'\$B'\$BA, A'\$B'\$B'\$B'\$A, \\ & A'\$B'\$B'\$B'\$A'\$, A'\$B'\$B'\$B'\$\$, A'\$B'\$B'\$\$\$, A'\$B'\$\$\$\$, \\ & A'\$\$\$\$\$, \$\$\$\$\$, \$\$\$\$\#, \$\$\$\#\#, \$\$\#\#\#, \$\#\#\#\#, \#\#\#\#\#, \\ & \#\#\#\#, \#\#\#, \#\#, \#\}. \end{aligned}$$

First, we prove that $L(G) \subseteq L(K)$. Let

$$S \Rightarrow^* uSvw \Rightarrow u'ABBBAv'w \Rightarrow u'v'w$$

be a derivation in G by productions from P and by $ABBBA \rightarrow \lambda$, where $u, u', v, v' \in \{A, B\}^*$ and $w \in T^*$. Then, $S \Rightarrow^* uSvw \Rightarrow u'ABBBAv'w$ is also a derivation in K by productions from P since any sentential form in this derivation is a word of $(\{S, A, B\} \cup T)^*$. By productions from R, we can continue the derivation to obtain $u'v'w$ in K too, i.e.,

$$\begin{aligned} S &\Rightarrow^* uSvw \Rightarrow u'ABBBAv'w \Rightarrow u'A'\$BBBAv'w \\ &\Rightarrow u'A'\$B'\$BBAv'w \Rightarrow u'A'\$B'\$B'\$BAv'w \Rightarrow u'A'\$B'\$B'\$B'\$Av'w \\ &\Rightarrow u'A'\$B'\$B'\$B'\$A'\$v'w \Rightarrow u'A'\$B'\$B'\$B'\$\$v'w \Rightarrow u'A'\$B'\$B'\$\$\$v'w \\ &\Rightarrow u'A'\$B'\$\$\$\$v'w \Rightarrow u'A'\$\$\$\$\$v'w \Rightarrow u'\$\$\$\$\$v'w \Rightarrow u'\$\$\$\$\#v'w \\ &\Rightarrow u'\$\$\$\#\#v'w \Rightarrow u'\$\$\#\#\#v'w \Rightarrow u'\$\#\#\#\#v'w \Rightarrow u'\#\#\#\#\#v'w \\ &\Rightarrow u'\#\#\#\#v'w \Rightarrow u'\#\#\#v'w \Rightarrow u'\#\#v'w \Rightarrow u'\#v'w \Rightarrow u'v'w \end{aligned}$$

where all sentential forms above are in M. The inclusions follows by induction.

Next, we prove that the inclusion $L(K) \subseteq L(G)$ is also held. Let D be a terminal derivation in K, i.e., $D : S \Rightarrow^* w, w \in T^*$. Let x be a sentential form containing S in this derivation.

Then, by construction of M, the sentential form x is in $(\{S, A, B\} \cup T)^*$, i.e., it cannot contain any nonterminal from $\{A', B', \$, \#\}$. It follows that any production from R is not applicable to x. Thus, the initial part of the derivation D is of the form $S \Rightarrow^* uSu' \Rightarrow uu'$, where $u \in \{A, B\}$, $u' \in (\{A, B\} \cup T)^*$, by productions from P. Moreover, since each sentential form derived from uu' must be in $\{A, B\}^*X\{A, B\}^*T^*$, u' is in the form of vw, where $v \in \{A, B\}^*$ and $w \in T^*$.

From the sentential form uvw, the applications of productions $A \to A'\$$ and $B \to B'\$$ are possible. But, by construction of M, only production $A \to A'\$$ is applicable, and the next sentential form must be in the form of $u'A'\$BBBAv'w$. Further, the applications of productions $A' \to \lambda$, $\$ \to \#$, $A \to A'\$$, $B \to B'\$$ are possible:

- if $A' \to \lambda$ is applied, the sentential form $u'\$BBBAv'w$ is obtained, which is not in M. Since each sentential form in a derivation must be in M, in this case, the derivation is blocked;
- if $\$ \to \#$ is applied, then, again, $u'A'\#BBBAv'w \notin M$;
- by construction of M, $A \to A'\$$ can only be applied to the occurrences of A which is in the subword $ABBBA$. But any word in M does not contain the subword $A'\$BBBA'\$$.
- by construction of M, $B \to B'\$$ must be applied to the occurrences of B in the subword $ABBBA$, in this case, it can only be applied to the occurrence of B following \$, which results in $u'A'\$B'\$BBAv'w \in M$.

By repeating the same arguments as above, we continue the derivation until the subword $A'\$B'\$B'\$B'\$A'\$$ is generated, i.e.,

$$S \Rightarrow^* uSvw \Rightarrow u'ABBBAv'w \Rightarrow^5 u'A'\$B'\$B'\$B'\$A'\$v'w \tag{2}$$

is the only valid derivation in K.

The derivation (2) can be continued by productions $A \to A'\$$, $B \to B'\$$, $A' \to \lambda$, $B' \to \lambda$ and $\$ \to \#$ as follows:

Case 1. Production $A \to A'\$$ or $B \to B'\$$ is applied to some occurrence of A or B, respectively, in u' or v', and the new sentential form

$$u''A'\$B'\$B'\$B'\$A'v'w \text{ or } u'A'\$B'\$B'\$B'\$A'v''w$$

is obtained. Since, by construction of M, words u'' and v'' can not contain nonterminals from $\{A', B', \$\}$, the obtained sentential form is not in M, thus, the derivation is blocked.

Case 2. Production $\$ \to \#$ is applied to some occurrence of \$ in the subword $A'\$B'\$B'\$B'\$A'\$$, and the new sentential form containing one of the subwords

$$A'\#B'\$B'\$B'\$A'\$,\ A'\$B'\#B'\$B'\$A'\$,\ A'\$B'\$B'\#B'\$A'\$,$$
$$A'\#B'\$B'\$B'\#A'\$,\ A'\$B'\$B'\$B'\$A'\#$$

is obtained. But every word in M does not contain any subword above, thus the derivation in this case is also blocked.

Case 3. It is not difficult to see that if production $B' \to \lambda$ is applied to any occurrence of B' in the subword $A'\$B'\$B'\$B'\$A'\$$, then the obtained sentential form is not in M, and the derivation is again blocked. The same situation happens if production $A' \to \lambda$ is applied to the first occurrence of A' in the subword $A'\$B'\$B'\$B'\$A'\$$. Thus, we can only apply the production to the last occurrence of A' in this subword, which results in the sentential form containing the subword $A'\$B'\$B'\$B'\$\$$.

The further only applicable sequence of productions is $B' \to \lambda$, $B' \to \lambda$, $B' \to \lambda$, $A' \to \lambda$, which are applied from the right to the left in the subword $A'\$B'\$B'\$B'\$\$$. As a result, the sentential form $u'\$\$\$\$\$v'w$ is obtained. Thus,

$$\begin{aligned} S \Rightarrow^* uSvw &\Rightarrow u'ABBBAv'w \\ &\Rightarrow^5 u'A'\$B'\$B'\$B'\$A'\$v'w \Rightarrow^5 u'\$\$\$\$\$v'w \end{aligned} \tag{3}$$

is the only valid derivation in K.

From (3), the applications of productions $A \to A'\$$, $B \to B'\$$ and $\$ \to \#$ are possible. Until all occurrences of \$ are erased, productions $A \to A'\$$ and $B \to B'\$$ are not applicable in K. Thus, we first replace all occurrence of \$ to # and, then, erase all occurrences of #, i.e.,

$$\begin{aligned} S \Rightarrow^* uSvw \Rightarrow u'ABBBAv'w &\Rightarrow^5 u'A'\$B'\$B'\$B'\$A'\$v'w \\ &\Rightarrow^5 u'\$\$\$\$\$v'w \Rightarrow^5 u'\#\#\#\#\#v'w \Rightarrow^5 u'v'w. \end{aligned} \tag{4}$$

Thus, from $u'ABBBAv'w$, only subword $ABBBA$ can be eliminated. Then,

$$S \Rightarrow^* uSvw \Rightarrow u'ABBBAv'w \Rightarrow u'v'w$$

is the corresponding derivation in G. By induction, the inclusion holds.

4 An Infinite Hierarchy

In this section we show that not using erasing rules in weakly conditional grammars leads to the infinite hierarchy of language families generated by these grammars.

Lemma 2. *Let for* $n \geq 1$,

$$L_n = a_1^+ \cup a_2^+ \cup \cdots \cup a_n^+.$$

Then

$$\mathrm{Var}_{w\mathcal{K}-\lambda}(L_n) = n + 2.$$

Proof. Let $K_n = (\{S, A_1, A_2, \ldots A_n\}, \{a_1, a_2, \ldots, a_n\}, P_n, S, M_n), n \geq 1$, be a weakly conditional grammar with

$$P_n = \{S \to A_i, A_i \to a_i A_i, A_i \to a_i : 1 \leq i \leq n\}$$

and

$$M_n = (\{S, A_1, A_2, \ldots A_n\} \cup \{a_1, a_2, \ldots a_n\})^*.$$

It is not difficult to see that $L(K_n) = L_n$. Since M_n can be generated by a regular grammar with one nonterminal, $\mathrm{Var}(K_n) = n + 2$. Consequently,

$$\mathrm{Var}_{w\mathcal{K}-\lambda}(L_n) \leq n + 2.$$

Next, we prove that $\mathrm{Var}_{w\mathcal{K}-\lambda}(L_n) \geq n + 2$.

Let $K = (N, \{a_1, a_2, \ldots, a_n\}, P, S, M)$ be a weakly conditional grammar generating the language L_n. Since K does not have erasing rules, for each terminal a_i, $1 \leq i \leq n$, it has to have a terminating rule of the form $A_i \to a_i^{r_i}$ for some positive integer r_i.

For each i with $1 \leq i \leq n$ and a sufficiently large positive integer l, we consider a derivation of a_i^l in K:

$$S \Rightarrow w_{i,1} \Rightarrow w_{i,2} \Rightarrow \cdots \Rightarrow w_{i,m_i} \Rightarrow a_i^l,$$

where $w_{i,m_i} = a_i^{k_i} A_i a_i^{k_i'} \in M$, $k_i, k_i' \in \mathbb{N}$. Then there is a rule $A_i \to a_i^{r_i} \in P$ such that $k_i + r_i + k_i' = l$.

Suppose that for some $1 \leq i < j \leq n$, $A_i = A_j$. Then, for two different terminals a_i and a_j, there are two terminating rules with the same left-hand side. Let $A_i \to a_i^{r_i'} \in P$ and $A_i \to a_j^{r_j''} \in P$ for some positive integers r_i' and r_j''. Then

$$S \Rightarrow w_{i,1} \Rightarrow w_{i,2} \Rightarrow \cdots \Rightarrow w_{i,m_i} = a_i^{k_i} A_i a_i^{k_i'} \Rightarrow a_i^{k_i} a_j^{r_j''} a_i^{k_i'}$$

is also allowed derivation, since $a_i^{k_i} A_i a_i^{k_i'} \in M$, but the word $a_i^{k_i} a_j^{r_j''} a_i^{k_i'}$ is not contained in L_n.

Hence, for all $1 \leq i, j \leq n$, $i \neq j$, $A_i \neq A_j$, i.e., $|N| \geq n$. Since for any grammar generating M, at least one nonterminal is needed, $\mathrm{Var}(K) \geq n + 1$.

Now let us assume that $\mathrm{Var}(K) = n + 1$. Then the following facts should be distinguished.

Fact 1. Any grammar G' generating M has to have exactly one nonterminal. If $\mathrm{Var}(G') \geq 2$, it immediately follows that $\mathrm{Var}(K) > n + 1$.

Fact 2. For each terminal a_i, $1 \leq i \leq n$, all terminating rules $A_i \to a_i^{r_i} \in P$, $r_i \in \mathbb{N}$, have the same left hand-side. If for some $1 \leq i \leq n$, there are two rules $A_i \to a_i^{r_i}$ and $A_i' \to a_i^{r_i'}$, $r_i, r_i' > 0$, with $A_i \neq A_i'$, then it immediately follows that $|N| > n$ and $\mathrm{Var}(K) > n + 1$.

Fact 3. $S = A_i$ for some $1 \leq i \leq n$. If $S \neq A_i$ for all $1 \leq i \leq n$, then, again $|N| > n$ and $\mathrm{Var}(K) > n+1$. Without loss of generality we assume that $S = A_1$.

Now we show that Fact 3 leads to contradiction, i.e., the weakly conditional grammar K with $n+1$ nonterminals generate words not contained in the language L_n.

Suppose that $G' = (\{B\}, \{A_1, A_2, \ldots, A_n\} \cup \{a_1, a_2, \ldots, a_n\}, P', B)$ is a regular grammar generating M. Due to the facts that M contains words of the form $a_1^q A_1 a_1^{q'}$ for sufficiently large positive integers q and/or q', and P' is a finite set, G' has rules of the form

– if $q, q' > 0$

$$\begin{aligned} &B \to a_1^{k_1} B, k_1 > 0, \\ &B \to a_1^{k_2} A_1 a_1^{k_3} B, k_2, k_3 \geq 0, \\ &B \to a_1^{k_4} B, k_4 > 0, \\ &B \to a_1^{k_5}, k_5 \geq 0; \end{aligned}$$

– if $q > 0$ and $q' = 0$

$$\begin{aligned} &B \to a_1^{k_1} B, k_1 > 0, \\ &B \to a_1^{k_2} A_1, k_2 \geq 0; \end{aligned}$$

– if $q = 0$ and $q' > 0$

$$\begin{aligned} &B \to A_1 a_1^{k_3} B, k_3 \geq 0, \\ &B \to a_1^{k_4} B, k_4 > 0, \\ &B \to a_1^{k_5}, k_5 \geq 0. \end{aligned}$$

Let for some $2 \leq j \leq n$, $A_1 \Rightarrow^* a_j^p A_j a_j^{p'}$, $p + p' > 0$, be a derivation in K. Then $a_j^p A_j a_j^{p'} \in M$.

Consider a derivation

$$A_1 \Rightarrow^* a_1^t A_1 a_1^{t'} \Rightarrow^* a_1^t a_j^p A_j a_j^{p'} a_1^{t'}$$

in G' such that $t+t' > 0$, $p+p' > 0$ and $t = km$, $t = k'm' + k''$ for some integers m, m'. Then

$$\begin{aligned} B \Rightarrow a_1^k B &\Rightarrow^* a_1^{km} B \\ &\Rightarrow^* a_1^{km} a_j^p A_j a_j^{p'} B \\ &\Rightarrow a_1^{km} a_j^p A_j a_j^{p'} a_1^{k'} B \\ &\Rightarrow^* a_1^{km} a_j^p A_j a_j^{p'} a_1^{k'm'} B \\ &\Rightarrow a_1^{km} a_j^p A_j a_j^{p'} a_1^{k'm'} a_1^{k''} \end{aligned}$$

is a derivation in G', i.e., $a_1^t a_j^p A_j a_j^{p'} a_1^{t'}$ is a word in M. It follows that the derivation

$$A_1 \Rightarrow^* a_1^t A_1 a_1^{t'} \Rightarrow^* a_1^t a_j^p A_j a_j^{p'} a_1^{t'} \Rightarrow a_1^t a_j^p a_j^{(r,j)} a_j^{p'} a_1^{t'}$$

is allowed in K but $a_1^t a_j^p a_j^{(r,j)} a_j^{p'} a_1^{t'} \notin L_n$. This contradicts our assumption $S = A_1$. Thus, $\mathrm{Var}(G') \geq n+1$ and

$$\mathrm{Var}_{w\mathcal{K}-\lambda}(L_n) \geq n+2.$$

Since for the languages $L_n = a_1^+ \cup a_2^+ \cup \cdots \cup a_n^+$, $n \geq 2$,

$$L_n \in \mathcal{L}_{n+2}(w\mathcal{K} - \lambda) - \mathcal{L}_{n+1}(w\mathcal{K} - \lambda),$$

we have the strict inclusions in Lemma 1 for the case $w\mathcal{K} - \lambda$, i.e.,

Theorem 2.

$$\mathcal{L}_2(w\mathcal{K} - \lambda) \subset \mathcal{L}_3(w\mathcal{K} - \lambda) \subset \cdots \subset \mathcal{L}_n(w\mathcal{K} - \lambda) \subset \cdots . \quad (5)$$

5 Conclusions

In this paper we have showed that every recursively enumerable language can be generated by a weakly conditional grammar with no more than ten nonterminals. But the optimality of this bound remains open. We should mention that as the nonterminal complexity of weakly conditional grammars, we have considered the number of the nonterminals of the underlying grammar and the control regular grammar. If we do not take into consideration the number of nonterminals of the control grammar, then the complexity bound is seven. Another interesting topic is to study the nonterminal complexity bounds for different subclasses of recursively enumerable languages.

Acknowledgments. This work has been supported by International Islamic University Malaysia via EDW B13-053-0938.

References

1. Friś, J.: Grammars with partial ordering of rules. Information and Control 12, 415–425 (1968)
2. Salomaa, A.: Formal languages. Academic press, New York (1973)
3. Kelemen, J.: Conditional grammars: Motivations, definitions, and some properties. In: Proc. Conf. Automata, Languages and Mathematical Sciences, pp. 110–123 (1984)
4. Meduna, A.: Generalized forbidding grammars. International Journal of Computer Mathematics 36(1-2), 31–38 (1990)

5. Meduna, A., Gopalaratnam, A.: On semi-conditional grammars with productions having either forbidding or permitting conditions. Acta Cybernetica 11(4), 307–323 (1994)
6. Navratil, E.: Context-free grammars with regular conditions. Kybernetika 2, 118–126 (1970)
7. Păun, G.: A variant of random context grammars: Semi-conditional grammars. Theoretical Computer Science 41, 1–17 (1985)
8. Král, J.: A note on grammars with regular restrictions. Kybernetika 9(3), 159–161 (1973)
9. Păun, G.: On the generative capacity of conditional grammars. Information and Control 43, 178–186 (1979)
10. Meduna, A., Švec, M.: Reduction of simple semi-conditional grammars with respect to the number of conditional productions. Acta Cybernetica 15, 353–360 (2002)
11. Meduna, A., Švec, M.: Descriptional complexity of generalized forbidding grammars. International Journal of Computer Mathematics 80(1), 11–17 (2003)
12. Vaszil, G.: On the descriptional complexity of some rewriting mechanisms regulated by context conditions. Theoretical Computer Science 330, 361–373 (2005)
13. Turaev, S., Dassow, J., Manea, F., Selamat, M.: Language classes generated by tree controlled grammars with bounded nonterminal complexity. Theoretical Computer Science 449, 134–144 (2012)
14. Turaev, S., Dassow, J., Selamat, M.H.: Language classes generated by tree controlled grammars with bounded nonterminal complexity. In: Holzer, M., Kurtib, M., Pighizzini, G. (eds.) DCFS 2011. LNCS, vol. 6808, pp. 289–300. Springer, Heidelberg (2011)
15. Turaev, S., Dassow, J., Selamat, M.: Nonterminal complexity of tree controlled grammars. Theoretical Computer Science 412, 5789–5795 (2011)
16. Dassow, J., Păun, G.: Regulated rewriting in formal language theory. Springer, Berlin (1989)
17. Rozenberg, G., Salomaa, A. (eds.): Handbook of formal languages, pp. 1–3. Springer (1997)
18. Geffert, V.: Context-free-like forms for phrase-structure grammars. In: Chytil, M.P., Koubek, V., Janiga, L. (eds.) MFCS 1988. LNCS, vol. 324, pp. 309–317. Springer, Heidelberg (1988)

Thai Grapheme-Phoneme Alignment: Many-to-Many Alignment with Discontinuous Patterns

Dittaya Wanvarie

Department of Mathematics and Computer Science
Faculty of Science, Chulalongkorn University, Bangkok, Thailand
Dittaya.W@chula.ac.th

Abstract. Grapheme-phoneme aligned data is crucial to the grapheme-to-phoneme conversion system. Although manual alignment is possible, the task is tedious and time-consuming. Therefore, unsupervised alignment algorithms are proposed to reduce this alignment cost. Several efficient algorithms rely on the assumption that patterns are continuous, but the assumption is not true for Thai. When applying these algorithms to Thai grapheme-to-phoneme alignment, some pre-processing steps for discontinuous patterns are necessary. We propose an algorithm to align Thai graphemes and phonemes which directly incorporates the discontinuous patterns. The experiments show that the precision of the proposed alignment algorithm substantially increases from the conventional alignment with only continuous patterns while the recall decreases from the original method. As a result, the proposed algorithm achieves similar $F1$ to the conventional algorithm.

1 Introduction

Language representation can be divided into 2 categories, written languages and spoken languages. Written languages were invented in order to record words in spoken languages. Therefore, we can convert spoken languages (sound) to written languages (text) and vice versa.

A grapheme is the smallest building block in a written language while a phoneme is the smallest unit in a spoken language. Both graphemes and phonemes may or may not contain any meaning since the smallest unit in language that contains meaning is a word or a morpheme, which usually consists of several graphemes or phonemes.

A writing system of modern languages is mainly divided into 3 categories, logographic, syllabic and alphabetic systems. Thai is in the alphabetic group whose characters or graphemes consisted of consonants, vowels, tonal marks, and some special symbols. Mapping from Thai graphemes to phonemes is not trivial because there are crossing alignment and discontinuous patterns. For example, as shown in Figure 1, a word “เกลี่ย"(equalize) in Thai is pronounced “/klia/”. The first, fourth and sixth characters, “เ ี / ย"are aligned to “/ia/” which is the

N.T. Nguyen et al. (Eds.): ACIIDS 2014, Part I, LNAI 8397, pp. 63–72, 2014.

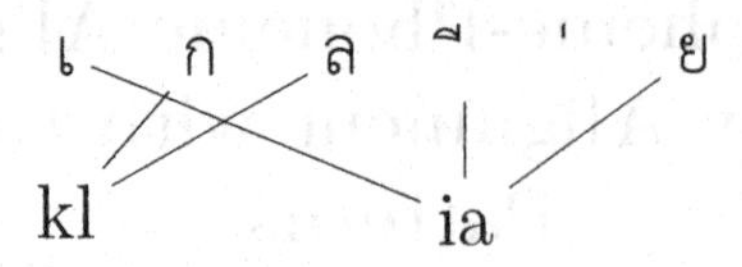

Fig. 1. An alignment of "เ - ก - ล ี - ่ - ย" and "/kl-ia/"

phoneme at the end of the phoneme sequence. The second and third characters "ก - ล" are mapped to "/kl/" at the beginning of the phoneme sequence. The fifth character is the tonal mark which is not mapped to any phoneme. The mapping relationship between graphemes and phonemes is many-to-many relationship.

In this paper, we propose an unsupervised many-to-many alignment that allows the alignment of discontinuous patterns and also allows the crossing alignment. We adopt the EM algorithm for the training and also modify the forward-backward score calculation and the sequence alignment computation in order to incorporate discontinuous patterns.

The organization of rest of this paper is as follows. Related works on grapheme-to-phoneme conversion and alignment methods are discussed in section 2. The EM algorithm is described in section 3. The modified forward and backward algorithm is explained in section 4. The experiments and analysis of the result are in section 5. Finally, we conclude our contribution and the future work in section 6.

2 Related Works

Early works on Thai grapheme-to-phoneme conversion are methods based on dictionary and pronunciation rules [1,2]. Both methods do not require any alignment information. Conventional statistical methods in grapheme-to-phoneme conversion utilize transcription corpora. In such corpora, each sentence or word is annotated with its pronunciation. The conversion system uses statistical scores of grapheme-phoneme pattern pairs to predict the candidate phoneme output sequence. However, the alignment in the corpora is usually in word-pronunciation level but not the grapheme-phoneme level. Hence, we need to align characters and phones in transcription corpora to obtain the grapheme-phoneme pattern scores.

Chotimongkol and Black proposed a set of hand-crafted preprocessing rules for front vowel patterns that cause discontinuous patterns [3]. On the contrary, Tarsaku et al. proposed a set of context-free rules that automatically extract alignment patterns. However, the extracted patterns are all continuous due to the restriction of the context-free grammar [4].

Another possible way to align graphemes to phonemes patterns is to ask an expert to annotate the corpus. However, the task is very tedious and time-consuming. Therefore, unsupervised alignment algorithms are proposed to reduce to annotation cost of the corpus [5,6].

Early methods in grapheme-phoneme alignment are restricted to 1-1 alignment because many languages usually have 1-1 mapping between a grapheme and a phoneme in a word. Expectation-Maximization (EM) algorithm is adopted to estimate the probability of the alignment [6]. Features for the alignment are either the existence or the number of occurrences of the pattern itself or the pattern together with its context such as in Hidden Markov Model [7]. The most probable alignment is calculated using dynamic programming. Subsequent works extend the 1-1 alignment to many-to-many alignment [5]. Alignment task can also be viewed as an optimization problem. The objective function is to achieve the alignment pairs with the highest probability or the least errors, subject to some constraints. Optimization algorithms such as Integer Programming can solve the grapheme-phoneme alignment task with considerable accuracy. Moreover, the ensemble method, which combine several techniques together, can achieve better alignment accuracy than the accuracy when using a single technique [8].

3 Expectation-Maximization Algorithm

Expectation-Maximization (EM) algorithm [9] is an algorithm to estimate the unknown or hidden parameters of a probabilistic model that best fit the data. In our task, we need to estimate the alignment score of each pattern pair. The desired score should maximize the likelihood of the real training data and the predicted model. EM algorithm is a hill-climbing, iterative method that iterates between 2 steps, the E-step that calculates the expected values of the dataset using the current model parameters and the M-step that re-calculates the parameter values using the expected value previously calculated in the E-step. The iterations continue until convergence, which indicates that the parameter values become stable. Let x and y be a grapheme pattern and a phoneme

Algorithm 1. EM algorithm

1: Initialize all $n(x, y)$ with their number of occurrences
2: Calculate $s_{new}(x, y)$
3: **repeat**
4: $\quad s(x, y) = s_{new}(x, y)$
5: $\quad$ **for** each (x, y) **do**
6: $\quad\quad n(x, y) = \Sigma_{(\mathbf{x},\mathbf{y})} p(x, y|(\mathbf{x}, \mathbf{y}))$
7: $\quad$ **end for**
8: $\quad$ Calculate $s_{new}(x, y)$
9: **until** $\Sigma_{(x,y)} (s_{new}(x, y) - s(x, y)) \leq \delta$

pattern, respectively. We adopt the EM algorithm to estimate the probability of each grapheme-phoneme alignment pair, (x, y). The algorithm is summarized in Algorithm 1. $n(x, y)$ is the estimated number of occurrences of pattern pair

(x, y). $s(x, y)$ and $s_{new}(x, y)$ are the alignment scores which are estimated using $n(x, y)$. Before starting any iterations, we need to estimate the score of each alignment pair. Although we do not have any alignment information beforehand, we set $n(x, y)$ to be the number of occurrences the the pair (x, y) in the training corpus.

Each iteration contains 2 steps, E-step and M-step. In the E-step, we estimate the number of occurrences, $n(x, y)$, of pattern pair (x, y) using its marginal probability in a sequence. A sequence pair $(\mathbf{x}, \mathbf{y})$ is a word-pronunciation pair in the dataset. $n(x, y)$ is summed over all $(\mathbf{x}, \mathbf{y})$ pairs. Subsequently, in the M-step, we re-calculate the alignment score $s(x, y)$ using the number of occurrences previously estimated in the E-step. The iterations continue until the difference between $s(x, y)$ of two consecutive iterations is smaller than a cut-off threshold δ.

The marginal probability of an alignment in a sequence is calculated using dynamic programming called the forward-backward algorithm. Let i be the position in the grapheme sequence and j be the position in the phoneme sequence. The forward score, $\alpha(i, j)$, is the sum of all prefix sequence scores to position (i, j) of the sequence. On the other hand, the backward score, $\beta(i, j)$, is the sum of all suffix sequence scores back to (i, j).

We calculate the forward and backward score by the following formula.

$$\alpha(i, j) = \sum_{(x,y)} \alpha(i - k, j - l) * s(x, y) \quad (1)$$

and

$$\beta(i, j) = \sum_{(x,y)} s(x, y) * \beta(i + k, j + l) \ . \quad (2)$$

k and l are the length of x and y, respectively. Dynamic programming is adopted to find all $\alpha(i, j)$ and $\beta(i, j)$. After all forward and backward scores are calculated, $\alpha(|\mathbf{x}|, |\mathbf{y}|)$ and $\beta(0, 0)$ represent the score of sequence pair $(\mathbf{x}, \mathbf{y})$ grapheme-phoneme alignment.

The marginal probability $p(x, y|(\mathbf{x}, \mathbf{y}))$ of pattern (x, y) starting at position (i, j) in sequence $(\mathbf{x}, \mathbf{y})$ is then calculated by the following equation,

$$p(x, y|(\mathbf{x}, \mathbf{y})) = \frac{\alpha(i, j) * s(x, y) * \beta(i + k, j + l)}{\alpha(|\mathbf{x}|, |\mathbf{y}|)} \ .$$

In the M-step, we re-calculate $s(x, y)$ using previously estimated count of occurrences in the E-step. Actually, the score can be the joint probability or the conditional probability. In this paper, we only perform the experiments using the conditional probability of the phonemes given the graphemes,

$$s(x, y) = p(y|x) = \frac{n(x, y)}{n(x)} \ .$$

$n(x)$ is the estimated number of grapheme x from all alignments in the training set.

4 Incorporating Discontinuous Patterns

The forward and backward score calculation in equation (1) and (2) is designed for continuous patterns (x, y). When some patterns are discontinuous, some prefix sequences may not yet be completely aligned.

We will explain using an example of alignment in Figure 1. If the alignment starts from the grapheme sequence, when the first, the fourth, and the fifth graphemes, "เ -ี -ย" are firstly aligned with the second phoneme, "/ia/", the first phoneme, "/kl/", is still left unaligned. On the other hand, if the alignment starts from the first phoneme, the first grapheme would still be unaligned. These unaligned positions will be aligned later when the alignment proceeds to next positions. Note that the forward score is the score of a completely aligned prefix sequence. We need to re-define the forward score for partially aligned prefix sequence. The same problems also occur when there is a crossing alignment even though all patterns are continuous. Therefore, we cannot directly calculate the forward and backward scores using equation (1) and (2) when there are discontinuous patterns or crossing alignments.

Although patterns may be discontinuous, the size of pattern can be fixed to a constant called a window size. A window size is the distance between the possible leftmost and the possible rightmost position in a pattern. For example, if the window size is 5, there are 32 possible patterns, both continuous and discontinuous, extending from the current position. We only align a new pattern at the position whose all previous graphemes are aligned and the new pattern does not overlap the partially aligned patterns. Since discontinuous patterns appear only in the grapheme sequence, we allow such patterns only in the grapheme sequence but restrict the phoneme patterns to be continuous. By using this pattern generation and alignment, we already incorporate the crossing alignment with no additional process.

However, most of the newly generated grapheme patterns are not valid patterns or useless. We need a large corpus to correctly estimate the number of occurrences of patterns. Otherwise, the estimated model will not be reliable. Since we have a limited size of corpus, we instead proposed a simple heuristic pruning method to eliminate such patterns in the pattern generation step. When two overlapping patterns appear to have similar number of occurrences, they tend to be merged into a large pattern. We simply remove the smaller grapheme pattern from the pattern list except the single character patterns and the continuous bigram character patterns.

By limiting the probable patterns in a window size, we can still use the forward and backward algorithm. We re-define the grapheme position in the algorithm from the position i to the position together with the extending pattern p.

We then modify the calculation of the forward score to

$$\alpha((m, n), j + l) = \alpha((i, p), j) * s(x, y) \ .$$

After adding pattern (x, y) to the prefix sequence ending at position $((i, p), j)$, m is the new position on a grapheme sequence whose all previous graphemes are already aligned and n is the new grapheme pattern extending from m. We could extend the prefix sequence at $((i,p),j)$ with (x,y) if and only if there is no overlapping pattern between the prefix sequence and (x,y).

Similarly, the backward calculation is defined as follows,

$$\beta((m,n),j+l) = \beta((i,p),j) * s(x,y) \ .$$

After adding pattern (x,y) to the suffix sequence starting at position $((i,p),j)$, m is the new position on a grapheme sequence whose all subsequent graphemes are already aligned and n is the new pattern extending from m to the beginning of the sequence.

5 Experiment and Result

5.1 Datasets and Pre-processing

The training part of the LOTUS corpus [10] contains 57525 words with their pronunciation counterparts. Words in the corpus are extracted from news text and general articles. The corpus consists of 4363 unique words. Most of them are frequent words in Thai. We leave duplicated words in the training set as is because we need to estimate the number of occurrences of each grapheme-phoneme pattern in the training set. The evaluation part of the corpus contains 8469 words with 1436 unique words. We remove duplicated words from the evaluation set and manually align characters in each word with their corresponding phonemes. We refer to a pattern pair in the test set as a gold pair or gold alignment.

We extract some corpus information from the annotated test set. The average number of characters per phone is 1.05. Since 527 out of 1436 words contain either discontinuous patterns or crossing alignment, the upper bound exact match accuracy of the system without any pre-processing for discontinuous patterns is approximately 63%.

A syllable in Thai consists of phones and tone. The tone is a level of sound that span the whole syllable while the phone is usually mapped to a group of consonants or vowels. The grapheme-phoneme alignment in this work will align only phonemes and graphemes. We leave the tone prediction task to future work. Therefore, we remove all tonal marks from the grapheme sequence in the pre-processing.

5.2 Parameters

There are 3 systems in the experiments.

- **Baseline** is a simple many-to-many alignment system with only continuous patterns.
- **Baseline-delX** is the same as the baseline system but allows some graphemes not to align with any phoneme.
- **Skip** is the baseline system with discontinuous patterns. This setting also allows some graphemes not to align with any phoneme.

We vary the window size of graphemes ($maxX$) and the window size of phonemes ($maxY$) in the baseline system from 2 to 5 characters, in order to capture whole syllable. We choose the window size from the best baseline settings for our proposed **Skip** systems.

The pruning threshold is the ratio between the number of occurrences of 2 grapheme patterns. If the ratio is greater than the threshold, we eliminate the smaller pattern. We perform the experiments with the threshold between 0.1 to 0.9.

5.3 Evaluation Metrics

We annotate the test set in order that aligned patterns are as small as possible. When the window size is large, several small patterns tend to be merged together and form a new large pattern. If the evaluation is the exact alignment, the number of alignment pairs that match the gold pairs would be very few even though the large pattern contains all correct small alignment pairs. Therefore, we propose to use the non-crossing-pattern precision and non-crossing-pattern recall to evaluate the alignment.We also evaluate the proposed algorithms using non-crossing-pattern $F1$ and the exact match accuracy.

When a span of a predicted pattern exactly matches a gold pattern of a group of gold patterns, the predicted pattern will be scored as correct. Note that the pattern pair is scored as correct if and only if both its grapheme and phoneme match the same gold pairs.

The numbers of a correctly predicted grapheme-phoneme pattern pairs are used to calculate the precision ($prec$) and recall (rec). The non-crossing-pattern precision is calculated by the following equation,

$$prec = \frac{cp}{np} .$$

cp is the number of non-crossing predicted pairs. np is the number of all predicted pattern pairs in the test set.

Similary, the numbers of correctly retrieved grapheme-phoneme pattern pairs are used to calculate the non-crossing-pattern recall, as shown in the following equation,

$$rec = \frac{cg}{g}.$$

cg is the number of non-crossing gold pairs while ng is the number of gold pairs.

$F1$ is the harmonic mean of precision and recall.

The last metric is the exact match accuracy. In a training set of nw words, if nm is the number of words whose all graphemes and phonemes are correctly segmented and aligned, the accuracy, Acc, is calculated by the following equation,

$$Acc = \frac{nm}{nw} .$$

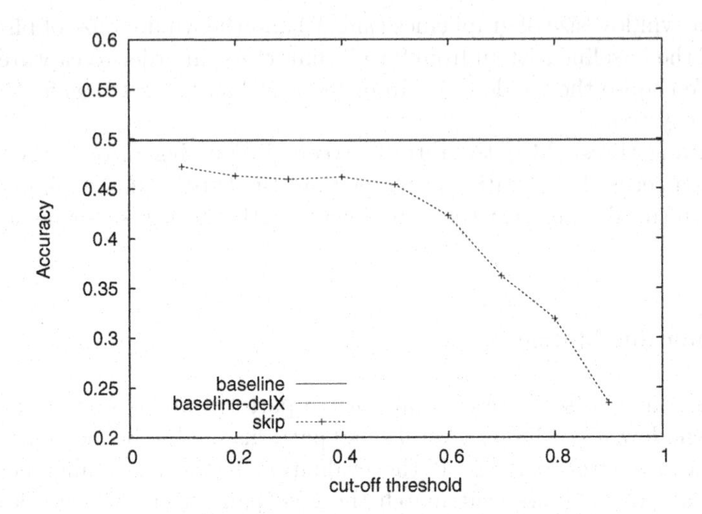

Fig. 2. Accuracy of all experiments

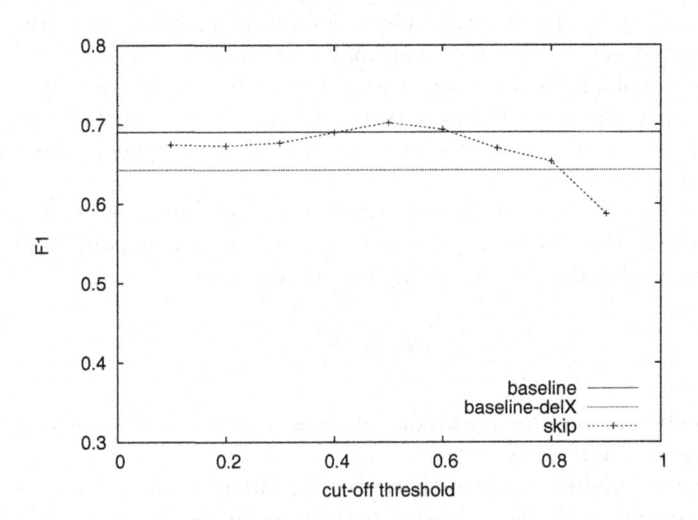

Fig. 3. $F1$ of all experiments

Table 1. Best precision, recall, $F1$, and accuracy of each system. Figures in parentheses are exact match precision, recall and $F1$, respectively.

System name	Parameters	Prec	Rec	F1	Acc
baseline	maxX=2, maxY=2	0.6923	0.6887	0.6905	0.4986
		(0.6924)	(0.6755)	(0.6839)	
baseline-delX	maxX=2, maxY=2, delX	0.6273	0.6588	0.6427	0.4993
		(0.6275)	(0.6432)	(0.6353)	
skip	maxX=5, maxY=2, skip, cut-off=0.5	0.7460	0.6637	**0.7024**	0.4540
		(0.7068)	(0.6462)	(0.6752)	

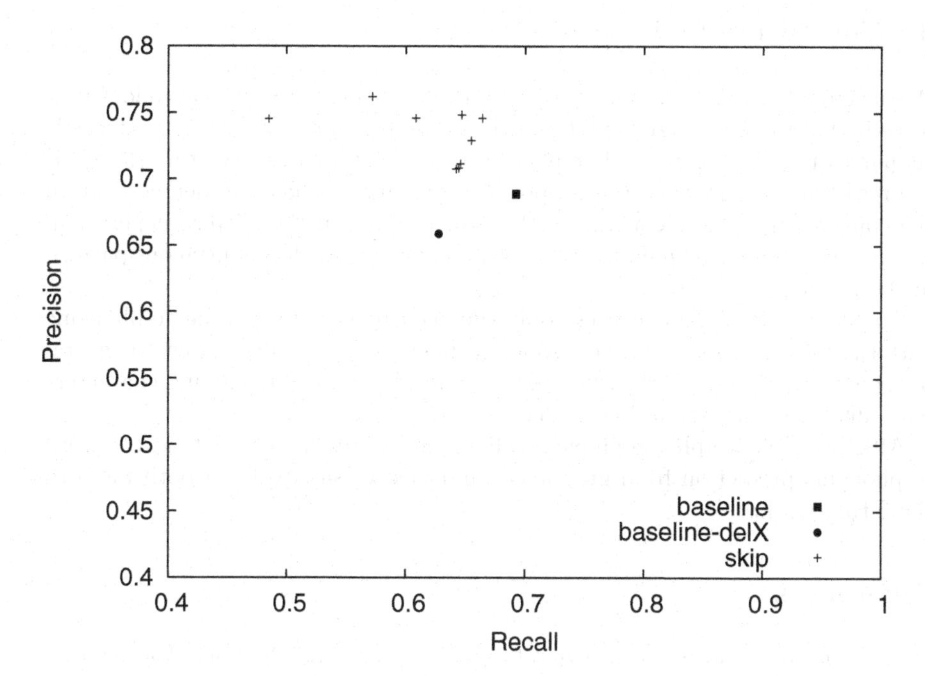

Fig. 4. Precision-Recall of proposed systems

5.4 Result and Discussion

Accuracy and $F1$ of all experiments is shown in Figure 2 and Figure 3, respectively. The exact match accuracy of the proposed method substantially decreased from the best baseline. We believe that the reason is from the heuristic pruning in the pattern generation. We did not distinguish the discontinuous and continuous patterns with exactly the same characters in the same order from one another. Although we have excluded the unigram and continuous bigram patterns from the pruning, some of them may appear as discontinuous patterns and are pruned. As a result, several patterns cannot be aligned because their alignment candidates are removed. From the result, we found that the percentage of unaligned patterns increases from 0.39% to 2.03%. The decrease in the recall may have the same cause.

In contrast, the precision of the proposed method is higher than the baseline as shown in Figure 4. Although the removed patterns may also be replaced by other incorrect patterns in other alignment, other predictions are correct thus increase the precision. We summarize the best settings for each system in Table 1. We choose the best system based on the best $F1$ because we should balance between precision and recall.

6 Conclusion and Future Work

We have proposed the alignment algorithm that incorporates discontinuous patterns by nature. The alignment $F1$ of the system with discontinuous patterns is similar to the conventional alignment with only continuous patterns. However, the recall and accuracy of the proposed alignment significantly decreases from the conventional alignment due to the over-pruning in the pattern generation. In our future work, we will distinguish the continuous and discontinuous pattern in the pruning.

Currently, the alignment takes only the occurrences of a grapheme-phoneme patterns into account. However, some graphemes can be translated to several phonemes depending on their context. We may add the surrounding characters into our alignment estimation model.

After we have grapheme-phoneme alignment scores, we can extend the work to phoneme prediction from graphemes and vice versa. We leave both tasks to the future work.

References

1. Luksaneeyanawin, S.: A thai text-to-speech system. In: Regional Workshop an Computer Processing of Asian Languages (1989)
2. Mittrapiyanuruk, P., Hansakunbuntheung, C., Tesprasit, V., Sornlertlamvanich, V.: Issues in thai text-to-speech synthesis: The nectec approach. In: NECTEC Annual Conference, pp. 483–495 (2000)
3. Chotimongkol, A., Black, A.W.: Statistically trained orthographic to sound models for Thai. In: INTERSPEECH, pp. 551–554. ISCA (2000)
4. Tarsaku, P., Sornlertlamvanich, V., Thongprasirt, R.: Grapheme-to-phoneme for thai. In: NLPRS (2001)
5. Jiampojamarn, S., Kondrak, G., Sherif, T.: Applying Many-to-Many Alignments and Hidden Markov Models to Letter-to-Phoneme Conversion. In: NAACL, pp. 372–379 (2007)
6. Daelemans, W., van den Bosch, A.: Language-Independent Data-Oriented Grapheme-to-Phoneme Conversion. In: Progress in Speech Processing, pp. 77–89. Springer (1997)
7. Och, F.J., Ney, H.: A systematic comparison of various statistical alignment models. Computational Linguistics 29(1), 19–51 (2003)
8. Jiampojamarn, S., Kondrak, G.: Letter-Phoneme Alignment: An Exploration. In: ACL, pp. 780–788 (2010)
9. Dempster, A.P., Laird, N.M., Rubin, D.B.: Maximum Likelihood from Incomplete Data via the EM Algorithm. Journal of the Royal Statistical Society: Series B (Methodological) 39, 1–38 (1977)
10. Kasuriya, S., Sornlertlamvanich, V., Cotsomrong, P., Kanokphara, S., Thatphithakkul, N.: Thai Speech Corpus for Thai Speech Recognition. In: The Oriental COCOSDA 2003, pp. 54–61 (2003)

A New Approach for Mining Top-Rank-*k* Erasable Itemsets

Giang Nguyen[1], Tuong Le[2,*], Bay Vo[2], and Bac Le[3]

[1] University of Technology, Ho Chi Minh City, Vietnam
nhgiang@hutech.edu.vn
[2] Faculty of Information Technology, Ton Duc Thang University, Ho Chi Minh City, Vietnam
tuonglecung@gmail.com, vdbay@it.tdt.edu.vn
[3] Faculty of Information Technology, University of Science, Ho Chi Minh City, Vietnam
lhbac@fit.hcmus.edu.vn

Abstract. Erasable itemset mining first introduced in 2009 is an interesting variation of pattern mining. The managers can use the erasable itemsets for planning production plan of the factory. Besides the problem of mining erasable itemsets, the problem of mining top-rank-*k* erasable itemsets is an interesting and practical problem. In this paper, we first propose a new structure, call dP-ID_List and two theorems associated with it. Then, an improved algorithm for mining top-rank-*k* erasable itemsets using dPID_List structure is developed. The effectiveness of the proposed method has been demonstrated by comparisons in terms of mining time and memory usage with VM algorithm for three datasets.

Keywords: data mining, erasable itemset, pattern mining, top-rank-*k*.

1 Introduction

Pattern mining is a well-established element of data mining. Frequent itemset mining [1] is the most popular which has been attracted many researchers with many methods such as the Apriori algorithm [2], the FP-tree-based algorithm [8], methods based on IT-tree [11, 14], hybrid methods [12-13], and so on.

An interesting variation of pattern mining, the problem of mining erasable itemsets [5] was first presented in 2009. In which, a factory produces many products created from a number of items. Each product brings an income to the factory. A financial resource is required to buy and store all items. However, in a financial crisis situation this factory has not enough money to purchase all necessary components as usual. This problem is to find the itemsets which can best be erased so as to minimize the loss to the factory's gain. Managers can then utilize the knowledge of these erasable itemsets to make a new production plan. There are many methods to solve this problem in recent years such as: META [5], VME [6], MERIT [4], dMERIT+ [9] and MEI [10].

* Corresponding Author.

N.T. Nguyen et al. (Eds.): ACIIDS 2014, Part I, LNAI 8397, pp. 73–82, 2014.

Besides, the problem of mining top-rank-*k* erasable itemsets [7] was presented in 2011, where *k* is the biggest rank value of all erasable itemsets to be mined. Currently there are two algorithms to solve this problem such as MIKE [7] and VM [3]. In which, VM is new and effective algorithm. However, VM uses PID_List structure and union PID_List strategy for mining top-rank-*k* erasable itemset. This makes the algorithm slow and requires a lot of memory usage. In this paper, we propose an enhanced algorithm of VM algorithm, called dVM, for mining top-rank-*k* erasable itemsets using a new structure, dPID_List.

The rest of the paper is organized as follows: section 2 presents related work, and then the proposed method is proposed in section 3. Experimental result is presented in section 4. The paper concluded in section 5.

2 Related Work

2.1 Top-Rank-*k* Erasable Itemset Mining

Let $I = \{i_1, i_2, ..., i_m\}$ be a set of all items, which are the abstract representations of components of products. A product dataset is denoted by $DB = \{P_1, P_2, ..., P_n\}$, where P_i $(1 \leq i \leq n)$ is a product presented in the form of ⟨*Items*, *Val*⟩, where *Items* are the items (or components) that constitute P_i and *Val* is the profit that the factory obtains by selling the product P_i. A set $X \subseteq I$ is also called an itemset, and an itemset with *k* items is called a *k*-itemset. An example dataset in Table 1 will be used throughout this article.

Table 1. An example dataset

Product	Items	Val (Million $)
P_1	a, b	1,000
P_2	a, b, c	200
P_3	b, c, e	150
P_4	b, d, e, f	50
P_5	c, d, e	100
P_6	d, e, f, h	200

Definition 1 (gain of an itemset). Let X $(\subseteq I)$ be an itemset. The gain of X is defined as:

$$g(X) = \sum_{\{P_k | X \cap P_k.Items \neq \emptyset\}} P_k.Val \tag{1}$$

Definition 2 (rank of an itemset). Given a product dataset *DB*, the rank of an itemset *X* is as follows:

$$R_X = |\{g(Y) | Y \subseteq I \text{ and } g(Y) \leq g(X)\}| \tag{2}$$

Definition 3 (top-rank-*k* erasable itemsets). An itemset X ($\subseteq I$) is called to be a top-rank-*k* erasable itemset if and only if $R_X \leq k$.

Given a transaction dataset *DB* and a threshold *k*, the problem of mining top-rank-*k* erasable itemsets is the task of finding the complete set of erasable itemsets whose ranks are no greater than *k*.

2.2 PID_List

In [3], Deng et al. presents the PID_List structure for mining top-rank-*k* erasable itemsets effectively. We summarize the basic concepts as follows:

Definition 4. The PID_List of 1-itemset $A \subseteq I$ is

$$PID\_List(A) = \bigcup_{\{P_k | A \cap P_k.Items \neq \emptyset\}} \langle P_k.ID, P_k.Val \rangle \tag{3}$$

Example 1. Let's consider *DB*, we have PID_List(*c*) = {⟨2, 200⟩, ⟨3, 150⟩, ⟨5, 100⟩} and PID_List(*d*) = {⟨4, 50⟩, ⟨5, 100⟩, ⟨6, 200⟩}.

Theorem 1. Let *XA* and *XB* are two erasable *k*-itemsets. Assume $PID\_List(XA)$ and $PID\_List(XB)$ are PID_Lists associated with *XA* and *XB* respectively. The PID_List of *XAB* is determined as follows:

$$PID\_List(XAB) = PID\_List(XA) \cup PID\_List(XB) \tag{4}$$

Example 2. According to *Example 1* and *Theorem 1*, we have PID_List(*cd*) = PID_List(*c*) ∪ PID_List(*d*) = {⟨2, 200⟩, ⟨3, 150⟩, ⟨5, 100⟩} ∪ {⟨4, 50⟩, ⟨5, 100⟩, ⟨6, 200⟩} = {⟨2, 200⟩, ⟨3, 150⟩, ⟨4, 50⟩, ⟨5, 100⟩, ⟨6, 200⟩}.

Theorem 2. The gain of an erasable itemset, *X*, can be computed as follows:

$$g(X) = \sum_{P_j \in PID\_List(X)} P_j.Val \tag{5}$$

Example 3. According to *Example 2* and *Theorem 2*, we have PID_List(*cd*) = {⟨2, 200⟩, ⟨3, 150⟩, ⟨4, 50⟩, ⟨5, 100⟩, ⟨6, 200⟩}; therefore $g(cd)$ = 200 + 150 + 50 + 100 + 200 = 700 dollars.

3 Mining Top-Rank-*k* Erasable Itemsets Using dPID_List

3.1 dPID_List

Definition 5 (dPID_List). Let *XA* and *XB* are two erasable *k*-itemsets. Assume $PID\_List(XA)$ and $PID\_List(XB)$ are PID_Lists associated with *XA* and *XB* respectively. The dPID_List of *XAB* is determined as follows:

$$dPID\_List(XAB) = PID\_List(XB) \setminus PID\_List(XA) \tag{6}$$

Example 4. According to *Example 1* and *Definition 4*, we have dPID_List(*cd*) = PID_List(*d*) \ PID_List(*c*) = {⟨4, 50⟩, ⟨5, 100⟩, ⟨6, 200⟩} \ {⟨2, 200⟩, ⟨3, 150⟩, ⟨5, 100⟩} = {⟨4, 50⟩, ⟨6, 200⟩}.

Theorem 3. Let *XA* and *XB* are two erasable *k*-itemsets. Assume $dPID\_List(XA)$ and $dPID\_List(XB)$ are dPID_Lists associated with *XA* and *XB* respectively. The dPID_List of *XAB* is also determined as follows:

$$dPID\_List(XAB) = dPID\_List(XB) \setminus dPID\_List(XA) \tag{7}$$

Proof. We have:

$$\begin{aligned} dPID\_List(XAB) &= PID\_List(XB) \setminus PID\_List(XA) \\ &= [PID\_List(X) \cup PID\_List(B)] \\ &\qquad \setminus [PID\_List(X) \cup PID\_List(A)] \\ &= [PID\_List(X) \cup [PID\_List(B) \setminus PID\_List(X)]] \\ &\qquad \setminus [PID\_List(X) \cup [PID\_List(A) \\ &\qquad \setminus PID\_List(X)]] \\ &= [PID\_List(B) \setminus PID\_List(X)] \setminus [PID\_List(A) \\ &\qquad \setminus PID\_List(X)] \\ &= dPID\_List(XB) \setminus dPID\_List(XA) \end{aligned}$$

Therefore, Theorem 3 is proved.

Theorem 4. Let *XA* and *XB* are two erasable *k*-itemsets. Assume $g(XA)$ is the gain of *XA*. The gain of *XAB* based on dPID_List as follows:

$$g(XAB) = g(XA) + \sum_{P_j \in dPID\_List(XAB)} P_j.Val \tag{8}$$

Proof. We have:

$$\begin{aligned} PID\_List(XAB) &= PID\_List(XA) \cup [PID\_List(XB) \setminus PID\_List(XA)] \\ &= PID\_List(XA) \cup dPID\_List(XAB) \end{aligned} \tag{a}$$

According to Definition 5, the gain of itemset XAB is determined as:

$$g(XAB) = \sum_{P_j \in PID\_List(X)} P_j.Val \tag{b}$$

From (a) and (b):

$$g(XAB) = \sum_{P_j \in [PID\_List(XA) \cup dPID\_List(XAB)]} P_j.Val$$

$$= \sum_{P_j \in PID\_List(XA)} P_j.Val + \sum_{P_j \in dPID\_List(XAB)} P_j.Val - \sum_{P_j \in [PID\_List(XA) \cap dPID\_List(XAB)]} P_j.Val \quad \text{(c)}$$

In addition:

$PID\_List(XA) \cap dPID\_List(XAB) = PID\_List(XA) \cap [PID\_List(XB) \setminus PID\_List(XA)] = \varnothing$

$$\Rightarrow \sum_{P_j \in [PID\_List(XA) \cap dPID\_List(XAB)]} P_j.Val = 0 \quad \text{(d)}$$

According to (c) and (d):

$$g(XAB) = \sum_{P_j \in PID\_List(XA)} P_j.Val + \sum_{P_j \in dPID\_List(XAB)} P_j.Val$$

$$= g(XA) + \sum_{P_j \in dPID\_List(XAB)} P_j.Val$$

Therefore, Theorem 4 is proved.

Example 5. According to *Example 4*, we have $g(c) = 450$ dollars and dPID_List(cd) = {⟨4, 50⟩, ⟨6, 200⟩}. Therefore, $g(cd) = g(c) + \sum_{j=1}^{n} dPID\_List(XAB)_j.Val = 450 + 50 + 200 = 700$ dollars.

3.2 dVM Algorithm

Based on dPID_List structure and its theorems, we develop an improved algorithm of VM algorithm, called dVM, for mining top-rank-*k* erasable itemsets in Fig. 1 with two main steps as follows:

1. Determining top-rank-*k* erasable itemset with their PID_List. Sort them in the gain ascending order.
2. Use the *l*-itemsets in the top-rank-*k* list to generate candidate (*l*+1)-itemsets. Each candidate which has the gain less than the largest value of gain of the top-rank-*k* list will be inserted into the top-rank-*k* list and remove the last element in top-rank-*k* if necessary. Repeat this step until no new itemsets can be inserted into top-rank-*k* list.

```
Input: a product dataset DB and a threshold k
Output: a top-rank-k list, which includes the complete
set of top-rank-k erasable itemsets.

Method:
The result Tab ← ∅
TR ← ∅ // the preserve temporary top-rank-k k-itemset
Scan DB to compute the gain and PID_List of each item
denoted by I1
```

```
Sort I1 in gain ascending order
For (j=0; j<m; j++) do
 If Tab = Ø then
   R.gain = I1[j].gain and R.list.add(I1[j])
   Add R to Tab and I1[j] to TR
 Else if Tab.last_tuple.gain = I1[j].gain
   Add I1[j] to Tab.last_tuple.list and I1[j] to TR
 Else if |Tab| < k then
   R.gain = I1[j].gain and R.list.add(I1[j])
   Add R to Tab and I1[j] to TR
While TR ≠ Ø do
 CR ← Candidate_Generation(TR);
 Sort CR in gain ascending order
 TR ← Ø
 Let x = 0 and l = 0
 While (l < |CR| and x < |Tab|) do
   If Tab[x].gain = CR[l].gain then
     Add CR[l] to Tab[x].list and CR[l] to TR
     l++;
   Else if Tab[x].gain > CR[l].gain then
     R.gain = CR[l].gain and R.list.add(CR[l])
     Insert R to TR into position which satisfy the gain
ascending order
     If |Tab| > k then
       Remove the last tuple in Tab
     Add CR[l] to TR
     l++;
   Else
     x++;
 If |Tab| < k then
   t ← min(k - |Tab|, |CR| - l + 1);
   For (x = l; x< l+t; x++) do
     R.gain = I1[j].gain and R.list.add(I1[j])
     Add R to Tab

Function Candidate_Generation(TR)
CR ← Ø
For each Cu ∈ TR do
 For each Cv ∈ TR (Cu ≠ Cv) do
   If Cu and Cv are in the same equivalence class then
     C = Cu ∪ Cv
     C.PID_List = Cv.PID_List \ Cu.PID_List; // Defini-
tion 4 and Theorem 3
     C.gain = Cu.gain + ∑_{Pj ∈ PID_List(Cv)} Pj.Val; // Theorem 4
     Add C to CR
Return CR
```

Fig. 1. dVM algorithm

4 Experimental Results

All experiments presented in this section were performed on a laptop with an Intel Core i3-3110M 2.4-GHz CPU and 4 GB of RAM. All the programs were coded in C# and .Net Framework Version 4.5.50709.

The experiments are conducted on datasets such as Chess, Mushroom and T10I4D100K which were downloaded from http://fimi.cs.helsinki.fi/data/. To make these datasets look like product datasets, a column was added to store the profit of products. To generate values for this column, a function denoted by $N(100, 50)$, for which the mean value is 100 and the variance is 50, was created. The features of these datasets are shown in Table 2.

Table 2. Features of datasets used in experiments

Dataset[1]	# of Products	# of Items
Chess	3,196	76
Mushroom	8,124	120
T10I4D100K	100,000	870

4.1 Memory Usage

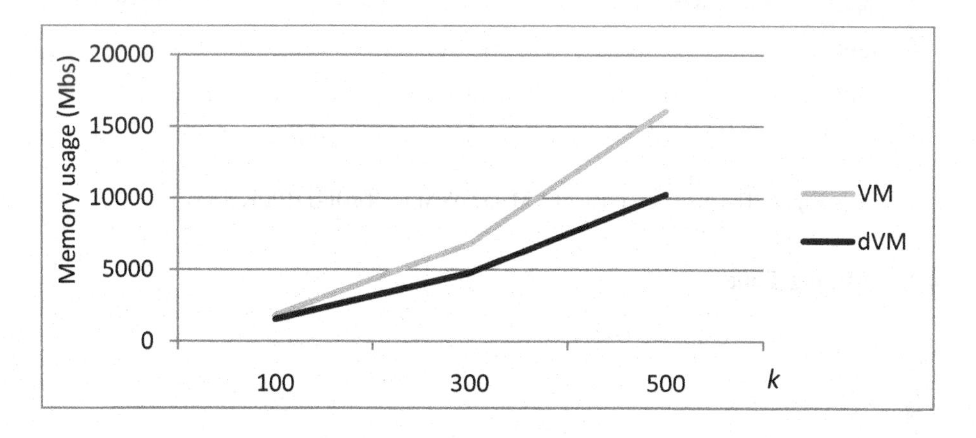

Fig. 2. The memory usage of VM and dVM for Chess dataset

Assume that each product identifiers, gains and items are represented as an integer (4 bytes in memory). We sum the memory usage of VM and dVM in the process of mining top-rank-*k* erasable itemsets. Then we obtained the charts in Figs. 2-4 which show that dVM is better than VM in terms of the memory usage. Especially, for T10I4D100K dataset, VM cannot run with $k = 500$, but dVM can do that.

[1] These datasets are available at `http://sdrv.ms/14eshVm`

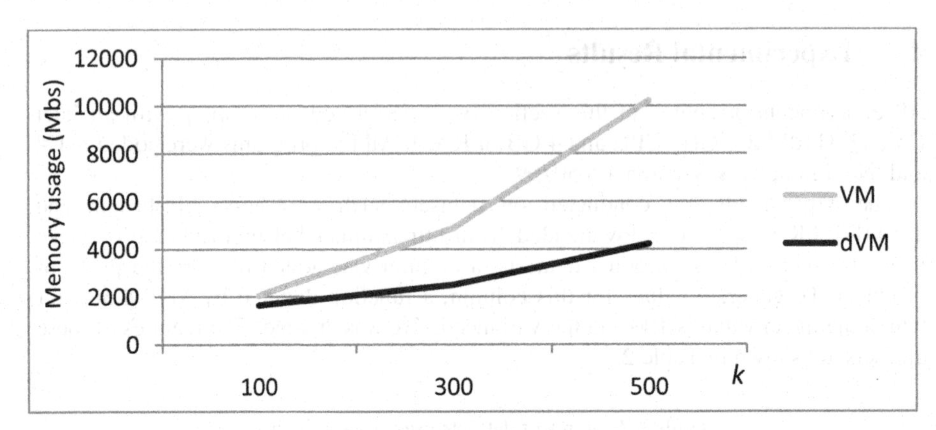

Fig. 3. The memory usage of VM and dVM for Mushroom dataset

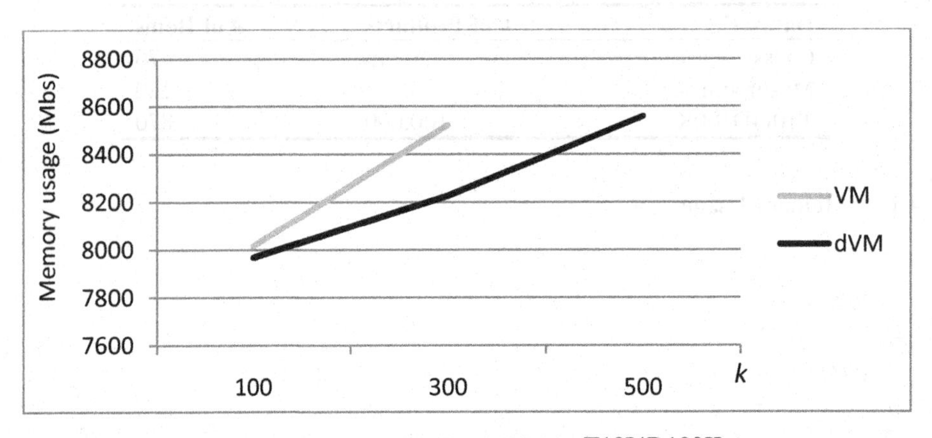

Fig. 4. The memory usage of VM and dVM for T10I4D100K dataset

4.2 Mining Time

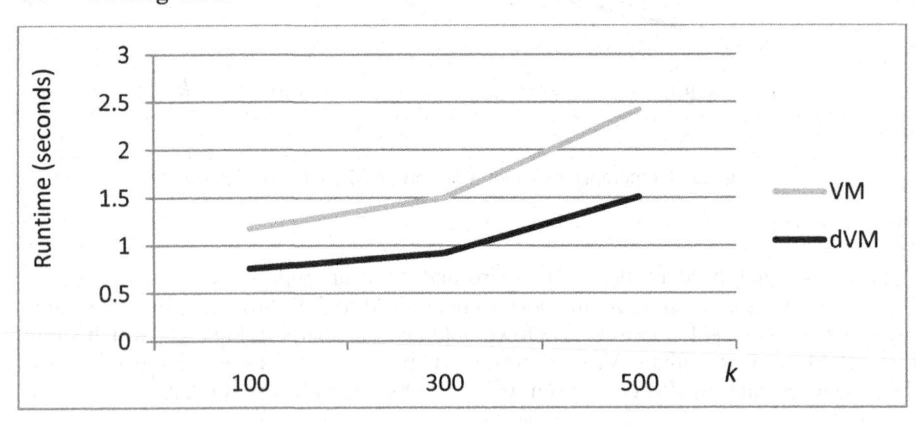

Fig. 5. The mining time of VM and dVM for Chess dataset

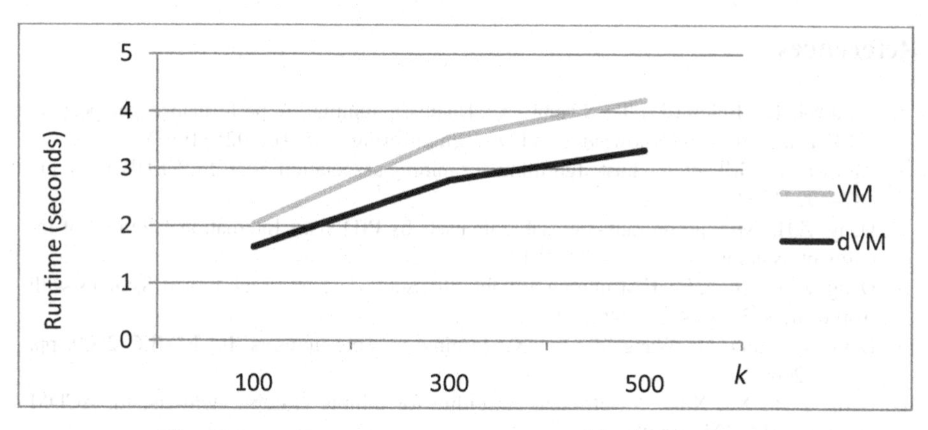

Fig. 6. The mining time of VM and dVM for Mushroom dataset

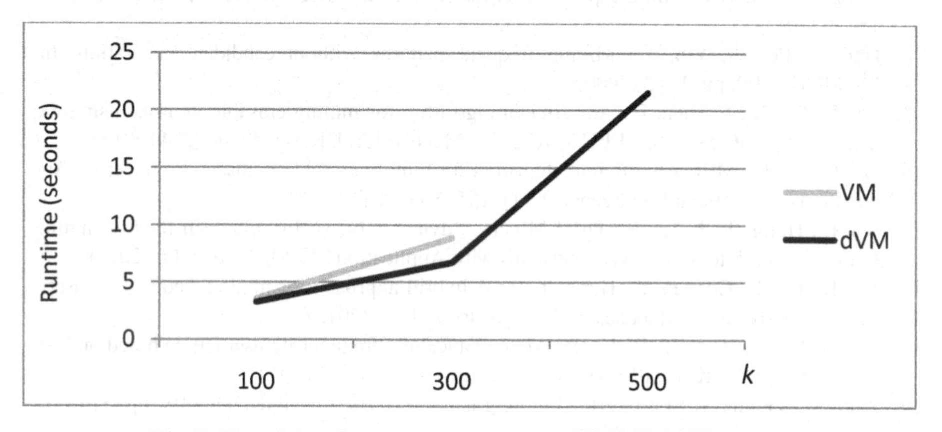

Fig. 7. The mining time of VM and dVM for T10I4D100K dataset

The experimental results presented in Figs. 5-7 show that the proposed algorithm, dVM, is better than VM algorithm in terms of the mining time. Therefore, using dPID_List is always better than using PID_List.

5 Conclusions and Future Work

In this paper, we proposed dPID_List structure to improve VM algorithm [3] in terms of mining time. Based on this structure and its theorems, dVM is proposed for mining top-rank-*k* erasable itemsets effectively. To show the effectiveness of dVM algorithm compared with VM algorithm, we conduct the experiments in terms of the mining time for three datasets. The experimental results show that dVM algorithm is better than VM algorithm. In future work, we will focus in mining erasable closed/maximal itemsets.

References

1. Agrawal, R., Imielinski, T., Swami, A.: Database mining: A performance perspective. IEEE Transactions on Knowledge and Data Engineering 5(6), 914–925 (1993)
2. Agrawal, R., Srikant, R.: Fast algorithms for mining association rules. In: VLDB 1994, pp. 487–499 (1994)
3. Deng, Z.H.: Mining top-rank-*k* erasable itemsets by PID_lists. International Journal of Intelligent Systems 28(4), 366–379 (2013)
4. Deng, Z.H., Xu, X.R.: Fast mining erasable itemsets using NC_sets. Expert Systems with Applications 39(4), 4453–4463 (2012)
5. Deng, Z., Fang, G., Wang, Z., Xu, X.: Mining erasable itemsets. In: ICMLC 2009, pp. 67–73 (2009)
6. Deng, Z.H., Xu, X.R.: An efficient algorithm for mining erasable itemsets. In: ACDM 2010, pp. 214–225 (2009)
7. Deng, Z., Xu, X.: Mining top-rank-*k* erasable itemsets. ICIC Express Letter 2011 5(1), 15–20 (2011)
8. Han, J., Pei, J., Yin, Y.: Mining frequent patterns without candidate generation. In: SIGMOD 2000, pp. 1–12 (2000)
9. Le, T., Vo, B., Coenen, F.: An efficient algorithm for mining erasable itemsets using the difference of NC-Sets. In: IEEE SMC 2013, Manchester, UK, pp. 2270–2274 (2013)
10. Le, T., Vo, B.: MEI: An efficient algorithm for mining erasable itemsets. Engineering Applications of Artificial Intelligence 27(1), 155–166 (2014)
11. Vo, B., Hong, T.-P., Le, B.: DBV-Miner: A dynamic bit-vector approach for fast mining frequent closed itemsets. Expert Systems with Applications 39(8), 7196–7206 (2012)
12. Vo, B., Le, T., Coenen, F., Hong, T.-P.: A hybrid approach for mining frequent itemsets. In: IEEE SMC 2013, Manchester, UK, pp. 4647–4651 (2013)
13. Vo, B., Le, T., Hong, T.-P., Le, B.: Maintenance of a frequent-itemset lattice based on Prelarge concept. In: KSE 2013, Ha Noi, Vietnam, pp. 295–305 (2013)
14. Zaki, M., Gouda, K.: Fast vertical mining using diffsets. In: SIGKDD 2003, pp. 326–335 (2003)

New Method for Extracting Keyword for the Social Actor

Mahyuddin K.M. Nasution*

Information Technology Department
Fakultas Ilmu Komputer dan Teknologi Informasi (Fasilkom-TI)
and
Centre of Information System
Universitas Sumatera Utara, Medan 20155 USU, Sumatera Utara, Indonesia
mahyunst@yahoo.com, mahyuddin@usu.ac.id

Abstract. In this paper we study the relationship between query and search engine by exploring some properties and also applying their relations to extract keyword for any social actor by proposing new method. The proposed approach based on considering the result of search engine in the singleton and doubleton. In this paper, we develop a novel method for extracting keyword automatically from Web with mirror shade concept (M2M). Results show the potential of the proposed approach, in experiment we get that the performance (recall and precision) of keyword depend on both weights (singleton and tfidf) and the distance of them.

Keywords: singleton, doubleton, searh engine, query, information retrieval.

1 Introduction

In the search space with a large repository such as Web, it is difficult to obtain accurate information about any social actor or social agent, that is endowed as an human agency which means recognising individually the attempt to grips the challenge for changing world around the agent to a good world. In this case, there are major obstacles that often accompanies the search engine capabilities such as ambiguity [2] and bias [5]. Therefore, it is always necessary corresponding keywords to pry information out of the heaps of data or documents in Web. In this paper we propose a new method for generating and selecting automatically the keyword for someone as an social actor in Web based on the principles of information retrieval (IR) and model of search engine.

* Please note that the LNCS Editorial assumes that all authors have used the western naming convention, with given names preceding surnames. This determines the structure of the names in the running heads and the author index.

N.T. Nguyen et al. (Eds.): ACIIDS 2014, Part I, LNAI 8397, pp. 83–92, 2014.

2 Problem Definition

We define some terminologies and the properties of a model of search engine [13,14,15,19].

T1 A *term* t_x consists of at least one or a set of words in a pattern, or $t_x = (w_1 w_2 \dots w_l)$, $l \leq k$, k is a number of parameters representing word w, l is the number of tokens (vocabularies) in t_x, $|t_x| = k$ is size of t_x.

T2 Let a set of web pages indexed by *search engine* be Ω, i.e., a set contains ordered pair of the terms t_{x_i} and the web pages ω_{x_j}, or (t_{x_i}, ω_{x_j}), $i = 1, \dots, I$, $j = 1, \dots, J$. The relation table that consists of two columns t_x and ω_x is a representation of (t_{x_i}, ω_{x_j}) where $\Omega_x = \{(t_x, \omega_x)_{ij}\} \subset \Omega$ or $\Omega_x = \{\omega_{x_1}, \dots, \omega_{x_j}\}$. The cardinality of Ω is denoted by $|\Omega|$.

T3 Let t_x is a search term, and $t_x \in \mathcal{S}$ where $\mathcal{S}$ is a set of singleton search term of search engine. A vector space $\Omega_x \subseteq \Omega$ is a singleton search engine event (*singleton space of event*) of web pages that contain an occurrence of $t_x \in \omega_x$. The cardinality of Ω_x is denoted by $|\Omega_x|$.

T4 Let t_x and t_y are two different search term, $t_x \neq t_y$, $t_x, t_y \in \mathcal{S}$, where $\mathcal{S}$ is a set of singleton search term of search engine. A doubleton search term is $\mathcal{D} = \{\{t_x, t_y\} : t_x, t_y \in \Sigma\}$ and its vector space denoted by $\Omega_x \cap \Omega_y$ is a double search engine event (*doubleton space of event*) of web pages that contain a co-occurrence of t_x and t_y such that $t_x, t_y \in \omega_x$ and $t_x, t_y \in \omega_y$, where $\Omega_x, \Omega_y, \Omega_x \cap \Omega_y \subseteq \Omega$.

For t_x and t_y are the search terms with conditions: $t_x \neq t_y$, $t_x \cap t_y \neq \emptyset$ and $|t_y| < |t_x|$. We have $\forall w_y \in t_y$, $w_y \in t_x$, $\exists w_x \in t_x$, $w_x \notin t_y \Rightarrow \forall w_y \in \omega_y$, $w_y \in \omega_x$, $\exists w_x \in \omega_x$, $\omega_x \notin \omega_y$ such that

$$t_x \cap t_y = t_y \text{ and } t_x \cup t_y = t_x \tag{1}$$

and

$$\omega_x \cap \omega_y = \omega_y \text{ and } \omega_x \cup \omega_y = \omega_x. \tag{2}$$

Similarly, let t_x and t_y are two search terms of the different queries, we have $\Omega_x \cap \Omega_y = \emptyset$ and clear that $\Omega_x \neq \Omega_y$ and $|\Omega_x \cap \Omega_y| = 0$, then

$$|\Omega_x \cup \Omega_y| = |\Omega_x| + |\Omega_y|. \tag{3}$$

Let $\Omega_x = \{(t_x, \omega_x)\}$, based on meaning Eq. (1) and Eq. (2), we have $\Omega_x = \{(t_x, \omega_x)\} = \{(t_x \cup t_y, \omega_x \cup \omega_y)\} = \{(t_x, \omega_x) \cup (t_y, \omega_y)\} = \{(t_x, \omega_x)\} \cup \{(t_y, \omega_y)\} = \Omega_x \cup \Omega_y$. Therefore, based on Eq. (3) we obtain $|\Omega_x| = |\Omega_x| + |\Omega_y|$. For different conditions we obtain another properties. Those properties are as follows.

P1 Let t_x and t_y are search term. If $t_x \neq t_y$, $t_x \cap t_y \neq \emptyset$ and $|t_y| < |t_x|$, then singleton search engine event of t_x and t_y is $\Omega_x = \Omega_x \cup \Omega_y$ or

$$|\Omega_x| = |\Omega_x| + |\Omega_y|, \tag{4}$$

where $\Omega_x, \Omega_y \subseteq \Omega$.

P2 If $t_y \neq t_z$ and $t_y \cap t_z = \emptyset$, then $|\Omega_y \cap \Omega_z| = 0$ and $|\Omega_y \cup \Omega_z| = |\Omega_y| + |\Omega_z|$.
P3 Let t_x and t_z are search terms. If $t_x \neq t_z$, $t_x \cap t_z = \emptyset$, and $\omega_x \cap \omega_z \neq \emptyset$, then $|\Omega_x| = |\Omega_z|$, $\Omega_x, \Omega_z \subseteq \Omega$.

Based on P1, $|\Omega_x \cap \Omega_y| = |\{(t_x, \omega_x)\} \cap \{(t_y, \omega_y)\}| = |\{(t_x \cap t_y, \omega_x \cap \omega_y)\}| = |\{(t_y, \omega_y)\}| = |\Omega_y|$ or

$$|\Omega_x \cap \Omega_y| = |\Omega_y| \tag{5}$$

Because $|\Omega_y| < |\Omega_x|$, we have $|\Omega_x \cap \Omega_y| < |\Omega_x|$. However, by P2, $|\Omega_x \cap \Omega_y| = |\{(t_x, \omega_x)\} \cap \{(t_y, \omega_y)\}| = |\{(t_x \cap t_y, \omega_x \cap \omega_y)\}| = \emptyset$. This means that

$$|\Omega_x \cap \Omega_y| < |\Omega_x| \wedge |\Omega_x \cap \Omega_y| < |\Omega_y|. \tag{6}$$

Based on P3, $|\Omega_x \cap \Omega_y| = |\{(t_x, \omega_x)\} \cap \{(t_y, \omega_y)\}| = |\{t_x \cap t_y, \omega_x \cap \omega_y)\}| = |\{(t_x, \omega_x)\}| = |\Omega_x|$ or

$$|\Omega_x \cap \Omega_y| = |\Omega_x| \tag{7}$$

Therefore, Eqs. (5), (6) and (7) clearly give $|\Omega_x \cap \Omega_y| \leq |\Omega_x| \leq |\Omega|$ or $|\Omega_x \cap \Omega_y| \leq |\Omega_y| \leq |\Omega|$. It has proved a theorem as follows.

Theorem 1. *Let t_x and t_y are search terms. If $t_x \neq t_y$, but $\{(t_x, \omega_x)\} \cap \{(t_y, \omega)\} \neq \emptyset$, then a doubleton search engine event of t_x and t_y is the $\Omega_x \cap \Omega_y$, $\Omega_x, \Omega_y \subseteq \Omega$, $|\Omega_x \cap \Omega_y| \leq |\Omega_x| \leq |\Omega|$ and $|\Omega_x \cap \Omega_y| \leq |\Omega_y| \leq |\Omega|$.*

Otherwise, let t_x and t_y are any search terms and we can derive a formula by starting from Eq. (7), based on Eq. (5) and then P1, i.e.,

$$|\Omega_x \cap \Omega_y| = |\Omega_x| + |\Omega_y| + |\Omega_x \cap \Omega_y| \tag{8}$$

and we know that $|\Omega_x| = |\Omega_x \cap \Omega_x|$ and $|\Omega_y| = |\Omega_y \cap \Omega_y|$, then Eq. (8) be

$$|\Omega_x \cap \Omega_y| = |\Omega_x \cap \Omega_y| + |\Omega_x \cap \Omega_x| + |\Omega_y \cap \Omega_y| \tag{9}$$

As information of any social actor the singleton and the doubleton are the basic of some properties of search engine statistically that related to the actor social. However, either a singleton or a doubleton depend on formulating a query, i.e. where and how the keyword there: Some of techniques for mining keyword from information sources have been proposed, for example is to estimate classifiers for labelling some messages by using simple [1] and more sophisticated [3,4] approaches. For features extraction several methods have been developed [7]. Some of them are the substring search method [20], model and prototype system [8], by using peer clustering [11], co-occurrence analysis [9], by using lexical chains [6], based on PageRank [21], laten sematic analysis [10], etc. In this case, the singleton [16] and the doubleton [17] are the necessary condition for gaining the information of social actor from Web because both singleton and doubleton contain bias and ambiguity, while other purpose requires a sufficient condition [18] so that the major obstacles can be reduced or eliminated. Some of the following formulas will be evidence against some of the approaches and methods that have been proposed by some researchers for extracting keyword.

3 The Proposed Approach

If the singleton is accompanied by a summary of the Web, then involvement of the singleton and doubleton in the computation generates descriptions (as keyword candidates) of an social actor as follows.

Definition 1. *Let t_a is a search term. $S = \{w_1, \ldots, w_{max}\}$ is a Web snippet (briefly snippet), $S \subset \omega_{a_i} \in \Omega$, where $max \leq 50$ words to the left and right of t_a that returned by search engines. $L = \{S_i : i = 1, \ldots, I\}$ is a list of snippets.*

We construct a relationship of actors-snippets-words based on frequency of words in Web pages as environments of an social actor as follows.

Definition 2. *A relationship between social actors, web snippets and words is defined as the mixture $p(a, S, w) = a \times S \times w$, $a \in A$, $S \in L \subseteq \Omega$, $w \in S$. A vector space of $P(a, S, w)$ is defined as $\mathbf{w} = \{w_i, \ldots, w_j\} = [\nu_i, \ldots, \nu_j]$, $\nu_i \geq \ldots \geq \nu_j$, where $w_i, \ldots, w_j$ are the unique words in S and $\nu_i, \ldots, \nu_j$ are the weights of word.*

Statistically, the task of relationship in Definition 2 is simply to gather and record information about words, features, and web pages where term weights reflect the relative importance of words in web pages. One of the most common type used in older retrieval models is known as *tf.idf* weighting [12] whereby we can generate the vector ν for each word/term w, and then this information is used for recognizing the different social actors in web pages based on clustering all words by using one of similarity measurements such as using Jaccard coeficient $jc = |\Omega_a \cap \Omega_b|/(|\Omega_a| + |\Omega_b| - |\Omega_a \cap \Omega_b|)$. For this purpose, we define the words undirected graph $G = (V, E)$ to describe the relations between words [12].

Definition 3. *Assume a sub-graph G', $G' \subset G$, G' is a* micro-cluster *satisfies the conditions as follows*

1. *There are a set of word $\mathbf{w} = \{w_x, \ldots, w_y\}$ whose vector space $[\nu_x, \ldots, \nu_y]$ and $\nu_x \geq \ldots \geq \nu_y \geq \alpha$, where α is a threshold.*
2. *There are an one-one function $f : \mathbf{w} \to V$ such that $f(w) = v$, $\forall w \in \mathbf{w} \exists v \in V$ where $v \in V$ is a vertex in G'.*
3. *There are an one-one function $\rho : \mathbf{w} \times \mathbf{w} \to E$ such that $\rho(w_x, w_y) = e$, $\forall w_x, w_y \in \mathbf{w}$, where ρ is a relation among words and $e \in E$ is a edge in G'.*

The micro-cluster is denoted by $G' = \langle V, E, \mathbf{w}, f, \rho, \alpha \rangle$.

A micro-cluster is maximal clique sub graph of t_o where the node represents word has the highest score in document. However, the collection of mentioned words do not exactly refer to the same social actors. To group the words into the appropriated cluster, we construct the trees of words. This based on an assumption that the words are that appear in same domains having closest relation. The tree is an optimal representation of relation in graph G.

Definition 4. *A tree T is an* optimal micro-cluster *if and only if T is a subgraph of micro cluster G', and is denoted by $T = \langle V_T, E_T, \mathbf{w}_T, f, \rho, \alpha \rangle$, where $V_T \subseteq V$, $E_T \subset E$, and $\mathbf{w}_T \subseteq \mathbf{w}$.*

In building the optimal micro-cluster, we save the strongest relations in T between a word and another in G' until T has no cycle. We introduce an intrusive word about the social actor, and there are at least one word of optimal micro-cluster has strongest relation with the intrusive word, and an optimal micro-cluster is a group of words refer to that social actor. However, the overlap keyword also exists in the same list. We define a strategy to select relevant words among all list candidates. In this case, there are a few potential words as keyword candidates.

Definition 5. *A vector space $\mathbf{s} = [|\mathbf{x}|, \ldots, |\mathbf{y}|]$ is a* mirror shade *of micro-cluster G' if there is an one-one function $g : \mathbf{w} \to \mathbf{s}$, where $\mathbf{x}, \ldots, \mathbf{y}$ are in event space. Let $\mathbf{z}$ is a vector whose greatest value in $\mathbf{s}$, the vector space in range of $[0, 1]$ is relatively defined as $\mathbf{s}_{[0,1]} = [|\mathbf{x}|/|\mathbf{z}|, \ldots, |\mathbf{y}|/|\mathbf{z}|] = [\mu_x, \ldots, \mu_y]$.*

We also can generate for example another vectors from $\Omega_i, \ldots, \Omega_j$ for words $w_i, \ldots, w_j$ respectively such that $[\mu_i, \ldots, \mu_j] = [\Omega_i, \ldots, \Omega_j]$ is a mirror shade of $[\nu_i, \ldots, \nu_j]$ from $tf.idf$.

Lemma 1. *Let $\mathbf{s}_T \subseteq \mathbf{s}$, then $\mathbf{s}_T$ is the mirror shade of an optimal micro-cluster T.*

Proof. Let $\mathbf{s}_T \subseteq \mathbf{s}$, based on Definition 5 we have $\mathbf{w}_T \subseteq \mathbf{w}$, i.e. $g(\mathbf{w}_T) = \mathbf{s}_T$ or because of g is one-one function, $g^{-1}(\mathbf{s}_T) = \mathbf{w}_T \subseteq \mathbf{w}$. Next, by applying Definition 3, $f(\mathbf{w}_T) = V_T$, or because of f is one-one function, $f^{-1}(V_T) = \mathbf{w}_T \subseteq \mathbf{w}$, and $s_T = g(\mathbf{w}_T) = g(f^{-1}(V_T)) = f^{-1}g(V_T)$ and we obtain $\rho(\mathbf{w}, \mathbf{w}) = \rho(f^{-1}(V), f^{-1}(V)) \subseteq E$, so $\rho(\mathbf{s}_T \times \mathbf{s}_T) = \rho(g(\mathbf{w}_T) \times g(\mathbf{w}_T)) = \rho(g(f^{-1}(V_T)) \times g(f^{-1}(V_T))) = \rho(f^{-1}g(V_T)) \times f^{-1}g(V_T))) = f^{-1}g\rho(V_T \times V_T) = f^{-1}g(\rho(V_T \times v_T))$ because of $f^{-1}g$ is also one-one function, this means that $V_T \subseteq V$ has $\mathbf{s}_T$ as a mirror shade of $\mathbf{w}_T$.

Lemma 2. *Let t_a, t_x, t_y are search terms, and $\Omega_a, \Omega_x, \Omega_y \subseteq \Omega$ are the singleton of them. If $\nu_x \geq \nu_y$, then $|\Omega_x \cap \Omega_a| \geq |\Omega_y \cap \Omega_a|$.*

Proof. By T2 and T3, $\Omega_a = \{(t_a, \omega_a)\}$, $\Omega_x = \{(t_x, \omega_x)\}$ and $\Omega_y = \{(t_y, \omega_y)\}$. Let L_a is a list of snippet S for query t_a or $L_a = \{(t_a, S)\}$, and by Definition 1 also $L_a = \{(t_x, S)\}$ or $L_a = \{(t_y, S)\}$. It applies that $\{(t_x, S)\} = \{(t_a, S)\} \cap \{(t_x, S)\} = \{(t_a, \omega_a)\} \cap \{(t_x, \omega_a)\} = \{(t_a, \omega_a)\} \cap \{(t_x, \omega_x)\} = \Omega_a \cap \Omega_x$, and similarly $\{(t_y, S)\} = \Omega_a \cap \Omega_y$, but by Definition 2 we obtain $\nu_x = |\{(t_x, S)\}|$ and $\nu_y = |\{(t_y, S)\}|$, and $\nu_x \geq \nu_y$ means that $|\{(t_a, S)\}| \geq |\{(t_y, S)\}|$ and $|\Omega_x \cap \Omega_a| \geq |\Omega_y \cap \Omega_a|$.

Lemma 2 declared that the words appeared frequently in certain snippets but rarely in the remaining of snippets are that words strongly associated with one of social actors only. If $\nu_x \geq \nu_y$ then $|\Omega_x \cap \Omega_a| \geq \ldots \geq |\Omega_y \cap \Omega_a|$. Otherwise, the words that do not appear frequently in Web pages except on only an social actor, then the words are strongly associated with that social actor. The last statement can be stated as follows.

Lemma 3. *Let t_a, t_x, t_y are search terms and $\Omega_a, \Omega_x, \Omega_y \subseteq \Omega$ are the singletons of them. Let $|\Omega_x| \geq |\Omega_y|$, if $|\Omega_x \cap \Omega_a| \leq |\Omega_y \cap \Omega_a|$, then $|\Omega_a - \Omega_x| \geq |\Omega_a - \Omega_y|$.*

Proof. By conditions of Theorem 1, we obtain $|\Omega_y| = |\{(t_y, \omega_y)\}| = |\{(t_y \cap t_a, \omega_y \cap \omega_a\})| = |\{(t_y, \omega_y)\} \cap \{(t_a, \omega_a)\}| = |\Omega_y \cap \Omega_a|$ or $|\Omega_y| = |\Omega_y \cap \Omega_a|$, and similarly we have also $|\Omega_x| = |\Omega_x \cap \Omega_a|$ such that $|\Omega_y| \leq |\Omega_x| \Rightarrow |\Omega_y \cap \Omega_a| \leq |\Omega_x \cap \Omega_a|$. It contradicts to the assumption that $|\Omega_x \cap \Omega_a| \leq |\Omega_y \cap \Omega_a|$, but do not contradict to $|\Omega_y \cap \Omega_a| \leq |\Omega_x|$. Similarly, it derived from the left side of $|\Omega_y| \leq |\Omega_x|$, because $|\Omega_y \cap \Omega_a| \leq |\Omega_y|$, then $|\Omega_x \cap \Omega_a| \leq |\Omega_x|$. Therefore, based on condition of P1 and Theorem 1, we obtain

$$
\begin{aligned}
|\Omega_x \cap \Omega_a| &\leq |\Omega_y \cap \Omega_a| \\
|(\Omega_x \cap \Omega_a)| - |\Omega_a| &\leq |(\Omega_y \cap \Omega_a)| - |\Omega_a| \\
|(\Omega_x \cap \Omega_a) - \Omega_a| &\leq |(\Omega_y \cap \Omega_a) - \Omega_a| \\
|\Omega_a - (\Omega_x \cap \Omega_a)| &\geq |\Omega_a - (\Omega_y \cap \Omega_a)| \\
|\neg(\Omega_x \cap \Omega_a)| &\geq |\neg(\Omega_y \cap \Omega_a)| \\
|(\Omega_x \cup \Omega_a) - (\Omega_x \cap \Omega_a)| &\geq |(\Omega_y \cup \Omega_a) - (\Omega_y \cap \Omega_a)|
\end{aligned}
$$

and for $\Omega_a = \Omega_a \cup \Omega_x = \Omega_a \cup \Omega_y$, $\Omega_x = \Omega_x \cap \Omega_a$ and $\Omega_y = \Omega_y \cap \Omega_a$, we obtain $|\Omega_a - \Omega_x| \geq |\Omega_a - \Omega_y|$.

Lemma 3 explains that distance between an social actor t_a and candidate words t_x and t_y can be used to select an appropriate keyword, or if $\mu_x \geq \mu_y$ then t_y is a priority word that is closest to t_a. Let $\nu \in [0,1]$ is a weight of word w and $\mu \in [0,1]$ is a vector in $\mathbf{s}_T$, there are three conditions of relation between ν and μ: (1) $\nu = \mu \Leftrightarrow \nu - \mu = 0$, (2) $\nu < \mu \Leftrightarrow \nu - \mu < 0$ (negative), and (3) $\nu > \mu \Leftrightarrow \nu - \mu > 0$ (positive).

Proposition 1. *If the internval $[0,1]$ divided by straight line into two areas: $[0, \frac{1}{2})$ and $[\frac{1}{2}, 1]$, then there are six patterns of conditions satisfying the relation between ν and μ, i.e., (1) $\nu \geq \mu$, $\nu \geq \frac{1}{2}$, $\mu \leq \frac{1}{2}$; (2) $\nu \geq \mu$, $\nu \geq \frac{1}{2}$, $\mu \geq \frac{1}{2}$; (3) $\nu \geq \mu$, $\nu \leq \frac{1}{2}$, $\mu \leq \frac{1}{2}$; (4) $\nu < \mu$, $\nu > \frac{1}{2}$, $\mu > \frac{1}{2}$; (5) $\nu < \mu$, $\nu < \frac{1}{2}$, $\mu < \frac{1}{2}$; and (6) $\nu < \mu$, $\nu < \frac{1}{2}$, $\mu > \frac{1}{2}$.*

Proof. Let us summarise the conditions of relation among ν and μ into (i) $\nu \geq \mu$ and (ii) $\nu < \mu$, and based on the condition $\nu \geq \mu$ from Lemma 2 and Lemma 3 we can determine the value in {TRUE,FALSE} for relation patterns between ν and μ: (1) If $\nu \geq \mu$, then $\nu \geq \frac{1}{2}$ and $\mu \leq \frac{1}{2}$ (TRUE); (2) If $\nu \geq \mu$, then $\nu \geq \frac{1}{2}$ and $\mu \geq \frac{1}{2}$ (TRUE); (3) If $\nu \geq \mu$, then $\nu \leq \frac{1}{2}$ and $\mu \leq \frac{1}{2}$ (TRUE); (4) If $\nu < \mu$, then $\nu > \frac{1}{2}$ and $\mu > \frac{1}{2}$ (TRUE); (5) If $\nu < \mu$, then $\nu < \frac{1}{2}$ and $\mu < \frac{1}{2}$ (TRUE); (6) If $\nu < \mu$, then $\nu < \frac{1}{2}$ and $\mu > \frac{1}{2}$ (TRUE); (7) If $\nu \geq \mu$, then $\nu \leq \frac{1}{2}$ and $\mu \geq \frac{1}{2}$ (FALSE); and (8) If $\nu < \mu$, then $\nu > \frac{1}{2}$ and $\mu < \frac{1}{2}$ (FALSE). Thus, there are only six patterns with TRUE value.

We can sort the candidate words by using six patterns of conditions for satisfying Proposition 1, and the selected word as keyword is a candidate word with a vector that satisfies one of six patterns.

Theorem 2. *Let T is an optimal micro-cluster containing the keyword candidates, then the suitable keyword is a keyword candidate with the highest value of vector space of $p(a, S, w)$ and lowest value of mirror shade, where the distance between two values is large enough.*

Proof. Let $\mathbf{w}_T$ are the keyword candidates in T. Each word in $\mathbf{w}_T$ has a value in a vector space of $p(a, S, w)$ and a value in a vector space of mirror shade. Based on Lemma 2, there is a word has a highest value of $p(a, S, w)$ in $[0, 1]$, while by Lemma 3 the mentioned word has a lowest value of mirror shade in $[0, 1]$, i.e., $\nu > \mu$ in $[0, 1]$. Therefore, only three of patterns on Proposition 1: (a) $\nu \geq \mu$, $\nu \geq \frac{1}{2}$, $\mu \geq \frac{1}{2}$; (b) $\nu \geq \mu$, $\nu \geq \frac{1}{2}$, $\mu \geq \frac{1}{2}$, and (c) $\nu \geq \mu$, $\nu \geq \frac{1}{2}$, $\mu \leq \frac{1}{2}$.
Let us define a distance between ν and μ: $\delta = \nu - \mu$. For first pattern with max values: $\delta = 1 - \frac{1}{2} = \frac{1}{2}$, with min values: $\delta = \frac{1}{2} - 0 = \frac{1}{2}$, and with max-min values: $\delta = 1 - 0 = 1$. For second pattern with max values: $\delta = 1 - 1 = 0$, with min values: $\delta = \frac{1}{2} - \frac{1}{2} = 0$, and with max-min values: $\delta = 1 - \frac{1}{2} = \frac{1}{2}$. Last pattern with max values: $\delta = \frac{1}{2} - \frac{1}{2} = 0$, with min values: $\delta = 0 - 0 = 0$, and with max-min values: $\delta = \frac{1}{2} - 0 = \frac{1}{2}$, where max value of ν and μ are respectively 1 and $\frac{1}{2}$, while min value of them are respectively $\frac{1}{2}$ and 0. Thus, one pattern gives the maximum value, i.e., $\delta = 1$. It means that there is a keyword candidate in $\mathbf{w}_T$ as an optimal keyword, where $\nu = max$ value and $\mu = min$ value, or $\nu - \mu = max - min$ values.

Three values of each word $w \in \mathbf{w}$ determine a relationship between $\mathbf{w}$ and any social actor. The last theorem expresses that the suitable keyword will provide to a query the enriched information with semantic relations of their contents, and this give more effectiveness retrieval of information. The effectiveness of using keywords dependent the query levels generally based on $\delta_x > \ldots > \delta_y$ if and only if t_x is suitable top keyword. This is an algorithm by using the micro-cluster and the mirror-shade (therefore we called it as MM method (M2M)) for generating keyword as follows.

generate(keyword)
INPUT : A set of social actors
OUTPUT : keyword(s) of each social actor
STEPS :

1. $\mathbf{w} = \{w_1, w_2, \ldots, w_n\} \leftarrow$ Collect words-(terms) per social actor from snippet.
2. $\{\nu_1, \nu_2, \ldots, \nu_n\} \leftarrow$ Generate vector $\forall w \in \mathbf{w}$ based on $tf.idf$.
3. $\{\mu_1, \mu_2, \ldots, \mu_n\} \leftarrow$ Generate vector for each hit count $w \in \mathbf{w}$ divided by highest hit count.
4. $G' \leftarrow$ Build a micro cluster using singleton and doubleton of W.
5. $T' \leftarrow$ Make optimal micro cluster.
6. If T' do not consist of trees, then collect and cut node with degree $deg > 1$ for seperating T' be trees.
7. Select a cluster from trees of T' by using a predefined stable attribute.
8. Take maximum δ from candidate keyword in a cluster.

Fig. 1. The optimal micro-cluster

4 Experiment

Let us consider information context of social actors that includes all relevant relationships with their interaction history, where Yahoo! search engines fall short of utilizing any specific information, especially micro-cluster information, and just therefore we use full text index search in web snippets. In experiment, we use maximum of 500 web snippets for search term t_a representing an actor, and we consider words where the TF.IDF value $> 0.3\times$ highest value of TF.IDF, or maximum number is 30 words. For example, Fig. 1 is a set of 30 words from web snippets for an actor = "Abdullah Mohd Zin". We test for 143 names, and we obtain 8 (5.59%) actors without a cluster of candidate words, 14 (9.09%) actors with only one cluster, and 122 (85.32%) persons have two or more keywords. In a case of "Abdulah Mohd Zin" we have trees of words as micro clusters of words. We can arrange the keyword candidates individual according to their proximity to the stable attribute "academic", i.e. a set of words in SK = {sciences, faculty, associate, economic, prof, environment, career, journal, network, university, report, relationship, context,...}. SK and δ maximum exactly determine that "network" be a keyword for actor "Abdullah Mohd Zin" as an academic (not a politician). In this case, first keyword is "computer", while second keyword is "university", and for dataset of "Abdullah Mohd Zin" with 143 files we obtain the recall and precision for the keywords are 59.44% & 58.74% and 58.62% & 53.85% respectively, see Fig. 2.

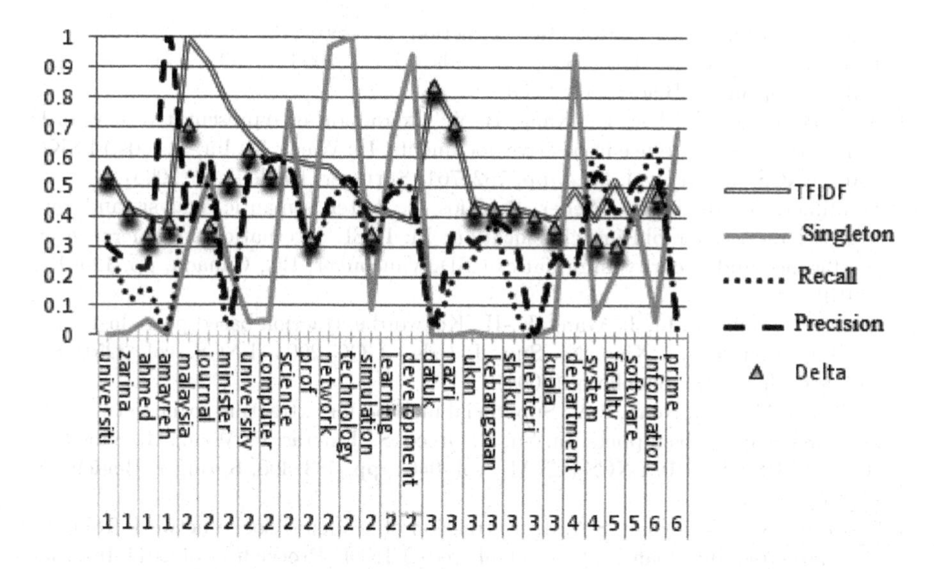

Fig. 2. Recall and precision of the optimal micro-cluster

5 Conclusion and Future Work

Studying to properties of relation between query and search engine gave the semantic meaning to the social actors. One of them is to provide keyword for any social actor or the clue about the social actor. The mirror-shade approach played a role to select top keyword from summary of web pages about the actor. Our near future work is to experiment and look into IR performance.

References

1. Abu-Nimeh, S., Nappa, D., Wang, X., Nair, S.: A comparison of machine learning techniques for phising detection. In: Proceedings of the Anti-Phising Working Groups 2nd Annual eCrime Researhers Summit, pp. 60–69 (2007)
2. Adriani, M.: Using statistical term similarity for sense disambiguation in cross-language information retrieval. Information Retrieval 2, 69–80 (2000)
3. Bergholz, A., Chang, J.-H., Paass, G., Reichartz, F., Strobel, S.: Improved phishing detection using model-based features. In: Proceedings of Fifth Conference on Email and Anti-Spam (2008)
4. Bergholz, A., Beer, J.D., Glahn, S., Moens, M.-F., Paass, G., Strobel, S.: New filtering approaches for phishing email. Journal of Computer Security (2009)
5. Buckley, C., Dimmick, D., Soboroff, I., Voorhees, E.: Bias and the limits of pooling for large collections. Information Retrieval 10, 491–508 (2007)
6. Ercan, G., Cicekli, I.: Using lexical chains for keyword extraction. Information Processing and Management 43, 1705–1714 (2007)
7. Fette, I., Sadeh, N., Tomasic, A.: Learning to detect phising emails. In: ACM Proceedings of the 16th International Conference on World Wide Web, pp. 649–656 (2007)

8. HaCohen-Kerner, Y.: Automatic extraction of keywords froom abstracts. In: Palade, V., Howlett, R.J., Jain, L. (eds.) KES 2003. LNCS, vol. 2773, pp. 843–849. Springer, Heidelberg (2003)
9. Kim, B.-M., Li, Q., Lee, K., Kang, B.-Y.: Extraction of representative keywords considering co-occurrence in positive documents. In: Wang, L., Jin, Y. (eds.) FSKD 2005. LNCS (LNAI), vol. 3614, pp. 752–761. Springer, Heidelberg (2005)
10. L'Huillier, G., Hevia, A., Weber, R., Ríos, S.: Laten semantic analysis and keyword extraction for phising classification. In: IEEE International Conference on Intelligence and Security Informatics (ISI), Vancouver, BC, Canada, pp. 129–131 (2010)
11. Liang, B., Tang, J., Li, J., Wang, K.-H.: Keyword extraction based peer clustering. In: Jin, H., Pan, Y., Xiao, N., Sun, J. (eds.) GCC 2004. LNCS, vol. 3251, pp. 827–830. Springer, Heidelberg (2004)
12. Nasution, M.K.M., Noah, S.A.: Superficial method for extracting social network for academic using web snippets. In: Yu, J., Greco, S., Lingras, P., Wang, G., Skowron, A. (eds.) RSKT 2010. LNCS (LNAI), vol. 6401, pp. 483–490. Springer, Heidelberg (2010)
13. Nasution, M.K.M., Noah, S.A.M.: Extraction of academic social network from online database. In: Noah, S.A.M., et al. (eds.) IEEE Proceeding of 2011 International Conference on Semantic Technology and Information Retrieval, Putrajaya, Malaysia, pp. 64–69. IEEE (2011)
14. Nasution, M.K.M., Noah, S.A.M., Saad, S.: Social network extraction: Superficial method and information retrieval. In: Proceeding of International Conference on Informatics for Development (ICID 2011), pp. c2-110-c2-115 (2011)
15. Nasution, M.K.M., Noah, S.A.M.: Information Retrieval Model: A Social Network Extraction Perspective. In: IEEE Proc. of CAMP 2012 (2012)
16. Nasution, M.K.M.: Simple search engine model: Adaptive properties. Cornell University Library (arXiv:1212.3906v1) (2012)
17. Nasution, M.K.M.: Simple search engine model: Adaptive properties for doubleton. Cornell University Library (arXiv:1212.4702v1) (2012)
18. Nasution, M.K.M.: Simple search engine model: Selective properties. Cornell University Library (arXiv:1303.3964v1) (2012)
19. Nasution, M.K.M. (Mahyuddin): Kaedah dangkal bagi pengekstrakan rangkaian sosial akademik dari Web, Ph.D. Dissertation, Universiti Kebangsaan Malaysia (2013) (in Malay)
20. Okada, M., Ando, K., Le, S.S., Hayashi, Y., Aoe, J.-I.: An efficient substring search method by using delayed keyword extraction. Information Processing and Management 37, 741–761 (2001)
21. Wang, J., Liu, J., Wang, C.: Keyword extraction based on PageRank. In: Zhou, Z.-H., Li, H., Yang, Q. (eds.) PAKDD 2007. LNCS (LNAI), vol. 4426, pp. 857–864. Springer, Heidelberg (2007)

Query Expansion Using Medical Subject Headings Terms in the Biomedical Documents

Ornuma Thesprasith and Chuleerat Jaruskulchai

Department of Computer Science, Faculty of Science, Kasetsart University, Bangkok, Thailand
ornuma.thesprasith@gmail.com, fscichj@ku.ac.th

Abstract. MEDLINE database is most resourceful of biomedical literatures. Lay users may get difficulty to formulate a query. Query expansion technique reformulates user query by adding more significant and related terms to original terms to retrieve more relevant results. Finding related terms are explored form external resources, collection and query context. Since each MEDLINE document is manually assigned with controlled vocabularies which is called MeSH (Medical Subject Headings). These controlled vocabularies may be beneficial for query expansion. This paper proposes pseudo-relevance feedback by using MeSH terms in documents for query expansion. Additionally, re-weighting scheme called RABAM-PRF (Rank-Based MeSH Pseudo-Relevance Feedback) for filtering misleading terms is studied. In experiment, we use Lucene to retrieve the OHSUMED collection as baseline. The proposed method improves retrieval performance in MAP, P@10, and B-pref. Furthermore, the experiment showed that not all MeSH terms should be included to the query.

Keywords: MeSH ontology, Re-weighting scheme, Pseudo-relevance feedback.

1 Introduction

In biomedical retrieval system, the primary resource is MEDLINE database[1], the bibliographic references of biomedical articles. This database is maintained by the U.S. National Library of Medicine (NLM). To catalogue and index literatures in MELDINE, the NLM's indexers use standard controlled vocabulary called MeSH[2] (Medical Subject Headings) to represent main concepts of each document. In general, some documents may use the same controlled vocabulary.

The NLM provides PubMed[3], the Boolean retrieval system, to access MEDLINE database. Expert users of PubMed such as professional in healthcare formulate queries using MeSH terms and Boolean operator to get better results because they are familiar with MeSH terms whereas lay users encounter difficulty to formulate effective queries. This gap between keywords in user's query and terms in documents is

---

[1] See http://www.nlm.nih.gov/pubs/factsheets/medline.html
[2] See http://www.nlm.nih.gov/mesh/meshhome.html
[3] See http:// www.ncbi.nlm.nih.gov/pubmed

N.T. Nguyen et al. (Eds.): ACIIDS 2014, Part I, LNAI 8397, pp. 93–102, 2014.

well-known problem of information retrieval (IR) community called "mismatch" problem. Query expansion techniques can solve this problem by adding more significant terms into original query to bridge the gap. The result from expanded query is more relevant documents. PubMed applies query expansion techniques by providing semi-automatic features for users to reformulate query with MeSH's synonyms and Boolean operators. Since characteristic of Boolean retrieval model is un-ranking model and equally-weighting scheme of original and added terms. This characteristic may cause to "query-drifting" problem, added terms leading to retrieval more irrelevant documents. Furthermore, PubMed retrieval system returns most recent documents whereas other MEDLINE retrieval systems [1, 2, 3] return most relevant documents to users using query expansion techniques.

Two frequently asked questions about query expansion technique are how to derive added terms and how to re-weighting expanded query. For the first question, obtaining added terms by finding co-occur terms or most related terms from three sources; external resource such as thesaurus or ontology, internal resource such as corpus or document collection, and query-context resource such as description in query set. For the second question, there are many re-weighting schemes and most of them are proposed based on Rocchio's formula [4] which associates with relevant feedback information.

Our query expansion approach use MeSH terms in pseudo-relevance feedback set. We propose a novel re-weighting scheme that using ranking of documents in pseudo-relevance feedback called RABAM-PRF (Rank-Based MeSH Pseudo-Relevance Feedback). The tunable parameter of this re-weighting scheme is dynamic based on ranking of pseudo-relevance documents. The ranking information prevents "query-drifting" problem effectively.

The rest of paper we address the related works in section 2 and then explain our proposed method in section 3. In section 4, we describe experimental design and then report results with discussion in section 5.

2 Related Works

Query expansion methods can be done by human. However, this method lack of experts to select related terms to be included in the original query thus a few work is done on manual query expansion [5]. On the other hand, automatic-based approaches can be divided into statistical-based method and ontology-based method [6]. Statistical-based approaches find related terms for a query expansion using local or global analysis method. The local analysis take results from initial retrieval into an account called pseudo-relevance set and known as "query-dependent" method.

Many researches based on pseudo-relevance are reported in [2],[7, 8, 9]. These are proposed to find candidate terms and several factors related to parameter setting. For example, top 200 documents were fixed to be pseudo-relevance set and investigated two techniques for finding related terms [7]; Latent Semantic Indexing (LSI) [10] and Association Rule (AR) [11]. Rather than using general terms, some researches mapped terms from pseudo-relevance set to corresponding concepts and the using LSI to find related concepts [7, 8],[12].

Rocchio's re-weighting formula is composed of three weighting components; terms in original query, terms in relevant documents, and terms in non-relevant documents. Each component associated with tunable parameters that are α, β, γ respectively. Pseudo-relevance feedback re-weighting method assumes that top documents are relevant and no information about non-relevant document thus ignores the third component by setting γ to 0. To improve retrieval performance, ranking mechanisms had been studied [2] and inverse document frequency of pseudo-relevance set had been applied [9].

Another finding related terms is global analysis methods which explore co-occur terms in different resources; a whole collection [7], clusters of documents from a whole collection [13], external resource such as encyclopedia [14]. The global analysis method is also known as "query-independent" method.

Query expansion researches based on ontology approaches have been extensively studied [3],[12],[15, 16]. The main external resources in biomedical field to find related terms are UMLS-Metathesaurus [17] and MeSH ontology. Both resources maintained by NLM. The UMLS-Metathesaurus contains a pair of several types of semantic relationships. Most common types of relationships have been selected in the experiment; narrowing relationship and synonymous terms by medical source. UMLS co-concept terms with same semantics type to the original query term have been [13].

The MeSH ontology contains more than 20,000 subject headings, known as MeSH descriptors. Each descriptor has a unique identifier and a short definition. One descriptor has many synonyms and related definition, known as entry terms. MeSH ontology with statistical analysis based on the frequency of concepts in top retrieval documents and relevant score of concept terms had been re-calculated [3]. There are several strategies to select MeSH ontology terms. Vahid et al [15] identified MeSH concepts in initial queries and used associated entry terms for expansion. They used number of related concepts and depth of concept in MeSH ontology to re-weighting expanded query. Furthermore, Vahid et al [16] used auxiliary frequency factor instead of direct frequency for re-weighting. This factor is composed of concept frequency, inverse frequency of concept, and number of concepts in documents.

Although various query expansion sources and re-weighting strategies are widely used, improving the retrieval accuracy of MEDLINE documents is still a challenging issue. Existing works combined variety methods and used external knowledge resources whereas our expansion method processes significant terms from internal resource. Since existing re-weighting scheme unsuitable with our candidate terms, we propose a novel re-weighting scheme, RABAM-PRF, as describe in the next section.

3 Method

3.1 OHSUMED Test Collection

One of reliable collection for biomedical information retrieval (IR) research is OHSUMED [18], and it is a subset of MEDLINE's references that collected from

1987 to 1991. It contains 348,566 references in total. Each document contains unique identifier (UI), title, MeSH terms, publication type, author, and source of reference. Only 75% of documents have an abstract. Test set of OHSUMED has 106 queries that mention about a patient such as gender or age and medical condition. Each query is formed by physicians and their results were judged by physicians from other groups to "definitely relevant", "possibly relevant", and "not relevant". Only 101 queries judged with "definitely relevant" and used in this research.

3.2 Medical Subject Headings (MeSH)

The proposed method utilizes MeSH terms within document collection as an internal resource which differ from existing works that used ontology as an external resource. Each MEDLINE document is assigned with medical controlled vocabularies about 9-10 MeSH terms. Main concept of a document is represented by 3-5 MeSH terms. Other MeSH terms in the document are additional features and used for specific purpose of NLM. For example, the "Check tag" type such as human, grant support, and publication type. These MeSH terms are uncommon used to formulate query by lay users because they are unlikely to relate with treatment or disease. Additionally, These MeSH terms are high frequency in collection and their distribution follows Zipf's law. Since MeSH terms can be single word or multiword, we take advantage of these multiword by separating them into individual word. Thus the term frequency in document is increased and we can use TF-IDF value to evaluate discriminating degree of MeSH terms as usual. To select most suitable MeSH terms for expansion process, we filter out MeSH terms with highest frequency in collection and low IDF as shown in Figure 1.

3.3 Proposed Expansion Method

Relevance feedback approach is a strategy to reformulate the query using a set of retrieval documents that are examined relevant documents by users. Then important terms or expression are added to original query. The relevance feedback process

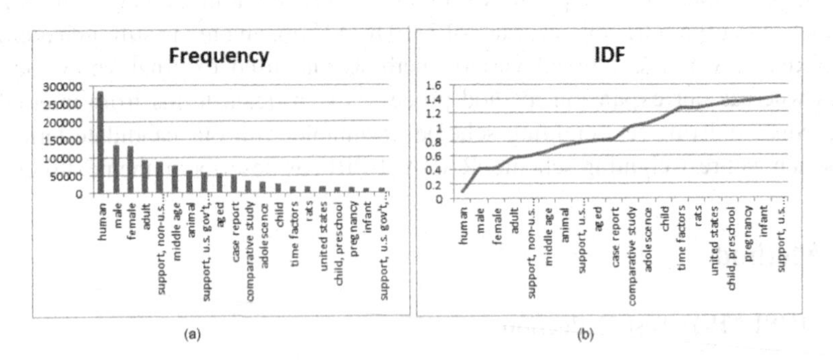

Fig. 1. Frequency and IDF of top 20 MeSH terms in OHSUMED collection

composes of two basic techniques; query expansion and term re-weighting. Since this process is required user to identify relevant documents, it is very time consuming.

The pseudo-relevance feedback is proposed and it takes the top K ranked documents as relevant set. Since each query term may affect the results with different improvement, probabilistic approach may be applied to re-weight query terms.

The Rocchio's re-weighting formula as show in equation (1) play an importance role in relevance feedback approach. The formula composes of three part; original weight, relevance-based weight, and non-relevance-based weight.

$$W'_Q = \alpha W_Q + \beta / |D_R| \times (\sum d_r) + \gamma / |D_N| \times (\sum d_n) \quad (1)$$

where α is tunable weight of initial query,
W_Q is weight of term in initial query,
β is tunable weight of relevant documents (d_r),
$|Dr|$ is number of relevant documents,
γ is tunable weight of non-relevant documents (d_n),
$|D_N|$ is number of non-relevant documents.

The pseudo-relevance re-weighting formula replaces the relevant part with pseudo-relevance part and ignores non-relevant part by setting γ to 0 as seen in [2],[9],[13] and the re-weighting scheme as follow.

$$W'_Q = \alpha W_Q + \beta \times W_{PRF} \quad (2)$$

where W_{PRF} is weight of term in pseudo-relevance documents.

There are many ways to set tunable parameters. For examples, balancing weight between original and pseudo-relevance feedback by setting α and β to 1.0 and W_{PRF} was computed using rank of candidate terms in pseudo-relevance set [2]. Alternatively, if the original query terms are more importance than the expanding terms then set value of α and β to 2.0 and 0.75 respectively, where W_{PRF} was computed using IDF of terms in pseudo-relevance set [9]. From the study [13] showed that query context is effective resource for finding expansion terms. This study mapped terms in query context with MeSH ontology. Since our method use only internal resource, we use terms in query context for re-weighting the original query terms. Furthermore, we believe that rank of documents in pseudo-relevance set have an importance role for weighting terms thus our re-weighting scheme called RABAM-PRF as follow.

$$W'_Q = \alpha_1 W_Q + \alpha_2 * W_{CONTEXT} + \beta * W_{MPRF} \quad (3)$$

where α_1 is tunable weight of initial query,
W_Q is weight of term in initial query, define as term-frequency (tf),
α_2 is tunable weight of query context,
$W_{CONTEXT}$ is weight of crossing terms in query context,
define as $1.0 + [tf_{CONTEXT} / \max(tf_{CONTEXT})]$,
β is tunable weight of pseudo-relevance documents,
W_{MPRF} is weight of MeSH terms crossing with tile in pseudo-relevance set,

define as $1.0 + [tf_{MPRF} / \max(tf_{MPRF})]$.

In the expansion process, we extract MeSH terms that matching with title terms in pseudo-relevance documents to be candidate MeSH set. We apply the RABAM-PRF with three variant schemes as follow.

The rank-based pseudo-relevance weighting scheme use rank information for dynamic tuning the β parameter by decreasing value when document is in lower rank. In this scheme, the candidate MeSH terms have 50-5o percent to be selected for expansion thus offset value is set to 0.5. To limit number of added terms, we set cutoff threshold weight at 1.0 thus the tunable weight of pseudo-relevance document as follow.

$$\beta = 0.5 + (1 - 0.2 \times r_k) \tag{4}$$

where r_k is rank of top retrieval document that term occur.

The IDF pseudo-relevance weighting scheme use document frequency and inverse document frequency in pseudo-relevance set for weighting terms. The W_{MPRF} in equation (3) is replaced with W_{IDF} which define as follow.

$$W_{IDF} = (DF_{PRF}/k * \log(N / DF_{PRF}) \tag{5}$$

where DF_{PRF} is number of document that candidate MeSH term occur.

The query-context boost weighting scheme uses crossing terms between title part and description part of each topic to be query context set and boosts up these terms with higher value than original query term. Since default of query term is 1.0 thus the tunable parameter of query-context is set to 1.75.

4 Experimental Design

Our study divided into three parts; the first experiment is testing MeSH terms usefulness in indexing construction process. The second experiment is investigating our variants re-weighting scheme compared with baseline. The last experiment is explored number of retrieved document for pseudo-relevance feedback approach.

Since main objective of assigning controlled vocabulary are to catalogue and index MEDLINE documents in NLM and have some specific purpose for NLM's services. The first experiment, we studied usefulness of MeSH in indexing construction process by setting up primary experiment to retrieve OHSUMED collection by comparing result from using MeSH term (IndexWithMeSH) and un-using MeSH terms (IndexWithoutMeSH) for indexing.

We used two sets of queries from TREC-9 Filtering Track for primary experiment. It composes of train and pre-test query set that contains 43 topics and 63 topics respectively. From the result of primary experiment shown in Figure 2, we choose better indexing construction method for consequent experiments. Because query characteristic of this collection infrequently use specific MeSH terms that already assigned in documents.

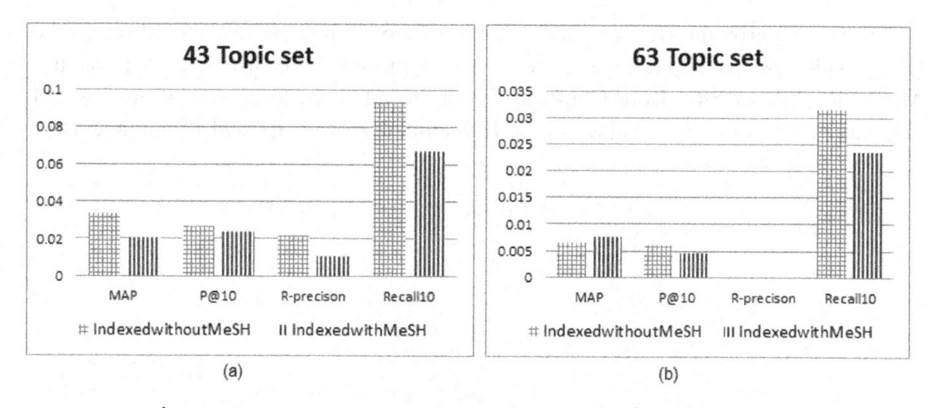

Fig. 2. Compare result of MeSH indexing (a) Retrieve 43 topic set and (b) Retrieve 63 topic set

Table 1. Experimental parameter testing

Expansion and weighting scheme	α_1	α_2	β_1	Wx
LUCENE_Baseline	1.0	0	0	W_Q
CrossTitleMeSH_RankBasedPRF	1.0	0	$(0.5 + [1- 0.2 * r_k])$	W_{MPRF}
CrossTitleMeSH_IDFPRF	2.0	0	0.75	W_{IDF}
CrossTitleMeSH_RankBasedPRF_BoostContext	1.0	1.75	$(0.5 + [1- 0.2 * r_k])$	$W_{CONTEXT}$

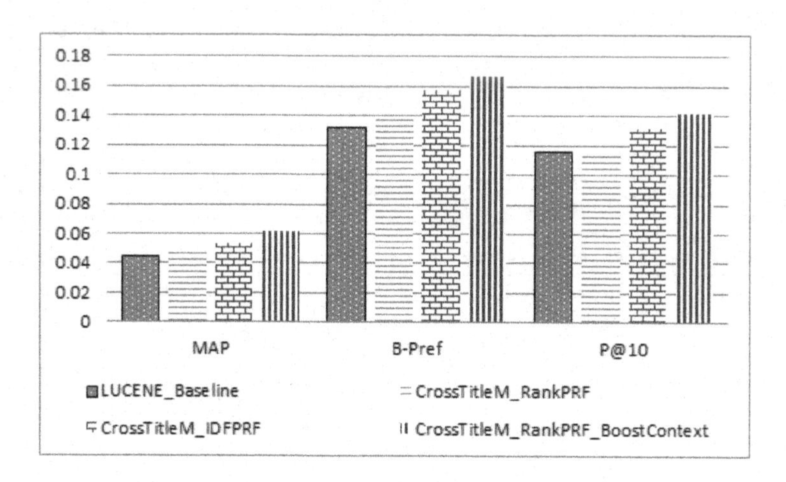

Fig. 3. Retrieval performance of 106-topic set (top k = 5) of four settings

The second experiment, we examine effectiveness of MeSH terms in query expansion process. Three query sets are 101 queries (from 106 queries) in the OHSUMED collection and two sets from TREC-9 Filtering Track as used in the primary experiment. Baseline is original title part of topics that retrieved by LUCENE. We compare with three variations of added methods and re-weighting scheme as shown in table 1.

The last experiment, we examined number of top documents that should return the best result for pseudo-relevance feedback approach. We use the "CrossTitle-MeSH_RankBasedPRF_BoostCon-text" testing to retrieved three sets of queries and compare results from three values of top K document sets (5, 10, and 15 documents).

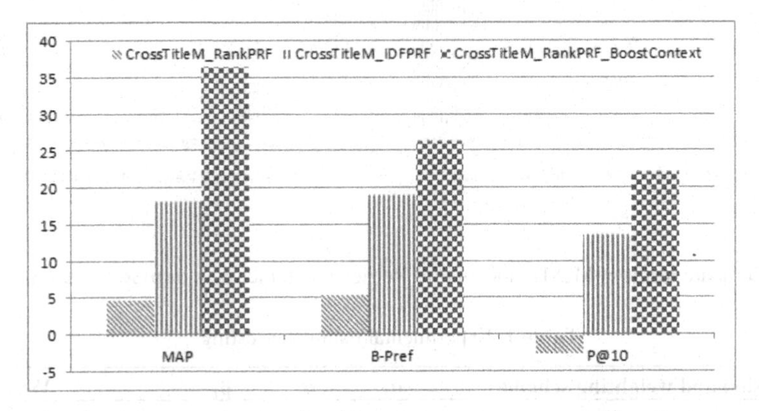

Fig. 4. Percentage of retrieval improvement compare with baseline

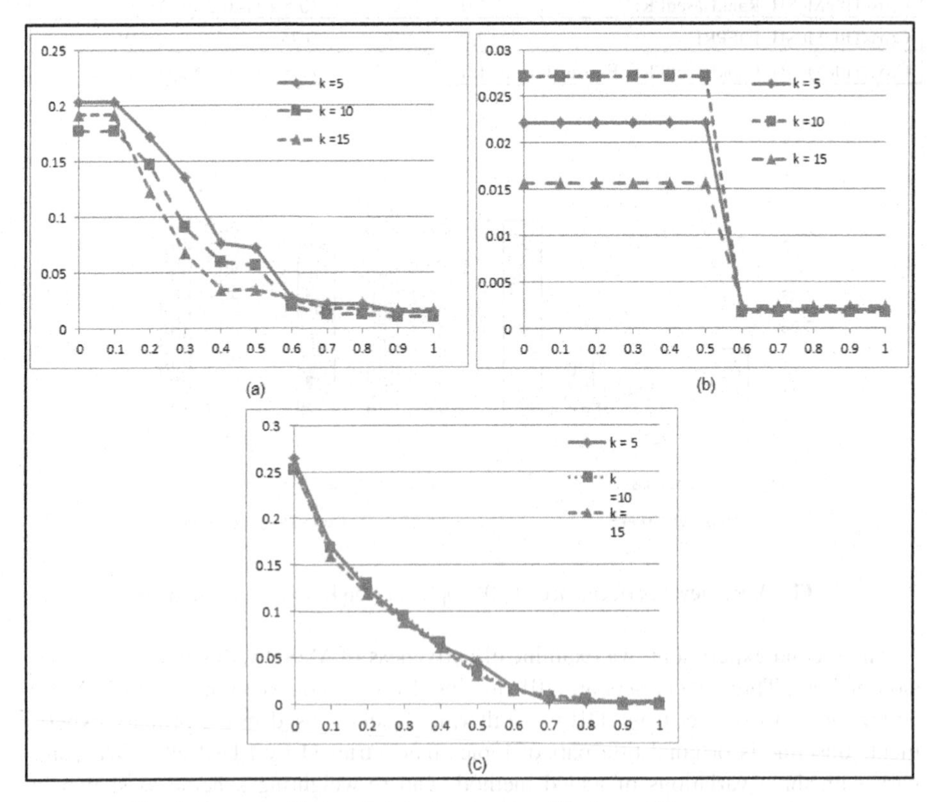

Fig. 5. Compare 11-Avg Point of variant top K values of three query sets

We used feature of LUCENE for boosting terms in query expression via caret mark "^". The weights of original and adding terms are setting after the caret.

We evaluated our experiments using program trec_eval [19]. We focus on MAP, b-pref and P@10 metric.

5 Discussion

Although MeSH terms are manually assigned to each MEDLINE document, directly using these terms for indexing process of ranking-based retrieval system without additional preprocess can degrade retrieval performance as seen in primary experiment. Because some manually assigned MeSH terms are less discrimination power and uncommon used in this query collection. However, carefully use MeSH terms for query expansion can improve retrieval performance as seen in Figure 2.

MeSH terms that crossing with title in pseudo-relevance set which is internal source for expansion show usefulness of MeSH terms in MEDLINE document. Results of using these terms along with our RABAM-PRF re-weighting scheme is better than baseline, the "CrossTitleMeSH_RankPRF" and "CrossTitleMeSH_IDFPRF" in Figure 3. To take advantage of all information in query context, we boosting query terms in title part that crossing with information need part and combine with MeSH terms in pseudo-relevance set, "CrossTitleMeSH_RankPRF_BoostContext" in Figure 4. The result shown that general term in query context more occur in relevant documents, boosting up these related terms to give better retrieval performance.

In Figure 5 shows 11-Average Point of results across three topic sets; (a) the 43-topic set, baseline retrieves of this query set fairly good (0.203) thus using MeSH terms from top five documents is the better than using term from lower rank; (b) the 63-topic set, baseline retrieves more irrelevant documents for this query set as seen from the very low value of 11-Average point (0.027). Terms from too few documents is not enough to push up possibility to find more relevant document thus using more documents. However, using fifteen documents is worse since the "query-drifting" problem; (c) the 106-topic set, baseline retrieves this query set fairly good (0.265) thus using terms from fewer documents of pseudo-relevance is better. Thus we recommend that using terms from few documents for expansion helping to increase precision and avoiding "query-drifting" problem.

Our proposed method can take advantage of MeSH terms in MEDLINE documents for query expansion with pseudo-relevance feedback approach effectively.

References

1. Fontaine, J.-F., Barbosa-Silva, A., Schaefer, M., Huska, M.R., Muro, E.M., Andrade-Navarro, M.A.: MedlineRanker: Flexible ranking of biomedical literature. Nucleic Acids Res. 37, W141-W146 (2009)
2. Yoo, S., Choi, J.: On the query reformulation technique for effective MEDLINE document retrieval. J. Biomed. Inform. 43, 686–693 (2010)

3. Jalali, V., Matash Borujerdi, M.: Information retrieval with concept-based pseudo-relevance feedback in MEDLINE. J. Knowl. Inf. Syst. 29, 237–248 (2011)
4. Rocchio, J.J.: Relevance feedback in information retrieval. In: Salton, G. (ed.) The SMART Retrieval System: Experiments in Automatic Document Processing, pp. 313–323. Prentice-Hall, Englewood Cliffs (1971)
5. William Hersh, S.P., Donohoe, L.: Assessing thesaurus-based query expansion using the UMLS Metathesaurus. AMIA, 344–348 (2000)
6. Baeza-Yates, R., Ribeiro-Neto, B.: Modern information retrieval. ACM press, New York (1999)
7. Xu, X., Zhu, W., Zhang, X., Hu, X., Song, I.-Y.: A comparison of local analysis, global analysis and ontology-based query expansion strategies for bio-medical literature search. In: IEEE International Conference on Systems, Man and Cybernetics, SMC 2006, pp. 3441–3446. IEEE (2006)
8. Xu, X., Zhang, X., Hu, X.: Using Two-Stage Concept-Based Singular Value Decomposition Technique as a Query Expansion Strategy. In: 21st International Conference on Advanced Information Networking and Applications Workshops, AINAW 2007, pp. 295–300 (2007)
9. Abdou, S., Savoy, J.: Searching in Medline: Query expansion and manual indexing evaluation. J. Infoproman. 44, 781–789 (2008)
10. Deerwester, S., Dumais, S.T., Furnas, G.W., Landauer, T.K., Harshman, R.: Indexing by Latent Semantic Analysis. JASIST 41, 391–407 (1990)
11. Witten, I.H., Frank, E., Hall, M.A.: Data Mining: Practical Machine Learning Tools and Techniques: Practical Machine Learning Tools and Techniques. Elsevier (2011)
12. Xu, X., Hu, X.: Cluster-based query expansion using language modeling in the biomedical domain. In: 2010 IEEE International Conference on International Conference on Bioinformatics and Biomedicine Workshops (BIBMW), pp. 185–188. IEEE (2010)
13. Zhu, W., Xu, X., Hu, X., Song, I.-Y., Allen, R.B.: Using UMLS-based Re-Weighting Terms as a Query Expansion Strategy. In: IEEE International Conference on Granular Computing, pp. 217–222. IEEE (2006)
14. Benjamin King, L.W., Provalor, I., Zhou, J.: Cengage Learning at TREC 2011 Medical Track. In: The 20th Text REtrieval Conference (TREC). National Institute for Standards and Technology (2011)
15. Jalali, V., Borujerdi, M.R.M.: The Effect of Using Domain Specific Ontologies in Query Expansion in Medical Field. In: IEEE Innovations in Information Technology (2008)
16. Jalali, V., Borujerdi, M.R.M.: A Hybrid Information Retrieval System for Medical Field Using MeSH Ontology. In: Prasad, S.K., Routray, S., Khurana, R., Sahni, S. (eds.) ICISTM 2009. CCIS, vol. 31, pp. 31–40. Springer, Heidelberg (2009)
17. Unified Medical Language Systems, http://www.nlm.nih.gov/research/umls
18. Hersh, W., Buckley, C., Leone, T.J., Hickam, D.: OHSUMED: An Interactive Retrieval Evaluation and New Large Test Collection for Research. In: Croft, B., Rijsbergen, C.J. (eds.) SIGIR 1994, pp. 192–201. Springer, London (1994)
19. Text REtrieval Conference. The trec eval Evaluation Package (2004), http://trec.nist.gov/trec_eval/

Improving Health Question Classification by Word Location Weights

Rey-Long Liu

Department of Medical Informatics, Tzu Chi University, Hualien, Taiwan
rlliutcu@mail.tcu.edu.tw

Abstract. Healthcare consumers often access the Internet to get health information related to specific health questions, which are often about several health categories such as the *cause*, *diagnosis*, and *process* (e.g., treatment) of disorders. Therefore, for a given health question *q*, a classifier should be developed to recognize the intended category (or categories) of *q* so that relevant information specifically for answering *q* can be retrieved. In this paper, we show that a Support Vector Machine (SVM) classifier can be trained to properly classify real-world Chinese health questions (CHQs), and more importantly by weighting the words in the CHQs based on their locations in the CHQs, the SVM classifier can be further improved significantly. The improved classifier can serve as a fundamental component to retrieve relevant health information from health information websites, as well as the collections of CHQs whose answers have been written by healthcare professionals so that healthcare consumers can get reliable health information, which is particularly essential in health promotion and disease management.

1 Introduction

The Internet has been a main channel to get health information services [2]. The critical success factor of the information services lies on the retrieval of medical information that needs to be both *relevant* (relevant to users' needs [4][16]) and *reliable* (as the information needs are often about disease treatment and health management [12], the information is often used to make health-related decisions [10], and inaccurate information may be even perceived as reliable information by readers [1]).

Therefore, given a health question *q* as a query, users require an intelligent retrieval system to retrieve both relevant and reliable information specifically for answering *q*. Two main sources of the information are (1) health information websites and (2) collections of health questions whose answers have been written by healthcare professionals so that healthcare consumers can get reliable health information. Table 1 lists several examples of Chinese health questions (CHQs) from KingNet[1], which is a health information provider on which many healthcare professionals keep answering health questions of healthcare consumers. The main difficulty of retrieving relevant information for CHQs lies on the fact that CHQs are often quite short, making it difficult to extract semantic evidences to identify relevant webpages and CHQs.

[1] Available at http://www.kingnet.com.tw

N.T. Nguyen et al. (Eds.): ACIIDS 2014, Part I, LNAI 8397, pp. 103–112, 2014.

Table 1. Example CHQs from KingNet, which is a health information provider on the Internet: The intended category (categories) of a CHQ can indicate a type of the fundamental semantics of the CHQ. A CHQ can even have no intended category and in that case the CHQ is actually asking about whatever categories.

CHQs	English Translation	Intended Category
(1) 何謂(what is)「過度換氣症」(hyperventilation syndrome)?	What is hyperventilation syndrome?	*None* (*whatever*)
(2) 兒童(children)常(often)吃(eat)山藥(yam)會不會(can or cannot)引發(cause)性早熟(precocious puberty)?	For children, will frequently eating yams cause precocious puberty?	*Cause*
(3) 失眠(insomnia)的原因(cause)與(and)診斷(diagnosis)	The cause and diagnosis of insomnia	*Cause* *Diagnosis*
(4) 安樂死 (Euthanasia)	Euthanasia	*None* (*whatever*)
(5) 嬰兒(infant)體溫(body temperature)太低(too low)怎麼辦(how to do)?	How to do for an infant whose body temperature is too low?	*Process*
(6) 如何(how to)克服(deal with)緊張(nervous)的情緒(mood)?	How to deal with a nervous mood?	*Process*

1.1 Problem Definition and Motivation

In response to the difficulty, we focus on a fundamental type of semantics: the intended health categories of health questions. Typical categories include the *cause*, *diagnosis*, and *process* (e.g., treatment) of diseases (or disorders). As illustrated in Table 1, the three categories (*cause*, *diagnosis*, and *process*) are respectively about the how a disorder is caused, diagnosed, and processed. Since healthcare consumers often care health promotion and disease management, recognition of the three categories has been shown to be helpful for the retrieval of relevant CHQs [9]. Two CHQs can be said to be relevant only if they are asking the same category of health information. The three categories were also noted as the key concepts to retrieve documents for clinical questions posted by healthcare professionals as well [7].

Note that a CHQ can have multiple intended categories (e.g., the 3rd example in Table 1). A CHQ can also be composed of a single disease name (e.g., euthanasia, ref. the 4th example in Table 1), indicating that the users are asking for all information about the disease. A CHQ can thus even have no intended category (e.g., the 1st and the 4th example in Table 1), and in this case the users are asking for *whatever* categories. A CHQ is an *in-space* CHQ if it belongs to some of the three categories. A CHQ is an *out-space* CHQ if it does not belong to any category.

Therefore, classification of both the in-space CHQs and the out-space CHQs is important for the recognition of the semantics of the CHQs. In this paper, we show that a Support Vector Machine (SVM) classifier can be trained to properly classify real-world CHQs, and more importantly, to further improve the SVM classifier we propose a technique WLW (Word Location Weight) that estimates the weights of the words in the CHQs based on their locations in the CHQs. The idea of WLW is based on an observation: those words (in a CHQ) that are more related to the intended category of the CHQ tend to appear at the beginning and end of the CHQ. For example, two words "如何" (how to) at the beginning of the 6th example in Table 1 provide a strong evidence indicating the CHQ might be about the *process* category. Similarly, the words "怎麼辦" (how to do) at the end of the 5th example in Table 1 indicate that the CHQ is likely to be about *process* of a health problem as well. Therefore, by properly amplifying the weights of the words at certain locations in a CHQ, the classifier can get helpful information to achieve better performance in classifying the CHQ.

We show that with the location weights estimated and encoded by WLW, the SVM classifier can be further improved significantly. The improved classifier can serve as a fundamental component to retrieve relevant health information from health information websites, as well as the collections of CHQs whose answers have been written by healthcare professionals so that healthcare consumers can get reliable health information, which is particularly essential in health promotion and disease management.

1.2 Main Challenges and Related Work

Main challenges of WLW include (1) *estimation* of the word location weights, and (2) *encoding* of the word location weights so that the underlying classifier can use the weights to improve CHQ classification. Obviously, without properly estimating and encoding the word location weights, the underlying classifier cannot be improved. To our knowledge, no previous studies tackled the challenges.

As noted above, the three basic categories (*cause*, *diagnosis*, and *process*) are a type of semantics that is particularly important in the domains of healthcare and medicine [7][9]. Previous studies often employ some kinds of semantic information to retrieve relevant questions as well. Typical kinds of semantic information included (1) question types of the questions (e.g., the categories of "what," "how," and "where," [15], and (2) syntactic or semantic structures of questions [3][13][14]. However, although the question type and the semantic structure of a question can indicate a kind of semantics of the question, it cannot indicate the basic intention of a CHQ, which is often about the cause, diagnosis, and process of a health problem. Moreover, many CHQs cannot fall into any specific question types, and as noted above a CHQ that does not fall into any category actually asks for the information for whatever categories.

One possible way to recognize the categories of a CHQ is to parse and analyze the CHQ. However, a good parser and analyzer for CHQs is often unavailable due to two reasons: (1) parsing Chinese questions is still a challenging task [6], and (2) CHQs are not always well-formed for parsing as they may even consist of a single term or multiple fragments[2]. Another way to recognize the categories was string matching using a set of *predefined* string patterns for each category [9]. However, the string patterns are difficult and costly to construct and maintain. Therefore, in this paper we employ a machine learning approach to build a classifier. We show that SVM can be a good

---

[2] We employed a parsing system (`http://parser.iis.sinica.edu.tw/`) to parse 38 CHQs and found that 34.2% of them cannot have a single and correct parse tree.

classifier for the classification of CHQs and it can be further improved significantly by properly encoding the word location weights into the learning and the testing processes.

WLW employs word locations (in a health question) as the evidence to improve text classifiers. Word locations are different from term proximity, which has been employed to improve text classifiers [8] and text rankers [11]. WLW cares the location of each individual word (rather than the proximity among multiple terms). It is based on an observation noted above and illustrated in Table 1: words at certain locations in a CHQ can be category-indicative. The observation is novel, and to our knowledge WLW is the first technique that improves question classification by the observation.

Section 2 presents WLW. To empirically evaluate WLW, section 3 reports a case study on the classification of hundreds of real-world CHQs. The result shows that WLW can significantly improve SVM, which is a state-of-the-art technique for text classification. The contribution is of technical significance to the enhancement of existing classifiers in recognizing the fundamental semantics of CHQs. It is also of practical significance to health promotion and disease management, since WLW can help information systems to identify relevant health information in webpages and those CHQs that have been edited by healthcare professionals.

2 Estimation and Encoding of Word Location Weights

WLW serves as a front-end processor for the underlying classifier (e.g., SVM). When training the classifier, WLW gets the training data as input (i.e., those CHQs that have been labeled with category labels), and outputs the weights of each word in each of the CHQs. The output is used to train the classifier. When testing the classifier (i.e., classifying a set of test CHQs), WLW gets the test CHQs, and outputs the weights of each word in each of the CHQs as well. The output is entered to the classifier, which classifies each CHQ into zero, one, or multiple categories. Therefore, WLW acts as a preprocessor in both the training and the testing phases of the underlying classifier.

The main mission of WLW is to properly amplify the weights of the words at certain locations in a CHQ. To achieve the mission, WLW has two main tasks: (1) *estimation* of the word location weights, and (2) *encoding* of the word location weights for the underlying classifier. For the former, given a location p in a question q, WLW employs Equation 1 and Equation 2 to respectively transform the locations into two weights LW_{front} and LW_{rear}.

$$LW_{front}(p,q) = \frac{1}{1 + (\#words\, before\; p\; in\; q)} \tag{1}$$

$$LW_{rear}(p,q) = \frac{1}{1 + (\#words\, after\; p\; in\; q)} \tag{2}$$

Therefore, LW_{front} and LW_{rear} are respectively the location weights estimated by using the *start* and the *end* of the question as the anchor. If p is near to the start (end) of q, its LW_{front} (LW_{rear}) will be large. Note that both LW_{front} and LW_{rear} of p can be large if q is quite short.

For the encoding of the location weights, we are concerned with the proper construction of the vector to be used as the training vector for the underlying classifier. Many text classifiers (e.g., SVM) require each text to be represented by a vector in which a word (or term) in the text corresponds to a dimension (feature) in the vector space. The corresponding value for the word in the vector can be the term frequency (TF) of the word in the text. Since WLW derives two location weights LW_{front} and LW_{rear} for each word (rather than one single value for the word), encoding of the two location weights into the vector space deserves proper design.

Among several alternatives for the encoding, WLW doubles the number of features in the vector by encoding both LW_{front} and LW_{rear} into the vector using Equation 3 and Equation 4.

$$Fvalue_{front}(w,q) = \sum_{p \in \{locations of\ w\ in\ q\}} (1 + LW_{front}(p,q)) \tag{3}$$

$$Fvalue_{rear}(w,q) = \sum_{p \in \{locations of\ w\ in\ q\}} (1 + LW_{rear}(p,q)) \tag{4}$$

Note that $Fvalue_{front}$ and $Fvalue_{rear}$ reduce to TF of w if the location weight is set to 0 on all positions (i.e., no location weights are employed). If w is near to the start (end) of q, its $Fvalue_{front}$ ($Fvalue_{rear}$) will be large. Both $Fvalue_{front}$ and $Fvalue_{rear}$ of w can be large if q is quite short.

It is interesting to note that the encoding strategy has two advantages. Firstly, both LW_{front} and LW_{rear} are encoded so that the learning-based classifier (e.g., SVM) can have more features to learn a proper model for question classification. This is particularly helpful, since for some categories and languages (e.g., Chinese and English), category-indicative words can tend to appear at the different locations (start or end) in health questions. By encoding both LW_{front} and LW_{rear} into the vector, the classifier can learn their relative importance using the trainng data (health questions with category labels). Moreover, the second advantage of the encoding strategy is that $Fvalue_{front}$ and $Fvalue_{rear}$ will approach 1.0 when LW_{front} and LW_{rear} happen to be quite small. The is helpful as well, since category-indicative words might still happen to be far away from the start or the end of the question. In that case, WLW actually sets the feature value to be the TF of the word, which is the traditional approach employed by the underlying classifier.

3 Empirical Evaluation

A case study on hundreds of CHQs had been conducted to empirically evaluate the contribution of WLW. It showed that WLW successfully helped SVM to have significantly better performance in classifying both the in-space and the out-space CHQs.

3.1 Experimental Data

The experimental CHQs were downloaded from KingNet[3], which is a Chinese healthcare information provider. For the experimental purpose, we manually label the categories of the CHQs. For the *cause* category (category 1), there were 313 CHQs; for the *diagnosis* category (category 2), there were 92 CHQs; and for the *process* category (category 3), there were 459 CHQs. Therefore, we had 864 *in-space* CHQs in the experiments. The category labels were also cross-checked to verify their validity.

Moreover, to measure how the classifiers performed in classifying *out-space* CHQs (those that belonged to none of the categories), we additionally collected 100 out-space CHQs from KingNet. As noted in Section 1, classification of out-space CHQs is particularly essential as an out-space CHQ actually asks for whatever categories of information about a health problem. Misclassifying an out-space CHQ into some category will make the intention of the CHQ too narrow and hence deteriorate the performance of retrieving relevant information for the CHQ.

To test the classifier under different settings of training and testing data, we conducted 5-fold cross validation: 4/5 of the in-space CHQs were used for training the classifier and the remaining 1/5 of the in-space CHQs were used for testing, and the experiment repeated five times so that each in-space CHQ was used as a test CHQ exactly once. With the 5-fold experiments, we can comprehensively evaluate the contribution of WLW.

3.2 Underlying Text Classifier

We employed SVM as the underlying classification technique. SVM is a popular technique in text classification, and previous studies have shown that SVM is one of the best classification techniques. To implement the SVM classifier, we employed SVM^{light} that is publicly available[4] [5] and has been tested in many previous studies.

3.3 Evaluation Criteria

We employed different criteria to evaluate the classification of in-space and out-space CHQs. For in-space CHQs, we employed F_1, which is a popular criterion measured by $2 \times P \times R/(P+R)$, where P (precision) is [total number of correct classifications / total number of classifications made] and R (recall) is [total number of correct classifica-

---

[3] The CHQs were collected from `http://www.kingnet.com.tw`, February 2012.

[4] SVM^{light} is available at `http://www.cs.cornell.edu/People/tj/svm%5Flight/old/svm_light_v5.00.html`. We employed all the default parameter settings of SVM^{light}, except that the parameter 'J' (about cost-factor, by which training errors on positive examples outweight errors on negative examples) was set to 10 (its default value is 1).

tions / total number of correct classifications that should be made]. There are two ways to compute average performance in F_1: *micro-averaged* F_1 and *macro-averaged* F_1. In measuring micro-averaged F_1, P and R are computed by viewing all categories as a system, and the resulting P and R are used to compute micro-averaged F_1. Macro-averaged F_1 is simply measured by averaging F_1 values on *individual* categories.

For the classification of out-space CHQs, we employed two evaluation criteria: *filtering ratio* (FR) and *average number of misclassifications* (AM). FR is [number of the out-space CHQs that are not classified into any category / number of the out-space CHQs] and AM is [number of misclassifications for the out-space CHQs / number of the out-space CHQs]. A system should filter out as many out-space CHQs as possible (i.e., higher FR) and avoid misclassifying the out-space CHQs into many categories (i.e. lower AM).

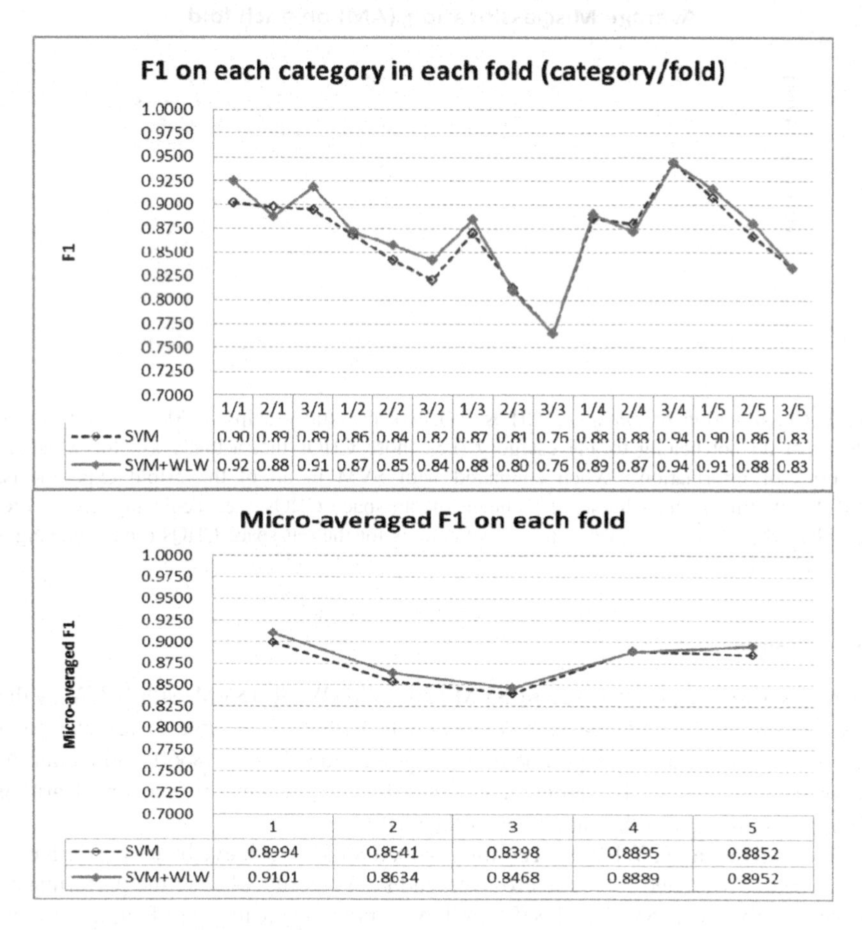

Fig. 1. Classification of *in-space* CHQs: WLW successfully helps SVM to achieve better performance in both macro-averaged F_1 and micro-averaged F_1. The performance differences between SVM and SVM+WLW are statistically significant.

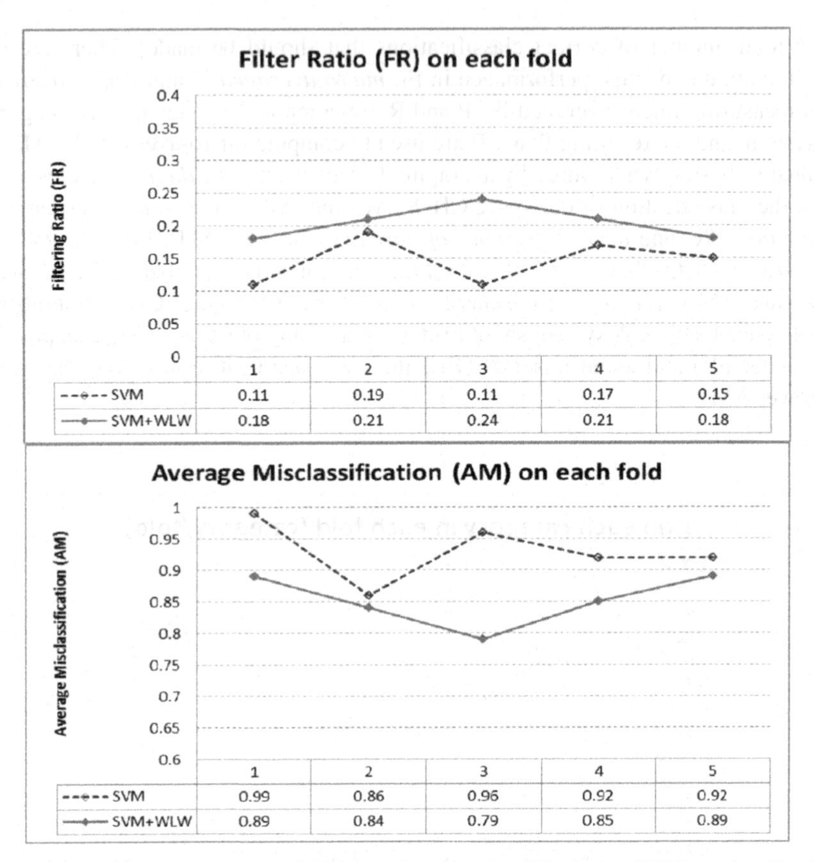

Fig. 2. Classification of *out-space* CHQs: WLW successfully helps SVM to classify those CHQs that do not belong to any category (i.e., those CHQs that actually ask for whatever categories of information). When compared with SVM in all of the 5-fold experiments, SVM+WLW filters out a higher percentage of out-space CHQs (i.e., achieving higher FR) and reduce the average number of misclassifications for the out-space CHQs (i.e., achieving a lower AM).

3.4 Results

Figure 1 shows the performance of SVM and SVM+WLW (SVM with WLW as the front-end processor) in classifying *in-space* CHQs. SVM was able to achieve good performance: in the 5-fold experiment, its macro-averaged F_1 is 0.8660, and its micro-averaged F_1 is 0.8736. Therefore, it is a feasible way to employ machine learning techniques to recognize the categories of the CHQs.

More interestingly, WLW successfully helped SVM to achieve better performance in both macro-averaged F_1 and micro-averaged F_1. To verify whether the performance differences (between SVM and SVM+WLW) were statistically significant, we conducted two-sided and paired t-test with 95% confidence level. The results showed that the performance differences were statistically significant. The results thus confirmed the contribution of WLW in the classification of in-space CHQs.

Figure 2 shows the performance of SVM and SVM+WLW in classifying *out-space* CHQs. Again, WLW helped SVM to achieve better performance in classifying out-space CHQs (i.e., achieving higher FR but lower AM). When compared with SVM in *each* of the 5-fold experiments, SVM+WLW successfully filtered out a higher percentage of out-space CHQs (i.e., achieving higher FR) and reduced the average number of misclassifications for the out-space CHQs (i.e., achieving a lower AM). On average, there is 39.73% promotion in FR (0.146 vs. 0.204) and 8.39% reduction in AM (0.93 vs. 0.852). Since out-space CHQs actually ask for whatever categories of information, the better performance in classifying them is essential for the retrieval of relevant information for the CHQs. Based on the results in Figure 1 and Figure 2, contributions of WLW to both the classification of in-space and out-space CHQs are confirmed.

4 Conclusion and Future Work

Healthcare consumers often require health information related to specific health questions, which are often about several health categories such as the *cause*, *diagnosis*, and *process* of disorders. Recognition of the intended categories of the health questions is thus essential in retrieving relevant information from websites and health questions whose reliable answers have been edited by healthcare professionals. We approach the category recognition problem by learning-based text classifiers, and show that a SVM classifier can be trained to properly classify real-world CHQs. To further improve the SVM classifier we propose a technique WLW that estimates the weights of the words in the CHQs based on their locations in the CHQs. With the location weights estimated and encoded by WLW, the SVM classifier can be further improved significantly. The contribution is of technical significance to the enhancement of existing classifiers in recognizing the fundamental semantics of CHQs. It is also of practical significance, as WLW can help to identify relevant and reliable information for health promotion and disease management.

Interesting future research directions include (1) application of WLW to English health questions, and (2) cross-lingual mapping of health questions. We expect that WLW can perform well in English, since locations of words in English health questions tend to indicate helpful categorical information as well. A preliminary analysis shows that category-indicative words in English health questions tend to appear at the beginning of the questions, since the questions are often started by interrogatives (e.g., "what" and "how"). As to the cross-lingual mapping of health questions, we expect that WLW can help to build the classifier for the mapping. Given a CHQ, the cross-lingual mapping aims at finding relevant English health questions so that healthcare consumers to get more relevant information.

Acknowledgment. This research was supported by the National Science Council of the Republic of China under the grant NSC 102-2221-E-320-007.

References

1. Abbas, J., Schwartz, D.G., Krause, R.: Emergency Medical Residents' Use of Google® for Answering Clinical Questions in the Emergency Room. In: Proc. of ASIST 2010 (2010)

2. Boguski, M.S.: Online Health Information Retrieval by Consumers and the Challenge of Personal Genomics. In: Willard, H.F., Ginsburg, G.S. (eds.) Genomic and Personalized Medicine, vol. 1-2 (2009)
3. Casellas, N., Casanovas, P., Vallbé, J.-J., Poblet, M., Blázquez, M., Contreras, J., López-Cobo, J.-M., Richard, V.: Semantic Enhancement for Legal Information Retrieval: IURISERVICE performance. In: Proceedings of ICAIL, Palo Alto, CA USA (2007)
4. Eysenbach, G., Köhler, C.: How do consumers search for and appraise health information on the world wide web? Qualitative study using focus groups, usability tests, and in-depth interviews. British Medical Journal 324, 573–577 (2002)
5. Joachims, T.: Making Large-Scale SVM Learning Practical. In: Schölkopf, B., Burges, C., Smola, A. (eds.) Advances in Kernel Methods - Support Vector Learning. MIT-Press (1999)
6. Lee, C.-W., Day, M.-Y., Sung, C.-L., Lee, Y.-H., Jiang, T.-J., Wu, C.-W., Shih, C.-W., Chen, Y.-R., Hsu, W.-L.: Boosting Chinese Question Answering with Two Lightweight Methods: ABSPs and SCO-QAT. ACM Trans. Asian Lang. Inform. Process Article 12 (2008)
7. Lin, J., Demner-Fushman, D.: The Role of Knowledge in Conceptual Retrieval: A Study in the Domain of Clinical Medicine. In: Proceedings of SIGIR 2006, Seattle, Washington, USA (2006)
8. Liu, R.-L.: A Passage Extractor for Classification of Disease Aspect Information. Journal of the American Society for Information Science and Technology 64(11), 2265–2277 (2013)
9. Liu, R.-L., Lin, S.-L.: A Conceptual Model for Retrieval of Chinese Frequently Asked Questions in Healthcare. In: Hou, Y., Nie, J.-Y., Sun, L., Wang, B., Zhang, P. (eds.) AIRS 2012. LNCS, vol. 7675, pp. 366–375. Springer, Heidelberg (2012)
10. Liszka, H.A., Steyer, T.E., Hueston, W.J.: Virtual Medical Care: How Are Our Patients Using Online Health Information? Journal of Community Health 31(5), 368–378 (2006)
11. Lv, Y., Zhai, C.: Positional Language Models for Information Retrieval. In: Proceedings of the 32nd Annual International ACM SIGIR Conference on Research and Development in Information Retrieval, pp. 299–306 (2009)
12. Shuyler, K.S., Knight, K.M.: What Are Patients Seeking When They Turn to the Internet? Qualitative Content Analysis of Questions Asked by Visitors to an Orthopaedics Web Site. Journal of Medical Internet Research 5(4), e24 (2003)
13. Wang, K., Ming, Z., Chua, T.-S.: A Syntactic Tree Matching Approach to Finding Similar Questions in Community-based QA Services. In: Proceedings of SIGIR 2009, Boston, Massachusetts, USA (2009)
14. Wu, C.-H., Yeh, J.-F., Lai, Y.-S.: Semantic Segment Extraction and Matching for Internet FAQ Retrieval. IEEE Transactions on Knowledge and Data Engineering 18(7) (2006)
15. Wu, C.-H., Yeh, J.-F., Chen, M.-J.: Domain-Specific FAQ Retrieval Using Independent Aspects. ACM Transactions on Asian Language Information Processing 4(1), 1–17 (2005)
16. Zeng, Q.T., Kogan, S., Plovnick, R.M., Crowell, J., Lacroix, E.-M., Greenes, R.A.: Positive attitudes and failed queries: an exploration of the conundrums of consumer health information retrieval. International Journal of Medical Informatics 73, 45–55 (2004)

Entity Recognition in Information Extraction

Novita Hanafiah[1,2] and Christoph Quix[3,4]

[1] King Mongkut's University of Technology North Bangkok, Thai-German Graduate School of Engineering, Bangkok, Thailand
[2] Bina Nusantara University, School of Computer Science, Jakarta, Indonesia
[3] Fraunhofer Institute for Applied Information Technology FIT, St. Augustin, Germany
[4] RWTH Aachen University, Information Systems and Databases, Aachen, Germany
nov.tau2@gmail.com, christoph.quix@fit.fraunhofer.de

Abstract. Detecting and resolving entities is an important step in information retrieval applications. Humans are able to recognize entities by context, but information extraction systems (IES) need to apply sophisticated algorithms to recognize an entity. The development and implementation of an entity recognition algorithm is described in this paper. The implemented system is integrated with an IES that derives triples from unstructured text. By doing so, the triples are more valuable in query answering because they refer to identified entities. By extracting the information from Wikipedia encyclopedia, a dictionary of entities and their contexts is built. The entity recognition computes a score for context similarity which is based on *cosine* similarity with a *tf-idf* weighting scheme and the string similarity. The implemented system shows a good accuracy on Wikipedia articles, is domain independent, and recognizes entities of arbitrary types.

1 Introduction

New technologies in the Internet provide new functions to publish and share data in a structured, semi-structured or non-structured form. Although (semi-)structured data formats such as RDF and XML have been proposed, unstructured text is still the dominant form. These articles can be interpreted by humans, but the huge amount of available documents requires a (pre-)processing of the documents by a machine, such that the human can focus only the relevant information. When a user does a keyword search, the search engine will retrieve some documents that are related to the input keywords. Due to the complexity of human language and the simple form of queries, keyword search fails sometimes to return the documents of interest. A deeper understanding of the semantics of a non-structured document is necessary to enable more complex queries over unstructured documents [9].

Semantic annotation is a technique that can provide more semantics to unstructured texts. Many entities in the real-world are described by the same term but indeed refer to different objects. For instance, the phrase 'Michael Jordan' may refer to the famous basketball player, Michael Jordan, but there is also an

N.T. Nguyen et al. (Eds.): ACIIDS 2014, Part I, LNAI 8397, pp. 113–122, 2014.

English football goalkeeper called Michael Jordan. Humans can differentiate to which 'Michael Jordan' that phrase refers by looking at the context.

Related Work: This problem is known as entity resolution, recognition, or disambiguation and various approaches have been proposed [12]. For example, the TBD approach (Taxonomy Based Disambiguation, [5]) resolves the disambiguation problem inside a taxonomy using the context of a term. The same concept is adopted in ESTER [1] for a semantic search engine application. ESTER makes use of the links in Wikipedia, as the occurrences of term as a link can be identified as an entity. Other methods leverage an encyclopedia [2,4] or especially Wikipedia [13,6]. Linked data is used for disambiguation using a graph-based approach in [8]. Another frequent source for entity disambiguation is YAGO [1] which is a knowledge base extracted from Wikipedia. For example, AIDA is a natural language processing system using YAGO as an entity catalog [12].

Contributions: This paper presents a method which applies the idea of the TBD algorithm to Wikipedia. We use Wikipedia as a source for information about entities, their notations and synonyms, and especially the usual context of entities. In all articles that have a link to another Wikipedia page, we collect the terms to the left and to the right of a link, count the frequencies of individual terms, and remember the most frequent terms to capture the ususal context of the referred entity. For the example above, references to the Basketball player have frequent terms like 'NBA', 'basketball' or 'Chicago', whereas the context of the football player have terms like 'goalkeeper', 'Farnborough', or 'English'. We use this data as training data in our approach. Entities are recognized using a cosine similarity scoring function to measure the context similarity between a given entity and the entities of the training data set. For example, if the context of a given entity has the terms 'NBA' and 'Chicago', it is more likely that it refers to a basketball player than to a football player. In addition, we include string similarity to increase the performance in terms of efficiency and effectivity. For example, 'Derrick Rose' and 'Michael Jordan' might have very similar contexts but their strings are different. Furthermore, we developed and evaluated several ranking techniques to rank the final result.

Recognizing the correct entity in the text is very useful in the information extraction area, which has the goal of extracting structured information (e.g., triple in form subject-predicate-object) from unstructured text. We integrated our approach into an information extraction system as the usefulness of triplets can be improved if the subject or the object refer to a recognized entity.

In the next section, we give a brief system overview and then explain the different phases of the entity recognition procedure: training, recognition, and mapping. Section 3 describes the evaluation results before we conclude the paper in section 4.

[1] `http://www.mpi-inf.mpg.de/yago-naga/yago/`

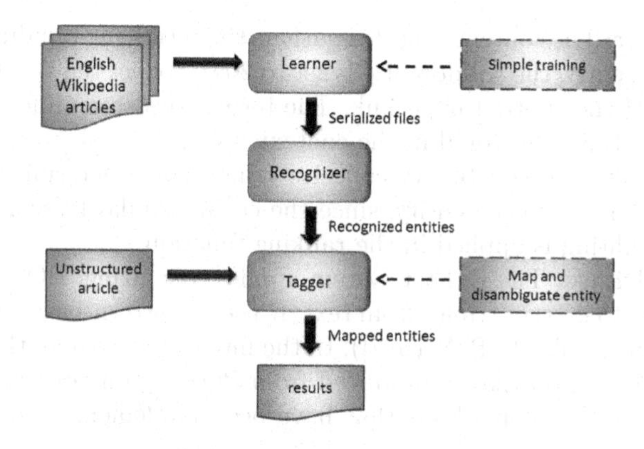

Fig. 1. Main system architecture

2 System Overview

The approach has 3 main phases, which are training, detection, and the matching processes (see fig. 1). In the training phase, the learner collects the entity names from the Wikipedia documents. Only nouns marked as hyperlinks are considered as entities, because the link points usually to the unique Wikipedia page defining that entity. In addition, the context of each entity's appearance is collected (e.g., a number of terms to the left and right of the appearance), as well as the terms used in the hyperlink to manage the name variations. The learner generates the trained files which are used in the disambiguation phase by the tagger. Next, the recognizer reads the input file and finds the entity which has to be annotated to only one actual entity (Recognizing Phase). The tagger assigns the meaning of the entity by applying the similarity ranking function (Mapping Phase).

In the next step, the main system is integrated with the information extraction system named ExFact which is currently developed at RWTH Aachen University. The integrated system aims to map the entities in the extracted triples (i.e., the subjects and objects) to the actual entities.

Training Phase: There are several components that are used in this phase:
Wikipedia files: Wikipedia has many features such as info-boxes, categories, and hyperlinks. The learner uses the extracted Wikipedia files as the data source for the training process and focus on hyperlinks. We use the *Wikipedia Extractor* [2] for this task.
Hyperlinks: To identify the occurrences of an entity in the documents, the system uses hyperlinks. Normally the title of a Wikipedia page is the same term as the actual term which occured in the text. Nevertheless, for entities having more than one meaning, the title's structure is written with the additional information inside the parentheses. For instance the title of novel as the story book is 'Novel' and the musician is 'Novel (musician)'.

[2] http://medialab.di.unipi.it/wiki/Wikipedia_Extractor

Context Collector: The context collector gathers the terms surrounding the entity occurrence. Currently, the window size is set to 20 to collect ten words to the left and right side of the spotted hyperlink. The terms pass stemming and filtering processes before they are stored in the context list.
Vocabulary collector: A vocabulary collector is needed to determine the corpus distribution of terms at each entity, since the cosine similarity scoring function with *tf- idf* weighting is applied in the ranking function.
Recognizing Phase: The ExFact system applies tokenizing, POS-tagging and phrase chunking to extract triples from text. It uses a pattern-based approach for triple extraction (similar to ReVerb [7]). In the integrated system, the recognizer exploits the information that is produced from POS-tagging of ExFact system. There are three different methods that have been implemented to find a good solution:
Standard named entity recognition: The ExFact system uses the OpenNLP library [3] to do the recognition of entities. This recognizer is harnessed in our integrated system to detect the entity occurrence in the text. The experiment shows that the recognizer works in general well, but the named entity recognition usually fails to identify a person name which consists of more than two words, e.g., 'Michael Jeffrey Jordan'. Additionally, in the matching phase we apply string similarity measurement. The captured entity name impacts the score of string similarity computation. Besides, the recognizer identified date, time and money, which can be ignored in the entity resolution process.
Entity hints: Entity hints are noun phrases which are recognized during information extraction. They are considered as hints because they are neither subject, object nor relation, but might deliver additional information. For example, in a sentence 'Albert Einstein was a German-born theoretical physicist who developed the general theory of relativity', the entity hints are 'Albert Einstein', 'a German-born theoretical physicist', 'who', and 'the general theory of relativity'.

By using this as a hint for the existence of entities in a sentence, we can identify more entities than the standard named entity recognizer of OpenNLP. However, it gives a big impact in the processing time. Moreover, there are too many general entities that are included, such as *time, offset, stamp.* The manual evaluation of the results shows that the ratio between the correct and incorrect mapping is 3:2, where the portion of general words is about 35%. This leads to an increasing number of incorrect taggings.
Pattern based recognition: A new pattern is constructed to apply the third method. The result of phrase chunker pattern extraction shows that the system can utilize the POS-tagging data to detect an entity. The patterns are

$$\text{(Noun Phrase)}^{+}\ \text{(for} \mid \text{of)?}\ \text{(Noun} \mid \text{Noun Phrase)}^{+}\ \text{and}$$
$$\text{(Noun} \mid \text{Noun Phrase)*}\ \text{(Noun)}\ \text{(Noun} \mid \text{Noun Phrase)}^{+}$$

The patterns are created by considering the possibilities of a noun and noun phrase to be unified each other. This approach reduces the amount of common entities such as 'white collar', 'chair', and 'book'. However, we find a deficiency of

[3] `http://opennlp.apache.org/`

this approach. Since we use the pattern which considers the occurrence of a noun or a noun phrase, we miss the entities which are built with an adjective. If we would add an additional regular expression to scoop the entities constructed by an adjective, we would get more general entities, such as 'Germanic Countries'.

Finally, we use the pattern based recognition in our evaluation since the approach of standard named entity recognition produced less entities and entity hints approach leads to generate many general entities.

Mapping Phase: The detected entities from the previous phase are annotated with the actual entity in the mapping phase, which has two main steps:

Candidate entity collection: The mapping process is started with selecting the candidate entities. The string similarity ranking function is applied to provide a list where the displayed name in the hypertext collection has a similarity with the detected entity names. To compute the similarity score, the system applies a token-based mechanism which is supported by the SecondString library [3]. After the score of each possibility is produced, the system ranks the score and get n candidates which have the highest similarity values. The parameter n is modified during the experiments to find the best value.

Disambiguation phase: Because the system provides a candidate entity list, the tagger maps the detected entities to only one entity. Therefore, the context similarity is computed by using the *cosine similarity* measurement with *tf-idf* weighting scheme. This measurement is one of the context similarity tested by [5] system that reached the highest precision. The n-best results are determined from the similarity function. However, the system needs to map the detected entity to the most similar actual entity in the n-best results. First, we use the context score to rank the n-best results. This approach has often failed to rank the actual entity to the first place. The analysis shows that some entities have the same context because they have the same profession. For example Michael Jordan and his son Jeffrey Jordan who are basketball player. The entity Jeffrey Jordan failed to rank to the first place because the terms that occurs in his father (Michael Jordan) has a higher score since Michael Jordan is the famous one. Therefore, the score is computed by using a combination of the string and context similarity is used.

$$score(s,e) = p \times Jaccard(s,e) + q \times SoftTFIDF(s,e) + r \times cosineSim(\boldsymbol{V_s}, \boldsymbol{V_e}) \quad (1)$$

The final score for the mapping of a spotted entity s to entity e is the sum of the string similarity score, which consist of *Jaccard* and *SoftTFIDF* scores [3], and the cosine similarity score. Entity s is obtained from the recognizing phase, while entity e is each entity inside the cadidate entity collection. The cosine similarity score denotes the similarity of the contexts which are expressed as vectors holding the frequencies of terms in the context of s and e. The weight of each score is adjusted by the parameters p, q, and r. The parameter value of p, q and r are the percentage of each similarity score that are used to get the best precision of final result. The total of those three parameters is 100 percent. As we rely more on the semantic context, the total coefficient of $p+q$ should be less than r.

We also applied here an n-gram and hashing algorithm to increase the performance of finding candidate entities. Firstly, we create n-grams for each entity during the training phase. The n-grams are stored in a hashtable which allows the quick retrieval of the entities which have a specific n-gram. Next, when the mapping process is started, the system generates the n-grams of the recognized entity. For each n-gram, we retrieve the corresponding entities from the hashtable. All entities found in this step are considered a candidate entity (i.e., a candidate entity shares at least one n-gram with the recognized entity). Only for these entities, we compute the detailed score.

3 Evaluation

To evaluate the approaches, the system is tested in a single computer with two Intel(R) Xeon(R) CPU running at 2.93GHz and 24GB RAM. From the provided RAM, we just used about 14GB to complete the training phase. There are two evaluation methods that we use: manual and automatic. The manual evaluation is supported by human effort. The results of the integrated system are evaluated by taking a random data sample. Alternatively, the system is evaluated by using an automatic technique which exploits the hyperlinks in the input files (in the same way as we do it for the training data). For the automatic evaluation, we use two different document sets: one document set is used for training (about one million documents) and another document set (about 5 articles which are disjoint from the first data set) is chosen for evaluation. The system calculates how many entities are evaluated and produce the amount of entities that are ranked correctly in the first place, second to fifth places, sixth to tenth places, and incorrectly mapped entities.

$$P@k = \frac{\sum_{e \in E} \begin{cases} 1 \text{ if } c(e) \in f(e_i, k) \\ 0 \text{ otherwise} \end{cases}}{|E|} \tag{2}$$

E is the set of all entities in the input text, $c(e)$ denotes the correct entity for an entity e, and $f(e, k)$ is the top-k result of the entity recognition process (i.e., the first k elements of the result list).

Data Set: The data set used for the training phase is a subset of the English Wikipedia articles. For the experiment, we exploit one million articles from Wikipedia database dump to be trained in the training phase.

As a comparison to the automatic evaluation, we choose five documents to check the system manually. We select the documents which have potential ambiguity problems, e.g., 'Michael Jordan' and 'Michael Man (director)'. Another document that might produce various entity types is chosen, such as the 'May Day' article. From 1089 entities which are recognized in the test data, we picked 100 samples randomly. We believe that the random sample data can represent the type distribution of the recognized entities. The random entities show that the number of location and the general entity are the smallest, about 10% for

Table 1. The result of context ranking function

Top-n candidates variants	Top 1	Top 2-5	Top 6-10	Incorrect	Total Entities	%Correct (Top1-5)	%Incorrect (Top 6-10 + incorrect)
20	463	321	82	223	1089	71.99	28.01
5	533	292	45	219	1089	**75.76**	**24.24**
3	565	256	36	232	1089	75.39	24.61
2	573	206	23	287	1089	71.53	28.47

Table 2. The result of combination scores of ranking function

Top-n candidates variants	Top 1	Top 2-5	Top 6-10	Incorrect	Total Entities	%Correct (Top1-5)	%Incorrect (Top 6-10 + incorrect)
25	703	191	26	169	1089	82.09	17.91
20	**706**	**189**	24	**170**	1089	**82.18**	**17.82**
15	**708**	**186**	24	**171**	1089	**82.09**	**17.91**
10	706	178	26	179	1089	81.17	18.83
5	693	170	21	205	1089	79.25	20.75

each, while entities which appear most are person and misc. The frequency of organization name occurs in the text is about 15%.

Results: To measure the effectiveness of mapping entities based on the context similarity, the system tested the scoring function by using the cosine similarity scoring function. The result can be seen in Tab. 1, where the highest precision is around 76% with $n = 5$, i.e., the system returns n candidate entities with the highest string similarity. From those candidates, the system ranks with the final scoring function and maps the entities. However, if we compare the top-1 result of context ranking function with the others string scores ranking function, we found that the number of entities in the context ranking function is lower than in the string scores ranking. By using the context score, the system failed to rank the correct actual entities in the first place. The contribution of correct entities in the top 2-5 push up the precision of this ranking method. Nonetheless, the incorrect results are lower than the others methods.

Therefore, we experimented with several ranking methods. The combined ranking function is one example. In Tab. 2, the results of combination score is presented with a weight of 0.8 for the cosine score and 0.1 for each string scores (*Jaccard* and *SoftTFIDF* similarity scores). From the tested ranking function, by using the string similarity scores, the system achieved about 74%, and using the context score, it is achieved about 75%. Nevertheless, by applying the combination of both scores, we reached about 82% precision.

The runtime of the process is quite high, especially when the system has to find the candidate entities. Thus, we tried to speed up the performance by applying n-gram algorithm and hashing as described in the section about the mapping phase. We test the system with 3-grams and 4-grams. The processing time is

Table 3. The comparison of *3*-grams and *4*-grams

Top-n candidates variants	Top 1	Top 2-5	Top 6-10	Incorrect	Total Entities	%Correct (Top1-5)	%Incorrect (Top 6-10 + incorrect)
			3-grams without normalizing				
25	719	200	25	145	1089	84.39	15.61
15	**720**	**200**	**24**	**145**	1089	**84.48**	**15.52**
5	723	188	26	152	1089	83.65	16.35
			4-grams without normalizing				
25	719	194	25	151	1089	83.84	16.16
15	**722**	**191**	**24**	**152**	1089	**83.84**	**16.16**
5	729	181	27	152	1089	83.56	16.44

rapidly reduced from 3.5 hours to around 25 minutes for 3-grams and about 8 minutes for 4-grams. Obviously, the mapping results are affected because the candidate list might be different when we applied the *n*-gram technique.

We experimented the *n*-gram methods by applying two different process. Firstly, we tested the *n*-gram with normalizing, like the previous method, where the hypertext list do not pass the normalizing process. This technique shows similar result with the previous method, that reached precision about 82% for *4*-gram testing. Secondly, we tested the method without applying the normalizing process for the hypertext list. The experiment shows that the *3*-grams produces higher precision for every variation of top *n* candidates number. The sample of the testing result are compared in Tab. 3. The *3*-grams approach achieves the precision almost 85% with the 720 entities are mapped correctly in the first place. On the other side, the highest precision is reached to 83% by *4*-grams approach. The overall of top 1 and top 2-5 entities in the *4*-grams is lower than *3*-grams. The limitation of candidate entities makes the *4*- grams approach do not perform as well as *3*-grams approach. However, comparing with the previous method, which do not apply the *n*-grams, the current results with *3*-grams are still a bit higher. This approach is evaluated with the same portion as combined scoring function in Tab. 2, which consist of 0.8 *cosine*, 0.1 *Jaccard* and 0.1 *SoftTFIDF* scores.

At the end, we test the system with several variation of context and similarity score portions as can be seen in Fig. 2. Even though the system has shown us the best result achieved with *3*-gram with the top 15 candidates entities parameters (Tab. 3), we tested some different parameters with *4*-grams due to the time consuming issues. The 4-grams test results are run with normalizing words, thus the score is a bit lower than without normalizing words. All kinds of tested variations scoring function demonstrate the same plot line style, the fewer the candidate is taken, the lower the accuracy. Nevertheless, by using more candidates, the level of accuracy also decreases, because sometimes the different entities may have the same context words to describe them. For instance, entities with almost the same name and they both are considered as candidate entities and by any chance they have the same profession, then this situation can be a factor of reducing accuracy. However, there is a peak point for each scoring function where they get the highest accuracy. From four variants scoring functions, Their peak points are with top 15 candidates entities. The highest accuracy is reached by the scoring function with 85% of cosine score, followed by 95% and 80% of cosine scores.

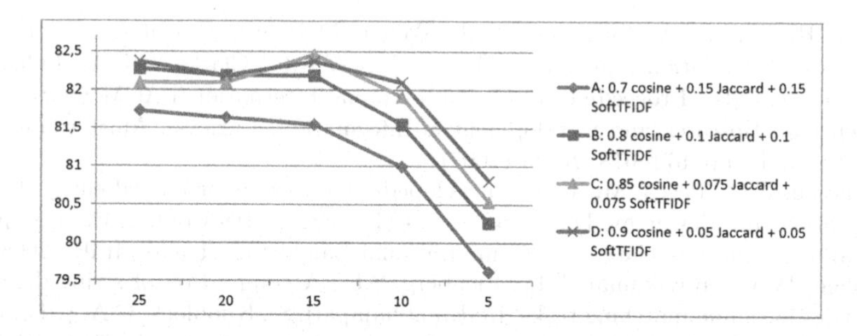

Fig. 2. The results of *4*-grams with scoring function variation

In summary, there are three kinds of combination scoring function where the differences are the portion for each applying scores. From the experiment, the system can reached a good accuracy by collecting 15 candidates hypertext to gather the candidate entities. The variations of combination scoring function present a good result with different amount of type of correct entities. Sometimes, the high precision can be reached because of the correct top 2-5 entities or the amount of entities in the first rank itself.

4 Conclusion

This paper presented an approach to annotate the extracted entities produced from fact extraction system to solve the ambiguity of those extracted entities. The recognition system applied the recognizer based on the patterns, whereas the similarity score combines context and string similarity. The experiment results show that the precision of the entity recognizer system can reach almost 85% by applying the combined score to rank the candidate entities. We evaluated different parameter settings and also introduced some optimizations to the original idea of entity recognition. To speed up the performance of entity recognition, the system applied the n-grams mechanism to replace the sequential process producing by *SoftTFIDF* mechanism. The n-grams method makes the system perform faster and produced the good result as the sequential method, even a bit higher. The system manage the variants name and recognize the entities which are not only a person name, but also scope the organization, location, and kind of common entities.

For future works, the system could also identify the entities which are constructed by an adjective, like "general relativity", without taking too much general entities, like "white collar". We also want to clean the training files from articles such as disambiguation pages. A further integration with the information extraction system is also planned, i.e., the score of the entity recognition should contribute to the score of an extracted triple.

References

1. Bast, H., Chitea, A., Suchanek, F.M., Weber, I.: Ester: efficient search on text, entities, and relations. In: Kraaij, W., de Vries, A.P., Clarke, C.L.A., Fuhr, N., Kando, N. (eds.) Proceedings of the 30th Annual International ACM SIGIR Conference on Research and Development in Information Retrieval, Amsterdam, The Netherlands, pp. 671–678. ACM (2007)
2. Bunescu, R.C., Pasca, M.: Using encyclopedic knowledge for named entity disambiguation. In: McCarthy, D., Wintner, S. (eds.) Proc. 11th Conf. of the European Chapter of the Association for Computational Linguistics, Trento, Italy (2006)
3. Cohen, W.W., Ravikumar, P.D., Fienberg, S.E.: A comparison of string distance metrics for name-matching tasks. In: Kambhampati, S., Knoblock, C.A. (eds.) Proceedings of IJCAI 2003 Workshop on Information Integration on the Web (IIWeb), Acapulco, Mexico, pp. 73–78 (2003)
4. Cucerzan, S.: Large-scale named entity disambiguation based on wikipedia data. In: Proceedings of the Joint Conference on Empirical Methods in Natural Language Processing and Computational Natural Language Learning, Prague, Czech Republic, pp. 708–716 (2007)
5. Dill, S., Eiron, N., Gibson, D., Gruhl, D., Guha, R., Jhingran, A., Kanungo, T., McCurley, K., Rajagopalan, S., Tomkins, A.: A case for automated large-scale semantic annotation. Web Semantics 1(1), 115–132 (2003)
6. Dredze, M., McNamee, P., Rao, D., Gerber, A., Finin, T.: Entity disambiguation for knowledge base population. In: Huang, C.R., Jurafsky, D. (eds.) Proc. 23rd International Conference on Computational Linguistics, Beijing, China, pp. 277–285. Tsinghua University Press (2010)
7. Fader, A., Soderland, S., Etzioni, O.: Identifying relations for open information extraction. In: Proc. Conference on Empirical Methods in Natural Language Processing (EMNLP), Edinburgh, UK, pp. 1535–1545 (2011)
8. Hakimov, S., Oto, S.A., Dogdu, E.: Named entity recognition and disambiguation using linked data and graph-based centrality scoring. In: Virgilio, R.D., Giunchiglia, F., Tanca, L. (eds.) Proc. 4th Intl. Workshop on Semantic Web Information Management (SWIM), Scottsdale, AZ. ACM (2012)
9. Halevy, A.Y., Etzioni, O., Doan, A., Ives, Z.G., Madhavan, J., McDowell, L., Tatarinov, I.: Crossing the structure chasm. In: Proc. 1st Biennal Conference on Innovative Data Systems Research (CIDR), Asilomar, CA, USA (2003)
10. In: Huang, C.R., Jurafsky, D. (eds.) Proc. 23rd International Conference on Computational Linguistics, Beijing, China. Tsinghua University Press (2010)
11. Nadeau, D., Sekine, S.: A survey of named entity recognition and classification. Lingvisticae Investigationes 30(1), 3–26 (2007)
12. Yosef, M.A., Hoart, J., Bordino, I., Spaniol, M., Weikum, G.: Aida: An online tool for accurate disambiguation of named entities in text and tables. PVLDB 4(12), 1450–1453 (2011)
13. Zhang, W., Su, J., Tan, C.L., Wang, W.: Entity linking leveraging automatically generated annotation. In: Huang, C.R., Jurafsky, D. (eds.) Proc. 23rd International Conference on Computational Linguistics, Beijing, China, pp. 1290–1298. Tsinghua University Press (2010)

Author Name Disambiguation by Using Deep Neural Network

Hung Nghiep Tran, Tin Huynh, and Tien Do

University of Information Technology - Vietnam,
Km 20, Hanoi Highway, Linh Trung Ward, Thu Duc District, HCMC, Vietnam
{nghiepth,tinhn,tiendv}@uit.edu.vn

Abstract. Author name ambiguity is one of the problems that decrease the quality and reliability of information retrieved from digital libraries. Existing methods have tried to solve this problem by predefining a feature set based on expert's knowledge for a specific dataset. In this paper, we propose a new approach which uses deep neural network to learn features automatically for solving author name ambiguity. Additionally, we propose the general system architecture for author name disambiguation on any dataset. We evaluate the proposed method on a dataset containing Vietnamese author names. The results show that this method significantly outperforms other methods that use predefined feature set. The proposed method achieves 99.31% in terms of accuracy. Prediction error rate decreases from 1.83% to 0.69%, i.e., it decreases by 1.14%, or 62.3% relatively compared with other methods that use predefined feature set (Table 3).

Keywords: Digital Library, Bibliographic Data, Author Name Disambiguation, Machine Learning, Feature Learning, Deep Neural Network.

1 Introduction

Author name ambiguity is a problem that occurs when a set of publication records contains ambiguous author names, i.e., the same author may appear under distinct names (synonymy), or distinct authors may have similar names (polysemy) [5]. This problem decreases the quality and reliability of information retrieved from digital libraries such as the impact of authors, the impact of organizations, etc. Therefore, author name disambiguation is a critical task in digital libraries.

There are two approaches to author name disambiguation: (1) grouping publication records of a same author by finding some similarity among them (author grouping methods) or (2) directly assigning them to their respective authors (author assignment methods) [5]. Both of them try to create, select and combine features based on the similarity of attributes (author names, keywords, etc.) by using some measures such as Jaccard, Jaro, etc., or some heuristics. However, most of those works are done manually based on experts' knowledge. Each predefined feature set could perform very well on a specific dataset that experts

N.T. Nguyen et al. (Eds.): ACIIDS 2014, Part I, LNAI 8397, pp. 123–132, 2014.

originally dealt with, but it could perform poorly on other datasets. To solve this problem, a method to learn features automatically from data is necessary.

Neural networks, which have many layers, are called deep neural networks (DNN). Recent researches [3, 10, 15] have shown their strong ability in feature learning in many tasks. Internal features learned by the DNN are relatively stable for variants in data if the training data are sufficiently representative [15]. This helps dealing with citation errors[1], which is an open challenge pointed out by Ferreira et al. [5]. Moreover, using neural network has the advantage that it would build a general model. This model could disambiguate author name incrementally when new publication records are incorporated into the dataset.

In this paper, we propose a new approach which uses deep neural network to learn features automatically for solving author name ambiguity. Additionally, we propose the general system architecture for author name disambiguation on any dataset. This system computes a representative for a dataset, and then uses a combination of many DNNs to learn features and disambiguate author names.

The remainder of the paper is organized as follows. Section 2 briefly presents related researches on author name disambiguation and feature learning using DNN. Section 3 will describe the proposed method and system architecture. In section 4, we will present the experiments, evaluation results and discussions. Finally, we conclude the paper and suggest future works in section 5.

2 Related Work

Ferreira et al. [5] did a brief survey of author name disambiguation methods. According to their survey, existing methods have tried to create, select and combine features based on the similarity of attributes by using some string-matching measures or some specific heuristic, such as the number of coauthor names in common, etc.

Bhattacharya and Getoor [1] proposed a combined similarity function defined on attributes and relational information. The method obtained a high F1 score around 0.99 in the CiteSeer collection, lower in the arXiv collection and only around 0.81 in the BioBase collection.

In another research, Torvik et al. [14] used a feature set resulting from the comparison between the common citation attributes along with medical subject headings, language, and affiliation of two references in MEDLINE dataset. In a subsequent work [13], Torvik and Smalheiser incorporated some features into their method to achieve better result.

In our previous research [9], we predefined a feature set to learn a similarity function specifically for Vietnamese author dataset, one of the most difficult case, and obtained around 0.98 of accuracy.

In those researches, the central task is predefining a feature set for a specific dataset. A good feature set will help improving accuracy on a specific dataset, but it need to be recalibrated for a new dataset. In this research, we aim at learning features automatically.

[1] Citation errors: errors in citation data which are sometimes impossible to detect.

DNN could be regarded as feedforward neural network with more than one layer [12]. Recently, many training and initialization schemes have been proposed in order to improve learning speed on such deep network, e.g., such as RBM [8], sparse auto-encoders [11], and normalized initialization [6]. Deep learning using DNN has been a popular method for automatic feature learning in many tasks.

Ciresan et al. [3] was very successful in using a big DNN to learn features in image recognition. They built a deep convolution neural network and trained such network by simple online back-propagation. Their models greatly outperformed previous methods on many well-known datasets such as MNIST [2], NORB [3], etc. without using complicated image pre-processing techniques.

Yu et al. [15] used a simple deep feedforward neural network to learn features in speech recognition. They proved the model's ability to extract discriminative internal features that are robust to variants in data. Their model outperformed state-of-the-art systems based on GMMs or shallow networks without the need for explicit model adaptation or feature normalization.

However, to the best of our knowledge, there has not been any research that attempts to learn feature automatically by using DNN for author name disambiguation. Therefore, in this research, we will explore that approach.

3 Our Approach

In this section, we describe our proposed method and the general system architecture for author name disambiguation on any dataset. This method uses a combination of many DNNs to learn features from a data representative, which could be computed automatically for any dataset.

3.1 Deep Neural Network

DNN is a popular method for automatic feature learning [12]. Some recent researches have successfully exploited feature learning using DNN to achieve state-of-the-art performance in many tasks [3,10,15]. There are many types of DNN. All of them are neural networks with many layers, but they are different in parameter initialization scheme, training algorithm, activation function, etc.

In this research, we use DNN with simple feedfoward architecture, a.k.a., multi-layer perceptron [12]. Figure 1 shows the general architecture of such network. The network has many layers stacked upon each other. Each neuron unit in each layer connects to every unit in the sequential layer. The number of units in the input layer corresponds with the number of basic features we use. Output layer contains two units which correspond with the case where two citations belong to the same author and otherwise, respectively.

If we denote the input and the ideal output of the DNN as x and y, respectively, a DNN can be interpreted as a directed graphical model that approximates the posterior probability $p_{y|x}(y = c|x)$ of a class c given an input x.

[2] `http://yann.lecun.com/exdb/mnist/`
[3] `http://www.cs.nyu.edu/~ylclab/data/norb-v1.0/`

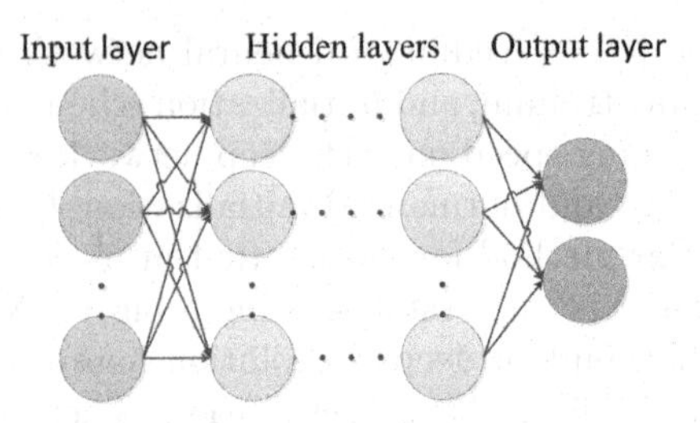

Fig. 1. Deep Neural Network Architecture

We consider a DNN with $(L-1)$ hidden layers, which is a stack of $(L+1)$ layers of log-linear models. Each layer l in the first L layers models the posterior probabilities of a hidden vectors h^l given the preceding layer's output v^l. If h^l consists of N^l hidden units, each denoted as h^l_j, the posterior probability can be expressed as

$$p^l(h^l|v^l) = \prod_{j=1}^{N^l} \frac{e^{z^l_j(v^l)\cdot h^l_j}}{\sum_{h^l_j} e^{z^l_j(v^l)\cdot h^l_j}}, 0 \leq l < L \tag{1}$$

where $z^l(v^l) = (W^l)^T v^l + a^l$, and W^l and a^l represent the weight matrix and bias vector in the $l-th$ layer, respectively.

Internal features will be learned in each hidden layer [15]. Each hidden unit's output represents an internal feature and each hidden layer's output composes an internal feature vector. Starting with the basic feature input ($v^0 = x$), the output of each layer becomes the input of the next one, i.e., $v^{l+1} = h^l$. The latter layer will learn more sophisticated features.

In the final layer, the class posterior probabilities are computed as a multinomial distribution

$$p_{y|x}(y=c|x) = p^L(y=c|v^L) = \frac{e^{z^L_c(v^L)}}{\sum_{c'} e^{z'^L_c(v^L)}} = softmax_c(v^L). \tag{2}$$

This type of DNN has the vanishing gradient issue when being trained with the traditional activation function $sigmoid(x) = 1/(1+e^{-x})$ and back-propagation algorithm. To address this issue, we use the activation function $softsign(x) = x/(1+|x|)$ and the adaptive resilient backpropagation algorithm together with some training techniques [6].

DNN's ability in a specific case is affected by the network's structure and its parameters. Network parameters could be learned by training using some optimization algorithms. Network structure includes two hyperparamters: the number of hidden layers and the number of hidden units in each layer. The number of units should be equal among hidden layers so that information could flow effectively [6]. In general, deeper and larger network will achieve better results. However, this makes training slower and is capable to yield overfitting.

Those two hyperparameters are usually chosen based on experiments on the validation set [3, 15]. In this research, we determine the optimum network size based on experiments using k-fold cross-validation. We begin with a small network, and then change the number of hidden layers and the number of hidden units respectively to create networks at larger sizes. The optimum network structure is the one with the highest average validation accuracy.

3.2 Data Representative

As we have shown in the previous subsection, DNN could learn internal features from data. In order to do this, data must be put into the input layer in a proper way. The input should be a good representative for data, i.e., it could describe details in data. We call that data representative the basic feature set.

There are many different ways to compute a data representative. One obvious way is to measure the similarity between all attributes of two publication records such as Author name, Affiliations, Coauthor, and Paper keyword, etc. using string-matching measures. We assume that the similarity between those attributes expresses how much two publications belong to the same author.

According to some surveys reviewing string-matching measures to identify the duplication [2, 4], there are three types of measure: (1) *Edit distance* such as Levenshtein, Monger-Elkan, Jaro, and Jaro-Winkler; (2) *Token-based* such as Jaccard and TF/IDF and (3) *Hybrid measures* such as Mogne-Elkan for comparing two long strings. We employ all three types of measure.

Each publication record has different attributes. In general case, we could apply computations to all available attributes, and use default value when they are unavailable. Therefore, how we build the data representative does not depend on a specific dataset.

In this research, we use these measures: Jaccard, Levenshtein, Jaro, Jaro-Winkler, Smith-Waterman, Mogne-Elkan. We apply them for these attributes: author name, co-author, affiliation, paper keyword, author interest keyword.

3.3 System Architecture

In this subsection, we describe the general system architecture for author name disambiguation. The system could run on any dataset without expert's modifications. Figure 2 shows the proposed system architecture. The system incorporates two components.

The first component takes the data and computes a data representative, i.e., a basic feature set ($v^0 = x$) to represent data. There could be many representatives for the same data, so this component could be implemented in many ways. In this research, we use string-matching measures to compute the data representative. The computations could be performed on any dataset automatically.

The second component takes the basic feature set as its input, and then learns features in its hidden layers to disambiguate author names. The last layer of DNN

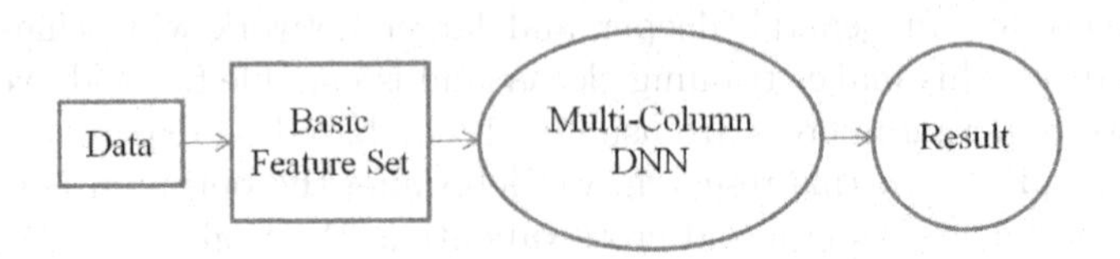

Fig. 2. System Architecture

computes the probabilities $p_{y|x}(y = c|x) = p^L(y = c|v^L)$ to determine whether two author instance names in a pair belong to the same author (when $p \geq 0.5$) or not (when $p < 0.5$).

In this system architecture, we use multi-column DNN technique, which is illustrated in figure 3, to improve the generalization capabilities of the system [3]. This technique is similar to an ensemble method known as bootstrap aggregating or bagging.

Specifically, we will train N DNNs simultaneously using data retrieved randomly from the training set in a manner similar to k-fold cross-validation. After training, we have N distinct DNNs. In testing phase, we will apply all those DNNs simultaneously to each item in a separate test set. Then, we will take the final result by averaging results from those DNNs as

$$P_{y|x}(y = c|x) = \overline{p_{y|x}(y = c|x)} = \frac{\sum_{n=1}^{N} p^n_{y|x}(y = c|x)}{N} \tag{3}$$

where p^n is the output of $n - th$ DNN.

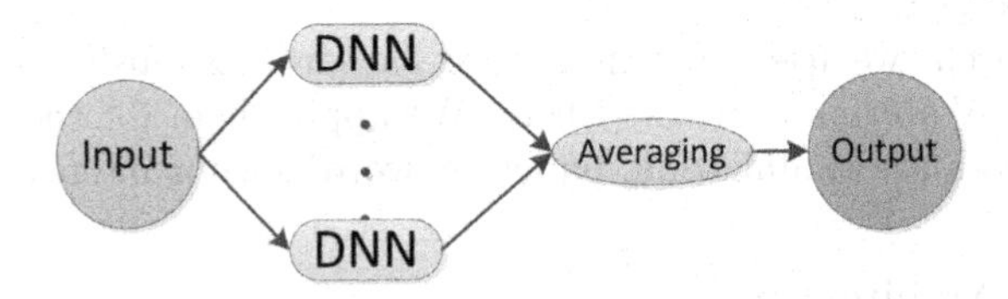

Fig. 3. Multi-Column Deep Neural Network

4 Experiment, Evaluation and Discussion

We evaluate the effect of using automatic feature learning by DNNs for the author name disambiguation problem on Vietnamese author dataset (very ambiguous cases [5]), which we collected from online digital libraries. This section presents our experiment settings, evaluation results, and discussions.

4.1 Dataset

Vietnamese Author Dataset:

In a previous work [9], we built a Vietnamese author dataset for checking author name ambiguity. Data was acquired from three online digital libraries that are ACM [4], IEEE Xplore [5], and MAS [6] by querying their search engine using names of 10 Vietnamese authors. For these authors, there are many different instance names, e.g., author 'Hoang Kiem' can have many instance names such as 'Hoang Kiem', 'Kiem Hoang', 'Hoang Van Kiem', 'Kiem Van Hoang', etc.

Query results are publications with different author instance names. We built the dataset by creating pairs of publications for each author. Based on our understanding about these authors, we manually labeled each pair with value 1 if ambiguous names in this pair actually were one person and value 0 otherwise.

In this research, we extend the dataset, so that, there are totally 30537 samples in the dataset. Table 1 shows the dataset details, the total number of pairs is counted without duplicated ones.

Table 1. The Vietnamese author dataset

Authors	**Number of pairs with label 0**	**Number of pairs with label 1**
Cao Hoang Tru	6522	409
Dinh Dien	6750	384
Duong Anh Duc	6757	406
Ha Quang Thuy	6812	351
Ho Tu Bao	6728	435
Hoang Kiem	6753	410
Le Dinh Duy	6725	387
Le Hoai Bac	6728	435
Nguyen Ngoc Thanh	6869	294
Phan Thi Tuoi	6728	435
Total	26591 (87.07%)	3946 (12.93%)

Dataset Preparation:

For each pair of publication records, we compute all basic features. So each pair of publication records is represented as a basic feature vector with label 1 or 0. We use this vector as the input for the DNN.

From the original dataset, we hold out 20% of the data as the test set for final performance evaluation. We use 5-fold cross-validation on the remaining 80% of the data to tune hyperparameters and avoid overfitting. Each split dataset contains almost equal percentage of random samples of one particular class. Those samples are picked randomly at uniform distribution.

[4] `http://dl.acm.org`
[5] `http://ieeexplore.ieee.org/Xplore/home.jsp`
[6] `http://academic.research.microsoft.com/`

4.2 Tuning Hyperparameters

On Vietnamese author dataset, we experiment with 5-fold cross-validation to choose network size. We begin with the smallest network size of 1 hidden layer and 10 hidden units. We use this network as the baseline for hyperparameters tuning and achieve average validation accuracy of 95.35%.

Then we increase the number of hidden layers and hidden units respectively and conduct experiments at many network sizes. Table 2 shows five network sizes (the number of hidden layers × the number of hidden units in each layer) with the highest accuracy. The network with 7 hidden layers and 50 hidden units in each layer achieves the highest average validation accuracy.

Table 2. Five network sizes with the highest accuracy

Network size	Average validation accuracy (%)
7 × 50	99.35
7 × 75	99.33
6 × 75	99.28
5 × 75	99.27
6 × 100	99.25

4.3 Evaluation

In our recent research [9], we proposed an approach based on a predefined feature set for Vietnamese author name and applied several classification models to that feature set. According to that research, k-NN, Random Forest, C4.5, SVM, and Naive Bayes, respectively, are the best suitable methods for the predefined feature set.

In this research, we compare the proposed method with those methods. The proposed method implements the system architecture that we have described. The DNNs use the hyperparameters that have been tuned. The other methods use the same settings and implementations as in our previous research [9].

Table 3 shows evaluation results in terms of accuracy and error on a separated test set. Results show that the proposed method significantly outperforms methods that use predefined feature set. The proposed method achieves 99.31% in terms of accuracy. Whereas, the best method that uses predefined feature set achieves 98.17% in terms of accuracy. Prediction error rate decreases from 1.83% to 0.69%, i.e., it decreases by 1.14%, or 62.3% relatively compared with other methods that use predefined feature set.

4.4 Discussion

Evaluation results clearly show benefits of learning features compared with predefining features in terms of accuracy. Moreover, automatic feature learning does not require expert's knowledge on specific dataset. DNN has been used to learn

Table 3. Evaluation results

Feature set	Method	Accuracy (%)	Error (%)
Predefined	k-NN	98.17	1.83
	Random Forest	98.13	1.87
	C4.5	98.02	1.98
	SVM	97.45	2.55
	Naive Bayes	96.68	3.32
Learned Automatically	DNN	**99.31**	**0.69**

features successfully. However, due to its high capability to learn complex features, it is prone to overfitting.

In this research, we use many techniques to reduce overfitting. We extend Vietnamese author dataset. We use k-fold cross-validation for hyperparameter tuning and early stopping. Moreover, our system architecture uses multi-column technique to have a lower variance result.

The type of DNN we use has the vanishing gradient issue when being trained with the traditional sigmoid activation function and back-propagation algorithm. One solution is choosing a good activation function [6]. The hyperbolic tangent function $tanh(x) = (1 - e^{-2x})/(1 + e^{-2x})$ is better than the traditional sigmoid function thanks to its zero mean. Whereas, the softsign function $softsign(x) = x/(1 + |x|)$ is better than $tanh(x)$ thanks to its smoother asymptotic behavior. The rectifier function $max(x, 0)$ is one of the best activation functions [6, 7].

The current DNN model is supervised, therefore, it needs labeled data, which is usually not easy to obtain. However, there are some techniques to pre-train DNN using unlabeled data, which are special kind of weight initialization methods, to improve performance much in terms of training time and accuracy, especially in case lack of labeled data.

The proposed method is prospective when data is integrated from heterogeneous sources, because in such case, it is difficult to predefine a feature set.

On the other hand, automatic feature learning using DNN has shown its ability in many complex tasks such as image recognition, where this method could recognize object just by using raw pixels [3]. Therefore, it is rational to think of creating such 'pixels' in author name disambiguation by encoding bibliographic data and use those 'pixels' as the basic data representative in the DNN.

5 Conclusion and Future Work

In this paper, we have proposed a new approach which uses deep neural network to learn features automatically for solving author name ambiguity. Additionally, we have proposed the general system architecture for author name disambiguation on any dataset.

We have evaluated the proposed method on a Vietnamese author dataset. The results show that this method significantly outperforms other methods that use predefined feature set. The proposed method achieves 99.31% in terms of accuracy. Prediction error rate decreases from 1.83% to 0.69%, i.e., it decreases by 1.14%, or 62.3% relatively compared with other methods that use predefined feature set.

The proposed method could be extended to solve some open challenges such as the lack of labeled training data, incremental and new-author disambiguation.

In the future, we will benchmark the proposed method on other datasets. We will also experiment with other activation functions and unsupervised pre-training methods on encoded bibliographic data.

Acknowledgments. This research is funded by University of Information Technology, VNU-HCMC under grant number C2012-07.

References

1. Bhattacharya, I., Getoor, L.: Collective entity resolution in relational data. ACM Trans. Knowl. Discov. Data 1(1) (March 2007)
2. Bilenko, M., Mooney, R., Cohen, W., Ravikumar, P., Fienberg, S.: Adaptive name matching in information integration. IEEE Intell. Sys. 18(5), 16–23 (2003)
3. Ciresan, D.C., Meier, U., Schmidhuber, J.: Multi-column deep neural networks for image classification. In: CVPR, pp. 3642–3649 (2012)
4. Cohen, W.W., Ravikumar, P.D., Fienberg, S.E.: A comparison of string distance metrics for name-matching tasks. In: IIWeb, pp. 73–78 (2003)
5. Ferreira, A.A., Gonçalves, M.A., Laender, A.H.: A brief survey of automatic methods for author name disambiguation. SIGMOD Rec. 41(2), 15–26 (2012)
6. Glorot, X., Bengio, Y.: Understanding the difficulty of training deep feedforward neural networks. JLMR - Proceedings Track 9, 249–256 (2010)
7. Glorot, X., Bordes, A., Bengio, Y.: Deep sparse rectifier neural networks. In: AISTATS, pp. 315–323 (2011)
8. Hinton, G.E., Osindero, S., Teh, Y.W.: A fast learning algorithm for deep belief nets. Neural Comput. 18(7), 1527–1554 (2006)
9. Huynh, T., Hoang, K., Do, T., Huynh, D.: Vietnamese author name disambiguation for integrating publications from heterogeneous sources. In: Selamat, A., Nguyen, N.T., Haron, H. (eds.) ACIIDS 2013, Part I. LNCS, vol. 7802, pp. 226–235. Springer, Heidelberg (2013)
10. Krizhevsky, A., Sutskever, I., Hinton, G.E.: Imagenet classification with deep convolutional neural networks. In: NIPS, vol. 25, pp. 1106–1114 (2012)
11. Ranzato, M., Boureau, Y.L., LeCun, Y.: Sparse feature learning for deep belief networks. In: NIPS (2007)
12. Rumelhart, D.E., Hinton, G.E., Williams, R.J.: Learning Internal Representations by Error Propagation. In: Parallel Distributed Processing: Explorations in the Microstructure of Cognition, vol. 1, pp. 318–362. MIT Press, Cambridge (1986)
13. Torvik, V.I., Smalheiser, N.R.: Author name disambiguation in medline. ACM Trans. Knowl. Discov. Data 3(3), 11:1–11:29 (2009)
14. Torvik, V.I., Weeber, M., Swanson, D.R., Smalheiser, N.R.: A probabilistic similarity metric for medline records: A model for author name disambiguation: Research articles. J. Am. Soc. Inf. Sci. Technol. 56(2), 140–158 (2005)
15. Yu, D., Seltzer, M.L., Li, J., Huang, J.T., Seide, F.: Feature learning in deep neural networks - A study on speech recognition tasks. CoRR abs/1301.3605 (2013)

Incremental Refinement of Linked Data: Ontology-Based Approach

Yusuke Saito[1], Boonsita Roengsamut[1], and Kazuhiro Kuwabara[2]

[1] Graduate School of Information Science and Engineering, Ritsumeikan University
[2] College of Information Science and Engineering, Ritsumeikan University
1-1-1 Noji-Higashi, Kusatsu, Shiga 525-8577 Japan

Abstract. This paper presents an approach to refining linked data using a domain ontology. Many linked data are available on the Internet, and since the volume is increasing, incrementally adding links between pieces of data is important. In this paper, we consider a Frequently Asked Questions (FAQs) database in the domain of rental apartments targeted to international students in Japan. This database is implemented using linked data and contains the relationships between part of the floor plan of the rental apartment and FAQs. To facilitate adding new questions, we propose a method that automatically derives a relationship between a new question and part of the floor plan using the domain ontology. Our experimental results with newly added questions indicate the effectiveness of our proposed approach.

Keywords: linked data, ontology, RDF.

1 Introduction

Many data are published on the Internet as linked data [1], forming the so-called Web of Data. In contrast to the typical document-based web, linked data are often automatically created or converted from existing data. For example, DBpedia [2], one well-known hub of linked data, is created from Wikipedia. The concept of linked data is used in many domains. For example, the BBS's media contents are interlinked using linked data [5]. A method has also been proposed to automatically interlink music datasets [8]. In medical domains, linked data principles maintain patient health records [7]. It is important to find a way to incrementally add meaningful data from existing data to enrich the Web of Data.

In this paper, we present an approach to introduce links in linked data using domain ontology. The example application we consider here is a multilingual application that has frequently asked questions (FAQs) concerning rental apartments [4]. This application, which is targeted at international students living in Japan, provides useful information to solve various problems they may encounter while living in such apartments. In this application, FAQ data are stored using an RDF database. To facilitate search for questions in the FAQ, its questions are linked with their relevant parts in the floor plan of a typical apartment. In this way, a question can be searched for by selecting part of the floor plan. In this

N.T. Nguyen et al. (Eds.): ACIIDS 2014, Part I, LNAI 8397, pp. 133–142, 2014.

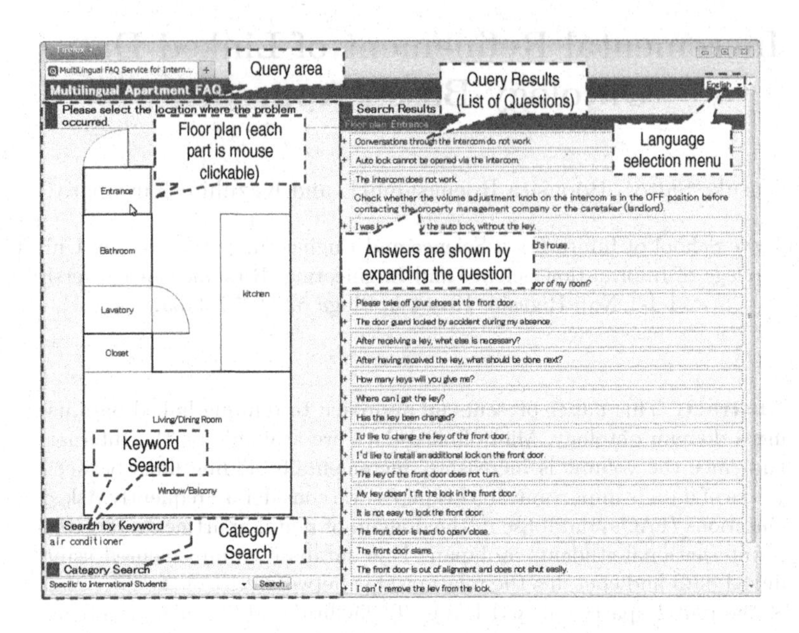

Fig. 1. Rental apartment FAQ system (screenshot)

paper, we propose a method to use the domain ontology of rental apartments to automatically derive relationships between a given question and a floor plan part with only simple natural language processing.

This paper is structured as follows. The next section describes the multilingual rental apartment FAQ system on which we implemented our proposed approach, and Section 3 describes the proposed method of deriving a new relationship in the FAQ data. Section 4 presents the results of the evaluation experiments, and Section 5 discusses some related works. The final section concludes our paper.

2 Rental Apartment FAQ System

2.1 Overview

Our multilingual rental apartment FAQ system contains frequently asked questions (FAQs) in the domain of rental apartments in four languages. With this system, a user can search for a question by selecting the part of the floor plan where the trouble occurs (Fig. 1). In this system, floor plans are categorized into the following seven parts: *entrance*, *bathroom*, *lavatory*, *kitchen*, *living/dining room*, *closet*, and *window/balcony*. This is a typical floor layout of an apartment for university students in Japan. When users select part of the floor plan, the questions related to that part will be shown. They can choose the question that is close to their problem.

2.2 Using an RDF

In this system, a question and its answers are stored in the RDF database. These data are defined by extracting them from spreadsheet data that contain sentences in four languages: Japanese, English, Chinese, and Korean. The original question and answers are written in Japanese, and the data in other languages are translated from Japanese by a human translator. To simplify extending the system, the data are separated into Japanese FAQs and their translations.

To allow search by selecting the relevant floor plan part, the association between a question and the floor plan part is defined manually. Links are made if the question contains a reference to a housing fixture or an appliance that is usually included in the designated part of the floor plan. For example, the statement, "The intercom does not work," contains a reference to the intercom. Since intercoms are often found at the entrance or its door, this statement is associated with the floor plan part of *entrance*.

When a new question (statement) is added, the association for it needs to be made manually. Here we consider a method that automatically derives the association between a statement and the floor plan part so that users do not need to specify which part of the floor plan is related when they add new questions. Additionally, when a user inputs a question, related questions can be found more easily by listing the questions that are associated with the same floor plan part.

3 Deriving a New Relationship

To make associations between a question and a floor plan, the system defines the ontology of an apartment that describes a typical floor plan and the housing fixtures and appliances contained in each part of it. In addition, the thesaurus is defined to handle variations of expressions in the question statements. Here we assume that a question is in Japanese.

3.1 Rental Apartment Ontology

The rental apartment ontology describes the relationship between the housing fixtures or appliances and each part of the apartment. As shown in Fig. 2, we consider the relationship between "entrance" and "intercom" since an "intercom" is often found at the "entrance." In this figure, `fp:entrance` is the URI that represents the *entrance* of the floor plan and `:part` indicates that the URI (`fp:entrance`) represents a floor plan[1].

Predicate `:related` means that the subject of the triple (RDF statement) is related to the triple's object. Thus, keyword `word:entrance` is related to keyword `word:intercom` where `word:entrance` and `word:intercom` represent keywords "entrance" and "intercom," respectively.

[1] The prefix `fp` is used to represent terms related to the floor plan, and `word` represents keywords. An empty prefix name is used for the default name space of this application.

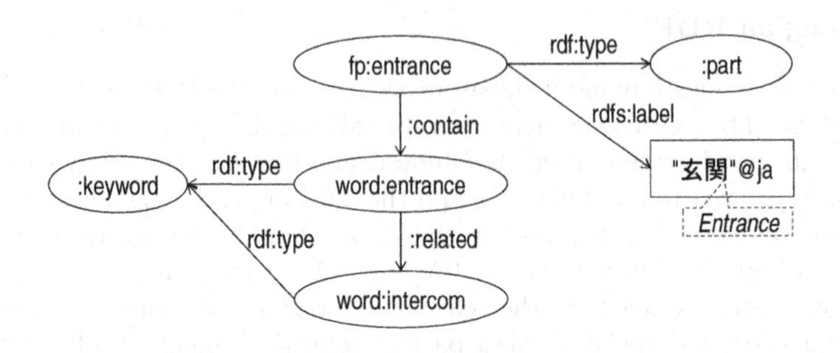

Fig. 2. Example of rental apartment ontology

Predicate `:contain` refers to the relationship between part of the floor plan and the keyword that represents a housing fixture or the part itself. Note that we distinguish between part of the floor plan and a keyword itself that represents it even if they use the same expression so that we can flexibly express relationships between keywords.

3.2 Thesaurus for Rental Apartment FAQ

There are cases where a keyword or a question can be linked with a different representation. For example, Japanese words "ドア" and "扉" are both translated into English as "door." "靴" ("shoe" in English) may share the same representation with "履物," which literally means "footwear."

To handle such cases, we introduced a thesaurus that is separate from the domain ontology of the rental apartment. It contains an entry that has the literal representation that is linked with a keyword in the rental apartment ontology by `skos:closeMatch`, which is defined in the Simple Knowledge Organization System (SKOS) [11]. This link means that two resources share almost the same meaning. Figure 3 shows an example, where the keyword (`word:intercom`) is connected to two thesaurus entries by `skos:closeMatch`. Each entry corresponds to a different expression that means the same thing. In this way, variations in question statements can be handled.

3.3 Keyword Extraction

To infer the association between a question and a floor plan part, we first extract the keywords from a question statement. We consider nouns in the questions to be the keywords. To extract the keywords from the question, a morphological analysis program for the Japanese language called kuromoji[2] is used. This program supports a user dictionary, which simplifies adding domain dependent expressions. The keyword extraction results are stored in the RDF database with a link from the question. Figure 4 shows an example, where `ques:1` is

[2] `http://www.atilika.org`

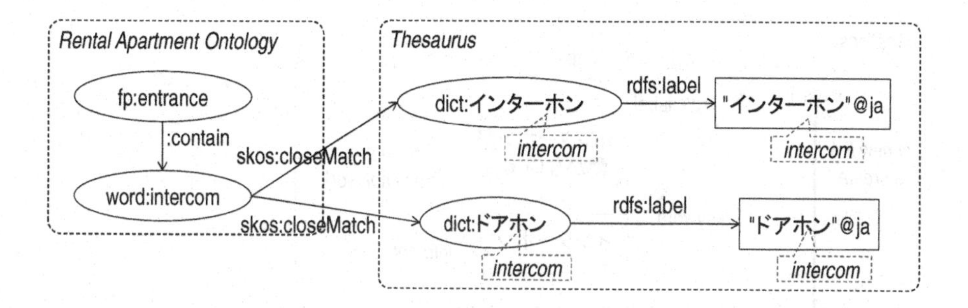

Fig. 3. Example thesaurus entry for rental apartment

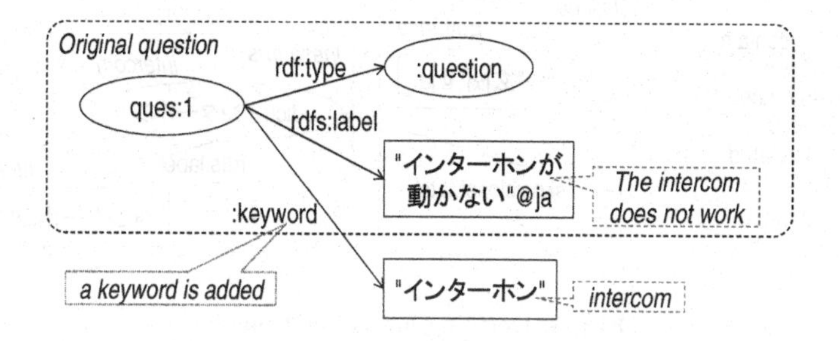

Fig. 4. Representing extracted keywords

the question's URI and `:question` indicates that this URI represents a question. Predicate `:keyword` refers to the keywords extracted from the question statement.

3.4 Example of Making an Association

As shown in Fig. 4, consider the following example: "The intercom does not work." As explained above, keyword "インターホン" ("intercom") is extracted from the question statement and stored in the question RDF database. Then the thesaurus entry with the same label as this keyword, `dict:インターホン` ("intercom"), is looked for (Fig. 3). This entry has a `skos:closeMatch` relationship with `word:intercom` in the rental apartment ontology (Fig. 2). Since `word:intercom` has links connected to the floor plan part (`fp:entrance`), the question (`ques:1`) can be linked with this part (Fig. 5).

3.5 Priority of Keywords

The extracted keywords might be related to multiple parts of the floor plan. For example, if a user inputs the statement: "トイレの水が止まらない" (which literally means "The water in the toilet keeps running"), keywords "water" and "toilet" will be extracted. The word "toilet" is related to the floor plan part

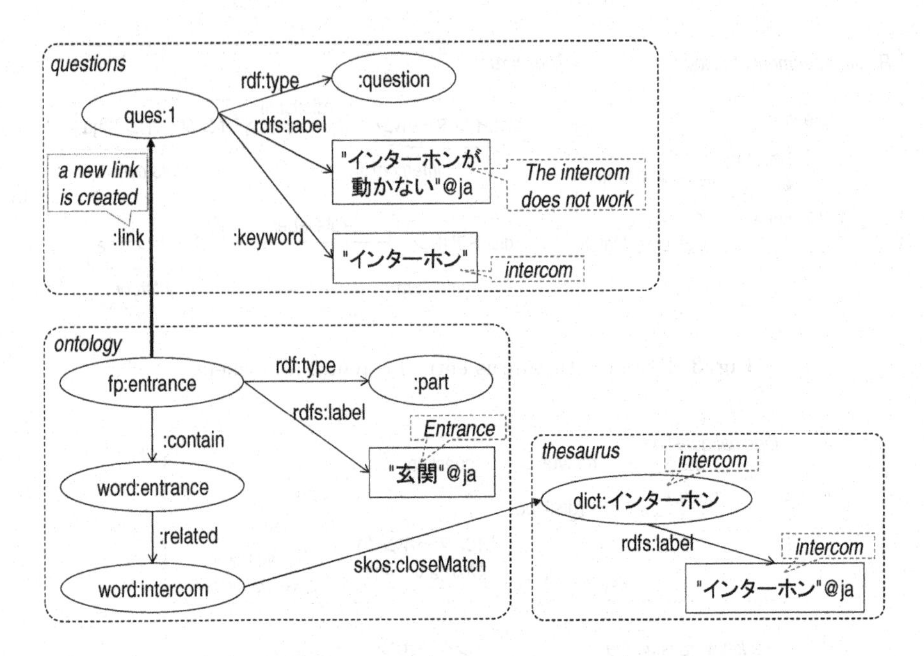

Fig. 5. Deriving an association

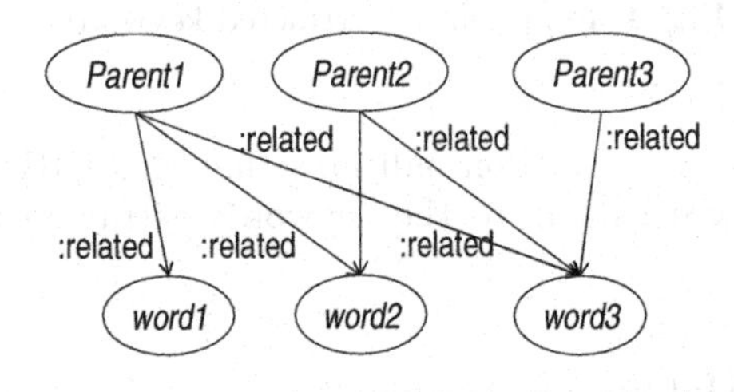

Fig. 6. Example settings for priority rules

bathroom and so is "water." However, there is also relevance between "water" and floor plan part *kitchen*. With accurate priority rules, the system will be able to determine that this question is associated with *bathroom*, not with *kitchen*.

For the above bathroom/kitchen example, we propose the following rules. Assume keywords *word1*, *word2*, and *word3* (Fig. 6). The node that is linked by `:related` is a *parent*. In this example, a parent node of *word1* is *parent1* and the parent nodes of *word2* are *parent1* and *parent2*. The parent nodes of *word3* are *parent1*, *parent2*, and *parent3*.

We put a higher priority on a node with fewer parent nodes. The heuristics behind this rule is that since a word with many parents tends to appear in various parts of the floor plan, it cannot be a representative word for part of the floor plan. In this example, the number of parent nodes of *word1*, *word2*,

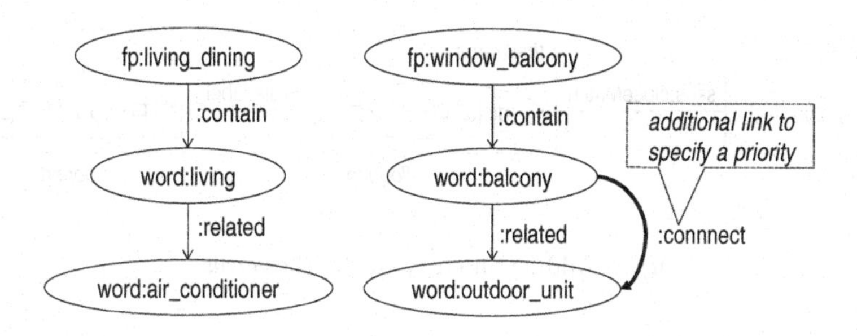

Fig. 7. Handling exceptions using `:connect` link

and *word3*, are 1, 2, and 3, respectively. Thus, *word1* will be given the highest priority, and the floor plan part will be selected that has a relationship with it.

3.6 Exception Handling

When a question includes multiple keywords that do not share the same parent node, determining the most relevant floor plan part is difficult. For example, the statement, "The air conditioner's outdoor unit is noisy," contains two keywords: "air conditioner" and "outdoor unit." "Air conditioner" has an association with the *living/dining* of the floor plan, and "outdoor unit" is related to its *window/balcony*.

If a word has only one relationship with part of the floor plan, the link (`:connect`) is put between the words. This link means that the word ("outdoor unit") takes precedence over any other words (Fig. 7).

4 Evaluation

We originally had 116 questions with manually defined associations. Using the above approach, we successfully derived the associations between the questions and the floor plan parts.

To evaluate our proposed approach, we made 121 new questions and applied our proposed approach to see if they were correctly linked. With the same program, 31 of 121 questions resulted in errors. We fixed 29 errors by adding more keywords to the thesaurus and ontology as well as the user dictionary of the morphological analysis program. The remaining two errors needed exception handling, as outlined in Section 3.6. Below we discuss how the errors were fixed.

Adding an Entry to the Thesaurus. Eleven of 31 errors were solved by simply expanding the thesaurus. For example, consider the following statement: "The flooring is damaged." If we consider "flooring" to be another representation of the term "floor," we just need to add an entry to the thesaurus and link it to the `word:floor` using `skos:closeMatch` (Fig. 8).

Fig. 8. Adding an entry to the thesaurus

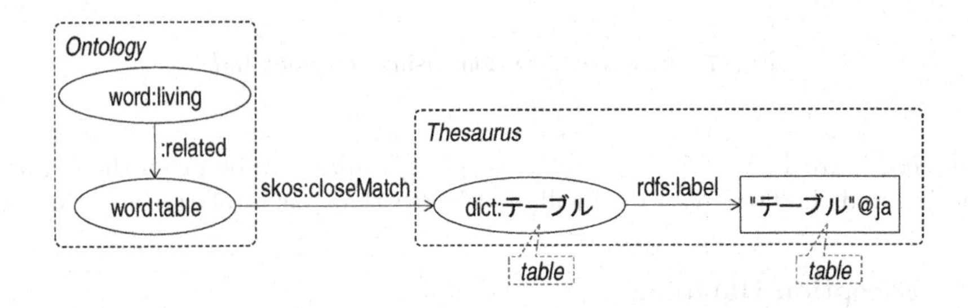

Fig. 9. Adding an entry to the ontology and the thesaurus

Adding an Entry to the Ontology. Five of 31 errors were fixed by adding words to the ontology and the thesaurus. For example, consider the following statement: "The legs of the table are broken." This can be solved by adding RDF statements concerning keyword "table," as shown in Fig. 9. Note that "leg" is ignored in this case since "table" is a more representative word from the viewpoint of the rental apartment ontology.

Adding an Entry to the User Dictionary. Thirteen of 31 statement errors occurred because the morphological analyzer did not recognize words. Such errors can be fixed by directly registering a keyword to the user's dictionary of the morphological analyzer. For example, the morphological analysis program failed to recognize "microwave oven" in "The microwave oven is broken." We solved this problem by simply adding a new keyword to the user's dictionary of the morphological analyzer and updating the thesaurus and ontology.

Using a Special Link. We fixed the remaining two statements that caused errors by treating them as exceptions, that is, by adding a special link `:connect` to denote a strong association between words. For example, the statement, "The smell from the garden of the room next door is awful," includes the keywords of "garden" and "room." Since the number of parent nodes of these keywords is the same, we resolved this issue by adding a special link (`:connect`) between `word:garden` and `word:balcony`.

5 Related Works

Our method proposed here links questions in FAQs and the relevant part of the floor plans of rental apartments. Keywords are extracted from the questions, and their matches are searched for in the rental apartment ontology. In this sense, this approach can be considered simplified ontology matching.

Ontology matching has been actively researched [9]. Most research topics in the field are concerned with monolingual ontology matching algorithms. However, we aim to bring ontology matching to another level by considering multilingual ontology matching.

There are several works on multilingual ontology matching. For example, a method of ontology alignment between Korean and Swedish ontologies was proposed [3]. Multilingual matching is also applied to increase the interoperability of financial information management, where machine learning techniques are used to rank the matches between two ontologies [10].

Our proposal focuses on developing a FAQ system for the rental apartment domain. Since the rental apartment's domain ontology is separated from the thesaurus that handles the variations of expressions in question statements, it is easier to extend our proposed approach to handle multiple languages by extending the thesaurus to include keyword representations in other languages.

6 Conclusion

In this paper, we proposed a mechanism to derive new links to associate a question in a FAQ database to a related floor plan part. With only simple natural language processing of morphological analysis, we derived links using the domain ontology and the thesaurus. Our experimental evaluation results indicate that the key to make correct associations is enriching the ontology and the thesaurus.

In this sense, refining the domain ontology is crucial. Currently we manually update it. In the future, we want to (semi-)automatically refine the ontology, for example, by a machine learning technique. In addition, since this application is targeted at particular users, we plan to exploit the power of a community of such users to refine the ontology and its mapping (such as [6]).

Currently we experimented with only question statements in Japanese. We plan to extend our proposed approach to deal with questions in other languages using a machine translation service.

The proposed approach here can possibly be applied to other domains where contents need to be linked to their related concept words. For example, in e-learning environments, learning materials may need to be linked to their related topics. We also plan to apply the proposed approach to such domains.

Acknowledgments. The authors thank Toshihiko Nakai for the multilingual FAQ and the glossary data for rental apartments.

References

1. Bizer, C., Heath, T., Berners-Lee, T.: Linked data - The story so far. International Journal on Semantic Web and Information Systems 5(3), 1–22 (2009)
2. Bizer, C., Lehmann, J., Kobilarov, G., Auer, S., Becker, C., Cyganiak, R., Hellmann, S.: DBpedia - A crystallization point for the web of data. Web Semantics: Science, Services and Agents on the World Wide Web 7(3), 154–165 (2009)
3. Jung, J.J., Håkansson, A., Hartung, R.: Indirect alignment between multilingual ontologies: A case study of Korean and Swedish ontologies. In: Håkansson, A., Nguyen, N.T., Hartung, R.L., Howlett, R.J., Jain, L.C. (eds.) KES-AMSTA 2009. LNCS (LNAI), vol. 5559, pp. 233–241. Springer, Heidelberg (2009)
4. Kinomura, S., Kuwabara, K.: Developing a multilingual application using linked data: A case study. In: Bădică, C., Nguyen, N.T., Brezovan, M. (eds.) ICCCI 2013. LNCS (LNAI), vol. 8083, pp. 120–129. Springer, Heidelberg (2013)
5. Kobilarov, G., Scott, T., Raimond, Y., Oliver, S., Sizemore, C., Smethurst, M., Bizer, C., Lee, R.: Media meets semantic web – how the BBC uses DBpedia and linked data to make connections. In: Aroyo, L., et al. (eds.) ESWC 2009. LNCS, vol. 5554, pp. 723–737. Springer, Heidelberg (2009)
6. Noy, N.F., Griffith, N., Musen, M.A.: Collecting community-based mappings in an ontology repository. In: Sheth, A.P., Staab, S., Dean, M., Paolucci, M., Maynard, D., Finin, T., Thirunarayan, K. (eds.) ISWC 2008. LNCS, vol. 5318, pp. 371–386. Springer, Heidelberg (2008)
7. Pathak, J., Kiefer, R.C., Chute, C.G.: Applying linked data principles to represent patient's electronic health records at mayo clinic: A case report. In: Proceedings of the 2nd ACM SIGHIT International Health Informatics Symposium (IHI 2012), pp. 455–464 (2012)
8. Raimond, Y., Sutton, C., Sandler, M.: Automatic interlinking of music datasets on the semantic web. In: Proceedings of the Linked Data on the Web Workshop (2008), `http://CEUR-WS.org/Vol-369/paper18.pdf`
9. Shvaiko, P., Euzenat, J.: Ontology matching: State of the art and future challenges. IEEE Trans. on Knowledge and Data Engineering 25(1), 158–176 (2013)
10. Spohr, D., Hollink, L., Cimiano, P.: A machine learning approach to multilingual and cross-lingual ontology matching. In: Aroyo, L., Welty, C., Alani, H., Taylor, J., Bernstein, A., Kagal, L., Noy, N., Blomqvist, E. (eds.) ISWC 2011, Part I. LNCS, vol. 7031, pp. 665–680. Springer, Heidelberg (2011)
11. W3C: SKOS simple knowledge organization system reference. W3C Recommendation (2009), `http://www.w3.org/TR/skos-reference`

Using Lexical Semantic Relation and Multi-attribute Structures for User Profile Adaptation

Agnieszka Indyka-Piasecka and Piotr Jacewicz

Institute of Informatics, Wrocław University of Technology, Poland
agnieszka.indyka-piasecka@pwr.wroc.pl

Abstract. This contribution presents a new approach to the representation of user interests and preferences at information retrieval process. The *adaptive user profile* includes both interests given explicitly by the user, as a query, and also preferences expressed by the valuation of relevance of retrieved documents, so to express field independent translation between terminology used by user and terminology accepted in some field of knowledge. Building, modifying, expanding by semantically related terms and using procedures for the profile are presented. Experiments concerning the profile, as a personalization mechanism of Web retrieval system, are presented and discussed.

1 Introduction

In today's World Wide Web reality, the common facts are: increasingly growing number of documents, high frequency of their modifications and, as consequence, the difficulty for users to find important and valuable information. These problems caused that much attention is paid to helping user in finding important information on Internet Information Retrieval (IR) systems. Individual characteristic and user needs are taken under consideration, what lead to system personalization. System personalization is usually achieved by introducing user model into information system. User model might include information about user preferences and interests, attitudes and goals [3], knowledge and beliefs [6], personal characteristics [7], or history of user interaction with a system [12]. User model is called user profile at the domain of IR. The profile represents user information needs, such as interests and preferences and can be used for ranking documents received from the IR system. Such ranking is usually created due to degree of similarity between user query and retrieved documents [8], [14]. In Information Filtering (IF) systems, user profile became the query during process of information filtering. Such profile represents user information needs, relatively stable over a period of time [1]. User profile also has been used for query expansion, based on explicit and implicit information obtained from the user [5].

The main issue, at the domain of user profile for IR, is a representation of user information needs and interests. Usually user interests are represented as a set of keywords or a n–dimensional vector of keywords, where every keyword's weight (at the vector) represents importance of keyword describing user interests [8], [9].

N.T. Nguyen et al. (Eds.): ACIIDS 2014, Part I, LNAI 8397, pp. 143–152, 2014.

The approaches with more sophisticated structures representing knowledge about user preferences are also described: stereotypes – the set of characteristics of a prototype user of users groups, sharing the same interests [4], or semantic net, which discriminates subject of user interests by underlining the main topic of interests [2].

The approaches to determine user profile can be also divided into a few groups. The first group includes methods where user interests are stated explicitly by the user in specially prepared forms or during answering standard questions [4], [8], or by an example piece of text, written by the user [11]. The second group can be these approaches, where user profile is based on the analysis of terms frequency in user queries directed to IR system [8]. There is an assumption for these methods that the interest of the user, represented by a term, is higher as the term is more frequent in the user query. Analysis of the queries with the use of genetic algorithms [13], reinforcement learning [18] or semantic nets [2] are extensions to this approach. The third group of approaches includes methods, where the user evaluates retrieved documents. From documents assessed as interesting by the user, additional index terms, describing user interests, are added to the user profile [4], [8].

Most of research, concerning user modeling for IR, acquire user information needs expressed directly by the user of the IR system. Serious difficulties of a user in expressing his real information need are frequently neglected. As well the fact is ignored that user usually does not know precisely, which words he should use to describe his interests and to receive valuable documents from IR system. User formulates his query with his *subjectively chosen words*, which may occur to be not very correct and popular at the retrieval domain. So the main idea of the *adaptive user profile*, presented below, is to join user subjective words with the objective description of retrieval domain. The connection between terminology used by the user and terminology accepted at some field of knowledge might be considered as a kind of *translation*, which describes the meaning of words used by the user in a context fixed by relevant documents. At the *adaptive user profile* the *translation* is described by assigning to the user *query pattern* a *subprofile*, created during the analysis of relevant documents of an answer. From relevant documents the objectively "proper" vocabulary of the retrieval domain could be identified.

We claim that user can express own preferences by retrieved documents relevance valuation. The valuation process is not a burdensome task for the user, while he only points out these documents, which he considers relevant and does not assign any relevance values.

2 User Profile

The IR system we define as four elements: set of documents D, user profiles P, set of queries Q and set of terms in dictionary T. The retrieval function is ω: $Q \rightarrow 2^D$. Retrieval function returns a set of documents, which is the answer to query q. Set $T = \{t_1, t_2, ..., t_n\}$ contains terms from all documents, which have been indexed by Web IR system and is called dictionary. By D_q we describe a set of relevant documents among retrieved documents D_q' for query q: $D_q' = \omega(q,D)$ and $D_q \subseteq D_q'$.

Therefore, the *user profile* $p \in P$ is a set of pairs:

$$p=\{\langle s_1, sp_1\rangle, \langle s_2, sp_2\rangle, \ldots, \langle s_l, sp_l\rangle\} \quad (1)$$

where: s_j – user query pattern, sp_j – user subprofile (user query pattern indicates one user subprofile univocally).

For profile p we define function π, which maps: user query q, the set of retrieved relevant documents D_q and the previous user profile p_{m-1}, into new user profile p_m. The function π determines the profile modifications. Thus, the profile is the following multi-attribute structure: $p_0 = \emptyset$, $p_m = \pi(q_m, D_q, p_{m-1})$. For the user profile we define a set of *user subprofiles SP* (presented below).

The user profile is created on the basis of *user verification of the documents* retrieved by Web IR system. During verification the user points out these documents which he considers relevant.

The user query pattern s_j is a Boolean statement, the same as user query q: $s_j=r_1\wedge r_2\wedge r_3\wedge\ldots\wedge r_n$, where r_i is a term: $r_i=t_i$, a negated term: $r_i=\neg t_i$ or logical one: $r_i=1$ (for terms which do not appear at the question). The user query pattern s_j is assigned to subprofile and is connected to only one subprofile.

The user subprofile $sp \in SP$ is defined as n-dimensional vector of terms weights, (the terms are from relevant documents): $sp_j^{(k)} = (w_{j,1}^{(k)}, w_{j,2}^{(k)}, w_{j,3}^{(k)}, \ldots, w_{j,n}^{(k)})$, where SP – set of subprofiles, n – number of terms in dictionary T: $n=|T|$, $w_{j,i}^{(k)}$ – weight of significant term tz_i in subprofile after k–th subprofile modification.

The terms from dictionary T are an indexing terms at Web IR system. These terms are indexing documents retrieved for query q and belong to relevant documents.

The weight of significant term tz_i *in subprofile* is calculated as following:

$$w_{j,i}^{(k)} = \frac{1}{k}((k-1)w_{j,i}^{(k-1)} + wz_i^{(k)}) \quad (2)$$

where: k – number of retrievals made with using the subprofile, i – index of term in the dictionary T, j – index of subprofile, $w_{j,i}^{(k)}$ – weight of significant term tz_i in subprofile after k–th modification of subprofile[1], subprofile is indicated by pattern s_j (i.e. after k–th document retrieval with the use of this subprofile), $wz_i^{(k)}$ – weight of significant term tz_i (3) (presented below) in k–th selection of term t_i.

3 Information Retrieval on Web IR System

Web IR system usually returns the answer at which only part of documents, found by the system, pertain to real user interest. We assume that asking user for pointing out these documents, among the answer, which are more interesting, we can come closer to the picture of user interest domain. In order to make selection process realistic, the user is making only *a binary choice*: interesting vs. non-interesting, i.e. points out some documents without any further assessments, e.g., how they pertain to his needs.

[1] The weight, called *cue validity*, is calculated according to a frequency of term tz_i in relevant documents retrieved by Web IR system at k–th retrieval and a frequency of this term in whole documents of the collection.

For the need of IR on the Web, a domain of user interest is represented and identified by keywords that originate and were extracted from the selected and relevant documents of the answer. An important part of the user interest representation process, proposed in this contribution, is an analysis of answer documents. The process is aimed to identify these terms which are representative for relevant documents and are key terms for user interest domain as well. The ultimate goal of the answer documents analysis is to identify vocabulary used at a certain knowledge domain of user interest and, considering only the content of these relevant documents, to construct user interest representation, i.e. the adaptive user profile.

3.1 Relevant Document's Terms Weighting

Each term from a document indexed in Web IR system is assigned the weight d_i according to the *tf–idf* schema [17]. The weights allow for *index terms* identification which describe properly the document content. A term t_i can occur in more than one relevant document. Thus the weight wz_i' is influenced also by the weights d_i of this term in each of the selected relevant documents of an answer.

3.2 Significant Terms Selection from Relevant Documents

Key terms, which are important for the domain of user interest, are automatically extracted from the relevant documents selected by the user. These terms are stored in the appropriate subprofile and next used during modification of a question asked by user to Web IR system.

Key terms chosen from the relevant documents selected are henceforth called *significant terms*. The proposed term selection method was inspired by the idea of discriminative terms [16], [19] and the cue validity factor [10]. Selected significant terms are introduced into a subprofile and then used to modify user query.

The proposed method of significant terms selection is performed on several steps. Assigning wz_i' weights to each term from relevant documents leads to set of significant terms tz_i from the relevant documents. A joint application of two term selection criteria is important novelty of our approach. In this method, weights of terms from the relevant documents together with the cue validity factor are combined to a two-step filter. A criterion obtained is a *weighted sum*. As an effect of combining two discussed methods of weighting terms from relevant documents, among all terms belonging to the relevant documents, only the terms pertaining to vocabulary used in the user interest domain are selected. A weight of term–candidate for inclusion into the set of significant terms is calculated as following:

$$wz_i = \alpha\, wz_i' + \beta\, cv_i \tag{3}$$

where values of factors α and β were chosen experimentally.

Selection of discriminative terms on the basis of a constant threshold is the technique often reported in literature and applied in IR, e.g. it was used for the authoritative collection of documents. According to this technique only terms with weight above some constant, pre-defined threshold are added to the selected group.

However collections of Web documents express substantially different properties. These collections are characterized by huge divergence of topics and large dynamics in time in relation to both: the number of terms and documents. In such collections, term significance, represented by term weight, is changing according to the modifications introduced into the collection, i.e. after adding new documents or altering already present ones. A typical method of discriminative terms identification, i.e. on the basis of the threshold expressed by the constant values, when applied to the Web collection would not produce an expected set of significant terms. So we propose multi-attribute criterion for significant terms identification. Values of such thresholds are constant, but defined on the basis of functions considering the dynamic of term weights in the Web collection. The process of significant terms selection is performed according to the algorithm described in [20].

Only terms occurring in all relevant documents can be included into a set of significant terms. The above criteria of terms selection have been aimed on finding only terms describing the domain of user interests and improving the number of relevant documents in the answer.

4 Modification of User Profile

The main idea is to join the user subjective words with the objective description of retrieval domain. The *adaptive user profile* expresses the connection between terminology used by the user and terminology accepted at some field of knowledge. This connection might be considered as a kind of translation describing the meaning of words used by the user in a context fixed by relevant documents. Translation is described by assigning to user query pattern s_j a subprofile ('translation') created during the process of significant terms tz_i selection from relevant documents of an answer. We assume the following designations: q – the user query, $D_q^{'}$ – the set of documents retrieved for the user query q, $D_q^{'} \subseteq D$, D_q – the set of documents pointed by the user as relevant documents among the documents retrieved for user query q, $D_q \subseteq D_q^{'}$.

As it was described above, user profile p_m is the representation of the user query q, the set of relevant documents D_q and the previous (former) user profile p_{m-1}. After every retrieval and verification of documents made by the user, the profile is modified. The modification is performed according to the following procedure: $p_0 = \varnothing$, $p_m = \pi(q_m, D_q, p_{m-1})$ where p_0 – the initial profile, this profile is empty, p_m – the profile after m–times the user has asked different queries and after each retrieval the analysis of relevant documents was made.

Traditionally, a user profile is represented by one n–dimensional vector of terms describing user interests. User interests change, and so should the profile. Usually changes of a profile are achieved by modifications of weights of the terms in the vector. After appearance of queries from various domains, modifications made for this profile can lead to an unpredictable state of the profile. By the unpredictable state we mean a disproportional increase of weights of some terms in the vector representing the profile that might not reflect a real increase of user interests in the domain represented by these terms. The weights of terms could be growing, because of high frequency of these terms at whole collection, regardless of the domain of actual retrieval.

The representation of a profile as single vector could also cause ambiguity during the use of such profile for query modification. At certain moment of user retrieval history, a query refers to only one domain of user interests. To use the profile mentioned above, a single vector form, we need a mechanism of choosing from single vector of terms, representing various interests of the user, only these terms that are related to a domain of current query. To obtain this information, knowledge about relationship between terms from a query and a profile, and between terms in profile is needed. In literature, this information is obtained from a co-occurrence matrix created for a collection of documents [15] or from a semantic net [9]. One of disadvantages of these approaches is that two mentioned above structures, namely a user profile and a structure representing term dependencies, should be maintained and managed for each user and also that creating the structure representing term relationships is difficult for so diverging and frequently changing environment as the Web.

The above-mentioned problems are not encountered at the adaptive user profile p created in this contribution. After singular retrieval, only weighs of terms from the subprofile identified by pattern s_j (identical to users' query) are modified, not all weighs of all terms in profile. Similarly, when the profile is used to modify user query, the direct translation between current query q and significant terms from the domain associated with the query is used. In profile p, between single user query pattern s_j and single subprofile sp_j a kind of mapping exists representing this translation.

The user profile is created during a period of time – during sequence of retrievals at Web IR system. There could appear a problem how many subprofiles should be kept in the user profile. We have decided that only subprofiles that are frequently used for query modifications should not be deleted. If a subprofile is frequently used, it is important for representing user interests.

The modification of the user subprofile sp is made when, from the set of relevant documents, pointed out from retrieved documents by the user, some significant term tz_i is determined [20]. The weight $w_{j,i}^{(k)}$ (2) is modified for these terms and only in one appropriate subprofile identified by the user query pattern s_j. During each retrieval cycle a modification is applicable to only one subprofile and for all significant terms tz_i obtain during the k–th selection of significant terms from the relevant documents retrieved for query q, which was asked k–th time. If the modification took place for significant terms tz_i for every subprofile in user profile, it would cause disfigurement of significant terms importance for single question.

The selected significant terms are further processed with the use of wordnet build for Polish language – plWordNet [21]. A wordnet is dictionary-like lexico-semantic resource that describes lexical meaning of words. Every wordnet, following the Princeton WordNet [22], includes network of synsets – sets of near-synonyms that are linked by lexico-semantic relations such as hypernymy, hyponymy, meronymy, holonymy etc. Synsets represent distinct lexical meanings.

We assume that the use of words lexically related to significant terms enriches the description of user interest domain. So at a subprofile of adaptive user profile, for each significant term we are selecting direct hypernyms, direct hyponyms and all synonyms to expand subprofile. If a word from newly selected set of hyponyms have already been among significant terms at subprofile, hyponyms of this word are not added to subprofile. Introducing further hyponyms for such word could affect unpredictable narrowing of modified query (presented below).

5 Application of User Profile

The user profile contains terms selected from relevant documents. These terms are good discriminators distinguishing relevant documents among the other documents of the collection and also represent the whole set of relevant documents.

The application of user profile p is performed during each retrieval for a user query q. One of the main problems is the selection of significant terms tz_i for query modification. Not all significant terms from a subprofile will be appropriate to modify the next user query, because the query becomes too long.

The user asks new query q_j to Web IR system, new pattern s_j and a subprofile identified by this pattern are added to the profile. The subprofile is created as a result of relevant documents analysis. If user asks next query q_k and this query is identical to previous query q_j, the given query is changed by user profile. The modified query is asked to Web IR system, retrieved documents are verified by the user and the subprofile in user profile is brought up to date. After each use of the same query as query q_j, the subprofile identified by the pattern s_j represents user interests, described at the beginning by the query q_j, even better. Each retrieval, with the use of the subprofile identified by the pattern s_j, leads to query narrowing, a decrease in the number of answer documents, an increase in the number of relevant documents.

Adaptive user profile can be used for query modification if pattern s_j , existing in the profile, is *identical* or *similar* to the current query q_i. For example, for queries: $q_a = t_1 \wedge t_2 \wedge t_3 \wedge t_4$, $q_b = t_1 \wedge \neg t_2$, following patterns: $s_1 = t_1 \wedge t_2 \wedge t_3 \wedge t_4$, $s_2 = t_1 \wedge \neg t_2$ are identical to queries q_a, q_b, respectively. For the same query q_a patterns: $s_2 = t_2 \wedge t_4$, $s_4 = t_1 \wedge t_2$, $s_5 = t_1 \wedge t_3$, $s_6 = t_2$ are similar to query q_a.

If pattern s_j is identical to current user query q_i, the query q_i is replaced by the best significant terms tz_i from subprofile identified by pattern s_j. The weights of these terms are over $\tau_{profile}$ – a dynamically calculated threshold. If user profile consists of more than one pattern similar to current user query q_i, all significant terms tz_i from all subprofiles identified by these patterns are taken under consideration. In such case significant terms weights from all subprofiles identified by similar patterns are summed. The n–dimensional vector $R=(r_1, r_2, \ldots, r_n)$ is created. The ranking list of all significant terms is created. If the weight r_i is over $\tau_{profile}$ dynamic threshold, significant term tz_i will be introduced to replace current user query q_i.

6 Experiments

The *adaptive user profile* was implemented as a part of Web IR system - the *Profiler*. User profile applied as a mechanism of retrieval personalisation, i.e. by user query modifications. User query modification is performed during user interaction with Web IR system (i.e. a verification of documents). After verification, the Profiler automatically asks the modified query to Web IR system and presents new answer to the user.

The experiments were performed into two directions. Firstly, we aimed to establish all parameters (i.e. thresholds) for the Profiler. Then we were verifying the usefulness of multi-attribute adaptive profile, i.e. our goal was to confirm the increase of number of relevant documents retrieved at every retrieval cycle and the decrease of number of document at following answers.

With the Netoskop search engine for Polish language and the Profiler (user profiling module) testing environment was prepared. The real users of Web IR system (group of 13 persons) established testing sets of documents consisting of relevant documents that represent their real interests domains. Three types of testing sets were used: *dense*, *loose* and *mixed sets of documents*. The dense sets of relevant documents consist of documents describing one domain of user interests; the strongly similar documents were assessed by users. The loose sets of documents consist of no similar to each other documents from different domains of user interests. The mixed sets of documents consist of subsets of closely related documents, identifying one domain of interests, and a number of documents from disjoint domains of user interests.

The experiments were arranged as a simulation of user behaviour during the information retrieval at Web IR system. The user interests could be from any domain of knowledge but the user query did not always reflects the vocabulary commonly used at this knowledge. Asking not very precise query by the user is simulated by the process of producing from dictionary T the number of 50 random queries. Each random query was asked to Web IR system. If in the answer there were relevant documents from testing sets of documents, the randomly generated query was modified – the significant terms replace the preliminary query. The modified query was automatically asked to Web IR system and another relevant documents were found from testing sets of documents. Thus the use of testing sets of documents simulates the process of answer verification made by the user.

Each stage of above described cyclic process is called *iteration*. Iterations were repeated until all relevant documents from the dense sets of relevant documents were found or no changes in number of relevant documents were observed (for the loose and mixed sets of relevant documents).

For every random query in experiments: the number of all retrieved documents D'_q, effectiveness $\%DR$ (the effectiveness is defined as a percent of relevant documents retrieved by the modified queries from the set of all relevant documents in the testing sets) and precision $Prec_m$ (as a standard precision in IR for the first m=10, 20, 30 documents in the answer) at every iteration were calculated. There are retrieval improvement measures for the proposed method.

Table 1. Measures of retrievals improvement made during experiments

	Percent of modified queries			Effectiveness $\%DR$			
	improvement in precision $Prec_m$	no improvement in precision $Prec_m$	partial improvement in precision $Prec_m$	75-100%	50-75%	25-50%	0-25 %
dense sets	82 %	12 %	6 %	54 %	10 %	18 %	18 %
loose sets	67 %	12 %	21 %	0 %	3 %	68 %	29 %
mixed sets	58 %	18 %	24 %	6 %	36 %	29 %	29 %

The retrievals made for dense sets of relevant documents confirmed the assumption that for most of the modified queries the retrieval results were better in comparison to preliminary query. For over 82% of preliminary queries, $Prec_m$ measure was increasing at every iteration of query modification (Table 1). The effectiveness $\%DR$ of the proposed method shows that over 75% of all relevant document from testing dense set was found for over 54% of asked queries. The number of all documents retrieved and returned as the answer diminishes with every iteration.

The experimental retrievals arranged for loose sets of relevant documents showed that all modified queries (from the same preliminary query) always focus on the same domain of user interests. This property was assured by the method of adaptive profile creation, modification and application and verified by experiments.

In case of loose sets of relevant documents, for more then 67% of preliminary queries, $Prec_m$ measure rose with each iteration of query modification. The number of all retrieved documents diminishes as well. For the rest of the questions at loose sets of relevant documents, the measured parameters were worse, because single documents were selected by Web IR system as the answer for each following modification of the query. No similar documents were found because the loose testing sets consist of no similar to each other documents taken from disjoint domains of user interests.

7 Conclusions and Future Work

The *adaptive user profile* is a new and universal approach to the representation of user interests and preferences. The profile includes both interests given explicitly by the user, as a query, and also preferences expressed by relevance valuation of retrieved documents. The achievement of the profile is the ability to express field independent translation between user terminology and terminology accepted in some field of knowledge. This universal translation is supposed to describe the meaning of words used by user (i.e. user query) in context fixed by the retrieved documents (i.e. user subprofile). In Web IR system the user is supported by the adaptive user profile, performing more precise document retrievals during every query reformulation. The query is modified in a way that, during next retrievals user receives the set of retrieved documents that definitely meet user information needs. In the future, some further experiments need to be done with a bigger group of WWW users, who will retrieve and assess the documents from the Web.

Acknowledgments. This contribution was partially supported by Polish Ministry of Science and Higher Education under grant no. S30080/I32.

References

1. Ambrosini, L., Cirillo, V., Micarelli, A.: A Hybrid Architecture for User-Adapted Information Filtering on the World Wide Web. In: Proc. of the 6th Int. Conf. on User Modeling, pp. 59–62. Springer (1997)

2. Asnicar, F., Tasso, C.: ifWeb: A Prototype of User Model-Based Intelligent Agent for Document Filtering and Navigation in the World Wide Web. In: Proc. of the Workshop Adaptive Systems and User Modeling on the World Wide Web, UM 1997. Springer (1997)
3. Billsus, D., Pazzani, M.: A Hybrid User Model for News Story Classification. In: Proc. of the 7th Int. Conf. on User Modeling, UM 1999, Banff, Canada, pp. 99–108. Springer (1999)
4. Benaki, E., Karkaletsis, A., Spyropoulos, D.: User Modeling in WWW: The UMIE Prototype. In: Proc. of the 6th Int. Conf. on User Modeling, pp. 55–58. Springer (1997)
5. Bhatia, S.J.: Selection of Search Terms Based on User Profile. Comm. of the ACM (1992)
6. Bull, S.: See Yourself Write: A Simple Student Model to Make Students Think. In: Proc. of the 6th Int. Conf. on User Modeling, pp. 315–326. Springer (1997)
7. Collins, J.A., Greer, J.E., Kumar, V.S., McCalla, G.I., Meagher, P., Tkatch, R.: Inspectable User Models for Just–In Time Workplace Training. In: Proc. of the 6th Int. Conf. on User Modeling, pp. 327–338. Springer (1997)
8. Daniłowicz, C.: Modelling of user preferences and needs in Boolean retrieval systems. Information Processing and Management 30(3), 363–378 (1994)
9. Davies, N.J., Weeks, R., Revett, M.C.: Information Agents for World Wide Web. In: Nwana, H.S., Azarmi, N. (eds.) Software Agents and Soft Computing: Towards Enhancing Machine Intelligence. LNCS (LNAI), vol. 1198, pp. 79–99. Springer, Heidelberg (1997)
10. Goldberg, J.L.: CDM: An Approach to Learning in Text Categorization. International Journal on Artificial Intelligence Tools 5(1 and 2), 229–253 (1996)
11. Indyka-Piasecka, A., Piasecki, M.: Adaptive Translation between User's Vocabulary and Internet Queries. In: Kłopotek, M.A., Wierzchoń, S.T., Trojanowski, K. (eds.) Proc. of the IIS IPWM 2003. ASC, vol. 22, pp. 149–157. Springer, Heidelberg (2003)
12. Danilowicz, C., Indyka-Piasecka, A.: Dynamic User Profiles Based on Boolean Formulas. In: Orchard, B., Yang, C., Ali, M. (eds.) IEA/AIE 2004. LNCS (LNAI), vol. 3029, pp. 779–787. Springer, Heidelberg (2004)
13. Jeapes, B.: Neural Intelligent Agents. Online & CDROM Rev. 20(5), 260–262 (1996)
14. Maglio, P.P., Barrett, R.: How to Build Modeling Agents to Support Web Searchers. In: Proc. of the 6th Int. Conf. on User Modeling, pp. 5–16. Springer (1997)
15. Moukas, A., Zachatia, G.: Evolving a Multi-agent Information Filtering Solution in Amalthaea. In: Proc. of the Conference on Agents, Agents 1997. ACM Press (1997)
16. Qiu, Y.: Automatic Query Expansion Based on a Similarity Thesaurus. PhD. Thesis (1996)
17. Salton, G., Bukley, C.: Term-Weighting Approaches in Automatic Text Retrieval. Information Processing & Management 24(5), 513–523 (1988)
18. Seo, Y.W., Zhang, B.T.: A Reinforcement Learning Agent for Personalised Information Filtering. In: Int. Conf. on the Intelligent User Interfaces, pp. 248–251. ACM (2000)
19. Voorhees, E.M.: Implementing Agglomerative Hierarchic Clustering Algorithms for Use in Document Retrieval. Inf. Processing & Management 22(6), 465–476 (1986)
20. Indyka-Piasecka, A.: Using multi-attribute structures and significance term evaluation for user profile adaptation. In: Jędrzejowicz, P., Nguyen, N.T., Hoang, K. (eds.) ICCCI 2011, Part I. LNCS, vol. 6922, pp. 336–345. Springer, Heidelberg (2011)
21. Piasecki, M., Szpakowicz, S., Broda, B.: A Wordnet from the Ground Up. Oficyna Wydawnicza Politechniki Wrocławskiej (2009)
22. Fellbaum, C. (ed.): WordNet – An Electronic Lexical Database. The MIT Press (1998)

Auto-Tagging Articles Using Latent Semantic Indexing and Ontology

Rittipol Rattanapanich and Gridaphat Sriharee

Department of Computer and Information Science
King Mongkut's University Technology of North Bangkok
1518 Pibulsongkram Road, Bangsur, Bangkok, Thailand 10800
rit_vr@hotmail.com, gridaphats@kmutnb.ac.th

Abstract. Tagging plays a crucial role in the success of social network and social collaboration. This paper proposes an auto-tagging methodology for articles using Latent Semantic Indexing (LSI) and ontology. The proposed methodology consists of pre-processing and tagging process. In pre-processing process, the LSI vector is created for article classification. The tagging process suggests some ontological tags. An accuracy evaluation of auto-tagging compared with manual-tagging is discussed. The experimental results show that the proposed auto-tagging methodology returns high accuracy and recall.

Keywords: Auto-tagging, Ontology, Latent Semantic Indexing.

1 Introduction

Tagging is implemented in internet forums, blogs, social collaboration (e.g. Wikipedia), and social networks (e.g. Flickr). Tagging is a mechanism for content retrieval and linking to relevant resources. The tag can be in-text keyword or out-of-text keyword and it can be labeled by word or phrase. In article sharing system, the tagged articles are retrieved regarding search keywords specified by users. The users may tag the article using their knowledge or content analyzer tool. As a number of articles are posted into the system, manual-tagging may take time because the administrator is required to read the content of article and specify relevant tags. Thus, auto-tagging is required and it is expected in returning accurate tags to the articles. The accurate tags should be suggested and those are conformable to the article detail.

This paper has the main contribution to propose the methodology of auto-tagging that is applied by IR concepts and ontology. The auto-tagging has pre-processing and tagging process. The pre-processing process uses LSI technique for article classification. The tagging process consists of classification process and tag selection process by which the cosine similarity and ontology are applied respectively. An evaluation of the proposed methodology is presented and discussed.

The remainders of the paper are as follows. Section 2 describes our motivation for tagging using ontology. Section 3 discusses some related works. Section 4 is the detail

N.T. Nguyen et al. (Eds.): ACIIDS 2014, Part I, LNAI 8397, pp. 153–162, 2014.

of the proposed methodology for auto-tagging. Section 5 presents the experiment result of an evaluation and Section 6 is a conclusion.

2 Motivation for Tagging Using Ontology

With the success of Web 2.0, tagging system is implemented on many online forums websites and social networks. The system supports the framework with different purposes. For example, tags are used to describe sharing resources, attract attention, self-presentation, and opinion expression. There are many web sites that use tagging as mechanism for resource/content retrieval for example, Delicious, Flickr, Blogger, Wordpress and Wikipedia. The tags are used as the linkage information to relevant resources. In social tagging system, the users specify tags to the published resources such as to images, news, and articles. The tags are represented in the support system and that are available for query.

The tags may be organized and managed as the part of folksonomy system and that may be simple terms or ontological terms [1]. The simple terms are represented by free-form texts specified by the users or system. The ontological terms are represented by the concepts defined in taxonomy. Tags are typically short textual labels, which provide an easy way to categorize, search, and browse the information they describe. Tags may be represented by a representation language and that is available for query. Linking resources with tags, retrieval across some application can be implemented.

Fig. 1 depicts an example of the tagged content from Wikipedia. Wikipedia defines a tag as a free-text keyword and tagging as an indexing process for assigning tags to resources. In this example, the content is annotated by keywords (see underlined terms): ข้าวหอมมะลิ (rice named in Thai), RTGS (refers to Royal Thai General System), long-grain, and rice. Some tags are proper names and some tags may have semantic relation for example, long-grain rice is a particular kind of rice.

> **Jasmine rice** (Thai: ข้าวหอมมะลิ; RTGS: Khao Hom Mali; Thai pronunciation: [kʰâ : w hɔ̌ : m malíʔ]), sometimes known as *Thai fragrant rice*, is a long-grain variety of rice that has a nutty aroma and a subtle *pandan*-like (*Pandanus amaryllifolius*-leaves) flavor caused by 2-acetyl-1-pyrroline.[1] Jasmine rice is originally from Thailand. It was named as Kao Horm Mali 105 variety (KDML105) by Sunthorn Seehanern, an official of the ministry of agriculture in the Chachoengsao Province of Thailand in 1954.[*citation needed*] The grains will cling when cooked, though it is less sticky than other rices as it has less amylopectin. It is also known as Thai Hom Mali. To harvest jasmine rice, the long stalks are cut and threshed. The rice can then be left in a hulled form and sold as brown rice or shucked and sold as white rice. Most Southeast Asians prefer the white variety of jasmine rice.

Fig. 1. An example of the content from Wikipedia

With regard to tag, there are many types of tags such as content-based tag, context-based tag, attribute tag, ownership tag and purpose tag [2]. The tagging system may provide a particular type. For content-based tags, the tags suggested to the article may be significant terms regarding term frequency. For example, information in Fig. 1 can be suggested with the tag *Rice* because the term rice has maximum frequency. Regarding semantic similarity, multiple tags may have the same meaning or may refer to the same thing. For example, *Jasmine Rice* and *Kao Horm Mali* (rice name in Thai with English spelling) are referred to the same thing. In this case, the users must rely on their own intuition to pick the appropriate tags for some contexts. In this paper, the auto-tagging methodology is proposed by which the tagging concerns both term frequency and semantic similarity.

3 Related Work

Auto-tagging is implemented in many research works. [3] proposed automatic in-text keyword tagging. Tagging system selects candidate keywords from the keyword dictionary by comparing the input document and whole terms in the dictionary. [4] proposed a tool to suggest tags for weblog. The tool finds similar tagged posts and suggests some set of the associated tags to a user for selection. [5] follows [4] to provide automatic tag suggestions for a blog post and focuses on performance of tag suggestion system. [6] presented interactive tag recommendation for Flickr by which the tags are retrieved from similar tagged pictures. The users are able to add more tags into the system. Regarding tagging using ontology, [7] discussed approaches for collaborative tagging at a semantic level. The tags are represented by some metadata languages and that are available for collaboration across the tagging systems. [1] proposed a formal representation model for tagging. The tagging ontology is represented by OWL.

This paper proposes a novel methodology for auto-tagging using ontological tags. The tagging system relies on both IR concepts and ontology. We do not focus on tag representation but apply the concept of ontology for tagging. There are some other works that proposed different methodologies and techniques. For example, [8] proposed a voting model where each feature in the resources votes for their favorite tags, and [9] proposed content-based similarity metrics for tagging. The approach of tagging with comparing the resource with the tagged resource, similar resource and dictionary (e.g. [3], [4] and [6]) may ignore semantic content of the resource itself but rely on the compared resource. Thus auto-tagging in this paper is focused on how to choose appropriate tags with semantic analysis of the resource content.

4 Auto-Tagging Methodology

Auto-tagging relies on the pre-processing process. The pre-processing process is a process for preparing dataset for use in runtime. Fig. 2 depicts the proposed methodology. It includes two main processes: pre-processing process and tagging process. The pre-processing process consists of LSI vector creating. The LSI vector is used for article classification in runtime. The article is assigned to a particular domain and the ontology of such domain is retrieved for tagging. The information from the LSI vector is also used for ontology building. The tag ontology is built for particular domain. The tag ontology is enhanced using information from dictionary. In this paper, there are 70 articles for the train dataset (used in classification step) and 140 articles for the test dataset for auto-tagging evaluation. Both datasets are articles collected from vcharkarn.com website. The articles are in Thai and categorized into four domains: food, tourism, sport and car. These domains are focused because most articles in the domains have complete content. The pre-processing process produces the LSI vector and this is used for article classification. The tagging process consists of two steps: article classification and tag selection. Article classification process assigns the article to the domain by cosine similarity computing. Later, the tagging process uses ontology of the assigned domain for tag selection. The tag selection process considers ontological tags by ontology weight computing. In this paper, an evaluation of auto-tagging accuracy is conducted to compare with manual- tagging.

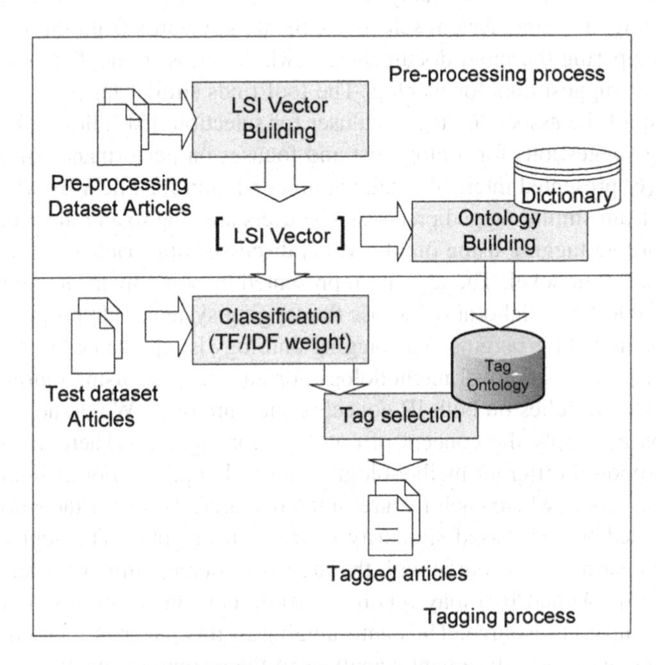

Fig. 2. The process of auto-tagging

4.1 Pre-processing Process

To prepare the data for use in runtime, the train dataset are used for LSI vector building. The pre-processing process has three steps as follows.

(i) We extract the text of the train dataset in part of title, abstract, and content. Lexitron dictionary [10] is adopted for use in the extraction. The train dataset is specified tags manually.

(ii) The extracted terms are considered their TF/IDF weight by:

$$TF\_IDFweight_{i,j} = tf_i \times IDF_i \quad (1)$$

Where $TF\_IDFweight_{i,j}$ is a TF-IDF weight of the term i in the domain j, tf_i is term frequency of the term i in articles of the domain j, $IDF_i = Log \frac{D}{df_i}$ by which D is a number of the domain of the train dataset, and df_i is a number of the domains that have the term i.

Table 1 is an example of LSI vector of terms across the domains: food, tourism, sport and car; indicated by 1, 2, 3 and 4 respectively. The LSI vector presents the term frequency, inverse document frequency and TF-IDF weight of terms in domains. For example, the term Oil is provided its TF-IDF weight with 1.201 and 94.824 for food and car domain respectively. The term may have its significance varied with the domains.

Table 1. An example of LSI Vector of terms

Terms	*TF1*	*TF2*	*TF3*	*TF4*	*df_i*	*D/ df_i*	*IDF_i*	*W1*	*W2*	*W3*	*W4*
Wheel	0	0	0	208	1	4	0.602	0	0	0	125.229
Geer	0	0	0	160	1	4	0.602	0	0	0	96.330
Car	0	3	0	174	2	2	0.301	0	0.903	0	52.379
Machine	0	0	0	190	1	4	0.602	0	0	0	114.391
Oil	4	0	0	315	2	2	0.301	1.204	0	0	94.824
CVT	0	0	0	71	1	4	0.602	0	0	0	42.746
Speed	0	0	0	81	1	4	0.602	0	0	0	48.767
Break	0	0	0	59	1	4	0.602	0	0	0	35.522
Drive	0	0	0	61	1	4	0.602	0	0	0	36.726
Mitsubishi	0	0	0	57	1	4	0.602	0	0	0	34.317
Steering Wheel	0	0	0	53	1	4	0.602	0	0	0	31.909
Mirage	0	0	0	51	1	4	0.602	0	0	0	30.705
Honda	0	0	0	49	1	4	0.602	0	0	0	29.501
Rate	3	0	0	41	2	2	0.301	0.903	0	0	12.342
Save	2	2	0	43	3	1.3	0.125	0.249	0.249	0	5.372
Nissan	0	0	0	45	1	4	0.602	0	0	0	27.093
Power	0	0	0	36	1	4	0.602	0	0	0	21.674
Gas	0	0	0	36	1	4	0.602	0	0	0	21.674
Air	0	0	0	35	1	4	0.602	0	0	0	21.070

(iii) We built the tag ontology by terms from LSI vector. The ontology is enhanced by adding concepts from the domain dictionary. The tags are organized into a hierarchy by considering on generalized and specialized relation regarding broader and narrower meaning. Fig. 3 is an example of some concepts defined in the tag ontology of food domain. The concepts are organized into hierarchy by considering on food classification. For example, food is categorized into vegetable, fruit, rice and beverage. The tag ontology is specified manually with human judgment.

4.2 Auto-Tagging Process

Tagging can be implemented after the article is classified its relevant category. The classification can support retrieving specific information within a set timeframe whereas choosing appropriate tags is important for finding resource quickly. In this paper, the auto-tagging is comprised of two processes: classification and tag selection with following detail.

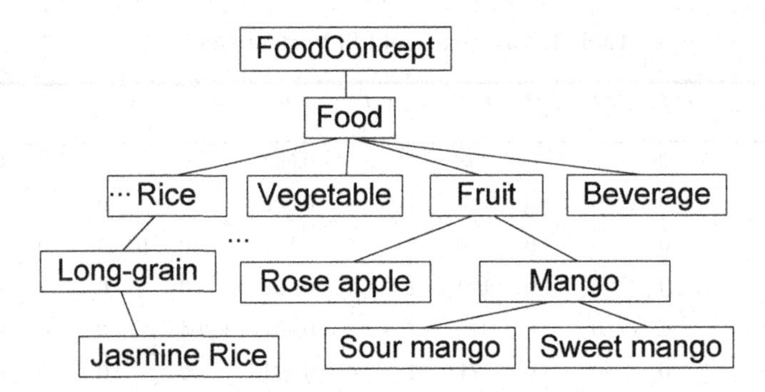

Fig. 3. Tag ontology of food domain

- *Article classification.* The article is classified into relevant domain using cosine similarity. The system computes cosine similarity and compares the article with the train dataset's articles. The article is assigned to the domain of the train article that has maximum cosine similarity. The cosine similarity function is computed by:

$$Similarity(A,D) = \frac{\sum_{i=1}^{n} w_{A_i} w_{D_i}}{\sqrt{\sum_{i=1}^{n} w^2_{A_i} \times \sum_{i=1}^{n} w^2_{D_i}}} \tag{2}$$

Where A is article, D is article in the train dataset, w_{A_i} is TF/IDF weight of term i in article A, w_{D_i} is TD/IDF weight of term i in article D.

- *Tag selection.* Tag selection has two steps as follows.
  (i) The extracted terms of article are matched to the terms in relevant tag ontology. The matched terms are considered to be suggested for tagging further.
  (ii) Ontology weight is computed to specify tag's significance. The tags are suggested with ranking regarding their significance.

In this paper, the ontology weight is computed with two conditions: with term frequency and without term frequency. We follow edge-based method (cf. [11]) and propose the computing as follows.

$$OntoWeight_{t,d} = \frac{N_t}{D_{N_t}} \quad (3)$$

$$OntoWeightTF_{t,i} = OntoWeight_{t,i} \times TF_t \quad (4)$$

Where N_t is a number of edges from root to tag *t*, D_{N_t} is a number of edges from root to the descendant node (that is the leaf node) of tag *t*, and TF_t is term frequency of tag *t* in domain *d*.

Fig. 4 depicts the suggested tags of the content from Fig. 1. With ontology weight score without TF, the suggested tags are *Jasmine Rice* (weight score = 1.00), *Long-grain* (0.75), and *Rice* (0.50) respectively in contrast, the suggested tags are *Rice* (weight score = 5.00, TF = 10), *Jasmine Rice* (weight score = 4.00, TF = 4) and *Long-grain* (weight score = 0.75, TF = 1) respectively.

Jasmine rice (Thai: ข้าวหอมมะลิ; RTGS: Khao Hom Mali; Thai pronunciation: [kʰâ ː w hɔ̌ ː m malíʔ]), sometimes known as *Thai fragrant rice*, is a long-grain variety of rice…
Ontology Weight: *Jasmine Rice, Long-grain, Rice*
Ontology Weight TF: *Rice, Jasmine Rice, Long-grain*

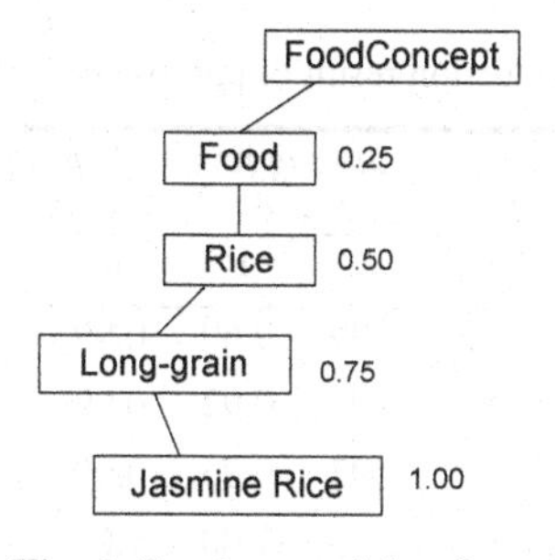

Fig. 4. Ontology weight of ontological tags

5 Evaluation

In this paper, we conduct two kinds of evaluation according to two purposes:

(i) To check whether the auto-tagging suggests tags that are similar to manual-tags or not.

(ii) To evaluate auto-tagging accuracy. The recall and precision are computed. The evaluation is conducted by comparing manual-tagging with auto-tagging.

Table 2 depicts auto-tagging accuracy. The suggested tags are compared with the manual-tags. For example, if the article has N manual-tags, the length of tag suggestions: N+1, N+2, N+3, N+4 and N+5 auto-tags, are evaluated. In this paper, the experiment used the test dataset for evaluation. With the proposed ontology weight methods, most tags with narrower meaning are tagged before the tags with broader meaning. Thus, manual-tags are matched to auto-tags when the length of tag-suggestion is increased. From Table 2, the length N+5 tag suggestion produces high tagging accuracy whereas the shorter length of tag suggestion has low accuracy.

Table 2. Evaluation result of purpose (i)

Tagging Methods	*N*	*N+1*	*N+2*	*N+3*	*N+4*	*N+5*
ontology	37.10	52.90	64.30	71.40	80.00	85.70
ontology x TF	67.10	81.40	84.00	87.10	88.60	94.30

Table 3 is experimental result of recall and precision evaluation. In this paper, the auto-tags and manual-tags are evaluated with different length of tag suggestion. The experiment is implemented by querying articles using a set of 10 keywords for each particular domain. The system retrieves the articles using such keywords. The accuracy is evaluated with ontology weight with term frequency and with different length of tag suggestion. The results show that the proposed auto-tagging accuracy returns the recall that is better than manual-tagging. However, precision may be low. The average recall and precision of auto-tagging from this experiment are 0.981 and 0.853 respectively. Also, the accuracy of the classification is 90%.

Table 3. Evaluation result of purpose (ii)

Domain	*Manual-tagging*		*Auto-tagging N = 4*		*Auto-tagging N = 6*		*Auto-tagging N = 8*		*Auto-tagging N = 10*	
	R	P	R	P	R	P	R	P	R	P
Car	0.93	0.79	0.85	0.58	0.99	0.60	0.99	0.66	1.00	0.65
Sport	0.98	0.95	0.97	0.92	0.99	0.92	0.99	0.93	1.00	0.93
Food	0.89	0.95	0.95	0.87	0.97	0.86	0.99	0.86	1.00	0.86
Tourism	1.00	1.00	1.00	1.00	1.00	1.00	1.00	1.00	1.00	1.00
Average	0.950	0.923	0.943	0.843	0.988	0.845	0.993	0.863	1.00	0.860

Note that R is recall value and P is precision value.

With the proposed auto-tagging methodology, the classification process supports retrieving specific information but tagging is focused on how to choose appropriate tags for the article. In this paper, classification is high because the test dataset (140 articles) is comprised of the train dataset (70 articles) that are used for classification. However, tag selection is not effected by such the train dataset. Because tagging is implemented by semantic analysis of the article's content by which TF weight and ontology weight are focused.

6 Conclusion

This paper proposed a methodology for auto-tagging using LSI and ontology. With experimental result, the tagging approach using ontology produces high accuracy and recall. Although, our experiment may use small dataset but the results show that auto-tagging using ontology is a suitable method to enhance recall of retrieval. In addition, tagging with semantic analysis is focused on tag significance in regard to their semantics and content of the article.

The proposed auto-tagging methodology includes classification process for retrieving information in regard to the tag ontology and tag selection is processed after. In this paper, the classification is conducted with a small dataset but using a large dataset may return high cost for computing and classification accuracy can be low accordingly. However, tagging mechanism is expected on how to choose appropriate tag whereas classification supports retrieving to specific information.

References

1. Knerr, T.: Tagging Ontology - Towards a Common Ontology for Folksonomies, `https://tagont.googlecode.com/files/TagOntPaper.pdf?` (retrieved November 4, 2013)
2. Gupta, M., Li, R., Yin, A., Han, J.: Survey on Social Tagging Techniques. ACM SIGKDD Explorations Newsletter 12(1), 58–72 (2010)
3. Kim, J., Jin, D., Kim, K., Choe, H.: Automatic In-Text Keyword Tagging based on Infor mation Retrieval. Journal of Information Processing Systems 5(3) (September 2009)
4. Mishne, G.: AutoTag: A Collaborative Approach to Automated Tag Assignment for Web log Posts. In: The 15th International World Wide Web Conference 2006, Edinburgh, Scotland (2006)
5. Sood, S.C., Owsley, S.H., Hammond, K.J., Birnbaum, L.: TagAssist: Automatic Tag Suggestion for Blog Posts. In: International Conference on Weblogs and Social Media, Boulder, Colorado, U.S.A., March 26-28 (2007)
6. Garg, N., Weber, I.: Personalized, Interactive Tag Recommendation for Flickr. In: The 8th ACM Recommender Systems Conference, Lausanne, Switzerland (2008)
7. Kim, H.L., Scerri, S., Breslin, J.G., Decker, S.: The State of the Art in Tag Ontologies: A Semantic Model for Tagging and Floksonomies. In: Proc. Int' l Conf. on Dublin Core and Metadata Applications (2008)

8. Si, X., Liu, Z., Li, P., Jiang, Q., Sun, M.: Content-based and Graph-based Tag Suggestion. In: Proceedings of ECML PKDD (The European Conference on Machine Learning and Principles and Practice of Knowledge Discovery in Databases) Discovery Challenge 2009, Bled, Slovenia (September 7, 2009)
9. Byde, A., Wan, H., Cayzer, S.: Personalized Tag Recommendations via Tagging and Content-based Similarity Metrics. In: International Conference on Weblogs and Social Media, Boulder, Colorado, U.S.A., March 26-28 (2007)
10. LEXITRON, http://www.lexitron.nectec.or.th/
11. Wu, Z., Palmer, M.: Verb semantics and lexical selection. In: Proceedings of the 32nd Annual Meeting of the Associations for Computational Linguistics (1994)

Evaluating Profile Convergence in Document Retrieval Systems

Bernadetta Maleszka and Ngoc Thanh Nguyen

Institute of Informatics, Wroclaw University of Technology,
Wybrzeze Wyspianskiego 27, 50-370, Wroclaw, Poland
Bernadetta.Maleszka@pwr.wroc.pl, Ngoc-Thanh.Nguyen@pwr.edu.pl

Abstract. In many document retrieval systems the user is not supported until sufficient information about him is collected. In some other systems randomly selected documents are recommended but they may not be relevant. To avoid so-called „cold-start problem" a method for determining a non-empty profile for a new user is presented in this paper. The experimental evaluations are usually performed using a few real users. This is a time- and cost-consuming method of evaluations, so we propose the methodology of experiments using simulations of user activities. The results were statistically analyzed and have shown that using the proposed method, the adaptation process allows to building a profile that is closer to user preference than in the situation when the first user profile is empty.

Keywords: user profile, profile convergence, evaluating retrieval systems.

1 Introduction

User profile is an important part of modern document retrieval systems. Existing user modeling methods are based on user preference, his current activities and activities of similar users. The problem occurs when a new user is coming to the system and any information about him are not gathered.

In many recommendation systems the user is not supported until sufficient information about him is collected. In some other systems randomly selected documents are recommended but they may not be relevant for him. The user can be discouraged when many non relevant documents are recommended to him. Developing a method for determining non-empty profile for a new user allows to omit the „cold start problem".

In this paper the problem of recommendation is considered. The object of recommendation is not a product or document as in most of such systems but the user profile. Based on first information about the user and knowledge about other users in document retrieval system, a profile for a new user can be predicted. The main assumption of proposed method for profile recommendation is the following: a new user is similar to other in respect to some demographic attributes. The method for determining a classifier based on usage data was presented in the previous works of the authors [9]. When a new user is coming to

N.T. Nguyen et al. (Eds.): ACIIDS 2014, Part I, LNAI 8397, pp. 163–172, 2014.

the system, he is asked to provide a few demographic attributes. Based on them, he is classified into group of users with the same values of these attributes and the centroid profile of this group is recommended for him. When user is interacting with the system, the history of his activities is collected and his profile is updated based on it.

A novel method for experimental evaluation is proposed. In many systems the recommendations are judged by a few users that were using the system for a short time. Obtained results may not be reliable and can not be analyzed using statistical tools. In this paper the methodology of system evaluation is described and an effectiveness measure is proposed.

The rest of the paper is organized as follows. In Section 2 we present the overview of classical approaches to evaluating effectiveness of document retrieval systems. The models of user preference and profile are presented in Section 3. Section 4 describes in detail a method for determining user profile. In Section 5 the way of simulating user activity, experimental evaluations are presented and obtained results are discussed. In the last Section 6 we gather the main conclusions and future works.

2 Related Works

The classical method for evaluating the recommendation system is Cranfield model [5]. In this model the testing set of documents, set of queries and information about the relevant documents are required. The effectiveness measures are based on precision and recall [8]. The most popular measures are F-measure or ROC curve [14]. Unfortunately, these measures are not adequate when the information about the relevance is unknown. Moreover the same document obtained as the result for the same query but different users can be useful for one user and useless for another. Authors of papers [13] and [15] noted that the effectiveness should take into account the satisfaction of the user not only the result documents.

Performance evaluation is time- and cost-consuming. For this reason any system should be evaluated based on at least 30 users. Each user should take part in evaluation many times in dependence of evaluation needs, parameters, etc. [3]. In practice, many systems are evaluated using only a few users and queries, e.g. authors of [2] present results for 13 users, each of whom ask about 50 queries; the system in paper [16] has 25 users that read some news for one month; Li [7] analyzed behaviour of 12 users that were interacting with his system for one week (one user asks on average 25 queries during this week).

A few researchers evaluate the system using a testing set, e.g. [1]. In this paper the methodology for experimental evaluations based on simulated users' behaviours is presented and the effectiveness measures are proposed.

3 User Preference and User Profile

User preference is a set of terms that the user is interested in. The preference directs the activities of the user. The system does not know this preference – it only observes his activities and based on this information builds the profile that is a model of the user. The aim of the system is to obtain the profile that is close to the preference. Non-empty profile for a new user will be proposed for the user when the system has no information about his preference. When user starts interaction with the system, his profile is updated based on his activities. We would like to show that the process of updating the profile that was empty at the beginning needs more time to be closer to his preference than when the profile is developed using proposed method and recommended to the user.

The ideal situation is when user profile contains the same terms and weights as in the preference. Unfortunately, the system can only observe user activities in order to create and update the profile. It should be closer to user preference when the system has more information about the user. The aim of experimental evaluation is to show that the distance between user preference and his profile (that was empty at the beginning) is bigger than the distance between user preference and his profile (that was recommended at the beginning). We may considered distance between user preference and profile as an effectiveness measure. The value of this measure becomes smaller and smaller when profile is updating so the profile converges to user preference.

3.1 User Preference

User preference contains a set of terms with weights that the user is interested in. The weights of these terms represents the degree of user interest in each term. Determining user preference is not a trivial problem even to this user. The system can collect information about user interests, e.g. by requiring him to fill the form during registration. The user can have many interest areas and he can not know which area is more important for him. Another problem is the fact that the user may not formulate the proper queries because he does not know the appropriate terminology in this area [6].

A more adequate way to obtain user preference is to ask the user which documents are relevant to him. Unfortunately, the user is required to look through a set of many documents. A test set of documents is often prepared to avoid this problem but such approach is not comfortable to the user [4].

User preference is the finite set of weighted terms:

$$Pref = \{(t_i, v_i) : t_i \in T_U \wedge v_i \in [0.5, 1), i = 1, 2, \dots, k\} \quad (1)$$

where t_i is a term, v_i is appropriate weight (degree of user interests in this term), and k is a number of terms in preference.

The set of terms T_U contains the terms that occur in user preference. The terms become from the set of all terms T ($T_U \cap T \neq \emptyset$). Number of terms can change between the blocks of sessions.

User preference is determined using the following procedure. A large collection of documents is considered (documents are described by weighted terms). The user chooses the set of relevant documents D_r. The precision and recall for this set is equal to 1. User preference is obtained based on this set of documents:

for each term the average weight is calculated when the term occurs in many documents or its weight is close to 1.

User preference is changing with time. In proposed system, we propose to change a part of relevant documents (e.g. 10% of D_r in each 5 blocks of sessions) and to recalculate the weights. New terms can also be added to the preference.

3.2 User Profile

The recommendation system uses information about user queries and relevant documents to build and update user profile. Let $D(s)$ be the set of queries and documents that user chooses as relevant in session s.

$$D(s) = \{(q_i^{(s)}, d_{i_j}^{(s)}) : i = 1, 2, \ldots, I; i_j = 1, 2, \ldots, i_J\} \quad (2)$$

where i is a subsequent number of the query in the session s and i_J is the quantity of documents that are relevant to query $q_i^{(s)}$.

User profile $UP(s)$ in session s is a set of weighted terms:

$$UP(s) = \{(t_j, w_p^{(s)}(t_j)) : t_j \in T_U^{profile} \wedge w_p^{(s)}(t_j) \in [0, 1), j = 1, 2, \ldots, p_s\}, \quad (3)$$

where t_j is a term from set of terms $T_U^{profile}$, $w_p^{(s)}(t_j)$ is a weight of this term in session s and p_s is the quantity of terms in user profile in session s.

User profile is updated based on user activities. After each session the average weight of term from user queries is calculated. If a new term occurs in user queries, the weight for this term is calculated. If the term is in user profile its weight is recalculated using the following equation:

$$w_p^{(s+1)}(t_j) = \eta \cdot w_p^{(s)}(t_j) + (1 - \eta) \cdot \Delta_1(w_d^{(s)}(t_j)), \quad (4)$$

where $w_p^{(s+1)}(t_j)$ is a weight of t_j term in user profile in the session $s+1$, η is a parameter that was tuned in experiments ($\eta = 0.3$) and $\Delta_1(w_d^{(s)}(t_j))$ is absolute change of degree of user interests in term t_j.

The absolute change is calculated based on the following equation:

$$\Delta_1(w_d^{(s)}(t_j)) = \begin{cases} w_d^{(s)}(t_j) - w_d^{(s-1)}(t_j), & \text{if } s > 1 \\ w_d^{(1)}(t_j), & \text{if } s = 1 \end{cases} \quad (5)$$

where $w_d^{(s)}(t_j)$ is the average weight of term t_j after the current session s and $w_d^{(s-1)}(t_j)$ is the average weight of term t_j after the previous session $s-1$.

User profile is updated after a few blocks of sessions (the block contains 5–10 sessions).

4 Profile Recommendation for a New User

The proposed idea of profile recommendation method is presented in Figure 1. It contains three steps. At the beginning the system has a set of users that have both demographic data and profiles. To create a set of similar users (users with similar interests), a clustering method (k-means algorithm) is performed based on the data in profiles (step 1). We assume that each profile reflects user preference. If two users are in the same group, it means that their preferences are similar.

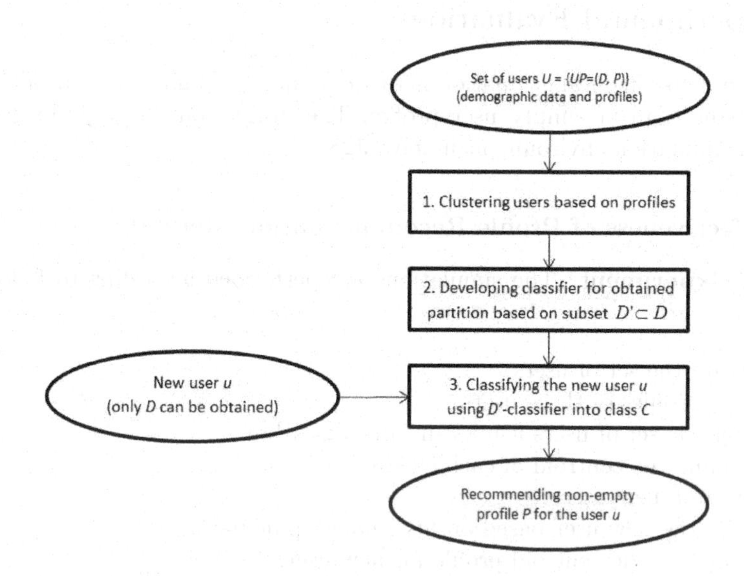

Fig. 1. Schema of profile recommendation based on a usage data classifier

The partition of users obtained in step 1 were performed based on their profiles. The second partition is determined based on demographic attributes (two users are in the same group if they have the same values of attributes). The aim of step 2 is to select a minimal set of attributes that determine a partition similar to partition obtained in step 1. The details of this method were presented in our previous paper [9]. Performed evaluations have shown that the most important sets of these attributes are as follows: {*gender, employment*}, {*gender, age*}, {*gender, education*}, {*gender, town, browser*} and {*gender, age, town*}.

The effectiveness of the system can be considered in the following way. We should check if a user was classified to the appropriate group (classifier based on demographic data is deterministic) – if profile updated based on user current activities is similar to the profiles of other users in this group. The profile that represents the group is recommended for a new user. When system has information about the user activities his profile is updated based on this information.

The profile for a new user will be recommended based on information about set of similar users. The idea of the algorithm is presented in [12,17]. When user registers to the system, he is asked to fulfill the questionnaire about his demographic data. Based on them, he is classified into group that has similar values of those data. Using knowledge about other users in this group, the centroid is determined and proposed to the new user. The profiles contain weighted terms. The centroid profile will be determined by calculating average value for each term from users in this group.

5 Experimental Evaluation

The aim of experimental evaluation is to show that non-empty user profile converges faster than an empty user profile. The evaluations were performed in prepared simulation environment in Java J2SE.

5.1 Effectiveness of Profile Recommendation Method

Plan of Experiment. The simulations are performed according to following steps:

1. Generate the set of users.
2. Update profiles of these users.
3. Cluster the set of users using k-means algorithm.
4. Determine the centroid of each cluster.
5. Generate a new user.
6. Classify the new user based on his demographic data.
7. Recommend the centroid profile for new user.
8. Update the profile of the new user in two cases:
   (a) When a beginning profile is empty (profile A).
   (b) When a beginning profile is recommended based on collective knowledge (profile B).
9. Calculate the distances between preference and profile in both cases.

The purpose of experiments was to show that the distance between user preference and profile A is bigger than the distance between user preference and profile B. It means that when a non-empty profile is being recommended for the new user, the profile converge to user preference faster than in the case when the profile is build and updated based only on single user activities.

Results Analysis. The experiments were performed for the following parameters: the set of users contains 100 users. The set was clustered using k-means algorithm for $k = 5$. The minimal set of demographic attributes were determined using the method described in [9]. In a single series of experiments a set of 100 new users was generated and non-empty profile was recommended to each new user. The distances were calculated after 20 updating steps. The results for 1249 users were collected.

The methodology of research was as follows. Let consider two random variables:

- X_1 – distances between preference and profile (that was empty at the beginning) after 20 updating steps
- X_2 – distances between preference and recommended profile (the centroid of the appropriate group) after 20 updating steps

The statistical analysis was performed for each pair of distances (differences between distances) for significance level $\alpha = 0.05$. The results are obtained using the same set of users' queries and relevant documents, so these variables are dependent. The a priori information about mean and standard deviation was unknown, so the first test was Lilliefors normality test. The null hypothesis is that the distribution of the differences is normal. Based on obtained p-value $p < 0.000001 < \alpha$, the null hypothesis is rejected. The conclusion is that the distribution of differences between distances is not consistent with normal distribution.

In such case the Wilcoxon test can be performed to check if the median of differences is equal to 0.

- $H_0 : \theta = 0$
- $H_1 : \theta \neq 0$

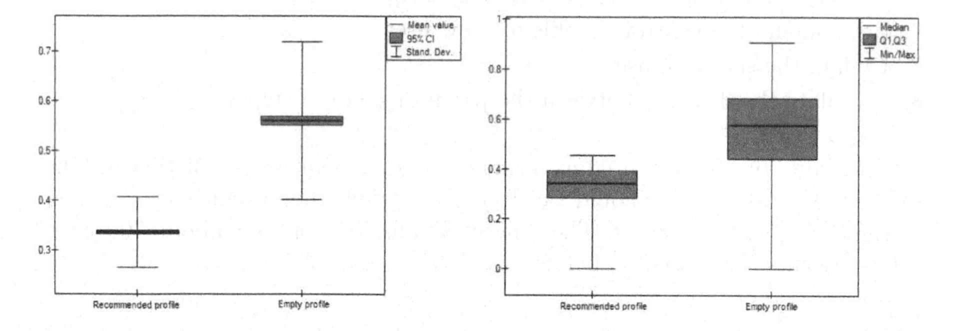

Fig. 2. The mean and median value for X_1 and X_2

The results of the Wilcoxon test was $p < 0.000001 < \alpha$, so the null hypothesis was rejected and the alternative hypothesis is accepted: the medians of both distribution are statistically different. The sum of the positive ranks is greater than the sum of the negative ranks, so the median of X_2 is smaller than the median of X_1.

The conclusion is the following: when a non-empty profile is being recommended for the new user, the profile converge to user preference faster than in the case when the profile is build and updated based only on single user activities.

5.2 The Effectiveness of Recommendation System

The effectiveness of this method of recommending a profile for a new user can be considered in the following way. A new user after few updating series becomes a source of knowledge for the system. The new user is added to the group that contains similar users. When many new users are added to the system (after many series of updating), the clustering procedure is necessary for the set of previous and new users. The system should also check if the demographic attributes are appropriate for the whole set of users. The minimal subset of attributes can change when the number of users is growing in the system.

The purpose of the experimental evaluations of the whole recommendation system is to show that with growing number of users, the usage-based classifier using demographic attributes will be better. In each iteration, the distances between partitions become smaller and smaller.

Plan of Experiment. The simulations are performed according to following steps:

1. Generate a set of 10 users.
2. Update profiles of these users.
3. Cluster the set of users using k-means algorithm.
4. Determine the centroid of each cluster.
5. Classify the new user based on his demographic data.
6. Recommend the centroid profile for new user.
7. Update the set of all users.
8. Calculate the distance between the partitions. Go to step 1.

The results of the presented experiment create a sequence of distances. The trend of these distances should be decreasing while the system collects more information about the users. The longest simulation contains about 70 cycles and the results are presented below.

Results Analysis. Simulations were performed according the presented plan in Java (J2SE) simulation environment. The parameters assumed were the same as in previous experiments. At the beginning a set of 100 users is generated. Based on theirs activities, the profiles are developed. Based on information about these users, the usage-based classifier is determined. In each iteration of experiments 10 new users are generated and classified into group of similar users. The profile that represents the appropriate group is recommended to each new user. Information about activities of all users are gathered (for one or a few updates). The procedure is reiterated for the current set of all users.

The experiments were performed for two independent series and the results are presented in Fig. 3 and in Fig. 4.

In each iteration 10 users were added, so the set of users is increasing. The longest series were performed for almost 70 iterations. At the beginning the system contains a small number of users and the distances between partitions are increasing. When the system contains about 100 users, the trends on both diagrams start decreasing.

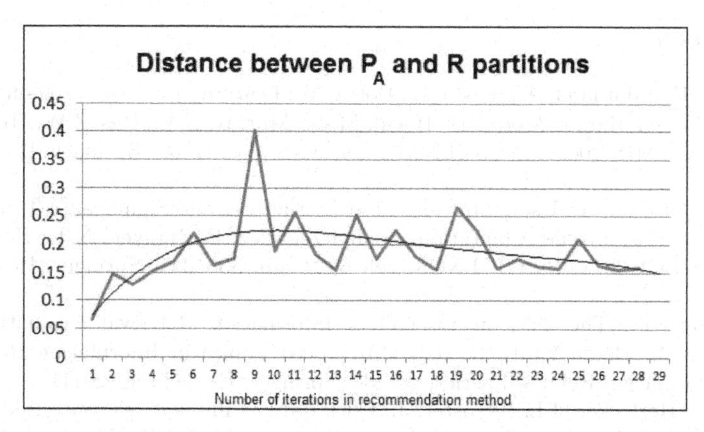

Fig. 3. Differences between partitions based on demographic data and profiles in subsequent iterations of experiments – series I

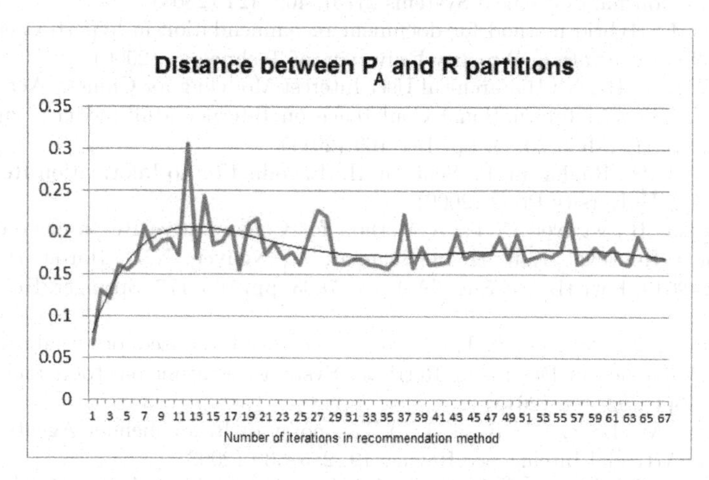

Fig. 4. Differences between partitions based on demographic data and profiles in subsequent iterations of experiments – series II

6 Summary and Future Works

In this paper the authors consider effectiveness of a method for recommending profile to a new user based only on knowledge about partial demographic data. A methodology of experimental evaluation was presented and simulations were performed. The main objective was to show that based on knowledge about other similar users, the system can classify a new user based on a subset of demographic data and recommend him a non-empty profile. The results have shown that the updating process is faster when the beginning profile is recommended to the user based on the collective knowledge about the similar users.

Acknowledgments. This research was partially supported by Polish Ministry of Science and Higher Education.

References

1. Galassi, U., Giordana, A., Saitta, L., Botta, M.: Learning Profiles Based on Hierarchical Hidden Markov Model. In: Hacid, M.-S., Murray, N.V., Raś, Z.W., Tsumoto, S. (eds.) ISMIS 2005. LNCS (LNAI), vol. 3488, pp. 47–55. Springer, Heidelberg (2005)
2. Indyka-Piasecka, A.: Using Multi-attribute Structures and Significance Term Evaluation for User Profile Adaptation. In: Jędrzejowicz, P., Nguyen, N.T., Hoang, K. (eds.) ICCCI 2011, Part I. LNCS, vol. 6922, pp. 336–345. Springer, Heidelberg (2011)
3. Ingwersen, P.: The User in Interactive Information Retrieval Evaluation. In: Melucci, M., Baeza-Yates, R. (eds.) Advanced Topics in Information Retrieval. The Information Retrieval Series, vol. 33, Springer, Heidelberg (2011)
4. Jung, S., Herlocker, J.L., Webster, J.: Click data as implicit relevance feedback in web search. Information Processing and Management 43, 791–807 (2007)
5. Kagolovsky, Y., Moehr, J.R.: Current Status of the Evaluation of Information Retrieval. Journal of Medical Systems 27(5), 409–424 (2003)
6. Kiewra, M.: Hybrid method for document recommendation in hypertext environment. PhD dissertation. Wroclaw University of Technology (2006)
7. Li, S., Wu, G., Hy, X.: Hierarchical User Interest Modeling for Chinese Web Pages. In: Proceedings of International Conference on Internet Multimedia Computing and Service (ICIMCS 2011), pp. 164–169 (2011)
8. Manning, C.D., Raghavan, P., Schütze, H.: Introduction to Information Retrieval. Cambridge University Press (2009)
9. Mianowska, B., Nguyen, N.T.: A Method for Collaborative Recommendation in Document Retrieval Systems. In: Selamat, A., Nguyen, N.T., Haron, H. (eds.) ACIIDS 2013, Part II. LNCS (LNAI), vol. 7803, pp. 168–177. Springer, Heidelberg (2013)
10. Mianowska, B., Nguyen, N.T.: Tuning User Profiles Based on Analyzing Dynamic Preference in Document Retrieval Systems. Multimedia Tools and Applications 65(1), 93–118 (2013)
11. Montaner, M., Lopez, B., Rosa, J.: A Taxonomy of Recommender Agents on the Internet. Artificial Intelligence Review 19, 285–330 (2003)
12. Nguyen, N.T.: Advanced Methods for Inconsistent Knowledge Management. Springer (2008)
13. Ren, F., Bracewell, D.B.: Advanced Information Retrieval. Electronic Notes in Theoretical Computer Science 225, 303–317 (2009)
14. Schein, A.I., Popescu, A., Ungar, L.H., Pennock, D.M.: Methods and Metrics for ColdStart Recommendations. In: Proceedings of the 25th Annual International ACM SIGIR Conference on Research and Development in Information Retrieval (SIGIR 2002), pp. 253–260 (2002)
15. Wang, Y.D., Forgionne, G.: Testing a Decision-Theoretic Approach to the Evaluation of Information Retrieval Systems. Journal of Information Science 34(6), 861–876 (2008)
16. Wolfe, S.R., Zhang, Y.: Interaction and Personalization of Criteria in Recommender Systems. In: De Bra, P., Kobsa, A., Chin, D. (eds.) UMAP 2010. LNCS, vol. 6075, pp. 183–194. Springer, Heidelberg (2010)
17. Hong, T.P., Liou, Y.L., Wang, S.L., Vo, B.: Feature selection and replacement by clustering attributes. Vietnam Journal of Computer Science (November 2013), doi:10.1007/s40595-013-0004-3

Using Non-Zero Dimensions and Lengths of Vectors for the Tanimoto Similarity Search among Real Valued Vectors

Marzena Kryszkiewicz

Institute of Computer Science, Warsaw University of Technology
Nowowiejska 15/19, 00-665 Warsaw, Poland
mkr@ii.pw.edu.pl

Abstract. The Tanimoto similarity measure finds numerous applications e.g. in chemical informatics, bioinformatics, information retrieval, text and web mining. Recently, two efficient methods for reducing the number of candidates for Tanimoto similar real valued vectors have been offered: the one using lengths of vectors and the other using their non-zero dimensions. In this paper, we offer new theoretical results on combined usage of lengths of real valued vectors and their non-zero dimensions for more efficient reduction of candidates for Tanimoto similar vectors. In particular, we derive more restrictive bounds on lengths of such candidate vectors.

Keywords: sparse high dimensional data, the Tanimoto similarity, chemical informatics, bioinformatics, text mining, data mining, information retrieval.

1 Introduction

The Tanimoto similarity measure finds numerous applications e.g. in chemical informatics, bioinformatics, information retrieval, text and web mining for finding similar vectors representing respective objects such as chemical substructures or documents. In the case of binary vectors, the Tanimoto similarity between two vectors equals the ratio of the number of dimensions with "1s" shared by both vectors to the number of dimensions with "1s" that occur in either vector. In this particular case of vectors, the Tanimoto similarity is equivalent to the Jaccard similarity, which is defined for pairs of sets as the ratio of the cardinality of the intersection of the two sets to the cardinality of their union. However, the Tanimoto similarity can be calculated for any non-zero real valued vectors and thus can be regarded as an extension of the Jaccard similarity.

The determination of Tanimoto similar vectors is challenging in the case of very large data sets with, say, thousands or tens of thousands of dimensions such as text data sets. While methods of efficient search of Jaccard similar binary vectors or binary vectors with weighted domains have been intensively explored (see e.g. [1, 2, 4, 6, 9]), there is no much work on efficient search of Tanimoto similar real valued vectors in the literature. In fact, the problem of the search of similar real valued vectors is often solved in an approximate way by transforming real valued vectors into binary vectors with even much larger number of dimensions [1, 2]. Recently,

N.T. Nguyen et al. (Eds.): ACIIDS 2014, Part I, LNAI 8397, pp. 173–182, 2014.

however, there were offered efficient exact methods for reducing the number of candidates for Tanimoto similar real valued vectors: i) by means of their lengths [3] and ii) by means of their non-zero dimensions [5]. The methods turned out efficient even for large sparse high dimensional data sets. In particular, the experiments carried out in [3, 5] showed that those methods returned on average from several (for lower values of the applied similarity threshold) by up to three orders of magnitude (for the similarity threshold close to 100%) less numerous candidate sets than the number of all vectors in a given data set. In this paper, we offer new theoretical results on combined usage of lengths of real valued vectors and their non-zero dimensions for more efficient reduction of candidates for Tanimoto similar vectors. In particular, we derive more restrictive bounds on lengths of such candidate vectors than in [3].

Our paper has the following layout. Section 2 provides basic notions and properties used in the paper. In Section 3, we introduce the Tanimoto similarity measure, its properties and the methods of determining candidates for Tanimoto similar vectors that were proposed in [3, 5]. In Section 4, we offer our new theoretical results on determining such candidates. Section 5 summarizes our work.

2 Basic Notions and Properties

In the paper, we consider *n*-dimensional vectors. A vector u will be also denoted as $[u_1, \ldots, u_n]$, where u_i is the value of the i-th dimension of u, $i = 1..n$. A vector all dimensions of which have zero values is called a *zero vector*. Otherwise, it is called a *non-zero vector*. By $NZD(u)$ we denote the set of those dimensions of vector u which have values different from 0; that is,

$$NZD(u) = \{i \in \{1, \ldots, n\} | \, u_i \neq 0\}.$$

Analogously, by $ZD(u)$ we denote the set of those dimensions of vector u which have zero values; that is,

$$ZD(u) = \{i \in \{1, \ldots, n\} | \, u_i = 0\}.$$

Table 1. Dense representation of an example set of vectors

Id	1	2	3	4	5	6	7	8	9
*v*1	3.0	4.0			3.0	5.0	3.0	6.0	
*v*2	4.0	8.0				6.0		5.0	
*v*3			5.0		2.0	2.0			
*v*4		10.0		8.0		10.0		4.0	
*v*5				1.0	4.0		7.0		
*v*6	1.0	2.0	1.0	4.0					5.0
*v*7			1.0	1.0			1.0	1.0	
*v*8				1.0					4.0
*v*9	1.0			2.0	5.0		4.0		2.0
*v*10			5.0						2.0

In Table 1, we present a set of vectors that will be used throughout this paper. In Table 2, we provide its alternative sparse representation, where each vector is represented only by its non-zero dimensions and their values. More precisely, each

Table 2. Sparse representation of the example set of vectors (extended by the information about lengths of the vectors)

Id	(non-zero dimension, value) pairs	length
*v*1	{(1, 3.0), (2, 4.0), (5, 3.0), (6, 5.0), (7, 3.0), (8, 6.0)}	10.20
*v*2	{(1, 4.0), (2, 8.0), (6, 6.0), (8, 5.0)}	11.87
*v*3	{(3, 5.0), (5, 2.0), (6, 2.0)}	5.74
*v*4	{(2, 10.0), (4, 8.0), (6, 10.0), (8, 4.0)}	16.73
*v*5	{(4, 1.0), (5, 4.0), (7, 7.0)}	8.12
*v*6	{(1, 1.0), (2, 2.0), (3, 1.0), (4, 4.0), (9, 5.0)}	6.86
*v*7	{(3, 1.0), (4, 1.0), (7, 1.0), (8, 1.0)}	2.00
*v*8	{(4, 1.0), (9, 4.0)}	4.12
*v*9	{(1, 1.0), (4, 2.0), (5, 5.0), (7, 4.0), (9, 2.0)}	7.07
*v*10	{(3, 5.0), (9, 2.0)}	5.39

vector is represented by a list of pairs, where the first element of a pair is a non-zero dimension of the vector and the second element – its value.

A *dot product* of vectors u and v is denoted by $u \cdot v$ and is defined as $\sum_{i=1..n} u_i v_i$. One may easily observe that the dot product of vectors u and v that have no common non-zero dimension equals 0 (see Property 1).

Property 1. Let u and v be non-zero vectors.

a) $u \cdot v = \sum_{i \in NZD(u)} u_i v_i = \sum_{i \in NZD(u) \cap NZD(v)} u_i v_i$.

b) If $NZD(u) \cap NZD(v) = \emptyset$, then $u \cdot v = 0$.

A *length of a vector u* is denoted by $| u |$ and is defined as $\sqrt{u \cdot u}$. A *normalized form of vector u* is denoted by u' and is defined as the ratio of u to its length $| u |$; that is,

$$u' = \frac{u}{| u |}.$$

Clearly, any i-*th* dimension u_i' of the normalized form of vector u equals $\frac{u_i}{| u |}$.

A vector u is defined as a *normalized vector* if $u = u'$. Obviously, the length of a normalized vector equals 1.

Property 2. Let u be a non-zero vector. Then:

a) $\sum_{i \in \{1,\dots,n\}} u_i'^2 = 1$.

b) $NZD(u) = NZD(u')$.

c) $ZD(u) = ZD(u')$.

It was shown in [5] how to calculate an optimistic estimation of the value of the dot product of two normalized vectors provided a subset of zero dimensions of one of the vectors is known. We recall this result as Lemma 1.

Lemma 1 [5]. Let u and v be non-zero vectors and $J \subseteq ZD(v)$. Then:

$$u' \cdot v' \leq \sqrt{1 - \sum_{i \in J} u_i'^2} .$$

3 The Tanimoto Similarity: Definition and Calculation Methods

3.1 Definition of the Tanimoto Similarity

The *Tanimoto similarity* between vectors u and v is denoted by $T(u, v)$ and is defined as follows,

$$T(u, v) = \frac{u \cdot v}{u \cdot u + v \cdot v - u \cdot v}.$$

In the sequel, non-zero vectors u and v for which $T(u, v) \geq \varepsilon$ will be called *Tanimoto ε-similar* (or briefly, *ε-similar*).

Property 3 [7]. Let u and v be non-zero vectors. Then, $T(u, v) \in \left[-\frac{1}{3}, 1\right]$.

Property 4. Let u and v be non-zero vectors. Then:

$$T(u, v) = \frac{u \cdot v}{|u|^2 + |v|^2 - u \cdot v} = \frac{u' \cdot v'}{\frac{|u|}{|v|} + \frac{|v|}{|u|} - u' \cdot v'}.$$

By Property 4, the Tanimoto similarity between two vectors can be perceived as a function of both the dot product of their normalized forms and the ratio of the length of one vector to the length of the other vector. Beneath, we present a property of the ratio of lengths of vectors.

Property 5. Let u and v be non-zero vectors. Then:

$$\frac{|u|}{|v|} + \frac{|v|}{|u|} \geq 2 \quad \text{and} \quad \frac{|u|}{|v|} + \frac{|v|}{|u|} - u' \cdot v' \geq 1.$$

Proof. Follows from the fact that $(|u| - |v|)^2 \geq 0$ and $u' \cdot v' \leq 1$ (by Lemma 1). □

3.2 Calculating Tanimoto Similar Vectors by Means of Inverted Indices

As follows from the definition of the Tanimoto similarity, $T(u, v)$ equals 0 if $u \cdot v = 0$. Hence, when looking for vectors v such that $T(u, v) \geq \varepsilon$, where $\varepsilon > 0$, vectors w that do not have any common non-zero dimension with vector u can be skipped as in their case $T(u, w) = u \cdot w = 0$. In Example 1, we show how to skip such vectors by means of *inverted indices* [8]. In a simple form, an inverted index stores for a dimension *dim* the list *I*(*dim*) of identifiers of all vectors for which *dim* is a non-zero dimension. In Table 3, we present the inverted indices that would be created for the set of vectors from Table 2.

Example 1. Let us assume that we are to find vectors v similar to vector $u = v1$ from Table 2 such that $T(u, v) \geq \varepsilon$, where $\varepsilon > 0$. To this end, it is sufficient to compare u only with vectors that have at least one non-zero dimension common with $NZD(u)$ =

{1, 2, 5, 6, 7, 8} (see Table 2); that is, with the vectors in the following set (see Table 3):

$$I(1) \cup I(2) \cup I(5) \cup I(6) \cup I(7) \cup I(8) = \{v1, v2, v3, v4, v5, v6, v7, v9\}.$$

Hence, the application of the inverted indices from Table 3 allowed us to reduce the set of vectors to be evaluated from 10 to 8. □

Table 3. Inverted indices for dimensions of vectors in the example set from Table 2

dim	*I(dim)*
1	<*v*1, *v*2, *v*6, *v*9>
2	<*v*1, *v*2, *v*4, *v*6>
3	<*v*3, *v*6, *v*7, *v*10>
4	<*v*4, *v*5, *v*6, *v*7, *v*8, *v*9>
5	<*v*1, *v*3, *v*5, *v*9>
6	<*v*1, *v*2, *v*3, *v*4>
7	<*v*1, *v*5, *v*7, *v*9>
8	<*v*1, *v*2, *v*4, *v*7>
9	<*v*6, *v*8, *v*9, *v*10>

Please note that in Example 1 the value of ε has not been taken into account when determining vectors potentially similar to the given vector u. In the remaining subsections of Section 3, we recall recently offered methods that use ε to make the reduction of candidates more efficient.

3.3 Using Bounds on Lengths of Tanimoto Similar Vectors

In this section, we recall the bounds on lengths of Tanimoto ε-similar vectors after [3] and illustrate their usefulness with an example.

Theorem 1 [3]. Let u and v be non-zero vectors and $\varepsilon \in (0, 1]$. Then:

$$(T(u,v) \geq \varepsilon) \Rightarrow \left(|v| \in \left[\frac{1}{\alpha}|u|, \alpha |u| \right] \right),$$

where $\alpha = \frac{1}{2}\left(\left(1+\frac{1}{\varepsilon}\right) + \sqrt{\left(1+\frac{1}{\varepsilon}\right)^2 - 4} \right)$.

Example 2. Let ε = 0.75 and u be vector $v1$ from Table 2, the length of which $|u| \approx 10.20$. We will use the bounds on vectors' lengths in order to reduce the number of candidates for vectors similar to u with respect to the given ε value. Then, by Theorem 1, $\alpha \approx 1.77$ and only vectors, the lengths of which belong to the interval $\left[\frac{1}{\alpha}|u|, \alpha |u| \right] \approx \left[\frac{1}{1.77} \times 10.20, 1.77 \times 10.20 \right] \approx [5.77, 18.03]$ have a chance to be sufficiently similar to u. Thus, only six out of ten vectors; namely, {$v1$, $v2$, $v4$, $v5$, $v6$, $v9$} can be ε-similar to u. □

3.4 Calculating Similar Vectors by Means of Vectors' Non-Zero Dimensions

In this section, we recall the method based on using non-zero dimensions in determining Tanimoto similar vectors, as offered in [5], and illustrate its usefulness with an example.

Theorem 2 [5]. Let u and v be non-zero vectors, $J \subseteq (NZD(u) \cap ZD(v))$ and $\varepsilon \in (0,1]$. Then:

$$(T(u,v) \geq \varepsilon) \Rightarrow \left(\sum_{i \in J} u_i'^2 \leq 1 - \left(\frac{2\varepsilon}{1+\varepsilon} \right)^2 \right).$$

Example 3. Let us consider vector $u = v1$ from Table 2. Then $NZD(u) = \{1, 2, 5, 6, 7, 8\}$. Let $J = \{5, 6\}$ and $\varepsilon = 0.75$. Then, $J \subseteq NZD(u)$ and $1 - \left(\frac{2\varepsilon}{1+\varepsilon} \right)^2 \approx 0.27$. We note that $\sum_{i \in J} u_i'^2 = \frac{u_5^2 + u_6^2}{|u|^2} = \frac{3.0^2 + 5.0^2}{|u|^2} \approx 0.33$. Hence, $\neg \left(\sum_{i \in J} u_i'^2 \leq 1 - \left(\frac{2\varepsilon}{1+\varepsilon} \right)^2 \right)$. Thus, by Theorem 2, all vectors v in Table 2 for which both dimension 5 and dimension 6 are zero dimensions are guaranteed not to be ε-similar to vector u. So, only the remaining vectors; that is, $I(5) \cup I(6) = \{v1, v2, v3, v4, v5, v9\}$ have such a chance (please, see Table 3 for inverted indices). □

3.5 Semi-Naive Calculation of Tanimoto Similar Vectors by Means of Both Bounds on Lengths and Non-Zero Dimensions

As suggested in [4, 5], a straightforward approach to enhancing the calculation of Tanimoto similar vectors could consist in applying both bounds on lengths of vectors and their non-zero dimensions in order to reduce the number of candidates.

Example 4. Let $\varepsilon = 0.75$ and u be vector $v1$ from Table 2. As shown in Example 2, applying the bounds on lengths of Tanimoto ε-similar vectors (by Theorem 1) resulted in restricting candidates to 6 vectors $\{v1, v2, v4, v5, v6, v9\}$. On the other hand, Example 3 showed that using non-zero dimensions 5 and 6 of u (by Theorem 2) resulted in another set of 6 candidate vectors $\{v1, v2, v3, v4, v5, v9\}$. The application of both the bounds on lengths and non-zero dimensions allows restricting the set of candidates to 5 vectors; namely, to: $\{v1, v2, v4, v5, v6, v9\} \cap \{v1, v2, v3, v4, v5, v9\} = \{v1, v2, v4, v5, v9\}$. □

4 New Approach to Calculating Tanimoto Similar Vectors

In this section, we propose and prove a new approach to searching Tanimoto ε-similar real valued vectors based on using both non-zero dimensions and lengths of vectors. We start with deriving an ε dependent upper bound on the sum of the ratio of lengths of any two non-zero vectors and its inverse provided the vectors are Tanimoto ε-similar.

Theorem 3. Let u and v be non-zero vectors, $L \subseteq (NZD(u) \cap ZD(v))$, $\varepsilon \in (0,1]$ and $\tau = \left(1+\frac{1}{\varepsilon}\right)\sqrt{1-\sum_{i\in L} u_i'^2}$. Then:

$$(T(u,v) \geq \varepsilon) \Rightarrow \left(\frac{|u|}{|v|}+\frac{|v|}{|u|} \leq \tau\right).$$

Proof. Let $T(u, v) \geq \varepsilon$. As $\varepsilon > 0$ and by Property 4, $T(u,v) = \frac{u' \cdot v'}{\frac{|u|}{|v|}+\frac{|v|}{|u|} - u' \cdot v'} \geq \varepsilon > 0$,

and by Property 5, $\frac{|u|}{|v|}+\frac{|v|}{|u|} - u' \cdot v' \geq 1$. Therefore, $\frac{|u|}{|v|}+\frac{|v|}{|u|} \leq \left(1+\frac{1}{\varepsilon}\right)(u' \cdot v')$.

Hence and by Lemma 1, $\frac{|u|}{|v|}+\frac{|v|}{|u|} \leq \left(1+\frac{1}{\varepsilon}\right)\sqrt{1-\sum_{i\in L} u_i'^2} = \tau$. □

Lemma 2. Let u and v be non-zero vectors, $\varepsilon \in (0,1]$, τ be a real number and $\beta = \frac{1}{2}\left(\tau + \sqrt{\tau^2 - 4}\right)$. Then:

$$\left(\frac{|u|}{|v|}+\frac{|v|}{|u|} \leq \tau\right) \Rightarrow \left((\tau \geq 2) \wedge (\beta \geq 1) \wedge \left(\frac{|v|}{|u|} \in \left[\frac{1}{\beta}, \beta\right]\right)\right).$$

Proof. Let $\frac{|u|}{|v|}+\frac{|v|}{|u|} \leq \tau$ and $k = \frac{|v|}{|u|}$. Hence and by Property 5, $2 \leq \frac{1}{k} + k \leq \tau$.

Thus, $\tau \geq 2$ (so, $\beta \geq 1$) and $k^2 - k\tau + 1 \leq 0$. As a result of solving the latter non-equation, we get $k \in [\beta', \beta]$, where $\beta' = \frac{1}{2}\left(\tau - \sqrt{\tau^2 - 4}\right)$. Now, we observe that $\beta\beta' = 1$, so $\beta' = \frac{1}{\beta}$. In consequence, $\frac{|v|}{|u|} = k \in \left[\frac{1}{\beta}, \beta\right]$. □

Theorem 4. Let u and v be non-zero vectors, $L \subseteq (NZD(u) \cap ZD(v))$, $\varepsilon \in (0,1]$, $\tau = \left(1+\frac{1}{\varepsilon}\right)\sqrt{1-\sum_{i\in L} u_i'^2}$ and $\beta = \frac{1}{2}\left(\tau + \sqrt{\tau^2 - 4}\right)$. Then:

$$(T(u,v) \geq \varepsilon) \Rightarrow \left((\tau \geq 2) \wedge \left(|v| \in \left[\frac{1}{\beta}|u|, \beta|u|\right]\right)\right).$$

Proof. Follows immediately from Theorem 3 and Lemma 2. □

Property 6. Let u be a non-zero vector, L be a subset of dimensions, $\varepsilon \in (0,1]$ and $\tau = \left(1+\frac{1}{\varepsilon}\right)\sqrt{1-\sum_{i\in L} u'^2_i}$. Then:

$$(\tau \geq 2) \Leftrightarrow \left(\sum_{i\in L} u'^2_i \leq 1-\left(\frac{2\varepsilon}{1+\varepsilon}\right)^2\right).$$

Theorem 5. Let u and v be non-zero vectors, $L \subseteq (NZD(u) \cap ZD(v))$, $\varepsilon \in (0,1]$ and $\beta = \frac{1}{2}\left(\tau + \sqrt{\tau^2 - 4}\right)$, where $\tau = \left(1+\frac{1}{\varepsilon}\right)\sqrt{1-\sum_{i\in L} u'^2_i}$. Then:

$$(T(u,v) \geq \varepsilon) \Rightarrow \left(\left(\sum_{i\in L} u'^2_i \leq 1-\left(\frac{2\varepsilon}{1+\varepsilon}\right)^2\right) \wedge \left(|v| \in \left[\frac{1}{\beta}|u|, \beta|u|\right]\right)\right).$$

Proof. Follows immediately from Theorem 4 and Property 6. □

Property 7. Let u be a non-zero vector, $L_1 \subset L_2 \subseteq NZD(u)$, $\varepsilon \in (0,1]$, $\tau_1 = \left(1+\frac{1}{\varepsilon}\right)\sqrt{1-\sum_{i\in L_1} u'^2_i}$, $\beta_1 = \frac{1}{2}\left(\tau_1 + \sqrt{\tau_1^2 - 4}\right)$, $\tau_2 = \left(1+\frac{1}{\varepsilon}\right)\sqrt{1-\sum_{i\in L_2} u'^2_i}$, $\beta_2 = \frac{1}{2}\left(\tau_2 + \sqrt{\tau_2^2 - 4}\right)$. Then, $\sum_{i\in L_1} u'^2_i < \sum_{i\in L_2} u'^2_i$ and $\left[\frac{1}{\beta_1}, \beta_1\right] \supset \left[\frac{1}{\beta_2}, \beta_2\right]$.

Proposition 1. Let u and v be non-zero vectors, $\tau = \left(1+\frac{1}{\varepsilon}\right)\sqrt{1-\sum_{i\in L} u'^2_i}$, $\beta = \frac{1}{2}\left(\tau + \sqrt{\tau^2 - 4}\right)$, $\alpha = \frac{1}{2}\left(\left(1+\frac{1}{\varepsilon}\right) + \sqrt{\left(1+\frac{1}{\varepsilon}\right)^2 - 4}\right)$ and $\varepsilon \in (0,1]$. Then:

$$(L = \emptyset) \Rightarrow \left((L \subseteq (NZD(u) \cap ZD(v))) \wedge \left(0 = \sum_{i\in L} u'^2_i \leq 1-\left(\frac{2\varepsilon}{1+\varepsilon}\right)^2\right) \wedge (\beta = \alpha)\right).$$

Proposition 2. Let u and v be non-zero vectors, $\varepsilon \in (0,1]$, J be a minimal subset of $NZD(u)$ such that $\neg\left(\sum_{i\in J} u'^2_i \leq 1-\left(\frac{2\varepsilon}{1+\varepsilon}\right)^2\right)$ and $\alpha = \frac{1}{2}\left(\left(1+\frac{1}{\varepsilon}\right) + \sqrt{\left(1+\frac{1}{\varepsilon}\right)^2 - 4}\right)$.

Let $L \subseteq ZD(v)$, $\tau = \left(1+\frac{1}{\varepsilon}\right)\sqrt{1-\sum_{i\in L} u'^2_i}$, $\beta = \frac{1}{2}\left(\tau + \sqrt{\tau^2 - 4}\right)$. Then:

$$(\emptyset \subset L \subset J) \Rightarrow \left((L \subset NZD(u)) \wedge \left(\sum_{i\in L} u'^2_i \leq 1-\left(\frac{2\varepsilon}{1+\varepsilon}\right)^2\right) \wedge \left(\left[\frac{1}{\alpha}, \alpha\right] \supset \left[\frac{1}{\beta}, \beta\right]\right)\right).$$

Proof. By Property 7 and Proposition 1. □

We will illustrate now the usefulness of Theorem 5 in determining candidates for vectors ε-similar to a given vector u by applying different subsets $L \subseteq NZD(u)$.

Example 5. Let $\varepsilon = 0.75$, u be vector $v1$ from Table 2 and $J = \{5, 6\}$. We will consider now using of Theorem 5 for the following subsets L of non-zero dimensions of u: a) $L = J = \{5, 6\}$, b) $L = \varnothing$, c) $L = \{5\}$, d) $L = \{6\}$.

Case $L = J = \{5, 6\}$: As we have already found in Example 3, $L = J = \{5, 6\}$ is a subset of $NZD(u)$ that does not fulfill the condition $\sum_{i \in L} u_i'^2 \le 1 - \left(\frac{2\varepsilon}{1+\varepsilon}\right)^2$ (in fact, $L = J = \{5, 6\}$ is a minimal subset of $NZD(u)$ having this property). Thus, by Theorem 5 (as well as by Theorem 2), each vector v for which both 5 and 6 are zero dimensions are not ε-similar to u . Hence, all vectors sufficiently similar to u are guaranteed to be found among $I(5) \cup I(6)$, where $I(5) = \{v1, v3, v5, v9\}$ and $I(6) = \{v1, v2, v3, v4\}$. So, an initial candidate set can be set to $C_1 = \{v1, v2, v3, v4, v5, v9\}$.

Case $L = \varnothing$: Then Theorem 5 is applicable to each vector v in C_1 as, in fact, $L = \varnothing$ is contained in zero dimensions' set as well as in non-zero dimensions' set of any vector. So, by Theorem 5, $\tau = \left(1+\frac{1}{\varepsilon}\right)\sqrt{1-\sum_{i \in \varnothing} u_i'^2} = \left(1+\frac{1}{\varepsilon}\right)$, $\beta = \frac{1}{2}\left(\tau + \sqrt{\tau^2 - 4}\right)$ $= \frac{1}{2}\left(\left(1+\frac{1}{\varepsilon}\right) + \sqrt{\left(1+\frac{1}{\varepsilon}\right)^2 - 4}\right) \approx 1.77$ and only vectors in C_1 the lengths of which belong to the interval $\left[\frac{1}{\beta}|u|, \beta|u|\right] \approx [5.77, 18.03]$ have a chance to be ε-similar to u (the same lengths' interval would have been found by applying Theorem 1). As vector $v3 \in C_1$ does not fulfill this length condition, it is not ε-similar to u, so the candidate set can be reduced to $C_2 = \{v1, v2, v4, v5, v9\}$ (this result has been also obtained in Example 4 by applying together Theorem 1 and Theorem 2).

Case $L = \{5\}$: Then Theorem 5 is applicable to each vector v in C_2 having 5 as its zero dimension; that is to vectors, $C_2 \setminus I\{5\} = \{v2, v4\}$. Thus, by Theorem 5, $\tau = \left(1+\frac{1}{\varepsilon}\right)\sqrt{1-\sum_{i \in \{5\}} u_i'^2}$, $\beta = \frac{1}{2}\left(\tau + \sqrt{\tau^2 - 4}\right) \approx 1.61$ and the vectors in $\{v2, v4\}$, have a chance to be ε-similar to u provided their lengths belong to the interval $\left[\frac{1}{\beta}|u|, \beta|u|\right] \approx [6.34, 16.40]$. Since vector $v4$ does not fulfill this length condition, it is not ε-similar to u, so the candidate set can be further reduced to $C_3 = \{v1, v2, v5, v9\}$.

Case $L = \{6\}$: Then Theorem 5 is applicable to each vector v in C_3 having 6 as its zero dimension; that is to vectors, $C_3 \setminus I\{6\} = \{v5, v9\}$. Thus, by Theorem 5, $\tau = \left(1+\frac{1}{\varepsilon}\right)\sqrt{1-\sum_{i \in \{6\}} u_i'^2}$, $\beta = \frac{1}{2}\left(\tau + \sqrt{\tau^2 - 4}\right) \approx 1.20$ and the vectors in $\{v5, v9\}$, have a chance to be ε-similar to u provided their lengths belong to the interval $\left[\frac{1}{\beta}|u|, \beta|u|\right] \approx [8.49,12.25]$. As neither vector $v5$ nor vector $v9$ fulfills this length condition, they are not ε-similar to u, and the candidate set can be finally reduced to $C_4 = \{v1, v2\}$.

Please note that Theorem 1 and Theorem 2 allowed us to reduce the number of candidates for vectors Tanimoto ε-similar to u from 10 vectors in the data set to 5, whereas Theorem 5 allowed us to reduce this number further to 2 vectors. □

Corollary 1. Theorem 5 allows reducing a set of candidates for Tanimoto ε-similar real valued vectors to not lesser degree than Theorem 1 and Theorem 2 used together.

5 Summary

Recently, we offered methods for reducing the number of candidates for Tanimoto ε-similar real valued vectors: i) by means of their lengths [3] and ii) by means of non-zero dimensions [5], which turned out efficient even for sparse high dimensional data sets. In this paper, we have offered new theoretical results on combined usage of both lengths of real valued vectors and their non-zero dimensions for more efficient reduction of candidates for Tanimoto similar vectors. In particular, we derived more restrictive bounds on lengths of candidate vectors. We have illustrated the usefulness of our new approach by means of an example.

References

1. Arasu, A., Ganti, V., Kaushik, R.: Efficient exact set-similarity joins. In: Proc. of VLDB 2006. ACM (2006)
2. Chaudhuri, S., Ganti, V., Kaushik, R.L.: A primitive operator for similarity joins in data cleaning. In: Proceedings of ICDE 2006. IEEE Computer Society (2006)
3. Kryszkiewicz, M.: Bounds on Lengths of Real Valued Vectors Similar with Regard to the Tanimoto Similarity. In: Selamat, A., Nguyen, N.T., Haron, H. (eds.) ACIIDS 2013, Part I. LNCS, vol. 7802, pp. 445–454. Springer, Heidelberg (2013)
4. Kryszkiewicz, M.: On Cosine and Tanimoto Near Duplicates Search among Vectors with Domains Consisting of Zero, a Positive Number and a Negative Number. In: Larsen, H.L., Martin-Bautista, M.J., Vila, M.A., Andreasen, T., Christiansen, H. (eds.) FQAS 2013. LNCS (LNAI), vol. 8132, pp. 531–542. Springer, Heidelberg (2013)
5. Kryszkiewicz, M.: Using Non-Zero Dimensions for the Cosine and Tanimoto Similarity Search among Real Valued Vectors. Fundamenta Informaticae 127, 307–323 (2013)
6. Rajaraman, A., Ullman, J.D.: Mining of Massive Datasets. Cambridge Univ. Press (2011)
7. Willett, P., Barnard, J.M., Downs, G.M.: Chemical similarity searching. J. Chem. Inf. Comput. Sci. 38(6), 983–996 (1998)
8. Witten, I.H., Moffat, A., Bell, T.C.: Managing Gigabytes: Compressing and Indexing Documents and Images. Morgan Kaufmann (1999)
9. Xiao, C., Wang, W., Lin, X., Yu, J.X.: Efficient similarity joins for near duplicate detection. In: Proc. of WWW Conference, pp. 131–140 (2008)

Finding the Cluster of Actors in Social Network Based on the Topic of Messages

Hoa Tran Quang, Hung Vo Ho Tien, Hoang Nguyen Le,
Thanh Ho Trung, and Phuc Do

University of Information Technology,
Vietnam National University – Ho Chi Minh City
{09520092,09520121}@aep.uit.edu.vn,
{nlhoang1203,hotrungthanh}@gmail.com, phucdo@uit.edu.vn

Abstract. Social Network, the most popular Internet service, has a miracle rapid increment of number of users in recent years. In this paper, we present how to use SOM network to cluster the actor based on vector. This vector is a distribution probability of topic that actor prefers. We use ART model to create the vector of interested topics. Moreover, we use Enron email corpus as a sample dataset to evaluate efficiency in SOM network. By experimenting on the dataset, we demonstrate that our proposed model can be used to extract well and meaningful cluster following the topics. We use F – measure method for this application for testing precision of SOM algorithm. As a result, from our sample tests, the F-measure cites the acceptable accuracy of the SOM method. Based on the result, application developers can use SOM to group the actors based on their interested topics.

Keywords: ART, Cluster, Social Network, SOM, Topic distribution.

1 Introduction

Social network is an environment that provides services for connecting members with the same interests, point of view or other purposes, regardless of age, gender, space and time. Beside connection, information exchange, communication, entertainment, etc. in the virtual world, social network is also an environment for people who work in online business, advertisement or politics, criminal investigation, and so forth. In this paper we will use SOM to detect groups of people or communities. Each actor in social network is represented by a vector which the components are the distribution probability of the interested topics of actor. Based on the components of this vector, we can know the interested degree of actor in each particular topic. We use the ART model [5][6] to create the vector. Then we use the SOM [3][4] to detect group of actors who have the common interested topics. The remaining of this paper is organized as follows 2) An Overview of SOM 3) Identify Interested Topics of Actors in Social Network by Using ART Model 4) Clustering Actors by using SOM 5) Experimental Result and Discussion 6) Conclusions and Future Work.

N.T. Nguyen et al. (Eds.): ACIIDS 2014, Part I, LNAI 8397, pp. 183–190, 2014.

2 An Overview of SOM

SOM or Kohonen network, an artificial neural network, which was introduced in 1982 by Tuevo Kohonen, a professor emeritus of the Academy of Finland [3]. It is known as self-organizing map (SOM), one of the relatively simple models of neuron networks.

Self-organizing is one of the attractive topics in neural network which can be trained to find out the rules, relationships and predicts the next results. Through the competitive learning process, neurons identify the equivalent input data. The main purpose of training SOM network is to identify a group of input vectors which have same type.

SOM network only includes an input vector and an output layer of neurons. Data input vector for the SOM network are training sample with size = n. The output layer includes nodes (neurons) which are arranged in grid (map). Each neuron contains weight vector with dimensions equal to dimensions of input vector as shown in Fig 1.

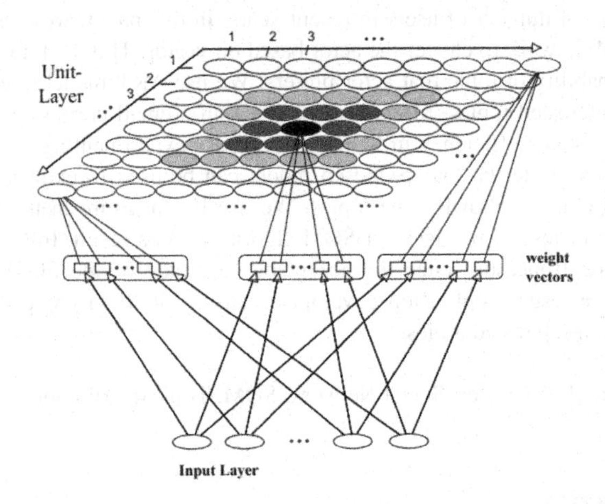

Fig. 1. SOM Structure

3 Identify Interested Topics of Actors in Social Network by Using ART Model

Topic: is a set of concepts (can be seen as process, object, event, property, etc.), they are identified by using ART model.

For example: topic 1 is "university education", author build the document by using relevant words such as:

- Bachelor's degree
- Scientific research
- Graduation thesis
- Others

Author – Recipient - Topic model (ART): is a Bayesian network that simultaneously models message content as well as the directed social network in which the messages are sent [5][6]. This model are based on the Latent Dirichlet Allocation (LDA) [2] and the Author – Topic model and add the important attribute that discovers topics based on the relationship between the senders and recipients[2]. ART model describes the interaction of each node by analyzing transferring information of each node in the network; a topic relates to author, recipient and discovers role of author and recipient in transferring information process.Hence, the identification of topics in ART model depends on the social network in which messages are sent and received. Each pair of sender and receiver has a distribution over topics and each topic has a distribution over words.

From the ART model, we base on the Enron email dataset which comprise message of actors on social network, the encoding process use this data as input in order to manipulate the output - a set of vectors of actors based on interested topics. Dimensionally reduction is an additional process to reduce dimensional space of the output vectors such as removing redundant topics. Each vector of actor based on interested topics is a vector representing the interested probability of the topics of each actor in a social network. Hence, the dimensionality of the vector is occasionally the number of interested topics.

During the analyzing process, the ART model will create three matrices: the distribution matrix of words according to topics T x V (Table 1), the distribution matrix of topics according to authors A x T (Fig. 2), and the distribution matrix of topics according to recipients R x T (Fig. 3) where A is the set of Authors, R is the set of Recipients, T is the set of Topics.

With distribution over topic 1 of sender A and recipient B is 0.3031 and a distribution over topic 1 of sender A and recipient C is 0.4102. Hence, the distribution over the topic 1 of sender A is 0.3031 + 0.4102 = 0.7133.

Table 1. The distribution of words in each topics

Topic #1		Topic #2		Topic #3	
attached	0.0704	love	0.1006	meeting	0.0594
comments	0.0573	wait	0.0385	invited	0.0221
letter	0.0322	tracy	0.0343	friday	0.0147
version	0.0274	miss	0.0257	scheduled	0.0123
draft	0.0263	great	0.0257	morning	0.0109
review	0.0251	meet	0.0193	reminder	0.0102
revised	0.0239	leave	0.0171	meet	0.0100

bill.rapp@enron.com: 39 21 31 38 12 28 2 43 20 48
danny.mccarty@enron.com: 25 16 0 20 48 3 12 40 35 9
drew.fossum@enron.com: 27 46 20 34 31 13 2 43 14 48
joe.stepenovitch@enron.com: 29 38 15 4 48 43 32 21 13 3
kevin.hyatt@enron.com: 41 11 2 4 31 17 12 19 45 38

Fig. 2. The list of topics of author

bill.rapp@enron.com: 40 38 12 19 44 28 41 24 39 43
danny.mccarty@enron.com: 32 0 49 44 4 46 33 20 2 16
drew.fossum@enron.com: 45 21 16 7 4 20 48 9 31 23
joe.stepenovitch@enron.com: 29 24 48 13 4 49 38 15 47 44
kevin.hyatt@enron.com: 27 22 17 12 20 46 10 9 49 0

Fig. 3. The list of topics of recipients

Thus, the output of the ART model is a pair of distribution of sender over a particular topic. Finally, we can easily calculate this output in order to figure out the interested topics that they most likely talk about as shown in Table 2.

Table 2. Example of vectors of actors

Actor	Topic 0	Topic 1	Topic 2
bill.rapp@enron.com	0.5006	0.4475	0.3872
danny.mccarty@enron.com	0.1298	0.5921	0.0327

4 Clustering Actors by Using SOM

To begin with using SOM for clustering, the encoding process is required to manipulate the actors and their messages in social network. After this process, we receive the set of vectors. In this set, each vector represents a specific actor and its dimensional value show the interested probability in a particular topic. These vectors will be the input vectors. They are the training data set of SOM. The training learning algorithm of SOM is as follows:

Step 1: Randomly initialize the weight of SOM output layer

Initialize $N_c(t)$ (the radius of neighboring area) and set time t = 1

Step 2: Present an input vector and normalize the input vector $\vec{v}_x$

Find the winning neuron by compute the Euclidean distance between the training vector and the neuron output. The winner is the neuron output, which has the smallest distance to the training vector.

Calculate the Euclidean distance from the input vector $\vec{v}_x$ to all weight vectors of all nodes in SOM output layer and choose the neuron with the minimum distance from input vector $\vec{v}_x$ to the weight vector (winner) as follows:

$$D_{i,j} = |\vec{v}_x - \vec{w}_{i,j}| = \sqrt{\sum_{k=0}^{n-1} \left(v_{x_k} - w_{(i,j)_k}\right)^2} \quad (1)$$

(Where: $D_{i,j}$ is the distance between the current training vector and the current neural vector at row i, column j in the SOM, $\vec{v}_x$ is the current training vector, and $\vec{w}_{i,j}$ is the current neural vector at row i, column j in SOM).

Step 3: Update the neuron's weight in the neighboring area of winning neuron.

$$w'_{(i,j)_k} = w_{(i,j)_k} + \alpha(t)h(r,t)\left(v_{x_k} - w_{(i,j)_k}\right) \tag{2}$$

(Where $w'_{(i,j)_k}$: The new (post-update) value of kth weight of the neuron at row i, column j, $w_{(i,j)_k}$: The current (pre-update) value of kth weight of the neuron at row i, column j; $\alpha(t)$: The learning rate at the current number of iteration;h(r,t): The result of topological neighborhood function with t is the current number of iterations, r *is the distance between the current neuron and the winning neuron).*

Step 4: Update t =t + 1, present new input vector and go back to step 2 until satisfying the convergence criteria or exceeding the maximum number of iterations.

Finally, we use the SOM output layer to find clusters of actors who has the same interested topics. Like the learning process, the cluster with the minimum of Euclidean distance is the cluster to which the vector belongs. The SOM output layer is shown in Fig 4 with many clusters.

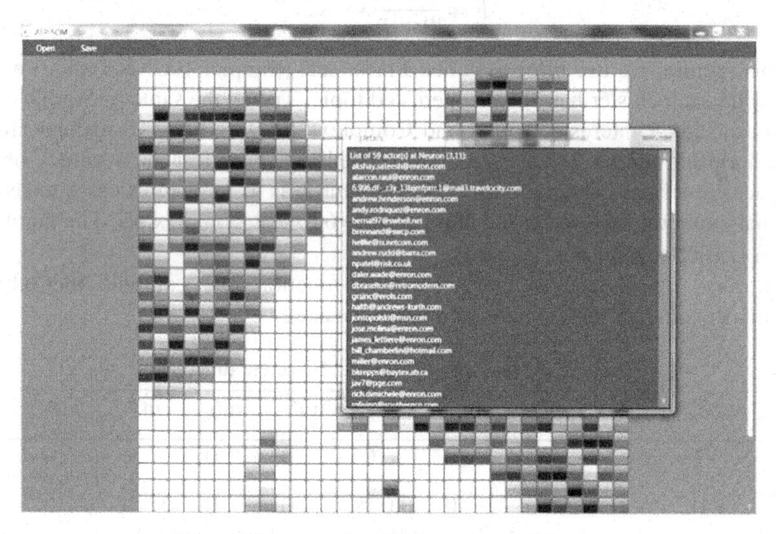

Fig. 4. The example of Kohonen output layer

5 Experimental Result and Discussion

We use Precision, Recall and F-Measure to compute the accuracy of the clustering result.

Assume that in set of actors we divide these actors into m clusters of actors by man-made. On the other hand, by using SOM, the set is split to k clusters.

Then at cluster $m_i$and cluster k_i we have:

- A: The set of actors belong to both clusters ($m_i \cap k_i$).
- B: The set of actors only belong to m_i($m_i \setminus k_i$)
- C: The set of actors only belong to k_i ($k_i \setminus m_i$)

Or $|m_i| = |A| + |B| = a + b$ and $|k_i| = |A| + |C| = a + c$

Thus, we have thefollowing measures:

Precision measure represents the ratio of the accuracy of a SOM cluster. If the ratio is 1, it means that all the vectors in cluster k_i are in cluster m_i, or $k_i \subset m_i$. We have:

$$P = \frac{a}{a + c} \tag{3}$$

Recall measure describes the ratio of the accuracy of a true positive actors clustered by SOM with the actual positive actors. If it is 1, $k_i \subset m_i$ or cluster m_i are in clusterk_i.We have:

$$R = \frac{a}{a + b} \tag{4}$$

According to Brew & Schulte im Walde (2002)[1], F-Measure, which is the combination of Precision and Recall, is used to compute the accuracy of the system. For the clustering system, this is the equation:

$$F = \frac{2PR}{P + R} \tag{5}$$

In this section, we use three assumed Enron sets (Simple: 100 vectors, 10 manual clusters and each cluster has only 1 interested topic; Triple: the same as Simple except each cluster has 3 interested topics;and Complex: 1000 vectors, 10 manual clusters and each cluster can have 1 or several interested topics) to test the accuracy of each topological functions. The number of output neuron is 4x4 for most cases except the additional situation we enlarge the number to 6x6 to test Complex set. The number of iteration is 1000 and$\varepsilon = 10^{-7}$

Each time we change the neighborhood function, we compute the table of Precision, Recall, then manipulate the total F-measure.

Table 3. F-Measure of Triple set (Discrete function)

	m0	m1	m2	m3	m4	m5	m6	m7	m8	m9
K0	1	0	0	0	0	0	0	0	0	0
K1	0	0	0	0	0.095	0	0	0.952	0	0
K2	0	0	0	0	0.947	0	0	0	0	0
K3	0	0	0	0	0	1	0	0	0	0
K4	0	0	0	0	0	0	0	0	0	1
K5	0	0.621	0.690	0	0	0	0	0	0	0
K6	0	0.182	0	0	0	0	0	0	0	0
K7	0	0	0	1	0	0	0	0	0	0
K8	0	0	0	0	0	0	0	0	1	0
K9	0	0	0	0	0	0	1	0	0	0
MAX	1	0.621	0.690	1	0.947	1	1	0.952	1	1

In this case, the total of maximal value of F-measure for vector clustering is:

$$1+0.621+0.690+1+0.947+1+1+0.952+1+1=9.21$$

Finally, the chart is sketched to provide a visual view of the accuracy of each situation:

From the result of our sample tests in Table 3 and Fig 5, the discrete function is appropriate h(r,t) for this situation with the total F-Measure is always greater than 8 (versus the number of manual clusters is 10). It means that the cluster of actor by using SOM is reliable.

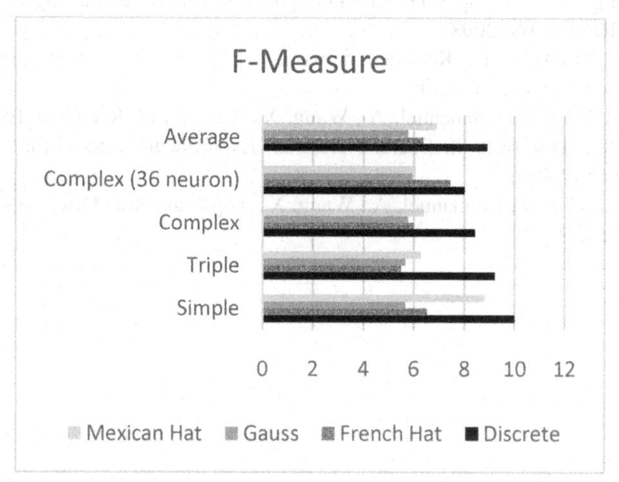

Fig. 5. F-Measure when Using a Specific Topological Function

6 Conclusions and Future Work

With the acceptable total F-measure value, the clustering actors' method by SOM has a potential to deploy in the actual social network system. To be a real system, there are several works need to be improved such as:

- Improving parallel processing and language support must be desirable. For improving parallel processing, the system should be able to process in parallel in training procedure. In addition, the improved algorithm should be able to solve the conflict of updating value in parallel. For improving language support, encoding process should be support several languages such as French, Chinese and so on. The solution for this issue is that we enlarge the encoding word dictionary for these languages.
- Developing a module to name the topics must be considered. The solution to develop this module is finding the distance of keywords with a specific measurement such as finding the least common hyponym of a couple of word in WordNet Dictionary.

Acknowledgement. This research is funded by Vietnam National University-HoChiMinh City (VNU-HCM) under grant number B2013-26-02.

References

1. Brew, C.: Schulte im Walde, Spectral Clustering for German Verbs. In: Proc. of the Conf. in Natural Language Processing, PA, USA, pp. 117–124 (2002)
2. Blei David, M.: Introduction to Probabilistic Topic Models. Princeton University (2007)
3. Phuc, D., Hung, M.X.: Using SOM based Graph Clustering for Extracting Main Ideas from Documents, RIVF 2008 (2008)
4. Kohonen, T., Honkela, T.: Kohonen network, `http://www.scholarpedia.org/article/Kohonen_network`
5. McCallum, A., Corrada-Emmanuel, A., Wang, X.: The Author-Recipient-Topic Model for Topic and Role Discovery in Social Networks: Experiments with Enron and Academic Email, Amherst (2005)
6. McCallum, A., Corrada-Emmanuel, A., Wang, X.: Topic and Role Discovery in Social networks, Amherst (2008)

Geodint: Towards Semantic Web-Based Geographic Data Integration

Tamás Matuszka and Attila Kiss

Eötvös Loránd University, Budapest, Hungary
{tomintt,kiss}@inf.elte.hu

Abstract. The main objective of data integration is to unify data from different sources and to provide a unified view to the users. The integration of heterogeneous data has some benefits both for companies and for research. However, finding the common schema and filtering the same element becomes difficult due to the heterogeneity. In this paper, a system is presented that is able to integrate geographic data from different sources using Semantic Web technologies. The problems that appear during the integration are also handled by the system. An ontology has been developed that stores the common attributes that are given after schema matching. To filter the inconsistent and duplicate elements, clustering and string similarity metrics have been used. The data given after integrating can be used among others for touristic purposes, for example it could provide data to an augmented reality browser.

Keywords: data integration, Semantic Web, ontology, entity resolution, Augmented Reality.

1 Introduction

The purpose of data integration is to combine and merge data from different data sources and to provide a unified query interface for the users. After the integration, the users do not have to worry about that the data came from different sources. The total amount of information could be regarded as a single dataset. There has long been a need to combine the heterogeneous data and to query it on a single interface. Data warehouses can be mentioned among the first solutions, which extract, transform and load the heterogeneous data to a unified schema. The problem with this approach is that the data warehouse is not always up-to-date. Later, the data integration has shifted towards the use of the mediated schema, which can obtain the information directly from the original database. This solution is enabled by mapping between the original schemas and the mediated schema. Due to the mediated schema, the query that is sent through a unified interface will transform into a form which complies with the original schemas. There are two approaches to this mediated schema solution. The first is the "Global As View (GAV) approach", which maps the entities of mediated schema to the original database schemas. The second is the "Local As

N.T. Nguyen et al. (Eds.): ACIIDS 2014, Part I, LNAI 8397, pp. 191–200, 2014.

View (LAV) approach", which maps the entities of the original data sources to the mediated schema [11]. Nowadays the preferred method is the ontology-based data integration that defines the schema and helps avoid the semantic problems [14]. Such semantic problem can be when the coordinates of a POI (Point of Interest) is stored in degree (e.g. 47.162494°) in the first database, and it is stored in degree-minute-second (e.g. 47° 9' 44.9" N) in the second database.

During the data integration a number of difficulties can be expected. Among these the most important is schema matching that gets two schemas as input and generates semantically correct schema mappings between them. Currently, the schema matching is typically done manually, usually with a graphical user interface. This method can be quite tedious, time-consuming and prone to errors [17]. Fortunately, nowadays there are some semi-automatic tools (e.g. COMA++ [2], Microsoft Biztalk) which can facilitate this process. Another similarly important problem is the entity resolution (also known as deduplication). This method is responsible for the identification and merging of the same real-word entities. A number of approaches have been developed for this problem, for example Swoosh approach [4], Karma [10].

The integration of geographic databases can play a particularly important role for location-based augmented reality browsers [13]. Augmented reality combines the real and virtual worlds in real-time. The location-based version takes advantage of the user's current geographical location and location-based information can be superimposed into the real life view. A typical example is when the user looks around with the mobile phone and could see the icons which represent restaurants located near in the real-life view. The current augmented reality browsers (e.g. Junaio[1], Layar[2]) use only one data source nowadays [12].

In this paper a system is presented that could handle the general disadvantages of data integration in case of geographic data. This system could be a basis of a location-based augmented reality browser. The main advantage of our system compared to previous existing that it can extract more information with the integration of data from different sources than individually. During the implementation different datasets were used (Facebook Places[3], Foursquare[4], Google Places[5], DBpedia [7], LinkedGeoData [1]) for data provision. The advantages of Semantic Web were used for the integration. To avoid semantic conflicts, an ontology storing the common schema has been developed, which can be easily extended with new data sources. Clustering and string similarity metrics were used for filtering duplications.

The rest of the paper is organized as follows. After the introductory Section 1, we outline the preliminary definitions in Section 2. Then, the details of our system is described in Section 3. Section 4 demonstrates the obtained results and the evaluation of the system. In Section 5 we present some applications that are

[1] `http://www.junaio.com/`
[2] `https://www.layar.com`
[3] `https://developers.facebook.com/docs/reference/api/search/`
[4] `https://foursquare.com`
[5] `https://developers.google.com/places/`

similar to our system. Finally, the conclusion and the future plans are described in Section 6.

2 Preliminaries

In this section, the concepts that are necessary for understanding are defined. We provide insight into the basic concepts of data integration, schema matching, entity resolution and Semantic Web.

A triple $\langle \mathcal{G}, \mathcal{S}, \mathcal{M} \rangle$ is called as *data integration system*, where $\mathcal{G}$ is the global schema, $\mathcal{S}$ is the set of source schemas and $\mathcal{M}$ is a mapping among the global schema and the heterogeneous source schemas. When a user sends a query, the request is executed over $\mathcal{G}$ by the system, and the $\mathcal{M}$ mapping is responsible for the mapping from this query to the schemas in $\mathcal{S}$.

Schema matching can be used for creating the $\mathcal{G}$ schema based on $\mathcal{S}$. The core of schema matching is the *match operator*. Before we could define the *match*, it is necessary to introduce the *mapping* between two schemas, S_1 and S_2. The *mapping* is a set of mapping elements. This method maps certain elements of S_1 to certain elements of S_2. The *match operator* is $f\colon S \times S \to M$ function, it gets two schemas as input and returns with the mapping between schemas. The given result is called *match result* [17].

During the data integration it is required to detect duplicates. Entity resolution gives a solution for this problem. It also uses a *match* function, which in this case is different from the one used for schema matching. Let E be the set of entities, the *match* is a $f\colon E \times E \to \mathit{Boolean}$ function, which will decide whether two entities are the same real-world entities or not (denoted $e_1 \approx e_2$, if $\mathit{match}(e_1, e_2) = \mathit{true}$, where $e_1, e_2 \in E$). A *merge* function $\mu\colon E \times E \to E$ merges two matching entities into one entity. The merge closure (denoted $\overline{I}$) can be obtained by executing all matching and merging on an instance I. If the *match* determines that e_1 and e_2 are same entity, then it just need to keep in the merge, which contains more useful information. For example, let e_1 = J. Smith and e_2 = John Smith. Then the latter has more information, denoted $e_1 \preceq e_2$. This theory can be extended to instances as well. Given this, we can now define the *entity resolution*. Let be I an instance, and $\overline{I}$ the merge closure of I. An *entity resolution* of I is the minimal set of records I', such that $I' \subseteq \overline{I}$ and $\overline{I} \preceq I'$ [4].

A possible way to manage the data available on the Internet is to use the Semantic Web [6]. The Semantic Web aims for creating a "web of data": a large distributed knowledge base, which contains the information of the World Wide Web in a format which is directly interpretable by computers. *Ontology* is recognized as one of the key technologies of the Semantic Web. An *ontology* is a structure $\mathcal{O} := (C, \leq_C, P, \sigma)$, where C and P are two disjoint sets. The elements of C and P are called *classes* and *properties*, respectively. A partial order $\leq_C$ on C is called class hierarchy and a function $\sigma\colon P \to C \times C$ is a signature of a property [18]. The Semantic Web stores the knowledge base as *RDF triples.* Let I, B, and L (IRIs, Blank Nodes, Literals) be pairwise disjoint sets. An *RDF triple* is a $(v_1, v_2, v_3 \in (I \cup B) \times I \times (I \cup B \cup L))$, where v_1 is the subject, v_2 is the predicate and v_3 is the object [15].

In this paper we present a *geographic data integration* system $\langle \mathcal{G}, \mathcal{S}, \mathcal{M} \rangle$, which can determine the $\mathcal{G}$ global schema from $\mathcal{S}$ in semi-automatic way with schema matching. The given common schema and the semantic relations are stored in an $\mathcal{O}$ ontology. The classes of ontology store the required types and the properties of ontology describe the relations among them. During the data integration, the system performs the *entity resolution* as well. In addition, the resulted data are transformed into RDF format by the system.

3 Geodint (Geographic Data Integration System)

In this section, we overview the details of our system and describe the used data sources and the creation of the common schema. After that, we show the ontology that is given after schema matching and the entity resolution method. Finally, we present the core algorithm of our system.

3.1 Applied Data Sources

Five data sources are used by the system. Three of them provide data through web services and two store the data in semantic databases.

The first is the Foursquare which is a social network with 40 million users. With this social network, the current position of the users could be shared. The most important element of Foursquare is the venue that are physically existing locations, where users can check-in. A venue has various attributes, but much of it is irrelevant in our case.

The second data source is the Facebook Places. Facebook, similarly to Foursquare, allows their users to share the current position. Due to this, numerous data can be obtained about touristic sights, restaurants, museums, etc.

Google also allows data provisioning about places which belong to different categories. For this purpose, the same database is used, what the Google Maps and Google+ Local use. The frequently updated database contains about 95 million POI-s.

There are several publicly available datasets in semantic form. These data can be queried with the SPARQL query language [16]. SPARQL formulates the queries as graph patterns, thus the query results can be calculated by matching the pattern against the data graph. The most well-known dataset is the DBpedia [7], which contains the knowledge of Wikipedia in semantic form. DBpedia contains the latitude and longitude coordinates of numerous places, therefore it can be used as geographical data source.

The last data source is the LinkedGeoData [1]. The goal of LinkedGeoData is to add a spatial dimension to Semantic Web. The spatial data is collected by OpenStreetMap[6] project and it is available as RDF format. The large spatial knowledge base contains about 20 billion triples and it is interlinked with DBpedia.

3.2 Schema Matching

To create a global schema from the schemas of different data sources, out of the existing tools, COMA++ have been used. COMA++ provides schema matching using various matching algorithms. For this, it provides a graphical user interface which allows many interaction to the users. Due to the generic data representation, the tool supports schemas (e.g. W3C XML Schema[7]) and ontologies (e.g. OWL[8]) [2].

Firstly, the schemas of Foursquare, Facebook and Google data was downloaded and converted to XSD schema. In case of semantic datasets, the corresponding ontologies were used for the schema matching. The useless attributes for an augmented reality browser were filtered in advance. With COMA++ we have determined the common schemas pairwise in semi-automatic way and then we got the global schema. The COMA++ recommends a matching between two schema and the users could confirm or discard the suggestions. An example schema matching can be viewed on Figure 1.

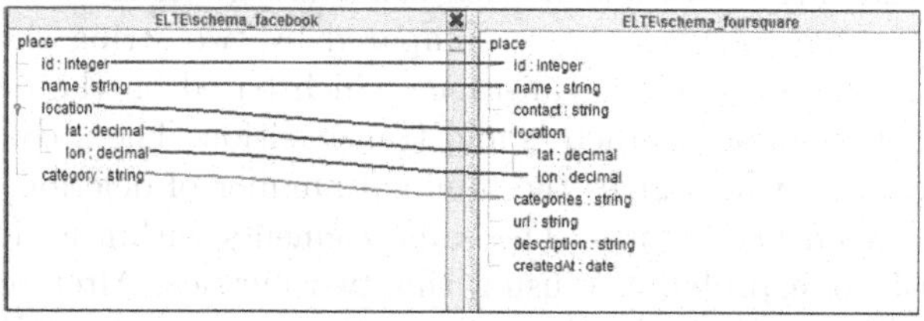

Fig. 1. Matching two schema with COMA++

3.3 Ontology to the Global Schema

After the semi-automatically executed global schema determination the given result was stored in an ontology. This ontology builds on the LinkedGeoData's ontology. We mapped the schemas of data sources to the classes of this ontology. This method fits one of the principles of Semantic Web, which has the purpose of reusing existing ontologies. We wanted to provide filtering by categories, thus these categories were also selected from this ontology. For this purpose, the ontology was extended with a *Category* class since the LinkedGeoData stores

[6] http://www.openstreetmap.org
[7] http://www.w3.org/XML/Schema
[8] http://www.w3.org/2004/OWL/

the different place types in the *amenity* class. The classes which are used as categories are also derived from this *Category* class. In addition, we had to create some classes and properties for the data sources. With the help of these classes and properties, the data-specific information (e.g. category matching) can be described. It was also required to create a POI class (it will be the type of the emergent result). The ontology was extended with some classes which could describe the attributes which correspond to the POI-s. These attributes were described with the properties of DBpedia ontology and the W3C Geo Vocabulary.

3.4 Entity Resolution and Result Generation

When we have the global schema, the actual data integration can be started. As we mentioned before, the filtering of duplicated elements plays important role. Our approach is based on clustering and string similarity metrics.

Firstly, a density clustering of the geographical data by coordinates was executed. For this purpose we used the DBSCAN [9] algorithm. According to density-based methods, the density within a cluster is much more than the density among the clusters. The DBSCAN algorithm uses two parameters. The first one is the radius (*eps*) and the second one is the threshold of the number of elements (*minpts*). The details of the algorithm can be viewed in [9]. The values of parameters was determined by empirical way. We set the value of *minpts* to 2 and the value of *eps* to 10 meter because of the inaccuracy of GPS.

The places that are located near each other are given after the clustering. After that, the names of places were compared by two string similarity metrics. The first is the Jaro-Winkler distance which based on the determination of the number of common characters and transposition. The second one is the Levenshtein-distance, which gives the minimal number of deletion, insertion or replacement between two strings. The same elements within a cluster can be determined with high probability using this two metrics. After removing and merging the duplicated elements, the given result is converted to RDF document by the system. This result can be queried easily with using SPARQL queries.

3.5 The Core Algorithm

Algorithm 1 describes the core algorithm of the system. The inputs are the coordinates where the user would like to search, the category and the radius. The system firstly creates the data source specific queries and collects the result and then converts it to POI type. After that, the clustering begins. The distance is determined for each POI, and if this distance is less than the radius, the two POI will be adjacent. Thereafter, the algorithm determines the elements which belong to a common cluster. Finally, it executes the string similarity search on the POI-s within a cluster. If the algorithm found two equivalent elements, then it merges them. Finally, it transforms the given, deduplicated result to RDF format.

Algorithm 1: The core algorithm of the system

```
Input: lat, lon, category, radius
Output: integrated RDF document

download the data from different sources and parse to POI
foreach POI do
 compute the distance among the other POI-s
 if distance ≤ radius then
  set the two POI-s to adjacent
 end
 determine whether it is core, boundary or outliner
 foreach adjacent POI-s do
  check whether the two POI-s are same according the Jaro-Winkler
  and Levenshtein distance or not
  if the two POI-s are same then
   merge them and add the given POI to the result
  end
 end
end
```

4 Results and Evaluation

For the evaluation of our system, various places have been selected in Budapest and Bangkok. The fundamental assumption was that the given result after data integration will be much wider than separately. Figure 2 shows the number of hits separately and collectively. On the x axis can be found the category of selected places. On this figure shows up that the number of the given unique places is much more after the integration than separately. It also shows that the semantic datasets do not contain any data in certain cases. However, in another case it exceeds the data coming from another type.

Figure 3 shows the number of aggregated hits, the number of unique hits and the number of the hits after the entity resolution. This figure demonstrates that the given result after the entity resolution approximates well the concrete results. It was found that the number of results can be sometimes more and sometimes less than the number of real unique elements. In case of the category *Cafe* in Budapest, a false positive hit was found. The "Cafe Illy" and "Cafe Vian" are the same according to our system, it cannot be filtered by the similarity search. There were false negative hits in few times, for example "Cafe Monaco" and "Cafe Monaco & Coctail Bar". These two places are different according to the system, while actually this is only one place. The number of false hits can be visited on Table 1.

We can say that our assumption was right based on the given result. The system, which is able to filter the duplicated elements have been developed. This system can be a basis of an augmented reality browser which can provide more data than the nowadays existing ones.

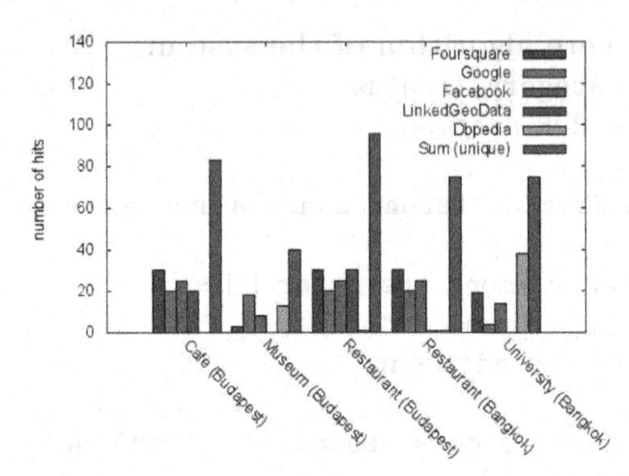

Fig. 2. Number of hits separately and collectively

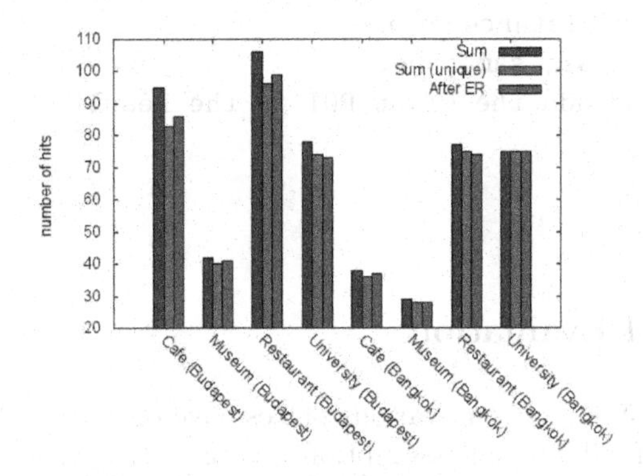

Fig. 3. Number of aggregated hits

Table 1. Number of false positive and false negative hits

Category	Sum	Sum (unique)	After ER	False positive	False negative
Cafe (Budapest)	95	83	86	1	4
Museum (Budapest)	42	40	41	0	1
Restaurant (Budapest)	106	96	99	0	3
University (Budapest)	78	74	73	2	1
Cafe (Bangkok)	38	36	37	0	1
Museum (Bangkok)	29	28	28	0	0
Restaurant (Bangkok)	77	75	74	1	0
University (Bangkok)	75	75	75	0	0

5 Related Work

A mobile application called csxPOI (collaborative, semantic, and context-aware points-of-interest) was presented in [8]. This application allows for their users to collaboratively create, share and modify Point of Interest. These POI-s represent real physical places. The properties of such places are stored in a collaboratively created ontology similarly to our solution. However, whereas our approach uses multiple data sources, their solution is based on POI-s created by the users.

In [3], the author shows a generic tool that provides automatic retrieval of the updates in geographic databases. The geographic data integration can be made easier with this method. This metod is based on data matching tools, which are similar to our entity resolution method.

Bennett presented a geoprocessing framework in [5]. This framework includes the basic principles of geographic information systems, modelbase management and computer simulation. All of these modules are integrated into an environment which could support the development of geopraphical models. The geographical data models include the spatial relations as well.

6 Conclusion

In this paper we presented a geographical data integration system which can be a base of an augmented reality browser. The global schema was created semi-automatically from different schemas using COMA++. The resulting schema was stored in an ontology. The system can be extended with other data sources easily. The deduplication is carried out in two steps. In the first step, a density clustering, namely DBSCAN is executed based on spatial dimensions. After that, string similarity search is executed among the elements within a cluster and the same elements are merged. The given results prove that we can get much wider information after integration than separately.

In the future, we will improve our system so that the ontology will be extended automatically after the schema matching. We would also like to examine whether storing the results in a triple store could speed up the system. In addition, we will use linguistic approaches (e.g. synonyms) apart from string similarity search.

Acknowledgments. This work was partially supported by the European Union and the European Social Fund through project FuturICT.hu (grant no.: TAMOP-4.2.2.C-11/1/KONV-2012-0013).

References

1. Auer, S., Lehmann, J., Hellmann, S.: Linkedgeodata: Adding a spatial dimension to the web of data. In: Bernstein, A., Karger, D.R., Heath, T., Feigenbaum, L., Maynard, D., Motta, E., Thirunarayan, K. (eds.) ISWC 2009. LNCS, vol. 5823, pp. 731–746. Springer, Heidelberg (2009)
2. Aumueller, D., Do, H.H., Massmann, S., Rahm, E.: Schema and ontology matching with COMA++. In: Proceedings of the 2005 ACM SIGMOD International Conference on Management of Data, pp. 906–908. ACM (2005)
3. Badard, T.: On the automatic retrieval of updates in geographic databases based on geographic data matching tools. Bulletin du Comit Franais de Cartographie (162), 34–40 (1999)
4. Benjelloun, O., Garcia-Molina, H., Menestrina, D., Su, Q., Whang, S.E., Widom, J.: Swoosh: A generic approach to entity resolution. The VLDB Journal The International Journal on Very Large Data Bases 18(1), 255–276 (2009)

5. Bennett, D.A.: A framework for the integration of geographical information systems and modelbase management. International Journal of Geographical Information Science 11(4), 337–357 (1997)
6. Berners-Lee, T., Hendler, J., Lassila, O.: The semantic web. Scientific American 284(5), 28–37 (2001)
7. Bizer, C., Lehmann, J., Kobilarov, G., Auer, S., Becker, C., Cyganiak, R., Hellmann, S.: DBpedia-A crystallization point for the Web of Data. Web Semantics: Science, Services and Agents on the World Wide Web 7(3), 154–165 (2009)
8. Braun, M., Scherp, A., Staab, S.: Collaborative creation of semantic points of interest as linked data on the mobile phone (2007)
9. Ester, M., Kriegel, H.P., Sander, J., Xu, X.: A density-based algorithm for discovering clusters in large spatial databases with noise. In: KDD, vol. 96, pp. 226–231 (1996)
10. Knoblock, C.A., et al.: Semi-automatically mapping structured sources into the semantic web. In: Simperl, E., Cimiano, P., Polleres, A., Corcho, O., Presutti, V. (eds.) ESWC 2012. LNCS, vol. 7295, pp. 375–390. Springer, Heidelberg (2012)
11. Lenzerini, M.: Data integration: A theoretical perspective. In: Proceedings of the Twenty-First ACM SIGMOD-SIGACT-SIGART Symposium on Principles of Database Systems, pp. 233–246. ACM (2002)
12. Matuszka, T.: Augmented Reality Supported by Semantic Web Technologies. In: Cimiano, P., Corcho, O., Presutti, V., Hollink, L., Rudolph, S. (eds.) ESWC 2013. LNCS, vol. 7882, pp. 682–686. Springer, Heidelberg (2013)
13. Matuszka, T., Gombos, G., Kiss, A.: A New Approach for Indoor Navigation Using Semantic Webtechnologies and Augmented Reality. In: Shumaker, R. (ed.) VAMR/HCII 2013, Part I. LNCS, vol. 8021, pp. 202–210. Springer, Heidelberg (2013)
14. Noy, N.F.: Semantic integration: A survey of ontology-based approaches. ACM Sigmod Record 33(4), 65–70 (2004)
15. Pérez, J., Arenas, M., Gutierrez, C.: Semantics and Complexity of SPARQL. In: Cruz, I., Decker, S., Allemang, D., Preist, C., Schwabe, D., Mika, P., Uschold, M., Aroyo, L.M. (eds.) ISWC 2006. LNCS, vol. 4273, pp. 30–43. Springer, Heidelberg (2006)
16. Prud'hommeaux, E., Seaborne, A.: SPARQL Query Language for RDF, `http://www.w3.org/TR/rdf-sparql-query/`
17. Rahm, E., Bernstein, P.A.: A survey of approaches to automatic schema matching. The VLDB Journal 10(4), 334–350 (2001)
18. Volz, R., Kleb, J., Mueller, W.: Towards Ontology-based Disambiguation of Geographical Identifiers. In: I3 (2007)

SPARQL – Compliant Semantic Search Engine with an Intuitive User Interface

Adam Styperek, Michal Ciesielczyk, and Andrzej Szwabe

Poznan University of Technology,
Maria Sklodowska-Curie Sq. 5, 60-965 Poznan, Poland
{adam.styperek,michal.ciesielczyk,andrzej.szwabe}@put.poznan.pl

Abstract. It is crucial to enable users of Linked Data to explore RDF-compliant knowledge bases in an intuitive and effective way. It is not reasonable to assume that a regular user posses any knowledge about the SPARQL nor about the ontology of the given knowledge base. This paper presents the Semantic Focused Crawler (SFC) system which features a graph-based querying interface that address this issue. As a result of the use of auto-complete recommendations within the SFC query builder interface, the user benefits from using the ontology irrespectively from the degree of the knowledge about semantic technologies he/she possesses. When compared to several widely-referenced alternative solutions in experiments performed with the use of 2011 QALD workshop questions, the presented system appears as achieving high query results accuracy and low complexity of the query formulation process.

Keywords: Semantic Web, Search, RDF, SPARQL, DBpedia.

1 Introduction

The idea of Semantic Web [1] has become popular in the last few years - partly thanks to the introduction of publicly available domain-generic knowledge bases like DBpedia [2] and YAGO [3]. These two Linked Data (LD) bases [4] contain data acquired from Wikipedia [5] and converted into the semantic form: an RDF graph [6] in the case of DBpedia and an extended RDF in the case of YAGO.

Obviously enough, users of such large knowledge bases should be provided with tools that enable them to query data in an effective and intuitive manner. However, as shown in this paper, the practical value of existing solutions to the problem is still rather limited. Such an observation motivates further efforts on developing user interfaces for semantic data stores that are intuitive but, at the same time, do not compromise the expressiveness of semantic queries.

1.1 Problem Statement

It seems reasonable not to assume that a regular user posses any knowledge about the SPARQL nor about the ontology of the given knowledge base. SPARQL, when used directly (i.e. in command line manner), is quite difficult for regular users [14,8]. Therefore it is not be surprising that alternative means for the realization of a semantic search

N.T. Nguyen et al. (Eds.): ACIIDS 2014, Part I, LNAI 8397, pp. 201–210, 2014.

engine user interface have been proposed [7,8,14]. The leading approaches to the problem are represented by the systems [7,8] that enable the users to create SPARQL-like queries without requiring any knowledge of the SPARQL syntax [11].

1.2 Concept of Semantic Focused Crawler

The search system presented in this paper is a component of the system referred to as Semantic Focused Crawler (SFC). SFC features a Web Crawler module that gathers and indexes selected data from the Internet. Each Web document collected by SFC is converted into an RDF graph consistent with the predefined ontology (OWL) [9], and subsequently stored using a software based on the JENA framework [10]. The ontology used in the SFC models elements of web pages such as a post, post's author, post's creation date, post's text, post's name, etc. It contains classes like *#Author, #Forum, #Post*, and properties such as *#hasAuthor, #hasTitle, #hasCreationDate*, etc. For example, when indexing an Internet forum, the Web Crawler may produce triples such as:

<:post_123><:hasAuthor><:user_123>.
<:post_123><:hasCreationDate>"2013-06-03T03:14:36+0100"^^xsd:dateTime .
<:post_123><:hasTitle>"Hello!".

The SFC is called semantic, because it is based on RDF [6], RDFS [12], OWL [9], SPARQL [11] standards which form the base of the Semantic Web [1]. The presented system is also able to load data from other RDF databases and third-party ontologies like DBPedia [2,15].

Additionally, the user is able to enhance the graph queries by using keyword-based search (applied to RDF literals). This function, available thanks to the use of the Apache Lucene [13], allows a user to perform free text searches within their semantic queries.

As a result of the use of auto-complete recommendations within the SFC query builder interface, the user benefits from using the ontology irrespectively from the degree of the knowledge about semantic technologies he/she possesses.

2 Related Work

Interfaces based on the Natural Language Processing (NLP) try to map every provided term to an ontology object, and thus they tend to be more user-friendly than graph-based query interfaces. On the other hand, due to natural language indeterminism, this kind of mapping is an error-prone process and it may cause irrelevant results [8]. As a consequence, solutions based on NLP are reported to provide less accurate results [8,14]. In this paper, so as not to obscure the introduced reasoning, we do not report our evaluation of NLP-based search engines.

Contrarily, each time a user issues a graph-based query, he/she has to specify a set of triple/fact templates that may be seen as simple sentences containing a subject, predicate and object. Such an approach provides users with the opportunity to query the system more precisely than while using the natural language [8,14]

GoRelations (GoR) [8] is a graph-based DBpedia [2] query interface allowing the user to type triple query patterns directly into a text-box field. Such queries are simpler to formulate than queries in the SPARQL and at the same time more precise than natural language queries. However, the GoRelations interface cannot be considered user-friendly for several reasons. First of all, the syntax is still quite strict an unintuitive. The user has to know to write each triple in a new line, separate each element of the triple by a comma, and specify the type of every new variable. Secondly, the user cannot create a query containing a variable predicate or an object of an unspecified type. Furthermore, GoRelations does not provide any recommendations about available types and relations, thus the user is required to be familiar with the underlying ontology. Although, the system maps the types provided by the user to the DBpedia ontology terms using statistic and semantic similarity components, it may result in greater inaccuracy of the results.

YAGO [3] is a knowledge base that integrates information from Wikipedia [5] and WordNet [16]. NAGA [7] is a search engine which operates on the data from the YAGO. It provides users with a web interface [7] in which they can create SPARQL-like queries. The user of NAGA is able to create a query which contains no more than five template facts. In this interface the user creates a query by completing 8 fields (fields are optional): Subject, Property, Object, two time fields, two location fields and Keywords field, where the Subject, the Property and the Object form a triple – a fundamental element of every created query. Similarly to GoRelations, NAGA does not automatically suggest subject and object values in the query builder. The user can choose a property from a provided list but it does not have to be relevant to the subject (domain) nor the object (range). For example, the user of NAGA or GoRelations may create a query like: *Zinedine Zidane hasCapital ?x*, which obviously does not make sense.

Table 1 presents the key features of the compared systems, such as their level of compliance with SPARQL. Only NAGA and SFC provides the user with results pagination function which may effectively support the user in the search results exploration. Only SFC supports SPARQL "*optional*" function which may be used as a logical disjunction. It should be also noted that none of the compared systems support the aggregation or ordering functions (probably due to scalability-related issues).

Table 1. Semantic search systems comparison

System	Domain of use	Keyword-based search	Ontology types and relation recommendations	Results pagination	Optional function	Filter function
NAGA	YAGO dataset	Yes	No	Yes	No	No
SFC	**Any domain**	**Yes**	**Yes**	**Yes**	**Yes**	**Yes**
GoR	DBPedia dataset	No	No	No	No	Yes

In this paper, we assumed that by reducing the amount of information the user has to type in, one makes the interface more intuitive. Therefore, in order to meet this requirement, the introduced system (as the only one) provides automatic recommendations of ontology entities improving its user-friendliness. Moreover, it is the only one of the compared graph-based semantic search systems that provides the user with use the functionality of full-text search on indexed literals and resource labels. In contrast, NAGA allows the user to filter the semantic search results only by using keywords, while GoRelations does not even support any function of such type [8].

3 Functionality of Semantic Search Engine User Interface

Figure 1 shows how SFC semantic search user interface looks like from the user's perspective.

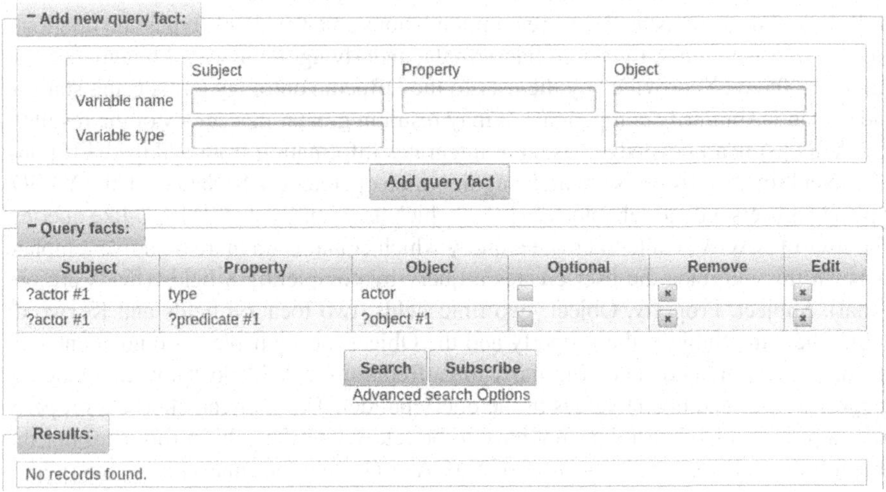

Fig. 1. Semantic search user interface

3.1 Example of Query Construction Process

The semantic search interface has five text fields which are used to construct a set of query triples. The more complex a query is, the more query triples it contains. Let X be the set of variables, matching any object or property in a query, $|X| = n$ and $x \in X$. For example, if the user wants to find names of the all models which are exactly as tall as *Claudia Schiffer*, the query should include 3 query triples:

Claudia Schiffer has in height x_1
x_2 *is a Model*
x_2 *has in height* x_1

Because of the fact that the user of Semantic Focused Crawler may define the variables' types he/she does not have to create all three query triples independently. Figure 2 presents the described query in the SFC system. There are still three query triples in the query but the user does not have to add all of them separately. It contains two variables, x_1 and x_2 (with defined type *Model*). For convenience, in SFC, all of the variables, by default, have names starting with a question mark (?) and ending with sequence numbers which are automatically complemented by the system and are used to distinguish variables of the same type.

The user of SFC may also define a variable property, which may be useful if the user wants to find all relations between two or more objects. For instance, to ask about a relation between Ryan Giggs and David Beckham the user may query DBpedia using a query presented in the Figure 3. Nevertheless, due to the fact that there is no direct relation between the footballers in the knowledge base, the user will receive an empty set as a result.

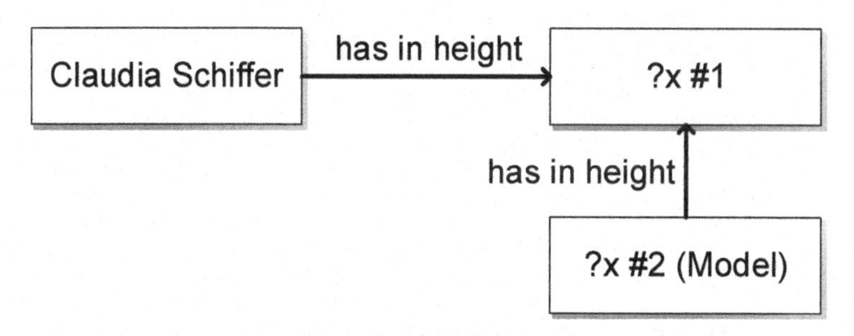

Fig. 2. Query example 1

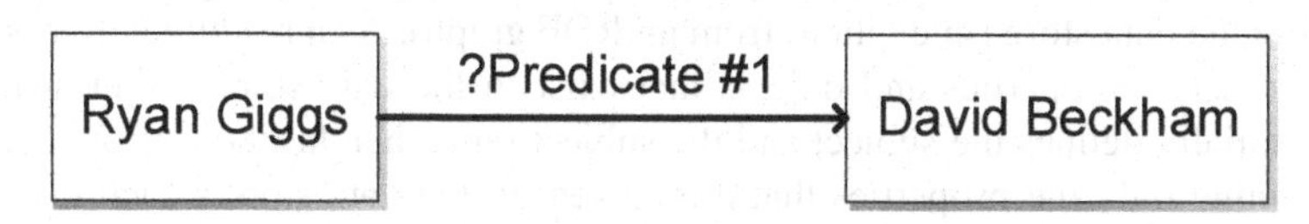

Fig. 3. Query example 2

However, if the user knows that they both played in Manchester United, then he/she may create another query that consists of two triples having a common variable (x_2):
Ryan Giggs x_1 x_2
David Beckham x_3 x_2

In SFC, all of the fields are optional, however at least one of them should be filled. For example, if the user types only the subject, then the system automatically creates variables for the property and the object. If the user wants to create a query similar to the example shown in the Figure 3, then he/she does not have to input the predicate variable. If he/she leaves the property field empty the system will automatically add a new variable property.

3.2 Recommendations of Object Types and Relations

All class types that may be recommended to the user are derived from the ontology. For example, if the user creates a new variable and he/she wants to specify its type, the system automatically provides applicable (i.e. relevant to the typed query) ontology type suggestions. For example, if the user types "*po*" the system will display a list which contains types like: *polo league, poker player, political party*. Figure 4 shows an example of that function. This solution allows to avoid mapping (which may always

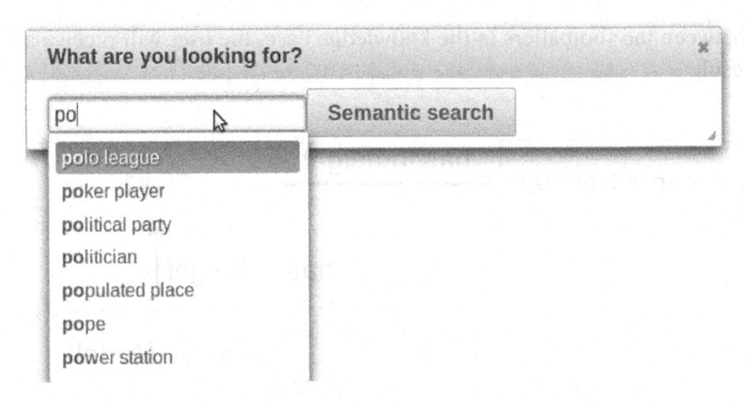

Fig. 4. Predicate and object recommendation example

introduce an error) terms from a user query to ontology classes. All the recommended types come from the ontology, and all recommended properties and class instances come from the data store (in explicit, from an RDF graph). As a result the user is only allowed to select properties and objects that exist in the data store. Additionally, if the user initially defines the subject and the subject type, then he/she is provided with list containing only the properties that are present in the data store - each occurring as a relation between the subject of this type and some object. For example, if the user defines a football player variable, then he/she will not receive prompts like $hasCapital$, which despite being present in the data store, are in fact used with other types of objects. In such a case the user will get only predicates related to the $FootballPlayer$ class instances, like $playsIn$, $weight$, $height$, etc.

3.3 Recommendations of Variables Names

Usually, in semantic search systems, the user is allowed to specify his/her own variable names. Although the user of the Semantic Focused Crawler is also able to do that, he/she is provided with some additional support. For instance, while looking for an information about an unspecified Facebook user, the SFC user may select the automatically suggested variable *?FacebookUser*. When the user starts typing, SFC displays useful suggestions of variables, similar to ontology type recommendations mentioned before. When the user selects a variable from the suggestions list, the variable type field will be automatically complemented. For example, when the user selects the *?FacebookUser#1* variable name, then the variable type field will be complemented with *Facebook user* type. The character '# ' (a special sign in the variable name) and the number after it are mandatory to enable the user to create other variables with the name starting with '*?FacebookUser*'. Naturally, the user is allowed use the same variables many times (the system also displays variables that were already used before). For instance, the query in the example shown in the Figure 5 represents the need to find a football player (along with his club) whose wife is an actress.

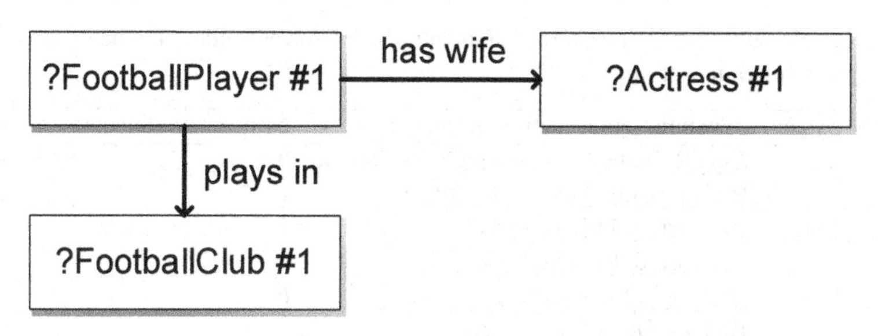

Fig. 5. Query example 3

4 System Evaluation

The comparison presented herein shows which of the evaluated semantic search systems return more relevant results, and tend to be more user-friendly. The comparison determining the user-friendliness has been performed by measuring how many elements the user has to type in order to construct every query.

4.1 Evaluation Methodology

The system presented in this paper was compared to other semantic search systems using the QALD-1 trainset [17] queries set which contains queries to the DBPedia knowledge base [2]. The QALD query set is a set of 100 questions and relevant answers for these questions [17]. In most cases the answers are represented as URIs of DBPedia resources or as a string. For evaluation purposes we selected a subset containing 28 queries, because some of the evaluated systems do not support aggregation, filter and ordering functions and are not able to represent some of the relations present in the QALD query set. Identifiers of these queries are: 2, 3, 6, 14, 15, 18, 22, 23, 25, 28, 29, 31, 35, 36, 38, 39, 42, 43, 44, 48, 54, 55, 56, 62, 73, 85, 87, 90.

Precision, Recall, and F-score values [18] have been measured to compare the systems accuracy. In this paper it is assumed that the precision will be calculated for the first n results, where n is equal to a number of the relevant answers from the QALD. Consequently, if there is only one relevant answer to the query "What is the official website of Tom Cruise" in the QALD, only the first result in all systems is considered.

A user interface may be considered more user-friendly when the user is forced to fill in less query elements. Therefore, in order to evaluate the user-friendliness of user interfaces, it is important to measure how many elements the user needs to type in to construct equivalent queries in all of the systems. In case of semantic search applications these elements are: resources, variables, variables' types, properties, and literals.

Table 2 presents an example query "*In which films did Julia Roberts as well as Richard Gere play?*" written in the form appropriate for each of the compared systems. The user of GoRelations has to specify types of all variables and resources and is obligated to type in every triple element. In contrast, in the same query scenario, the user of SFC is not required to manually add the subject variable (x_1) in the first query line and to specify all the object types. Consequently, he/she has to add only five elements instead of nine, while in NAGA the user has to add variable x_1 in the both query lines.

Table 2. Example query representations with the number of corresponding elements the user has to type in (underlined in every query)

System	Semantic represenation of the query	Number of elements
GoR	Julia Roberts/Actor, starredIn, x_1/Movie Richard Gere/Actor, starredIn, x_1	**9**
SFC	x_1, starred, Julia Roberts x_1, starred, Richard Gere	**5**
NAGA	Julia Roberts, actedIn, x_1 Richard Gere, actedIn, x_1	**6**

4.2 Analysis of the Results

The results of the system accuracy evaluation are presented in the Table 3. Each of these values is the average of the individual results obtained for each query.

Table 3. Systems comparison results

System	Precision	Recall	F-Score	Average number of elements required to create queries
SFC	**0.961**	**0.906**	**0.927**	**2.786**
GoR	0.693	0.672	0.682	5.714
NAGA	0.536	0.506	0.520	3.607

It should be noted that one of the key factors affecting the search results is the use of different knowledge bases. However, the comparison should still be regarded as objective for two reasons. Initially, all of the systems contain data necessary to generate all relevant answers for each of the selected queries. Secondly, all of the compared systems allow the user to create all of the tested queries and to issue them to the system.

According to the results, a user of Semantic Focused Crawler does not have to type in as many query elements as the users of other systems. On average, less than three elements have to be used to construct each query. In case of GoRelations almost six elements per query have to be used, and in NAGA more than three elements. As a result, the SFC user may create queries without typing redundant information, like types of query variables. This feature, unique among other semantic search systems [8,7], enables to reduce significantly the amount of data that the user has to type in to construct a query.

The presented results show that a system featuring a graph-based, user-friendly querying interface, may still be very useful in terms of semantic expressiveness of queries. As shown in the experiments, Semantic Focused Crawler outperforms the other systems in both the accuracy test and the query formulation complexity test. The introduced system enabled us to achieve higher precision and recall mainly due to the fact that it avoids mapping of the query terms to ontology classes and properties (which tend to be an error-prone process causing irrelevant results). It may be concluded that the proposed system provides the user with more useful responses to queries than the other compared systems do, and features the querying interface requiring the user to type a smaller number of query elements.

5 Conclusions

Many of existing semantic search systems provide graph-based querying interfaces. However, these user interfaces are still not very self-explaining and simple to use. As shown in the experimental results, the system presented in this paper may be seen as effectively combining graph-based querying systems usability and user-friendliness that so far could be observed only in natural language-based querying systems.

The system presented in this paper allows the user to create graph-based queries in a simpler way than it is done in the other graph-based querying systems. The key outcomes of the work reported in the paper include:

- Keyword-based search provided as a function of the graph-based search interface
- Query triples construction accompanied by an intuitive variable naming function, which makes the process simpler than in the case of using semantic search systems presented in the literature [8,7]
- Automatic type and property suggestions that additionally supporting the user in the query construction process.

It should be stressed that the presented system has achieved the highest score in the semantic search accuracy test performed with the use of 2011 QALD workshop questions [17]. Moreover, it has also achieved the lowest average query formulation complexity in a test performed with the use of 2011 QALD workshop questions.

Acknowledgements. This work is supported by research project O ROB 0025 01 financed by The National Center for Research and Development on purpose of the defence and the security of the country, Contract 0025/R/D1/2011/01 from 12-12-2011 and by the Polish National Science Center under grant DEC-2011/01/D/ST6/06788.

References

1. Lassila, O., Berners-Lee, T., Hendler, J.: The Semantic Web. Scientific American (2001)
2. Bizer, C., Soren, A., Kobilarov, G., Lehmannm, J., Cyganiak, R.: DBpedia - Querying Wikipedia like a Database. In: 16 International World Wide Web Conference: Developers Track (2007)
3. Weikum, G., Suchanek, F., Kasenci, G.: YAGO: A Core of Semantic Knowledge Unifying WordNet and Wikipedia. In: 16 International World Wide Web Conference (2007)
4. Berners-Lee, T., Bizer, C., Heath, T.: Linked Data - The Story So Far. In: WebDB Workshop at SIGMOD 2010 (2010)
5. Wikipedia, `http://www.wikipedia.org/`
6. Resource Description Framework (RDF), `http://www.w3.org/TR/RDF/`
7. Kasneci, G., Suchanek, F., Ifrim, G., Ramanath, M., Weikum, G.: NAGA: Searching and Ranking Knowledge. In: 24th IEEE International Conference on Data Engineering, ICDE 2008 (2008)
8. Han, L., Finin, T., Joshi, A.: GoRelations: An Intuitive Query System for DBpedia. In: Pan, J.Z., Chen, H., Kim, H.-G., Li, J., Wu, Z., Horrocks, I., Mizoguchi, R., Wu, Z. (eds.) JIST 2011. LNCS, vol. 7185, pp. 334–341. Springer, Heidelberg (2012)
9. Web Ontology language Document Overview, 2nd edn., `http://www.w3.org/TR/owl2-overview/`

10. Carroll, J., Dickinson, I., Dollin, C.: Jena: Implementing the Semantic Web Recommendations. World Wide Web, pp. 74–83 (2004)
11. SPARQL Query Language for RDF, http://www.w3.org/TR/rdf-sparql-query/
12. Resource Description Framework Schema (RDFS), http://www.w3.org/TR/rdf-schema/
13. Apache Lucene, http://lucene.apache.org/core/
14. Damljanovic, D., Agatonovic, M., Cunningham, H.: Natural Language Interfaces to Ontologies: Combining Syntactic Analysis and Ontology-based Lookup Through the User Interaction. In: Aroyo, L., Antoniou, G., Hyvönen, E., ten Teije, A., Stuckenschmidt, H., Cabral, L., Tudorache, T. (eds.) ESWC 2010, Part I. LNCS, vol. 6088, pp. 106–120. Springer, Heidelberg (2010)
15. Wu, F., Weld, D.: Automatically Refining the Wikipedia Infobox Ontology. In: 17 World Wide Web Conference (WWW 2008), pp. 645–644 (2008)
16. WordNet, http://www.wordnet.princeton.edu
17. Question Answering over Linked Data, http://greententacle.techfak.uni-bielefeld.de/~cunger/qald/
18. Goutte, C., Gaussier, É.: A Probabilistic Interpretation of Precision, Recall and F-Score with Implication for Evaluation. In: Losada, D.E., Fernández-Luna, J.M. (eds.) ECIR 2005. LNCS, vol. 3408, pp. 345–359. Springer, Heidelberg (2005)

A General Model for Mutual Ranking Systems

Vu Le Anh[1,5], Hai Vo Hoang[2,*], Kien Le Trung[3], Hieu Le Trung[4], and Jason J. Jung[1,5]

[1] Nguyen Tat Thanh University, Ho Chi Minh City, Vietnam
[2] Information Technology College, Ho Chi Minh City, Vietnam
[3] Hue University of Sciences, Hue City, Vietnam
[4] Duy Tan University, Da Nang City, Vietnam
[5] Big IoT BK Project Team, Yeungnam University, Gyeongsan, Korea

Abstract. Ranking has been applied in many domains using recommendation systems such as search engine, e-commerce, and so on. We will introduce and study N-linear mutual ranking, which can rank n classes of objects at once. The ranking scores of these classes are dependent to the others. For instance, PageRank by Google is a 2-linear mutual ranking, which ranks the webpages and links at once. Particularly, we focus to N-star ranking model and demonstrate it in ranking conference and journal problems. We have conducted the experiments for the models in which the citations are not considered. The experimental results are based on the DBLP dataset, which contains more than one million papers, authors and thousands of conferences and journals in computer science. Finally, N-star ranking is a very strong ranking algorithm can be applied in many real-world problems.

Keywords: N-star ranking, Markov chain, PageRank, Academic ranking, Conference ranking, Ranking algorithms, Prolific ranking, Recommendation systems, Bibliographical database, DBLP.

1 Introduction

Ranking is an interesting but difficult problem on many information processing systems. With a large amount of information, the systems need to adapt efficient ranking schemes to sort out (or to select) only the information which are highly relevant to the users' contexts. Particularly, in the context of *bibliometrics*, a set of given entities can be quantified to compare several evaluation indicators (e.g., popularity and reputation). For example, impact factors (IF) of international journals can be measured by taking into account how many times the papers in the corresponding journals have been cited.

However, there are some problems on the ranking schemes of traditional systems (i.e., web searching and impact factor). These systems are based on simple links among only the target entities (e.g., web pages and papers). We assume that this type of referencing process is not enough for measuring the evaluation indicators more precisely.

In this work, we formulate a mathematical model of N-star ranking scheme, which can represent a generalized referencing process. The N-star ranking is based on N-linear mutual ranking system in which N classes are ranked at once. Especially, as a

* Corresponding Author: Phone/Fax: +84 938 642 717. vohoanghai2@gmail.com

N.T. Nguyen et al. (Eds.): ACIIDS 2014, Part I, LNAI 8397, pp. 211–220, 2014.

real-world problem, conference ranking has been regarded as an interesting issue on academic communities. The conference publication (e.g., proceedings) is related with various factors such as authors (and their affiliation), papers (and their citations), and conferences.

The remainder of this paper is as follows. In Sect. 2, we describe the backgrounds of the N-linear mutual ranking (e.g., the rank scoring and conceptual modeling on relationships between classes). Sect. 3 defines the N-star ranking as a N-linear mutual ranking system in which there exists a core class. We study two N-star ranking models for the author, publication and conference ranking problem in different contexts in Sect. 4. Also, Sect. 5 shows the experiments for the simple N-star ranking model of authors, publications and conference ranking. In Sect. 6, we exhibit the related work for comparison with the proposed N-star ranking model. Finally, Sect. 7 draws a conclusion of this work.

2 Backgrounds

2.1 N-linear Mutual Ranking System

The couple $(\mathcal{A}, R)$ is called a *ranking system* if (i) $\mathcal{A} = \{a_1, \ldots, a_n\}$ is a finite set, and (ii) $R : \mathcal{A} \to [0; +\infty)$. $\mathcal{A}$ is called a *class*, $a \in \mathcal{A}$ is called an *object of the class* $\mathcal{A}$, and R is called a *score* of $\mathcal{A}$. R is *positive* if $R(a) > 0 \ \forall a \in \mathcal{A}$. $n = |\mathcal{A}|$ is the *size* of $\mathcal{A}$.

Definition 1. *$\Omega = \{(\mathcal{A}_i, R_i)\}_{i=1}^N$ is called a N-*linear mutual ranking system *described by a system $\{\alpha_{ij}, \beta_i, I_i, W_{ij}\}$ if $(\mathcal{A}_i, R_i)$ is a ranking system and $\alpha_{ij}, \beta_i \in [0; +\infty)$, $I_i = (t_u^{(i)})_{n_i}$ is a n_i-dimensional normalized nonnegative real number vector, $W_{ij} = (\omega_{kl}^{(ij)})_{n_i \times n_j}$ is a nonnegative real number and normalized columns matrix such that*

$$\sum_j \alpha_{ij} + \beta_i = 1 \qquad \text{and} \qquad R_i = \sum_{j=1}^{N} \alpha_{ij} W_{ij} R_j + \beta_i I_i, \quad i = 1, \ldots, N.$$

$\{\alpha_{ij}, \beta_i, I_i, W_{ij}\}$ is called a *linear constraint system* of Ω.

Let a_{iu}, a_{jv} be objects in $\mathcal{A}_i, \mathcal{A}_j$ respectively. Suppose $\mathcal{C}^*(a_{iu}, a_{jv}) = \alpha_{ij}\omega_{uv}^{(ij)}$. From the definitions, we have:

$$R_i(a_{iu}) = \sum_{j=1}^{N} \sum_{v=1}^{n_j} \alpha_{ij}\omega_{uv}^{(ij)} R_j(a_{jv}) + \beta_i t_u^{(i)} = \sum_{j=1}^{N} \sum_{v=1}^{n_j} \mathcal{C}^*(a_{iu}, a_{jv}) R_j(a_{jv}) + \beta_i t_u^{(i)}$$

a_{jv} is called *affect to* a_{iu} (denoted by $a_{jv} \to a_{iu}$) if $\mathcal{C}^*(a_{iu}, a_{jv}) > 0$. Class $\mathcal{A}_i$ is called *total affect and reflect directly to* class $\mathcal{A}_j$ (denoted by $\mathcal{A}_i \to \mathcal{A}_j$) if $\forall a_{jv} \in \mathcal{A}_j$: $\exists a_{iu_1}, a_{iu_2} \in \mathcal{A}_i$: $a_{jv} \to a_{iu_2} \wedge a_{iu_1} \to a_{jv}$.

Definition 2. *Class $\mathcal{A}_i$ is called* total affect and reflect to *class $\mathcal{A}_j$, denoted by $\mathcal{A}_i \rightsquigarrow \mathcal{A}_j$, if $\mathcal{A}_i \to \mathcal{A}_j$ or $\exists \mathcal{A}_k : \mathcal{A}_i \to \mathcal{A}_k \wedge \mathcal{A}_k \rightsquigarrow \mathcal{A}_j$.*

2.2 PageRank

We rewrite the PageRank into a 2-linear mutual ranking system as follows:

$\mathcal{W} = \mathcal{A}_1$ is the class representing for the set of webpages. $\mathcal{L} = \mathcal{A}_2$ is the class representing for hyperlinks. For each hyperlink $l \in \mathcal{L}$ from web $u \in \mathcal{W}$ to web $v \in \mathcal{W}$, we denote $u = in(l)$ and $v = out(l)$. For each $v \in \mathcal{W}$, we denote: $IN(v) = \{l \in \mathcal{L} | v = out(l)\}$ and $N_{out}(v) = |\{l \in \mathcal{L} | v = in(l)\}|$.

PageRank[1] determined the ranking system of webpages by the following formula: $\forall v \in \mathcal{W}$,

$$R_w(v) = d \sum_{l \in IN(v), u = in(l)} \frac{R_w(u)}{N_{out}(u)} + \frac{1-d}{|\mathcal{W}|} \tag{1}$$

where $d \in (0, 1)$ is a constant.

Suppose $W_{21} = (\delta_{kt})_{|\mathcal{L}| \times |\mathcal{W}|}$ is a matrix in which $\delta_{kt} = \frac{1}{N_{out}(w_t)}$ if $l_k \in \mathcal{L} \wedge w_t = in(l_k) \in \mathcal{W}$, otherwise 0. W_{21} is a nonnegative real number and normalized columns matrix. Suppose $W_{12} = (\gamma_{tk})_{|\mathcal{W}| \times |\mathcal{L}|}$ is a matrix in which $\gamma_{tk} = 1$ if $l_k \in L \wedge w_t = out(l_k) \in \mathcal{W}$, otherwise 0. We construct a 2-linear mutual ranking system on two classes $\mathcal{W}$ and $\mathcal{L}$ as follows: Let $\bar{R}_w$ and $\bar{R}_l$ be scores on the classes $\mathcal{W}$, $\mathcal{L}$ respectively. They are satisfied:

$$\bar{R}_l = W_{21} \bar{R}_w \qquad and \qquad \bar{R}_w = d W_{12} \bar{R}_l + (1-d) I_{|\mathcal{W}|}, \tag{2}$$

where $I_{|\mathcal{W}|}$ denotes the $|\mathcal{W}|$-dimensional vector in which all its elements are $1/|\mathcal{W}|$. It is not difficult to see that (2) confirms: for all web $v \in \mathcal{W}$,

$$\bar{R}_w(v) = d \sum_{l \in IN(v), u = in(l)} \frac{\bar{R}_w(u)}{N_{out}(u)} + \frac{1-d}{|\mathcal{W}|}.$$

Since the equation (1) has the unique solution which is the PageRank score (see in [1]), $\bar{R}_w$ is the PageRank score R_w. Vice versa, if R_w is a solution of (2), R_w should be $\bar{R}_w$. Thus, the PageRank score R_w is totally determined by the equation (2), or in other words, PageRank can be presented as the two-linear ranking system described by (2).

Note that, since for each link $l \in \mathcal{L}$, let $u = in(l)$ and $v = out(l)$ then web u affects to link l, $(u \to l)$ and link l affects to web v, $(l \to v)$. Therefore, the class $\mathcal{W}$ total affect and reflect directly to the class $\mathcal{L}$, $\mathcal{W} \leadsto \mathcal{L}$.

3 N-star Ranking Model

Definition 3. *Let $\Omega = \{(\mathcal{A}_i, R_i)\}_{i=1}^N$ be a N-linear mutual ranking system. Ω is called a N-*star ranking *if*

1. $\exists i : \mathcal{A}_i : (\beta_i > 0) \wedge (I_i$ *is positive* $) \wedge (\forall \mathcal{A}_j (j \neq i) : \mathcal{A}_i \leadsto \mathcal{A}_j)$.
2. $\forall j \neq i : \alpha_{j1} = 1$.

$\mathcal{A}_i$ is called a core *of the system Ω.*

If $\Omega = \{(\mathcal{A}_i, R_i)\}_{i=1}^N$ is a N-star ranking system described by a linear constraint system $\{\alpha_{ij}, \beta_i, I_i, W_{ij}\}$, $\{\alpha_{ij}, \beta_i, I_i, W_{ij}\}$ is called *N-star constraint system* of Ω.

Proposition 1. *PageRank is a 2-star ranking system.*

Proof. Because $1 - d > 0$ and $\mathcal{W} \rightsquigarrow \mathcal{L}$, $\mathcal{W}$ is the core of PageRank. The second condition is clear since $\alpha_{21} = 1$. Hence, PageRank is a 2-star ranking system.

The ranking scores are determined by the N-star constraint system and the classes.

Proposition 2. *Suppose the classes $\{\mathcal{A}_i\}_{i=1}^N$ and the N-star constraint system $\{\alpha_{ij}, \beta_i, I_i, W_{ij}\}$ are given. There exists a unique $\{R_i\}_{i=1}^N$ in which R_i is a score on $\mathcal{A}_i$, such that $\Omega = \{(\mathcal{A}_i, R_i)\}_{i=1}^N$ is a N-star ranking described by $\{\alpha_{ij}, \beta_i, I_i, W_{ij}\}$ and for all i, $\sum_{a\in\mathcal{A}_i} R_i(a) = 1$.*

Proof. Assuming without loss of generality that $\mathcal{A}_1$ is a core of a N-star ranking system described by $\{\alpha_{ij}, \beta_i, I_i, W_{ij}\}$, the sequence of scores $R_1, \ldots, R_N$ satisfy the following equations:

$$R_1 = WR_1 \qquad \textit{and} \qquad R_i = W_{i1}R_1 \qquad \forall i = 2, \ldots, N, \tag{3}$$

where

$$W = \alpha_{11}W_{11} + \alpha_{12}W_{12}W_{21} + \cdots + \alpha_{1N}W_{1N}W_{N1} + \beta_1\mathbf{I}_1 \tag{4}$$

and $\mathbf{I}_1$ is the $(n_1 \times n_1)$-matrix which its columns are I_1. It is not difficult to infer that because W_{1i} and W_{i1} are transition matrices (i.e. nonnegative and normalized columns matrices), the new square matrix $W_{1i}W_{i1}$ is a stochastic matrix (i.e. a transition and square matrix). The matrices W_{11} and $\mathbf{I}_1$ are also stochastic matrices. Since $\mathbf{I}_1$ has positive entries, $\beta_1 > 0$ (because $\mathcal{A}_1$ is the core), and $\sum_j \alpha_{1j} + \beta_1 = 1$, the matrix W is also a stochastic matrix with positive entries. The Perron-Frobenius theorem (see in [2,3]) confirms that there exists a unique score R_1 with $\sum_{a\in\mathcal{A}_1} R_1(a) = 1$ such that

$$R_1 = WR_1.$$

From (3), the unique existence of R_1 infers the unique existences of $R_2, \ldots, R_N$. Moreover, since $W_{21}, \ldots, W_{N1}$ are normalized columns and $\sum_{a\in\mathcal{A}_1} R_1(a) = 1$, we have $\sum_{a\in\mathcal{A}_i} R_i(a) = 1$ for all $i = 2, \ldots, N$. The proposition is proven.

The ranking scores are computed by following algorithm:

Algorithm : Finding the sequence scores $\{R_i\}_{i=1}^N$

Input : $\alpha_{ij},\ \beta_i,\ W_{ij}, I_i$

Output : $\{R_i\}_{i=1}^N$

1. **begin**
2. Check the N-star ranking model with the core $\mathcal{A}_1$
3. Let
$$W \leftarrow \alpha_{11}W_{11} + \sum_{i=2}^{N} \alpha_{1i}W_{1i}W_{i1} + \beta_1\mathbf{I}_1$$
4. Initialize $R_1^{(0)}$: uniform distribution, $k = 0$
5. **repeat**

6. $k = k + 1$
7. Update $R_1^{(k)} \leftarrow W R_1^{(k-1)}$
8. **until** $\|R_1^{(k)} - R_1^{(k-1)}\| \leq$ a stopping criterion
9. Let $R_1 = R_1^{(k)}$ and $R_i = W_{i1} R_1, \ i = 2, \ldots, N$
10. **end**

4 Ranking Authors, Publications and Conferences

In this section, we apply the N-star ranking model for constructing a model to evaluate authors, publications and conferences (journals) in the world of science. Concretely, we consider a four-star ranking model corresponding with four ranking systems: $(\mathcal{A}, R_a)$ - $\mathcal{A}$ is a set of all scientists which has publications, $(\mathcal{P}, R_p)$ - $\mathcal{P}$ is a set of all publications, $(\mathcal{C}, R_c)$ - $\mathcal{C}$ is a set of all sciential conferences and sciential journals, and $(\mathcal{L}, R_l)$ - $\mathcal{L}$ is a set of all citations between publications. R_a, R_p, R_c and R_l are the scores for each classes $\mathcal{A}$, $\mathcal{P}$, $\mathcal{C}$ and $\mathcal{L}$, respectively.

For each citation $l \in \mathcal{L}$, $u = in(l)$ and $v = out(l)$ if l is from $u \in \mathcal{P}$ to $v \in \mathcal{P}$. For each publication $v \in \mathcal{P}$, we denote: $IN(v) = \{l \in \mathcal{L} | v = out(l)\}$; $OUT(v) = \{l \in \mathcal{L} | v = in(l)\}$ $N_{out}(v) = |OUT(v)|$. If a publication cites no where, we assume that it cites to all publications; $C(v) \in \mathcal{C}$ is the conference of v; $A(v) \subseteq \mathcal{A}$ is the set of authors of v. For each author $a \in \mathcal{A}$, $P(a) = \{v \in \mathcal{P} | a \in A(v)\}$ is a set of publications of a. For each conference $c \in \mathcal{C}$, $P_c(c) = \{v \in \mathcal{P} | c = C(v)\}$ is a set of publications published in c.

The 4-star ranking system model for ranking authors, publications, conferences and citations is constructed based on some following ideas:

1. The score of an author depends only on his publications, and each publications affects to all of its authors:
$\forall a \in \mathcal{A}, p \in \mathcal{P}$:
$$R_a(a) = \sum_{p' \in P(a)} \mathcal{C}^*(a, p') R_p(p') \ \ and \ \ \sum_{a' \in A(p)} \mathcal{C}^*(a', p) = 1 \tag{5}$$
If *a publication affects equally to its authors* (a), (5) is rewritten as follows:
$\forall a \in \mathcal{A}, p \in \mathcal{P}, a' \in A(p)$:
$$\mathcal{C}^*(a', p) = \frac{1}{|A(p)|} \ \ and \ \ R_a(a) = \sum_{p' \in P(a)} \frac{R_p(p')}{|A(p')|} \tag{6}$$
2. The score of a conference depends only on its publications:
$$\forall c \in \mathcal{C} : R_c(c) = \sum_{p' \in P_c(c)} R_p(p'). \tag{7}$$
3. The score of a citation depends on the citing publication, and each publications affects to all of its citations:
$\forall l \in \mathcal{L}, p \in \mathcal{P}, p' = in(l)$:
$$R_l(l) = \mathcal{C}^*(l, p') R_p(p') \ \ and \ \ \sum_{l' \in OUT(p)} \mathcal{C}^*(l', p) = 1 \tag{8}$$

If *a publication affects equally to its citations* (b), (8) is rewritten as follows:
$\forall l \in \mathcal{L}, p \in \mathcal{P}, p' = in(l), l' \in OUT(p):$

$$\mathcal{C}^*(l', p) = \frac{1}{N_{out}(p)} \quad and \quad R_l(l) = \frac{R_p(p')}{|N_{out}(p')|} \tag{9}$$

4. The score of a publication depends on its citations, its authors and its conference and randomly finding by some reader. Each conference affects to all of its publications. Each author affects to all of its publications. Hence :
$\forall p \in \mathcal{P}, c \in \mathcal{C}, a \in \mathcal{A}, c' = C(p):$

$$R_p(p) = \alpha_1 \sum_{l' \in IN(p)} R_l(l') + \alpha_2 \sum_{a' \in A(p)} \frac{\mathcal{C}^*(p, a')}{\alpha_2} R_a(a') + \alpha_3 \frac{\mathcal{C}^*(p, c')}{\alpha_3} R_c(c') + \beta_p I_p, \tag{10}$$

$$\sum_{p' \in P(a)} \mathcal{C}^*(p', a) = \alpha_2 \quad and \quad \sum_{p' \in P_c(c)} \mathcal{C}^*(p', c) = \alpha_3 \tag{11}$$

where $\alpha_1, \alpha_2, \alpha_3, \beta_p > 0$ and $\alpha_1 + \alpha_2 + \alpha_3 + \beta_p = 1$, I_p is a $|\mathcal{P}|$-dimensional normalized uniform random vector.
If *a conference affects equally to its publications* (c) and *an author affects equally to its publications* (d), the equation (10) and (11) are rewritten as follows:
$\forall p \in \mathcal{P}, c \in \mathcal{C}, a \in \mathcal{A}, c' = C(p), p' \in P(a), p'' \in P_c(c):$

$$\mathcal{C}^*(p', a) = \frac{\alpha_2}{|P(a)|} \quad and \quad \mathcal{C}^*(p'', c) = \frac{\alpha_3}{|P_c(c)|} \tag{12}$$

$$R_p(p) = \alpha_1 \sum_{l' \in IN(p)} R_l(l') + \alpha_2 \sum_{a' \in A(p)} \frac{R_a(a')}{|P(a')|} + \alpha_3 \frac{R_c(c')}{|P_c(c')|} + \beta_p I_p, \tag{13}$$

The model which is described by equations (5), (7), (8), (10) and (11) is called *the general 4-star ranking model for the ranking authors, publications and conferences problem.* And the model which is described by equations (6), (7), (9), (12) and (13) is called *the simple 4-star ranking model for the ranking authors, publications and conferences problem.* Both of the general and simple 4-star ranking models are really N-star ranking in which the publication class is the central.

5 Experiments

5.1 Dataset, NPC and SD4R Model

Datasets. We exploited the DBLP data set[1], which provides bibliographic information on major computer science journals and proceedings, to conduct experiment for the simple 4-star ranking model for the ranking authors, publications and conferences problem on fields of computer science. A publication can be both in a conference and another journal, too. To avoid this gap, we keep only the publications related to conferences and

[1] `http://dblp.uni-trier.de` accessed on May 2013.

ignore the journals's ones. We has built program to parse DBLP dataset in XML format to extract the authors, title, and publication venue information from the guides [5,6]. We mention our readers that DBLP has no information about the citations between publications. Finally, we generate two datasets D_1 and D_2 for the experiments. D_1 contains all authors, publications ($\|\mathcal{P}\| = 1253997, \|\mathcal{A}\| = 845295, \|\mathcal{C}\| = 3351$). D_2 contains only the authors, and the publications belong to *not small* conferences which have over 300 publications ($\|\mathcal{P}\| = 1046030, \|\mathcal{A}\| = 753896, \|\mathcal{C}\| = 949$).

NPC Model. Because there is no information about the citations between publications, we adopt a *naive model based on publication counting (NPC)* for ranking author and conference. In NPC model, for each author $a \in \mathcal{A}$ $R'_a(a) ::= |P(a)|$ and for each conference $c \in \mathcal{C}$ $R'_c(c) ::= |P_c(c)|$.

SD4R Model. We implemented a *simple DBLP 4-star ranking (SD4R) model* with central publication class follow adapted versions of equations (6), (7), (9), (12) and (13). Because we do not have any information about the citations and for the simplicity, we omitted equations (9), (12) and set $\alpha_1 = 0, \alpha_2 = \alpha_3$ and $\beta_p = 1 - 2\alpha_2$. The equation (13) are rewritten as follows :$\forall p \in \mathcal{P}, c' = C(p)$

$$R_p(p) = \alpha_2 \sum_{a' \in A(p)} \frac{R_a(a')}{|P(a')|} + \alpha_2 \frac{R_c(c')}{|P_c(c')|} + (1 - 2\alpha_2) I_p, \qquad (14)$$

5.2 Experimental Results

We scale the rank scores of SD4R such that:$\sum_{a \in \mathcal{A}} R_a(a) = \sum_{a \in \mathcal{A}} R'_a(a)$ and $\sum_{c \in \mathcal{C}} R_c(c) = \sum_{c \in \mathcal{C}} R'_c(c)$. For each author $a \in \mathcal{A}$ and conference $c \in \mathcal{C}$, we measure the difference between ranking scores of two models: (i)$\Delta(a) = R_a(a) - R'_a(a)$, $\%\Delta(a) = \frac{\Delta(a)}{R'_a(a)}$ and (ii)$\Delta(c) = R_c(c) - R'_c(c)$, $\%\Delta(c) = \frac{\Delta(c)}{R'_c(c)}$. By examining the Δ and $\%\Delta$ functions, we found following interesting results.

- *SD4R ranking scores reflect how hot the conferences are.* Table 1 has show the top 5 conferences most increasing and decreasing the ranking score over D_2. All *increasing* conferences are young, annual events and get hot topics now. Their topics are about remote sensing (IGARSS - IEEE International Geoscience and Remote Sensing Symposium, from 2005), computer human interaction (CHI - Computer Human Interaction, from 1990), medical image computing (MICCAI - Medical Image Computing and Computer-Assisted Intervention, from 1998), robot (IROS - International Conference on Intelligent RObots and Systems, from 1998), solid-state circuits (ISSCC - International Solid-State Circuits Conference, from 2009)... The top 5 *decreasing* conferences are held for over a long time or biennial events, in local community (GI-Jahrestagung - language spoken is Germany, from 1972) or with less interesting topics such as artificial intelligence (IJCAI - International Joint Conference on Artificial Intelligence, biennial from 1969), image processing(ICIP - International Conference on Image Processing, from 1994; IFIP - International Federation for Information Processing, biennial from 1959), computational linguistics (ACL - Meeting of the Association for Computational Linguistics, from 1979),...

Table 1. Top 5 conferences most differences of ranking value of conferences using SD4R vs. NPC over D_2; (a) increasing and (b) decreasing ranking values

Conference name	nPubs	SD4R val	DIFF val	Δ %
IGARSS	9691	10173	482	4.97
CHI	8737	9218	481	5.51
MICCAI	3778	4246	468	12.39
ISSCC	1185	1552	367	30.97
IROS	7906	8218	312	3.95

Conference name	nPubs	SD4R val	DIFF val	Δ %
IJCAI	5635	5321	-314	5.57
ICIP	15125	14821	-304	2.01
IFIP	2796	2529	-267	9.55
GI	4349	4132	-217	4.99
ACL	3489	3278	-211	6.05

- *SD4R affects much on authors and little on conferences.* The average of $|\%\Delta|$ of the conferences is 2.7%. And the average of $|\%\Delta|$ of the authors is 23%. It implies that SD4R does not change the conference ranking scores but do change the author ranking scores.
- *SD4R reflects the contribution of the author betters than the NPC.* Table 2 has show the top 5 authors most increasing and decreasing the ranking score over D_2. We observe that all top *decreasing* authors are have a large number of publications. Their publications have a large number of co-authors. All top *increasing* authors are really key-person in their research topics.

Table 2. Top 5 authors most differences of ranking value of authors using SD4R vs. NPC over D_2; (a) increasing and (b) decreasing ranking values

Author name	nPubs	SD4R val	DIFF val	Δ %
Debenham, J.K.	132	247	115	87.3
Tsumoto, S.	204	307	103	50.5
Yama kami, T.	55	156	101	183
Hancock, E.R.	396	485	89	22.6
Kamimura, R.	55	139	84	153

Author name	nPubs	SD4R val	DIFF val	Δ %
Gao, W.	488	367	-121	-121
Barolli, L.	283	200	-83	-83
Takanishi, A.	166	90	-76	-76
Catthoor, F.	241	168	-73	-73
Liu, W.	426	354	-72	-72

6 Related Work

Link-based ranking has been applied in many application in various domains. In this section, we want to describe such applications and compare them.

- Webpages: One of the most well-known example is PageRank utilized by Google search engine. Since the hyperlink structure among the webpages are easily represented as a web graph, the PageRank of each webpage can be measured (see Sect. 2.2 for more detail).
- Named entities: In Natural Language Processing (NLP) communities, named entity recognition is an important problem. Ranking scheme has been applied to solve the problem. Collins [11] proposes a ranking method based on a maximum-entropy tagger. Also, Vercoustre et al. [12] has presented how to apply the ranking method to Wikipedia.
- Scientific articles: In bibliometrics, the scientific articles (e.g., research papers, technical reports, and so on) are evaluated with respect to the quality (e.g., novelty and originality) as well as academic influence to the communities (e.g., impact) by relaying on the citations (e.g., references and quotation) [13].

- Researchers: Also, Researchers has been ranked by citation analysis (e.g., how many papers has he/she published, how many times have his/her papers cited, and so on). More interestingly, H-index (Hirsch index) has been designed to measure both the productivity and impact of the published work of the researchers.
- Complex system: The complex system ranking has been already explored using a different formalism for ranking or classification in heterogeneous networks [14,15]. The poprank model [14] introduce the Popularity Propagation Factor to express the relationship between classes. Their model is based on the markov chain model which can be applied in the N-linear mutual ranking systems. The quantium ranking [15] is based on quantium navigation. Their formula is come from the quantium theory and quite different to ours.

So far, conferences have been ranked by subjective opinions and consensus among well known experts in a domain. Such lists in computer science area are compiled here[2]. In this work, we have proposed a novel conference ranking framework to integrate all possible evidence.

7 Conclusion and Future Works

We have introduced and studied N-star ranking for mutual ranking systems. The mutual relationships between ranking objects are described by a system of linear equations, N-mutual ranking system. A N-linear mutual ranking system is a N-star ranking systems if it has a core class which affects and reflects all other classes in the system. The rank scores of the N-star ranking system are unique and computed by a Markov chain. We have pointed out that PageRank is a 2-star ranking. It has two classes: the webpages (a core class) and links.

We have introduced and studied a general and a simple 4-star ranking models for ranking authors, publications, conferences. A general model is a generic one. In a simple model, we consider each publication, author, conference, citation is equally. We have conducted the experiments for the models in which the citations are not considered. The experimental results are based on the DBLP dataset. By comparing the difference between the SD4R vs. NPC models, we have shown that our ranking system can reflect how hot the conference are and record the contribution of the authors better than the naive ranking system. Moreover, our ranking system makes a big change on author ranking.

As future work, we are planning to *i*) get the citations between the publication to upgrade the quality of our ranking system, *ii*) study how to combine a N-star ranking systems with a given ranking systems, *iii*) investigate the time series in N-star ranking and the trend prediction problem, and *iv*) apply N-star ranking systems in various ranking problems, e.g., business ranking, event ranking, and so on.

References

1. Brin, S., Page, L.: The anatomy of a large-scale hypertextual web search engine. In: Proceedings of the 7th International World Wide Web Conference, pp. 107–117 (1998)

[2] http://intelligent.pe.kr/ConfRank.txt

2. Keener, J.: The Perron-Frobenius theorem and the ranking of football teams. SIAM Review 35(1), 80–93 (1993)
3. Kien, L.T., Hieu, L.T., Hung, T.L., Vu, L.A.: MpageRank: The Stability of Web Graph. Vietnam Journal of Mathematics 37, 475–489 (2009)
4. Sidiropoulos, A., Katsaros, D., Manolopoulos, Y.: Generalized Hirsch h-index for disclosing latent facts in citation networks. Scientometrics 72(2), 253–280 (2006)
5. Ley, M., Reuther, P.: Maintaining an Online Bibliographical Database: The Problem of Data Quality. EGC 2006, 5–10 (2006)
6. Ley, M.: DBLP - Some Lessons Learned. PVLDB 2(2), 1493–1500 (2009)
7. Furukawa, T., Okamoto, S., Matsuo, Y., Ishizuka, M.: Prediction of social bookmarking based on a behavior transition model. In: Proceedings of the 2010 ACM Symposium on Applied Computing, pp. 1741–1747 (2010)
8. Rendle, S., Freudenthaler, C., Thieme, L.S.: Factorizing personalized Markov chains for next-basket recommendation. In: Proceedings of the 19th International Conference on World Wide Web, pp. 811–820 (2010)
9. Freudenthaler, C., Rendle, S., Thieme, L.S.: Factorizing Markov Models for Categorical Time Series Prediction. In: Proceedings of ICNAAM, pp. 405–409 (2011)
10. Microsoft Corporation: Microsoft Academic Search (June 26, 2013), `http://academic.research.microsoft.com/`
11. Collins, M.: Ranking algorithms for named-entity extraction: Boosting and the voted perceptron. In: Proceedings of the 40th Annual Meeting on Association for Computational Linguistics, pp. 489–496 (2002)
12. Vercoustre, A.-M., Thom, J.A., Pehcevski, J.: Entity ranking in Wikipedia. In: Proceedings of the 2008 ACM Symposium on Applied Computing (SAC 2008), pp. 1101–1106 (2008)
13. Cronin, B.: Bibliometrics and beyond: Some thoughts on web-based citation analysis. Journal of Information Science 27(1), 1–7 (2001)
14. Nie, Z., Zhang, Y., Wen, J., Ma, W.: Object-Level Ranking: Bringing Order to Web Objects, Study of the eXplicit Control Protocol (XCP). IEEE Infocom (2005)
15. Snchez-Burillo, E., Duch, J., Gmez-Gardenes, J., Zueco, D.: Quantum Navigation and Ranking in Complex Networks Nature online journal (2012)

Automated Interestingness Measure Selection for Exhibition Recommender Systems

Kok Keong Bong[1,2], Matthias Joest[2], Christoph Quix[3], and Toni Anwar[1,4]

[1] The Sirindhorn International Thai German Graduate School of Engineering, King Mongkut's University of Technology North Bangkok, 1518 Pibulsongkram Road, Bangsue, 10800 Bangkok, Thailand
[2] Heidelberg Mobil International GmbH Industriestrasse 41, D-69190 Walldorf, Germany
[3] Information Systems, RWTH Aachen University, Germany and Fraunhofer FIT, St. Augustin, Germany
[4] Faculty of Computing, Universiti Teknologi Malaysia (UTM), 81310 Johor Bahru, Johor, Malaysia

Abstract. Exhibition guide system contain various information pertaining to exhibitors, products and events that are happening during the exhibitions. The system would be more useful if it is augmented with a recommender system. Our recommender system would recommend users a list of interesting exhibitors based on associations that mined from the web server logs. The recommendations are ranked based on various Objective Interestingness Measures (OIMs) that quantify the interestingness of an association. Due to data sparsity, some OIMs cannot provide distinct values for different rules and hamper the ranking process. In mobile applications, the ranking of recommendations is crucial because of the low real estate in mobile device screen sizes. We show that our system is able to select an OIM (from 50 OIMs) that would perform better than the regular Support-Confidence OIM. Our system is tested using data from exhibitions held in Germany.

Keywords: Association Rule Mining, Objective Interestingness Measures, Clustering.

1 Introduction

The exhibition recommender domain presents great challenges for recommender systems that are based on user interaction because the exhibition system is only being used a few days before and during an exhibition. This leads to sparse user navigation data. Running association rule mining on sparse dataset would produce many association rules that have the same Support or Confidence values. This hinders the effectiveness of the recommender system to rank recommendations. So, there is a need for an OIM that could provide distinct values for different association rules/recommendations. The recommendations that have

N.T. Nguyen et al. (Eds.): ACIIDS 2014, Part I, LNAI 8397, pp. 221–231, 2014.

the same interestingness value would require extra rank parameter or more complex OIMs.

The OIMs also act to filter out possibly irrelevant rules in the face of huge amounts of data and huge amounts of association rules mined. The survey in [7] noted that the roles of OIMs in association rule mining consists of: (1) pruning search space so as to enable efficient rule mining, (2) rule ranking (3) post-processing to uncover only interesting rules. The primary methods used to utilize OIMs are the selection of OIM and the aggregation of OIMs.

In this paper, we explored the former method as there are currently more than 50 OIMs reported to be used in various literatures. We tried to answer the question of when Support-Confidence values fail to provide distinct valuations for association rules, how do we select the best OIM to be used? In addition, we provide mobile guides to various exhibitions (of different domains like book exhibition, car exhibition, electronics exhibitions and industrial exhibitions) throughout the year and we require a mechanism that can automatically choose an OIM for each exhibition. The recommender system should perform considerably good (equal or better than the standard Support-Confidence OIMs in terms of precision and recall).

2 Related Work

Given a transactional database, the FPGrowth or Apriori algorithm would mine the association rules that satisfy the regular Support-Confidence OIM threshold. The Support OIM is defined as the probability of the association rule occuring (e.g. how many customers bought egg, milk and bread together?). The Confidence OIM is the probability that the consequent occurs in a particular transaction given that the antecedent has occurred. In a supermarket example, it is interpreted as "if a customer has bought egg and milk, what are the probability that the same customer would also buy bread?".

In an empirical study of OIM contributions for building a classfier, Jalali-Heravi and Osmar [9] utilized a total of 53 OIMs for building classifiers for 20 datasets from the machine learning repository at University of California, Irvine (UCI). The results were reported as which OIMs perform better in the rule pruning and selection. They concluded that within their study, there are no OIMs that perform better than other OIMs consistently across different datasets and different phases of association rule mining. So, it is important that data miners are exposed to the methods of which OIMs could be better utilized, as opposed to customizing equations by hand to fit their problem.

Data miners can decide on which OIMs to use based on their experience and assumption about the association rules that they would like to find. There are many ways to select the appropriate OIMs: manually examining the OIMs [17], looking for specific properties in the dataset [12], using decision aids [5,10], clustering the OIMs based on experiments [2,8], employing meta learning mechanisms to predict user selection [1,6], or even by applying brute force analysis [15]. However, most selection methods suffer from the requirement of human participation, in which a human expert is required to either use their experience or define some properties which they would want in an OIM. This is a huge problem for us because we provide various cross-domain exhibition guide systems throughout the year and it would be ideal to automate such selection process to ensure low resource requirement and also high recommender precision-recall. Thus, we attempt to select an OIM based on correlations between OIMs.

Earlier works [11,13,16] focused on interestingness measures as a determinant of type of rule and proceed to perform correlation analysis on these values. The correlation mechanism is either through individual or combinations of the Pearson's, Spearman's or Kendall's correlation coefficients. Performing these analysis would require a long processing time as interestingness values need to be calculated for each association rule and correlations need to be done all those values. The only sampling method that we have found is [11] that sampled rules based on the most amount of ranking conflicts among the OIMs.

3 Idea and Methodology

Our contribution is two-fold. Firstly, we propose a novel method that is based on the fact that OIMs are an interplay of variables (which are the values of the contingency table values). In view of a sparse dataset, we argue that some contingency table cell values will not have a wide range of values (or not as important as a feature in calculating interestingness) and any OIM that uses such cell will not be able to properly differentiate association rules.

As an example, there are OIMs that uncovers association rules that have low number of counter examples (the contingency table value, $N_{a\bar{b}}$) but if counter example values are not an important criteria from the dataset (i.e. because most rules in the dataset also have the same low number of counter examples), then these OIMs will not be able to uncover this type of association rules. Instead, they will give values that correlate with other OIMs that do not specifically uncover these types of rules.

Thus, these OIMs will tend to correlate together in groups and differing distribution of contingency table cells will provide different groups of OIM. We proposed that these correlation groups represent a kind of voting mechanism. In principle, an OIM from this group is selected as the OIM to be used in our recommendations. The groups are extracted per domain and so, our methods would work across different domain of exhibitions.

Secondly, we improved on our algorithm's computational performance by proposing a sampling method before the OIMs are correlated. Thus, before we start correlating the OIMs, we cluster the rules based on their contingency table values. The clustering would produce groups of rules that are for example: "contain high number of counter examples" or "contain low number of null transactions". These clusters are actually the type of rules that can be found in the dataset. From the clusters, we could select representative rules. These sampled rules would speed up correlations process and limit the amount of outlier rules that might skew the correlations.

4 Choosing the Best OIM through Correlations

In this project, we implemented a set of 50 OIMs (listed in Table 1). Due to space constraints, readers can obtain the formulas used in the calculation of the OIMs through the works of [3,4,14]. The association rule mining algorithm was obtained from the KNIME tool while the correlations were done using the R package. After association rule mining is performed on the web server logs, the system would perform K-Means clustering on all the association rules based on 4 cells in the contingency table: N_a, N_b, $N_{a\bar{b}}$ (counter example, CE) and $N_{\overline{ab}}$ (null transaction, NT). We are not required to consider other cells in the contingency table because the contingency table has the property where given 4 unrelated cells, one can calculate the values for others cells in the contingency table. In other words, the cells other than the 4 mentioned earlier are related to the 4 mentioned earlier and clustering using all cells or just 4 cells would not yield different results.

After the clustering has been done, we would choose two rules from each cluster: the rule closest to the center point and the rule that has the median distance from the center point. Using a cluster size of ten, the previous step gives us 20 association rules. We can then perform Kendall's Tau correlation analysis using the 20 association rules' OIM values. This correlation analysis will uncover groups of OIMs that have the same ranking for different association rules. The final selected OIM is selected from the largest group of similar OIMs that have the most number of distinct values. In summary, the workflow for this system is (1) perform clustering on the rules contingency table values, (2) calculate correlations between interestingness values of the rule and finally (3) choosing the representative measure (of the largest group) as the one having the most discriminant values. In our opinion, the most appropriate measure should have the following 2 criteria: in line with majority other OIMs and has the most discriminant values. This process is illustrated in Figure 1.

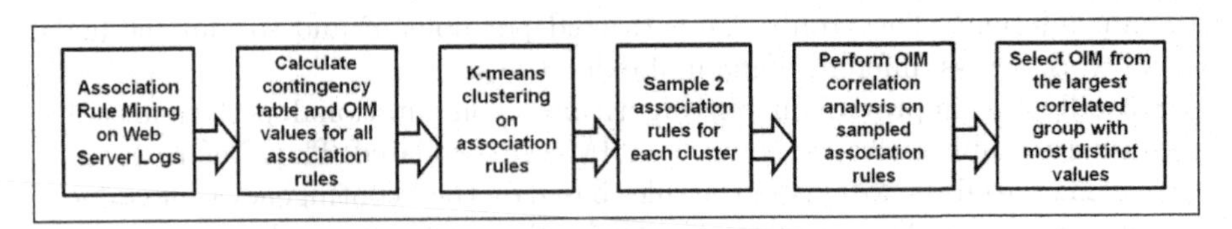

Fig. 1. Flowchart of proposed selection method

Table 1. List of OIMs used for the automated selection

Accuracy	Geometric Implication Intensity (GII)	Odds Ratio
Bayes Factor	IPEE	One-Way Support
Causal Confirmed Confidence	Implication Index	Putative Causal (PC) Dependency
Causal Confidence	Info Gain	Piatetsky Shapiro (PS)
Causal Confirm	J Measure	Pavillion
Causal Support	Jaccard	Phi Coefficient
Certainty Factor	Kappa	Relative Risk
Collective Strength	Klosgen	Rule Interest
Confidence	Kulczynski	SebagSchoenauer
Conviction	Laplace	Specificity
Cosine	Least Contradiction	Support
Dependency	Lerman	Two Way Support Var
Descriptive Confirmed Confidence	Leverage	Two-Way Support
Descriptive Confirm	Lift	YuleQ
Example Contra Example	Loevinger	YuleY
F Measure	Mutual Info	Zhang
Gini Index	Odds Multiplier	

Table 2. Exhibition Dataset Properties

Exhibition	Number of days	Average Web Views	Average Recommendations	Average Test Cases
CeBIT 2013	6	4,901	146,504	3,951
Hannover Messe 2013	5	2,576	238,744	2,753

5 System Evaluation

5.1 Dataset Utilized

We have utilized the web server logs from the CeBIT 2013 (electronics exhibition) and Hannover Messe 2013 (HM2013, industrial products exhibition) web application. Association rule mining is applied to the web server logs and the resultant associations are ranked and provided as recommendations. For evaluations, we compared our recommendations to the exhibitors that the user had bookmarked. Due to the fact that a user only bookmarks an exhibitor that he or she likes, these test cases would provide a real-life evaluation of our recommender systems. Table 2 shows the amount of visitor views that we have uncovered from the web server logs and the amount of bookmarks that we have to test our recommendation system.

5.2 Evaluation Protocol

For each day of the exhibition, association rule mining was ran on the data from historical exhibition days to retrieve the recommendations. A recommendation is considered good if a user who has bookmarked an exhibitor also bookmarked the recommended exhibitor. To better gauge the performance of our selection method, we would compare our selection method with OIMs selected using the following methods:

- **Simple**: Calculate the 50 OIM values for all association rules, take the OIM that had the most number of distinct values.

- **Extreme Association Rule (AR) Sampling**: Correlations are ran on rules sampled using the "Extreme" method, which are using only rules with highest value for individual contingency table values.
- **Previous-Day**: The best performing OIM based on previous day test is used for the current day.

The goal is to determine which of the above methods is capable of selecting the OIM(s) that perform better than the Support-Confidence framework in both precision and recall measures. Any OIM that only performs better in terms of precision (or recall) are not counted as better performing. The next section highlights the OIMs that were better than Support-Confidence OIM and the relative performance of each selection method (and the OIM selected from the respective method).

6 Results

The performance of the recommender system is illustrated in Table 3, 4, 5 and 6. The results presented indicated that there is no OIM that performed consistently better than the Support-Confidence OIM in all the days tested. This places more importance in a context-based selection of OIM. Context in this paper means selecting an OIM based on the contingency table value distribution. We compared our method with intuitive methods like Support-Confidence, Extreme selection and Previous-Day selection. OIMs like Support-Confidence often denote popularity-based recommendations and in exhibitions, recommending popular items would definitely perform well. Previous-Day OIM selection is also a viable solution to the problem because the historical dataset data set used could be seen as a learning model to predict current day recommendations.

As a definition, a test scenario is (for either Top-3 or Top-5), the performance of the selection method on a particular exhibition day. If we defined every "BETTER" performance as having two points, "EQUAL" as having one point and "WORSE" performance as having zero point, our method would have scored 27 points while the closest competitor is the Previous-Day method with 24 points. The bottom two performers are the Extreme method with 11 points and the Simple method with 4 points. In addition, our selection method only performed worse than Support-Confidence in two of the test scenario, while the Previous-Day selection performed worse in seven of the test scenarios. From the tables as well, we could observe that out of the 20 test scenarios, the Support-Confidence OIM only perform as the best OIM in only 2 test scenario. Thus, if the recommender system uses only Support-Confidence, then it will fail to realize better performance in most of the exhibition days. As a short note, the Simple method highlights that choosing the OIM that gives the most distinct values is not an appropriate method but if we merge distinctness with a group-voting mechanism (like in our work), we can achieve better performance.

Table 7 and 8 shows the average precision across the different exhibition days for CeBIT 2013 and HM 2013 respectively. The first four columns denote the same benchmark methods as shown in Table 3, 4, 5 and 6. The last two columns

Table 3. Recommender Systems Top-3 Performance in CeBIT 2013

Exhibition Day	Better OIMs	Our work	Simple	Extreme	Previous-Day
1	Causal Confirmed Confidence, Causal Confidence, Certainty Factor, Conviction, Dependency, Gini Index, Least Contradiction, Leverage, Loevinger, PC Dependency, PS, Pavillion	EQUAL (Confidence)	WORSE (Implication Index)	WORSE (Phi Coefficient)	BETTER (Least Contradiction)
2	Descriptive Confirmed Confidence, Desc Confirm, Example Contra Example, Gini Index, Laplace, Least Contradiction, Sebag Schoenauer	BETTER (Descriptive Confirmed Confidence)	WORSE (Implication Index)	WORSE (Leverage)	BETTER (Least Contradiction)
3	Certainty Factor, Conviction, Dependency, Gini Index, Leverage, Loevinger, PC Dependency, PS, Pavillion, Rule Interest	BETTER (Pavillion)	WORSE (Implication Index)	BETTER (Conviction)	EQUAL (Descriptive Confirmed Confidence)
4	Causal Confirmed Confidence, Causal Confidence, Certainty Factor, Conviction, Dependency, Example Contra Example, Gini Index, Leverage, Loevinger, PC Dependency, PS, Pavillion	BETTER (Example Contra Example)	WORSE (Implication Index)	EQUAL (Descriptive Confirmed Confidence)	BETTER (Gini Index)
5	Causal Confirmed Confidence, Causal Confidence, Certainty Factor, Conviction, Dependency, Gini Index, Leverage, Loevinger, PC Dependency, PS, Pavillion, Rule Interest	BETTER (PC Dependency)	WORSE (Implication Index)	WORSE (Phi Coefficient)	BETTER (Causal Confirmed Confidence)
6	Implication Index, Least Contradiction	EQUAL (Leverage, PS)	BETTER (Implication Index)	EQUAL (Conviction)	EQUAL (Conviction)

Table 4. Recommender Systems Top-5 Performance in CeBIT 2013

Exhibition Day	Better OIMs	Our work	Simple	Extreme	Previous-Day
1	Least Contradiction	EQUAL (Confidence)	WORSE (Implication Index)	WORSE (Phi Coefficient)	BETTER (Least Contradiction)
2	Descriptive Confirmed Confidence, Least Contradiction	EQUAL (Confidence)	WORSE (Implication Index)	WORSE (Leverage)	BETTER (Least Contradiction)
3	Descriptive Confirmed Confidence, Gini Index, Least Contradiction, Leverage, PS	EQUAL (Pavillion)	WORSE (Implication Index)	WORSE (Conviction)	BETTER (Least Contradiction)
4	Causal Confirmed Confidence, Causal Confidence, Certainty Factor, Conviction, Dependency, Leverage, Loevinger, PC Dependency, PS, Pavillion, Rule Interest	EQUAL (Confidence)	WORSE (Implication Index)	WORSE (Phi Coefficient)	WORSE (Least Contradiction)
5	Certainty Factor, Conviction, Dependency, Gini Index, Loevinger, PC Dependency, Pavillion, Rule Interest	BETTER (PC Dependency)	WORSE (Implication Index)	WORSE (Phi Coefficient)	WORSE (Causal Confirmed Confidence)
6	Implication Index, Least Contradiction	EQUAL (Leverage, PS)	BETTER (Implication Index)	EQUAL (Conviction)	EQUAL (Dependency)

which are labelled as "Support/Conf." and "Optimal". The former lists down the precision values in the scenario where we only used the Support and Confidence metric to sort recommendations while the latter illustrates the scenario where one managed to select the best performing metric in each exhibition days. Those two columns are grouped under the "Context" column to provide context into our analysis. In this perspective, we observe that our method and the closest competitor ("Previous-Day" method) performed closely to the "Optimal" precisions although we noted that the "Previous-Day" method has a higher average precision at 3.97% in CeBIT 2013 than our method (with precision 3.77%).

Table 5. Recommender Systems Top-3 Performance in Hannover Messe 2013

Exhibition Day	Better OIMs	Our work	Simple	Extreme	Previous-Day
1	Certainty Factor, Conviction, Example Contra Example, Least Contradiction, Loevinger, Rule Interest	EQUAL (Confidence)	WORSE (Implication Index)	WORSE (Collective Strength)	WORSE (Dependency)
2	Certainty Factor, Conviction, Dependency, Descriptive Confirmed Confidence, Descriptive Confirm, GII, Loevinger, One-Way Support, PC Dependency, Pavillion, Sebag Schoenauer, Two-Way Support	EQUAL (Confidence)	WORSE (Implication Index)	WORSE (Collective Strength)	WORSE (Rule Interest)
3	None	WORSE (PC Dependency)	WORSE (Implication Index)	WORSE (Conviction)	WORSE (Dependency)
4	Certainty Factor, Conviction, Dependency, Leverage, Loevinger, One-Way Support, PC Dependency, PS, Pavillion, Rule Interest, Two-Way Support	EQUAL (Confidence)	(Implication Index)	BETTER (Conviction)	EQUAL (Confidence)
5	Certainty Factor, Collective Strength, Conviction, Dependency, Descriptive Confirm, F Measure, GII, J Measure, Jaccard, Kappa, Loevinger, One-Way Support, PC Dependency, Pavillion, Sebag Schoenauer, Two-Way Support	BETTER (PC Dependency)	WORSE (Implication Index)	BETTER (Conviction)	BETTER (Certainty Factor, PC Dependency)

Table 6. Recommender Systems Top-5 Performance in Hannover Messe 2013

Exhibition Day	Better OIMs	Our work	Simple	Extreme	Previous-Day
1	Least Contradiction	EQUAL (Confidence)	WORSE (Implication Index)	WORSE (Collective Strength)	BETTER (Least Contradiction)
2	Dependency, One-Way Support, PC Dependency, Pavillion, Two-Way Support	EQUAL (Confidence)	WORSE (Implication Index)	WORSE (Collective Strength)	EQUAL (Least Contradiction)
3	Causal Confidence, Example Contra Example, Laplace	WORSE (PC Dependency)	WORSE (Implication Index)	WORSE (Conviction)	WORSE (Dependency)
4	None	EQUAL (Confidence)	WORSE (Implication Index)	WORSE (Conviction)	WORSE (Causal Confidence, Example Contra Example, Laplace)
5	Conviction, Dependency, F Measure, Gini Index, GII, J Measure, Jaccard, Kappa, Laplace, Leverage, Loevinger, One-Way Support, PC Dependency, PS, Pavillion, Rule Interest, Sebag Schoenauer, Two-Way Support	BETTER (PC Dependency)	WORSE (Implication Index)	BETTER (Conviction)	EQUAL (Confidence)

Table 7. Average precision values for CeBIT 2013

	Benchmarks				Context	
	Work	Simple	Extreme	Previous-Day	Support/Conf.	Optimal
Top 3	4.20	2.31	3.77	4.47	4.11	4.53
Top 5	3.35	1.99	2.66	3.47	3.29	3.61
Average	3.77	2.15	3.22	3.97	3.70	4.07

Table 8. Average precision values for Hannover Messe 2013

	Benchmarks				Context	
	Work	Simple	Extreme	Previous-Day	Support/Conf.	Optimal
Top 3	2.01	1.00	1.95	1.99	1.98	2.10
Top 5	1.66	0.98	1.55	1.63	1.62	1.69
Average	1.83	0.99	1.75	1.81	1.80	1.90

In HM 2013, our work (with precision 1.83%) outperformed the "Previous-Day" method (with precision 1.81%).

Due to the sampling methods that we have proposed, the amount of time required to select a proper OIM was reduced significantly. The speedup comes from the bottleneck of performing Kendall's Tau correlations on the association rules. In our experiments, performing Kendall's Tau correlation on 20 association rules took less than 1 minute while the same correlation done on 12,000 association rules would took more than 60 minutes. Referring to Table 2, our dataset could have up to 130,000 association rules per exhibition day. Thus the sampling method enabled our recommender system to execute efficiently and enabled us to perform more selections at a lower time granularity (for example, per hour instead of the current experiments' daily selection). This enables us to refresh our OIM selection frequently. It is noted that all association rule mining and test simulation are performed on a workstation that utilizes a quad-core Central Processing Unit (CPU) running at 2.00 GHz, 16GB Random Access Memory (RAM) and a 256 GB Solid State Drive (SSD) storage.

7 Improvement Opportunities

To accomodate automation, we have decided to choose an OIM that comes from the largest group of OIMs that agree with each other. Given such decision, there exists a danger of biased selection of OIMs. If our collection contains many OIM implementations (e.g. more than half of the OIM list) that is derived with the Support-Confidence framework, theoretically, the selection method will always group them together. This would mean that the largest group will always be the same.

Due to the fact that association rule mining algorithms produce rules that are based on the Support-Confidence measure, it is impossible that by using other measures we can achieve a much higher precision-recall than the maximal calculated. This is because all the OIMs function at a post processing stage where rules are ranked and filtered accordingly. Thus, all other OIMs used in this case study can only perform better than Support-Confidence in ranking of recommendations/rules. This aligns with our problem where due to the data sparsity, the Support-Confidence measures are not able to provide a high amount of distinct values for all the rules in order to rank them for recommendations.

8 Conclusion

The problem in which this paper addresses is the automated selection of an appropriate interestingness measure for a sparse dataset. Our application context is in exhibition guide system where the server log datasets only grow at the start of the exhibition and cease to be useful after exhibition ends. Given the short lived dataset, conventional OIMs like Support and Confidence will always return association rules that have the same values. This poses great challenge if the association rules are to be used for recommendations, as the OIM to be used must have high precision-recall in its recommendations and also need to be able to provide differing valuations for different rules. Therefore, it is crucial that the system is able to select an OIM that could rank the association rules distinctly in a high precision-recall manner.

In this work, we have opted to follow the idea of: an interesting rule, is the one that satisfies many different interestingness measures (OIMs). If there are many measures that rank a particular association rule highly, then that rule is taken to be interesting. Though in a normal sense, optimal rule mining is done by trading off some values of OIMs for other OIMs to achieve optimal interestingness value, we propose here a unique perspective in which OIMs correlate to form groups depending to the distribution of the contingency table values of the association rules. The correlated groups mean that the OIMs in the group agree that a particular association rule is interesting. That and the criteria of being discriminant are the criteria we used when selecting the appropriate interestingness measure. We also proposed the clustering of association rules based on contingency table distribution to sample a smaller ruleset for correlations. This relieves us from the computational cost of computing the Kendall's Tau.

By using the aforementioned architecture, we showed that the selection method we proposed are able to perform better than Support-Confidence in the real life data set of exhibitions. The method allowed the recommender system to achieve higher precision-recall with the sparse data of CeBIT 2013 and Hannover Messe 2013 exhibitions. It also outperformed other intuitive selection methods while at the same time required no human intervention and have more efficient computation through the sampling process.

References

1. Abe, H., Ohsaki, M., Tsumoto, S., Yamaguchi, T.: Evaluating a rule evaluation support method with learning models based on objective rule evaluation indices - A case study with a meningitis data mining result. In: Proceedings of the 5th IEEE International Conference on Hybrid Intelligent Systems, HIS 2005, pp. 169–174. IEEE Computer Society, Washington, DC (2005), http://dx.doi.org/10.1109/ICHIS.2005.37
2. Belohlavek, R., Grissa, D., Guillaume, S., Nguifo, E.M., Outrata, J.: Boolean factors as a means of clustering of interestingness measures of association rules. In: Proceedings of the 8th International Conference on Concept Lattices and Their Applications (2011)
3. Bonchi, F., Lucchese, C.: Pushing tougher constraints in frequent pattern mining. In: Ho, T.-B., Cheung, D., Liu, H. (eds.) PAKDD 2005. LNCS (LNAI), vol. 3518, pp. 114–124. Springer, Heidelberg (2005), http://dx.doi.org/10.1007/11430919_15
4. Bong, K.K.: A Framework for Objective Interestingness Measures Selection. Master's thesis, The Sirindhorn International Thai German Graduate School of Engineering, King Mongkut's University of Technology North Bangkok (2012)
5. Delpisheh, E., Zhang, J.Z.: Evaluating association rules by quantitative pairwise property comparisons. In: Proceedings of the IEEE International Conference on Data Mining Workshops, ICDMW 2010, pp. 927–934. IEEE Computer Society, Washington, DC (2010), http://dx.doi.org/10.1109/ICDMW.2010.145

6. Delpisheh, E., Zhang, J.Z.: A dynamic composite approach for evaluating association rules. In: The 7th International Conference on Natural Computation (ICNC), pp. 1893–1898 (2011)
7. Geng, L., Hamilton, H.J.: Interestingness measures for data mining: A survey. ACM Computing Survey 38 (September 2006), http://doi.acm.org/10.1145/1132960.1132963
8. Grissa, D., Guillaume, S., Nguifo, E.M.: Combining clustering techniques and formal concept analysis to characterize interestingness measures. Computing Research Repository (CoRR) abs/1008.3629 (2010)
9. Jalali-Heravi, M., Zaïane, O.R.: A study on interestingness measures for associative classifiers. In: Proceedings of the 2010 ACM Symposium on Applied Computing, SAC 2010, pp. 1039–1046. ACM, New York (2010), http://doi.acm.org/10.1145/1774088.1774306
10. Lenca, P., Meyer, P., Vaillant, B., Lallich, S.: On selecting interestingness measures for association rules: User oriented description and multiple criteria decision aid. European Journal of Operational Research 184(2), 610–626 (2008)
11. Tan, P.N., Kumar, V., Srivastava, J.: Selecting the right interestingness measure for association patterns. In: Proceedings of the 8th ACM SIGKDD International Conference on Knowledge Discovery and Data Mining, KDD 2002, pp. 32–41. ACM, New York (2002), http://doi.acm.org/10.1145/775047.775053
12. Vaillant, B., Lallich, S., Lenca, P.: Modeling of the counter-examples and association rules interestingness measures behavior. In: Crone, S.F., Lessmann, S., Stahlbock, R. (eds.) The 2nd International Conference on Data Mining (DMIN), pp. 132–137. CSREA Press (2006), http://dblp.uni-trier.de/db/conf/dmin/dmin2006.html#VaillantLL06
13. Vaillant, B., Lenca, P., Lallich, S.: A clustering of interestingness measures. In: Suzuki, E., Arikawa, S. (eds.) DS 2004. LNCS (LNAI), vol. 3245, pp. 290–297. Springer, Heidelberg (2004)
14. Wu, J., Zhu, S., Xiong, H., Chen, J., Zhu, J.: Adapting the right measures for pattern discovery: A unified view. The IEEE Transactions on Systems, Man, and Cybernetics, Part B: Cybernetics PP(99), 1–12 (2012)
15. Xianneng, L., Mabu, S., Huiyu, Z., Shimada, K., Hirasawa, K.: Analysis of various interestingness measures in classification rule mining for traffic prediction. In: Proceedings of The Society of Instrument and Control Engineers (SICE) Annual Conference, pp. 1969–1974 (August 2010)
16. Xuan-Hiep, H., Guillet, F., Briand, H.: Arqat: An exploratory analysis tool for interestingness measures. In: International Symposium on Applied Stochastic Models and Data Analysis (2005)
17. Zhang, L., Yu, D.L., Wang, Y.G., Zhang, Q.M.: Selecting an appropriate interestingness measure to evaluate the correlation between chinese medicine syndrome elements and symptoms. Chinese Journal of Integrative Medicine, 1–7 (2011), http://dx.doi.org/10.1007/s11655-011-0859-z, doi:10.1007/s11655-011-0859-z

Equivalent Transformation in an Extended Space for Solving Query-Answering Problems

Kiyoshi Akama[1] and Ekawit Nantajeewarawat[2]

[1] Information Initiative Center, Hokkaido University, Hokkaido, Japan
akama@iic.hokudai.ac.jp
[2] Computer Science Program, Sirindhorn International Institute of Technology
Thammasat University, Pathumthani, Thailand
ekawit@siit.tu.ac.th

Abstract. A query-answering problem (QA problem) is concerned with finding all ground instances of a query atomic formula that are logical consequences of a given logical formula describing the background knowledge of the problem. Based on the equivalent transformation (ET) principle, we propose a general framework for solving QA problems on first-order logic. To solve such a QA problem, the first-order formula representing its background knowledge is converted by meaning-preserving Skolemization into a set of clauses typically containing global existential quantifications of function variables. The obtained clause set is then transformed successively using ET rules until the answer set of the original problem can be readily derived. Many ET rules are demonstrated, including rules for unfolding clauses, for resolution, for dealing with function variables, and for erasing independent satisfiable atomic formulas. Application of the proposed framework is illustrated.

Keywords: Equivalent transformation; query-answering problems; unfolding; Skolemization; extended clauses.

1 Introduction

Query-answering problems (QA problems) are an important class of problems in the Semantic Web. Given a logical formula K, representing background knowledge, and an atomic formula (atom) q, representing a query, a QA problem is to find the set of all ground instances of q that are logical consequences of K. It is characteristically an "all-answers finding" problem, i.e., all ground instances of the query atom satisfying the requirement must be found.

Most studies on solving QA problems on the Semantic Web, e.g., [6,7,8], can be characterized as specific approaches, since they restrict logical formulas under consideration to certain specific subclasses of first-order logic, with a belief that more efficient computation can be achieved by such restriction. These specific approaches have been centered around proof problems—existing methods for solving proof problems, e.g., tableau-based methods and resolution-based methods, have been adapted to address some specific classes of QA problems.

N.T. Nguyen et al. (Eds.): ACIIDS 2014, Part I, LNAI 8397, pp. 232–241, 2014.

By contrast, we take a general approach to dealing with QA problems. Our approach is based on the equivalent transformation (ET) principle. A given QA problem is transformed equivalently into simpler forms until its answer set can be readily obtained. To complete this ET-based strategy, when there is no transformation path to proceed equivalently, a new space is devised and a new transformation path is made in the new space. For this purpose, meaning-preserving Skolemization has been developed in [1] together with a new extended space, called the $\mathrm{ECLS_F}$ space, over the set of all first-order logical formulas. This extended space includes function variables, which are variables ranging over function constants. Since function constants are mappings from tuples of ground terms to ground terms, atoms with function variables are regarded as "second-order" atoms.

In this paper, we present a framework for solving QA problems using ET and provide many ET rules. With meaning-preserving Skolemization and ET-based problem solving, our approach deals with a QA problem as follows: First, a logical formula representing background knowledge is converted into a first-order formula. The obtained formula is then converted into a set of extended clauses in the $\mathrm{ECLS_F}$ space using meaning-preserving Skolemization. Next, ET rules are applied for problem transformation in the $\mathrm{ECLS_F}$ space. ET rules on $\mathrm{ECLS_F}$ are demonstrated, e.g., ET rules for unfolding, for removing useless definite clauses, for resolution, for dealing with atoms with function variables, and for erasing independent satisfiable atoms.

To begin with, Section 2 introduces QA problems, meaning-preserving Skolemization, and the $\mathrm{ECLS_F}$ space. Section 3 presents our ET-based procedure for solving QA problems. Section 4 gives ET rules on the $\mathrm{ECLS_F}$ space. Section 5 illustrates application of our procedure. Section 6 provides conclusions.

2 QA Problems on an Extended Space

A *query-answering problem* (*QA problem*) is a pair $\langle K, q\rangle$, where K is a first-order formula, representing background knowledge, and q is a usual atom, representing a query. The answer to a QA problem $\langle K, q\rangle$, denoted by $\mathit{answer}(K, q)$, is the set of all ground instances of q that are logical consequences of K. As shown in [3], $\mathit{answer}(K, q)$ can be equivalently defined as

$$\mathit{answer}(K, q) = (\bigcap \mathit{Models}(K)) \cap \mathit{rep}(q), \tag{1}$$

where $\mathit{Models}(K)$ denotes the set of all models of K and $\mathit{rep}(q)$ the set of all ground instances of q.

To solve a QA problem $\langle K, q\rangle$ on first-order logic, the first-order formula K is usually converted into a conjunctive normal form. The conversion involves removal of existential quantifications by Skolemization, i.e., by replacement of an existentially quantified variable with a Skolem term determined by its relevant quantification structure. The classical Skolemization, however, does not preserve the logical meaning of a formula—the formula resulting from Skolemization is not necessarily equivalent to the original one [5]. In [1], a theory for extending the space of first-order formulas was developed and how meaning-preserving Skolemization can be achieved in the obtained extended space was shown. The basic idea of meaning-preserving Skolemization is to use existentially quantified function variables instead of usual Skolem functions. Clauses and QA problems on the extended space are introduced below.

Given any n-ary function constant or n-ary function variable f, an expression $\mathit{func}(f, t_1, \ldots, t_n, t_{n+1})$, where the t_i are usual terms, is considered as an atom of a new type, called a *func-atom*. When f is a function constant and the t_i are all ground, the truth value of this atom is true iff $f(t_1, \ldots, t_n) = t_{n+1}$.

A *clause* C in the extended space is a formula of the form

$$a_1, \ldots, a_m \leftarrow b_1, \ldots, b_n, \mathbf{f}_1, \ldots, \mathbf{f}_o,$$

where (i) $a_1, \ldots, a_m$ are usual atoms, (ii) each of $b_1, \ldots, b_n$ is a usual atom or a constraint atom, and (iii) $\mathbf{f}_1, \ldots, \mathbf{f}_o$ are *func*-atoms. The sets $\{a_1, \ldots, a_m\}$ and $\{b_1, \ldots, b_n, \mathbf{f}_1, \ldots, \mathbf{f}_o\}$ are called the *left-hand side* and the *right-hand side*, respectively, of the clause C, denoted by $\mathit{lhs}(C)$ and $\mathit{rhs}(C)$, respectively. When $m = 0$, C is called a *negative clause*. When $m = 1$, C is called a *definite clause*, the only atom in $\mathit{lhs}(C)$ is called the *head* of C, denoted by $\mathit{head}(C)$, and the set $\mathit{rhs}(C)$ is also called the *body* of C, denoted by $\mathit{body}(C)$. When $m > 1$, C is called a *multi-head clause*. All usual variables in a clause are universally quantified and their scope is restricted to the clause itself.

By meaning-preserving Skolemization [1], a first-order formula can be transformed equivalently into a set of clauses in the extended space. Function variables in the resulting clause set are all existentially quantified and their scope covers entirely all clauses in the set.

The set of all clause sets in the extended space is called the $\mathrm{ECLS_F}$ space. A QA problem $\langle \mathit{Cs}, q\rangle$ such that Cs is a clause set in $\mathrm{ECLS_F}$ and q is a usual atom is called a *QA problem on* $\mathrm{ECLS_F}$.

3 Solving QA Problems

Using the notation introduced in Section 3.1, our ET-based procedure is presented in Section 3.2.

3.1 Triples for Representation and Transformation of QA Problems

Let $\mathcal{A}$ be the set of all usual atoms and for any atom $a \in \mathcal{A}$, let $\mathit{rep}(a)$ denote the set of all ground instances of a. For more flexible representation and transformation, we represent a QA problem on $\mathrm{ECLS_F}$ as a triple $\langle \mathit{Cs}, q, \pi\rangle$, instead of a pair $\langle \mathit{Cs}, q\rangle$, where Cs is a set of clauses, $q \in \mathcal{A}$, and π is a partial mapping from $\mathcal{A}$ to $\mathcal{A}$ such that the range of π contains all instances of q. The answer to the QA problem $\langle \mathit{Cs}, q, \pi\rangle$, denoted by $\mathit{answer}(\mathit{Cs}, q, \pi)$, is defined by

$$\mathit{answer}(\mathit{Cs}, q, \pi) = \pi((\bigcap \mathit{Models}(\mathit{Cs})) \cap \mathit{rep}(q)), \qquad (2)$$

where $Models(Cs)$ is the set of all models of Cs. A QA problem $\langle Cs, q\rangle$ can thus be transformed equivalently into the triple form $\langle Cs, q, id\rangle$, where id is the identity mapping, since $answer(Cs, q) = answer(Cs, q, id)$ (cf. Equations (1) and (2)).

3.2 A Procedure for Solving QA Problems by ET

Assume that a QA problem $\langle K, q\rangle$ is given, where K is a first-order formula and q is a usual atom. To solve this problem using ET, perform the following steps:

1. Transform K by meaning-preserving Skolemization [1] into a clause set Cs in the $\mathrm{ECLS_F}$ space.
2. Change the QA problem $\langle Cs, q\rangle$ into the triple form $\langle Cs, q, id\rangle$, where id is the identity mapping.
3. Successively transform the QA problem $\langle Cs, q, id\rangle$ in the $\mathrm{ECLS_F}$ space using unfolding and other ET rules (see Section 4).
4. Assume that the transformation yields a QA problem $\langle Cs', q', \phi\rangle$. Then:
   (a) If Cs' is not satisfiable, then output $rep(\phi(q'))$ as the answer set.
   (b) If Cs' is a set of unit clauses the head of which are instances of q', then output the answer set

$$\bigcup_{C \in Cs'} rep(\phi(head(C))).$$

   (c) Otherwise stop with failure.

The obtained answer set is always correct since all transformation steps used in the procedure are answer-preserving.

4 ET Rules on $\mathrm{ECLS_F}$

ET rules for transforming QA problems on $\mathrm{ECLS_F}$ are provided below. Throughout this section, let $\langle Cs, q, \pi\rangle$ be a given QA problem on $\mathrm{ECLS_F}$.

4.1 An ET Rule for Inclusion of Query Information

To include a query atom in clause transformation, $\langle Cs, q, \pi\rangle$ can be transformed as follows:

1. Let p be the predicate of q. Introduce a new predicate p' that appears in neither Cs nor q. (Assume that the set $\mathcal{A}$ of all usual atoms is large enough for introducing such a predicate p'.)
2. Let ϕ be a partial mapping from $\mathcal{A}$ to $\mathcal{A}$ defined as follows: For any p-atom b, let $\phi(b)$ be the atom obtained from b by replacing the predicate p with p'. For any p'-atom b', let $\phi^{-1}(b')$ be the atom obtained from b' by replacing the predicate p' with p.
3. Transform $\langle Cs, q, \pi\rangle$ into $\langle Cs \cup \{(\phi(q) \leftarrow q)\}, \phi(q), \pi \circ \phi^{-1}\rangle$, where $\pi \circ \phi^{-1}$ is the composition of π with ϕ^{-1}.

4.2 ET Rules for Unfolding (Unf) and Definite-Clause Removal (RmD)

Usual unfolding transformation uses a set of definite clauses to unfold a definite clause at an atom in its body. Unfolding for extended clauses on $\mathrm{ECLS_F}$ is the same as usual unfolding transformation except that (i) a set D of definite clauses defining a predicate is determined from the whole clause set Cs and (ii) D is used to unfold a clause in $Cs - D$ (possibly negative or multi-head) at a usual atom in its right-hand side. Function variables are not changed by the most general unifier used in an unfolding operation, and *func*-atoms in a clause right-hand side are treated in the same way as constraint atoms.

After unfolding using a set D of definite clauses, the number of atoms to be unfolded may decrease. If no atom in the remaining clauses in $Cs - D$ can be further unfolded using D, the set D can be removed. The precise definition of unfolding and that of definite-clause removal on $\mathrm{ECLS_F}$ are given in [2].

4.3 An ET Rule for Resolution (Reso)

Resolution for extended clauses on $\mathrm{ECLS_F}$ is the same as the resolution for usual clauses except the possible existence of *func*-atoms. Only usual variables in *func*-atoms are changed by the most general unifier in use; function variables are not changed. More precisely, assume that:

1. $C_1, C_2 \in Cs$.
2. $lhs(C_1) = A_1$ and $rhs(C_1) = B_1 \cup \{b\}$.
3. $lhs(C_2) = A_2 \cup \{a\}$ and $rhs(C_2) = B_2$.
4. ρ is a renaming substitution such that C_1 and $C_2\rho$ have no common variable.
5. θ is the most general unifier of b and $a\rho$.
6. C_3 is a clause such that $lhs(C_3) = A_1\theta \cup A_2\rho\theta$ and $rhs(C_3) = B_1\theta \cup B_2\rho\theta$.

Then $\langle Cs, q, \pi\rangle$ can be transformed into $\langle Cs \cup \{C_3\}, q, \pi\rangle$.

4.4 Specialization with Respect to a Left-Hand-Side Atom (Spec)

A clause C in Cs can be specialized by finite ground substitutions determined by an atom in its left-hand side as follows:

1. Find an atom b in $lhs(C)$ and a finite set $\{b_1, \ldots, b_n\}$ of atoms such that $\mathcal{M}(\textsc{Split}(Cs)) \cap rep(b) \subseteq rep(b_1) \cup \cdots \cup rep(b_n)$, where
   - $\textsc{Split}(Cs) = \bigcup\{\textsc{Split}(C) \mid C \in Cs\}$,
   - for any clause $C \in Cs$, $\textsc{Split}(C)$ is the set of all definite clauses C' such that $head(C') \in lhs(C)$ and $body(C') = rhs(C)$, and
   - $\mathcal{M}(\textsc{Split}(Cs))$ is the least model of $\textsc{Split}(Cs)$.
2. Let $\Theta = \{mgu(b, b_i) \mid (1 \leq i \leq n)\ \&\ (b \text{ and } b_i \text{ are unifiable})\}$, where for any $i \in \{1, \ldots, n\}$, $mgu(b, b_i)$ denotes the most general unifier of b and b_i if they are unifiable.
3. Transform $\langle Cs, q, \pi\rangle$ into $\langle (Cs - \{C\}) \cup \{C\theta \mid \theta \in \Theta\}, q, \pi\rangle$.

4.5 Other ET Rules on ECLS_F

Elimination of Isolated *func*-Atoms (EIF). A *func*-atom $func(f, t_1, \ldots, t_n, v)$, where v is a usual variable, is said to be *isolated* in a clause C iff there is only one occurrence of v in C. Assume that:

1. $C \in Cs$ and C contains a *func*-atom $\mathbf{f}$ that is isolated in C.
2. C' is the clause obtained from C by removing $\mathbf{f}$.

Then $\langle Cs, q, \pi \rangle$ can be transformed into $\langle (Cs - \{C\}) \cup \{C'\}, q, \pi \rangle$.

Elimination of Subsumed Clauses (ESUB). A clause C_1 is said to *subsume* a clause C_2 iff there exists a substitution θ for usual variables such that $lhs(C_1)\theta \subseteq lhs(C_2)$ and $rhs(C_1)\theta \subseteq rhs(C_2)$. If Cs contains clauses C_1 and C_2 such that C_1 subsumes C_2, then $\langle Cs, q, \pi \rangle$ can be transformed into $\langle Cs - \{C_2\}, q, \pi \rangle$.

Elimination of Valid Clauses (EVAD). A clause is *valid* iff all of its ground instances are true. Given a clause C, if some atom in $rhs(C)$ belongs to $lhs(C)$, then C is valid. A valid clause can be removed.

Erasing Independent Satisfiable Atoms (EIS). Let $C \in Cs$. Assume that:

1. $B \subseteq rhs(C)$ such that B and $lhs(C) \cup (rhs(C) - B)$ have no common variable.
2. There exists a ground substitution θ such that each atom in $B\theta$ is true with respect to $Cs - \{C\}$.
3. C' is the clause obtained from C by removing all atoms in B from its right-hand side.

Then $\langle Cs, q, \pi \rangle$ can be transformed into $\langle (Cs - \{C\}) \cup \{C'\}, q, \pi \rangle$.

Side-Change Transformation (SCH). Assume that p is a predicate occurring in Cs such that p does not appear in the query atom q. Cs can be transformed by changing the clause sides of p-atoms as follows: First, determine a new predicate *notp* for p. Next, move all p-atoms in each clause to their opposite side in the same clause (i.e., from the left-hand side to the right-hand side and vice versa) with their predicates being changed from p to *notp*. Side-change transformation is useful for decreasing the number of atoms in multi-head clauses in Cs when (i) every negative clause in Cs has at most one p-atom in its right-hand side and (ii) every non-negative clause in Cs has more p-atoms in its left-hand side than those in its right-hand side.

Elimination of Positive Independent Clauses (EPI). A set of clauses is positive iff it contains no negative clause. Given a set $\tilde{Cs}$ of clauses, let $\text{LeftPred}(\tilde{Cs})$ and $\text{RightPred}(\tilde{Cs})$ be defined by:

- $\text{LeftPred}(\tilde{Cs}) = \{p \mid (C \in \tilde{Cs})$ & (p is a predicate appearing in $lhs(C))\}$
- $\text{RightPred}(\tilde{Cs}) = \{p \mid (C \in \tilde{Cs})$ & (p is a predicate appearing in $rhs(C))\}$

A subset Cs' of Cs is independent with respect to the query atom q iff the following conditions hold:

1. LEFTPRED(Cs') and LEFTPRED($Cs - Cs'$) are disjoint.
2. LEFTPRED(Cs') and RIGHTPRED($Cs - Cs'$) are disjoint.
3. The predicate of q does not belong to LEFTPRED(Cs').

If Cs' is a positive independent subset of Cs with respect to the query atom q, then $\langle Cs, q, \pi \rangle$ can be transformed into $\langle Cs - Cs', q, \pi \rangle$.

5 Example

Consider the knowledge base in Fig. 1, taken from [6], where (i) the two columns refer to the structural component and the relational component, (ii) the two rows refer to the intensional level and the extensional level, and (iii) x, y, and z are variables. The structural component is described using the description logic $\mathcal{ALC}$ [4]. The intensional part of the relational component is described using an extension of Horn clauses, where class membership constraints are specified after the symbol '&'. The query to be considered is to find every pair of a student s and a professor p such that s may do his/her thesis with p.

Based on standard translation from $\mathcal{ALC}$ to first-order logic [4], the knowledge base in Fig. 1 can be converted into a first-order formula K and this problem can then be formalized as the QA problem $\langle K, mayDoTh(x, y) \rangle$ on first-order logic, where x and y are variables. Application of the procedure in Section 3.2 to solve this QA problem is shown below. By meaning-preserving transformation [1], K is transformed into a set of extended clauses in the ECLS_F space. The resulting clause set Cs consists of the clauses C_1–C_{25} in Fig. 2, where f is a unary function variable. The QA problem $\langle Cs, mayDoTh(x, y), id \rangle$ is then successively transformed by applying the ET rules in Section 4 as follows, with reference to the clauses C_{26}–C_{36} in Fig. 3 and the clauses C_{37}–C_{53} in Fig. 4:

(UNF) By unfolding using the definitions of St ($\{C_{17}\}$), FP ($\{C_4, C_{13}, C_{15}\}$), $expert$ ($\{C_{24}, C_{25}\}$), Tp ($\{C_{19}, C_{20}\}$), $exam$ ($\{C_{21}\}$), $subject$ ($\{C_{22}, C_{23}\}$), and $curr$ ($\{C_{10}\}$), the clauses C_1, C_2, C_{11}, and C_{12} are transformed into C_{26}–C_{36}.

Structural	Relational
$NFP = FP \sqcap \neg\exists teach.Co$ $FP \sqsubseteq FM$ $AC \sqcup BC = Co$ $AC \sqcap BC \sqsubseteq \bot$	$curr(x, z) \leftarrow exam(x, y), subject(y, z)$ $\quad\quad \& \; x{:}\, St,\ y{:}\, Co,\ z{:}\, Tp$ $mayDoTh(x, y) \leftarrow curr(x, z), expert(y, z)$ $\quad\quad \& \; x{:}\, St,\ z{:}\, Tp,\ y{:}\, FP \sqcap \exists teach.AC$ $mayDoTh(x, y) \leftarrow \& \; x{:}\, St,\ y{:}\, NFP$
$john{:}\, FP,\ \ teach(john, ai)$ $mary{:}\, FP \sqcap \forall teach.AC$ $paul{:}\, St,\ \ ai{:}\, AC,\ \ kr{:}\, Tp,\ \ lp{:}\, Tp$	$exam(paul, ai)$ $subject(ai, kr),\ \ subject(ai, lp)$ $expert(john, kr),\ \ expert(mary, lp)$

Fig. 1. A knowledge base

C_1: $Co(x), NFP(y) \leftarrow FP(y), func(f, y, x)$
C_2: $teach(y, x), NFP(y) \leftarrow FP(y), func(f, y, x)$
C_3: $\leftarrow NFP(x), teach(x, y), Co(y)$

C_4: $FP(x) \leftarrow NFP(x)$ C_5: $FM(x) \leftarrow FP(x)$ C_6: $AC(x), BC(x) \leftarrow Co(x)$
C_7: $Co(x) \leftarrow AC(x)$ C_8: $Co(x) \leftarrow BC(x)$ C_9: $\leftarrow AC(x), BC(x)$

C_{10}: $curr(x, z) \leftarrow exam(x, y), subject(y, z), St(x), Co(y), Tp(z)$
C_{11}: $mayDoTh(x, y) \leftarrow curr(x, z), expert(y, z), St(x), Tp(z), FP(y), AC(w), teach(y, w)$
C_{12}: $mayDoTh(x, y) \leftarrow St(x), NFP(y)$

C_{13}: $FP(john) \leftarrow$ C_{14}: $teach(john, ai) \leftarrow$ C_{15}: $FP(mary) \leftarrow$
C_{16}: $AC(x) \leftarrow teach(mary, x)$ C_{17}: $St(paul) \leftarrow$ C_{18}: $AC(ai) \leftarrow$
C_{19}: $Tp(kr) \leftarrow$ C_{20}: $Tp(lp) \leftarrow$ C_{21}: $exam(paul, ai) \leftarrow$
C_{22}: $subject(ai, kr) \leftarrow$ C_{23}: $subject(ai, lp) \leftarrow$ C_{24}: $expert(john, kr) \leftarrow$
C_{25}: $expert(mary, lp) \leftarrow$

Fig. 2. Representing the knowledge base in Fig. 1 using clauses in the ECLS_F space

C_{26}: $Co(x), NFP(john) \leftarrow func(f, john, x)$ (Replaced by SCH)
C_{27}: $Co(x), NFP(mary) \leftarrow func(f, mary, x)$ (Replaced by SCH)
C_{28}: $Co(x), NFP(y) \leftarrow func(f, y, x), NFP(y)$ (Removed by EVAD)
C_{29}: $teach(john, x), NFP(john) \leftarrow func(f, john, x)$ (Replaced by SCH)
C_{30}: $teach(mary, x), NFP(mary) \leftarrow func(f, mary, x)$ (Replaced by SCH)
C_{31}: $teach(x, y), NFP(x) \leftarrow func(f, x, y), NFP(x)$ (Removed by EVAD)
C_{32}: $mayDoTh(paul, x) \leftarrow NFP(x)$ (Replaced by SPEC)
C_{33}: $mayDoTh(paul, john) \leftarrow AC(x), teach(john, x), Co(ai)$ (Replaced by EIS)
C_{34}: $mayDoTh(paul, mary) \leftarrow AC(x), teach(mary, x), Co(ai)$ (Replaced by EIS)
C_{35}: $mayDoTh(paul, john) \leftarrow AC(x), teach(john, x), NFP(john), Co(ai)$ (Removed by ESUB)
C_{36}: $mayDoTh(paul, mary) \leftarrow AC(x), teach(mary, x), NFP(mary), Co(ai)$ (Removed by ESUB)

Fig. 3. Clauses obtained by unfolding

(RMD) By definite-clause removal, the definitions of *St*, *FP*, *expert*, *Tp*, *exam*, *subject*, and *curr*, which are used by unfolding above, along with the definition of *FM* ($\{C_5\}$) are removed.

(ESUB) C_{35} and C_{36} are subsumed by C_{33} and C_{34}, respectively, and are removed.

(EIS) By C_7 and C_{18}, $Co(ai)$ is true. It is thus removed from the bodies of C_{33} and C_{34}. The resulting clauses are C_{37} and C_{38}.

(EIS) By C_{14} and C_{18}, the conjunction of $AC(x)$ and $teach(john, x)$ is satisfiable. C_{37} is then replaced with C_{39}.

(EVAD) Since C_{28} and C_{31} are valid, they are removed.

(SPEC) Since *NFP* only occurs in C_{26}, C_{27}, C_{29}, and C_{30}, the variable x in C_{32} is instantiated into *john* or *mary*, yielding C_{40} and C_{41}.

(ESUB) C_{40} is subsumed by C_{39} and is removed.

C_{37}:	$mayDoTh(paul, john) \leftarrow AC(x), teach(john, x)$	(Replaced by EIS)
C_{38}:	$mayDoTh(paul, mary) \leftarrow AC(x), teach(mary, x)$	(Removed by ESub)
C_{39}:	$mayDoTh(paul, john) \leftarrow$	
C_{40}:	$mayDoTh(paul, john) \leftarrow NFP(john)$	(Removed by ESub)
C_{41}:	$mayDoTh(paul, mary) \leftarrow NFP(mary)$	(Removed by ESub)
C_{42}:	$mayDoTh(paul, mary) \leftarrow teach(mary, x)$	(Removed by ESub)
C_{43}:	$mayDoTh(paul, mary), NFP(mary) \leftarrow func(f, mary, x)$	(Removed by ESub)
C_{44}:	$mayDoTh(paul, mary) \leftarrow func(f, mary, x)$	(Replaced by EIF)
C_{45}:	$mayDoTh(paul, mary) \leftarrow$	
C_{46}:	$notNFP(x) \leftarrow teach(x, y), Co(y)$	(Removed by EPI)
C_{47}:	$Co(x) \leftarrow func(f, john, x), notNFP(john)$	(Removed by EPI)
C_{48}:	$Co(x) \leftarrow func(f, mary, x), notNFP(mary)$	(Removed by EPI)
C_{49}:	$teach(john, x) \leftarrow func(f, john, x), notNFP(john)$	(Removed by EPI)
C_{50}:	$teach(mary, x) \leftarrow func(f, mary, x), notNFP(mary)$	(Removed by EPI)
C_{51}:	$AC(x) \leftarrow Co(x), notBC(x)$	(Removed by EPI)
C_{52}:	$Co(x), notBC(x) \leftarrow$	(Removed by EPI)
C_{53}:	$notBC(x) \leftarrow AC(x)$	(Removed by EPI)

Fig. 4. Clauses generated in the transformation process after unfolding

(Reso) By applying the resolution rule to C_{38} and C_{16} and eliminating a duplicate atom in the resolvent, C_{42} is derived.
(Reso) Applying the resolution rule to C_{42} and C_{30} yields the resolvent C_{43}.
(Reso) By applying the resolution rule to C_{43} and C_{41} and eliminating a duplicate atom in the resolvent, C_{44} is derived.
(EIF) Since $func(f, mary, x)$ is isolated in C_{44}, it is removed, resulting in C_{45}.
(ESub) C_{38}, C_{41}, C_{42}, and C_{43} are subsumed by C_{45}, and they are removed.
(SCh) By applying side-change transformation for *NFP*, the clauses C_3, C_{26}, C_{27}, C_{29}, and C_{30} are changed, respectively, into C_{46}, C_{47}, C_{48}, C_{49}, and C_{50}, where for any term t, $notNFP(t)$ corresponds to the negation of $NFP(t)$.
(SCh) By applying side-change transformation for *BC*, the clauses C_6, C_8, and C_9 are changed, respectively, into C_{51}, C_{52}, and C_{53}, where for any term t, $notBC(t)$ corresponds to the negation of $BC(t)$.
(EPI) The set of all remaining clauses can be partitioned into $Cs' = \{C_{39}, C_{45}\}$ and $Cs'' = \{C_7, C_{14}, C_{16}, C_{18}, C_{46}, C_{47}, C_{48}, C_{49}, C_{50}, C_{51}, C_{52}, C_{53}\}$. Since Cs' and Cs'' have no common predicate, they are independent. Since Cs'' has no negative clause, it is removed.

The final clause set is $Cs' = \{C_{39}, C_{45}\}$. From Cs', the answer set $\{mayDoTh(paul, john), mayDoTh(paul, mary)\}$ is directly derived.

6 Conclusions

The equivalent transformation (ET) principle provides a basis for solving a very large class of QA problems. Our proposed ET-based procedure for solving QA problems is a state-transition procedure in which a state is a QA problem and application of an ET rule results in state transition. Using ET, a given QA problem is transformed equivalently into simpler forms until its answer set can be readily obtained. To find computation paths toward solving QA problems on first-order logic, meaning-preserving Skolemization [1] has been invented together with a new extended space, called the $\mathrm{ECLS_F}$ space, over the set of all first-order logical formulas. The extended space includes function variables. As its characteristic features, a QA problem on $\mathrm{ECLS_F}$ involves clauses possibly with (i) multiple atoms in their left-hand sides and/or (ii) *func*-atoms containing function variables and/or function constants in their right-hand sides.

To cope with QA problems with these features, usual unfolding and resolution are not sufficient. Many additional ET rules are required. For dealing with *func*-atoms and function variables, ET rules for elimination of isolated *func*-atoms are provided. For reduction of a problem's size, ET rules for removing useless definite clauses, removing subsumed clauses and valid clauses, erasing independent satisfiable atoms from clause right-hand sides, and eliminating positive independent clauses are illustrated. Each transition step preserves the answer set of a given input problem and therefore the correctness of the proposed procedure is guaranteed.

References

1. Akama, K., Nantajeewarawat, E.: Meaning-Preserving Skolemization. In: 3rd International Conference on Knowledge Engineering and Ontology Development, Paris, France, pp. 322–327 (2011)
2. Akama, K., Nantajeewarawat, E.: Unfolding-Based Simplification of Query-Answering Problems in an Extended Clause Space. International Journal of Innovative Computing, Information and Control 9, 3515–3526 (2013)
3. Akama, K., Nantajeewarawat, E.: Embedding Proof Problems into Query-Answering Problems and Problem Solving by Equivalent Transformation. In: 5th International Conference on Knowledge Engineering and Ontology Development, Vilamoura, Portugal, pp. 253–260 (2013)
4. Baader, F., Calvanese, D., McGuinness, D.L., Nardi, D., Patel-Schneider, P.F.: The Description Logic Handbook, 2nd edn. Cambridge University Press (2007)
5. Chang, C.-L., Lee, R.C.-T.: Symbolic Logic and Mechanical Theorem Proving. Academic Press (1973)
6. Donini, F.M., Lenzerini, M., Nardi, D., Schaerf, A.: $\mathcal{AL}$-log: Integrating Datalog and Description Logics. Journal of Intelligent and Cooperative Information Systems 10, 227–252 (1998)
7. Motik, B., Sattler, U., Studer, R.: Query Answering for OWL-DL with Rules. Journal of Web Semantics 3, 41–60 (2005)
8. Tessaris, S.: Questions and Answers: Reasoning and Querying in Description Logic. PhD Thesis, Department of Computer Science, The University of Manchester, UK (2001)

Knowledge Generalization during Hierarchical Structures Integration

Marcin Maleszka

Wroclaw University of Technology, Wyb. Wyspianskiego 27, 50-370 Wroclaw
Marcin.Maleszka@pwr.wroc.pl

Abstract. Hierarchical data structures are common in modern applications. Tree integration is one of the tools that is not fully researched in this scope. Therefore in this paper we define a complex tree to model common hierarchical structures. Complex tree integration aim is determined by specific integration criteria. In this paper we define and analyze a criterion measuring generalization of knowledge – upper semantic precision. We analyze the criterion in terms of simpler syntactic criteria and describe an extended example of an information retrieval system using this criterion.

Keywords: knowledge generalization, knowledge precission, tree integration, integration criteria, integration algorithms.

1 Introduction

Hierarchical and graph-based structures are nowadays common both in theoretical and practical applications. Thesaurus and ontology are commonly used in many research papers and the XML format is widely used in various computer systems. With the spread of hierarchical structures, the need for more efficient or complex tools for processing them is increasing. This paper aim is to provide such tools for integrating knowledge stored in hierarchical format, both in the theoretical and practical aspect.

In our previous work we proposed a criteria-based approach to hierarchical data integration [8]. We introduced a notion of criteria that help describe the result of integration in terms of the input structures, e.g. how many vertices are in the integrated tree, in relation to how many were in the input trees. The focus of that research was entirely of structural and data level and we did not consider any semantic relations.

We have also proposed an information retrieval system that makes use of hierarchical structure integration [9]. In this work we integrated user profiles to improve recommendations. We used several previously defined structural criteria and proposed two new ones that dealt with the structure on the knowledge level. *Reliability* was based on the knowledge about user queries (how user queries influence the tree structure of the profile), while *Conflict Solving* was used to generalize user interests.

N.T. Nguyen et al. (Eds.): ACIIDS 2014, Part I, LNAI 8397, pp. 242–250, 2014.

In this paper we show our first proposition of a fully semantic integration criterion. We introduce *Upper Semantic Precision* as a method to model the generalization aspects of the integration procedure, using the modified measure of semantic similarity in trees. With this criterion we may describe both the integrated structure and the input ones (in terms of their diversity). We also present general descriptions of how such criterion may be used in an information retrieval system.

This paper is organized as follows: in Section 2 we present similar research into criterial approach to integration and into semantic similarity in trees; in Section 3 we define the required mathematical structures in order to define and analyze the criterion; in Section 4 we describe the information retrieval system using such information; we conclude this paper with some final remarks and future work aspects in Section 5.

2 Related Works

Criteria-based approach to hierarchical structures integration was proposed several times in the literature, mostly in relation to XML documents. In several works, Passi and Madria [5,10] propose general criteria for integrated global schema. They describe completess and correctness (integrated tree should include and correctly represent all concepts from input schemas and the integrated schema should represent the union of input schema domains), minimility (if an element occurs multiple times in different input trees, it should occur only once in the integrated tree) and understandability (integrated tree should be readable by the final user; if multiple structures satisfy other criteria, then the most readable one should be selected). One may note that these criteria are not formally defined and they may be treated as only general guidelines for designing integration algorithms.

Minimality criterion described as above was further analysed in [2] in a general case. Authors introduce multiple possible relations between concepts, e.g. semantic equivalence, semantic similarity, identity. User input is required to determine the proper relation, but further calculations are made using a specific algorithm. As each relation has some measure of similarity (e.g. identity is 1, semantic equivalence is close to 1), minimality may be calculated and used to eliminate redundant concepts.

Significant research in XML schema integration criteria wa done in [3,4,11]. In the latter one, significant technical criteria are analysed, including schema vs. instance, element vs. structure, language vs. constraint, etc. These were not criteria describing the result, as previously, but classifications used to categorize different practical systems for XML integration. In [4] the same authors focus more on describing the result of integration, proposing four groups of criteria (input, output, effort, quality measures). Some of the proposed solutions are very similar to those discussed in our previous papers, e.g. schema similarity (a measure of the relation between the number of correctly mapped elements in input trees to the total number of elements), precision and recall (based on information retrieval measures, the relation between the number of correctly mapped nodes to all mapped nodes, and the number of nodes correctly mapped by the automatic systems to the number of nodes mapped by the user). These types of criteria may be used to improve integration algorithms.

Measures of knowledge generalization and knowledge similarity have also been previously proposed in literature. In [12] the author discusses a similarity measure between concepts in a thesaurus based on the notion of *information content*. The similarity is then equal the amount of information in *the most informative subsumer*. A more complex approach proposed in [6,7] calculates the relation between information content of the common ancestor and the compared terms. They also extend the approach to a non-tree ontology and use fuzzy membership values.

3 Knowledge Generalization

In previous papers the author has introduced the complex tree structure to facilitate integration of hierarchically structured data. In this paper the knowledge aspect of integration is considered, but the same structure may be used to represent it. Thus in this section we will describe the complex tree and the integration task for it, before describing a measure of knowledge generalization and defining the criterion for knowledge generalization.

3.1 Complex Tree Integration

To define the complex tree we must first define its substructure - denrite:

Definition 1. ***Dendrite*** *is a rooted tree (consistent, acyclic graph with a single distinct node – root)* $D = (W, E)$*, where:*

- W *is a finite set of nodes.*
- E *is a set of edges* $E \subseteq W \times W$.

We also introduce the following notations: A is a set of attributes. $l \in A$ is a special attribute called label (it is not the unequivocal node identifier of graph theory, labels may be repeated in a complex tree). V_a is the domain of attribute $a \in A$. $V = \bigcup_{a \in A} V_a$ is the set of all values of all attributes. Due to these notations one may say that (A, V) is a representation of real world.

Let now T be a finite set of node types. Each node of the same type represents a real world object of the same class (e.g. tags in XML).

Definition 2. ***Complex Tree*** *is a five* $CT = (W, E, T_{CT}, A_{CT}, V_{CT})$*, where:*

- (W, E) *is a dendrite.*
- $T_{CT} : W \to T$ *is a function that assigns a single type to each node.*
- $A_{CT} : W \to 2^A$*, where* $\forall_{w \in W} l \in A_{CT}(w)$ *is a function that assigns each node a set of attributes (each node has at least one attribute – label).*

– $V_{CT} : W \times A \rightarrow V$ *is a partial function that assigns to nodes and attributes some value such that:*

$$\forall_{w \in W} \forall_{a \in A_{CT}(w)} V_{CT}(w, a) \in V_a .$$

From this definition, each node $w \in W$ representing a real world object has the following form $w = \big(T_{CT}(w), A_{CT}(w), \{V_{CT}(w, a) : a \in A_{CT}\}\big)$. An object is described by a label $V_{CT}(w, l)$, type $T_{CT}(w)$, a set of attributes $A_{CT}(w)$ and their values $V_{CT}(w, a), a \in A_{CT}(w)$.

Let now $\Pi_k(B)$ be a set of every k-element subset with repetitions of some set B and 2^B be a powerset of B. Let also:

$$\Pi(B) = \bigcup_{k \in \mathbf{N}} \Pi_k(B) \tag{1}$$

Therefore $\Pi(B)$ is a set of all non-empty, finite subsets with repetitions of some set B.

We may now define a complex tree integration function, based on the set of all possible complex trees $\overline{CT}$:

Definition 3. *We use the name complex tree* ***integration function*** *for every such function* I:

$$I : \Pi(\overline{CT}) \rightarrow 2^{\overline{CT}}, \tag{2}$$

such that for every complex tree $CT^* \in I(\Pi(\overline{CT}))$ *one or more integration criterion* $K_j, j \in \{1, \ldots, k\}$ *is met.*

We use the notation $\mathbf{CT}^*$ for a set of complex trees that is the result of such integration function.

We defined some integration criteria in [8]. Each integrated tree may be described by a normalized measure $M_j(CT^* | CT_1, CT_2, \ldots, CT_N)$. Criterion K_j is met, if the corresponding measure $M_j = 1$.

3.2 Measure of Knowledge Generalization

In this paper we use a distance function for knowledge generalization d – an extension of Resnick's information-based approach [12] to semantic relatedness, as described in [1]. The idea of the distance function is to find the common information content of two nodes in some external ontology (when applied to hierarchical structures: the common parent in the tree). Here, we focus on the generalization aspect of the similarity (relatedness) calculation and reverse it in order to use the distance aspect.

Consequently, the knowledge generalization distance has the following properties for the hierarchical structure:

– The maximal distance between two nodes is the distance between the root of the tree, and the most specialized leaf:

$$\max_{v_1, v_2 \in V} d(v_1, v_2) = \max_{l \in L} d(v_{\text{root}}, l) \tag{3}$$

where L is the set of all leafs in the tree

- Distance is never equal 0 for different nodes. It is positive for generalization and negative for specialization:

$$\forall_{v_1,v_2 \in V} d(v_1, v_2) = 0 \Leftrightarrow v_1 = v_2 \quad (4)$$

$$\forall_{v_1,v_2 \in V} d(v_1, v_2) > 0 \Leftrightarrow v_1 \text{ is ancestor of } v_2 \quad (5)$$

$$\forall_{v_1,v_2 \in V} d(v_1, v_2) < 0 \Leftrightarrow v_2 \text{ is ancestor of } v_1 \quad (6)$$

- Consequently, the distance is not simmetrical (it is not a metric). Distance between more general and less general node is opposite to the distance between less general and more general node.

$$d(v_1, v_2) = -d(v_2, v_1) \quad (7)$$

- Distance is transitive on a single path.

$$d(v_1, v_3) = d(v_1, v_2) + d(v_2, v_3) \text{ if } v_1 \text{ is parent of } v_2 \text{ and } v_2 \text{ is parent of } v_3 \quad (8)$$

One may also observe that due to this function, the second Key Assumption (K2) proposed in our previous research ([9]) becomes invalid when this measure is used. The reason for this is that the information content of a more specialized term includes the information of the more general one.

3.3 Knowledge Integration Criterion

With the distance function for knowledge generalization defined, we can now propose the criterion of upper semantic precision, or generalization precision. It is possible to measure this criterion using the methodology we proposed in our previous works [8]:

Definition 4. ***Upper Semantic Precision*** *is a criterion measure equal to the opposite of average distance between each node in the integrated tree and their corresponding specialized nodes in the input trees.*

$$U_{SP}(CT_1, \ldots, CT_n) = \frac{\sum_{v \in V^*} \sum_{w \in V_1 \cup \ldots \cup V_n} B(v, w) \cdot (1 - d(v, w))}{\sum_{v \in V^*} \sum_{w \in V_1 \cup \ldots \cup V_n} B(v, w)} \quad (9)$$

where:

$$B(v, w) = \begin{cases} 1 & \textit{if } v \textit{ is a generalization of } w, \\ 0 & \textit{otherwise.} \end{cases}$$

Additionally, the distance $d(v, w)$ is calculated using an external hierarchical thesaurus, not the input or integrated trees.

For such measure, the Upper Semantic Precision criterion is met only if the measure is maximized (equal to 1).

3.4 Criterion Analysis

In this section we provide several observed properties of the proposed criterion.

Theorem 1. *If Data Completeness is maximum then Upper Semantic Precision is maximum.*

Proof. Data Completeness is only equal 1 (maximum value) if all nodes from input trees exist in the output tree [8]. Therefore, for each node in the output tree, the only corresponding nodes are themselves. From properties of the distance function, the distance is equal 0, therefore from (9) the criterion measure is equal 1.

Theorem 2. *Upper Semantic Precision is minimal if and only if all leaf nodes are generalized to a single common root and other nodes are discarded.*

Proof. Assume that all leaf nodes from input trees are generalized to a single common root in the integrated tree and no other nodes in the output tree exist. From (3) this means that for all leaf nodes w, the distance to the corresponding node w' is maximum. Consequently, as all nodes that have corresponding ones have maximum distance, the USP measure is minimal.

Let now assume that the criterion measure is minimal. This only occurs if all the distances are maximum (if there are no corresponding nodes in trees, the value of the measure is undefined: $U_{SP} = \frac{0}{0}$). From (3) the distance is maximum between the root and leaf nodes. Therefore only leaf nodes should be generalized and all should generalize to a common single node.

Note. Upper Semantic Precision is only equal 0, if this single node is infinitely distant in the hierarchy (ontology) used to measure the distance.

Theorem 3. *For the same distance measure, if 1-Optimality is maximum then Upper Semantic Similarity is maximum.*

Proof. 1-Optimality criterion is maximum if the distance from the integrated tree to the input trees is minimal [8]:

$$O_1(CT^*|CT_1,\ldots,CT_N) = \frac{min_{CT\in\overline{CT}}(\sum_{i=1}^{N} d(CT,CT_i))}{\sum_{i=1}^{N} d(CT^*,CT_i)} \tag{10}$$

If this distance is minimal, then from (9) the USP measure is maximum.

Observation 1. *Integration of similar input trees leads to high Upper Semantic Precision, while integration of diverse input trees leads to low Upper Semantic Precision.*

Trees are similar, if nodes (in applications: terms or concepts) in them are similar. This means that the generalization required to find a common ancestor in the hierarchy is small (the distance is small), which in turn means that USP measure is high. Similarly, if the trees are very different, then a lot of generalization is required. The distances will then be large and the USP measure will be lower.

Observation 2. *Low values of Upper Semantic Precision lead to high value of General (syntactic) Precision.*

The measure of Precision measures repetitions of nodes. Low values of USP measure require that similar or identical concepts are generalized to a common ancestor (for very low USP even only slightly similar concepts are generalized to the common one). This means that few or no repetitions will occur in the integrated tree. The reverse is not true - high values of USP may also occur for high values of General Precision.

4 Integration and Generalization Algorithms in a Knowledge Based System

As a practical example for this criterion, we propose a following collaborative recommendation system. Each user is described by a profile that models his preferences and is used to make recommendations or personalize search results. Similar users are grouped together, and each new user is assigned to a group of similar ones (using demographical data). This new user has no personal profile when first using the system so a centroid profile of his group (integrated profile) is proposed to him. This helps avoid the so-called „cold-start problem". There are two generalization algorithms in the proposed system. One is responsible for generalizing knowledge about a single user into a hierarchy. The other works to combine such hierarchies from different users, in order to find a centroid of the group. The latter is also an extension of our previous work in [8,9]. We may use Upper Semantic Precision to describe this centroid profile – if it generalizes the group knowledge more or less.

Algorithm 1. Creating a hierarchical user profile

Data: User logs
Result: Hierarchical user profile
BEGIN
Transform use logs into data nodes of the tree (leafs).
Group similar data nodes.
Find a generalization of grouped nodes and add it as a parent node.
while *the nodes do not constitute a single tree* **do**
  Group similar knowledge nodes.
  Find a generalization of grouped nodes and add it as a parent node.
end
Delete such nodes v_k in the tree, that are further from the root than a given threshold ϵ_U: $d(v_root, v_k) > \epsilon_U$.
END

The algorithms are very similar, as their main task is finding similar nodes and generalizing them. The main difference is that for a single profile the algorithm

Algorithm 2. Creating a centroid profile

Data: Group of n hierarchical user profiles
Result: Centroid profile
BEGIN
Create a root for the centroid profile
while *distance from root to leaves in centroid is smaller than given threshold* ϵ_C*:* $d(v_root, v_k) > \epsilon_C$ **do**
  Find similar nodes at the current level of the tree.
  Find a generalization of grouped nodes and add it as child nodes.
end
END

is bottom-up: from single nodes (data) to a full tree (knowledge structure); while for centroid profiles the algorithm is top-down: from the root of the tree (common general knowledge) to deeper nodes (common specific knowledge).

As the generalization is most dependent on user data (similar or different) and the used thesaurus, those two factors influence the USP measure greatly. We may therefore determine the USP measure based on the selected thesaurus. Alternatively, we may select the thesaurus, depending on the required value of USP measure. We may also use the thresholds ϵ_U and ϵ_C to slightly influence the USP measure.

5 Conclusions

In this paper we proposed the *Upper Semantic Precision* criterion for hierarchically structured knowledge integration. This criterion may be used to describe both the integrated tree (in terms of the generalization that occured during integration) and the input trees (in terms of their diversity). The proposed criterion may be used, among others, to create a more efficient information retrieval system.

Knowledge-based criteria for integration of hierarchical structures are a novel research topic. There are several similar aspects with currently used ontology alignment methods and this will be used to define further criteria, as well as to help analyze them. In our future work we aim to focus especially on the practical aspects of criteria-based integration, ie. using them in a real-world system.

Acknowledgment. This research was co-financed by Polish Ministry of Science and Higher Education.

References

1. Budanitsky, A., Hirst, G.: Evaluating wordnet-based measures of lexical semantic relatedness. Computational Linguistics 32(1), 13–47 (2006)

2. Comyn-Wattiau, I., Bouzeghoub, M.: Constraint confrontation: An important step in view integration. In: Loucopoulos, P. (ed.) CAiSE 1992. LNCS, vol. 593, pp. 507–523. Springer, Heidelberg (1992)
3. Do, H.H.: Schema Matching and Mapping-based Data Integration. Ph.D. Thesis, University of Leipzig (2006)
4. Do, H.-H., Melnik, S., Rahm, E.: Comparison of Schema Matching Evaluations. In: Chaudhri, A.B., Jeckle, M., Rahm, E., Unland, R. (eds.) NODe-WS 2002. LNCS, vol. 2593, pp. 221–237. Springer, Heidelberg (2003)
5. Madria, S., Passi, K., Bhowmick, S.: An XML Schema integration and query mechanism system. Data and Knowledge Engineering 65, 266–303 (2008)
6. Maguitman, A.G., et al.: Using topic ontologies and semantic similarity data to evaluate topical search. XXXVI Conferencia Latinoamericana de Informática. Centro Latinoamericano de Estudios en Informática, Facultad Politécnica-Universidad Nacional de Asunción and Universidad Autónoma de Asunción, Asunción, Paraguay (2010)
7. Maguitman, A.G., et al.: Algorithmic detection of semantic similarity. In: Proceedings of the 14th International Conference on World Wide Web, pp. 107–116. ACM (2005)
8. Maleszka, M., Nguyen, N.T.: A Method for Complex Hierarchical Data Integration. Cybernetics and Systems 42(5), 358–378 (2011)
9. Maleszka, M., Mianowska, B., Nguyen, N.T.: A method for collaborative recommendation using knowledge integration tools and hierarchical structure of user profiles. Knowledge-Based Systems 47, 1–13 (2013)
10. Passi, K., Lane, L., Madria, S., Sakamuri, B.C., Mohania, M., Bhowmick, S.: A Model for XML Schema Integration. In: Bauknecht, K., Tjoa, A.M., Quirchmayr, G. (eds.) EC-Web 2002. LNCS, vol. 2455, pp. 193–202. Springer, Heidelberg (2002)
11. Rahm, E., Bernstein, P.A.: A survey of approaches to automatic schema matching. The VLDB Journal 10, 334–350 (2001)
12. Resnick, P.: Semantic Similarity in a Taxonomy: An Information-Based Measure and its Application to Problems of Ambiguity in Natural Language. Journal of Artificial Intelligence Research 11, 95–130 (1999)

Design and Implementation of an Adaptive Tourist Recommendation System

Leila Etaati and David Sundaram

Department of Information Systems and Operations Management,
University of Auckland, Auckland, New Zealand
{l.etaati,d.sundaram}@auckland.ac.nz

Abstract. Recommendation Systems and Adaptive Systems have been introduced in travel applications in order to support travellers in their decision-making processes. These systems should respond to the unexpected changes during travel. In this case, they need to sense the changes holistically before, during, and after the travel. In addition, they should also be adapted to the specifications and conditions of the traveller. For example, there is a need to consider all aspects of the traveller's needs, such as personal, cultural, and social. Similarly, the information about accommodations, flights, cities, activities and countries should be gathered through different sources. Furthermore, these systems need to learn from travellers' feedback to improve the quality of recommendations. However, the majority of travel applications do not satisfy the above requirements. To address these problems and issues, we propose and implement a travel process that is supported by an adaptive tourist recommendation framework, architecture, and system.

Keywords: Adaptive System, Holistic, Integration, Adaptive Tourist Recommendation System.

1 Introduction

Most tourism literature describes a traveler as a visitor who visits a country or city other than his/her hometown for a period of less than 12 months. They face different decision-making challenges during their travel, and are frequently overwhelmed by a plethora of questions and travel information [1]. Before and during travel, they think about how to obtain personalized and current information about events, accommodation, and activities. Moreover, the traveller has to extract travel information from different sources such as travel websites, friends, and experienced travellers, which is not personalized to the needs of the traveller. During travel, they struggle to find out how to be notified about upcoming events. In addition, travellers' preferences, and information related to destination, accommodation, flight and so forth change rapidly. As a result, how to adapt to these changes becomes another vital element in travel planning before and during travel.

Taking all of the above problems into account, we tackle these requirements by proposing an *Adaptive Tourist Recommendation System* (ATRS), which can alter

N.T. Nguyen et al. (Eds.): ACIIDS 2014, Part I, LNAI 8397, pp. 251–260, 2014.

its behaviour to *adapt* to the new situation. One way to enhance the adaptation process is to provide a *holistic* view regarding the traveller's specifications and situations, travel products, travel information and travel process. Moreover, to reach a holistic perspective, there is a need to *integrate* travel information, travel process and travel applications.

This paper is organized as follows: Section 2 introduces adaptive tourist recommendation systems. Then Section 3 illustrates the current situation of ATRS and discusses what is missing. Further, in Section 4 we propose ATRS concepts, processes, and system frameworks that address the research lacuna. Finally in Section 5 we illustrate our proposed ATRS architecture and describe our prototypical implementation.

2 Adaptive Tourist Recommendation Systems

In this section, we will summarise the recent ATRS applications and we will identify which ATRS aspects are not completely covered.

2.1 Current ATRS

ATRS provide recommendations based on travellers needs and interests, their devices (e.g. mobile, PC, PDA) and their locations. Michael and Jean [2] proposed several solutions to cope with travel dynamic packaging. They considered two scenarios in travel planning: the first one just recommends specific travel services such as accommodation or flights; whilst the other recommends travel package that are a combination of flights, accommodation, POI and so forth.

Cena et al [3] proposed an ATRS which supports different forms of adaptation. For example, adapting to different devices (Laptop, and smartphone and web access) and adapting to the user's preferences and experiences. In the research done by Mahmood et al [4], they coped with existing problems within the dynamic packaging travel planning by offering a new adaptive recommendation system that allows users to select travel components such as hotels, events, and attractions, and put them in the new plan. In addition, their system was a kind of interactive system that was conversational and followed two processes. The first process was seeking inspiration for a list of products for users to encourage them to buy travel products; and the second process was that the travellers do some query searching themselves.

Coelho et al [5] proposed an ATRS that can adapt to different traveller preferences. They got information about the travellers in various ways such as considering domain independent data or domain dependant data. As a result, they got different types of data by diverse methods (e.g. Likelihood Matrix, Stereotypes, Socialization, Psychological Model, User Explicit Knowledge, and Retrieval).

2.2 ATRS Problems and Issues

To achieve success in all business domains, it is crucial to adapt to new situations and respond quickly to unexpected changes [6]. Likewise, ATRS needs to cater to different users' requirements and situations before, during, and after travel.

According to the integral theory of Wilber [7], all of the reality in the world is composed of whole/parts, not just wholes nor parts (holistic perspectives). Consequently, the first step to be adaptive before and during travel is about holistically sensing the status and specification of traveller and travel elements (e.g. accommodation, transportation, and so forth). Moreover, ATRS should collaborate with other systems to collect information [8]. The collected data should be recent, and should encompass all aspects of traveller and travel elements [9].

However, current ATRS just rely on a single and poor representation of the travellers' specifications and assumptions. They do not take into consideration all the aspects of the traveller, which is composed of the individual's interior and exterior aspects, as well as social and cultural realms (Fig 1).

Upper Left (Individual Interior Subjective) Purpose of Travel, Traveler Values, Traveler implicit preference and interest on travel elements	Upper Right (Individual Exterior Objective) Traveler previous experiences, Traveler's explicit preferences and interests, Traveler activity on social networks, Traveler location and health situation, Traveler demographic attributes
Lower Left (Cultural) Traveler nationality, Traveler language, Traveler shared values	Lower Right (Social) Friends, Social groups, economic aspects

Fig. 1. All-Quadrants, ALL-Levels (AQAL), Traveler Dimension, Adopted from Integral Theory [7]

Information underlying the dimensions of travel elements is another important ATRS aspect. The general travel elements are about accommodation, places, types of transportation, restaurants, cafes, shopping centres etc. However, different types of travel have different travel elements; for example, medical travel information is composed of general travel elements information, as well as pharmacies and hospital information. Nonetheless, most of the current ATRS just focus on general travel elements information.

During travel, information about the travel elements can be changed. Current ATRS mostly focus on the static aspect of travel element, and they rarely consider the ephemeral aspect. For example, each accommodation has some static information such as the room size and accommodation location. In a similar way, it has some ephemeral aspects such as the room price, which can change with seasons and economic conditions.

Another important issue that has been less covered in ATRS literature is about the travel process. Before and during the travel, there are various travel processes such as *Recommendation on POI, Recommendation on Accommodation,* and *Recommendation on Transportation.* To have a holistic view concerning the travel elements and to be able to dynamically plan and respond to the changes, it is crucial to take all the pre-defined travel processes (Fig. 3), as well as travel linkage processes (Fig. 2), into consideration. During the travel, there is a need to modify the ATRS behaviour to fit the current context. Feigh [8] introduced the four primary ways to be adaptive to changes. The first one is about the modification of function allocation by dividing functions or tasks between people and machines. The second one is modification of task scheduling (timing, prioritization and duration); and the third way concerns modification of the interaction by providing different styles, amounts of data, and interface features. The last one concerns modification of the content, based on quantity, quality, and abstraction. However, most ATRS just support one approach.

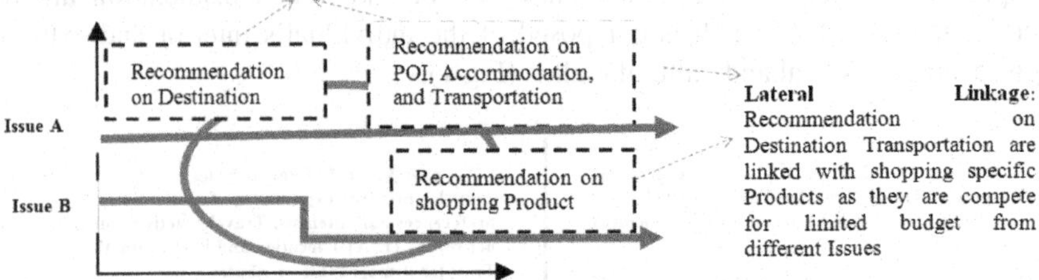

Fig. 2. Travel process linkage adopted from Langley [13]

During and after travel, it is crucial to learn from feedback. Learning and feedback loops represent a key feature of ATRS [11]. There are different types of learning namely: *Single Loop Learning* (SLL) and *Double Loop Learning* (DLL). SLL is mostly about solving problems based on the existing assumptions and pre-defined rules. In contrast, DLL goes a step further and examines the existing rules and improves them [12].

3 Proposed ATRS Concepts and Conceptual Frameworks

Taking into account the existing problems mentioned in the discussion above, in this section we propose the main ATRS concepts, processes and frameworks which can cope with these problems.

3.1 Traveler, Travel Elements and Travel Processes

Bearing in mind that changing the ATRS elements' context can trigger the system to adapt to the new situations, changing travel status, recent information about the travel elements, and the traveler, can cause the system to change its behavior.

In figure 3, we proposed the main travel status and the travel processes. Some of the important processes in this activity diagram are concerned with gathering information holistically. For example, in the *Get Traveler Information* process, it is crucial to take into account the main four traveler's dimensions adopted from the integral theory, namely: *Individual, Social, Behavioral and Cultural* (section3.2); figure 2 shows these dimensions briefly.

3.2 Main ATRS Concepts

As it can be seen from figure 4, there are four main ways to alter ATRS behavior, namely: modification of ATRS information, modification of ATRS stakeholders' interaction, modification of ATRS process task scheduling and modification of function allocation.

Moreover, there is a need to take into account the main triggers for these adaptations. For example:

- Changing the ATRS state which is more concerned with the travel process (Fig. 3).
- Environment's status such as weather conditions, social ceremonies and so forth.
- Traveler's status such as traveler's location and preferences, his/her health situation.
- Finally the need for adaptation can be identified by comparing the traveler's expectations and what the ATRS has already suggested to him/her.

To reach an acceptable level of adaptation, it is necessary to have a holistic understanding of both cause and effect [9]. Bear in mind that all of the ATRS elements' information is not located in a single place (section 1) so it is difficult to reach a holistic perspective. One solution to overcome this problem is to obtain information from other systems by employing integration and collaboration techniques [14]. To satisfy the above high-level ATRS concepts, the ATRS framework has been presented in figure 5. Moreover, taking into account the ATRS as a type of Decision Support System (DSS), it is important to consider all important DSS's elements (i.e. model, solver, and visualization). Consider ATRS as a subset of recommendation systems; it should encompass user profile (traveler profile), as well as item profile (travel element profile). In addition, there are three main external services which are responsible for collaborating with external applications and websites (Fig. 5).

4 Proposed ATRS Architecture and Implementation

4.1 ATRS Architecture

The ATRS architecture proposed in figure 5. ATRS can be presented at four different layers: Presentation, Application, Database and External Services Layers. The *Presentation Layer* is mainly responsible for establishing a connection between the main stakeholders. The *Application Layer* is the core of the ATRS which contains seven main components (Fig. 5). *ATRS Models* are the main component of the *Application Layer* which consists of both *Specific ATRS Models* and *General Models*. *Specific ATRS Models* support the main ATRS aspects: Holism, Adaptation, Integration and Collaboration. For example, a traveler logs into the ATRS; *Travel Workflow Management Model* identifies the traveler's status (Fig. 3) and then executes the other required ATRS models. *Travel Process Linkage* supports the process integration concepts (Fig. 2). *Traveler Information Gathering Model* (Fig. 3) and *Travel Elements Information Gathering Model* are for handling the holistic aspects, which are connected with the *External Collaboration Engine* to gather information. *Travel Adaptation Model* is responsible for managing adaptation approaches, as well as adaptation triggers such as how to adapt to the different devices and identify the current status of the traveler (Fig. 4).

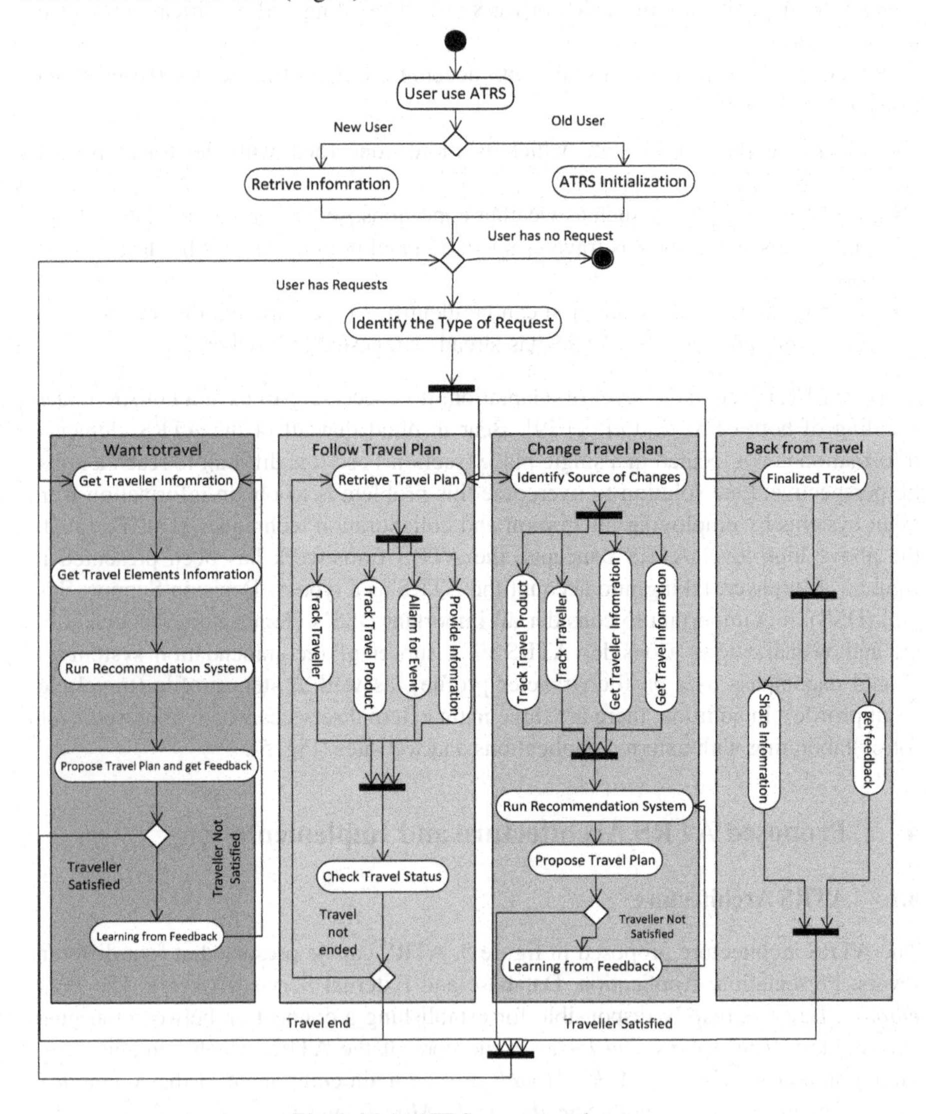

Fig. 3. Main ATRS process and states

General Models are not restricted to use in ATRS. For instance, *Recommendation Models* can be used in all types of recommendation systems. The *Rule Based Model* identifies which recommendation algorithm should be applied. In addition, the *Rule Based Model* employs *Single Loop Learning* and *Double Loop Learning Models* to increase the ATRS' future performance.

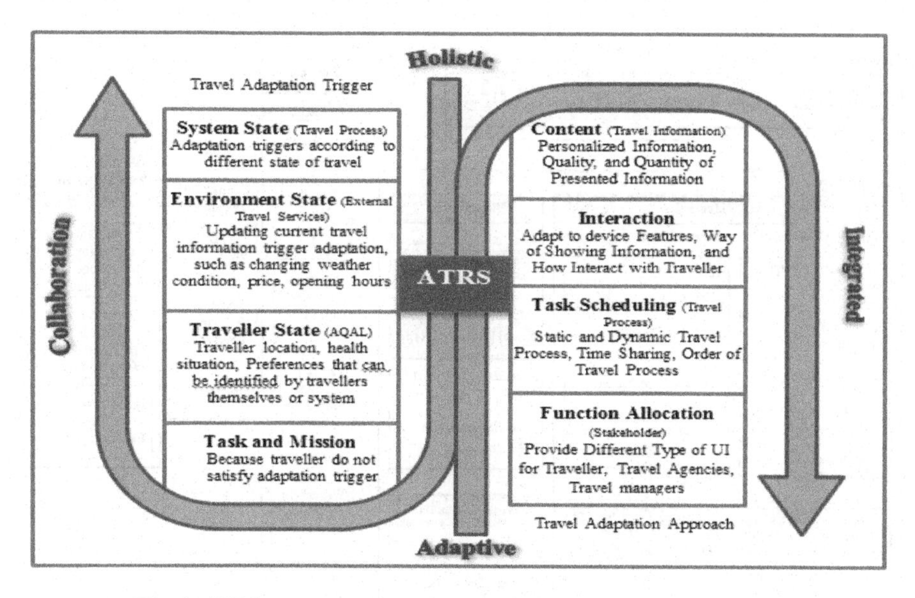

Fig. 4. ATRS concepts: adaptation approaches, triggers for adaptation

ATRS scenarios contain one or several executed *ATRS Models* that can be saved by *ATRS Management Systems* in the *Scenario Database*. The *ATRS Management System* is responsible for maintaining, modifying and retrieving the ATRS solvers, models, traveler profile, travel elements profile, travel process, and scenarios. Finally, the *External Collaboration Engine* is mainly responsible for establishing a connection between the ATRS and external travel websites and applications.

4.2 Two Scenarios Illustrating Integration and Holism

In this section we demonstrate two different scenarios which describe the ATRS functionality. The first scenario is about the Travel Agency that wants to initiate and customize ATRS for a traveler. In the first step (Fig. 6), the Travel Agency logs in (Fig. 6:1) and retrieves current Traveler Profile and Travel Elements Profile (Fig. 6:2), and sets the traveler's status to Want to Travel (Fig. 6:3). Then, he identifies the main websites which the ATRS should search through and capture the implicit, as well as explicit, information about the traveler and travel elements (Fig. 6:4). Moreover, the Travel Agency identifies which methods should be applied to gather information (Fig. 6:5). For example, he/she may use API as a tightly coupled application integration approach which assists the ATRS to obtain some general information (Fig. 6:6). The Travel Agency also may choose a WebCrawler which is less tightly coupled than the previous one. In this way, the ATRS can get the websites' addresses from a traveler or Travel Agency, and starts to search through them (Fig. 6:7). The last approach is more loosely coupled than the others, which just gets favorite websites' links from the traveler and then shows them to the traveler in a single page during the travel process (Fig. 6:8). The second scenario is when the traveler wants to get recommendations on different travel elements with different priorities, which are about the travel interrelated decisions (Fig.2).

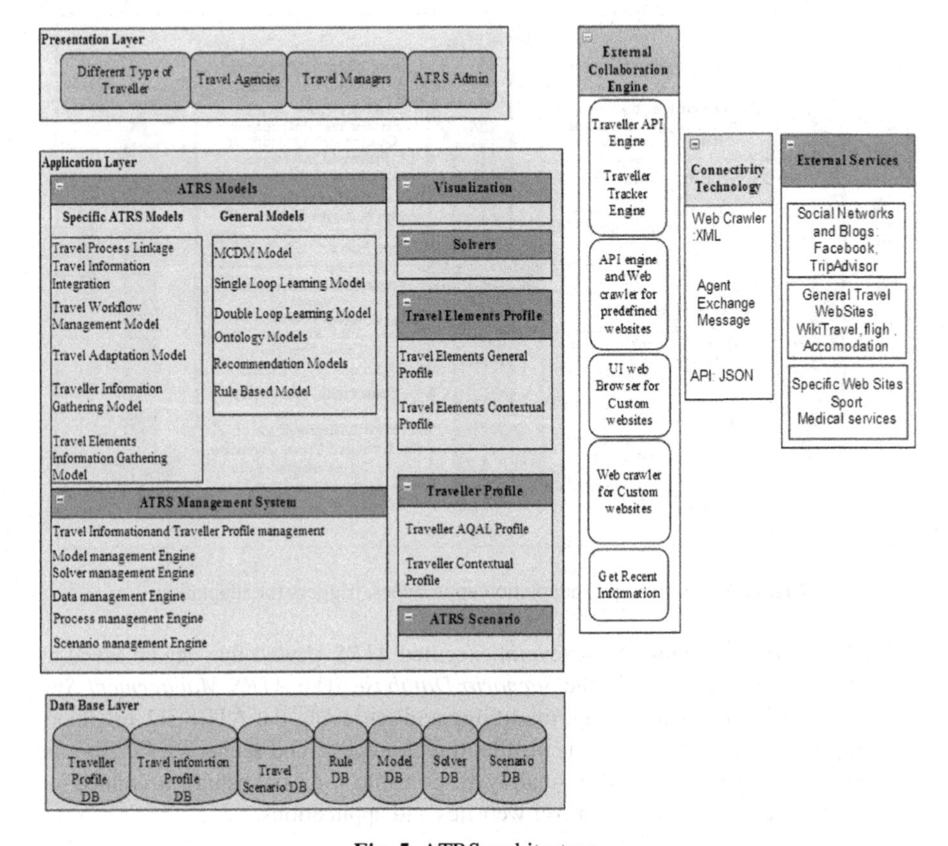

Fig. 5. ATRS architecture

First, the traveler should identify the priority of travel elements by AHP engine (Fig. 7:1), and then, as it can be seen from figure 7:2, the interrelated decisions (Fig.2). First, the traveler should identify the priority of travel elements by AHP engine (Fig. 7:1). Then, as it can be seen from figure 7:2 the traveler got recommendations first on *Destination* then *Activities* and finally on *Transportation*.

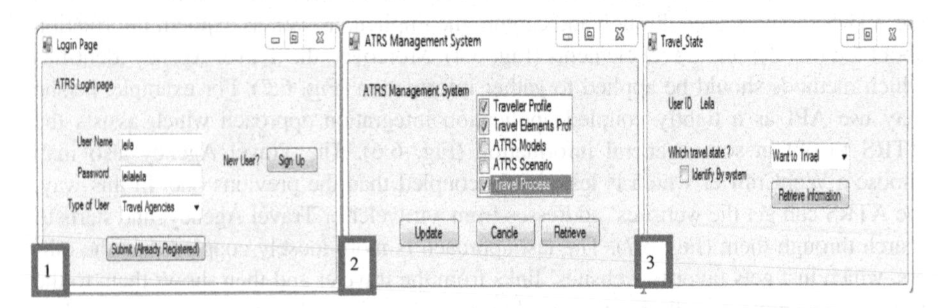

Fig. 6. First Scenario, Travel Agencies Interaction

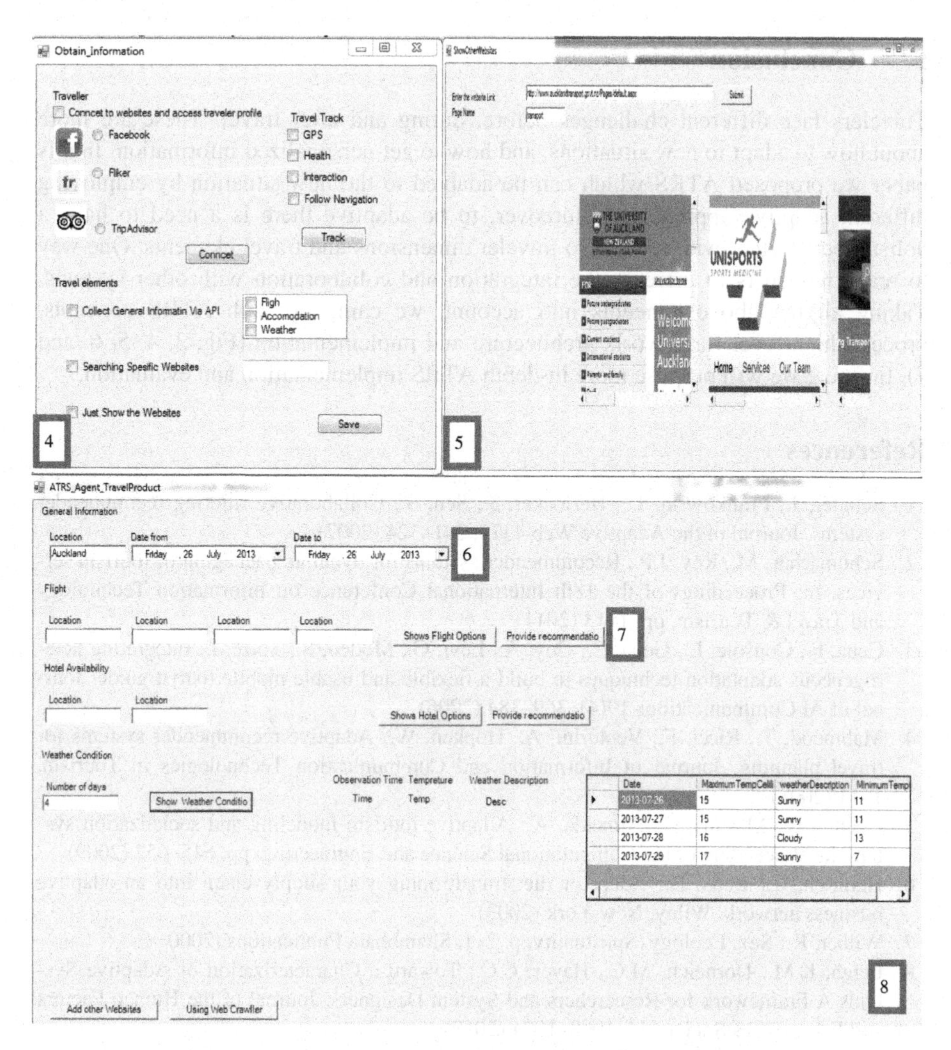

Fig. 6. *(Continue)* First Scenario, Travel Agencies Interaction

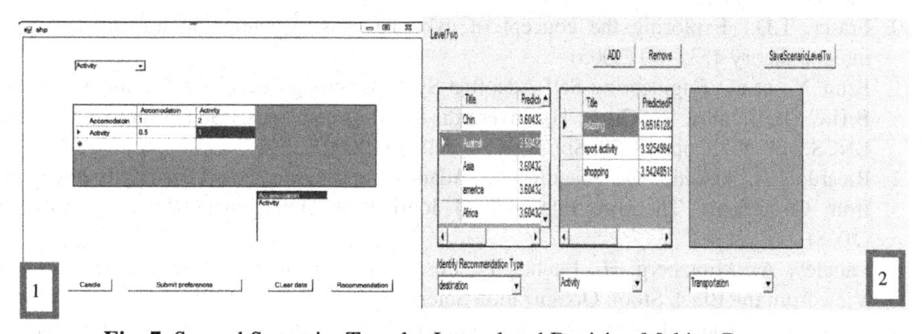

Fig. 7. Second Scenario, Traveler Interrelated Decision Making Process

5 Conclusion

Travelers face different challenges before, during and after travel. These are more about how to adapt to new situations, and how to get personalized information. In this paper we proposed ATRS which can be adapted to the new situation by employing different adaptive approaches. Moreover, to be adaptive there is a need to have a holistic perspective with regards to traveler dimensions and travel elements. One way to reach holism is to increase the integration and collaboration with other systems. Taking all the above concepts into account, we came up with ATRS concepts, process, high-level framework, architecture and implementation (Fig. 3, 4, 5, 6, and 7). In future we will propose more in-depth ATRS implementation and evaluation.

References

1. Schafer, J., Frankowski, D., Herlocker, J., Sen, S.: Collaborative filtering recommender systems. Journal of the Adaptive Web 4321, 291–324 (2007)
2. Schumacher, M., Rey, J.P.: Recommender systems for dynamic packaging of tourism services. In: Proceedings of the 18th International Conference on Information Technology and Travel & Tourism, pp. 1–13 (2011)
3. Cena, F., Console, L., Gena, C., Goy, A., Levi, G., Modeo, S., Torre, I.: Integrating heterogeneous adaptation techniques to build a flexible and usable mobile tourist guide. Journal of AI Communications 19(4), 369–384 (2006)
4. Mahmood, T., Ricci, F., Venturini, A., Höpken, W.: Adaptive recommender systems for travel planning. Journal of Information and Communication Technologies in Tourism, 1–11 (2008)
5. Coelho, B., Martins, C., Almeida, A.: Adaptive tourism modeling and socialization system. In: Proceeding of the Computational Science and Engineering, pp. 645–652 (2009)
6. Heinrich, C., Betts, B.: Adapt or die: transforming your supply chain into an adaptive business network. Wiley, New York (2003)
7. Wilber, K.: Sex, Ecology, Spirituality, p. 211. Shambhala Publications (2000)
8. Feigh, K.M., Dorneich, M.C., Hayes, C.C.: Toward a Characterization of Adaptive Systems A Framework for Researchers and System Designers. Journal of the Human Factors and Ergonomics Society 54, 1008–1024 (2012)
9. O'Brien, K., Hochachka, G.: Integral adaptation to climate change. Journal of Integral Theory and Practice 5, 89–102 (2010)
10. Fraser, T.D.: Exploring the concept of cybercartography using the holonic tenets of integral theory 4, 35–60 (2005)
11. Brun, Y., et al.: Engineering Self-Adaptive Systems through Feedback Loops. In: Cheng, B.H.C., de Lemos, R., Giese, H., Inverardi, P., Magee, J. (eds.) Self-Adaptive Systems. LNCS, vol. 5525, pp. 48–70. Springer, Heidelberg (2009)
12. Ricardo, C., Antonio, G., Joaquín, A.: Adaptive and Generative Learning:Implications from Complexity Theories. International Journal of Management Reviews, 114–128 (2008)
13. Langley, A., Mintzberg, H., Pitcher, P., Posada, E.: Opening up Decision Making: the View from the Black Stool. Organization Science 6(3), 260–279 (1995)
14. Bénaben, F., Touzi, J., Rajsiri, V., Pingaud, H.: Collaborative information system design. In: Proceeding Conference of the Association Information and Management, pp. 281–296, Luxembourg (2006)

Improving Efficiency of PromoRank Algorithm Using Dimensionality Reduction

Metawat Kavilkrue and Pruet Boonma*

Faculty of Engineering, Chiang Mai University, Chiang Mai, Thailand
comengi49@gmail.com, pruet@eng.cmu.ac.th

Abstract. Promotion plays a crucial role in online marketing, which can be used in post-sale recommendation, developing brand, customer support, etc. It is often desirable to find markets or sale channels where an object, e.g., a product, person or service, can be promoted efficiently. Since the object may not be highly ranked in the global property space, PromoRank algorithm promotes a given object by discovering promotive subspace in which the target is top rank. However, the computation complexity of PromoRank is exponential to the dimension of the space. This paper proposes to use dimensionality reduction algorithms, such as PCA, in order to reduce the dimension size and, as a consequence, improve the performance of PromoRank. Evaluation results show that the dimensionality reduction algorithm can reduce the execution time of PromoRank up to 25% in large data sets while the ranking result is mostly maintained.

1 Introduction

Online marketing becomes an important tool for business and organization [1]. Big companies like Google or Amazon relies heavily on online marketing operations, such as, online advertising, recommendation or promotion. For instance, when a customer buys a book from Amazon, he/she will be provided with *recommendation* on similar books. These recommendation are automatically generated from customers' buying/browsing history and also the target promotion from books' publisher.

Ranking is a technique to carry out promotion. It is used widely, for instance, in many bookstores, where top selling books are shown on the front of the stores. This can accelerate those books selling because people are tend to believe that, because so many other customers already bought these books, they should be good. This is also applied to many other fields in business as well, e.g., American Top Forty, Fortune 500, or NASDAQ-100. Because, the number of top ranking is limited, only those who are the best on every dimensions can be in the list. Nevertheless, there are many cases that when consider only a subset of the dimensions, some interesting objects can be found. Consider following two examples:

Example 1. (Product Promotion) It is impossible that Donald Knuth's The Art of Computer Programming series can be in the top list of all books in Amazon store. However, when consider only computer science books with readership toward college

* Corresponding Author.

N.T. Nguyen et al. (Eds.): ACIIDS 2014, Part I, LNAI 8397, pp. 261–270, 2014.

Table 1. Example of multidimensional data

City	Year	Object	Score
NY	2012	O_1	0.9
SF	2012	O_1	0.2
SF	2012	O_2	0.8
SF	2011	O_2	0.7
NY	2011	O_2	0.5
NY	2012	O_3	0.4
SF	2012	O_3	0.8

Table 2. Target object O_1's subspaces and its ranks

Subspace	Rank	Object Count
{*}	3	3
{City=NY}	1	3
{City=SF}	3	3
{Year=2012}	2	3
{City=NY, Year=2012}	1	2
{City=SF, Year=2012}	3	3

students, this book will be ranked on the top list. So, this book series should be promoted only in that category and readership.

Example 2. (Person Promotion) a CEO wants to promote a salesman into a manager; however, when consider data in all category for the past few years, the salesman is only ranked third among all three salesman. On the other hand, when consider only the sales in New York, the salesman is the first rank. So, the CEO can promote the salesman to take care of the New York office.

From the examples, the data space is breakdown into subspace, e.g., instead of all cities, only New York is consider. As a consequence, the target object, e.g, Knuth's book or the salesman, can be the top rank, i.e., top-R, in only some of the subspace. This subspace where the target object is the top rank is called *promotive subspace*.

Table 1 shows a concrete example of a multi-dimensional data set. From the table, there are one object dimension, *Object*, with three target objects, O_1, O_2, O_3. There are two subspace dimensions, *City and Year*, and a score dimension, *Score*. Consider O_1 as the target object to promote, Table 2 lists O_1's 6 subspaces and the corresponding rank and object count in each subspace. The rank is derived from the sum-aggregate score of all objects in the subspace. For example, in {NY, 2012}, O_1 ranks 1st because the score of $O_1 : 0.9 > O_2 : 0.5 > O_3 : 0.4$. Object count is the number of objects in that subspace. Thus {NY, 2012} is a promotive subspace of O_1.

Thus, given a target object, the goal is to find subspace with large promotiveness, i.e., subspace where the target object is top-R. For example, observe that O_1, which is ranked third in all dimensions ({*}), should be promote in subspace {NY} because it is the first rank. In the other word, {NY} is a promotive subspace of O_1. The problem of finding large promotiveness subspace is formally defined in section 2.

PromoRank [2] proposes to use subspace ranking for promoting a target object by finding a subspace where the target object is in Top-R ranking. Section 3 discusses PromoRank in detail. However, the computation complexity of PromoRank is exponential to the dimension size, this paper proposes to use dimensionality reduction algorithms, such as principal component analysis (PCA) and factor analysis (FA), to reduce the number of dimensions before performing PromoRank. This approach is explained in Section 4. Dimensionality reduction in recommendation system, in generally, has been studied for many years, e.g., in [3–5]; however, this work is the first attempt to apply dimensionality reduction technique to subspace ranking, in general, and PromoRank, in particular. The evaluation results, in Section 5, show that the dimensionality reduction can improve the performance of PromoRank, i.e., reduce the execution time, with small impact on Top-R ranking result.

2 Problem Definition

Consider a d-dimensional data set $\mathcal{D}$ with the size of n, each tuple in $\mathcal{D}$ has d **subspace dimension** $\mathcal{A} = \{A_1, A_2, ..., A_d\}$, **object dimension** I_o and **score dimension** I_s. Let $\text{dom}(I_o) = O$ is the complete set of objects and $\text{dom}(I_s) = \mathbb{R}^+$. Let $S = \{a_1, a_2, ..., a_d\}$, where $a_i \in A_i$ or $a_i = *$ ($*$ refers to any value) is a **subspace** of $\mathcal{A}$. In Table 1, $O = \{O_1, O_2, O_3\}$, $\mathcal{A} = \{A_{\text{city}}, A_{\text{year}}$. An example of S is {City=NY, Year=2012}

As a consequence, S induces a projection of $\mathcal{D}_S \subseteq \mathcal{D}$ and a subspace of object $O_S \subseteq O$. For example, when $S = \{City = NY, Year = 2012\}$, $O_S = \{O_1, O_3\}$.

For a d-dimensional data, all subspaces can be group into 2^d cuboids. Thus, S belongs to a d'-dimensional cuboid $\mathcal{A}'$ denoted by $A'_1 A'_2 ... A'_{d'}$, iff S has non-star values in these d' dimensions and star values in the other $d - d'$ dimensions.

Then, for a given **target object** $t_q \in O$, $\mathcal{S}_q = \{S_q | t_q \in O_{S_q}\}$ is the set of **target subspace** where t_q occurs. For example, O_1 in Table 2 has 6 target subspaces, as in Table 1, subspace {2011} is not a target subspace because O_1 does not occur in it.

There are many ways to measure the promotiveness of objects in each subspace. One way to measure the promotiveness is percentile-rank, calculated from inverse of the rank of the target object in the subspace times distinct object count, i.e.,

$$\mathsf{P} = \mathsf{Rank}^{-1} \cdot \mathsf{ObjCount}. \tag{1}$$

For example, in Table 2, promotiveness of subspace {City=SF} is $\frac{1}{3} \cdot 3 = 1$ while the promotiveness of subspace {CIy=NY, Year=2013} is $\frac{1}{1} \cdot 2 = 2$.

Finally, the definition of the promotion query problem is, *given data set $\mathcal{D}$, target object t_q, and promotiveness measure* P, *find the top-R subspaces with the largest promotiveness values.*

This promotion query is a challenging problem because it has a combinatorial nature, the number of combination of subspace with multiple dimensions can increase exponentially. The brute-force approach that enumerates all subspaces and compute the promotiveness in each subspace is prohibitive.

3 PromoRank Algorithm

To address this challenge, **promotion analysis through ranking (PromoRank)** [2] utilizes the concept of subspace ranking, i.e., ranking in only a selected dimensions. PromoRank consists of two phases: aggregation and partition. Algorithm 1, shows the pseudo-code of PromoRank [2].

Algorithm 1. PromoRank($t_q, S, \mathcal{D}, O_S, d_0$)

```
Input: target object t_q, subspace S, data set D, object set in current
   subspace O_S, current partition dimension d_0
Output: Top-R promotive subspaces Results
 1: Results ← ∅
 2: if |D| < minsup ∨ t_q ∉ O then
 3:    return
 4: end if
 5: Compute Rank and P
 6: Enqueue (S, P) to Results
 7: for d' ← d_0 + 1 to d do
 8:    Sort D based on d'-th dimension
 9:    for all value v in d'-th dimension do
10:       S' ← S ∪ {d' : v}
11:       PromoRank(S', D_S', O_S', d')
12:    end for
13: end for
```

In aggregation phase, if the size of data set is not less than a threshold (*minsup*) and t_q is in the given current subspace, then, promotiveness P of a subspace S is computed and kept in *Results* priority queue. From Algorithm 1, Rank and P of the target object are computed for the input subspace S (Line 5). In particular, Rank can be measured from the rank of the target object in the subspace and P is calculated using Equation 1. Then, S and P are inserted into the priority queue, where P is the key (Line 6). This priority queue maintains the top-R results.

In partition phase, the input data is iteratively processed for an addition dimension (d'). Then, for each distinct value on the d'-th dimension, a new subspace is defined and processed recursively. In particular, the input data $\mathcal{D}$ is sorted according to the d'-dimension (Line 8). Then, $\mathcal{D}$ can be projected into multiple partition, corresponding to the distinct values on the d'-t dimension. A new subspace S' is defined for each partition (Line 10). Then, PromoRank recursively computes over subspace S' (Line 9).

At each recursion, the aggregation phase runs in $O(|\mathcal{D}| + |O|)$ and the partition phase runs in $O(|\mathcal{D}|)$. Given that there are d dimension, the number of recursion will be 2^d. Thus, the computational complexity of this algorithm is $O(2^d(|\mathcal{D}| + |O|))$.

From the time complexity, computes all subspaces could be excessive time for large data sets. However, only subspaces where the target object has top ranks are interested by users, so PromoRank proposes to use two optimization techniques, namely, subspace pruning and object pruning.

In subspace pruning, because users are interested in only top-R promotive subspaces. Thus, unpromising subspaces, where promotiveness are lower than a bound, are pruned out. In object pruning, on the other hand, a minimal bound for the score is set for each target subspace, then objects which the score are lower than the bound are removed. Both approaches reduce the number of S or the size of O.

4 Dimensionality Reduction for PromoRank

In order to further improve PromoRank, this paper proposes to reduce the number of dimensions (d) of the data set. From the computational complexity of PromoRank, reduce dimensions should impact the performance greatly [6,7]. Moreover, this approach can be performed as a pre-processing for PromoRank; thus, it can be combined with the pruning approaches.

Given a d-dimensional data set $\mathcal{D}$ with subspace dimension $\mathcal{A}$, a dimensionality reduction algorithm, such as PCA and FA, reduces the number of dimension to d^*, such that $d^* < d$, and a reduced data set $\mathcal{D}^*$ is produced with subspace dimension $\mathcal{A}^*$. Please note that, it does not necessary that $\mathcal{A}^* \subset \mathcal{A}$ because the dimensionality reduction algorithm might generate a new dimension for $\mathcal{A}^*$. In other words, there might exists a subspace $S^* = \{a_1^*, a_2^*, ..., a_{d^*}^*\}$ from $\mathcal{A}^*$ where $a_i^* \notin A_j$ for any $A_j \in \mathcal{A}$.

Thus, the top-R promotive subspace from PromoRank with original data set might be different from the top-R promotive subspace with reduced data set; then, they cannot be compared directly. In order to handle this, a simple mapping scheme is proposed based on the relationship between the original dimensions and reduced dimensions. Suppose that two original subspace dimensions, A_i and A_j, are reduced to a new subspace dimension A_k^*. As a consequence, for a top-R promotive subspace with contains $a_k^* \in A_k^*$, it will be compared with a subspace that has $a_i \in A_i$ and/or $a_j \in A_i$; together with the common other subspace dimensions. For example, let's assume that the original dimensions are {City, Country, Year}, then, after a dimensional reduction algorithm is performed on the data set, the new dimensions are {Location, Year} where *Location* is reduced from *City* and *Country*. Thus, if PromoRank considers a subspace {location=Lanna} where *Lanna* is reduced from *Chiang Mai* and *Thailand*, then, the subspace {location=Lanna} will be compared with the subspace {City=Chaing Mai}, {Country=Thailand} and {City=Chiang Mai, Country=Thailand}.

Nevertheless, performing dimensionality reduction algorithms to reduce the number of dimensions incurs extra computational cost to the PromoRank workflow. However, dimensionality reduction algorithms such as PCA and FA have lower computational complexity than PromoRank. For instance, PCA that use Cyclic Jacobi's method has complexity of $O(d^3 + d^2n)$ [8]. The polynomial complexity of PCA is much lower than the exponential complexity of PromoRank. The experimental results in Section 5.5 confirms that the extra computational cost from a dimensionality reduction algorithm is lower than the performance gain from reducing dimensions.

5 Experimental Evaluation

To evaluate the proposed approach, an experimental evaluation with three data sets, namely, **NBA** [9], **Top-100 US private collage** [10] and **Market analysis** [11], was carried out. A Java version of PromoRank was developed and tested on a computer with an Intel Core2 Duo 3GHz processor and 4GB of memory. The pruning optimization of PromoRank was disabled in the evaluation to remove the impact on the result.

The dimensionality reduction algorithm used in this evaluation is principal component analysis (PCA) and factor analysis (FA). They reduces the number of variables, i.e., dimensions, by measuring the correlation among them. Then, the variables that highly correlate with the others are removed or combined into new variables. The evaluation was performed on three parts; first part compares the Top-R promotive subspace of dimensional-reduced data sets and original data sets with PromoRank. For all data set, only top-3 promotive subspaces of each target object are considered. The result of this part is presented in Section 5.2, 5.1 and 5.3 for NBA, Top-100 US private college and Market Analysis data set, respectively. This part uses PCA for reducing dimensions of the data set. The second part, presented in Section 5.4, substitutes PCA with FA to verify that the proposed approache is independent from a particular dimensionality reduction algorithm. The last part, presented in Section 5.5, investigates the performance improvement from the dimensionality reduction.

5.1 NBA Data Set with PCA

This data set consists of 4,051 tuples with 12 subspace dimensions, namely *First Name, Last Name, Year, Career Stage, Position, Team, Games, Minutes, Assists, Block, Turnover* and *Coach*. The result from PCA dictates that 6 dimensions, *Game, Minutes, Assists, Block, TurnOver* and *Coach* can be removed. Thus, after dimensional reduction, the reduced data set contains only 6 subspace dimensions, namely, *First Name, Last Name, Year, Career Stage, Position* and *Team*.

Table 3 compares ranks of Top-3 promotive subspaces from the original data set and reduced data set of three target objects. From the table, even 6 dimensions are removed, the Top-3 promotive subspaces do not change in this data set. The result show that, by using a dimensionality reduction algorithm to reduce the number of correlated subspace dimension, there is no impact on the ranking of Top-3 promotive subspaces.

5.2 Top-100 US Private College Data Set with PCA

This data set consists of 100 tuples with 8 subspace dimensions, namely *State, Enrollment, Admission Rate, Admission Ratio, Student/faculty Ratio, 4yrs Grad Rate, 6yrs Grad Rate* and *Quality Rank*. In contrast with NBA data set in Section 5.1, Top-100 US Private College data set and the Stock Market data set in the next section contains quantitative data, e.g., *6yrs Grad Rate*, which cannot be used in PromoRank directly. From Section 2, the subspace dimension has to be a set. Thus, in order to process quantitative data, they have to be converted to categorical data first. In this paper, the number of categories is set to ten. Each quantitative data will be assigned to a category based on its value. For example, the value of *4yrs Grad Rate* is from 0% to 100%; therefore, a data will be assigned to first category when its value is from 0% to 10%, to second category when its value is more than 10% to 20% and so on.

The result from PCA dictates that there are two new principle components, namely, *Grad Rate* and *Ratio*. *Grad Rate* strongly correlates, i.e., has low variance, with *4yrs*

Table 3. Subspace ranking of NBA data set with PCA

Target	Original Data Set		Reduced Data Set	
Object	Top-3 Promotive Subspaces	Rank	Top-3 Promotive Subspaces	Rank
Kareem	{*}	1	{*}	1
Abdul-	{Position=Center}	1	{Position=Center}	1
Jabbar	{Team=LA Lakers, Year=1978}	2	{Team=LA Lakers, Year=1978}	1
Karl	{*}	2	{*}	2
Malone	{Position=Forward}	1	{Position=Forward}	1
	{Team=Utah Jazz}	1	{Team=Utah Jazz}	1
Michael	{*}	3	{*}	3
Jordan	{Position=Guard}	1	{Position=Guard}	1
	{Team=Chicago Bulls}	1	{Team=Chicago Bulls}	1

Grad Rate and *6yrs Grad Rate*. *Ratio*, on the other hand, strongly correlates with *Admission Ratio* and *Admission Rate*. Thus, after dimensional reduction, the reduced data set contains only 8 subspace dimensions, namely, *State, Enrollment, Student/faculty Ratio, Quality Rank, Grad Rate* and *Ratio*.

Table 4 compares ranks of Top-3 promotive subspaces from the original data set and reduced data set of three target objects. From the table, when compare ranks of the two data, original and reduced, there is only one difference, marked by a star in the table. With Williams College as the target object, when the subspace of original data is { *Admission-Rate=30%, 6yrs Grad Rate=100%* }, the target object is ranked third in this subspace. On the reduced data, the comparable subspace is { *Rate=95%, Grad Rate=95%*} , the target object is ranked first in the subspace. First of all, these two subspaces are compared because *Rate* is the principle component of *Admission-Rate* and *Grad-Rate* is the principle component of *6yrs Grad Rate*, according to PCA. Because the subspace is changing, it is possible that , when compared with the other object in *O*, the rank of the target object in the reduced subspace can be different from the original subspace. However, this mismatch is infrequently happened. Therefore, the result shows that the ranking of Top-3 promotive subspace is mostly maintained even after the dimensionality reduction is performed on the data.

5.3 Stock Market Data Set with PCA

This data set consists of 5,891 tuples with 23 subspace dimensions, namely *Company Name, Industry Name, Ticket Symbol, SIC Code, Exchange Code, Size Class, Stock Price, Price/Piece, Trading Volume, Market Price, Market Cap, Total Debt, Cash, FYE Date, Current PE, Trailing PE, Firm Value, Enterprise Value, PEG Ratio, PS Ratio, Outstanding, Revenues* and *Payout Ratio*. Similar to the previous data set, this data set is converted to categorical data with ten categories for each subspace dimension.

The result from PCA dictates that a subspace dimension, *Price/Piece* can be removed, and there are two new principle components, namely, *Price* and *Forward PE*.

Table 4. Subspace ranking of Top-100 US Private College data set with PCA

Target Object	Original Data Set		Reduced Data Set	
	Top-3 Promotive Subspaces	Rank	Top-3 Promotive Subspaces	Rank
California	{*}	1	{*}	1
Institute of	{State=CA}	1	{State=CA}	1
Technology	{6yrs Grad Rate=90%}	1	{Grad Rate=85%}	1
Rice	{*}	2	{*}	2
University	{6yrs Grad Rate=90%}	2	{Grad Rate=85%}	2
	{Enrollment=2, 4yrs Grad Rate=70%}	1	{Enrollment=2, Grad Rate=85%}	1
Williams	{*}	3	{*}	3
College	{4yrs Grad Rate=90%, 6yrs Grad Rate=100%}	1	{Grad Rate=95%}	1
	{Admission Rate=30%, 6yrs Grad Rate=100%}	3*	{Rate=95%, Grad Rate=95%}	1*

Price strongly correlates with *Stock Price* and *Market Price*. *Forward PE*, on the other hand, strongly correlates with *Current PE* and *Trailing PE*. Thus, after dimensional reduction, the reduced data set contains 20 subspace dimensions, namely, *Company Name, Industry Name, Ticket Symbol, SIC Code, Exchange Code, Size Class, Trading Volume, Market Cap, Total Debt, Cash, FYE Date, Firm Value, Enterprise Value, PEG Ratio, PS Ratio, Outstanding, Revenues, Payout Ratio, Price* and *Forward PE*.

Table 5 compares ranks of Top-3 promotive subspaces from the original data set and reduced data set of three target objects. From the table, when compare ranks of the two data, original and reduced, there is only one difference, marked by a star in the table. With AppTech Corp as the target object, when the subspace of original data is {Market Price=$0, Stock Price=$0}, the target object is ranked fifth in this subspace. On the reduced data, the comparable subspace is {Price=$0} , the target object is ranked first in the subspace. Similar to the previous section, these two subspaces are compared because *Price* is the principle component of *Market Price* and *Stock Price*, according to PCA. Because the subspace is changing, it is possible that, when compared with the other object in O, the rank of the target object in the reduced subspace can be different from the original subspace. However, this mismatch is infrequently happened. Therefore, the result shows that the ranking of Top-3 promotive subspace is mostly maintained even after the dimensionality reduction is performed on the data.

5.4 Top-100 US Private College Data Set with FA

To verify that this approach is not limit to any particular dimensionality reduction algorithm, factor analysis (FA) is used instead of PCA in the same evaluation setup as in Section 5.2. The result from FA dictates that a dimension, i.e., *Admission Ratio*, can be removed; therefore, only 7 dimensions remain. Table 6 compares ranks of Top-3 promotive subspaces from the original data set and reduced data set of three target objects. From the table, when compare ranks of the two data, original and reduced, there is only a difference, marked by the star in the table. With Yale University as the target object, when the subspace of original data is { *4yrs Grad Rate=90%, 6yrs Grad Rate=100%* },

Table 5. Subspace ranking of Stock Market data set with PCA

Target	Original Data Set		Reduced Data Set	
Object	Top-3 Promotive Subspaces	Rank	Top-3 Promotive Subspaces	Rank
Bank of	{*}	1	{*}	1
America	{Stock Price=$6, Market Price=$6}	1	{Price=$6}	1
	{Size Class=10, FYE Date=31/12/2010}	1	{Size Class=10, FYE Date=31/12/2010}	1
Greenshift	{*}	2	{*}	2
Corp	{Stock Price=$1, Market Price=$1}	1	{Price=$1}	1
	{Size Class=8, FYE Date=31/12/2010}	2	{Size Class=8, FYE Date=31/12/2010}	2
AppTech	{*}	3	{*}	3
Corp	{Market Price=$0, Stock Price=$0}	5*	{Price=$0}	1*
	{Size Class=4}	1	{Size Class=4}	1

Table 6. Subspace ranking of Top-100 US Private College data set with FA

Target Object	Top-3 Promotive Subspaces	Original Data Set's Ranking	Reduced Data Set's Ranking
Williams	{*}	1	1
College	{State = MA}	1	1
	{4yrs Grad Rate=90%}	1	1
Massachusetts	{*}	2	2
Institute of	{State = MA}	2	2
Technology	{4yrs Grad Rate=80%}	2	2
Yale	{*}	3	3
University	{State = CT}	1	1
	{4yrs Grad Rate=90%, 6yrs Grad Rate=100%}	2*	1*

the target object is ranked second in this subspace. On the reduced data, the target object is ranked first in the subspaces. Because of a dimension is removed, the ranking is changed as well. However, this small mismatch is infrequently happened. Therefore, the result shows the proposed approach is independent from the dimensionality algorithms in used.

5.5 Performance Improvement

To evaluate the performance improvement, the time to perform PromoRank was measured. Table 7 shows the comparison between the execution time of PromoRank on the original data set and the execution time of PCA to produce a reduced data set plus the execution time of PromoRank on the reduced data set. The execution time result with FA, i.e., Section 5.4, is the same as of the result with PCA. The result shows that the usage of dimensionality reduction algorithm such as PCA can improve performance of PromoRank, for about 20-25% for a large data set.

Table 7. Performance comparison

Date Set	Execution Time	
	Original Data Set	Reduced Data Set
Top-100 Private College	2 seconds	1 second
NBA	20.5 minutes	15.2 minutes
Stock Market	124 minutes	99 minutes

6 Conclusion

In this work, a dimensionality reduction algorithm, i.e., PCA, is introduced to reduce the size of data set in order to improve the performance, i.e., execution time, of PromoRank algorithm. The results confirm that the dimensionality reduction algorithm can reduce the execution time of PromoRank up to 25% while mostly maintains the ranking result. Following extensions of this work is planned: first, different dimensional reduction algorithms will be compared on different data sets to investigate the nature of algorithms that compatible with nature of data set. This will lead to a framework where a suitable dimensional reduction algorithm will be applied for each data set. Second, a better subspace mapping scheme, from reduced data set to original data set, will be developed.

References

1. Kotler, P., Keller, K.: Marketing Management. Prentice Hall (2008)
2. Wu, T., Xin, D., Mei, Q., Han, J.: Promotion analysis in multi-dimensional space. In: International Conference on Very Large Databases, Lyon, France, VLDB Endowment (2009)
3. Symeonidis, P., Nanopoulos, A., Manolopoulos, Y.: Tag recommendations based on tensor dimensionality erduction. In: ACM Conference on Recommender Systems, Lausanne, Switzerland. ACM (2008)
4. Kamishima, T., Akaho, S.: Dimension reduction for supervised ordering. In: International Conference on Data Mining, Hongkong. IEEE Press (2006)
5. Ahn, H.J., Kim, J.W.: Feature reduction for product recommendation in internet shopping malls. International Journal of Electronic Business 4(5), 432–444 (2006)
6. Fodor, I.: A survey of dimension reduction techniques. Technical report, Center for Applied Scientific Computing, Lawrence Livermore National Research Laboratory (2002)
7. Ailon, N., Chazelle, B.: Faster dimension deduction. Commun. ACM 53(2), 97–104 (2010)
8. Golub, G., van Loan, C.: Matrix Computations. Johnm Hopkins University Press (1996)
9. DatabaseSports.com, http://www.basketballreference.com
10. Drexel University. http://mathforum.org/workshops/sum96/data.collections/datalibrary/data.set6.html
11. StataCorp LP, http://www.stata.com

A Framework to Provide Personalization in Learning Management Systems through a Recommender System Approach

Hazra Imran, Quang Hoang, Ting-Wen Chang, Kinshuk, and Sabine Graf

Athabasca University, Edmonton, Canada
{hazraimran,tingwenchang,kinshuk,sabineg}@athabascau.ca,
hoangdangquang@yahoo.com

Abstract. Personalization in learning management systems (LMS) occurs when such systems tailor the learning experience of learners such that it fits to their profiles, which helps in increasing their performance within the course and the quality of learning. A learner's profile can, for example, consist of his/her learning styles, goals, existing knowledge, ability and interests. Generally, traditional LMSs do not take into account the learners' profile and present the course content in a static way to every learner. To support personalization in LMS, recommender systems can be used to recommend appropriate learning objects to learners, not only based on their individual profile but also based on what worked well for learners with a similar profile. In this paper, we propose a framework to integrate a recommender system approach into LMS. The proposed framework is designed with the goal of presenting a flexible integration model which can provide personalization by automatically suggesting learning objects to learners based on their current situation as well as successful learning experiences of learners with similar profiles in a similar situation. Such advanced personalization can help learners in many ways such as reducing the learning time without negative impact on their marks, improving learning performance as well as increasing the level of satisfaction.

Keywords: Personalization, E-Learning, Learning Management Systems, Recommender System.

1 Introduction

With the advancement in technology, e-learning is becoming more and more popular. E-learning can comprise either fully online or blended courses. While in fully online courses, everything is delivered in an online mode, blended courses have an online and a face-to-face component. To facilitate the delivery and organization of e-learning, especially in large-scale educational institutions, learning management systems (LMSs) are typically used. According to Szabo [1], a "Learning Management System is the infrastructure that delivers and manages instructional content, identifies and assesses individual and organizational learning or training goals, tracks the progress towards meeting those goals, and collects and presents data for supervising

N.T. Nguyen et al. (Eds.): ACIIDS 2014, Part I, LNAI 8397, pp. 271–280, 2014.

the learning process of an organization as a whole". Courses in LMSs typically consist of learning objects (LOs). LOs can be defined as "any entity, digital or non-digital, that may be used for learning, education or training" [2]. Generally, LMSs deliver the same kind of course structure and LOs to each learner [3, 4]. This is coined as "one size fits all" approach. But, each learner has different characteristics, and therefore, a "one size fits all" approach does not support most learners particularly well. One of the possible ways to support each learner individually based on his/her characteristics is the use of personalization. Personalization in LMS refers to the functionality which enables the system to uniquely address a learner's needs and characteristics such as levels of expertise, prior knowledge, cognitive abilities, skills, interests, preferences and learning styles [5] so as to improve a learner's satisfaction and performance within the course. Personalization in the form of recommendations for resources and learning materials is an area that has gained significant interest from researchers recently. Recommendations exhibit prominent social behavior in day-to-day life [6]. In real life, people seek and trust the recommendations of others in making decisions. Reflecting this societal behavior, recommender systems are increasingly being adopted in different fields in order to support users in their decision making processes and help them in making wise choices with less effort. Many online companies, such as Amazon [7] and Netflix [8] are using recommender systems to offer users personalized information to help them in their decisions [9]. Such successful integration of recommender systems in e-commerce has prompted researchers to explore similar benefits in the e-learning domain [10, 11] since the integration of recommender systems in e-learning has high potential for achieving advanced personalization.

This paper presents a novel framework that integrates a recommender system into a LMS in order to provide personalization to learners based on their situations and successful learning experiences of other learners with similar characteristics in similar situations. The proposed framework is designed to:

- Integrate a recommender system into LMSs
- Consider a learner's profile consisting of characteristics like learning style, expertise level, prior knowledge and performance to provide advanced personalization.
- Form a neighborhood of learners based on their profile and discover associations among learning objects (through association rule mining) that led to successful learning experiences of other similar learners in similar situations.
- Create a personalized list of recommendations of learning objects to be presented to an individual learner in situations where members of his/her neighborhood benefitted from the suggested learning objects.

An important feature of the proposed framework is the approach to find similar learners for building a neighborhood of learners. The approach is advanced as it is considering different characteristics of the learners such as their learning styles, prior knowledge, expertise level and performance within the course. Accordingly, we get more similar learners in a neighborhood, which enables our approach to generate more suitable recommendations that fit to the learners' situations more accurately.

The paper is organized as follows. The next section begins with an analysis of the state of the art in the field of providing personalization through recommender systems using data mining technique. Section 3 describes the proposed framework and its main components. Section 4 concludes the paper and discusses future research directions.

2 Related Work

Recommender systems use behavior or opinions of a group with similar characteristics/behavior to help individual users in making decision from vast available choices. Recently, some recommender systems have been applied in the e-learning domain. In this section, such works are described based on two directions: First, we discuss research works that focus on providing recommendations based on learners' activities in a course. These works use association rule mining to find associations among the activities done by learners and then recommendations are provided accordingly to the individual learner. In these works, recommendations are based on learners' activities in a course rather than learner characteristics, needs and/or profiles. Second, we describe research works that provide recommendations based on similar learners who have similar characteristics. These works either used clustering techniques based on learners' characteristics to create groups or compute similarity between the learners based on the ratings they provided. Subsequently, recommendations were provided based on what worked well for similar learners.

Research work falling under the first group used association rule mining to find rules based on which recommendations were provided to learners. For example, Zaiane [12] built a recommender agent that provides recommendations of learning activities within a course based on learner access histories. Khribi, Jemni and Nasraoui [13] developed a recommender system based on learners' recent navigation histories, and similarities and dissimilarities among the contents of the learning materials. The first group of research works considers the web usage data of the learners in a course as well as associations between the activities of learners in a course. These works focus on grouping learners based on their activities. Our work is different from these works as we are finding similar learners based on their characteristics (e.g., learning styles, skills, prior knowledge and performance) rather than activities, which has potential to allow for a more accurate grouping since we are considering the underlying reason for learners' behavior (e.g., not much background knowledge, a certain learning style) rather than just the actions themselves.

The second group of research works finds similar learners and then recommendations are provided based on the information from these similar learners. For example, Tang and McCalla [14] proposed an evolving web-based learning system that finds the relevant content from the web. They use a clustering technique to cluster the learners (based on their learning interests) to calculate learners' similarities for content recommendation. Tai, Wu and Li [15] proposed a course recommender system by using self-organizing maps and data mining techniques. Self-organizing maps were used to classify learners based on similar interests into groups. Then a data mining technique was used to elicit the rules of the best learning path for each group of learners. Kerkiri, Manitsaris and Mavridou [16] proposed a framework that uses reputation metadata in a recommender system. Reputation is the cumulative scale of user opinions regarding persons, products, and ideas. The system describes the learning resources based on learning object metadata and the leaners profile based on PAPI [17]. The registered learners were asked to provide information for their profile including qualifications, skills, licenses etc. The similarity between the learners is calculated by using the Pearson's r correlation coefficient. The learners were asked to provide ratings to learning resources, which is termed as reputation metadata. Having all the information about learners and learning resources (metadata and reputation metadata), collaborative filtering was applied to recommend personalized learning resources. An experiment showed that the use of reputation metadata augmented learners' satisfaction by retrieving learning materials which were evaluated positively by learners. Yang, Sun, Wang, and Jin [18] proposed a personalized recommendation algorithm for curriculum resources based on semantic web technology using a domain ontology. The algorithm first collects curriculum resources of interest in terms of user evaluation and user browsing behavior. Yang et al. [18] assume that "different users evaluate different core concepts, according to domain knowledge, as there is a certain similarity between core concepts, so there are similarities between the user's interests". Therefore, similarity among users can be computed from similarity between core concepts. The users were asked to provide ratings to the learning resources. The similarity among learners was computed based on their ratings. Then the interest degree of users is calculated for each interest category of the nearest neighbors and finally recommendations were provided based on interest of the nearest neighborhood.

The research works in the second group provide recommendations based on similar learners. However, these works mainly used the learner interest and ratings from the learner as the parameter for creating groups. In the e-learning domain, we generally do not have ratings for the content. If a learner is asked to provide ratings for each learning object in a course, it puts a lot of effort on the learner. In our work, we aim at providing automatic recommendations without requiring any additional effort from learners. Instead of using ratings, information about whether or not a certain learning object was helpful for a particular learner is retrieved from his/her navigation and behavior in the course as well as his/her performance. Furthermore, our work is different in that it considers students' characteristics, including their learning styles, expertise level and prior knowledge, together with their performance in the course. By identifying similar learners based on multiple characteristics, we expect to place a learner together with learners who learn in a very similar fashion, leading to more accurate recommendations.

3 Framework for Integrating a Recommender System Approach into LMSs

This section describes the proposed framework for integrating a recommender system approach into LMSs. The aim of the framework is to enable LMSs to provide recommendations to learners based on the successful learning experience of other similar learners. The framework is illustrated in Figure 1. The modules in this framework are designed in such a way that they are not dependent on the LMS and hence, can be integrated easily into different LMSs with minimum required changes. In the following subsections, the modules are described in more detail.

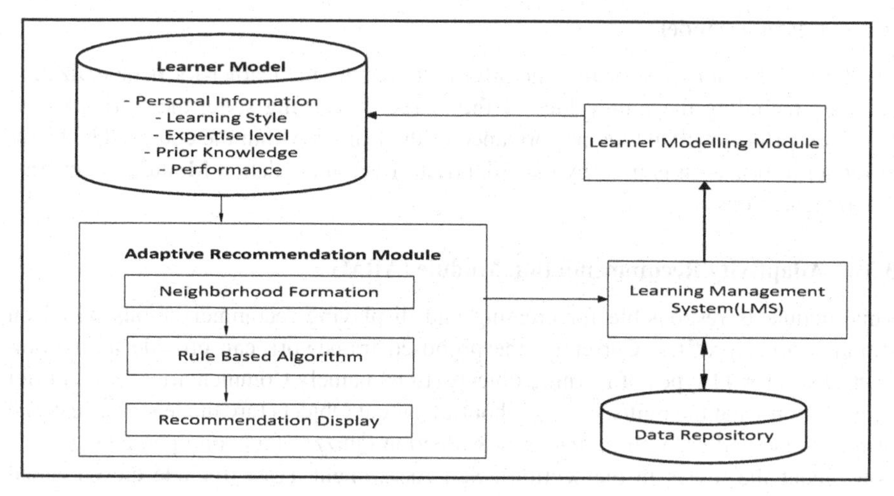

Fig. 1. Architecture of proposed framework

3.1 Learner Modelling Module

The Learner Modelling Module aims to generate the Learner Model. The Learner Model contains information gathered from the learner, i.e. personal information (first name and last name), previous knowledge (related to the course), expertise level (i.e., Beginner, Intermediate or Expert), learning styles and performance. When learners register in the LMS through a registration form, the Learner Model is initialized. During the registration, learners provide personal information such as first name and last name, which is stored in the Learner Model. Furthermore, they are asked about their prior knowledge and expertise level. In addition, the Learner Model aims at gathering information about the learning styles of learners. Every learner learns in a different and unique way, as each one has their own preferences, need and approaches toward learning. These individual differences are coined as learning styles. According to Dunn, Dunn, and Freeley [19], learning styles can be defined as "unique manners in which learners begin to concentrate on, process, absorb, and retain new and difficult information". To identify the learning styles, the Learner Modelling Module uses a well-investigated and commonly used questionnaire, called Index of Learning Styles (ILS)[20] developed by Felder and Solomon, which identifies the preferences of learning in four dimensions based on the Felder-Silverman Learning Style Model [21]. These four dimensions are: active/reflective, sensing/intuition, visual/verbal and sequential/global. At the time of registration, a learner is asked to fill out the ILS questionnaire, consisting of 44 questions. Based on a learner's responses, the result is calculated as four values between +11 to -11 indicating the preference on each of the four learning style dimensions. These four values are stored in the learner model and are used as the identified learning styles of learners. Performance data describe a learner's performance in the course units. The performance data are gathered from the learner's performance on assignments and quizzes within each unit.

3.2 Learner Model

The Learner Model aims to store the information about the learner for personalization purpose, including the four values of the learner's learning style, their prior knowledge, expertise level and the performance of the learner within the course. The learner model information is used by the Adaptivity Recommendation Module to generate recommendations.

3.3 Adaptivity Recommendation Module (ARM)

This module is responsible for creating and displaying recommendations based on similar learner profiles. Currently, the proposed framework can provide the recommendations for 11 types of learning objects (LOs) namely Commentaries (give a brief overview on what the unit/section is about), Content Objects (are the learning material of the course and are rich in content), Reflection Quizzes (contain open-ended questions about the topics in the section), Self-Assessment Tests (include closed-ended questions about the topics in a section), Discussion Forums (allow learners to ask question and join/initiate a discussions with their peers and instructor), Additional Reading Materials (provide additional sources of reading materials about the topics in a section), Animations (explain the concepts of a section in an animated multimedia format), Exercises (allow learners to practice their knowledge and skills), Examples (illustrate the theoretical concepts in a more concrete way), Real-Life Applications (demonstrate how the learned material can be applied in a real-life situations) and Conclusions (summarize the topics learned in a section).

ARM has information about learners' behavior through accessing log data tracked by the LMS, which include what learning objects have been visited by each learner and how much time he/she spent on each learning object. This is information that every LMS typically tracks. In order to provide recommendations, ARM finds the neighbors of a learner who have similar characteristics. We are making the assumption that since learners within a neighborhood are similar to each other, successful learning experiences of one learner can be beneficial to other similar learners. The overall aim of ARM is to provide recommendations of learning objects to the learner in a situation where the learner is visiting different learning objects than other similar learners. For example, a learner may be advised to consult some unread material that other similar learners have read before attempting a particular reflection quiz. ARM has three main steps: neighborhood formation, rule generation and recommendation display. Each step is discussed in the next subsections in more detail.

Neighborhood formation. In ARM, we assume that if a learner visits particular LOs and performed well in the course, the learner had a successful learning experience. Accordingly, those LOs might be helpful to other similar learners who have not yet visited those LOs. These other similar learners build the neighborhood of a learner and are learners with similar characteristics (i.e., learning styles, prior knowledge, expertise level and performance). The purpose of the neighborhood formation step is to find such other similar learners. There were two main requirements for our algorithm to build a neighborhood: (1) the number of learners in the neighborhood of a particular learner should not be predefined but flexible and (2) the neighborhood should include the data points (learners) that are close to another. Based on the above stated requirements, we choose a neighborhood approach for finding similar learners. Such neighborhood approach does not demand the number of neighborhoods or neighbors as input a priori and can use a distance measure to place a learner only together with learners who have very similar characteristics.

To find the neighborhood, we use an algorithm that describes each learner, L_i $(I = 1,\ldots,m)$ as a vector and compute similarities between learners based on the commonly used distance measure, Euclidean distance. As mentioned before, we are using different characteristics of learners including learning styles, expertise level, prior knowledge and performance. Each characteristic has a different scale of values. To ensure the equal impact of each characteristic, we normalize the data between 0 to1. Once the characteristics values are normalized, Euclidean distance is used to compute the similarity between learners based on their characteristics. Euclidean distance (L_i, L_j) is the distance between the vectors representing two learners. The formula to calculate the Euclidean distance between two learners is shown in Formula (1).

$$Euclidean\_distance\ (L_i, L_j) = \sqrt{\sum_{k=1}^{n} (L_{ik} - L_{jk})^2}\ , \qquad (1)$$

where L_{ik} denotes the characteristic k of learner i.

In order to calculate the neighborhood of a learner, a threshold t is used as radius for the neighborhood. Accordingly, for a learner L_i, we consider every other learner L_j (j=1 … m and j != i) as a member of the neighborhood if $Euclidean\_distance\ (L_i, L_j) <= t$. To determine a suitable value for a threshold t, we assume that two learners can be considered as similar if the difference between each characteristic is on average equal or lower than 0.25 (on a scale from 0 to 1). Accordingly, the Euclidean distance between two such learners would be 0.66. Therefore, we consider 0.66 as threshold to calculate the neighborhood.

Rule Generation. In order to generate recommendations, some data processing is needed. The learning objects visited by learners are recorded and are converted into transactions consisting of learner ID and all learning objects visited by the learner within the course. Table 1 shows an example of such transactions.

Table 1. Example of Transaction

Learner ID	Learning objects visited by the learner
1	{Content Object1, Example1}

After pre-processing, association rule mining algorithm [22] is applied to the transaction within the neighborhood to discover associations between the learning objects among similar learners. In the following, an example of rules, resulting from the association rule mining algorithm is presented:

R1 : {Content Object1, Forum1} → {Self-Assessment Test1}

According to R1, learning objects, Content Object1 and Forum1 are associated with Self-Assessment Test1. That mean, before performing Self-Assessment Test1, learners have visited Content Object1 and Forum1. To provide recommendations to a learner, ARM consults the association rules to check for mismatches between the learning objects visited by the current learner and the learning objects visited by the learners within the neighborhood. For example, suppose the current learner has not visited Forum1 yet and he/she is trying to attempt Self-Assessment Test1, but other similar learners in his/her neighborhood have visited Forum1 before completing Self-Assessment Test1 successfully. In such case, the recommendation to be provided (to the current learner) is to visit Forum1 before Self-Assessment Test1. Such recommendations are then passed to the recommendation display for being presented to the learner.

Recommendation Display. In this step, the personalized recommendations are displayed to the learner in an informative, precise and simple way. Recommendations include links to the recommended learning objects so that the learners can go to these learning objects easily. A learner can either click on the links or choose to close the recommendation. As and when the learner clicks on any recommended learning object, the learning object pops up and other recommendations (if any) are saved so that the learner can visit them later on. Figure 2 shows an example of a recommendation for a learner. In the example, when the learner tries to attempt Reflection Quiz1, the recommender system recommends two learning objects namely, Forum1 and Example1. The learner may choose to visit Forum1 first by clicking on the respective link. In this case, Forum1 pops up and Example1 is saved as further recommendation. When the learner tries to attempt Reflection Quiz1 again, then the other recommendation of Example1 is displayed if Example1 has not been visited by the learner already. If the learner clicks the Continue button, he/she can proceed without the recommendation with Reflection Quiz1.

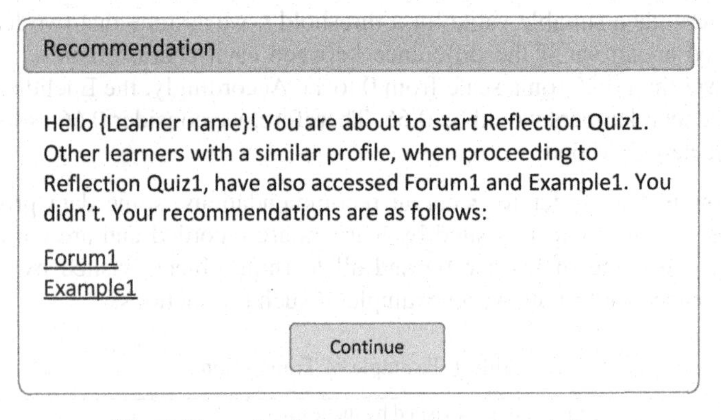

Fig. 2. Example of personalized recommendation

4 Conclusion

This paper introduces a framework to integrate a recommender system approach into learning management systems, enabling these systems to provide recommendations of learning objects to learners based on successful learning experiences of similar learners. The recommendation mechanism uses association rule mining and a neighborhood algorithm. The main contributions of the work are: First, to find similar learners, our framework does not consider ratings given by learners as done in most of the traditional recommender systems. Instead, it uses different characteristics/attributes of learners like learning style, previous knowledge, expertise levels, and performance to identify highly similar learners. Second, recommendations are provided to learners for appropriate learning objects based on what worked well for other learners with similar characteristics in similar situations. Third, in most of the previous works similar learners are found by using a clustering approach. In our work, we consciously decided against a clustering algorithm. Clustering algorithms typically aim at assigning each learner to a group/cluster. This leads to several relevant drawbacks such as the risk of creating clusters that include data points (or learners) that are actually not too close, the risk of getting different clusters when running the same clustering algorithm again, meaning that the clustering algorithm does not always group the nearest data points (or learners), or the need for a predefined number of clusters. Since our aim is to find learners who are close to a particular learner, a neighborhood approach is more accurate and free of the abovementioned drawbacks. By using such neighborhood approach, we expect to place a learner only together with learners who learn in a very similar way, and use the experience of similar learners to provide accurate recommendations. Fourth, while most other works focus on using a recommender system in a particular e-learning systems, the aim of our work is to integrate a recommender system into any LMS. LMSs are commonly used by educational institutions and by enhancing LMSs with personalized functionality to provide individual recommendations, teachers can continue using the systems that they are already using for online learning and learners are receiving additionally some personalized support. The provided recommendations can help learners to better navigate the course (by suggesting learning objects that could improve their performance within the course) as well as improve their learning performance and satisfaction. Currently, we are providing recommendations of learning objects within a course. As a future work, we will extend the framework to additionally provide recommendations from the web.

Acknowledgment. The authors are grateful to MITACS for their partial financial support through the ELEVATE program. The authors acknowledge the support of Athabasca University, NSERC, iCORE, Xerox, and the research related gift funding by Mr. A. Markin.

References

1. Szabo, M.: CMI Theory and Practice: Historical Roots of Learning Management Systems. In: Proceedings of World Conference on E-Learning in Corporate, Government, Healthcare, and Higher Education, pp. 929–936 (2002)
2. IEEE Learning Technology Standardization Committee. Draft standard for learning object metadata (IEEE 1484.12.1-2002). New York, NY (2002)
3. Brusilovsky, P., Miller, P.: Course Delivery Systems for the Virtual University. In: Tschang, F.T., Della Senta, T. (eds.) Access to Knowledge: New Information Technologies and the Emergence of the Virtual University, pp. 167–206. Elsevier (2001)

4. Shishehci, S., Banihashem, S.Y., Zin, N.A.M., Noah, S.A.M.: Review of Personalized Recommendation Techniques for Learners in learning management system Systems. In: Proc. of the Int. Conf. on Semantic Technology and Information Retrieval, pp. 277–281. IEEE Press (2011)
5. Huang, M.J., Huang, H.S., Chen, M.Y.: Constructing a Personalized Learning Management System based on Genetic Algorithm and Case-Based Reasoning Approach. Expert Systems with Applications 33(3), 551–564 (2007)
6. Tseng, C.: Cluster-based Collaborative Filtering Recommendation Approach. Master's Thesis, National Sun Yatsen University (2003)
7. Amazon, `http://www.amazon.com/`
8. Netflix, `http://www.netflix.com/`
9. Linden, G., Smith, B., York, J.: Amazon.com Recommendations: Item-to-Item Collaborative Filtering. Internet Computing 7(1), 76–80 (2003)
10. Capuano, N., Iannone, R., Gaeta, M., Miranda, S., Ritrovato, P., Salerno, S.: A Recommender System for Learning Goals. In: Lytras, M.D., Ruan, D., Tennyson, R.D., Ordonez De Pablos, P., García Peñalvo, F.J., Rusu, L. (eds.) WSKS 2011. CCIS, vol. 278, pp. 515–521. Springer, Heidelberg (2013)
11. Manouselis, N., Drachsler, H., Verbert, K., Duval, E.: Recommender Systems for Learning. Springer Briefs in Electrical and Computer Engineering. Springer (2012)
12. Zaïane, O.: Building a Recommender Agent for e-Learning Systems. In: Proceedings of the International Conference in Education, Auckland, New Zealand, pp. 55–59 (2002)
13. Khribi, M.K., Jemni, M., Nasraoui, O.: Automatic Recommendations for E-Learning Personalization based on Web Usage Mining Techniques and Information Retrieval. In: Proc. of the Int. Conf. on Advanced Learning Technologies, pp. 241–245. IEEE Press (2008)
14. Tang, T., McCalla, G.: Smart Recommendation for an Evolving Learning Management System: Architecture and Experiment. International Journal on Learning Management System 4(1), 105–129 (2005)
15. Tai, D.W., Wu, H., Li, P.: Effective Learning Management System Recommendation System based on Self-Organizing Maps and Association Mining. The Electronic Library 26, 329–344 (2008)
16. Kerkiri, T., Manitsaris, A., Mavridou, A.: Reputation Metadata for Recommending Personalized E-Learning Resources. In: Proceedings of the Second International Workshop on Semantic Media Adaptation and Personalization, pp. 110–115. IEEE Press (2007)
17. IEEE Learning Technology Standards Committee, `http://www.ieeeltsc.org/`
18. Yang, Q., Sun, J., Wang, J., Jin, Z.: Semantic Web-Based Personalized Recommendation System of Courses Knowledge Research. In: Proceedings of the International Conference on Intelligent Computing and Cognitive Informatics, pp. 214–217. IEEE Press (2009)
19. Dunn, R., Dunn, K., Freeley, M.E.: Practical Applications of the Research: Responding to Students' Learning Styles – Step One. Illinois State Research and Development Journal 21(1), 1–21 (1984)
20. Felder, R.M., Soloman, B.A.: Index of Learning Styles Questionnaire. NorthCarolina State University (1996), `http://www.engr.ncsu.edu/learningstyles/ilsweb.html`
21. Felder, R.M., Silverman, L.K.: Learning and Teaching Styles in Engineering Education. Engineering Education 78(7), 674–681 (1988), Preceded by a preface in 2002, `http://www4.ncsu.edu/unity/lockers/users/f/felder/public/Papers/LS-1988.pdf`
22. Agrawal, R., Imieliński, T., Swami, A.: Mining Association Rules between Sets of Items in Large Databases. In: Proceedings of ACM SIGMOD International Conference on Management of Data, pp. 207–216. ACM Press (1993)

Agent-Based Modelling the Evacuation of Endangered Areas

František Čapkovič*

Institute of Informatics, Slovak Academy of Sciences, Bratislava, Slovakia
Frantisek.Capkovic@savba.sk

Abstract. The evacuation process from endangered areas (EA) in crisis situations is modelled by means of simple agents (gate-ways equipped by sensors). Timed Petri nets (TPN) and first-order hybrid Petri nets (FOHPN) are utilized here to model the EA structure as well as the agents and their cooperation. Rooms, other spaces to be evacuated (corridors) and safe spaces out of EA (where people are evacuated) are modelled by TPN places and FOHPN continuous places. Gate-ways are modelled by TPN subnets and by FOHPN continuous transitions. While the supervisor for the TPN gate-ways can be synthesized by means of place/transition Petri nets (P/T PN), the blocks of FOHPN discrete places and transitions are used to affect the gate-ways. Depending on the immediate throughput of the gate-ways the escape time behaviour is found in the process of simulation. This paper is a free continuation of [4] where the problem was solved solely by means of P/T PN.

Keywords: Agents, evacuation, first-order hybrid Petri nets, modelling, place/transition Petri nets, simulation, timed Petri nets.

1 Introduction

In crisis situations the evacuation of people from endangered areas (EA) is the overriding and principal matter. There are many methods how to model the evacuation processes - see e.g. [9]. In [4] the place/transition Petri nets (P/T PN) [14, 15] were used at agent-based modelling the evacuation process. Here, alternatively, timed Petri nets (TPN) viewed e.g. in [18] and first-order hybrid Petri nets (FOHPN) defined and/or used in [1–3, 5, 7, 8, 16] will be tested in such a role. Namely, while P/T PN do not include time and yield only the successive discrete marking of their places, TPN and FOHPN yield the time behaviour of the marking which depict the escape of people from EA. TPN offer stepped time functions while FOHPN offer real continuous ones. Crisis situations cannot be completely controlled anyway. They can only be partially influenced. To manage the evacuation process from EA, flexible escape strategies are required. Namely, it is necessary to find safety and free escape routes [13]. Suitable models utilized in the simulation process, especially the models representing the EA structure,

* Partially supported by the grants VEGA 2/0039/13 and APVV-0261-10.

N.T. Nguyen et al. (Eds.): ACIIDS 2014, Part I, LNAI 8397, pp. 281–290, 2014.

are able to yield useful information. TPN and FOHPN used here allow to model the EA structure and to analyze the evacuation process. Namely, their places directly model the spaces to be evacuated and marking the places models people, while their transitions models the gate-ways. In TPN the transitions are components of blocks modelling the gate-ways. In FOHPN the gate-ways are modelled by continuous transitions affected by discrete PN subnets. Because only the gate-ways play an active role in EA they can be regarded as simple *agents.* Although such agents have not own goals (in fact they are rather reactive subsystems), they are able to cooperate each other. In TPN they cooperate by means of a suitable supervisor, in FOHPN by means of discrete PN subnets. The cooperation favourably affects the evacuation dynamics. Supervisors synthesized and utilized in [4] were able to ensure the intended agent cooperation. The supervised evacuation of EA given in Fig. 1 left was modelled there by P/T PN. The models of the one-way door and two-ways door [4, 13] are displayed in Fig. 1 right. Supervisors given in Fig. 2, displaying the P/T PN model of EA, were synthesized in [4] by means of supervision principles known in discrete event systems control theory [10–12]. While the Supervisor 1 monitors number of people leaving the rooms, the Supervisor 2 determines the priority of leaving the Corridor through the Exit 1. However, when e.g. the preferential Exit 1 is busy the escape across the Room 2 or the Room 4 is possible too. Details about the supervisors synthesis can be found in [4]. In contrast to P/T PN, TPN yield the

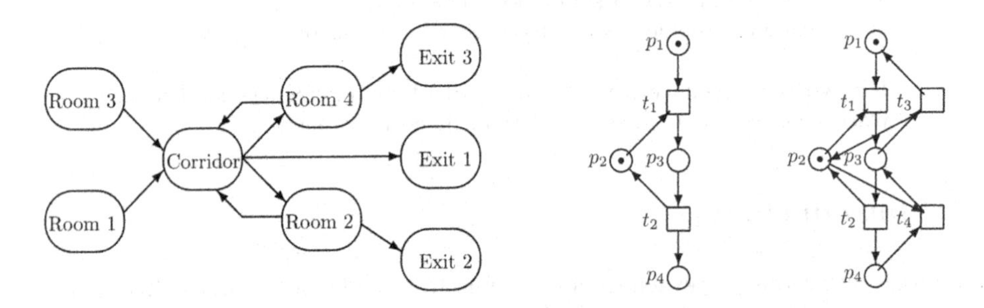

Fig. 1. The scheme of EA to be evacuated (left). P/T PN-based model of two kinds of gate-ways - the non-reversible way and the reversible (bidirectional) one (right).

time behaviour of marking (stepped functions), expressing the escape of people from the rooms and corridors to the safety spaces. FOHPN offer similar information but in the form of piecewise-linear real functions - flows. The FOHPN continuous part corresponds directly to the EA structure (the places model the rooms and the transitions model the gate-ways), while their discrete part makes possible to affect the throughput of the gate-ways. Instead of the supervisor(s), the P/T PN or TPN modules are used to influence the continuous transitions, and consequently, also the escape. First of all TPN and FOHPN will be concisely introduced. Then, the relation between the TPN model and P/T PN one [4] will be explained. The supervisors used in the P/T PN model will be utilized in TPN

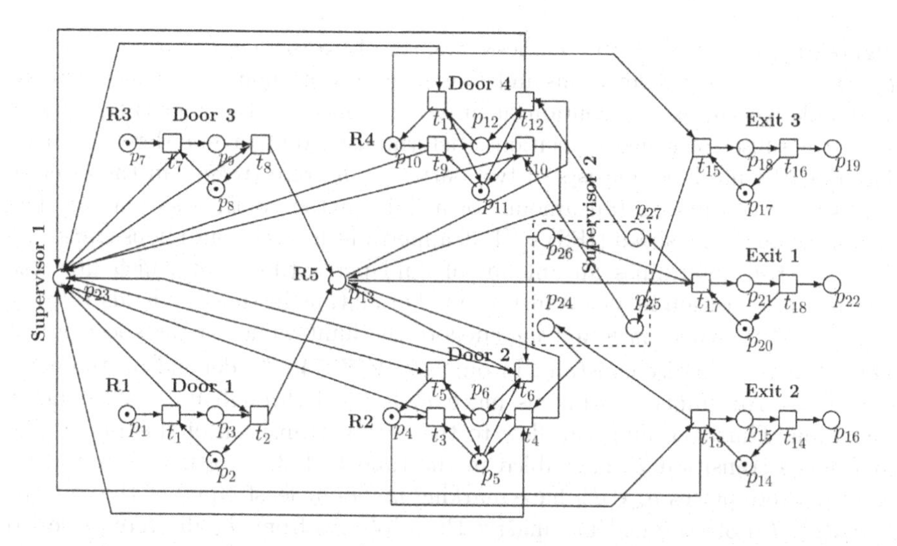

Fig. 2. The P/T PN-based model of EA. Two supervisors influence the evacuation.

too. Time specifications will be assigned to the TPN model. Next, the FOHPN model will be presented and described. Finally, simulation results obtained by the HYPENS tool [16] in MATLAB will be displayed for both kinds of models. They will show emptying the rooms and corridor and filling the safety spaces.

2 Timed Petri Nets and Hybrid Petri Nets

P/T PN transitions, places, arcs and tokens do not depend on time. In TPN [18] time specifications are defined. Here, we will use TPN assigning time exclusively to P/T PN transitions (delays in deterministic case or a kind of probability distributions of timing in non-deterministic case) because they model doorways (being *bottle-necks*) playing the most important role at the throughput of the escape routes. Hybrid Petri nets (HPN) [6] are an extension of standard PN. They model the coexistence of discrete and continuous variables. HPN have two groups of places and transitions - discrete and continuous. Thus, there are three kinds of directed arcs here: (i) between discrete places and discrete transitions; (ii) between continuous places and continuous transitions; (iii) between discrete places and continuous transitions as well as between the continuous places and discrete transitions. The discrete places and transitions handle discrete tokens, while the continuous places and transitions handle continuous variables. FOHPN are a simplified kind of HPN. Here, they can model the escape from EA (flows are flocks of people). Rules defining mutual interactions between their groups of places, transitions and arcs are defined in details in [1–3, 7, 16]. Duplicating the FOHPN comprehensive definition here is needles. It is sufficient to give a basic idea about the FOHPN principle and function. The set of places $P = P_d \cup P_c$, where P_d is a set of discrete places and P_c is a set of continuous places (figured by double concentric circles). The set of transitions $T = T_d \cup T_c$, where T_d is a set of discrete transitions and T_c is a set of continuous transitions (figured by double rectangles). T_d contains a subset of immediate (no-timed) transitions and/or a subset of timed transitions (deterministic and/or non-deterministic). The FOHPN marking consists of two parts: (i) discrete (tokens in the discrete places); (ii) continuous (an amount of a fluid in the continuous places). The instantaneous firing speed (IFS) v_j [1–3] appertains to each continuous transition $T_j \in T_c$. IFS determines an amount of fluid per a time unit which fires the continuous transition in a time instance τ. For all time instances τ holds $V_j^{min} \leq v_j(\tau) \leq V_j^{max}$, where *min* and *max* denote the limit values of the speed $v_j(\tau)$. Thus, IFS is piecewise-constant. In our case V_j^{min}, V_j^{max} depend on the gateways sizes. An empty continuous place P_i is filled through its enabled input transition. Thus, the fluid can flow to the output transition of this place. The continuous transition T_j is enabled in the time τ [1–3, 7, 16] if and only if its input discrete places $p_k \in P_d$ have marking $m_k(\tau)$ at least equal to the element $Pre_{dc}(p_k, T_j)$ of the incidence matrix $\mathbf{Pre}_{dc}$ of arcs from the discrete places to the continuous transitions and all of its input continuous places $P_i \in P_c$ satisfy the condition that their markings $M_i(\tau) \geq 0$ - i.e. the places P_i are filled. If all of the input continuous places of the transition T_j have nonzero marking then T_j is strongly enabled, otherwise T_j is weakly enabled. The continuous transition T_j is disabled if some of its input places are not filled. Namely, T_j cannot take more fluid from any empty input continuous place than the amount entering the place from other transitions. This corresponds to the principle of mass conservation.

3 Problem Formulation and Simulation Results

Let us model EA by both the TPN and the FOHPN. To obtain the TPN model it is sufficient only to add the time specifications to transitions in the P/T PN model [4]. The model structure retains the form of P/T PN in Fig. 2. However, the FOHPN model is completely different from the P/T PN model. It has the structure given in Fig. 4 - the continuous PN affected by the discrete PN modules.

3.1 Using the TPN Model of the Endangered Area

Consider the supervised P/T PN-based model [4] given in Fig. 2. Assign time specifications to its transitions in order to obtain the TPN model. The time delays in the deterministic case and the parameters of the probability distribution of timing in the non-deterministic one depend on the kind of doorways in the real EA, especially on their throughput (determined primarily by the doorways size). Namely, the transitions model the doorways being the *bottle-necks* at the escape of people from the rooms and the corridor to the safety spaces and affect the time behaviour of the escape. For simulation we have chosen the universal PN tool HYPENS [16, 17] for MATLAB. Consider the EA initial state (before evacuation): 15 people in the Room 1, 17 people in the Room 2, 43 people in the Room 3, 55 people in the Room 4 and 5 people in the Corridor. Let the

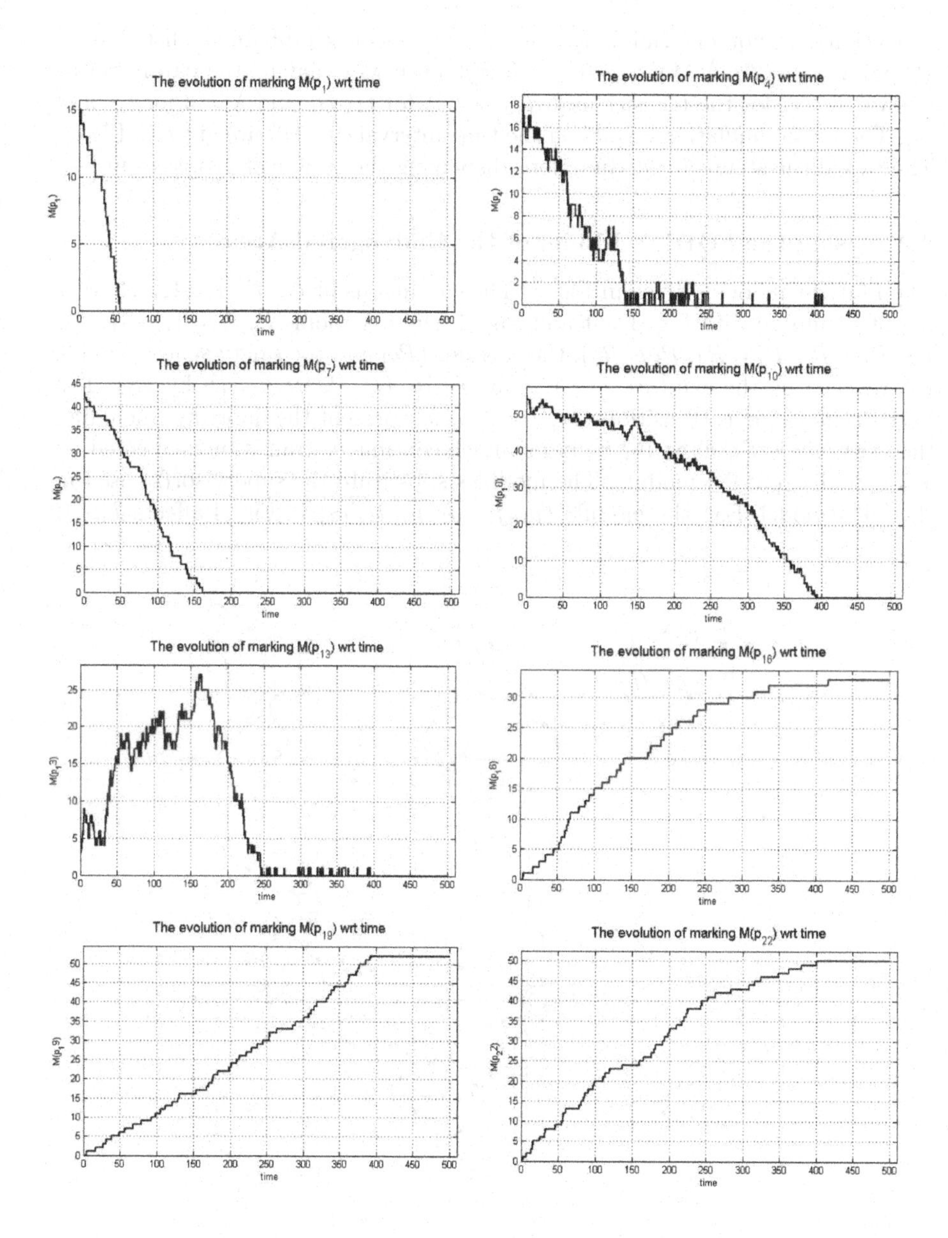

Fig. 3. The escape dynamics expressed by the evolution of markings $m(p_i)$ of the TPN places (compare with Fig. 2): p_1, p_4, p_7, p_{10} (rooms), p_{13} (corridor) and p_{16}, p_{19}, p_{22} (safety spaces) in time at the exponential probability distribution of timing the TPN transitions. The markings express people while the stepped functions show how people escape from the rooms and from the corridor as well as how the numbers of people accrue in the particular safety spaces.

transitions are non-deterministic, with the exponential probability distribution of timing $f_x = \lambda.e^{-\lambda.x}$ for $x \geq 0$ and $f_x = 0$ otherwise, having the parameter as follows: $\lambda = 2$ for the transitions t_1-t_{12}, $\lambda = 4$ for t_{13}-t_{16} and $\lambda = 3$ for t_{17} and t_{18}. Then, the simulation results on the time interval (0, 500) are given in Fig. 3. They yield the stepped time functions displaying the escape of people from EA.

3.2 Using the FOHPN Model of the Endangered Area

The FOHPN model is given in Fig. 4. The continuous places P_i, $i = 1, ..., 8$, represent (compare with Fig. 1 left and Fig. 2) the EA rooms R_j, $j = 1, ...4$ (P_1 is R_3, P_2 is R_1, P_3 is R_4, P_4 is R_2), the corridor (P_5) and the safety spaces outside EA reached by the exits E_i, $i = 1, ..., 3$ (P_6 by E_1, P_8 by E_2, P_7 by E_3). The FOHPN continuous transitions T_j, $j = 1, ..., 9$ represent the doorways including the exits (E_1 is T_7, E_2 is T_9, E_3 is T_8). Each continuous transition is affected by a simple discrete PN module. The modules substitute the supervisor(s) used in the TPN model. E.g. the module $\{p_1, p_2, p_3, p_4, t_1, t_2\}$ in Fig. 4 affects T_1: (i)

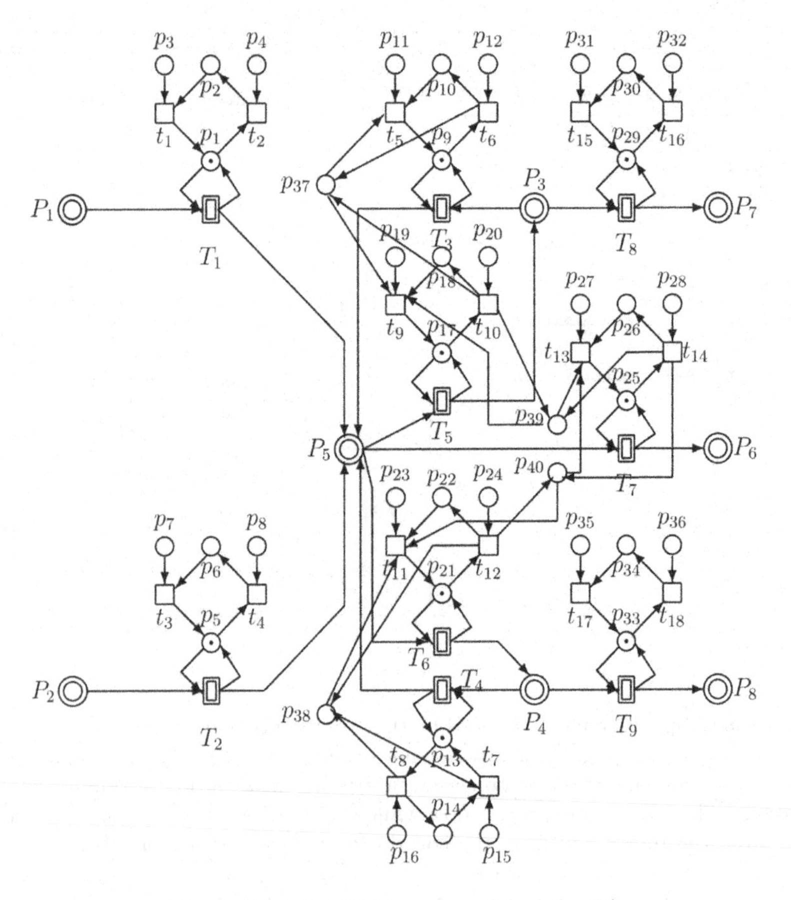

Fig. 4. The FOHPN-based model of the EA

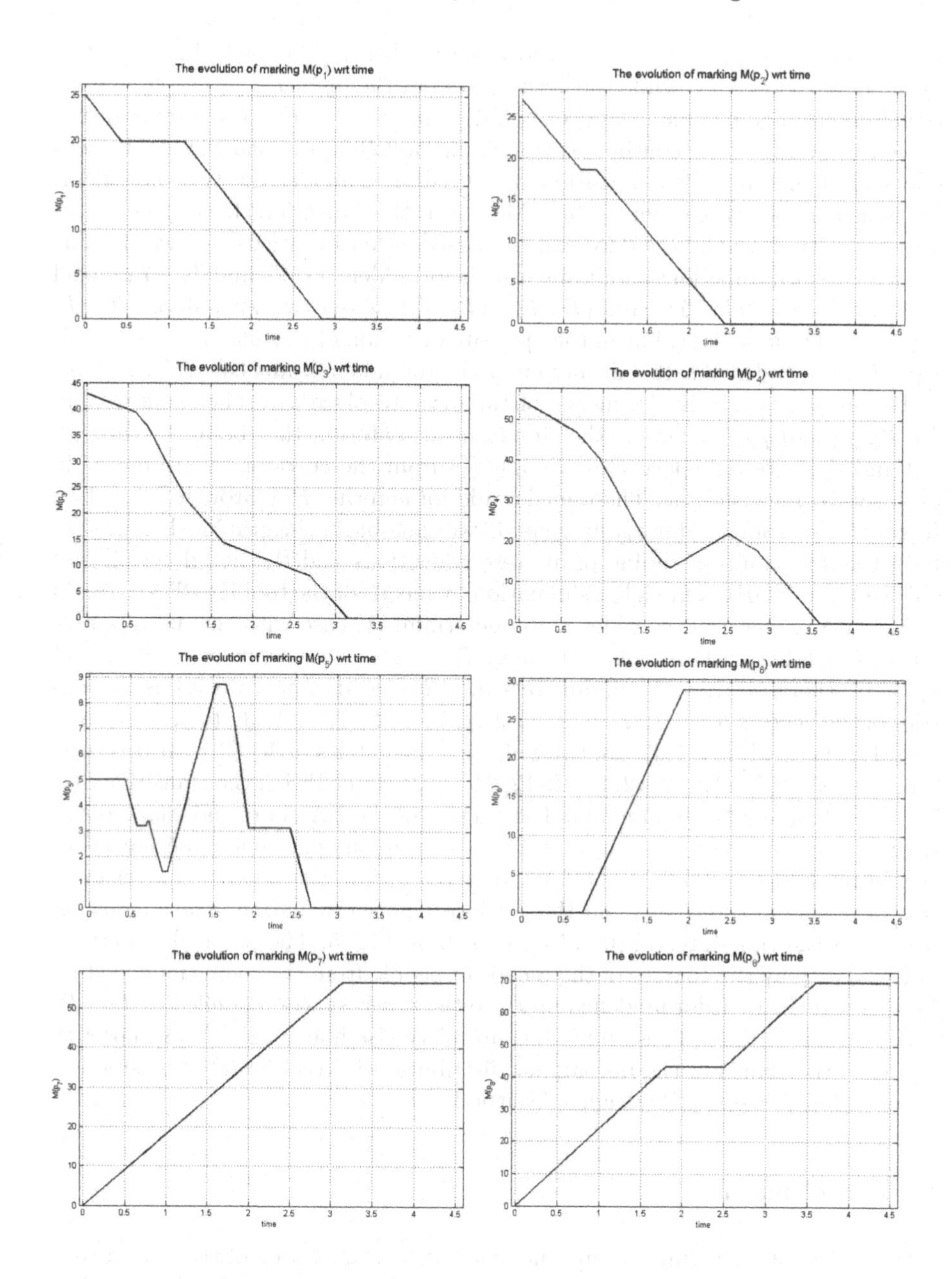

Fig. 5. The escape dynamics expressed by the evolution of markings $M(P_i)$ of the FOHPN continuous places $P_1 - P_8$ in time at the discrete uniform probability distribution of timing the FOHPN discrete transitions. The piecewise-linear real functions show the flows of people escaping from the rooms modelled by $P_1 - P_4$ and from the corridor modelled by P_5 as well as the flows of people filling the particular safety spaces modelled by $P_6 - P_8$.

if p_1 is active then T_1 is open; (ii) if p_2 is active then T_1 is closed; (iii) the active p_3 opens the closed T_1, while the active p_4 closes the open T_1; (iv) t_1, t_2 may be either deterministic (time delays) or non-deterministic (with a probability distribution of timing). The continuous transition modelling the one-way door needs one such a module. Two transitions are needful to model the two-ways door. Each of them needs the own module. Moreover, the transitions have to be mutually exclusive. Namely, the door can be passed either outwards or inwards. The passage in both directions simultaneously is forbidden. T_1, T_2 and $T_7 - T_9$ model one-way doors while the pairs (T_3, T_5), (T_4, T_6) model two-ways doors. T_3, T_5 represent the same door, but in the opposite direction of passing. To avoid a collapse (at the endeavour to exit and enter the room simultaneously), T_3, T_5 have to be mutually exclusive by means of the discrete place p_{37}. The same is valid for T_4, T_6 and p_{38}. Likewise, the simultaneous entering the room P_3 (through T_5) and the external space P_6 (through T_7) from the corridor P_5 are mutually exclusive by means of p_{39}. The same is valid for entering P_4 (through T_6) and P_6 (through T_7) from P_5 and p_{40}. In general, we can obtain the graphical results for both the deterministic timing of discrete transitions and the non-deterministic one using the FOHPN model at simulation by means of the tool HYPENS [16,17] in MATLAB. However, consider a non-deterministic case with the discrete uniform probability distribution of timing: $f_x = 1/(b - a)$ when $x \in (a, b)$ and $f_x = 0$ otherwise. Let the parameters a, b for 18 discrete transitions are entries of the vectors $\mathbf{a} = (.1, .2, .1, .2, .1, .2, .1, .2, .1, .2, .1, .2, .1, .3, .1, .4, .1, .4)$, $\mathbf{b} = (1, 1, 1, 1, 1, 1, 1, 1, 1, 1, 1, 1, 1, 3, 1, 4, 1, 4)$. Let the entries of $\mathbf{V}^{min} = (0, 0, 0, 0, 0, 0, 0, 0, 0)$, $\mathbf{V}^{max} = (12, 12, 12, 12, 12, 12, 24, 18, 24)$ limit IFS of the transitions T_j. These parameters depend on a kind and a size of the EA doors and characterize their throughput. Let the initial states of continuous places are the entries of the vector $\mathbf{M0}_C = (25\ 27\ 43\ 55\ 5\ 0\ 0\ 0)$ representing the number and dislocation of people in the rooms and the corridor before evacuation. The simulation results on the time interval (0, 4.5) are given in Fig. 5. The particular pictures in Fig. 5 display the courses of the escape of people from the rooms modelled by $P_1 - P_4$ and the corridor modelled by P_5 to the safety spaces modelled by $P_6 - P_8$ with respect to time. These functions are piecewise-linear. HYPENS allows to use also other kinds of the probability distribution for the discrete transitions - exponential, Poisson's, Rayleigh's, Weitbull's.

4 Conclusion

The problem of modelling EA and the model-based simulation of the EA evacuation were examined. As it can bee seen in Fig.1 left the Room 2 has the own Exit 2 and the Room 4 has the own Exit 3 to safety spaces. Moreover, they also have two-ways doors to the Corridor. Thus, e.g. when the direct Exit 1 from the Corridor is busy, the escape of people can be realized across these rooms too. Two alternative models were used. Firstly, the TPN model of EA was proposed, arising from the supervised P/T PN model presented in [4]. The doorways (one-way and reversible) were considered to be simple *agents* able to cooperate by means of the supervisor(s). Although the cooperation is forced (especially by the Supervisor 2), at simulation it helps to master the escape of people from EA. The TPN model conserves the structure of the supervised P/T PN model [4] and adds the time specifications to the transitions - the time delays to deterministic transitions or a kind of the probability distribution of timing to non-deterministic ones - depending on the kind and size of doorways. Namely, the transitions model the doorways being the *bottle-necks* of the escape. Thus, they have the most important influence on the escape. Secondly, the FOHPN model of EA was built. It models EA structure by the continuous PN (the first part of FOHPN). Here, the continuous places represent the rooms, the corridor and the safety spaces (where the people escape tends). The continuous transitions (modelling the doorways) are affected by means of discrete PN modules (the second part of FOHPN). Thus, the flow of people escaping from EA is modelled by the evolution of marking the FOHPN continuous places in time. The discrete PN modules are either simple P/T PN subnets or TPN subnets with deterministic or non-deterministic timing of their transitions. The parameters of the discrete transitions timing as well as IFS of continuous transitions have a coherence with the kind and size of the real EA doorways. In effect, the discrete PN modules substitute the supervisors applied in TPN. In both approaches (TPN and FOHPN) the model-based simulation of the evacuation process was performed in MATLAB by means of the universal simulation tool HYPENS [16,17]. The TPN simulation results and the FOHPN ones were presented, respectively, in Fig. 3 and Fig. 5 where the marking evaluation of particular relevant places with respect to time were displayed. In comparison with the P/T-PN-based approach [4] the TPN-based approach as well as the FOHPN-based one seem to be more suitable for the simulation of the evacuation process. Namely, they yield time behaviour of the escape while P/T PN does not. This allows to test the throughput of particular escape routes in time. At comparison both approaches (TPN and FOHPN) each other we can state that TPN approach has a higher accuracy (its marking exactly express human individuals) while FOHPN offer the piecewise-linear functions (something like *tangents* or *envelopes* of the courses of the actual continuous marking). In general, the simulation results achieved by such models can be useful not only at modelling the evacuation but also at the development of new buildings.

Acknowledgments. The author thanks for the support both the Slovak Grant Agency for Science VEGA under grant # 2/0039/13 and the Slovak Grant Agency APVV under grant APVV-0261-10.

References

1. Balduzzi, F., Giua, A., Menga, G.: First-Order Hybrid Petri Nets: A Model for Optimization and Control. IEEE Trans. on Robotics and Automation 16, 382–399 (2000)
2. Balduzzi, F., Giua, A., Seatzu, C.: Modelling and Simulation of Manufacturing Systems Using First-Order Hybrid Petri Nets. International Journal of Production Research 39, 255–282 (2001)

3. Balduzzi, F., Di Febbraro, A., Giua, A., Seatzu, C.: Decidability Results in First-Order Hybrid Petri Nets. Discrete Event Dynamic Systems 11, 41–58 (2001)
4. Čapkovič, F.: Supervision of Agents Modelling Evacuation at Crisis Situations. In: Jezic, G., Kusek, M., Nguyen, N.-T., Howlett, R.J., Jain, L.C. (eds.) KES-AMSTA 2012. LNCS (LNAI), vol. 7327, pp. 24–33. Springer, Heidelberg (2012)
5. Dotoli, M., Fanti, M., Giua, A., Seatzu, C.: First-Order Hybrid Petri Nets. An Application to Distributed Manufacturing Systems. In: Nonlinear Analysis: Hybrid Systems, vol. 2, pp. 408–430 (2008)
6. David, R., Alla, H.: On Hybrid Petri Nets. Discrete Event Dynamic Systems: Theory and Applications 11, 9–40 (2001)
7. Dotoli, M., Fanti, M., Giua, A., Seatzu, C.: Modeling Systems by Hybrid Petri Nets: an Application to Supply Chains. In: Kordic, V. (ed.) Petri Net Theory and Applications, ch. 5, pp. 91–109. I-Tech Education and Publishing, Vienna (2008)
8. Dotoli, M., Fanti, M., Iacobellis, G., Mangini, A.M.: A First-Order Hybrid Petri Net Model for Supply Chain Management. IEEE Transactions on Automation Science and Engineering 6, 744–758 (2009)
9. Hofman, U., Veichtlbauer, A., Miloucheva, T.: Dynamic Evacuation Architecture Using Context-Aware Policy Management. International Journal of Computer Science and Applications 6, 38–49 (2009)
10. Iordache, M.V., Antsaklis, P.J.: Supervision Based on Place Invariants: A Survey. Discrete Event Dynamic Systems 16, 451–492 (2006)
11. Iordache, M.V., Antsaklis, P.J.: Supervisory Control of Concurrent Systems: A Petri Net Structural Approach. Birkhäuser, Boston (2006)
12. Iordache, M.V.: Methods for the Supervisory Control of Concurrent Systems Based on Petri Nets Abstraction. Ph.D. Dissertation, University of Notre Dame, Notre Dame, Indiana, USA (2003)
13. Lino, P., Maione, G.: Applying a Discrete Event System Approach to Problems of Collective Motion in Emergency Situations. In: Klingsch, W.W.F., Rogsch, C., Schadschneider, A., Schreckenberg, M. (eds.) Pedestrian and Evacuation Dynamics 2008, pp. 465–477. Springer, Heidelberg (2010)
14. Murata, T.: Petri Nets: Properties, Analysis and Applications. Proceedings of the IEEE 77, 541–580 (1989)
15. Peterson, J.L.: Petri Nets Theory and the Modelling of Systems. Prentice-Hall Inc., Englewood Cliffs (1981)
16. Sessego, F., Giua, A., Seatzu, C.: HYPENS: A Matlab Tool for Timed Discrete, Continuous and Hybrid Petri Nets. In: van Hee, K.M., Valk, R. (eds.) PETRI NETS 2008. LNCS, vol. 5062, pp. 419–428. Springer, Heidelberg (2008)
17. Sessego, F., Giua, A., Seatzu, C.: HYPENS Manual (2008), http://www.diee.unica.it/automatica/hypens/Manual_HYPENS.pdf
18. Popova-Zeugmann, L.: Time Petri Nets: Theory, Tools and Applications, Part 1, Part 2 (2008), http://www2.informatik.hu-berlin.de/ popova/1-part-short.pdf, http://www2.informatik.hu-berlin.de/~popova/2-part-short.pdf

DPI: Dual Private Indexes for Outsourced Databases

Yi Tang[1,2], Fang Liu[1,2], and Liqing Huang[1,2]

[1] School of Mathematics and Information Science
Guangzhou University, Guangzhou 510006, China
[2] Key Laboratory of Mathematics and Interdisciplinary Sciences of Guangdong
Higher Education Institutes
Guangzhou University, Guangzhou 510006, China

Abstract. Designing secure and efficient indexes to support selective queries over encrypted data at server side is necessary for outsourced databases. On one hand, the indexes must be associated with plain values in order to locate tuples precisely. On the other hand, the indexes may open a door to leak sensitive information, especially when the indexes are combined with selective encryption. In this paper, we propose DPI, a dual private index for outsourced databases. According to DPI, two types of indexes, incorporated with user-specified random salts, are defined for each attribute. The generalization-based index is used to support server side query over encrypted data and protect data from link inferences, and the value-based index is adaptively used at client side to reduce extra decryption costs. We have conducted some experiments to validate our proposed method.

Keywords: Outsourced Database, Private Index, Computation Cost.

1 Introduction

Outsourcing databases to a third party is a typical paradigm of cloud storages. In order to protect the privacy of database owners, it needs to store data encrypted on server side. It also needs secure and efficient indexes to support SQL queries over encrypted databases.

Early efforts on outsourced databases [3] are focused on translating the client-side plain queries into corresponding server-side encrypted versions, and tuples are encrypted in row by row by a symmetric encryption method with a single key [1][3]. The selective encryption methods use different keys to encrypt different data portions such as tuples or attributes [4]. To avoid users from managing too many keys, the keys can be derived from user hierarchy [2] and the symmetric methods can be changed into some asymmetric methods such as the attribute-based encryption (ABE) method [8].

DAS model is a well-known model for outsourcing databases [3]. According to this model, data are encrypted tuple by tuple. Auxiliary attributes are introduced to store index tag values for data indexing. These index tags represent

N.T. Nguyen et al. (Eds.): ACIIDS 2014, Part I, LNAI 8397, pp. 291–300, 2014.

index mapping images of plain attribute values. The preimage of each image can be a single value or a set of values, and hence the indexes are classified into value-based and bucket-based [6]. However, the index tags only shadow real attribute values. They are inextricably linked to the original plain values. The combination of index and selective encryption may introduce potential disclosures of confidential information [1][5][6][7].

An adversary user can exploit the inconsistencies among the encrypted tuples with equal index tag values and non-equal but intersected access control lists (ACLs) to infer some real attribute values that he cannot access. Those leaked tuples can be from tuples at rest [6] or tuples from query results [5]. To defend against this type of attacks, the user-specified random salts are introduced to preserve the consistencies with ACLs [6], and the generalization-based indexes are used to prevent the inferences from query results [5]. However, the generalization-based indexes would also introduce tuples which do not satisfy the SQL query conditions, and hence increase more decryption costs at client side.

In this paper, we try to reduce the computation costs at client side by combining the salted value-based and generalization-based index. We proposed DPI, a dual private index strategy to define private indexes. We use two private indexes for each attribute. The generalization-based index is used to support server side query over encrypted data to protect data from link inferences, and the value-based index is adaptively used at client side to locate tuples precisely to reduce extra decryption costs.

The rest of this paper is structured as follows. In Section 2, we give our basic models and overview the index-based and generalization-based index for encrypted data. In Section 3, we introduce the dual private index strategy for indexing attributes and propose an adaptive method to translate client queries. In Section 4, we conduct some experiments to validate our proposed method. And finally, the conclusion is drawn in Section 5.

2 Background

2.1 The Outsourced Databases

The typical outsourced database model is as demonstrated in Fig. 1. Three parties, the users, the proxy, and the outsourced database service provider, are involved in the scenario where the users and the proxy are on client side and the provider on server side. A general assumption is as follows. The proxy is trust and secure and the user can access any outsourced data based on his access rights. The service provider is honest but curious, sometimes a bit greedy. This means that the provider can provide services he claims to be able to provide and he may leak some stored encrypted tuples out to others for curiosity or benefits.

We consider the DAS model described in [3]. According to this model, the plain relation $R = (\mathsf{A}_1, \mathsf{A}_2, ..., \mathsf{A}_n)$ is stored as corresponding encrypted relation $R^s = (etuple, \mathsf{A}_1^s, \mathsf{A}_2^s, ..., \mathsf{A}_n^s)$ on server where the attribute *etuple* denotes the encrypted tuple in R and A_i^s denotes the index for A_i in R^s with $i = 1, 2, ..., n$.

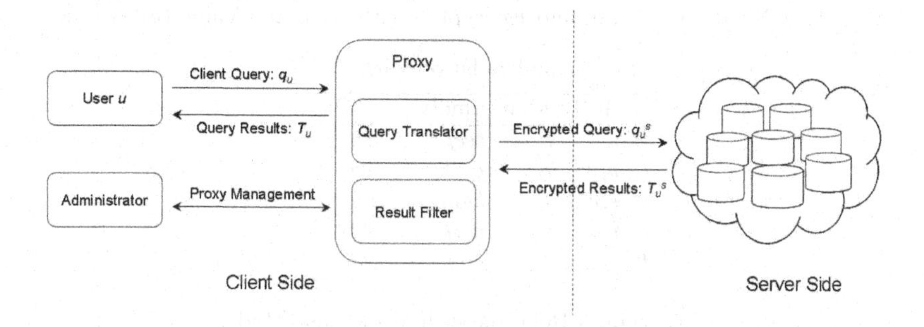

Fig. 1. The Outsourced Databases Scenario

When user u initiates a query q_u, a proxy at client side will translate the plain q_u into an encrypted query q_u^s and deliver it to the remote server. After executing q_u^s over encrypted data on server side, the server replies to client with encrypted tuple set T_u^s. The proxy then decrypts tuples in T_u^s, refilters the decrypted tuples according to the conditions in q_u, and returns satisfied tuples to user u.

The attribute A_i^s in R^s denotes the index for attribute A_i in R. The index values are generally decided by a user-specified function $\mathsf{id}()$. Two types of indexes can be used in outsourced databases scenario, the value-based index and the bucket-based index. For an attribute value a, its corresponding value-based index is denoted as $a^s = \mathsf{id}(a)$, while the bucket-based index is denoted as $a^s = \mathsf{id}(bucket_a)$ with $a \in bucket_a$. With the auxiliary attribute for index, the query initiated by user u at client side, q_u: **select** * **from** R **where** $R.\mathsf{A} = a$, will be translated into an encrypted version, q_u^s: **select** *etuple* **from** R^s **where** $R^s.\mathsf{A}^s = a^s$.

Table 1(a) and 1(b) shows an original relation with an access control list and corresponding encryption version on server side, respectively. The attribute indexes are value-based. For example, the tuple t_1, $\langle 7, beef, 22\rangle$ in Table 1(a) is stored as the tuple t_1^s, $\langle e_1, \mathsf{id}_u(7), \mathsf{id}_u(beef), \mathsf{id}_u(22)\rangle$ in Table 1(b), where $e_1 = encrypted(\langle 7, beef, 22\rangle)$, the encrypted version of t_1, and $\mathsf{id}_u()$ is an index mapping function, privately used by user u, for attributes.

2.2 Linking in Tuple Sets with Selective Views

It seems secure that the tuple encryption protects all plain values in a tuple. Without access rights, an adversary user cannot get the decryption key to decrypt the encrypted tuple. However, the value-based index scheme naturally implies that the equal index values means the equal plain attribute values. This implication could be used to launch a link attack when an adversary user gets some encrypted tuples he cannot access [6].

For example, considering the tuples t_1^s and t_2^s in Table 1(b), we have $t_1^s.\mathsf{Mon}^s = t_2^s.\mathsf{Mon}^s$, and $t_1.ACL \neq t_2.ACL \wedge t_1.ACL \cap t_2.ACL \neq \phi$. As a result, if user v takes t_1^s, he can infer that $t_1^s.\mathsf{Mon}^s$ represent the index value for $\mathsf{Mon} = 7$ although he

Table 1. A Relation in Plain and Encrypted with ACL and Value-based Index

(a) Original Relation with ACL

	ACL	Mon	Cmdty	Qty
t_1	u	7	*beef*	22
t_2	u,v	7	*pork*	25
t_3	v,w	7	*ham*	25
t_4	u,v	8	*pork*	23
t_5	u	9	*beef*	23

(b) Encrypted Relation with Value-based Index

	etuple	Mon^s	Cmdty^s	Qty^s
t_1^s	e_1	$\mathsf{id}_u(7)$	$\mathsf{id}_u(beef)$	$\mathsf{id}_u(22)$
t_2^s	e_2	$\mathsf{id}_u(7)\mathsf{id}_v(7)$	$\mathsf{id}_u(pork)\mathsf{id}_v(pork)$	$\mathsf{id}_u(25)\mathsf{id}_v(25)$
t_3^s	e_3	$\mathsf{id}_v(7)\mathsf{id}_w(7)$	$\mathsf{id}_v(ham)\mathsf{id}_w(ham)$	$\mathsf{id}_v(25)\mathsf{id}_w(25)$
t_4^s	e_4	$\mathsf{id}_u(8)\mathsf{id}_v(8)$	$\mathsf{id}_u(pork)\mathsf{id}_v(pork)$	$\mathsf{id}_u(23)\mathsf{id}_v(23)$
t_5^s	e_5	$\mathsf{id}_u(9)$	$\mathsf{id}_u(beef)$	$\mathsf{id}_u(23)$

(c) Encrypted Relation with Salted Value-based Indexes

	etuple	Mon^s	Cmdty^s	Qty^s
t_1^s	e_1	$\mathsf{id}_u(7, S_{u1})$	$\mathsf{id}_u(beef, S_{u1})$	$\mathsf{id}_u(22, S_{u1})$
t_2^s	e_2	$\mathsf{id}_u(7, S_{u2})\mathsf{id}_v(7, S_{v1})$	$\mathsf{id}_u(pork, S_{u1})\mathsf{id}_v(pork, S_{v1})$	$\mathsf{id}_u(25, S_{u1})\mathsf{id}_v(25, S_{v1})$
t_3^s	e_3	$\mathsf{id}_v(7, S_{v2})\mathsf{id}_w(7, S_{w1})$	$\mathsf{id}_v(ham, S_{v1})\mathsf{id}_w(ham, S_{w1})$	$\mathsf{id}_v(25, S_{v2})\mathsf{id}_w(25, S_{w1})$
t_4^s	e_4	$\mathsf{id}_u(8, S_{u1})\mathsf{id}_v(8, S_{v1})$	$\mathsf{id}_u(pork, S_{u1})\mathsf{id}_v(pork, S_{v1})$	$\mathsf{id}_u(23, S_{u1})\mathsf{id}_v(23, S_{v1})$
t_5^s	e_5	$\mathsf{id}_u(9, S_{u1})$	$\mathsf{id}_u(beef, S_{u1})$	$\mathsf{id}_u(23, S_{u2})$

(d) Encrypted Relation with Salted Generalization-based Index Mon_g^s

	etuple	Mon_g^s
t_1^s	e_1	$\mathsf{id}_u([7, 9], \mathrm{S}_{u1})$
t_2^s	e_2	$\mathsf{id}_u([7, 9], \mathrm{S}_{u2})\mathsf{id}_v([7, 9], \mathrm{S}_{v1})$
t_3^s	e_3	$\mathsf{id}_v([7, 9], \mathrm{S}_{v2})\mathsf{id}_w([7, 9], \mathrm{S}_{w1})$
t_4^s	e_4	$\mathsf{id}_u([7, 9], \mathrm{S}_{u2})\mathsf{id}_v([7, 9], \mathrm{S}_{v1})$
t_5^s	e_5	$\mathsf{id}_u([7, 9], \mathrm{S}_{u3})$

has no right to access t_1. This is because that he knows that $t_2.\mathsf{Mon} = 7$, and one of the index tag values in $t_2^s.\mathsf{Mon}^s$ is just the same as $t_1^s.\mathsf{Mon}^s$.

This kind of inference is based on the explicit equality among index tag values. A simple defence method is to introduce random *salts* in constructing indexes to destroy the equalities of index tag values [6].

Table 1(c) demonstrates an encrypted table with salted indexes. For the tuples t_1^s and t_2^s, user u can construct the index tags for attribute Mon^s with different salts. The two salts, S_{u1} and S_{u2}, make $t_1^s.\mathsf{Mon}^s \neq t_2^s.\mathsf{Mon}^s$. Though user v can still realize that both $\mathsf{id}_u(7,S_{u2})$ and $\mathsf{id}_v(7, S_{v1})$ represent the index with $\mathsf{Mon} = 7$, he cannot infer any other tuples are with $\mathsf{Mon} = 7$ except the tuples he can access.

2.3 The Generalization-Based Indexes

In [5], we address another kind of inference attacks, the collusion-based inference, based on the implicit equality in query results.

As a collusion-based inference instance for Table 1, we consider a specified single-attribute-single-value query, **select** $*$ **from** R **where** $R.\mathsf{Mon} = 7$, initiated by user u and v, respectively. The two returned sets, $T_u \triangleq \{t_1^s, t_2^s\}$ and $T_v \triangleq \{t_2^s, t_3^s\}$, means that any tuples in the two sets are with $\mathsf{Mon} = 7$. Suppose that user v colludes with server s, i.e., s can leak the query results from other users, especially for those intersected with v's query results, to v. When s sends T_u to v, v can definitely determines that the tag value represented by $t_1^s.\mathsf{Mon}^s$ corresponds to the index of 7 although he cannot be granted to access tuple t_1 and t_1 is still encrypted.

We have proposed a private generalization-based index to defend against the collusion-based inference attack [5]. A generalization-based index mapping maps a set of attribute values into a single index tag value. With the generalization mapping $gm_{\mathsf{A}}()$, the original query, say **select** * **from** R **where** $R.\mathsf{A} = a$, will be translated into the generalized version, **select** * **from** R **where** $gm_{\mathsf{A}}(R.\mathsf{A}) = g$ where $g = gm_{\mathsf{A}}(a)$. Let T_a and T_g be the query results for original and generalized query, respectively, it is obviously that $T_a \subseteq T_g$. This makes the real results be hidden in a larger tuple set.

We have developed a method to construct secure and private generalization-based indexes [5]. We design a combined measure, (k, α), to measure the collusion-based inference resistance for a certain attribute generalization index mapping, where α is defined on the information entropy which measures the privacy protected in T_g and k measures the attribute value diversities in query results. We also introduce the notion of salt into the generalization-based indexes to prevent an adversary user obtain more information.

Table 1(d) demonstrates a salted generalization-based index over attribute Mon. The generalization index mapping over attribute Mon is defined as gm_{Mon} : $\{7, 8, 9\} \rightarrow \{[7, 9]\}$ and we can show that the mapping gm_{Mon} is (4, 0.25)-secure. More details are referred in [5].

3 Dual Private Index

In this section, we introduce the notion of dual private index. The dual private index means that each attribute is attached with two indexes, the value-based private index and the generalization-based private index. To defend against collusion-based attacks, the generalized index is necessary. To reduce extra decryption cost introduced by the generalized indexes, an adaptive procedure will be executed to determine whether or not refiltering the returned encrypted tuples on value-based indexes.

3.1 Time Cost for Post-query

Consider the query q_u: **select** $*$ **from** R **where** $R.\mathsf{A} = a$, initiated by user u. Once it is translated into an encrypted generalization version and executed on server, the server will return total m_g tuples with only m_a tuples are matched with $R.\mathsf{A} = a$. Let n_g^* denote the maximum number of salts for all generalization images over attribute A and n_a^* denote the maximum number of salts for all attributes values over A. The procedures for the proxy with different indexes can be described as follows.

- *Case 1: the generalization-based index only*
  The q_u is translated into **select** *etuple* **from** R^s **where** $R^s.\mathsf{A}_g^s = \vee_{i=1}^{n_g^*} \mathsf{id}_u(g, S_{u_i})$ and sent to server. After receiving query results, decrypts m_g tuples and then filters out m_a tuples.
- *Case 2: the dual index*
  The q_u is translated into **select** *etuple*, A^s **from** R^s **where** $R^s.\mathsf{A}_g^s = \vee_{i=1}^{n_g^*} \mathsf{id}_u$ (g, S_{u_i}) and sent to server. After receiving query results, filters the returned m_g tuples with condition $\mathsf{A}^s = \vee_{i=1}^{n_a^*} \mathsf{id}_u(a, S_{u_i})$, and then decrypts the m_a filtered tuples.

The main difference for the above two cases lies in the operation order between the refiltering and decrypting at client side. For the former, it needs m_g decryptions. For the latter, it needs to generate n_a^* index tags to filer encrypted tuples and then execute m_a decryptions. Intuitively, decryption is a time-consuming computation, and the index tag generation is relatively simpler and faster. If we adopt the dual index, the total computation cost may be reduced.

Let t_{dec} be the time for decrypting a tuple and t_{idx} be the time for generating a value-based attribute index value at client side. Both operations need time cost $n_g^* \cdot t_{idx}$ to generate generalization indexes to support querying on server. However, when the query results are returned, the time costs for post-query, i.e., filtering out matched tuples, are different. With only the generalization-based index, the time cost of post-query is $time_g \triangleq m_g \cdot t_{dec}$ while the cost is $time_d \triangleq m_a \cdot t_{dec} + n_a^* \cdot t_{idx}$ with dual indexes. It is obviously that the minimum time cost is $time_a \triangleq m_a \cdot t_{dec}$ which is with value-based index.

For the example in Table 1(d), we consider the queries initiated by user u. Suppose he decides to initiate query, q_u: **select** $*$ **from** R **where** $R.\mathsf{Mon} = mon$, where the parameter *mon* can be chosen as *7*, *8*, and *9*, respectively. The post-query time costs are listed in Table 2. We find that with the generalization-based index only, all of the three cases need 4 decryptions to decide matched tuples. While with the dual index, there needs at most 2 decryptions and 2 index generations. Since the size of the tuple set for post-query is limited, the search cost can be ignored. And hence, in those cases, if $time_d < time_g$, we can reduce post-query time cost by adopting dual indexes.

Given a query, to compare the size of $time_d$ and $time_g$ is impossible because we cannot predict how many matched tuples are in query results. We consider a probability based method. We introduce a parameter p to denote the probability of adopting generalization-based index method to translate queries for next query. Correspondingly, the probability of adopting dual index method for next query is $1 - p$.

Table 2. Time Cost Estimation for User u

mon	$time_g$	$time_d$	$time_a$
7	$4 \cdot t_{dec}$	$2 \cdot t_{idx} + 2 \cdot t_{dec}$	$2 \cdot t_{dec}$
8	$4 \cdot t_{dec}$	$2 \cdot t_{idx} + t_{dec}$	t_{dec}
9	$4 \cdot t_{dec}$	$2 \cdot t_{idx} + t_{dec}$	t_{dec}

```
procedure pEstimator()
  if 0 < p < 1
    if m_a + r · n*_a ≥ m_g and p < 0.98
      p = p + 0.02
    else
      if m_a + r · n*_a < m_g and p > 0.02
        p = p - 0.02
      endif
    endif
  endif
  return p
endprocedure
```

We design an adaptive procedure as above to decide the parameter p. We compute a ratio r where $r \approx \frac{t_{dec}}{t_{idx}}$. Note that r can be decided by the encryption method and the index generator. For example, if we use *AES* cipher as the encryption tool and *md*5 hash function as the index generator, with both cryptological functions are provided by the Crypto library in JDK 6, the r can be set as 7. When running current post-query procedure at client side, the value p is changed on the comparison between $m_a + r \cdot n_a^*$ and m_g.

3.2 Multiple Index Values

Note that in our proposed method, multiple index values may be defined for a same index attribute values. This kind of data is not directly supported in the relational data model. We can logically represent them with two separate tables, one is for encrypted relation and the other is for attribute indexes [6].

Table 3 demonstrates the different tables stored on server for attribute Mon. By joining the separate tables, the original encrypted relation with dual private indexes over attribute Mon can be simply obtained at server side. And at the same time, the query initiated by user u, q_u: **select** $*$ **from** R **where** $R.\mathsf{Mon} = 7$, will be translated into correspond encrypted version, $q_{u,g}^s$: **select** *etuple*,Mon^s **from** R^s `join` R^s_{Mon} on $R^s.\mathsf{tid} = R^s_{\mathsf{Mon}}.\mathsf{tid}$ **where** $R^s_{\mathsf{Mon}}.\mathsf{Mon}^s_g = \mathsf{id}_u([7, 9], S_{u1}) \vee \mathsf{id}_u([7, 9], S_{u2}) \vee \mathsf{id}_u([7, 9], S_{u3})$.

To support our proposed dual private index for an attribute A, each user u has the following knowledge:

1. the function $\mathsf{id}_u()$ used to generate two types of indexes;
2. the method used to generate random salts;

Table 3. Encrypted and Index Relations for Table 1(d)

(a) R^s

tid	*etuple*
t_1^s	e_1
t_2^s	e_2
t_3^s	e_3
t_4^s	e_4
t_5^s	e_5

(b) R^s_{Mon}

tid	Mon^s	Mon^s_g
t_1^s	$\mathsf{id}_u(7, S_{u1})$	$\mathsf{id}_u([7, 9], S_{u1})$
t_2^s	$\mathsf{id}_u(7, S_{u2})$	$\mathsf{id}_u([7, 9], S_{u2})$
t_2^s	$\mathsf{id}_v(7, S_{v1})$	$\mathsf{id}_v([7, 9], S_{v1})$
t_3^s	$\mathsf{id}_v(7, S_{v2})$	$\mathsf{id}_v([7, 9], S_{v2})$
t_3^s	$\mathsf{id}_w(7, S_{w1})$	$\mathsf{id}_w([7, 9], S_{w1})$
t_4^s	$\mathsf{id}_u(8, S_{u1})$	$\mathsf{id}_u([7, 9], S_{u2})$
t_4^s	$\mathsf{id}_v(8, S_{v1})$	$\mathsf{id}_v([7, 9], S_{v1})$
t_5^s	$\mathsf{id}_u(9, S_{u1})$	$\mathsf{id}_u([7, 9], S_{u3})$

3. the maximum number of salts $n^*_{g,\mathsf{A}}$ for generalization values over A;
4. the maximum number of salts $n^*_{a,\mathsf{A}}$ for values over A;
5. the query translation probability p_{A} for attribute A.

4 Experiments and Discussion

4.1 The Datasets

To evaluate the behavior of our proposed method, we need two types of materials for experiments, the data tuples and the authorized users for tuples. We first generate a relational table with 800,000 tuples following the TPC-H benchmark specifications, and then randomly select 13,000 tuples to construct a tables with three attributes. To show the relations with the different sizes of attribute domains on constructed indexes, we define the domain of these attributes are with 10,000, 9,999, and 1,000 distinct integers, respectively. For the authorized users for tuples, we extract the authors coauthored with Professor Jiawei Han from the DBLP repository. In particular, we extract the top 140 most productive authors and obtain 204 author sets from the repository. We view these author sets as the authorized user sets, i.e., the *ACL* lists for tuples.

To construct the satisfied generalization based index, we adopt two strategy, the increasing-generalization-level strategy and the adding-noise-tuples strategy, with probability β and $1-\beta$, respectively, to construct generalization mapping. With constructed (2, 1.5)-secure generalization mapping, we obtain 3 generated datasets with different βs for testing.

4.2 The Results

To determine the p for that user, we random generate single-attribute-single-value equality query and estimate the probability p. Since our experiments are on JDK 6 and use *AES* cipher and *md*5 hash function for encryption and index construction, respectively, we set r as 7 and initiate p as 0.5. Fig. 2 demonstrates the dependency between the query translation estimator probability p and query

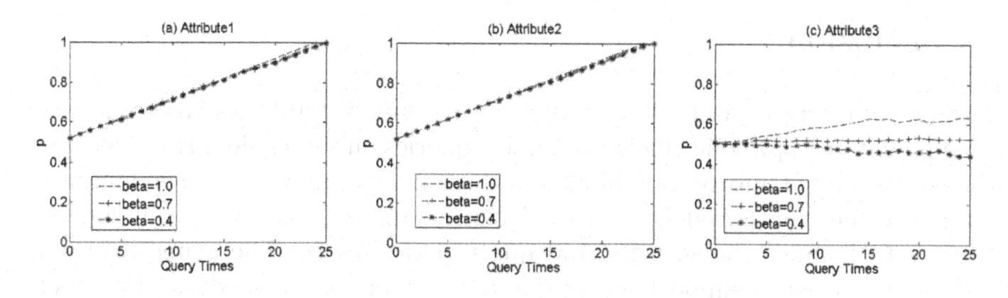

Fig. 2. Query Translation Probability p and Query Times

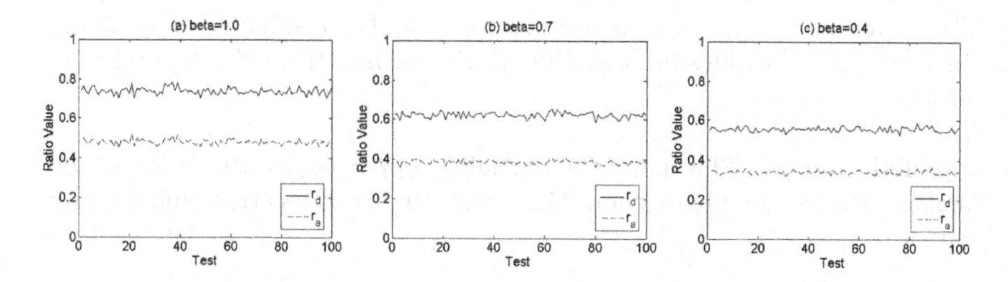

Fig. 3. Comparison between r_d and r_a for Queries over Attribute3

times for a random chosen user. The cases for the first two attributes are shown in Fig. 2(a) and Fig. 2(b). Both figures show that the p is monotonically increasing to 1 for all three βs. It means that with similar domain size and similar value distribution, the required value of p is similar. The case $p \to 1$ implies that the generalization-based indexes are preferred in queries for the first two attributes. The case for Attribute3 is different. For different βs, the probability ps are different. Fig. 2(c) shows that after limited query times, the value of p becomes stable and is trapped in a limited scope. When $\beta = 1$, the probability adopting dual index in queries is about 0.38, while for the case $\beta = 0.7$ and $\beta = 0.4$, the probability is about 0.5 and 0.42, respectively.

Fig. 3 demonstrates the time costs for post-query of our proposed dual index method. In this experiments, we random generate single-value equality queries over Attribute3 and decide query translation method based on precomputed p. For each query, we compute $time_g$, $time_d$, and $time_a$ (discussed in section 3.1), respectively. And then compute $r_d = time_d/time_g$ and $r_a = time_a/time_g$. Figure 3 shows that our proposed dual index method can save 25% to 40% time cost comparing to the generalization-based index. Comparing to the value-based index, the dual index can defend against collusion-based inference with 50% time costs increased for post-query.

From the view point of the sizes of returned query results, the dual private indexes introduce more traffic comparing to the generalization-based index only strategy. This is because that the dual index strategy needs extra value-based index tags to perform post-query filtering. However, the increased volume of data can be controlled by limiting the number of adopted value-based indexes via analyzing the query conditions in detail.

5 Conclusion

Ensuring data privacy is fundamental for outsourced databases. Indexes over encrypted data can provide efficient selective queries on server side. However, the plaintext associated information hidden in those indexes may introduce inference attacks. We have proposed DPI, a dual private index strategy for outsourced databases to balance the security and query performance. The generalization-based index is used to support server side query over encrypted data to protect data from link inferences, and the value-based index is used in client side to locate tuples precisely to reduce extra decryption cost. We also define an adaptive query translation method to reduce extra decryption cost. The conducted experiments on single-attribute-single-value equality queries demonstrate the feasibility of our proposed method.

Acknowledgments. This paper is partially supported by the Science and Technology Project of Guangzhou Municipal Higher Education under grant 2012A022.

References

1. Damiani, E., Vimercati, S., Jajodia, P.S., Samarati, P.: Balancing Confidentiality and Efficiency in Untrusted Relational DBMSs. In: Proceedings of ACM CCS 2003, pp. 93–10 (2003)
2. Damiani, E., Vimercati, S., Foresti, S., Jajodia, S., Paraboschi, S., Samarati, P.: Key Management for Multi-user Encrypted Databases. In: Proceedings of StorageSS 2005, pp. 74–83 (2005)
3. Hacigumus, H., Iyer, B., Li, C., Mehrotra, S.: Executing SQL over Encrypted Data in the Database-Service-Provider Model. In: Proceedings of ACM SIGMOD 2002, pp. 216–227 (2002)
4. Miklau, G., Suciu, D.: Controlling Access to Published Data Using Cryptography. In: Proceedings of VLDB 2003, pp. 898–909 (2003)
5. Tang, Y., Liu, F., Huang, L.: Generalization-Based Private Indexes for Outsourced Databases. In: Meng, W., Feng, L., Bressan, S., Winiwarter, W., Song, W. (eds.) DASFAA 2013, Part I. LNCS, vol. 7825, pp. 161–175. Springer, Heidelberg (2013)
6. Vimercati, S., Foresti, S., Jajodia, S., Paraboschi, S., Samarati, P.: Private Data Indexes for Selective Access to Outsourced Data. In: Prodeedings of WPES 2011, pp. 69–80 (2011)
7. Wang, H., Lakshmanan, L.: Efficient Secure Query Evaluation over Encrypted XML Databases. In: Proceedings of VLDB 2006, pp. 127–138 (2006)
8. Yu, S., Wang, C., Ren, K., Lou, W.: Achieving Secure, Scalable, and Fine-grained Data Access Control in Cloud Computing. In: Proceedings of INFOCOM 2010, pp. 534–542 (2010)

Anomaly SQL SELECT-Statement Detection Using Entropy Analysis

Thanunchai Threepak and Akkradach Watcharapupong

Department of Computer Engineering, Faculty of Engineering
King Mongkut's Institute of Technology Ladkrabang, Bangkok, Thailand
ktthanun@kmitl.ac.th, akkra_watch@yahoo.com

Abstract. Database systems are often intruded because they store valuable information and can be accessed through Internet web applications which sometimes are not developed with security in mind. Attackers can inject some crafted inputs to those programs that work on database systems so that some unexpected results occur. We analyze the database system log files, focus on query statements (SQL SELECT statements), using the Shannon entropy to detect such anomaly attempts that would change conditional entropy significantly. Our experiment shows that the proposed anomaly detection using entropy analysis is effective.

Keywords: Database Security, SQL Injection, Anomaly Detection, Entropy Analysis.

1 Introduction

In every information system, database servers are the important subsystems that collect, manage, and provide raw data for other applications. There is a query mechanism in every database system which provides other applications the ability to submit and retrieve the required data. The SQL language is used as the standard query language because of its flexibility and effectiveness. For administration and security purpose, database systems have query log files to collect every statements which operate in database systems.

Database systems are the target of hacker because they mostly contain important information. Attacks on database systems primarily target on internal information disclosure and privilege gaining on the systems. SQL injection is a database attack that inserts crafted string to alter normal query statements then process malicious functions. This attack can damage the database system and cause privacy problems. As a result, database administrators always need an intrusion detection system to detect attacks.

Intrusion detection systems are used to identify intrusion behavior in the systems. They help system administrators to detect dangerous behaviors and alert related staffs as quick as possible. Intrusion detection systems are divided by detection schemes into two areas, namely the misuse detection, and the anomaly detection. The misuse

N.T. Nguyen et al. (Eds.): ACIIDS 2014, Part I, LNAI 8397, pp. 301–309, 2014.

detection uses the collection of attack patterns to identify malicious events using pattern matching technique. In another way, anomaly detection uses the collection of normal behaviors to distinguish normal events in the system from intrusions.

There are many techniques that are used to implement anomaly detection systems. Fuzzy logic, genetic algorithm, hidden Markov model, Bayesian methodology, artificial neural network, self organizing map, support vector machine, principle component analysis, singular value decomposition, data mining, and statistical methods [1], [2] are examples of techniques which are used to discover system profiles and detect anomalies.

Entropy based detection is one of interesting techniques that are applied in many researches. Entropy analysis is the process which converts data to entropy domain and make decision by some specified criteria. This scheme is applied in many research topics such as anomaly detection in network applications [3], anomaly detection in network traffic [3-7], detecting computer worm using entropy analysis [8], [9], anomaly detection in stream [10], in space shuttle engine [11], and in computer network behavior [12].

In this paper, we present the entropy analysis method for detecting database attack. Our approach uses entropy analysis to analyze the database query log file by using relationships that query statements from intruders change the conditional entropy significantly.

Details of each sequence are described in next section. Section 2 describes principle of entropy. Section 3 specifies database operations and attacks. Section 4 explains the proposed anomaly detection procedure using conditional entropy. Section 5 shows the experimental results and Section 6 is the conclusion and future works.

2 Entropy Calculation

Shannon entropy [13] is the widely used method to measure uncertainty value in the system. In principle, the entropy, H(X), of continuous random variable X with probability density function p(x) is given by equation (1).

$$\mathrm{H}(X) = -\int \mathrm{p}(x)\log_2(\mathrm{p}(x))\,dx \tag{1}$$

For discrete domain, the entropy value of discrete variable X with density probability p(x) is specified by equation (2).

$$\mathrm{H}(X) = -\sum_{x\in X} \mathrm{p}(x)\log(\mathrm{p}(x)) \tag{2}$$

Entropy of discrete variable X is maximum entropy when X varies to N different values, {x1, x2, .. xN}, and each value has equal probability, 1/N. Maximum entropy is formulated in equation (3). From differential mathematics, the differential ratio of maximum entropy equals to 1/(N*ln(10)).

$$H(X) = -\sum_{i=1}^{N} \frac{1}{N} \log\left(\frac{1}{N}\right) = \log(N) \tag{3}$$

The theory of entropy analysis is extending into other arrangements. Conditional entropy is one of variation that measures the complexity of relation on two random variables.

$$H(Y|X) = \sum_{x \in X} p(x) H(Y|X = x) \tag{4}$$

$$H(Y|X = x) = -\sum_{y \in Y} p(y|x) \log(p(y|x)) \tag{5}$$

The conditional entropy, H(Y|X), of the random variable Y when given random variables X is written as (4), and H(Y|X=x) is the entropy of all Y conditions on a fix $x \in X$, shown in equation (5). Where, p(x) is probability of variable x and p(y|x) is conditional probability that variable y occur when variable x is known to happen.

3 Database Operations and Attacks

Application programs access information in databases using function calls. They set up connections at first then pass commands which mostly are the SQL SELECT-statements as those function's arguments to database services. These query statements are composed by predefined phrases and variant words which are string or not strongly typed variables from application programs as follows:

```
dbconn = connectDB("myDB")
var = getValue("ExtInputVar")
sqlstmt = "SELECT * FROM tb WHERE col='"+ var +"';"
result = SendSQL(dbconn, sqlstmt)
```

Attackers or malicious tools may put crafted character streams as application inputs that transfer to such variables (*var*) which are from external input sources then combined to the SQL statements (*sqlstmt*) then processed by database systems and threats occur finally. They inject through user input or client application both directly (e.g., HTML forms) and indirectly (e.g., HTTP variables). These are also known as the SQL injection attacks [14]. Such as:

Case 1: Put ' OR '1'='1 into *var*. The *sqlstmt* will be as follows which makes its where-clause always true.

```
SELECT * FROM tb WHERE col='' OR '1'='1';
```

Case 2: Put x';DROP TABLE tb into *var*. The *sqlstmt* will be as follows which purges the tb table if it has enough privilege.

```
SELECT * FROM tb WHERE col='x';DROP TABLE tb;
```

Attacks mostly try to find something (e.g., raw information, database schemata, and DBMS configurations) or try to perform some negative actions (e.g., data manipulation, privilege escalation, and command execution).

However some SQL injection attacks may use the character encoding evasions (e.g., Unicode, URL, and Hex encoding) and the character meaning evasions (e.g., multiple comments). Such as:

Case 3: Put x'; %44%52%4F%50%20%54%41%42%4C%45%20%74%62 into *var*. The *sqlstmt* will be as follows which purges the tb table if it has enough privilege.

```
SELECT * FROM tb WHERE col='x';DROP TABLE tb;
```

Case 4: Put x';D/*foo*/ROP TA/*bar*/BLE tb into *var*. The *sqlstmt* will be as follows which purges the tb table if it has enough privilege.

```
SELECT * FROM tb WHERE col='x';DROP TABLE tb;
```

4 Conditional Entropy for Anomaly Detection

Every application has its own patterns to acquire data from each database tables. This makes every query statements to each tables have repeated and correlated patterns. When uncommonly events occur, either by rare function calls or database attacks, query patterns will increase their complexity significantly. The conditional entropy value of some parameters in query string conditions on other parameters are used as query complexity. Anomaly query strings are detected when conditional entropy value increase significantly.

To detect anomaly characteristics, the change on maximum entropy is used as primary detection level. When we compute the entropy of this situation, we will get the maximum entropy. Because the maximum entropy is increased when new anomaly log records are read and included in entropy calculation, the change of the maximum entropy can be applied to identify which log record means intrusion.

With normal queries, the conditional entropy differentiation rate in general case should never be greater than the change rate of maximum entropy. When normal query statements occur, only conditional probability values are increased, thus bring conditional entropy to a small change. In the same scenario, a new entropy term will be added to the maximum entropy value. This causes the conditional entropy to increase less than the maximum entropy in normal operations.

However with anomaly queries, the conditional entropy differentiation rate of new log records, either rare SQL patterns or command injection attacks, will change the conditional entropy higher than change on the maximum entropy. New log patterns produce new entropy term and increase the probability in conditional entropy calculation; only increase entropy term in maximum entropy calculation.

Although the change in maximum entropy value can be used as anomaly detection line, our chosen safety factor reduces false positive rate in the operation by using 2 times of maximum entropy changes as the anomaly inspection level. When the change on conditional entropy is greater than inspection level then that query should be identified as an anomaly.

Our process to detect anomaly query statements is summarized in the following pseudocode:

Procedure: Anomaly Query String Detection Algorithm
Pre-Condition: $S_0=\phi$, $W_0=\phi$, $H_0(\phi|\phi)=0$
Algorithm:

for each ith SELECT-statement of target table in query log
  $S_i = S_{i-1} \cup \{ \alpha \mid \alpha$ is parameter of SELECT commands$\}$
  $W_i = W_{i-1} \cup \{ \beta \mid \beta$ is expression parameter of where-clause, logic declaration, and logic expression parameter$\}$
  Calculate conditional entropy : $H_i(W_i|S_i)$
  $\Delta H_i(W_i|S_i) = H_i(W_i|S_i) - H_{i-1}(W_{i-1}|S_{i-1})$
  Anomaly detected when $\Delta H_i(W_i|S_i) > 2/(i * \ln(10))$

After choosing a database table to detect anomaly behaviors, our process starts to read each line in the query log file and choose only SELECT-statement records that are referred to the specified table name. Then, splits each SELECT-statement record, find SELECT and WHERE parameters and insert them into set *S* and set *W*, respectively. Some example log records are as follows:

```
1: select * from table1 where p1=34 order by id asc;
2: select count(*) from table1 where p1=12;
3: select count(*) from table2 where p2=10;
4: select * from table1 where p3=13 and p4=33;
5: select * from table1 where p2=10 and p3=0;
```

When reading the first line, screen SELECT and WHERE parameters and insert them into *S* and *W*, *S* equals to {"*"} and *W* equals to {"p1"}. After processing the second line, *S* equals to {"*", "count(*)"} and *W* equals to {"p1"}. After processing the entire records, *S* equals to {"*", "count(*)"} and *W* equals to {"p1", "p2", "p3 and p4", "p2 and p3"}. The number of each pairs (SELECT parameters, WHERE parameters) occurred are remembered for conditional entropy calculation.

Next, computes the conditional entropy of W condition on S, H(W|S), as shown in equation (6) and equation (7). Occurrences of the pairs (SELECT parameters, WHERE parameters) and number of all selected records are used to analyze probability p(S), as shown in equation (8), and conditional probability p(W|S), as shown in equation (9).

$$\mathrm{H}(W|S) = \sum_{s \in S} \mathrm{p}(s)\mathrm{H}(W|S = s) \tag{6}$$

$$\mathrm{H}(W|S = s) = -\sum_{w \in W} \mathrm{p}(w|s)\log(\mathrm{p}(w|s)) \tag{7}$$

$$\mathrm{p}(s) = \frac{number\ of\ select\ statement\ with\ s\ parameter}{number\ of\ all\ select\ statement} \tag{8}$$

$$\mathrm{p}(w|s) = \frac{number\ of\ select\ s\ with\ where\ w\ statements}{number\ of\ select\ s\ statement} \tag{9}$$

After the calculation, any query statements are detected as anomaly whenever their difference of conditional entropy values are greater than 2 times of maximum entropy differential ratio (i.e., $2/(n*\ln(10))$ where n is the number of SELECT-statements on concerning table).

5 Experimental Results

A database log file from an anonymous commercial website that operates 24-hours a day is used in our experiment. It contains 3,526,362 query records of 197 operation days. After choosing table name to inspect anomaly signal by our proposed method, the SELECT-statements are screened and there are 52,837 log records of which that query in targeted table name.

Our experiment is expressed in two cases, to detect rare requests and to detect attack query statements.

5.1 Rare SQL Requests Detection

To detect rare SQL statements, we use the proposed method on the experiment log file. Analysis process reads each filtered record and calculates conditional entropy of WHERE parameters condition on SELECT parameters. Entropy calculation of log record series from start to the end of file produces series of entropy value as shown in figure 1.

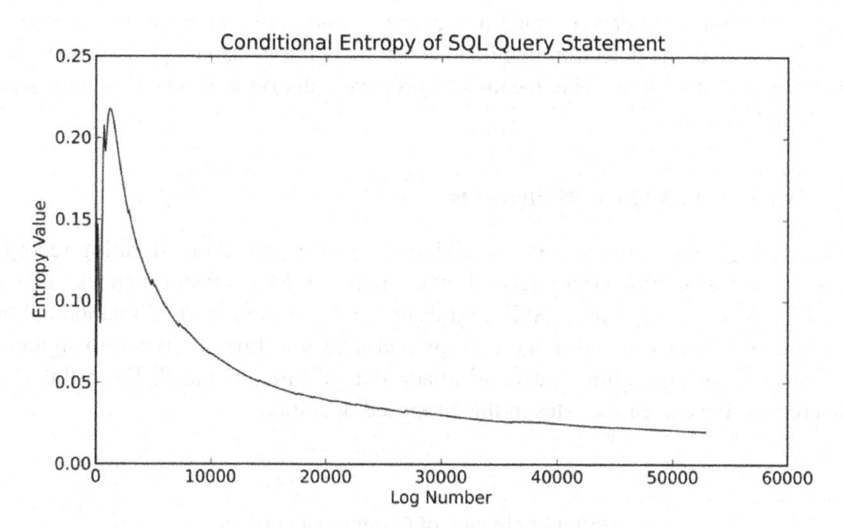

Fig. 1. Conditional entropy value of each target table's SELECT-statements

The characteristic of conditional entropy value of WHERE parameters condition on SELECT parameters of log record is a sharp rise in the beginning to the highest point and slides down to a saturated level. At start period, low numbers of SELECT-statement bring high value of each p(s) and p(w|s) and cause rapid rising. After highest point, conditional entropy reduce slowly because higher number of SELECT-statements reduces each p(s) and p(w|s) in the calculation.

Database attacks typically create new combination of SELECT-statement and WHERE-clause in log records, conditional entropy of WHERE-clause condition on SELECT-statement will increase significantly.

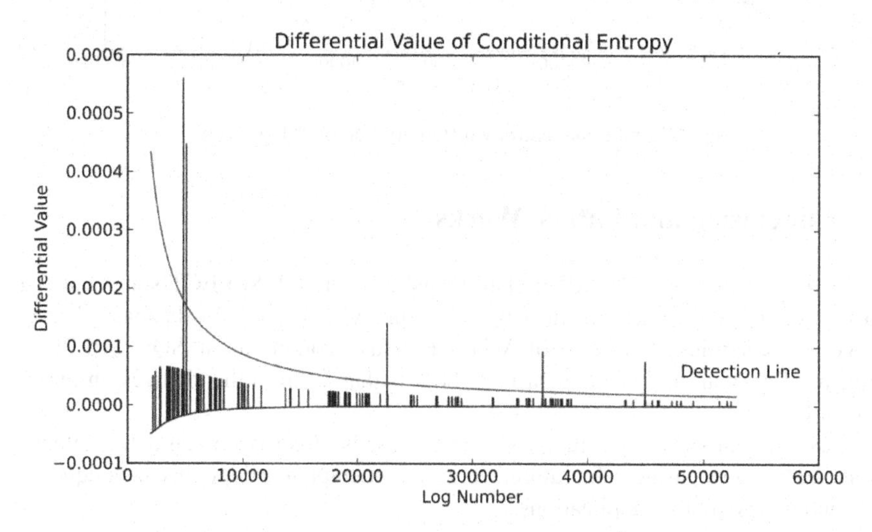

Fig. 2. Conditional entropy differentiation of ith log record

In figure 2, the difference of condition entropy values are plot. Conditional entropy that is greater than the detection line denotes anomaly query statement. This operation found 14 rare log records. This means our proposed algorithm has 0.03% false positive rate in this test.

5.2 Detect Attack Query Statements

To detect attack query statements, we add attack phases into 20 random log records. The attack command including general attack phase in SQL injection attack such as "or %1% == %1%" or "; drop table". After that, we use the proposed method to calculate the difference of conditional entropy and detection line, as shown in figure 3. The result of our algorithm detects all attack that we are simulated. From this test, there are zero false negative rates in this proposed operation.

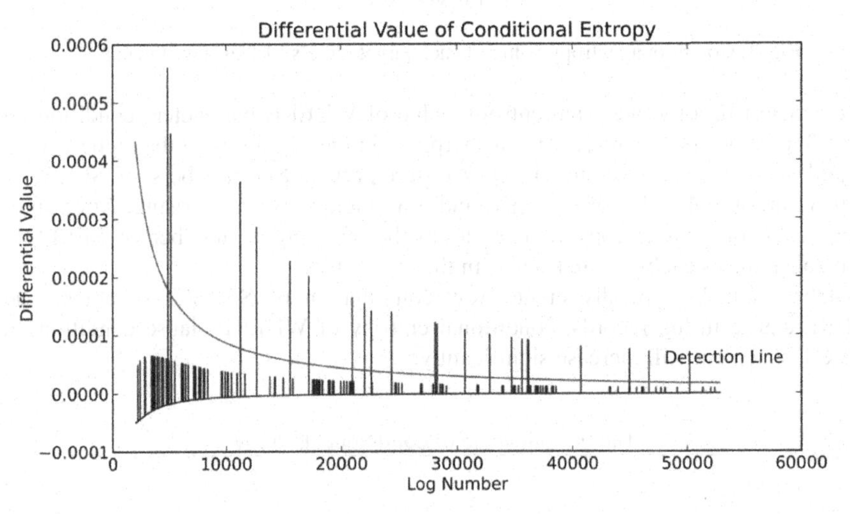

Fig. 3. Conditional entropy differentiation of ith log record

6 Conclusion and Future Works

Our conditional entropy of WHERE-clause condition on SQL SELECT-statement is a highly accurate procedure to investigate anomaly behavior in database query. Changes of conditional entropy of WHERE-clause condition on SQL SELECT-statement of normal query statement must be less than 2 times the changes on maximum entropy.

Our experiment shows that the proposed process is effective; it can detect attacks and anomaly query strings. In future work, we attempt to adopt this technique to detect anomalies in other applications.

References

1. Kabiri, P., Ghorbani, A.A.: Research on Intrusion Detection and Response: A Survey. I. J. Network Security, 84–102 (2005)
2. Patcha, A., Park, J.M.: An overview of anomaly detection techniques: existing solutions and latest technological trends. Elsevier Computer Networks 51(12), 3448–3470 (2007)
3. Lee, W., Xiang, D.: Information-Theoretic Measures for Anomaly Detection. In: IEEE Symposium on Security and Privacy (2001)
4. Nyalkalkar, K., Sinha, S., Bailey, M., Jahanian, F.: A Comparative Study of Two Network-based Anomaly Detection Methods. In: INFOCOM, Shanghai, China (2011)
5. Gu, Y., McCallum, A., Towsley, D.: Detecting anomalies in network traffic using maximum entropy estimation. In: IMC 2005, pp. 1–6. ACM, New York (2005)
6. Tellenbach, B., Burkhart, M., Sornette, D., Maillart, T.: Beyond Shannon: Characterizing Internet Traffic with Generalized Entropy Metrics. In: Moon, S.B., Teixeira, R., Uhlig, S. (eds.) PAM 2009. LNCS, vol. 5448, pp. 239–248. Springer, Heidelberg (2009)
7. Nychis, G., Sekar, V., Andersen, D.G., Kim, H., Zhang, H.: An empirical evaluation of entropy-based traffic anomaly detection. In: IMC (2008)
8. Wagner, A., Plattner, B.: Entropy based worm and anomaly detection in fast IP networks. In: IEEE International Workshops on Enabling Technologies: Infrastructure for Collaborative Enterprise, WET ICE (2005)
9. Kopylova, Y., Buell, D., Huang, C.T., Janies, J.: Mutual Information Applied to Anomaly Detection. Journal of Communications and Networks 10(1), 89–97 (2008)
10. Arackaparambil, C., Bratus, S., Brody, J., Shubina, A.: Distributed Monitoring of Conditional Entropy for Anomaly Detection in Streams. In: IEEE Workshop on Scalable Stream Processing Systems (SSPS), Atlanta (2010)
11. Agogino, A., Tumer, K.: Entropy Based Anomaly Detection Applied to Space Shuttle Main Engines. In: IEEE Aerospace Conference (2006)
12. Winter, P., Lampesberger, H., Zeilinger, M., Hermann, E.: On detecting abrupt changes in network entropy time series. In: De Decker, B., Lapon, J., Naessens, V., Uhl, A. (eds.) CMS 2011. LNCS, vol. 7025, pp. 194–205. Springer, Heidelberg (2011)
13. Shannon, C.: Prediction and entropy of printed English. Bell Systems Tech. Jour. 30, 50–64 (1951)
14. OWASP, https://www.owasp.org/index.php/Top_10_2013-Top_10

Deriving Composite Periodic Patterns from Database Audit Trails

Marcin Zimniak[1], Janusz R. Getta[2], and Wolfgang Benn[1]

[1] Faculty of Computer Science, TU Chemnitz, Germany
{marcin.zimniak,benn}@cs.tu-chemnitz.de
[2] School of Computer Science and Software Engineering, University of Wollongong, Australia
jrg@uow.edu.au

Abstract. Information about the periodic changes of intensity and structure of database workloads plays an important role in performance tuning of functional components of database systems. Discovering the patterns in workload information such as audit trails, traces of user applications, sequences of dynamic performance views, etc. is a complex and time consuming task. This work investigates a new approach to analysis of information included in the database audit trails. In particular, it describes the transformations of information included in the audit trails into a format that can be used for discovering the periodic patterns in database workloads. It presents an algorithm thatthe fluctuations finds elementary periodic patterns through nested iterations over a four dimensional space of execution plans of SQL statements and positional parameters of the patterns. Finally, it shows the composition rules for the derivations of complex periodic patterns from the elementary and other complex patterns.

Keywords: elementary periodic pattern, composite periodic pattern, database audit trail, automated performance tuning, online database design.

1 Introduction

It is a well-known fact that database workloads periodically change in time. The oscillations are caused by the recurrent invocations of database applications that access and change data on behalf of database users performing the typical real world processes. Discovering the patterns in a continuously changing database workload may, to large extent, allow for anticipation of the future operations on data containers and it can be used for automated performance tuning and online database design [1]. In a typical scenario, information about a period of low database workload and about data containers to be accessed in the future allows for appropriate restructuring of data containers to speed up their processing.

Information about the structures and characteristics of the past database workloads can be collected from a database management system in a form of audit trails, traces from processing of SQL statements, sequences of snapshots of

N.T. Nguyen et al. (Eds.): ACIIDS 2014, Part I, LNAI 8397, pp. 310–321, 2014.

internal systems structures and dynamic performance views, etc. Due to a large number of concurrently processed user applications a structure of workload trace is very complex and irregular with no simple access patterns to data containers and computations of database operations. This is why direct discovery of complex periodic patterns in a database workload is a difficult and time consuming task. On the other hand, many of user applications are processed in regular ways because the same applications implement the real world processes, which due to business or legal reasons must be systematically performed by the individuals and organizations. This paper shows how to discover the complex periodic patterns from the database audit trails.

A solution proposed in this work finds the simple periodic patterns created through processing of simple components of user applications and later on "derives" the complex periodic patterns from the elementary ones. The outcomes from such derivations can be compared with the original workload history in order to eliminate a "noise" created by the random processing of accidental applications. We assume an environment of relational database system where processing of user applications is traced and recorded in the anonymized *audit trails*. A selected audit trail contains information about the scopes of user applications and about SQL statements processed by the applications. A statement `EXPLAIN PLAN` is used to process SQL statements within the environments of the relevant relational schemas and to create precise specifications of execution plans and estimations of the processing costs. Analysis of execution plans represented as syntax trees of extended relational expressions leads to elimination of the common subtrees as the common fragments of the plans. The results are saved in the "compressed" *syntax tree table*. Further reduction of syntax tree table leads to elimination of subtrees whose frequencies of execution are the same as the syntax trees that belong to. At the end of this process we obtain a *reduced syntax tree table*. An audit trail passes through "data cleaning" phase and later on a predefined set of time units is used to create a workload histogram of an audit trail. A reduced syntax tree table and workload histograms of individual syntax trees are used by an algorithm that discovers elementary period patterns. In the final stage, the elementary periodic patterns are combined to generate the composite periodic patterns.

The paper is organized in the following way. The next section refers to the previous works in the related research areas. Section 3 defines an environment of relational database and the concepts of audit trail, time units, syntax tree table, and reduced syntax tree table. A concept of periodic pattern in a database workload is defined in Section 4 and discovering of elementary periodic patterns is explained in Section 5. Section 6 shows how to derive the composite periodic patterns from the elementary and other composite patterns. Section 7 concludes the paper.

2 Related Work

Data mining techniques that inspired the works on periodic patterns came from the works on mining association rules [2] and later on from mining frequent episodes [3] and its extensions on mining complex events [4].

The problem seems to be very similar to a typical periodicity mining in time series [5], where analysis is performed on the long sequences of elementary data items discretized into a number of ranges and associated with the timestamps. In our case, input data is a sequence complex data processing statements, like for example SQL statements and due to its internal structure cannot be treated in the same way as analysis of elementary data elements in time series or genetic sequences. The complex data processing statement form a lattice whose elements are syntax trees of the statements with a partial order determined by an inclusion relationship on syntax trees [6].

The recent approaches, which addressed full periodicy, partial periodicity, perfect and imperfect periodicity [7], and asynchronous periodicity [8] are all based on fixed size and adjacent time units and fixed length of discovered patterns. In our case, the cycles are pretty well determined by the real world events and because of that it may have variable size and non contiguous structure.

Our problem is also similar to a problem of mining cyclic association rules [9] where an objective is to find the periodic executions of the largest sets of items that have enough support. However, in our case the largest sets of operations do not necessarily mean the highest workload and sometimes a single periodically processed application significantly contributes to a database workload. Invocation of operation on data along the various points in time can be easily described by temporal predicates within a formal scope of Temporal Programming Logic and temporal deductive database systems [10]. The reviews of data mining techniques based on analysis of ordered set of operations on data performed by the user applications ara available in [11], [12]. The model of periodicity considered in this paper is a significant extension of the model proposed in [13] with the new concepts of overlapping periodic patterns and derivations of periodic patterns.

3 Database Processing Model

We consider a typical relational database system where the relational model of data is used to represent data containers. Let x be a nonempty set of attribute names later on called as a *schema* and let $dom(a)$ denotes a domain of attribute $a \in x$. A *tuple* t defined over a schema x is a full mapping $t : x \rightarrow \cup_{a \in x} dom(a)$ and such that $\forall a \in x$, $t(a) \in dom(a)$. A *relational table* r created on a schema x is a set of tuples over a schema x.

Query processor transforms SQL statements submitted by the user applications into the query execution plans formulated as the expressions of extended relational algebra. The operations of extended relational algebra include the implementation dependent variants of operations of standard relational algebra such as *selection*, *projection*, *join*, *antijoin*, *set operations*, and operations of *grouping*, *sorting*, and *aggregate functions*. Due to the different implementation techniques, the operations included in the basic system of relational algebra, e.g. *selection* or *join* contribute to an number of different elementary operations depending on their implementations, e.g. *index based selection*, *full scan selection*, *hash based join*, *index based join*, etc.

3.1 Audit Trail

SQL statements submitted by the *user applications* $a_1, \ldots, a_n$ are recorded in an *application trace*. A *trace of an application* a_i is a finite sequence of pairs $<c_i{:}t_{c_i}$, $s_{i_1}{:}t_{i_1}$, ..., $s_{i_n}{:}t_{i_n}$, $d_i{:}t_{d_i}>$ where c_i is a *connect* statement, t_{c_i} is a timestamp when the statement has been processed, each s_{i_j} is SQL statement with a timestamps t_{i_j} attached, and d_i is a *disconnect* statement with its timestamp t_{d_i}. Processing of an application a_i starts from processing of a connect statement c_i, the processing of SQL statements s_{i_j}, and it finally ends with processing of a disconnect statement d_i.

An *audit trail* is a sequence of interleaved trails of user applications. For example, a sequence $<c_i{:}t_{c_i}, s_{i_1}{:}t_{i_1}, c_j{:}t_{c_j}, s_{j_1}{:}t_{j_1}, s_{i_2}{:}t_{i_2}, d_i{:}t_{d_i}\ d_j{:}t_{d_j}>$ is a sample audit trail from the processing of applications a_i, and a_j. In practice SQL statements can be easily extracted from an audit trail and `EXPLAIN PLAN` statement can be used in the contexts of respective user schemas to transform the statements into the syntax trees of query execution plans over a set of operations of extended relational algebra. A complete information about syntax trees obtained from a database audit trail is kept in a *syntax tree table*.

3.2 Syntax Tree Table

Let s_i and s_j be SQL statements included in an audit trail and let T_{s_i}, T_{s_j} be the syntax trees obtained by application of `EXPLAIN PLAN` statement to the statements. The codes of operations of extended relational algebra are used as the labels of nonleaf nodes and the names of data containers processed by the operations are the labels of leaf nodes in a syntax tree.

If there exists a nonleaf node n in a syntax tree T_{s_j} such that a subtree with a root node n is the same as a syntax tree T_{s_i} then we say that a syntax tree T_{s_i} is *included in or equal to* a syntax tree T_{s_j}, and we denote by $T_{s_i} \subseteq T_{s_j}$.

A *syntax tree table* contains a complete and compressed information about the syntax trees of SQL statements extracted from an audit trail. A syntax tree table is a set of tuples $<tree, operation, left, right, workload, timestamps>$ where *tree* is a unique identifier of a syntax tree, *operation* is a code of extended relational algebra operation at the root of syntax tree identified by $tree$, $left$ and $right$ are the identifiers of left and right argument of syntax tree identified by $tree$ or the names of relational tables, $workload$ is an average workload required to process a syntax tree, and $timestamps$ is a set of all timestamps when a syntax tree $tree$ was processed by a database system.

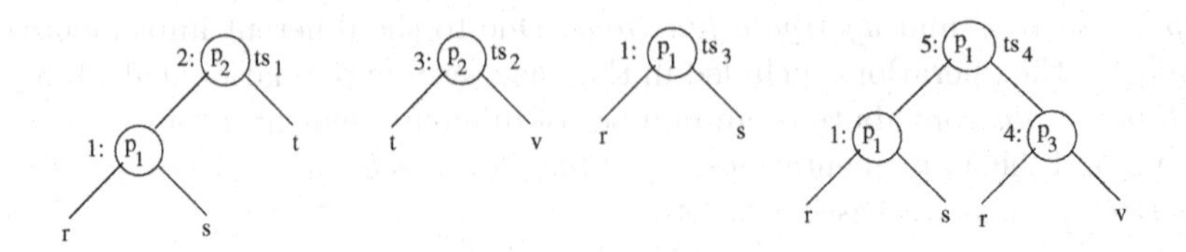

Fig. 1. A sequence of syntax trees

We say that a subtree t_{leaf} is a *leaf level subtree* of a syntax tree T_s if $t_{leaf} \subseteq T_s$ and both arguments of an operation in a root node of t_{leaf} are data containers. A syntax tree table is created in the following way.

(1) In the first step we create an empty syntax tree table.

(2) Next, we iterate over all statements included in an audit trail starting from the first statement in the trail. All connect and disconnect statements are ignored. We collect the next statement s from the trail and we create its syntax tree T_s. If no more statements are available in the trail then the process of creating a syntax tree table stops.

(2.1) We iterate from left to right over all leaf level subtrees in T_s. Let t_{leaf} be the next leaf level subtree in T_s.

(2.1.1) With the current leaf level subtree t_{leaf} we search a syntax tree table for a tuple that has a value of *code* equal to an operation code in the root node of t_{leaf} and $left$ equal to the left argument of t_{leaf} and $right$ equal to the right argument of t_{leaf}

(2.1.2) If a tuple is found then it means that a subtree the same as t_{leaf} has been already recorded in the table and we append a timestamp of t_{leaf} to *timestamps* in the tuple found.

(2.1.3) Otherwise, we append a new tuple to a syntax tree table. The new tuple obtains automatically generated identifier *tree*, code of operation in the root of t_{leaf} becomes a value of *code* in the tuple. The values of left and right arguments of t_{leaf} become the values of $left$ and $right$, and these are either the names of relational tables or the identifiers of syntax trees already recorded in the table. An average workload needed to process the syntax tree t_{leaf} is estimated using a workload need to process its subtrees an information about an operation in the root of t_{leaf}. Finally, a timestamp of t_{leaf} is appended to a set *timestamps* in the tuple.

(2.1.4) A subtree t_{leaf} is removed from T_s such that a root node of t_{leaf} is replaced either with an identifier *tree* found in step (2.1.2) or with a new identifier *tree* created in a step (2.1.3).

(2.1.5) If T_s still has at least one leaf level subtree then return to step (2.1) otherwise processing of T_s is completed and we return to a step (2).

As a simple example consider a sequence of syntax trees processed at the timestamps t_1, t_2, t_3, and t_4 given in Fig. 1 above where p_1, p_2, and p_3 are the codes of operations. The respective syntax tree table is given below.

Table 1. A sample syntax tree table

tree	*operation*	*left*	*right*	*workload*	*timestamps*
1	p_1	r	s	w_1	$\{ts_1, ts_3, ts_4\}$
2	p_2	1	t	w_2	$\{ts_1\}$
3	p_2	t	v	w_3	$\{ts_2, ts_4\}$
4	p_3	r	v	w_4	$\{ts_4\}$
5	p_1	1	4	w_5	$\{ts_4\}$

3.3 Time Units

Let $<t_{start}, t_{end}>$ be a period of time over which an audit trail is recorded. The period is divided into a contiguous sequence of disjoint and fixed size *elementary time units* $<t_e^{(i)}, \tau_e>$ where $t_e^{(i)}$ for $i = 1, \ldots, n$ is a timestamp when an elementary time unit starts and τ_e is a length of the unit. Elementary time units are distributed over $<t_{start}, t_{end}>$ such that $t_{start} = t_e^{(1)}$ and $t_e^{(i+1)} = t_e^{(i)} + \tau_e$ and $t_e^{(n)} + \tau_e = t_{end}$.

A *time unit* is a pair $<t, \tau>$ where t is a start point of a unit and τ is a length of the unit. A time unit consists of one or more consecutive elementary time units. A nonempty sequence U of n disjoint time units $<t^{(i)}, \tau^{(i)}>$ $i = 1, \ldots, n$ over $<t_{start}, t_{end}>$ is any sequence of time units that satisfies the following properties: $t_{start} \leq t^{(1)}$ and $t^{(i)} + \tau^{(i)} \leq t^{(i+1)}$ and $t^{(n)} + \tau_{(n)} \leq t_{end}$.

As a simple example consider an audit trail that starts on $t_{01:01:2007:0:00am}$ and ends on $t_{31:01:2007:12:00pm}$. Then, a sequence of disjoint time units called as *morning tea time* consists of the following units $<t_{01:01:2007:10:30am}, 30mins>$, $<t_{02:01:2007:10:30am}, 30mins>$, ..., $<t_{31:01:2007:10:30am}, 30mins>$.

3.4 Workload Histogram

Let $|U|$ denotes the total number of time units in U. Then $U[n]$ denotes the n-th time unit in u where n changes from 1 to $|U|$.

A multiset M is defined as a pair $<S, f>$ where S is a set of values and $f : S \rightarrow N^+$ is a function that determines multiplicity of each element in S and N^+ is a set of positive integers [6]. In the rest of this paper we shall denote a multiset $<\{T_1, \ldots, T_m\}, f>$ where $f(T_i) = k_i$ for $i = 1, \ldots, m$ as $(T_1^{k_1} \ldots T_m^{k_m})$. We shall denote an empty multiset $<\emptyset, f>$ as $\emptyset$ and we shall abbreviate a multiset (T^k) to T^k and T^1 to T.

A *workload histogram of a syntax tree* T is a sequence W_T of $|U|$ multisets of syntax trees such that $W_T[i] =<\{T\}, f_i>$ and $f_i(T) = |T.timestamp(i)|$ $\forall i = 1, \ldots, |U|$ i.e. $f_i(T)$ is equal to the total number of times a syntax tree T was processed in the i-th time unit $U[i]$. A workload histogram can be created from information about time units in U and the values in a column *timestamps* in a syntax tree table.

Let $\mathbf{T}$ be a set of all syntax trees obtained from an audit trail A and recorded in a syntax tree table. A *workload histogram of an audit trail* A is denoted by W_A

and $W_A[i] = \biguplus_{T \in \mathbf{T}} W_T[i], \forall i = 1, \ldots, |U|$, i.e. it is a sum of workload histograms of all syntax trees included in a syntax tree table.

4 Periodic Patterns

A *periodic pattern* is a tuple $<\mathcal{T}, U, b, p, e>$ where $\mathcal{T}$ is a nonempty sequence of multisets of syntax trees, U is a sequence of disjoint time units that partitions the audit trail into disjoint sequences of SQL statements, $b \geq 1$ is a number of time unit in U where the repetitions of $\mathcal{T}$ start, $p \geq 1$ is the total number of time units after which processing of $\mathcal{T}$ is repeated in every processing cycle, $e > b$ is a number of time unit in U where the processing of $\mathcal{T}$ is performed for the last time. A sequence of multisets $\mathcal{T}$ may contain one or more empty multisets. The positional parameters b, p, and e of a periodic pattern must satisfy a property $(e-b) mod\ p = 0$. A value $c = \frac{e-b}{p} + 1$ is called as the *total number of cycles* in the periodic pattern.

Let $\mathcal{T}_{ext}$ be a sequence of multisets of syntax trees obtained from $\mathcal{T}$ and extended on the right with $e - b$ empty multisets. Then, a *workload histogram of a periodic pattern* $<\mathcal{T}, U, b, p, e>$ with the total number of cycles $c = \frac{e-b}{p} + 1$ is a sequence $W_{\mathcal{T}}$ of $e - b + |\mathcal{T}|$ multisets of syntax trees such that $W_{\mathcal{T}}[i] = \mathcal{T}_{ext}[g(i)] \uplus \mathcal{T}_{ext}[g(i-p)] \uplus \mathcal{T}_{ext}[g(i-2*p)] \uplus \ldots \mathcal{T}_{ext}[g(i-(c-1)*p)]$ for $i = 1, \ldots, e - b + |\mathcal{T}|$ where a function g is computed such that `if` $x > 0$ `then` $g(x) = x$ `else` $g(x) = x + e - b + |\mathcal{T}|$.

For example, $<\emptyset TV, U, 1, 1, 3>$ is a periodic pattern where processing of a sequence of syntax trees $\emptyset TV$ starts in the time units 1, 2, and 3 and its workload histogram is a sequence of multisets $\emptyset T(VT)(VT)V$. The periodic pattern has 3 cycles.

Let $|W_{\mathcal{T}}|$ be the total number of elements in $W_{\mathcal{T}}$. Let v be the total number of elements in $W_{\mathcal{T}}$ such that $W_{\mathcal{T}}[i] \sqsubseteq W_A[b + i - 1]$ for $i = 1, \ldots, e - b + |\mathcal{T}|$. Then, we say that a periodic pattern $<\mathcal{T}, U, b, p, e>$ is *valid in an audit trail A with a support* $0 < \sigma \leq 1$ if $W_{\mathcal{T}}[1] \sqsubseteq W_A[b]$ and $W_{\mathcal{T}}[e - b + |\mathcal{T}|] \sqsubseteq W_A[e + |\mathcal{T}|]$ and $\sigma \leq v/|W_{\mathcal{T}}|$.

For example, a periodic pattern $<(T^2V)\emptyset W, U, 2, 3, 8>$ has a workload histogram $(T^2V)\emptyset W(T^2V)\emptyset W(T^2V)\emptyset W$. The pattern is valid in an audit trail A with support $\sigma = 1$ if every element of its workload histogram is included in a workload histogram W_A from position 2 to position 10.

5 Discovering Elementary Periodic Patterns

We say that a periodic pattern $<\mathcal{T}, U, b, p, e>$ is an *elementary periodic pattern* when a sequence of multisets $\mathcal{T}$ consists of one multiset that consists only of k identical elements, i.e. $\mathcal{T} = T^k$.

Discovery of elementary periodic patterns can be done over a number of dimensions such as syntax trees of all statements included in an audit trail, all possible partitions of audit time into time units in U, all workload levels expressed as multiplicity coefficients in multisets, and the dimensions of positional parameters b, p, and e. In this work, we assume that a sequence of time units U does not change. In the first step we perform "data cleaning" stage on the contents of workload histogram W_A. A *threshold workload* w is used to eliminate from a workload histogram W_A of an audit trail A created over a sequence of time units U all elements of multisets such that their workload is below a threshold value w in all time units. All elements representing a syntax tree T are removed from all multisets in a workload histogram W_A if the total contribution to a workload of an element $T^{max(k)}$ in W_A, i.e. the largest number of times a syntax tree T was processed in a time unit measured as $k * T.workload$ is less than the threshold workload w and does not exist at least two multisets in a workload histogram W_A such it contains T and its total workload is greater or equal w. In the other words, "data cleaning" removes from the multisets in a workload histogram all elements representing syntax trees whose processing has no significant impact on overall workload and whose processing does not contribute to any periodic patterns.

5.1 Reduced Syntax Tree Table

Let $\mathbf{T}$ be a set of syntax trees that consists of all syntax trees of statements in an audit trail. Let T_ϵ be an *empty syntax tree* and let T_π be a syntax tree obtained from concatenation of all syntax trees from syntax tree table, which are not included in any other syntax tree. Then, discovering elementary periodic patters in an audit trail is performed over a lattice $< \mathbf{T}, \subseteq>$ implemented as a syntax tree table with a minimum T_ϵ and maximum T_π and partial order $\subseteq$ representing inclusion of syntax trees. The following three rules can be used to reduce the total number of iterations over the syntax trees. Let A be an audit trail.

(1) If a elementary periodic pattern $<T_i^k, U, b, p, e>$ is valid in A then for any syntax tree T_j such that $T_j \subseteq T_i$ an elementary periodic pattern $<T_j^k, U, b, p, e>$ is valid in A.
(2) If a an elementary periodic patterns $<T_i^k, U, b, p, e>$ is not valid in A then for any syntax tree T_j such that $T_i \subseteq T_j$ an elementary periodic pattern $<T_j^k, U, b, p, e>$ is not valid in A.
(3) If an elementary periodic patterns $<T_i, U, b, p, e>$ is not valid in A then for any syntax tree $T_j \sqsubseteq T_i$ and not shared with any other subtree an elementary periodic pattern $<T_j^k, U, b, p, e>$ is not valid in A.

The rules listed above reduce a syntax tree table to a simple table of pairs $<tree, timestamps>$ where *tree* is an identifier of a syntax tree that suppose to be verified against periodic patterns and *timestamps* is a set of timestamps when the processing of a syntax tree identified by *tree* occurred in an audit trail. A *reduced syntax trees table* includes identifiers of all sub-lattices determined by the rules (1)-(3) above. The table contains only information about the syntax trees of the statements from an audit trail and about subtrees shared by two or more syntax trees. For example, a syntax tree table given in Table 1 reduces to a set of pairs $\{<1, \{ts_1, ts_3, ts_4\}>, <2, \{ts_1\}>, <3, \{ts_2, ts_4\}>, <5, \{ts_4\}>\}$.

5.2 Iterations

Discovering an elementary periodic pattern <T^k, U, b, p, e> for a given set of time units U, and a given value of support parameter $0 < \sigma \leq 1$ is performed through the nested iterations over the syntax trees included in a reduced syntax tree table and the iterations over the positional parameters b, p, and e. At the beginning all syntax trees in a reduced syntax tree table are marked as "not processed yet" and a set $\mathcal{P}$ of periodic patterns that occur in an audit trial A is set to empty. At each level the iterations are performed in the following way.

(1) At the outermost level we pick a syntax tree T from a reduced syntax tree table such that it is not included in any other "not processed yet" syntax tree. If such tree does not exist then the iterations are completed. Otherwise, we create a workload histogram W_T for T.

(1.1) At the first inner level the iterations are performed over the values of positional parameter b. The parameter b iterates over an increasing sequence of numbers $1, 2, 3, \ldots, |W_T| - 1$. Let b_c be the current value of parameter b. If $W_T[b_c] = \emptyset$ then a value of b_c is increased by one and the same condition is tested again. If no more iterations over the values of parameter b are possible then we move to a step (1.2) below.

(1.1.1) At the next inner level the iterations are performed over the values of parameter e for a fixed value b_c set at outer level. A parameter e iterates over a decreasing sequence of numbers $|W_T|, |W_T| - 1 \ldots, b_c + 2, b_c + 1$. Let e_c be the current value of parameter e. If $W_T[e_c] = \emptyset$ then we take the next value of parameter e the same condition is tested again. If no more iterations over the values of parameter e are possible we return to level(1.1).

(1.1.1.1) At the lowest level the iterations are performed over an increasing sequence of values of parameter p such that $(e_c - b_c) mod\ p = 0$ and $b_c + p < e_c$. If no more iterations over the values of parameter p are possible we return to level (1.1.1). Otherwise, we set the current value of parameter p to p_c.

(1.1.1.2) Next, we create a candidate elementary periodic pattern <$T^{min(k)}$, U, b_c, e_c, p_c> where $min(k)$ denotes the smallest value of k in all instances of element T^k in workload histogram W_T in a range $[b_c, e_c]$.

(1.1.1.3) We use a histogram W_T to check whether the candidate pattern is valid in an audit trail with a given support σ. We compute $c = \frac{e_c - b_c}{p} + 1$ and v as the total number of times $T^{min(k)}$ is valid in W_T at the positions b_c,$b_c + p_c$,...,e_c. Then, the candidate pattern <T^k, U, b_c, p_c, e_c> is valid in an audit trail with a support σ when $T^{min(k)} \sqsubseteq W_T[b_c]$ and $T^{min(k)} \sqsubseteq W_T[e_c]$ and $\sigma \leq v/c$.

If the candidate pattern is not valid in an audit trail then we return to step (1.1.1.1) to collect the next value of parameter p.

(1.1.1.4) If the candidate pattern is valid in an audit trail then we append <$T^{min(k)}$, U, b_c, p_c, e_c> to a set $\mathcal{P}$. Then we modify the entries of histogram W_T such that $W_T[i] := W_T[i] - T^{min(k)}$ for all $\forall i = b_c, b_c +$

$p_c, \ldots, e_c$. Next we return to step (1.1.1.1) to collect the next value of parameter p.

(1.2) At the end of iterations over the positional parameters we are left with the single elements in a workload histogram W_T, which are not attached to any periodic pattern in $\mathcal{P}$. If there exists a periodic pattern $< T^k, U, b, e, p > \in \mathcal{P}$ and an element $W_T[n] = T^m$ such that $n \geq e + p$ and $m \geq k$ then we split the pattern into $< T^k, U, b, e - p, p >$ and $< T^k, U, e, n, n - e >$ and we modify histogram $W_T[n] := W_T[n] - T^{min(k)}$. Splitting of periodic patterns is repeated until no more single elements in W_T can be used. When finished we mark a syntax tree T as "processed" in a reduced syntax tree table and we return to step (1) above.

It is important to note, that support of all periodic patterns in $\mathcal{P}$ is equal to 1.

6 Deriving Sequential Periodic Patterns

We shall call a periodic pattern $<\mathcal{T}, U, b, p, e>$ such that $p \geq |\mathcal{T}|$ as a *sequential periodic pattern*. A binary operation of *composition of periodic patterns*, denoted by $\oplus$, derives a new sequential periodic pattern from the pairs of sequential periodic patterns. Consider the sequential periodic patterns $\mathcal{P}_i = <\mathcal{T}_i, U, b_i, p_i, e_i>$ and $\mathcal{P}_j = <\mathcal{T}_j, U, b_j, p_j, e_j>$ where $p_i \geq |\mathcal{T}_i|$ and $p_j \geq |\mathcal{T}_j|$ and such that both supports $\sigma_i = \sigma_j = 1$. For the composition operation we consider the following three cases.

Case I (*Vertical composition*)
In this case the scopes of arguments either overlap or one is included in the other. Then, it is possible to compose the arguments into a sequential periodic pattern when $min(e_i, e_j) - max(b_i, b_j) + 1 \geq 2 * LCM(|\mathcal{T}_i| + p_i - 1, |\mathcal{T}_j| + p_j - 1)$ where LCM means the Least Common Multiply. The result of $\mathcal{P}_i \oplus \mathcal{P}_j$ is a sequential periodic pattern $P_{ij} = <\mathcal{T}_{ij}, U, b_{ij}, e_{ij}, p_{ij}>$ that satisfies the following conditions:

(i) $b_{ij} = max(b_i, b_j)$,
(ii) $p_{ij} = LCM(|\mathcal{T}_i| + p_i - 1, |\mathcal{T}_j| + p_j - 1)$,
(iii) If $e_i = min(e_i, e_j)$, then $c_{ij} = \lfloor (e_i - b_{ij})/p_{ij} \rfloor + \lfloor (|\mathcal{T}_i| + p_i - 1)/p_{ij} \rfloor$ else $c_{ij} = \lfloor (e_j - b_{ij})/p_{ij} \rfloor + \lfloor (|\mathcal{T}_i| + p_j - 1)/p_{ij} \rfloor$; $e_{ij} = b_{ij} + (c_{ij} - 1) * p_{ij}$
(iv) If $b_i = max(b_i, b_j)$ then $\mathcal{T}_{ij}[k] = W_{\mathcal{T}_i}[k] \uplus W_{\mathcal{T}_j}[k + b_i - b_j]$ for $k = 1, \ldots, e_{ij} - b_{ij}$ else $\mathcal{T}_{ij}[k] = W_{\mathcal{T}_j}[k] \uplus W_{\mathcal{T}_i}[k + b_j - b_i]$ for $k = 1, \ldots, e_{ij} - b_{ij}$ where $W_{\mathcal{T}_i}$ and $W_{\mathcal{T}_j}$ are the workload histograms of $\mathcal{T}_j$ and $\mathcal{T}_j$ and operation $\uplus$ is a sum of multisets.

As a simple example consider a composition of the sequential periodic patterns $<T^2, U, 1, 3, 13>$ and $<V, U, 3, 2, 13>$. As the initial conditions hold, the result is a sequential periodic pattern $<VT^2V\emptyset(VT^2), U, 3, 2, 9>$.
Case II (*Horizontal composition*)
The arguments are disjoint and $\mathcal{T}_i = \mathcal{T}_j$ and $p_i = p_j$ and $\exists n \in \mathcal{N}$ $max(b_i, b_j) - min(e_i, e_j) + 1 = n * p_i$. The result of $\mathcal{P}_i \oplus \mathcal{P}_j$ is a sequential periodic pattern $P_{ij} = <\mathcal{T}_{ij}, U, b_{ij}, e_{ij}, p_{ij}>$ of composition satisfies the following conditions:

(i) $b_{ij} = min(b_i, b_j)$,
(ii) $p_{ij} = p_i$,
(iii) $e_{ij} = max(e_i, e_j)$,
(iv) $\mathcal{T}_{ij} = \mathcal{T}_i$, and
(v) support $\sigma_{ij} = (|W_{\mathcal{T}_i}| + |W_{\mathcal{T}_j}|)/(|W_{\mathcal{T}_i}| + |W_{\mathcal{T}_j}| + n)$.

For instance, the result of composition of disjoint periodic patterns $<T^2, U, 5, 1, 7>$ and $<T^2, U, 10, 1, 11>$ both with a support equal to 1, is a sequential periodic pattern $<T^2, U, 5, 1, 12>$ with a support $\sigma_{ij} = \frac{5}{7}$.

Case III (*"Echo" composition*)
The arguments are disjoint and $p_i = p_j$. The result of $\mathcal{P}_i \oplus \mathcal{P}_j$ is a sequential periodic pattern $P_{ij} =<\mathcal{T}_{ij}, U, b_{ij}, e_{ij}, p_{ij}>$ of composition satisfies the following conditions:

(i) $b_{ij} = min(b_i, b_j)$,
(ii) $p_{ij} = p_i$,
(iii) $e_{ij} = min(c_i, c_j) * (|\mathcal{T}_j| + p_i - 1)$,
(iv) $\mathcal{T}_{ij}[k] = \mathcal{T}_i[k]$ for $k = 1, \ldots, |\mathcal{T}_i|$,
$\mathcal{T}_{ij}[k] = \emptyset$ for $k = |\mathcal{T}_i| + 1, \ldots, max(b_i, b_j) - min(b_i, b_j)$, and
$\mathcal{T}_{ij}[k] = \mathcal{T}_j[k]$ for $k = max(b_i, b_j) - min(b_i, b_j) + 1, \ldots, max(b_i, b_j) - min(b_i, b_j) + |\mathcal{T}_j|$.

For instance, the result of composition of disjoint periodic patterns $<TV^2, U, 1, 3, 7>$ and $<VVT, U, 11, 3, 14>$ is a sequential periodic pattern $<TV^2\emptyset\emptyset\emptyset\emptyset\emptyset\emptyset\emptyset VVT, U, 1, 3, 4>$.

7 Summary and Conclusions

Discovering the complex periodic patterns in the database audit trails is a difficult and time consuming task. An approach investigated in this papers shows that it is easier to find the elementary periodic patterns and to compose them into the complex ones instead of directly searching for all complex patterns. A data preparation stage of the process starts from defining the time units and partitioning an audit trail over time units. Next, the syntax trees are obtained from an audit trail, and then the trees are compressed, reduced, and not important ones are eliminated. The discovery stage consists of finding elementary periodic patterns and applying the composition rule to derive the required complex periodic patterns. The computational complexity of search for elementary periodic patterns is approximately $O(k * n^3)$ where $0 < k < 1/8$ and n is the total number of partitions in an audit trail. Complexity of search over syntax trees is hard to estimate as it depends on the total number of access methods to relational tables, complexity of SQL statements, and a level of sharing common components among SQL statements.

The periodicity of database workload provides very usefull information for many functional components of a database management system, like for example transaction scheduler, data buffer cache controller, automated performance tuner, etc. Usually, such functional components need only partial information about the behavior of selected database applications. An approach in which only some of elementary periodic patterns are discovered and later on composed is more practical as it targets only the specific SQL statements in an audit trail. Another advantage of the proposed approach is the possibility to use the discovered periodic patterns to model future workload after the old applications are replaced with the new ones or the new applications are added to a system. It is also easier to reconcile the new audit trails with the collections of periodic patterns discovered from the previous audit trails than to integrate the complete trails.

References

1. Bruno, N. (ed.): Automated Physical Database Design and Tuning. CRC Press Taylor and Francis Group (2011)
2. Agrawal, R., Imielinski, T., Swami, A.: Mining association rules between sets of items in large databases. In: Proceedings of The 1993 ACM SIGMOD Intl. Conf. on Management of Data, pp. 207–216 (1993)
3. Mannila, H., Toivonen, H., Verkamo, A.I.: Discovery of frequent episodes in event sequences. Data Mining and Knowledge Discovery 1, 259–289 (1997)
4. Wojciechowski, M.: Discovering frequent episodes in sequences of complex events. In: Proceedings of Challenges, Enlarged Fourth East-European Conference on Advances in Databases and Information Systems (ADBIS-DASFAA), pp. 205–214 (2000)
5. Rasheeed, F., Alshalalfa, M., Alhajj, R.: Efficient periodicity mining in time series databases using suffix trees. IEEE Transactions on Knowledge and Data Engineering 23(1), 79–94 (2011)
6. Simovici, D.A., Djeraba, C.: Mathematical tools for data mining: Set theory, partial orders, combinatorics. In: Advanced Information and Knowledge Processing. Springer, London (2008)
7. Huang, K.Y., Chang, C.H.: SMCA: A general model for mining asynchronous periodic patterns in temporal databases. IEEE Transactions on Knowledge and Data Engineering 17(6), 774–785 (2005)
8. Yang, J., Wang, W., Yu, P.S.: Mining asynchronous periodic patterns in time series data. IEEE Trans. on Knowl. and Data Eng. 15(3), 613–628 (2003)
9. Özden, B., Ramaswamy, S., Silberschatz, A.: Cyclic association rules. In: Proceedings of the Fourteenth International Conference on Data Engineering, pp. 412–421 (1998)
10. Baudinet, M., Chomicki, J., Wolper, P.: Temporal deductive databases (1992)
11. Laxman, S., Sastry, P.S.: A survey of temporal data mining. Sadhana, Academy Proceedings in Engineering Sciences 31(2), 173–198 (2006)
12. Roddick, J.F., Society, I.C., Spiliopoulou, M., Society, I.C.: A survey of temporal knowledge discovery paradigms and methods. IEEE Transactions on Knowledge and Data Engineering 14, 750–767 (2002)
13. Zimniak, M., Getta, J., Benn, W.: Discovering periodic patterns in database audit trails. In: Proceedings of International Conference on Interdisciplinary Research Theory and Technology, pp. 365–367 (2013)

Comparison of Stability Models in Incremental Development

Alisa Sangpuwong and Pornsiri Muenchaisri

Department of Computer Engineering, Faculty of Engineering,
Chulalongkorn University, Bangkok, Thailand
Alisa.S@Student.chula.ac.th, Pornsiri.mu@chula.ac.th

Abstract. There are many stability models that are developed with a different of factors, indicators, and methods. The objective of this paper is to compare models for estimating class logical stability of software design in incremental development from class diagrams and sequence diagrams. The models are developed with different methods such as multiple regression analysis (MRA), principle component analysis (PCA), and design logical ripple effect analysis (DLREA). The empirical result shows that the models are acceptable for estimating stability. Then we compare and discuss the results to help developers make decision when selecting and using the methods for developing the stability estimation models.

Keywords: Stability Estimation, Stability Model, Class Logical Stability, Incremental Development, Software Design Stability.

1 Introduction

Stability models are used for indicating of software quality which resist to the effects of changes [1]. Software stability is divided into 2 types: logical stability is concerned with design structure and performance stability is concerned with design behavior [2]. Based on previous research [3], we focus only logical stability which is the logical ripple effect of a change in one class causes an inconsistency in other classes that are related to the changed area.

Incremental software development is a process that divides work into increments. When the current increment[1] is added to the previous increment[2], the new design structure[3] may be affected by the changes as well as its stability. Therefore, stability should be evaluated after adding each increment so that developers may decide whether to re-design the new design structure or not.

Researchers have been performed on evaluation of software stability at design phase. If the stability is reduced, the design models should be modified. Otherwise, modification will take place at maintenance phase which are more difficult and more

[1] Current increment is modules of program that will be added to the previous increment.

[2] Previous increment is modules of program that was already developed.

[3] New design structure is the result of adding the current increment to the previous increment.

N.T. Nguyen et al. (Eds.): ACIIDS 2014, Part I, LNAI 8397, pp. 322–331, 2014.

expensive. Some researchers propose estimation models which are developed by using metrics [2, 4, 5, 6] and by defining new specification of variables in the models [1, 7, 8]. Our previous research [3] is focused on methodology for construction of stability models by using the design logical ripple effect analysis (DLREA). The model is acceptable for estimating stability in incremental development but it has not been compared with other models that are developed by different methods. The comparison of estimating stability models will help developers to determine when selecting methods for developing models. So we compare and discuss the models which are developed by different methods.

In this paper, we construct and compare three models. The first model is constructed from metrics using multiple regression analysis (MRA). The second model is constructed from metrics using principal component analysis (PCA). The third model is defined with a new specification using design logical ripple effect analysis (DLREA). MRA is general method for constructing several models and studying the relationship between a dependent variable and two or more independent variables. PCA [9] is a method for the extracting components in order to reduce the number of variables. DLREA [3] is method for constructing model by analyzing the design logical ripple effect to define the model specifications.

This paper is organized as follows. Section 2 reviews related work. Section 3 proposes the process of construction stability estimating models. Comparison the stability estimation models are presented in Section 4. Finally, this paper is concluded in Section 5.

2 Related Works

Stability is an important attribute for measuring software quality. So our previous research [3] has proposed a model for estimating the stability of software design in incremental development. Most researchers have been performed on stability models from design models at design phase and code at maintenance phase.

Measurement models are used for measuring software logical stability from code at maintenance phase. Researchers [8] propose model to compute the logical stability of class in object-oriented programming. The metrics are calculated from source code and identified factors that affect the stability of a class. They specify and explain the method to define the factors for develop model for estimating stability. In addition, Yau and Collofello [1, 7] propose a model to calculate the stability of the logical structure of the program based on the modular structure chart.

Estimation models are used for estimating software logical stability from design models at design phase. Researchers have many methods such as multiple regression analysis (MRA) [4], analogy [5], artificial neural networks (ANN) [6], and design logical ripple effect analysis (DLREA) [3] that are proposed for applying diagram metrics in models. The logical stability estimating at the design phase can help designer and developer to decide to re-design structure model before developing that reduce maintenance cost. This concept will be used in this paper for constructing models.

3 Process of Construction Stability Estimating Models

In this section, we compare models for estimating class logical stability in incremental software development at design phase. The models are constructed by using three methods consist of MRA, PCA and DLREA. The process is consists of preparing Java source code from open-source programs, calculating the logical stability of classes from java source code, converting Java source code to class diagram and sequence diagram, developing software stability estimation models, validating the models and comparing the models, as shown in Fig. 1.

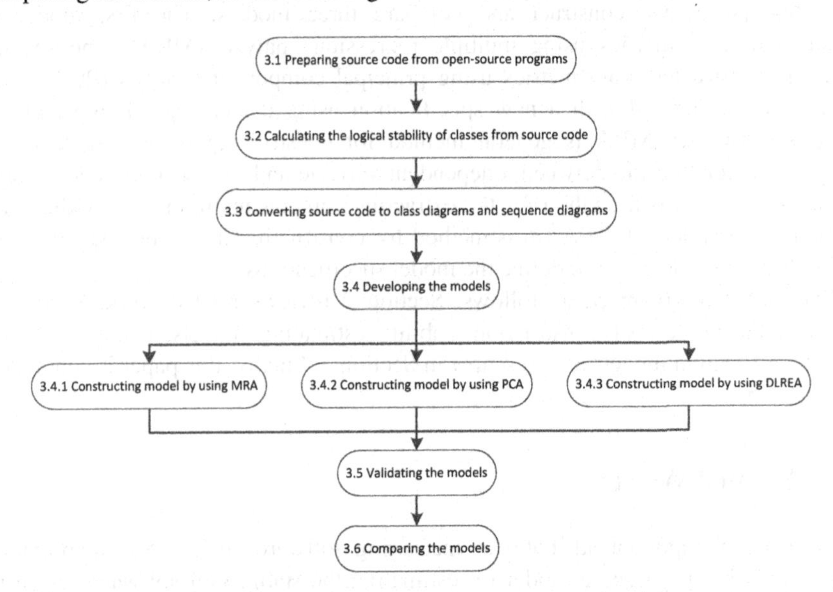

Fig. 1. Overview of the process to compare the models

3.1 Preparing Source Code from Open-Source Programs

Java programs are selected from open-source programs [10, 11] based on version number and focused on major version that the main functions are changed and identified in each version. The result of this procedure consists of the research set and validation set of the Java source code.

3.2 Calculating the Logical Stability of Classes from Source Code

Class logical stability is calculated for each class in each version by assuming a test case from changes at the system level and the class level that can occur in each class. The results as the number of all possible changes and the number of time which class received effect from changes. Changes and affected are collected from previous researches [4, 5, 6] for use to calculating class logical stability. Then calculate the stability based on the Equation (1) by Elish and Rine [2].

$$CLS_i = 1 - \frac{NTE_i}{TNC} \quad (1)$$

Where CLS_i is the Class logical stability of class i when i=1,…,n; NTE_i is the number of time which class i received effect from changes when i=1,…,n and TNC is total number of changes. The logical stability values of classes from this step are used to construct and validate the model.

3.3 Converting Source Code to Class Diagrams and Sequence Diagrams

Java programs are converted to a class diagram and sequence diagrams of every method. Software tool that used in this experiment is MagicDraw UML [12] for converting diagrams from Java source code, and then diagrams are saved in XML file format.

3.4 Developing Models

We compare three models for estimating the stability of the object-oriented design in incremental development from class diagram and sequence diagram. The models are constructed by using different methods: MRA, PCA, and DLREA.

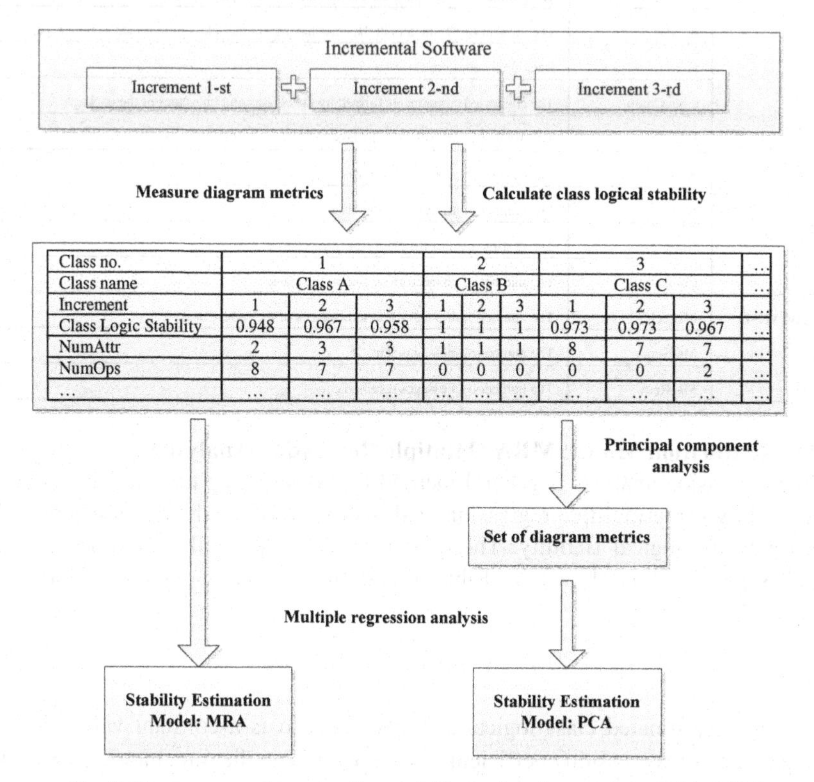

Fig. 2. The process of construction stability estimating models: PCA, MRA

Table 1. Diagram metrics description

Diagram	Metric	Description
Static view: Structure	NOC	The number of children of the class (UML Generalization).
	NOP	The number of parent of a class
	OpsInh	The number of inherited operations.
	AttrInh	The number of inherited attributes.
	NumDesc	The number of descendants of the class (UML Generalization).
	NumAnc	The number of ancestors of the class.
	DIT	The depth of the class in the inheritance hierarchy.
	CLD	The longest path from the class to a leaf node in the inheritance hierarchy below the class.
	IFImpl	The number of interfaces the class implements.
	Assoc_out	The number of associated elements via outgoing associations.
	Assoc_In	The number of associated elements via incoming associations.
	NumAss_User	The number of association of a class in class diagram as user
	NumAss_Provider	The number of association of a class in class diagram as provider
	EC_Attr	The number of times the class is externally used as attribute type.
	IC_Attr	The number of attributes in the class having another class or interface as their type.
	EC_Par	The number of times the class is externally used as parameter type.
	IC_Par	The number of parameters in the class having another class or interface as their type.
	H	Relational cohesion.
	ConnComp	The connected components formed by the classes and interfaces of the package.
	NumOps	The number of operations in a class.
	NumPubOps	The number of public operations in a class.
	NumAttr	The number of attributes in the class.
	Nesting	The nesting level of the class (for inner classes).
	Connectors	The number of connectors owned by the class.
Dynamic view: Behavior	MsgSelf	The number of messages sent to instances of the same class.
	MsgSent	The number of messages sent.
	MsgRecv	The number of messages received.

Stability Estimation Model: MRA (Multiple Regression Analysis)

Multiple regression analysis is general method for constructing model. This model is constructed by using multiple regression analysis to determine the relationships of 27 metrics with the logical stability. The diagram metrics are collected from previous researches [4, 5, 6] is shown in Table 1 and the results of model as following Equation (2).

$$y = a + b_1x_1 + b_2x_2 + \ldots + b_ix_i + e \tag{2}$$

Where y is an estimated class logical stability value; a is a constant value; b is an estimated value of regression coefficient (i = 1,..,n); x is an diagram metrics (i= 1,..,n); e is an error term. The process of construction model is shown as Fig. 2.

Stability Estimation Model: PCA (Principal Component Analysis)
This model is constructed by using principal component analysis to determine the relationships of 27 metrics with the logical stability. PCA is reducing the dimensionality of a data set so the result is a set of diagram metrics. Then we use multiple regression analysis for estimated value of regression coefficient of metrics to construct the model which the form following Equation (2).

Stability Estimation Model: DLREA (Design Logical Ripple Effect Analysis)
DLREA is used for construction this model by analyzing the design logical ripple effect to define specification of the model. Then calibrate the model by adjusting coefficients of the model. Some variables may be introduced into the model that depends on necessity. The conceptual of this model that uses class diagrams and sequence diagrams for analyzing the design logical ripple effect as shown in Fig. 3.

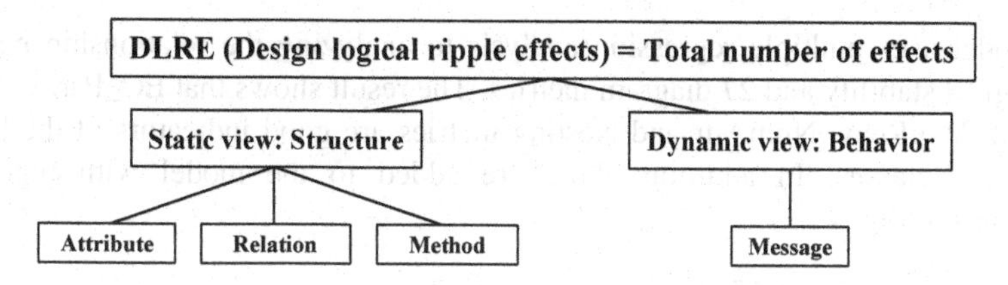

Fig. 3. The conceptual of construction stability estimating model: DLREA

3.5 Validating the Models

The evaluation process is similar to the process of constructing the models in steps 3.2 and 3.3. The logical stability of design is estimated from the class diagram and sequence diagram based on the models obtained from step 3.4. The models are evaluated estimation accuracy by compare the actual values with the estimates and validated the model with relative error and l-level prediction that can be explained as follows: if PRED(0.15) = 0.80, this means that 80% of the estimated value will deviate within 15% of the actual value. L-level prediction generally accepts estimation model is PRED(0.25) $\geq$ 0.75, which means that the deviation is not more than 25% at least 75% of all estimates.

3.6 Comparison Criteria

We compare estimation accuracy value of the stability estimation models for comparing the methods that use to constructing the models. So the best model has the highest estimation accuracy, this means that the estimation value from this model is nearest the actual value. Moreover, the result is used to determine for selecting the methods to develop the stability estimation models.

4 Comparison the Stability Estimation Models

The experimental subjects for comparing the models were twenty Java software projects from open source project hosting web sites (CodePlex, SourceForge).

The selected projects have at least two versions that record version history consisting of new features, changes, fixes and removals. The total number of classes of selected projects in each version is shown in Table 2 which is divided into 2 sets: research set and validation set.

Table 2. Set of java source code

Java source code	Number of programs	Total number of classes		
		version 1st	version 2nd	version 3rd
research set	5	48	133	75
validation set	15	550	1871	981

4.1 Constructing the Model by Using Multiple Regression Analysis (MRA)

This model uses multiple regression analysis to analyzing the relationship between class logical stability and 27 diagram metrics. The result shows that EC_Par, EC_Attr, MsgSent, MsgRecv, NumAttr and Nesting metrics are good indicators of the logical stability of classes. In addition, there are added to the model with regression coefficients in Fig. 4.

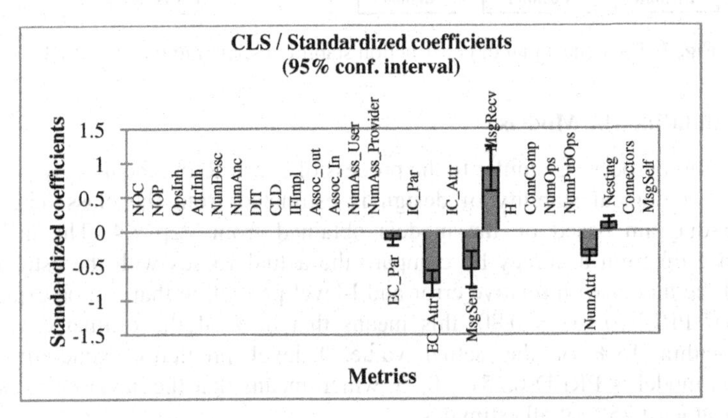

Fig. 4. Standardized coefficients of the 27 diagram metrics

4.2 Constructing the Model by Using Principal Component Analysis (PCA)

The result shows that NOC, NOP, OpsInh, AttrInh, NumDesc, NumAnc, DIT, CLD, IFImpl, EC_Par, IC_Attr, NumOps and NumPubOps metrics are correlated with the class logical stability. The descriptions of diagram metrics are shown in Table 1. Principal component analysis is reduces the number of metrics from 27 metrics to 13 metrics and using multiple regression analysis for constructing model. As shown in Fig. 5, EC_Par, NumOps and NumPubOps metrics were found to be good indicators of the logical stability of classes.

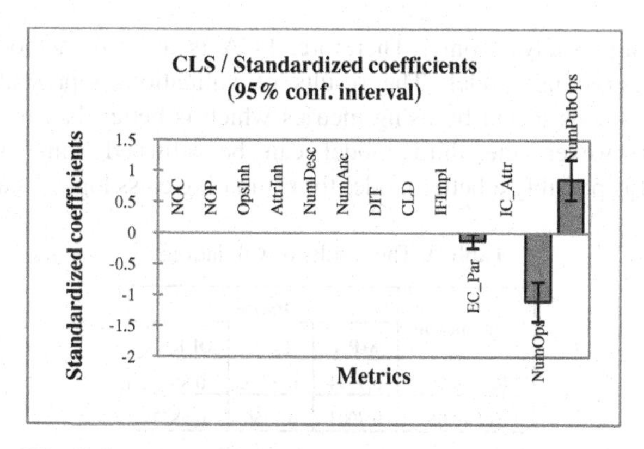

Fig. 5. Standardized coefficients of the 13 diagram metrics

4.3 Constructing the Model by Using Design Logical Ripple Effect Analysis (DLREA)

The design logical ripple effect of object-oriented design is total number of effects from attributes, relations, methods and messages. After adding variables to model, it is calibrated by adjusting coefficients, and validated by estimating the stability in experiment. The result of analyzing the design logical ripple effect is shown as following Equation (3).

$$DS_x = 1/(1+DLRE_x)$$

$$DLRE_x = TA_x + TA_{xy} + TR_x + \sum_{y \in J_{xy}} TR_{xy} + TM_x + \sum_{y \in J_{xy}} TM_{xy} + \sum_{y \in J_{xy}} TS_{xy} + \sum_{y \in J_{xy}} TS'_{xy} \qquad (3)$$

Where DS_x is the design stability of class x; $DLRE_x$ is the design logical ripple effect of class x; x is a focus class in the program; y is other class in the program that related to class x; J_x is the set of super class of class x; J_{xy} is the set of classes that relate class x as an actor e.g. x is association class of y; TA is the total number of effects from attributes; TR is the total number of effects from classes; TM is the total number of effects from methods and TS is the total number of messages.

4.4 Comparing the models

In the Table 3, the first model is constructed by using MRA shows PRED(0.25) is 95.94%, the second model is constructed by using PCA shows PRED(0.25) is 97.63% and the third model is constructed by using DLREA shows PRED(0.25) is 85.71%. Three models for estimating class logical stability of design are accepted because the prediction performed at 0.25 level of significance (75% confidence level) have not less than 0.75.

The second model uses principal component analysis to determine the relationship of metrics by reduce the metrics that without relationship with stability and then uses multiple regression analysis to develop the model that more effective than uses multiple regression analysis only. Therefore, PCA is a good method to selection metrics for constructing model. The results of validations represent concept for stability model is constructed by using metrics which is better than by defining new specification. However, the third model can be adjusted some variables and coefficients that is possibly a better model for estimating class logical stability.

Table 3. The results of validations

L-prediction	Result		
	MRA	PCA	DLREA
PRED(0.25)	0.9594	0.9763	0.8571
PRED(0.05)	0.7903	0.7956	0.6857

In Table 4 show comparison of estimating class logical stability (CLS) of three models. We compare estimating CLS of all classes of current increment, next increment and change of CLS when adding next increment. This set of classes is java program from open-source project hosting web site [11]. There are 2 versions which the number of classes in first version (current increment) is 22 classes and second version (current + next increment) is 67 classes.

Table 4. Comparison of estimating CLS of three models

No.	Class Name	Current Increment			+ Next Increment			Change of CLS		
		MRA	PCA	DLREA	MRA	PCA	DLREA	MRA	PCA	DLREA
1	ExtraReturnsRemover	0.9628	0.9595	0.9651	0.9491	0.9494	0.9412	-0.0136	-0.0100	-0.0238
2	HTMLReplacer	0.9628	0.9595	0.9651	0.9628	0.9595	0.9576	0	0	-0.0075
3	LetterPulse	0.9518	0.9585	0.9563	0.9518	0.9576	0.9442	0	-0.0009	-0.0121
4	LetterPulseDialog	0.9437	0.9604	0.9542	0.9437	0.9604	0.9542	0	0	0
5	NonPrintingChars	0.9628	0.9604	0.9701	0.9628	0.9604	0.9625	0	0	-0.0070
6	Search	0.9546	0.9500	0.9143	0.9546	0.9428	0.9068	0	-0.0072	-0.0075
7	FindDialog	0.9110	0.9538	0.9437	0.9001	0.9481	0.9225	-0.0109	-0.0057	-0.0212
8	LibTTx	0.9628	0.9395	0.8807	0.9600	0.9247	0.8493	-0.0027	-0.0148	-0.0314
9	EndsWithFileFilter	0.9600	0.9633	0.9982	0.9600	0.9633	0.9982	0	0	0
...	...	...	...	...	...	...	...	...	...	...
65	Getopt	-	-	-	0.9082	0.9375	0.9353	-	-	-
66	LongOpt	-	-	-	0.9409	0.9604	0.9576	-	-	-
67	GetoptDemo	-	-	-	0.9628	0.9642	1	-	-	-

5 Conclusions

This paper compares three stability models that are constructed with difference methods. The first model is constructed from metrics using multiple regression analysis (MRA). The second model is constructed from metrics using principal component analysis (PCA). The third is defined with a new specification using design logical ripple effect analysis (DLREA). The principal component analysis is the best of three methods for constructing the stability estimation model because it has the highest estimation accuracy.

At the design phase of incremental development, the new design structure may be affected by the changes when the current increment is added to the previous increment. This study will help software developers and designers to determine to re-design the new design structure in the design phase of incremental development. The data in this paper is collected from different domains. In the future, we plan to construct and compare models in each domain.

References

1. Yau, S.S., Collofello, J.S.: Some Stability Measures for Software Maintenance. IEEE Transactions on Software Engineering SE-6(6), 545–552 (1980)
2. Elish, M.O., Rine, D.: Investigation of metrics for object-oriented design logical stability. In: Proceeding of the Seventh European Conference on Software Maintenance and Reengineering, pp. 193–200. IEEE Computer Society, Washington, DC (2003)
3. Sangpuwong, A., Muenchaisri, P.: Estimating Stability of Software Design in Incremental Development. In: The 7th International Conference on Software, Knowledge, Information Management and Applications, Chiang Mai, Thailand (2013)
4. Rangsiyawath, S., Muenchaisri, P.: Estimating Software Logical Stability from Class Diagram and Sequence Diagram. In: International Joint Conference on Computer Science & Software Engineering (2007)
5. Cheewaviriyanon, C., Muenchaisri, P.: Estimating Software Logical Stability Using Analogy from Class Diagram and Sequence Diagram. In: The 2009 Joint Conference on Computer Science and Software Engineering, Phuket, Thailand (2009)
6. Nimol, D., Muenchaisri, P.: Estimating Software Logical Stability using ANN from Class diagram and Sequence diagram. The Information Technology Journal of IT faculty of KMUTNB, 58–63 (2011)
7. Yau, S.S., Collofello, J.S.: Design Stability Measures for Software Maintenance. IEEE Transactions on Software Engineering SE-11(9), 849–856 (1985)
8. Alshayeb, M., Naji, M., Elish, M.O., Al-Ghamdi, J.: Towards measuring object-oriented class stability. IET Software, The Institution of Engineering and Technology 5(4), 415–424 (2011)
9. Norman, E.F., Shari, L.P.: Software Metrics A Rigorous and Practical Approach, 2nd edn. PWS Publishing Company (1997)
10. CodePlex Project Hosting for Open Source Software, `http://www.codeplex.com/`
11. SourceForge Open Source Applications and Software Directory, `http://sourceforge.net/`
12. MagicDraw Software Package, `http://www.nomagic.com/products/magicdraw.html`

An Approach of Finding Maximal Submeshes for Task Allocation Algorithms in Mesh Structures

Radosław J. Jarecki[1], Iwona Poźniak-Koszałka[2], Leszek Koszałka[2], and Andrzej Kasprzak[2]

[1] Control Theory and Applications Center
Coventry University,
Coventry, the United Kingdom
[2] Department of Systems and Computer Networks
Wroclaw University of Technology
Wroclaw, Poland
dzidmail@gmail.com,
{iwona.pozniak-koszalka,leszek.koszalka,
andrzej.kasprzak}@pwr.wroc.pl

Abstract. This paper concerns the problem of finding efficient task allocation algorithms in mesh structures. An allocation algorithm, called Window-Based-Best-Fit with Validated-Submeshes (WBBFVS), has been created. The core of this algorithm lays in the way of finding maximal submeshes what is an important part of the task allocation algorithms based on the stack approach. The new way of finding maximal submeshes avoids post-checking submeshes' redundancy thanks to the proposed initial validation. The elimination of redundancy may increase the speed of the algorithms' performance. The three considered algorithms have been evaluated on the basis of the results of simulation experiments made using the designed and implemented experimentation system. The obtained results of investigation show that the WBBFVS algorithm seems to be very promising.

Keywords: mesh structure, task allocation algorithm, efficiency, maximal submeshes, simulation.

1 Introduction

The given problem concerns the task allocation in the 2D mesh networks. Two-dimensional topologies are simple and popular within the large-scale integrated systems [1]. For two-dimensional meshes it is assumed that processors communicate only with the other adjacent processors. Processors are interpreted as the nodes in mesh topology and can have only up to four neighbors. Incoming tasks define rectangular configuration of nodes they want to use within the mesh structure. Hence, the tasks need to be allocated to the submeshes with proper sizes without overlapping.

There are many algorithms for solving such an allocation problem, e.g., well-known algorithm Busy List (BL) which is described in [2], and several algorithms based on stack ideas [3], called Stack Based Algorithms (SBA-family), including WSBA and WBBF presented in [4].

N.T. Nguyen et al. (Eds.): ACIIDS 2014, Part I, LNAI 8397, pp. 332–341, 2014.

The WBBF (Windows Based Best Fit) algorithm is a synergy of the WSBA (Window-Stack-Based-Allocation). The authors of [5] provide brief description and present the results of the comparison of these two algorithms. In general, the both algorithms required using a list of all maximal free submeshes in order to allocate tasks. A maximal free submesh is one that cannot be extended in any direction. Consequently, each not busy node can be included in one or in more maximal free submeshes. The BL algorithm is based on Best Fit idea [2], in particular on the concept of border values. This concept has been also adapted to WBBF (see [5] and [6]).

There are available a few papers which concern finding maximal submeshes, e.g., [1], [2], [4], and [7]. The main contribution of this paper consists in creating a new algorithm that finds all the maximal submeshes within any 2D mesh. The proposed way of finding maximal submeshes can be applied essentially to any task allocation algorithm which needs to perform such an operation. Moreover, an experimentation system has been designed and implemented to give possibilities for comparative experiments and to conclude about efficiency of the allocation algorithms. The comparison between three algorithms WBBF, BL, and the proposed WBBFVS is presented.

The rest of the paper is organized as follows. In Section 2, the problems of task allocation and of finding maximal submeshes are described briefly. Section 3 is devoted to presentation of known ideas in finding maximal submeshes, but Section 4 describes the newly proposed approach in detail. Section 5 contains a short description of the created experimentation system and the obtained results of the comparative study of efficiency of the considered allocation algorithms. As the measures of efficiency, the averaged fragmentation and the averaged processing time have been introduced. The final remarks and plans for the further research appear in Section 6.

2 Problem Statement

The problem of task allocation within a 2D mesh structure may be stated as follows:

Given: (i) the rectangular (or squared) mesh with the length L and the width W; (ii) the set contained *n* rectangular tasks with known sizes (l_k x w_k), k=1, 2, …,*n*., and with known execution times.

To find: the allocation of the all tasks within the mesh.

Such that: to minimize the fragmentation expressed by the formula (1)

$$F = \frac{[L\,W - \sum_{k=1}^{n} l_k w_k]}{L\,W} \quad (1)$$

The fragmentation describes processors utilization and should be as low as possible to avoid waste of CPU time.

The task allocation algorithm for solving such a problem should fulfill another criterion – to minimize the processing time needed for allocating all tasks. For the creation of the allocation algorithm an important issue (especially for the SBA [3] family of algorithms) is finding maximal submeshes in an efficient way. This problem can be described in the following way: During allocating a next given task within a given 2D mesh with some busy area we need to find all the maximal submeshes – a potential places for the task being allocated. A busy area is composed of the nodes just taken by the previously allocated tasks. A maximal submesh is such submesh that cannot be extended more in any direction [4]. For illustration, an example is shown in Fig. 1. In figure the numbers assigned to the nodes inform about connections to the busy nodes and to the border of the mesh. Finding maximal submeshes is a repeating process. After adding a task or removing a task (when the task has been just executed) the list of the recent maximal submeshes within the mesh is prepared.

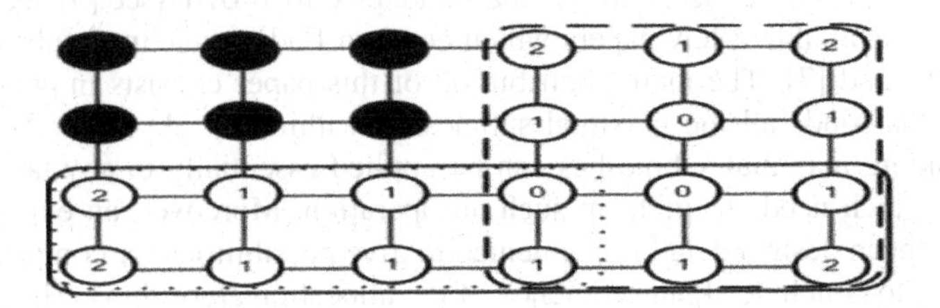

Fig. 1. Example of two potential maximal submeshes (busy area is blackened)

3 Finding Maximal Submeshes – Known Approach

The procedure of finding maximal submeshes [4] consists of the following steps:

Step 1. Searching for *bases* of the maximal submeshes. The *bases* are called *base nodes* and refer to the top left node of a submesh.

Comment: According to the WBBF [4], these nodes have to be not busy and have the wall or a busy node on the left or/and on the top side. However, checking all such nodes does not cover all the cases in which such submesh can be built. In Fig. 2 a unique maximal submesh that would not be generated by this algorithm is shown.

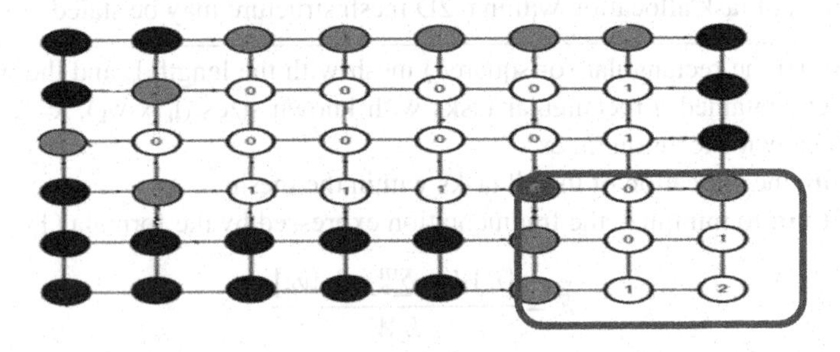

Fig. 2. WBBF action - the grey nodes are considered as base nodes; the black nodes are busy ones. The top-left node (row=4, column=6) is omitted.

Step. 2. For each base node the submeshes are generated. An additional condition has been added to ensure generating submeshes which can bind incoming tasks. The procedure used by WBBF produces many redundant submeshes (Fig. 3). A term *redundant* describes submesh which is found by an algorithm; however, it is also a submesh of another bigger submesh found. The redundant submeshes are not maximal, so they have to be removed.

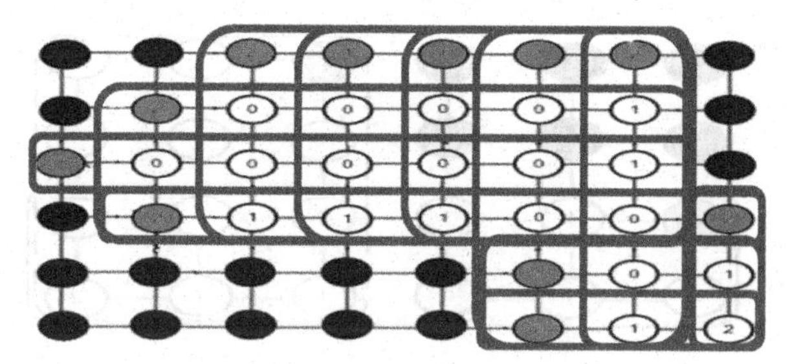

Fig. 3. Redundant submeshes

Step. 3. A list of maximal submeshes is ready for the further use (Fig. 4). It can be observed, that one maximal submesh, mentioned in *Step 1*, is still missing.

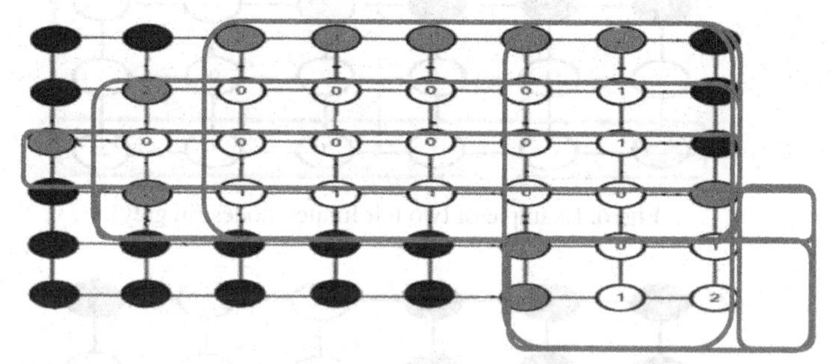

Fig. 4. A set of maximal submeshes - still missing one submesh

4 Finding Maximal Submeshes – A New Approach

In contrast to the way described in the previous section, the WBBFVS algorithm evaluates redundancy of submeshes before their creation. The scheme is as follows:

Step 1. Searching for the bases of the maximal submeshes (using top left node principle). As the base nodes the following nodes are chosen:

1.1. The nodes on corners. They have the wall or a busy node on the top and on the left side (Fig. 5). These nodes are a base for non-redundant maximal submeshes.

1.2. The nodes with busy the left neighbor node or the nodes having the wall on the left. An example is shown in Fig. 6. For these nodes, there is taken an action to check whether these nodes guarantee creating the redundant maximal submeshes. In the case of the left limited nodes, we have to check whether in the upper row the distance to the right border (the wall or a busy node) is closer than in the current row. If it is true, all the maximally spanned submeshes produced a maximal submesh with this node. In Fig. 6, a submesh composed of all nodes in bottom row is redundant. The submesh composed of nine nodes (down–right) is maximal.

1.3. The nodes with a busy top neighbor node or the nodes having a wall on the top (Fig. 7). For these nodes there is also taken an action to check whether they guarantee to produce the redundant maximal submeshes, i.e., checking if in the column on the left the distance to the bottom border (to the wall or a busy node) is closer than in the current column. If it is true, all the maximally spanned submeshes produced by this node (and longer than mentioned (left) distance), are the maximal submeshes.

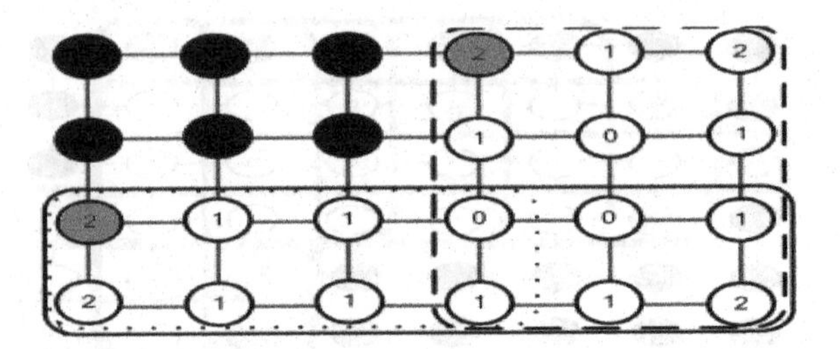

Fig. 5. Example of nodes on corners (in grey)

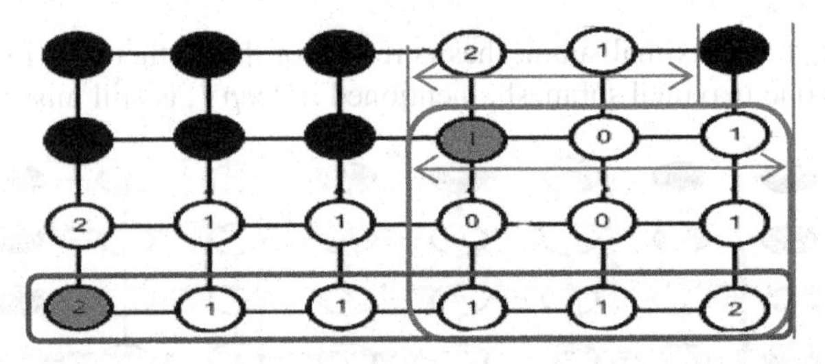

Fig. 6. Example of two left limited nodes (in grey)

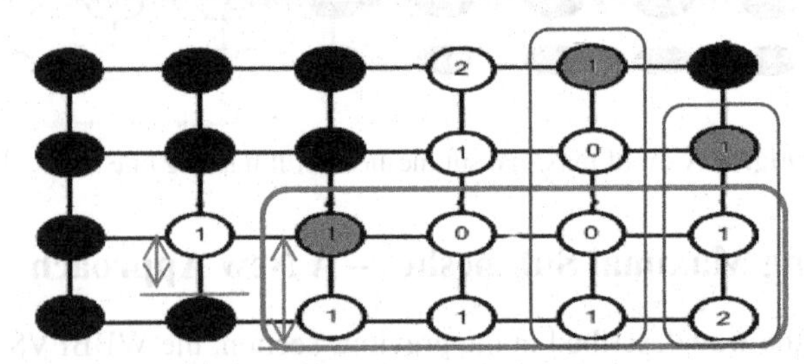

Fig. 7. Example of two top limited nodes. Vertical submeshes are redundant. Horizontal is maximal.

1.4. The floating nodes which do not have any left or top busy neighbor or the wall (Fig. 8). Finding them is more difficult because checking the conditions is done in two dimensions. (Note: finding this kind of nodes takes statistically the same CPU time as the rest of algorithm). To find such a node we have to check whether configuration of adjacent column and row meets following conditions: (i) the distance from upper row to right border (the wall or a busy node) is closer than from the current row; (ii) the distance from the column on the left to the bottom border (the wall or a busy node) is closer than in the current column. If both conditions are satisfied then all the maximally spanned submeshes produced by this node are the maximal submeshes. (Note: to save CPU time, if the first condition is not satisfied, the second one should not be checked).

Step 2. For every base node found, one or more maximal submeshes are being created. All these unique maximal submeshes are stored in the list for the further usage.

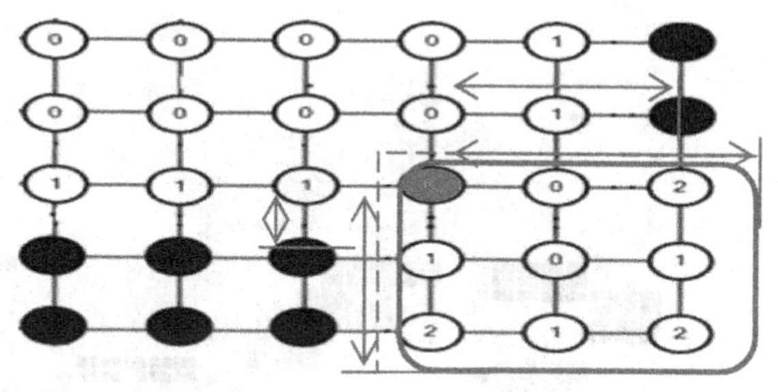

Fig. 8. Example of a base node which was found by the intersection of the virtual lines. A distance conditions is denoted by arrows.

5 Investigation

The investigations have been made using the designed and implemented experimentation system with a simulator program compiled from the C# source code, provided by co-author of this paper, and described in details in [2]. Originally the simulator was designed to compare BL and WBBF algorithms. For purpose of this paper, the experimentation system was developed. It was added ability to benchmark WBBFVS algorithm. This system can run the real time experiments and can generate statistics about fragmentation rate and computational (processing) time of each tested algorithm.
The available inputs are:

- The task allocation problem parameters (i) mesh size; (ii) the total number of tasks, (iii) the sizes of tasks, (iv) the execution times (duration) of tasks;
- The task allocation algorithm used, including BL, WBBF, WBBFVS;
- The number of tests (single experiments).

The available outputs are:

- The graphical representation of the allocation process, the queue of incoming tasks and the progress bar (see screenshot in Fig. 9);
- The computed indices of performance for the task allocation algorithms, including the fragmentation F defined by (1) and the processing time T.

For the purposes of the analysis of the obtained results the averaged values of the indices were computed, including the averaged fragmentation defined by (2), and the averaged processing time defined by (3), where N is the total number of tasks.

$$AvgF = \frac{\sum_{j=1}^{N} F_j}{N} \tag{2}$$

$$AvgT = \frac{\sum_{j=1}^{N} T_j}{N} \tag{3}$$

The complex experiments to compare the algorithms were made. The values of the task sizes and the task execution times were generated from the fixed (designed) ranges at random. The series of experiments were carried on 3.2 GHz single core for each algorithm for the same instances (the sets of inputs). The values of the task size and the task execution time were generated from the fixed ranges at random.

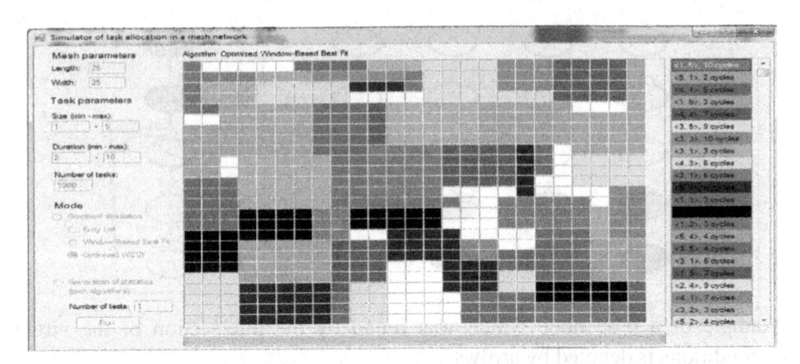

Fig. 9. Visualization module of the experimentation system [2]

The results of three comparative experiments are presented and discussed:

- The comparison of WBBF and WBBFVS on the basis of AvgF.
- The comparison of BL and WBBFVS on the basis of AvgF.
- The comparison of BL and WBBF and WBBFVS on the basis of AvgT.

***Experiment #1*. Fragmentation - WBBF vs. WBBFVS.**

The exemplary results are shown in Fig. 10. It may be observed that the input problem parameters have impact on the obtained results. For the sizes of tasks in the range from 1 to 10, the averaged fragmentations were almost the same. However, for the case with larger sizes of tasks to be allocated (in the range from 1 to 15) it can be observed that the created algorithm WBBFVS outperformed WBBF - having the averaged fragmentation lower by 3 %.

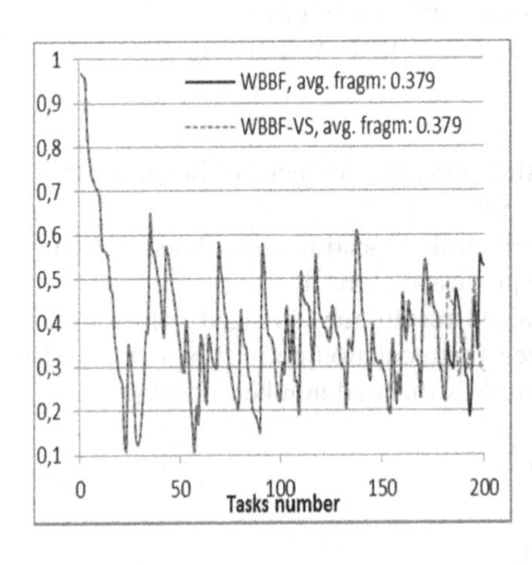

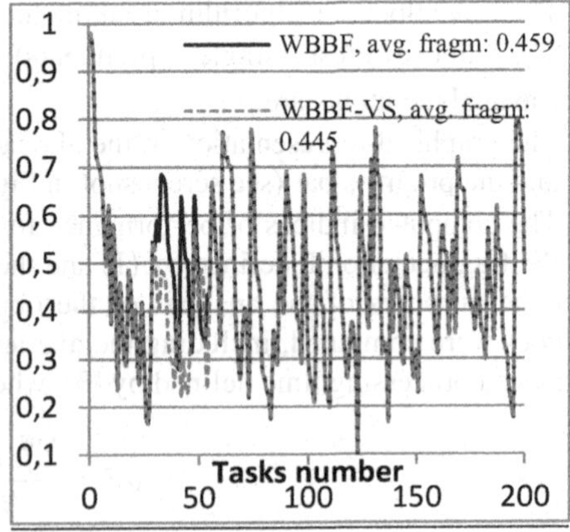

Fig. 10. Fragmentation for mesh 25x25, tasks times 2-10, task sizes 1-10 (left), 1-15 (right)

***Experiment #2*. Fragmentation - BL vs. WBBFVS.**

The obtained results are given in Table 1. It can be taken into consideration that these two algorithms allocate tasks in very different way but the average fragmentation AvgF (where F is defined by (1)) was almost the same for the smaller meshes – the both algorithms produced a similar allocation pattern. However, for bigger meshes, the averaged results for WBBFVS algorithm were better by around 6 %.

Table 1. Comparison of BL and WBBFVS

Input parameters			Outputs	
Mesh size	Task size	Task time	AvgF for BL	AvgF for WBBFVS
25 x 25	1 - 20	2 - 10	0.501	0.500
50 x 50	1 - 20	2 - 10	0.396	0.374

Experiment #3. **Processing Time – WBBFVS vs. BL vs. WBBF**

In Table 2, the results of 7 series of simulation experiments carried out with the experimentation system are presented.

In Table 2, the exemplary, interesting cases are presented. It is a part of the three-stage experiment in which three stages were corresponded to the variables regarded as the problem parameters. The mesh size was on the three levels (N_I=3): 25x25, 50x50, 100x100; the tasks size ranges were also on the three levels (N_{II}=3): 1-10, 5-10, 1-40;

Table 2. The averaged processing times for the tested algorithms

	AvgT WBBFVS	*AvgT BL*	*AvgT WBBF*
Mesh 100x100 Task size: 5-10 Task time: 2-10 No. of tasks: 500	1788	16414 9.2 x slower	381887 213 x slower
Mesh 50x50 Task size: 1-40 Task time: 2-10 No. of tasks: 500	331	1376 4.2 x slower	19803 60 x slower
Mesh 50x50 Task size: 5-10 Task time: 2-10 No. of tasks: 100	78	96 1.2 x slower	2954 38 x slower
Mesh 50x50 Task size: 5-10 Task time: 2-10 No. of tasks: 5000	2031	2087 1.1 x slower	90411 45 x slower
Mesh 25x25 Task size: 5-10 Task time: 2-10 No. of tasks: 100	12.8	15 1.2 x slower	174 13,6 x slower
Mesh 25x25 Task size: 5-10 Task time: 2-10 No. of tasks: 1000	203	187 1.09 x faster	2196 15 x slower
Mesh 50x50 Task size: 1-10 Task time: 2-10 No. of tasks: 500	261	246 1.06 x faster	12268 47 x slower

and the total number of tasks was on four levels (N_{III}=4): 100, 500, 1000, 5000. The execution time (task time) was in the same range 2-10. The whole experiment was composed of $N = N_I \times N_{II} \times N_{III} = 36$ single series of experiments.

As we can see in Table 2, WBBFVS has the best (the fastest) average processing time in five of the seven different experiments. WBBFVS is remarkable faster than WBBF (for instance, it is 203 times faster for 100 x100 mesh size), and in the most cases it is also faster than BL. On the basis of the detailed results obtained during the experiment in which 5000 tasks were allocated, it was observed that the algorithms' speed depends also on the sizes of the tasks.

When analyzing the results of the other multistage experiment we may found that WBBFVS performed faster (up to 9 times more) than BL, especially for tasks varying in size in the ranges 1-20 and 1-24 (see Fig. 11) as well as in the ranges 1-40 and 1-49. This relationship could be modeled (in the future) using mathematical formulas.

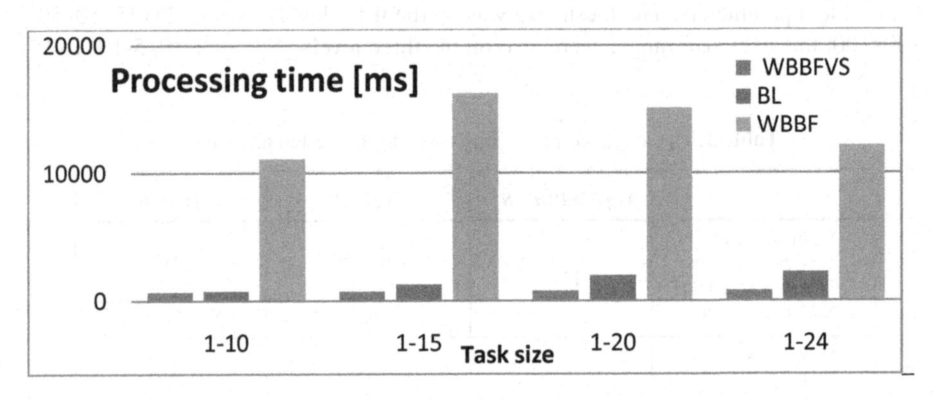

Fig. 11. Processing time on mesh 25x25, with 5000 tasks, for tasks time 1-30

It may be observed from Fig. 11 that for many ranges of tasks' size the WBBFVS outperformed the other considered algorithms.

6 Conclusion

The created WBBFSV task allocation algorithm with the new implemented way of finding maximal submeshes gives a real boost. The algorithm is characterized with the slightly better fragmentation than BL and WBBF; moreover, it is 10 to 200 times faster than the previous implementation of the WBBF algorithm.

The presented experimentation system is serving as a tool to aid teaching graduate students and preparing projects in computer science and telecommunications areas in Faculty of Electronics, Wroclaw University of Technology.

In our opinion, the further research in the considered area should concentrate on (i) practical applications of the allocation algorithms following those presented in [9]; (ii) further development of the experimentation system by implementing other allocation algorithms, e.g., algorithms based on evolutionary ideas [10]; (iii) preparing new modules of the system to ensure designing multistage experiments [4] in the automatic way.

Acknowledgement. This work was supported by the statutory funds of the Department of Systems and Computer Networks, Faculty of Electronics, Wroclaw University of Technology, Wroclaw, Poland.

References

1. Yoo, B.S., Das, C.R.: A fast and efficient processor allocation scheme for mesh-connected multicomputers. IEEE Transactions on Computers 51(1), 46–60 (2002)
2. Linkowski, T., Koszalka, L., Kasprzak, A.: Window-list-based best fit task allocation in mesh. In: Information Systems Architecture and Technology: System Analysis in Decision Aided Problems, OWPWr, pp. 379–388 (2009)
3. Sharma, D., Pradhan, D.K.: Submesh allocation in mesh multicomputers using busy-list: A best-fit approach with complete recognition capability. Journal of Parallel and Distributed Computing 36, 106–118 (1996)
4. Kaminski, R.T., Koszalka, L., Pozniak-Koszalka, I., Kasprzak, A.: Evaluation and comparison of task allocation algorithms for mesh networks. In: Proc. 9th International Conference on Networks, pp. 104–108. IEEE Computer Society Press (2010)
5. Koszalka, L.: Static and dynamic allocation algorithms in mesh structured networks. In: Madria, S.K., Claypool, K.T., Kannan, R., Uppuluri, P., Gore, M.M. (eds.) ICDCIT 2006. LNCS, vol. 4317, pp. 89–101. Springer, Heidelberg (2006)
6. Papadimitriou, C.: Calculation Complexity. WNT, Warsaw (2002)
7. Seong-Moo, Y., Hee, Y.Y., Hyun, S.C.: Fault-free maximal submeshes in faulty torus-connected multicomputers. Informatica 28(3), 289–296 (2004)
8. Bani-Mohammad, S., Ababneh, I., Hamdan, M.: Comparative performance evaluation of non-contiguous allocation algorithms in 2D mesh-connected multicomputers. In: CIT, pp. 2933–2939. IEEE Computer Society (2010)
9. Zydek, D., Selvaraj, H., Koszalka, L., Pozniak-Koszalka, I.: Evaluation scheme for NoC-based CMP with integrated processor management system. International Journal of Electronics and Telecommunications 56(2), 157–167 (2010)
10. Kmiecik, W., Wojcikowski, M., Koszalka, L., Kasprzak, A.: Task allocation in mesh connected processors with local search meta-heuristic algorithms. In: Nguyen, N.T., Le, M.T., Świątek, J. (eds.) Intelligent Information and Database Systems. LNCS (LNAI), vol. 5991, pp. 215–224. Springer, Heidelberg (2010)

A GA-Based Approach for Resource Consolidation of Virtual Machines in Clouds

I-Hsun Chuang[1],Yu-Ting Tsai[1], Mong-Fong Horng[2], Yau-Hwang Kuo[1], and Jang-Pong Hsu[3]

[1] Department of Computer Science and Information Engineering, National Cheng Kung University, Tainan, Taiwan (ROC)
[2] Department of Electronics Engineering, National Kaohsiung University of Applied Sciences, Kaohsiung, Taiwan (ROC)
[3] Advance Multimedia Internet Technology, Inc. Tainan, Taiwan (ROC)
mfhorng@ieee.org

Abstract. In cloud computing, infrastructure as a service (IaaS) is a growing market that enables users to access cloud resources in the convenient, on-demand manner. The IaaS can provide user to rent the resources of cloud computing and virtual machines (VMs) through virtualization technology. Because different VMs may demand different amounts of resources, an important problem that must be addressed effectively in the cloud is how to decide the mapping adaptively in order to satisfy the resource needs of VMs. The mapping problem solution is called virtual machine placement policy (VMPP). However, VM will change the requirement of resources according to the workload of application VM. Thus, it's necessary to apply resource consolidation technology to satisfy dynamically resource on demand. In this thesis, we present a two-phase approach for resource consolidation to minimize resource consumption. In the first phase, we use a genetic algorithm to find a reconfiguration plan. In the second phase, we propose a mechanism to find a way to migrate VMs such that the number of active nodes and the overall migration cost could be minimized. Finally, the experimental results show that we obtain well-consolidating active nodes than other existing approaches.

Keywords: Resource consolidation, VM migration, Cloud computing, Genetic algorithm.

1 Introduction

Cloud computing is a key concept to greatly improve the living quality for people. Cloud computing [1] changes the ways companies managing their computing in a cost-effective manner. A growing number of companies, like Google, Microsoft, Amazon, Yahoo and others, attend to build cloud computing environments. In a cloud environment, those companies handle the huge amounts of data using distributed computing over hundreds of nodes. The popular IaaS clouds must be able to automatically configure the deployment, control the computing resources to the needs of the virtual machines (VMs), memory allocated or CPU allocated to a virtual machine, etc. Through

N.T. Nguyen et al. (Eds.): ACIIDS 2014, Part I, LNAI 8397, pp. 342–351, 2014.

hardware virtualization, there are more flexibility and scalability from traditional data processing of being burdensome and private to being shared, pay-per-use on a short-term basis [4].

Resource consolidation is a way of reducing the number of active nodes used in a cloud cluster. Because different VMs may demand different amounts of resources due to the diverse set of applications, an important problem that must be addressed effectively in the cloud is how to decide the mapping adaptively in order to satisfy the resource needs of VMs while minimizing the number of active nodes used. The purpose of consolidation is envisioned to achieve not only lower hardware cost but also efficient resource utilization of an active node. Consolidation technology is applied to many research areas, such as power consumption and network. In whether power consumption or network, the issues are about to utilize radio resource more efficiently to reduce the consumption of resources in systems and increase the overall performance. On the other hand, in several proposed consolidation approaches [2] [7] focused on how to calculate a new configuration and ignore the higher migration time of VMs to impact on the performance of system. In this paper, we have two parts to solve the virtual machine placement problem and virtual machine migration problem. First, different from existing consolidation approach, we propose a GA-based approach, finding quickly near optimal solution, to solve virtual machine placement policy (VMPP). And then, virtual machine migration policy (VMMP) is formulated a migration cost model to find the minimal cost of VMPP solution. Finally, some experiments are made to evaluate the performance of the GA-based approach. The experimental results present that the approach we purpose is achieving minimizing the number of active nodes.

The paper is organized as follows. In Section 2, related work is reviewed. The architecture and resource consolidation approach of proposed system are described in Section 3. Numeric results are shown in Section 4. Finally, we conclude this work.

2 Related Work

Virtualization simplifies service management and reduces the server consumption to save on hardware, management of the infrastructure in data centers. That's the reason why virtualization becomes a key technology. Through virtualized technique, no matter what time a user demand changes, VMs can be resized and migrated to other physical servers if necessary. However, a problem to decide when, how, and which virtual machines (VMs) have to be consolidated into a single physical server in data center, called server consolidation, is produced. In recent years, resource consolidation approaches which can be found in the fields of resource management in many related work involve VM migration, which has a direct impacted on service response time. This section is a brief summary of the approaches in our work.

In [8], a genetic algorithm (GA) and a reconfiguration algorithm were proposed to optimize the resource consumptions in the cloud cluster systems. They attempt to maximize a global utility function and optimize the system state at the low transition overhead. In their method scheme, the VMs are moldable which exposes a new dimension that should be addressed when designing consolidation strategies. Example for the rule based knowledge management approach, and the knapsack modeling method. In earlier approaches such as [2, 5-6], they focused on the migrations of VMs as a whole and heuristic methods were developed to solve resource replacement problem. However, the produced solution cannot be guaranteed to be optimal.

A genetic algorithm (GA) is used to find approximate solutions to NP problems through application of the principles of evolutionary biology to computer science. There are a lot of discussions on the advantages and costs of the genetic algorithm and how to model cloud into a function. We can utilize a more efficient mechanism than [8] and [5] by combining the reconfiguration, proposed in [8], and the performance model, in [5]. The aim to reduce the number of migrations is minimized with minor penalty on the number of physical servers in supposed cloud systems. There are some ways to migrate Virtual Machines in clouds, like live migration and quick migration. Live migration of virtual machines is an important technology in modern Cloud environments to allow serving customers while migrating. Many methods have been proposed to improve the live migration time. They formulate the memory transfer during migration of a VM and to improve the warm-up processes of VM migration with introducing a page selection method for the memory pages.

3 Resource Consolidation for Virtual Machines in Clouds

A. System Architecture Framework

In IaaS cloud environment, duo to the QoS requirements that change over time, such as types of resources, inflexible price, and usage time, the new generation blueprint of Cloud Computing architecture is depending on demand-based market and flexible infrastructure [3]. As shown in Fig.1, it illustrates the hierarchical architecture of resource consolidation in IaaS Clouds proposed in this thesis. There are nine components as follows.

1. Request (Brokering services/ Applications): Customers or Service Brokers, are agents between cloud providers and cloud consumers.
2. Interface: It offers customers or Service Brokers the choice of services based on cost saving function or others through Web impact on the Cloud provider interaction with consumers.
3. Resource Allocator: The efficient resource management is depending on the following components.
4. Service Manager: Analyzing the service requirements of a request and allocate it resources according to service-level agreement (SLA).
5. VM Monitor: The VM Monitor is used to view information about the availability of VMs and their resources utilization, e.g., CPU utilization, memory utilization, network utilization, disk utilization.
6. Accounting: It accounts records for VM resource usage and price based on resource usage.
7. Resource Manager: It is a resource allocation decision that plans the best way for using available resources to map VMs to PMs.

8. Virtualization Layer: Virtual machines offer numerous advantages of flexibility, snapshot feature, high availability, and scalability to assist developers in development more convenient.
9. Physical Infrastructure: Physical Infrastructure, such as clusters, or data centers, is composed of different types of physical nodes.

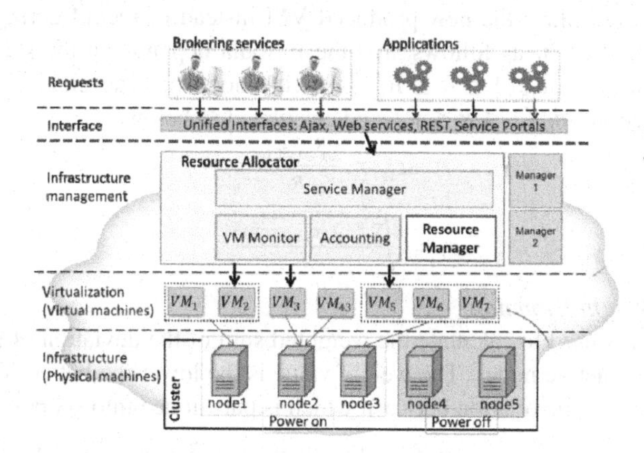

Fig. 1. Hierarchical System Architecture

A. System Model

The problem of VM-to-PM assignment within the limit of cloud environment is formulated by three models. System Profile of Cloud environment is proposed to define the essential components of a cluster, including the resource requirements of VMs and the relation between physical nodes. The Cloud environment is composed of a cluster of N active nodes, n_i (where $i = 1,2, \ldots, N$) is denoted an active node. $i$is denoted the identity of the active node. In a homogeneous environment, sc_{ik} is normalized as a percentage of available resources of a type of resource $r_k$in an active node n_i. Each element fc_{ik} in v_j consists of two units, $fc_{k,min}$and $fc_{k,max}$. $fc_{k,min}$is denoted the minimal free resources of a type of resource r_k and $fc_{k,max}$ is the maximal free resources of a type of resource r_k. Thus, the value of fc_{ik} is between $fc_{k,min}$ and $fc_{k,max}$.

Modeling of virtual machines is given as follows, $v_j$for $j = 1,2, \ldots, M$ denotes a virtual machine created by Cloud environment.jis denoted the identity of the VM, and there are M VMs in the Cloud environment. Each element r_{jk} (where $k = 1,2, \ldots, R$) allocates to v_j is the type k of the demanded resources of v_j . And r_{jk} is normalized as a percentage of the total resources. k is denoted the type of resource, and there are R types demanded resources provided by Cloud environment. Each element r_{jk} in v_j consists of two units, $r_{k,min}$and $r_{k,max}$. Moreover, $r_{k,min}$ and $r_{k,max}$ are normalizedas a percentage of the total resources in an active node. $r_{k,min}$ is denoted the minimal requirement of the resource type k when created a VM in Cloud environment and

$r_{k,max}$ is the peak demand of the resource type k when configuring an individual VM. Thus, the range of r_{jk} value is between $r_{k,min}$ and $r_{k,max}$.

We use weight function to model available resources of system resources. The weight function is in the case where an active node has balanced spare CPU cycles and memory. Many types of resources in an active node are balanced in order to have enough resources to place the new produced VM instead of reconfiguring the system status. We define that σ_i as follows, and the formula depends on the standard deviation of the variables, $sc_{ik}(1 \leq k \leq R)$, in active node n_i. The element $\overline{sc_i^s}$ is denoted the average of $sc_{ik}(1 \leq k \leq R)$ and calculated as below.

$$\overline{sc_i^s} = \frac{\sum_{k=1}^{R} sc_{ik}}{R} \tag{1}$$

$$\sigma_i = \sqrt{\frac{\sum_{k=1}^{R}(sc_{ik} - \overline{sc_i^s})^2}{R}} \tag{2}$$

Definition 1: Weight Function

Weight function is used to calculate the weighted sum of the deviation of the available resource r_k in all active nodes. The weight value is the lower the better. The weight is determined based on the relation between which is partitioned into six bound:

$$\mathbf{W}(\sigma_i, \overline{sc_i^s}) = \begin{cases} \mathbf{w_i} (\mathbf{i} - \mathbf{1}) \times \mathbf{20\%} \times \overline{sc_i^s} \leq \sigma_i < i \times \mathbf{20\%} \times \overline{sc_i^s} \\ \mathbf{w_0} \sigma_i \geq \overline{sc_i^s} \end{cases} \tag{3}$$

Modeling of Migration Cost Profile:

$f(p)$ denotes the estimated cost of a reconfiguration plan p is the sum of the costs of migrations of each preceding step. An element S_n denotes the total number of migration steps. $f(s)$ is denoted the largest live migration time of VM that is migrated in step s. An element s is denoted the identity of migrated step. $T(v_j)$ is denoted the live migration time of VM v_j that is migrated in step s. An element G_n is denoted the total number of migrated VMs. Assume that a VM is allocated r_m for memory. And the source node is connected to the destination node with a link with bandwidth b, delay t_d, and error rate err. A element $\overline{e_c}$ denotes that the average compression rate and VM is modifying the memory with rate of e_w. The overhead for a single migration and the delay incurred for preceding migrations. Therefore, we refer the factors in [9] to define the migration cost function $T(v_j)$ as follows.

Definition 2 : Migration Cost Function

We define that $C = f(p)$, and the formula depends on the amount of memory allocated to the migrated VMs.

$$f(p) = \sum_{s=1}^{S_n} f(s) \tag{4}$$

$$f(s) = max\left(T(v_j)\right), 1 \leq j \leq G_n \tag{5}$$

$$T(v_j) = t_d - \frac{ln\,(1 - \frac{r_m \times \overline{e_c} \times e_w \times (1+err)}{b})}{e_w} \tag{6}$$

B. A Resource Consolidation Approach

In this section, the resource consolidation approach takes into account both the problem of efficiently allocating the VMs to the available nodes and the problem of how to migrate the VMs to these nodes. We try to find a condition which is a compromise between the minimal number of active nodes and the minimal migration overhead of the VMs to these nodes. Our purpose is to minimize the number of active nodes used and the cost of migrations. The approach has two parts: The Virtual Machine Placement Policy (VMPP) and The Virtual Machine Migration Policy (VMMP). The VMPP invoke GA to discover available resources of active nodes and include VMMP. The proposed VMMP is composed of two parts. First, we select ten solutions in the order by fitness value from VMPP to calculate the cost of migrations. And the second, the solution with the lowest migration cost is selected as the new configuration of VM-to-PM mapping.

System Flow Chart

The proposed approach is composed of two parts, VMPP and VMMP. We would first describe the flow chart of our system and the executed situations of VMPP. And then, we would describe the details of the VMPP procedure as well as the VMMP. As shown in Fig. 2, it is the flow chart of the proposed approach. There are two situations to execute VMPP. First, when a new VM which is serving requests is produced, the system would monitor whether the available resource of active node is enough to accommodate a new VM.

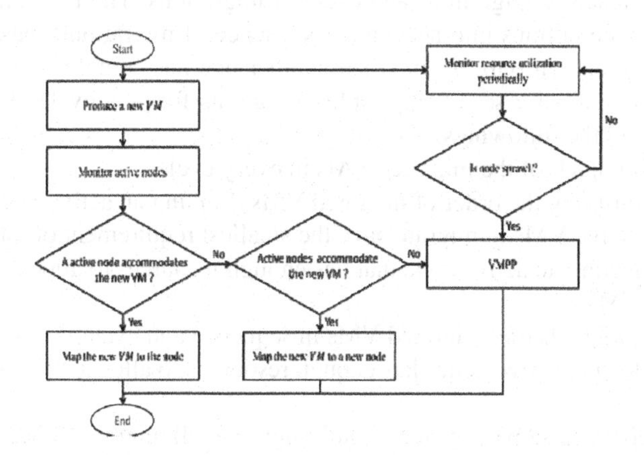

Fig. 2. Flow Chart of the proposed approach

If there is an active node the available resources of which are enough, the new VM will be mapped to the active node. If there is not an active node the available resources of which are enough, system will calculate the total available resources across all active nodes. If total available resources are not enough to accommodate the new VM, the new VM will be mapped to an inactive node and the inactive node will turn to active node. If not, the VMPP will be executed to invoke GA. Second, the system would monitor the every resource utilization of active node periodically. If there is a situation of node sprawl, the VMPP will be executed to invoke GA.

The Virtual Machine Placement Policy

In this consolidation strategy, to admit the minimum number of active nodes is mainly taken into account. We refer the utility function in [8] and improving the algorithm more efficient. We assume an element $A[i,j,k]$ is denoted the percentage of the resource r_k allocated to VM v_j in node n_i in the array. The variables sc_{ik} can reflect the convergence level r available resources across N nodes. $\overline{sc_k^a}$ is denoted the average of $sc_{ik}(1 \le i \le N)$ and calculated as below.

$$\overline{sc_k^a} = \frac{\sum_{i=1}^{N} sc_{ik}}{N} \tag{7}$$

The objective function is formulated as bellow.

$$\sum_{k=1}^{R}\sum_{i=1}^{N}\frac{\left(sc_{ik}-\overline{sc_k^a}\right)^2}{W(\sigma_i, sc_i^s)} \tag{8}$$

A GA is invoked for resource consolidation and many relatives and variants are based on the use of local information. And the function value and the derivatives with respect to the parameters optimized are used to take a step in an appropriate direction towards a local maximum or minimum. The objective of the Virtual Machine Migration Policy (VMMP) is to construct a reconfiguration plan for possible configurations that use the solutions determined by the VMPP, and choose the one with the lowest estimated reconfiguration cost.We assume two kinds of constraints on migrations, sequence of migrations and cycle of migrations. The migration algorithm takes the bellow conditions into account. A sequence of migrations constraint occurs when one migration can only begin when another one has completed and a cycle of migrations occurs when a set of infeasible migrations forms a cycle. The migration policy steps are in the followings:

- Step1: Finding all of the migrated VMs in every cycle.
- Step2: Identifying the order of migrated VMs. Finding an active node in each cycle where the VMs to migrate have the smallest requirement of total memory.
- Step3: Selecting an active node that has enough resources to allocate to these migrated VMs.
- Step4: Finding feasibly migrated VMs in sequence and cycle.
- Step5: Selecting active nodes has enough resources to allocate to these migrated VMs.
- Step6: Repeating step4 and step 5 until migrating all migrated VMs.
- Step7: Estimating migrated cost

This approach is implemented on Depth-First Search. The goal is to try to do as many migrations in parallel as possible, so that each migration will take place with the minimum possible delay. Although the overhead can be heavy during the migration time, the migration time is fairly short, and has little impact on the overall performance calculated by (4).

4 Experimental Evaluation

In this chapter, some experiments are designed and made to evaluate the performance of a GA-based resource consolidation approach. The experimental results are also compared with three existing strategies, and Entropy [5], First Fit Decreasing (FFD) heuristic and Best Fit Decreasing (BFD) method. Experimental environment is illustrated as follows. To set up a whole generic Cloud computing environment (i.e. IaaS) needs to consider many phases and integrate large-scale devices. However, it is very difficult to establish a real Cloud computing environment. Therefore, we narrow the scope of the experiment and simplify the experimental environment by a way of simulations to evaluate the proposed heuristics algorithms. In this experiment, we put $err = 0.1\%$, $b = 1$ Gigabyte/second, $t_d = 0.5$ second, $\bar{e_c} = 50\%$ and $e_w = 4\%$. Assume every node is composed of 1024 Gigabytes memory and 4-core CPU. Besides, we set $R = 2$, $k = 1$ as the demanded CPU and $k = 2$ is denoted the demanded memory of VM. Randomly producing $r_k, 1 \leq k \leq R$. On the other hand, we assume that experiment on single server sprawl situation so take low resource utilization situation as the case study and the fixed bandwidth in a WAN environment. In Fig. 3, x-axis denotes the number of VMs (M) and the number of active nodes (N) and the ratio is 2.

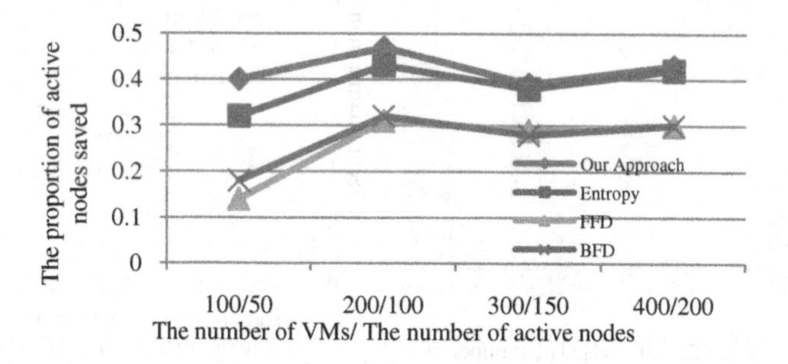

Fig. 3. Saving performance of active nodes

And y-axis denotes the proportion of active nodes saved. Each line represents the results of four resource consolidation. The blue line represents the results of our approach. The red line represents the results of Entropy. The green line represents the results of BFD. The purple line represents the results of FFD. Our approach limits the time below one minute. Entropy limits the time below one minute. As in Fig. 3, we can find that the slope of curve in our scheme is larger than others. The BFD and FFD are lower because they consider only where to allocate the VMs. And the Entropy is lower than our approach because entropy algorithm is using Depth-First Search. When the big amount of data is input, the computation time is larger.

In Fig. 4, x-axis denotes the number of VMs(M) and the number of active nodes (N) and the ratio is 2. And y-axis denotes migration time in seconds. The blue block represents the results of our approach. The red block represents the results of Entropy. The green block represents the results of BFD. The purple block represents the results of FFD. And y-axis denotes the number of migrated steps. The blue block represents the results of our approach. The red block represents the results of Entropy. The green block represents the results of BFD. The purple block represents the results of FFD. Our approach limits the time below one minute. Entropy limits the time below one minute. In Fig. 5, the method of migration of four methods is implemented by our approach to solve VMMP. The FFD and BFD are larger than others because it only considers the size of VMs. And FFD and BFD cause larger number of migrated steps as shown in Fig. 4. We can find that the slope of curve in our scheme is almost lower than others when the number of VMs increases in Fig. 4.

In Fig. 5, x-axis denotes the number of VMs (M) and the number of active nodes and the ratio is 2. And y-axis denotes the cost of migrations. Each line represents the results of four resource consolidation. The blue line represents the results of our approach. The red line represents the results of Entropy. The green line represents the results of BFD. The purple line represents the results of FFD. In Fig. 5, we can find that the slope of curve in our scheme is larger than others. The BFD and FFD are lower because they only consider size of VMs without considering resource utility.

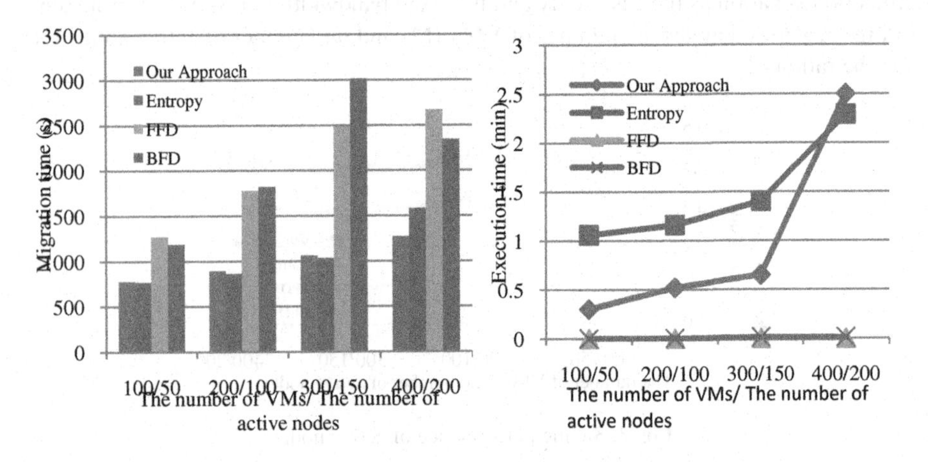

Fig. 4. Migration time in seconds

Fig. 5. Execution time in minutes

5 Conclusion

We provide an efficient resource consolidation method based on genetic algorithm for an important and complex problem encountered in cloud computing environments, called virtual machine placement problem (VMPP) and virtual machine migration problem (VMMP). First, in this work, we consider the resource utility by modeling the VMPP and VMMP. Thus, the problems can be easily solved through genetic algorithm and efficiently maintained the overall performance. In addition, our approach has been done that the architecture and preliminary design of proposed approach for further system implementation in Cloud environment. Finally, the experimental results show that we obtain a better consolidation of active nodes than other existing approaches.

Acknowledgement. The authors would like to express their sincere appreciation for the financial support from National Science Council, Ministry of Economic Affairs and Southern Taiwan Science Park (STSP) under the reseach grants of 102-2218-E-151 -005 - , 101-2221-E-006 -259 -MY3, 102CC02 and 101-EC-17-A-02-S1-222.

References

[1] Armbrust, M., Fox, A., Griffith, R., Joseph, A.D., Katz, R., Konwinski, A., Lee, G., Patterson, D., Rabkin, A., Stoica, I., Zaharia, M.: A view of cloud computing. Communications of the ACM 53(4), 50–58 (2009)

[2] Bobroff, N., Kochut, A., Beaty, K.: Dynamic placement of virtual machines for managing SLA violations. In: Proceeding of IEEE International Symposium on Integrated Network Management, Germany, pp. 119–128 (2007)

[3] Duncan, D., Chu, X., Vecchiola, C., Buyya, R.: The structure of the new ITfrontier: Cloud computing part II (2009), `http://texdexter.wordpress.com/2009/12/21/cloud-computing`

[4] Han, Y.: On the clouds: A new way of computing. Information Technology & Libraries. Chicago 29(2), 87–92 (2010)

[5] Hermenier, F., Lorca, X., Menaud, J.-M., Muller, G., Lawall, J.: Entropy: A consolidation manager for clusters. In: Proceeding of ACM SIGPLAN/SIGOPS International Conference on Virtual Execution Environments, March 11-13 (2009)

[6] Khanna, G., Beaty, K., Kar, G., Kochut, A.: Application performance management in virtualized server environments. In: IEEE Symposium on Network Operations and Management, pp. 373–381 (2006)

[7] Grit, L., Irwin, D., Yumerefendi, A., Chase, J.: Virtual Machine Hosting for Networked Clusters: Building the Foundations for Autonomic Orchestration. In: Proceedings of International Conference on Virtualization Technology in Distributed Computing, p. 7 (November 2006)

[8] He, L., Zou, D., Zhang, Z., Yang, K., Jin, H., Jarvis, S.A.: Optimizing resource consumptions in clouds. Grid Computing (2011)

[9] Moghaddam, F.F., Cheriet, M.: Decreasing live virtual machine migration down-time using a memory page selection based on memory change. In: Proceeding of International Conference on Sensing and Control (ICNSC), April 10-12, pp. 355–359 (2010)

Problems of SUMO-Like Ontology Usage in Domain Modelling

Bogumiła Hnatkowska, Zbigniew Huzar, Iwona Dubielewicz, and Lech Tuzinkiewicz

Wrocław University of Technology, Institute of Informatics, Poland
{bogumila.hnatkowska,zbigniew.huzar,iwona.dubielewicz, lech.tuzinkiewicz}@pwr.wroc.pl

Abstract. Ontologies are increasingly used, especially in the early stages of software development. It is widely believed that the use of ontologies has positive impact on the quality of the final software product. The aim of the paper is an introductory analysis of possible mapping between SUMO-like ontologies and domain models expressed in UML as a commonly accepted modelling language in software development. The main contribution of the paper is identification of basic problems within this mapping.

Keywords: SUMO ontology, domain modelling, UML, MDA.

1 Introduction

The notion of ontology is used in different contexts. In philosophy [5], ontology is the study of the nature of being, the basic categories of being and their relations. In software engineering [12], ontology is understood as a representation of domain knowledge in the form of a set of concepts within the domain, the properties of the concepts and interrelations of those concepts.

Ontologies are more and more used in software development [1], especially at the initial stages of the development, i.e. business and conceptual modelling. It is generally believed that the use of ontology in these stages improves the quality of communication between the participants of the development process, can reduce the time needed to determine user's requirements, and thus affects the cooperation [7]. The cumulative effect of the above mentioned elements has influenced on the quality of the final software product.

The motivation for this paper is the problem how to use the knowledge included in specific ontologies (which are based on high-level ontology) in modelling of application domain. The problem is worth considering as the application of ontologies brings many benefits. Among others, ontology [11], [12]: delivers a common vocabulary for communication among stakeholders; may be used as a conceptual schema of a relational data-base; enables reusing of a knowledge base.

In order to more specifically describe our considerations, we have concentrated on SUMO (Suggested Upper-Merged Ontology) as one from numerous high-level

N.T. Nguyen et al. (Eds.): ACIIDS 2014, Part I, LNAI 8397, pp. 352–363, 2014.

ontologies (Cyc, UMBEL, and General Formal Ontology etc.) [11]. We have chosen SUMO because there are a large number of specific SUMO ontologies. Specifically, the aim of the paper is an introductory analysis of possible mapping between SUMO-like ontologies and domain models expressed in UML being a commonly accepted modelling language in software development. The main contribution of the paper is identification of basic problems in the mapping between SUMO and UML. Due to the size limitation, we restrict our discussion to the mapping of structural aspects only. This structural aspect is modelled in UML in the form of class and object diagrams.

First, in Section 2, we present basic notions related to ontologies and their classification. In Section 3 we discuss the general relationship between ontologies and domain models represented in UML. In the core Section 4 the most important problems on the way of transformation of SUMO models into UML models are discussed. The consideration is illustrated by simple examples coming from SUMO. Finally, Section 5 concludes the paper.

2 Ontology Components

In further, we apply the definition of ontology proposed in [6]: *An ontology is a formal (machine-readable), explicit (consensual knowledge) specification (concepts, properties, relations, functions, constraints, axioms are explicitly defined) of a shared conceptualization (abstract model of some phenomenon in the world).* It is coherent with the definition given in [12]: *ontology formally represents knowledge as a set of concepts within a domain, using a shared vocabulary to denote the types, properties and interrelationships of those concepts.* According to both definitions, ontology can include the following components:

- individuals: instances or objects (the basic or "ground level" objects);
- classes: sets, collections, concepts, types of objects, or kinds of things;
- attributes: aspects, properties, features, characteristics, or parameters that objects (and classes) can have;
- relations: ways in which classes and individuals can be related to each other;
- function terms: complex structures formed from certain relations that can be used in place of an individual term in a statement;
- restrictions: formally stated descriptions of what must be true in order for some assertion to be accepted as an input;
- rules: statements in the form of an "if-then" (antecedent-consequent) sentence that describe the logical inferences that can be drawn from an assertion in a particular form;
- axioms: assertions (including rules) in a logical form that together comprise the overall theory that the ontology describes in its domain of application.

Nowadays, there are many attempts to define formally ontologies of different level of abstractions – from high-level ontologies up to specific low-level ontologies related to specific domains, e.g. medicine, banking, communication etc. In general, ontologies are grouped in three categories [2]: high-level (e.g. Penman Upper Level, Cyc), mid-level (e.g. Knowledge Representation Ontology, Linguistic Ontologies, and Engineering Ontologies), and low-level (or domain specific) (e.g. Modelling Enterprise-Enterprise, TOVE) ontologies. Low-level ontologies are developed on the base of higher-level ontologies; they use notions defined in higher-level ontologies.

Among the high-level ontologies SUMO ontology is the most widespread. The SUMO was created by merging publicly available ontological content into a single, comprehensive, and cohesive structure. The ontology is formally defined in declarative language SUO-KIF (Standard Upper Ontology Knowledge Interchange Format) [3], [4].

3 Models in Object-Oriented Software Development

Model-based software development is a modern paradigm of computer system design. Model-driven architecture (MDA) is an expression of this paradigm [8]. In MDA the software development cycle is perceived as sequential elaboration of Computer Independent Model (CIM), Platform Independent Model (PIM), and finally Platform Specific Model (PSM) which is the specification of the software code. The CIM model, considered as the starting point to software development, strongly bases on deep understanding and precise description of the domain for which a future computer system is intended. Domain knowledge is represented by a corresponding ontology. We have observed that ontologies play essential role in business modelling. As mentioned in the Introduction, the presence of ontology has a positive impact on the quality of the domain models.

Object-orientation is the second paradigm in modern software development. Unified Modelling Language (UML) is a basic general-purpose modelling language in the field of software engineering [10]. There are many software development methodologies supported by UML tools.

Primarily, UML is used for object-oriented modelling. A specific viewpoint in which we observe an interesting domain is the starting point to modelling. The viewpoint encourages looking at the given domain through the observed entities grouped in classes, their properties and relationships between them.

The notion of a class in UML, as opposed to some high-level ontologies, does not allow instances of the class to be collective entities. Two kinds of properties for the entities are discriminated: structural, called attributes, and behavioural, called operations. The classes with attributes may be compared to concepts introduced by upper-level ontologies, however in business modelling the attributes of classes are usually assumed to be the concepts that represent sets of values. The values may be simple (elementary) or (hierarchically) composite. Operations are considered as functions over some sets of values.

The set of observed entities from an interesting domain with similar properties and behaviours are grouped into sets called classes. The class C is defined as a pair $<Att_C, Op_C>$, where: Att_C is a set of attributes; each attribute with the name has its data type; Op_C is a set of operations.

A given class C represents a set of entities from the interesting domain $Dom(C)$. The entities belonging to the domain of a given class are called class instances, or business objects, or shortly objects. The attributes of an entity are characterized by their valuation. Note that instances of classes have identity in contrast to instances of data types that have no identity.

Following the general description of ontology components, we try to give more precise, partially formal description of domain models that are expressed in UML.

A domain model *DM* is defined as:

$$DM = \langle Cl, Ins, RIns, RTCl, RAss, RTAss, Con \rangle$$

where:

- Cl is a set of business classes; each class $C \in Cl$ represents a set of its instances (business objects) that are real or abstract entities; an instance of a class has its identity and is discriminated from other instances of the class;
- Ins is a set of selected instances from the set of classes Cl;
- $RIns$ is an instantiation relation of the signature $RIns \subseteq Ins \times Cl$; the relation assigns each instance a class (a type of instance);
- $RTCl$ is a partially ordered taxonomic relation of the signature $RTCl \subseteq Cl \times Cl$; if $\langle C_1,C_2 \rangle \in RTCl$ it means that the superclass C_2 is more general then the subclass C_1, i.e. $Dom(C_1) \subseteq Dom(C_2)$;
- $RAss$ is a family of relations of the following signatures $Ass(C_1,\ldots,C_n) \subseteq Dom(C_1) \times \ldots \times Dom(C_n)$ for $C_1,\ldots,C_n \in Cl$, and $n \geq 2$; relations in the family are called associations; the most often used are binary associations;
- $RTAss$ is a partially ordered taxonomic relation of the signature $RTAss \subseteq RAss \times RAss$; if R_1 and R_2 are relations of the same signature, then $\langle R_1,R_2 \rangle \in RAss$ means that the relation R_2 is more general then the relation R_1, i.e. $R_1 \subseteq R_2$;
- Con is a set of constrains; a constraint is represented in the form of a formula in first-order logics; the terms of the logics are instances of classes Cl, classes' attributes and instances of relations $RTCl$ and $RTAss$.

The structure of the domain model follows the components listed for ontology: classes with attributes, instances, some relations. Restrictions, rules and axioms are represented by constraints. Including constraints to the definition of domain model is justified by the fact that constraints are in common use in business modelling, e.g. in form of business rules. Constrains relate to some aspect of business domain that is intended to assert business structure or to control or influence the behaviour of the business [9]. They may be expressed in the Object Constraint Language (OCL) which is an inherent part of the UML standard. The OCL is a formal language for describing constraints applying to UML models.

4 Problems of Mapping between SUMO and UML

A UML business model, which is the first step in a software process development, describes usually a fragment of an application domain for which a future software system is to be built. In further, we assume that specific ontologies related to the application domain may describe this domain only partially. It means that only some domain concepts have their counterparts in the ontology. Therefore, most often, the specific ontology may support construction of the UML business model only partially. The Fig. 1 is a graphic representation of the discussed relationships under assumption that only one specific ontology is considered.

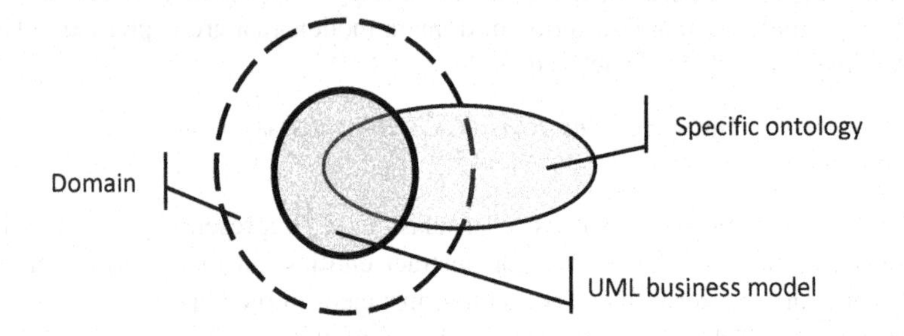

Fig. 1. Domain ontology-based modelling approach

The Fig. 1 reminds us that we faced with two problems. The first one deals with moving the knowledge included in the specific ontology to the UML business model or with verification of the business model against the ontology. The second one is how to extend the specific ontology using the elaborated UML business model. In further, our consideration is related to the first problem only; more accurately, we discuss how to transform basic components of SUMO-like ontology into UML.

The aim of this subchapter is to identify what elements of SUMO can be directly translated to the UML diagrams (mainly class diagrams) and how, and to identify potential mapping problems. To make the discussion clear we illustrate our ideas with examples coming from specific SUMO ontologies.

Class diagram is the main UML diagram used for describing entities, entities' properties, and relationships among them. The way of a class diagram instantiation can be constrained with the use of OCL which is able to express first-order logic constructs (SUMO allows to express also higher !ogic rules). The instances of UML classes and associations are represented in UML object diagrams.

SUMO entities form multi-level structure which main elements are presented in Fig. 2 as UML class diagram. Every SUMO element can be related to many others with the use of generalization relationship (can have many children and many parents). The generalization relationship between SUMO entities is expressed with *subclass* binary predicate, e.g. "(subclass Abstract Entity)". The domain for this binary predicate is defined as: *SetOrClass* x *SetOrClass* (Merge.kif) [11], where *SetOrClass* represents any subclass of *Entity* (with the *Entity* itself).

SUMO entities are partitioned into two basic groups: *Physical* and *Abstract*. Physical entities differ from abstract ones as they need to have location in space and time.

Entities that can be directly mapped to UML class diagram elements are filled in white in Fig. 2. From physical entities we are especially interested in structural elements, i.e. *Objects*. Among abstract entities, we are especially interested in *Classes*, *Attributes*, *Quantities*, and *Relations*. Other kinds of entities could be expressed with UML notions (e.g. *Graph*, *GraphElement*) but the translation is not strict.

Relations are further specialized, what is shown in Fig. 3 with the use of UML class diagram.

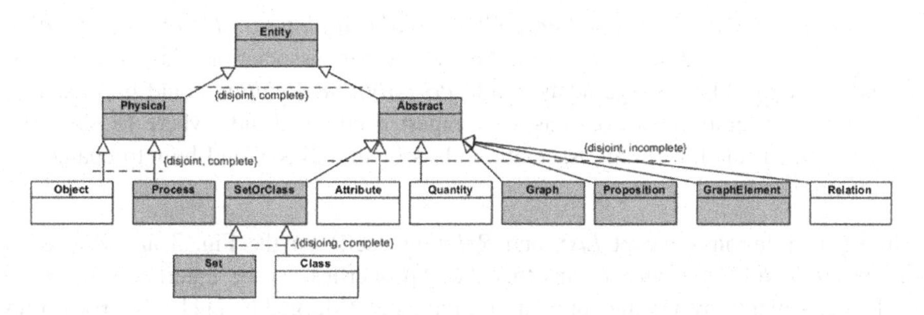

Fig. 2. Part of SUMO subclasses hierarchy tree expressed in UML; prepared on the base of Merge.kif [11][1]

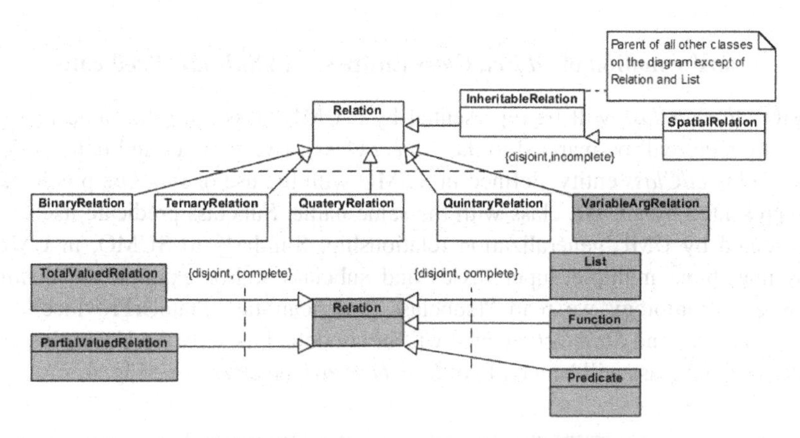

Fig. 3. Part of hierarchy of SUMO relations expressed in UML; prepared on the base of Merge.kif [11]

SUMO Relations are grouped into three generalization sets:

- The first generalization set takes into account "arity" of relations and include *BinaryRelation*, *TernaryRelation*, …, *VariableArgRelation*. It is incomplete in UML sense. Most of relations from this group (except *VariableArgRelation*) can be transformed to UML (the details are described below).
- The second generalization set serves for denoting sets of ordered n-tuples with the use of *Function*, and *Predicate*, as well as particular ordered n-tuples (*List*). SUMO predicates in UML are expressed at instance level, represented by UML object diagrams, not at conceptual level, represented by UML class diagrams. SUMO function can be expressed in OCL or as a UML class operation; however, to rewrite SUMO functions to UML collaboration with a domain expert is needed.
- The third generalization set allows defining relations which take from 1 to n-1 entities as arguments and assign a value to the last n-th relation argument (*TotalValuedRelation*), and distinguishes them from functions, which produce a value of

[1] All UML diagrams were prepared with Visual Paradigm tool.

a specific range for *n* arguments (*PartialValuedRelation*). *TotalValuedRelation* could be modelled in UML as a function in an association class linking the classes being relation arguments, while *PartialValuedRelation* could be modelled as a class operation. In both cases an expert needs to decide where to place the transformed relation (in an association class or in a class), and how to change its signature.

Almost all relations (except *List*, and *Relation* itself) on the Fig. 3 are subclasses of *InheritableRelation* what means that they "properties can be inherited downward in the class hierarchy via the subrelation predicate" (Merge.kif) [11]. The interesting inheritable relation is *SpatialRelation* including metrological and topological relations.

4.1 Transformation of *Object, Class* Entities, and *Subclass* Predicate

SUMO *Object/Class* will be represented by a UML class with the same name (e.g. SUMO *Object* will be mapped to the *Object* class). Every direct and indirect descendant of *Object/Class* entity, defined in SUMO with the use of *subclass* predicate, will be represented by a UML class with the same name. Subclass predicate itself will be represented by UML generalization relationship. Similarly to SUMO, in UML one class may have multiple superclasses and subclasses. For example, in *Countries-AndRegions* ontology we read "(subclass AmericanState StateOrProvince)". Both, *AmericanState*, and *StateOrProvince* will be modelled as separate UML classes, and *AmericanState* class will be a child of *StateOrProvince* class – see Fig. 4.

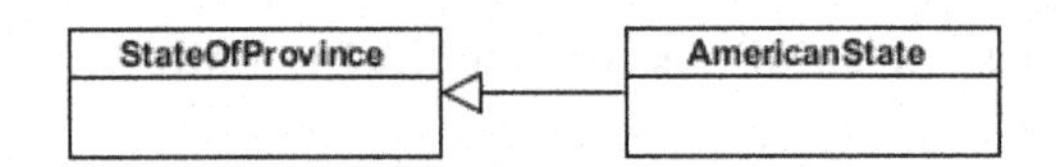

Fig. 4. Exemplary transformation of *subclass* predicate

4.2 Transformation of *Quantities*

Quantity represents "how many" or "how much" with the usage of two subclasses: *Number*, and *PhysicalQuantity*. Both can be represented by UML data types, e.g. primitive types, enumerations, or complex types with own properties and operations.

4.3 Transformation of *Instance* Predicate

In SUMO we deal with instances of different entities, among them we find objects and relations. Both are treated in SUMO in the same way, and defined as "an object is an instance of a set or class if it is included in that set or class" (Merge.kif) [11]. But in UML instances of SUMO objects are represented differently than instances of SUMO relations.

An instance of SUMO object will be translated to an instance of UML object diagram, e.g. SUMO expression: "(instance California AmericanState)" from *RegionsAndCountries* ontology could be transformed to a UML object California: AmericanState. An instance of SUMO relation has its domain defined in terms of SUMO classes. So, it can be represented at conceptual level as a UML association – for details see next subchapter.

4.4 Transformation of *BinaryRelation* … *QuintaryRelation* Relations, and *Subrelation* Predicate

A SUMO relation instance with a fixed number of arguments will be represented by a UML association having the name of relation and the same number of ends as the relation. Fortunately, UML allows defining n-ary associations. To make UML diagram more readable, the association ends can be labelled in the way that corresponds to the definition of SUMO relation domain, e.g. end1, end2, etc.

For example SUMO specification of property predicate looks as follows: "(domain property 1 Entity) (domain property 2 Attribute)(instance property BinaryPredicate)" (Merge.kif) [11]. This specification will be transformed to a UML binary association linking *Entity* and *Attribute* classes, and represented in a UML class diagram – see Fig. 5. The association end near to *Entity* end will be labelled with end1, and the opposite end next to *Attribute* class will be labelled with end2. The multiplicity at both ends will be defined as "*". Possible constraints on multiplicities can be concluded from SUMO rules but it needs a domain expert collaboration. The fact, that property is an instance of *BinaryPredicate* can be marked with the use of UML tagged value.

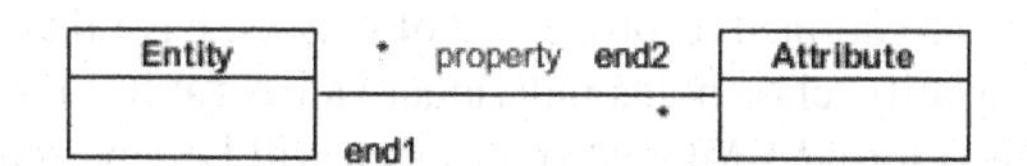

Fig. 5. Exemplary transformation of *property* predicate

Below there is an example of translating an instance of a ternary relation coming from *Food* ontology [11]: "(instance ingredientAmount TernaryRelation) (domainSubclass ingredientAmount 1 SelfConnectedObject) (domainSubclass ingredientAmount 2 SelfConnectedObject) (domain ingredientAmount 3 PhysicalQuantity)". This SUMO relation will be represented in UML as ternary association named *ingredientAmount* with 3 ends: (a) with *SelfConnectedObject* class labelled *end1*, (b) with *SelfConnectedObject* labelled *end2*, (c) with *PhysicalQuantity* class labelled *end3*. All ends with the multiplicity "*".

All above mentioned SUMO relations are *InheritableRelations.* It means that it is possible to define their subrelations, e.g. "(subrelation superficialPart part)". The domain of subrelation can remain the same as for parent or be refined. In the case of *superficialPart* the domain is unchanged. Additionally, *superficialPart* is also an instance of *IrreflexiveRelation* and *TransitiveRelation*. These facts can be modelled in UML by tagged values.

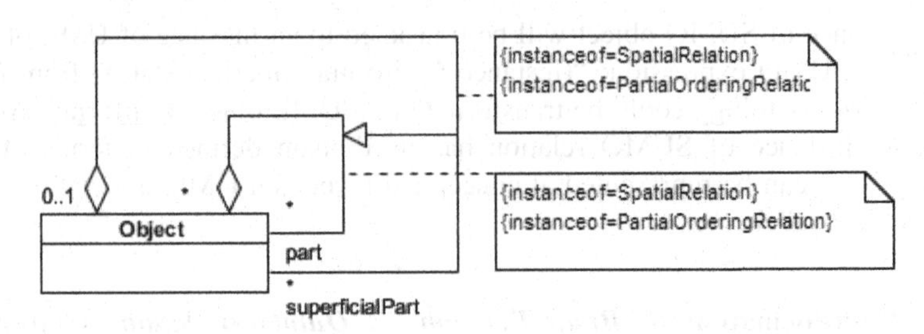

Fig. 6. Exemplary transformation of *subrelation* predicate

Subrelation predicate will be represented as a UML association linking UML classes corresponding to the subrelation domain. This association will inherit from the association representing the parent of subrelation – see Fig. 6 for an example.

4.5 Transformation of *Attribute* Entity, and *Property* and *Attribute* Predicates

Attributes in SUMO are defined as "qualities which we cannot or choose not to reify into subclasses of objects" (Merge.kif) [11]. They are partitioned into *InternalAttributes, RelationalAttributes*, and *PerceptualAttributes*, and further specialized, e.g. *PhysicalAttribute* is a subclass of *InternalAttribute.*

In SUMO an instance of an attribute is bound to an instance of entity with the use of *property* binary predicate. It is also possible to bind an instance of attribute to an instance of *Object* with the use of *attribute* binary predicate, which is a subrelation of *property* predicate, e.g. "(attribute MyLittleRedWagon Red)".

In UML attributes are defined as properties of classes, have their names and types. Every instance of a specific class holds individual values for its class attributes.

An attribute defined for a SUMO entity instance will be represented by a UML attribute in the class representing the type of the entity instance. However, the transformation from SUMO to UML can't be done automatically, as we need to establish both the name of an attribute and its type. For example transformation of: "(attribute MyCar Red) (instance MyCar Car)(instance Red Color)" can resulted in one UML class called *Car* with one attribute *CarColor: Color*. The valuation of the *CarColor* attribute (Red) will be represented together with an entity instance *MyCar* on a UML object diagram.

Opposite to UML, attributes are not very popular in SUMO. There are many ontologies not defining attributes at all.

4.6 Transformation of *Part* Predicate

In SUMO, two Objects can be related with so called part relation. SUMO expression "(part x y)" means that x is a part of y. The relation is an instance of *SpatialRelation*, and "is the parent of other metrological relations" (Merge.kif) [11]. The relation is *ReflexiveRelation* meaning that every object is a part of itself.

Part relation will be represented by UML aggregation. The association end representing parts will be labelled "part". Selected subrelations of *part* relation, depending of their nature, will be represented by UML composition (e.g. component relation).

4.7 Transformation of Axioms

Axioms form most of SUMO low-level ontologies. They typically represent facts from the domain expressed at instance level, e.g. "(part Germany Europe)" in *CountriesAndRegions*. Such axiom can be represented in a UML object diagram as a link between *Germany* and *Europe* instances.

4.8 Transformation of Rules

Rules are very important part of each SUMO ontology. They introduce constraints on instances and their relationships. They can be considered as a specific business rules in a domain ontology. An exemplary rule coming from Merge.kif is given below:

```
(<=>
 (properPart ?OBJ1 ?OBJ2)
 (and
    (part ?OBJ1 ?OBJ2)
    (not
         (part ?OBJ2 ?OBJ1))))
```

We can read this rule as "obj1 is a proper part of obj2 if and only if obj1 is a part of obj2 and obj2 is not a part of obj1".

Such rules can't be directly expressed in UML, but they can be translated into OCL by an experienced software engineer.

4.9 SUMO to UML Mapping Problems by Example

As was mentioned at the beginning of this chapter, SUMO ontology is very big (over 25.000 terms and 80.000 axioms) [11] what brings many problems, when we try to represent specific ontology in software development process. The main problem is that any low-level ontology in SUMO can freely use concepts from other ontologies, especially upper-level and mid-level.

For example, upper-level ontology defines part binary relation between instances of two *Objects* (Merge.kif). This relation is used in *CountriesAndRegions* ontology for instances of *Object* subclasses, e.g. "(part Paris France)", where *Paris* is an instance of *EuropeanCity*, and *France* is an instance of *EuropeanNation*. The inheritance hierarchy of SUMO entities involved in this example is very deep (almost all classes are defined in Merge.kif) – see Fig. 7 – and difficult to interpret, as part relation can be defined between two instances of descendants of *Object* class, provided they fulfil constrains expressed by rules. What interesting, in *CountriesAndRegions* ontology there is no rule expressing the fact that *EuropeanCity* is a part of *EuropeanNation*. Instead we found two rules, expressing the facts that *EuropeanCity* is a part of *Europe* (instance of *Continent* class), and that *EuropeanNation* is a part of *Europe*.

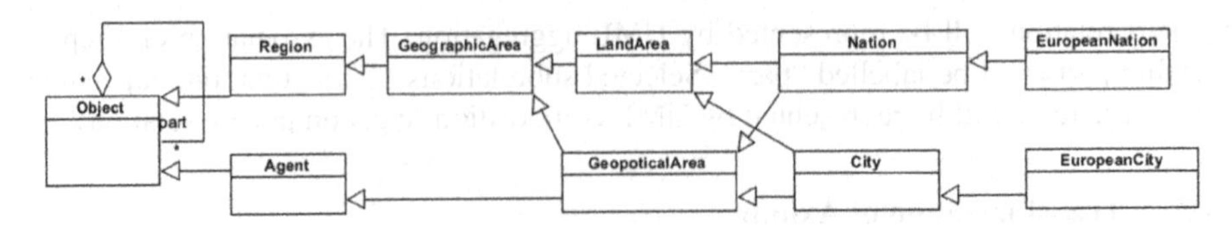

Fig. 7. Complexity of SUMO ontology by simple example

The result of direct transformation of SUMO specific ontology to a UML class diagram seems to be unacceptable. The resulting class diagram leaves to many freedom of its possible instantiation. For example, it is possible to associate two city instances with a part relationship.

The important relationships from specific ontology should be defined directly and clearly in UML model. These relationships, of course, can refine relationships defined at general level, and play the role of SUMO rules, constraining the way of possible UML class diagram instantiation. For example, the natural way of modelling in UML the facts that *City* is a part of *Nation* or that *EuropeanCity* is a part of *European-Nation* is to use UML redefinitions of UML properties – see Fig. 8.

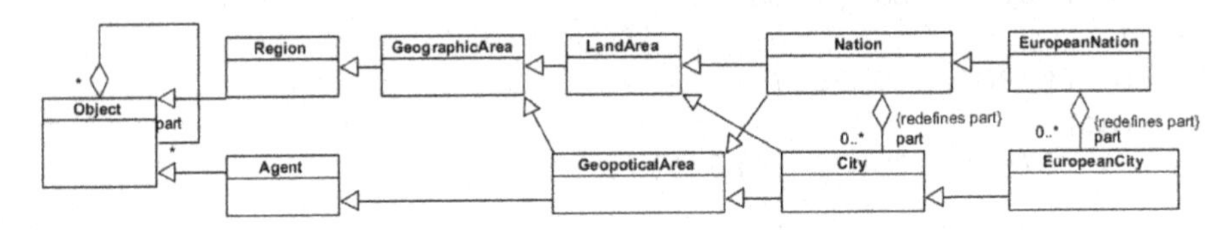

Fig. 8. Usage of UML redefinitions to improve the model

5 Conclusions

Nowadays, model-driven software development intensively exploits UML-like languages. So, a pragmatic question arises: how the knowledge included in different ontologies may be effectively extracted and presented in the form of UML models.

The use of ontologies in software development process allows both to support the steps involved in modelling a part of reality, as well as, to verify existing models to detect inconsistencies, duplicates or non-compliance of domain concepts.

In the paper we presented preliminary analysis of possible mapping problems between SUMO-like ontologies and domain models expressed in UML. The analysis was informally illustrated by simple examples. On the base of the analysis carried out, we have concluded that mapping from SUMO to UML models may be only partially automated. The reason is a different modelling approach in SUMO and UML. In contrast to UML, which operates separately on classes (class diagrams) and objects (object diagrams), SUMO uses classes together with objects, which play the role of specific constraints. In result, we can automatically transform some part of SUMO ontology into a UML class diagram, but the diagram has to be supplemented by constraints, which have to be deduced from an analysis of classes and objects in SUMO. This analysis may involve a domain expert.

SUMO can be used to create models at the level of the CIM and PIM levels of MDA approach.

Other potential uses of ontologies in the modelling include:

- selection of the conceptual models from a set of proposed (candidate) models in the context of compliance with the modelled area (an ontology),
- generation of conceptual models based on a specific ontology by projecting and mapping ontology elements in the context of defined application functional requirements.

  The problem of the mapping between SUMO and UML is also interesting in the context of such issues as:
- creation of an ontology based on a set of verified data models (specified at different levels of abstraction),
- selection of the most complaint ontology or ontologies for a considered legacy model.

References

1. Gašević, D., Djurić, D., Devedžić, V.: Model Driven Arcitecture and Ontology Development. Springer (2006)
2. Gómez-Pérez, A.: Ontological Engineering, http://www.imamu.edu.sa/topics/IT/IT%206/Ontological%20Engineering.pdf
3. Genesereth, M., Fikes, R.: KIF: Knowledge Interchange Format. Version 3.0. Reference Manual. Report Logic-92-1. Computer Science Department. Stanford University, CA (1992)
4. Pease, A.: Standard Upper Ontology Knowledge Interchange Format (2004), http://ontolog.cim3.net/file/resource/reference/SIGMA-kee/suo-kif.pdf
5. Random House Kernerman Webster's College Dictionary. Random House, Inc. (2010)
6. Studer, R., Benjamins, V., Fensel, D.: Knowledge Engineering: Principles and Methods. Data and Knowledge Engineering 25, 161–197 (1998)
7. Said Cherfi, S., Ayad, S., Comyn-Wattiau, I.: Improving Business Process Model Quality Using Domain Ontologies. J. Data Semant. 2, 75–87 (2013)
8. MDA (2001), http://www.omg.org/mda/specs.htm
9. OCL 2.3 (2012), http://www.omg.org/spec/OCL/
10. UML 2.4 (2011), http://www.omg.org/spec/UML/
11. http://www.ontologyportal.org/
12. http://en.wikipedia.org/wiki/Ontology/

Implementation of Emotional-Aware Computer Systems Using Typical Input Devices

Kaveh Bakhtiyari[1,2], Mona Taghavi[1], and Hafizah Husain[1]

[1] Department of Electrical, Electronics and Systems Engineering,
Faculty of Engineering and Built Environment,
Universiti Kebangsaan Malaysia (The National University of Malaysia)
43600 UKM Bangi, Selangor Darul Ehsan, Malaysia
[2] Department of Computer & Cognitive Science, Faculty of Engineering
University of Duisburg-Essen
47048 Duisburg, North Rhine-Westphalia (NRW), Germany
academic@bakhtiyari.com, mona@siswa.ukm.edu.my,
hafizah@eng.ukm.my

Abstract. Emotions play an important role in human interactions. Human Emotions Recognition (HER - Affective Computing) is an innovative method for detecting user's emotions to determine proper responses and recommendations in Human-Computer Interaction (HCI). This paper discusses an intelligent approach to recognize human emotions by using the usual input devices such as keyboard, mouse and touch screen displays. This research is compared with the other usual methods like processing the facial expressions, human voice, body gestures and digital signal processing in Electroencephalography (EEG) machines for an emotional-aware system. The Emotional Intelligence system is trained in a supervised mode by Artificial Neural Network (ANN) and Support Vector Machine (SVM) techniques. The result shows 93.20% in accuracy which is around 5% more than the existing methods. It is a significant contribution to show new directions of future research in this topical area of emotion recognition, which is useful in recommender systems.

Keywords: human emotion recognition, keyboard keystroke dynamics, mouse movement, touch-screen, human computer interaction, affective computing.

1 Introduction

Emotional Intelligence can recognize human emotions and respond to the user accordingly. Human Emotions Recognition (HER) might be used in different categories such as e-learning, game, adaptive user interfaces, etc. Emotion is one of the features which promotes human-computer interactions, and plays a significant role in making trust. If a recommender system can recognize the user's emotion, it would produce responses relevant to the user's emotional state. Consequently, it attracts the user's attention and loyalty. Emotion is a way of interaction which is about the message owner features. Emotions are discussed by three parameters. The first

N.T. Nguyen et al. (Eds.): ACIIDS 2014, Part I, LNAI 8397, pp. 364–374, 2014.

parameter is *Arousal* which shows the energy of feeling. The second parameter is *Valence*. Valence presents whether the feeling is a pleasure (positive) or displeasure (negative) in case of the energy. The third is *Dominance* which shows the strength of the feelings.

A classification of emotions is proposed by Plutchik, which is used as a standard classification in 8 emotions of *Acceptance, Fear, Surprise, Sadness, Disgust, Anger, Anticipation, Joy* [1].

HER has been done by various methods and techniques to achieve this goal. However, there are some challenges in different areas, which make it an open research topic to work through [1]. The first challenge is achieving a higher accuracy in emotion recognition with a reliable precision (lower false positive rate) [2]. Still the available techniques are not reliable and accurate enough to be employed in real applications. New methods and hypotheses can be applied to gain better results with a higher performance, thus new techniques are being introduced. The second challenge is the real time processing [3-5].

This research tends to perform a solution for human emotions recognition to address those mentioned problems. We have analysed the users' inputs on common input devices such as keyboard keystroke dynamics, mouse movements, and touch-screen interactions. Chang has tracked the individual pattern of keyboard, mouse and mobile device usage [6]. He showed that the users' patterns are unique, and it can be applied for security purposes. In addition, a hybrid analysis by combination of few input devices tries to perform a better performance in HER.

2 Critiques

The first issue is the recognition accuracy. Only facial expression recognition could achieve the highest recognition accuracy of 90% lately in 2012 by using image processing techniques. The other methods have still less performance which are not reliable in business applications. Facial expression recognition gained one of the best accuracies in emotions recognition. However, for real time processing, it works worse than the other methods, because in fact image processing techniques are time and resource consuming. Natural Language Processing (NLP) and common devices can be used for real time applications, but the resulted recognition accuracies in recent research are not satisfying. Some methods such as using EEG machines' signals are expensive and still those machines are not available to be used in a daily usage. There are other methods for HER such as using a microphone, camera and other input devices. However because of the security and privacy issues, many computers may not use microphones and cameras. These challenges cause to have the limited number of applications for facial expression recognition, body gestures recognition and voice processing.

3 Methodology

This paper is presenting a methodology based on a software prototype which records the data from user's inputs on mouse, keyboard and touch screen. Following this method, a prototype application has been designed and developed to collect the required data from computer users' interactions. The keyboard keystroke dynamics, mouse movements and touch screen interactions of 50 users with various cultural backgrounds were collected while they were using the system. These users were mostly settled physically in Malaysia, Germany and Iran. Every 4 hours for a month, users were asked to answer a question about their current emotions. Then the collected data were used in RapidMiner to be trained using the SVM technique for classification. For evaluation of the mouse and touch-screen interaction, the methodology presented by Schuller [7] has been used which collects all the mouse movements and mouse keystrokes.

A key question at the beginning was the selection of appropriate emotions which in this study should be considered. First, the seven universal emotions by Paul Ekman [2] have been used as a basis. Then emotions were clustered, and all investigations in this work are concentrated on the following four emotion categories:

- Neutral (includes above all the emotion happiness and as perceived normal mood)
- Fright (Afraid) (includes above all helplessness, confusion and surprise)
- Sadness (primarily sadness, anger and resentment)
- Nervousness (including nervous, fatigue and light-headedness)

3.1 Keyboard

Keystroke dynamics are habitual rhythm patterns by way of typing a word [5]. It has three major parts as *representation*, *extraction* and *classification*. Representation shows the input values as the words. When the user is typing, actually he is representing his identification. The next step is features extraction that the system extracts and defines the features as a fingerprint and records them in a database. The last section is a classification that matches the extracted features of a new user with the existing features in the database to identify him/her. Now, this research is using the similar method but there are differences by using the novel training algorithms to identify the emotions rather than the identification of the users.

There are three major and important features in keystroke dynamics: 1) *Key down-to-down*; 2) *Key down-to-up*; and 3) *Key up-to-down*.

The Keystroke Features were selected from the timing differences of single keystrokes, digraphs (two-letter combinations) and tri-graphs (three-letter combinations).

3.2 Mouse

It seems reasonable to divide the mouse movements in two different sections. The first section is the movement of the mouse without using the left mouse button pressed. The second section is where the mouse button is pressed.

This curve is then transformed into a 2-dimensional coordinate system. The ideal line corresponds to the x-axis of the coordinate system, and therefore y-axis describes a measure of the local variation of the mouse movement from the ideal line. Since these distance values have lost the absolute commitment to its original screen position, it can already measure global properties of the local mouse movement to be made. For example, the sum over all possible distance values states how much the mouse was moving on entire place above or below the ideal line. The properties which were studied are:

- The length of the racing line from start to end point
- The sum over all distance values
- The zero crossings
- Maximum deviation of the values
- Average of the individual values
- Standard deviation
- Variance
- Correlation function of the curve
- First order and second derivatives
- Min. and max. of the values
- Average amount over all values
- Standard deviation,
- Variance
- Autocorrelation function

Time Properties

In parallel with the above discussed features, the time intervals, which register with the result of a new (x,y) point are analysed. It should not be forgotten that only a change in the x or y coordinate of a new data value is read. This elapsed time between two consecutive points together is not only the total time the mouse moves, but the specified values describe information about the movement individually. It explains the time between jerky and slow; and also it can be used to distinguish verse breaks in existing movements very well. However, a complete overview is firstly presented of all the examined given features in Figure 1. This figure shows a possible sequence of values of time intervals, from which the main features are very well seen. This figure presents the time between the pressed keys. For instance, the time between the first and second character is the minimum in comparison with the time between 19th and 20th characters which is the maximum spent time for typing 25 characters. This figure is only a demonstration of a sample registered mouse keystroke time.

It is similar to the local variation, made and analysed with a number of time delta values. Then first two statements about the time relationships are possible:

✓ Total time of motion by summing over all values
✓ Average time distance between two points or the average required time.

However, when a change occurs to the location coordinates of the mouse movements, averaging is performed on the "Standard deviation of individual values" and "Variance of these values". Finally, the derived variables of mouse movements and keystrokes are:

- Correlation function
- First derivative
- Second derivative with the corresponding analysis

More precise statements are possible to be described. However, the formation of a distribution function of these values and the derived properties of this distribution function lead to the catalogue of the properties.

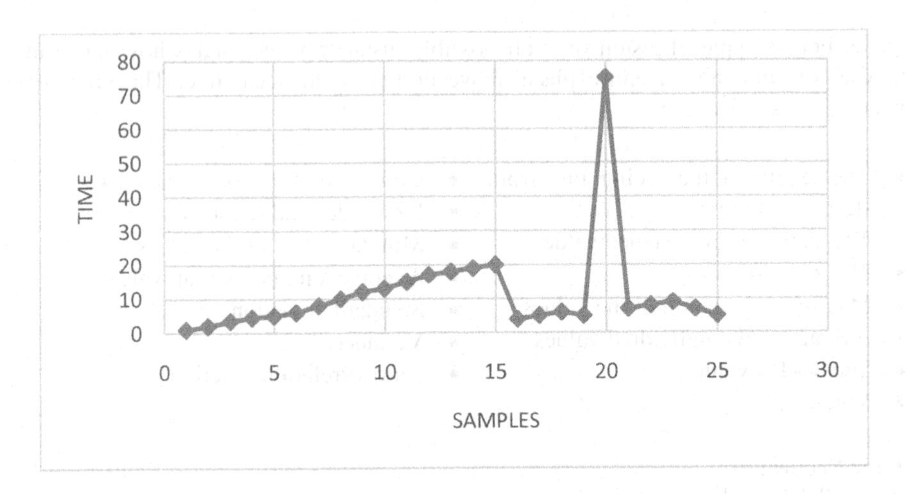

Fig. 1. Elapsed time between the modified coordinates of mouse movements

3.3 Touch Screen

The touch screen is only able to determine points in x, y and z coordinates. User interaction with touch screen monitors only result the changes in these coordinate values [7]. These values along with the time interval of changes are collected and prepared to measure the other important features such as velocity and the movement's details. Some other companies have introduced some new technologies which make the touch-screen monitor to react according to the user's eyes, hands and behaviours. All of these advanced technologies are the combination of image processing with the touch screen displays and AI techniques; and they are not directly related to touch-screen monitors. The most significant expansion was therefore to complement the additional available z-component (pressure strength), which has been evaluated in parallel. Thus, analogous reads the (x,y) coordinates of an initial set of z-values, where they open up a value range between 0 and 255. Straight from the emerging contours of the first and second derivatives as well as the correlation function, some additional values can be used to interpret better. These values include the average, minimum, maximum, standard deviation and variance of the first and second derivatives and the correlation function. By considering this number of features on touch-screen monitors, all the values are obtained from the Cartesian coordinate system. However, a three-dimensional coordinate space is presented. This can also offer a transformation in spherical coordinates (r, α, β).

4 Evaluation

This section demonstrates the diagnosis of the research based on the theories and methods of research methodology.

4.1 Evaluation Criteria

Evaluation of the system is based on the emotions recognition methods and machine learning techniques which have been used in the HER system. There are several criteria to evaluate and measure the performance of the system. These criteria are mainly composed of Classification/Recognition Accuracy, False Positive Rate, and Computational/Process Time. *Classification / Recognition Accuracy* shows how precise a system is able to recognize the emotions. It mostly focuses on the output of machine learning techniques. This criterion is measured by the machine learning classification methods. Generally, for this purpose, from 60% to 80% of the data would be trained, and then the rest of 20% to 40% of the remained data would be tested. *False Positive Alarm* gives some false classified emotions. These emotions are recognized but they are not matched with the actual recorded emotions. *Computational Time* is a classification procedure time to be applied to the collected data set. Different classifiers follow different algorithms, and they have different time complexities.

4.2 Data Analysis

Keyboard

The recognition performance is determined by using Support Vector Machine (SVM) as a classifier in term of classification accuracy and false positive rates. The number of mistakes (backspace + delete key) was calculated. There are many different methods to correct the mistakes, but it was not possible to catch all of the possible correction scenarios as keystrokes were collected from different computer applications. Outliers for all of the features that involved multiple keys were calculated to remove these pauses (e.g. digraph latency). Pauses were removed by considering the mean and standard deviation for all keystroke dynamic features, which they were 12 standard deviations greater than the mean for each individual participant [8].

This process has been considered in the prototype application while recording and collecting the data from users. The Kappa statistic indicates how much the classification rate was a true reflection of the model or how much chance/probability could be attributed to be succeeded.

Table 1. Keyboard keystroke dynamics classification of human emotions

Emotion	Accuracy %	Kappa	False Positive %
Nervousness	85.20	0.67	10.26%
Relaxation / Neutral	79.40	0.55	17.60%
Sadness	87.10	0.76	9.36%
Fright	91.24	0.68	4.35%
Anger	83.90	0.53	12.86%

Table 1 shows the classification results of human emotions based on the keyboard keystroke dynamics with their Kappa values and False Positive Rates. Fright emotion has the strongest classification of 91.24% with the least value of 4.35% for false positive rate.

Mouse

In the first detection process, features were selected. The best result was on Neutral emotion but in the other emotions, the outcome is less than 40%. Then in the second phase, the features were selected according to the Schuller [7] features, and it is far better than the first result as shown in Table 2. Despite in the *fright* (*afraid*) emotion the resulted percentage is weak; but in the other two emotions of *pensive* and *annoyed*, the results are much improved.

The collected data from our volunteers are evaluated and then analysed with our emotions. Table 2 summarizes the evaluation of all collected data set of 4 set of emotions. This evaluation was done based on the 2003 collected data vectors [9, 10]. The overall average of the correctly classified emotions is 0.866 with a mean variance of 0.075.

The correctly classified emotions by RapidMiner are the values at the junction of the same detected emotion with the intended emotion as shown in bold. The other values are called as false positive alarms, which are classified incorrectly. The increased classification rate for neutral emotion can be explained easily. The test subjects were accumulated primarily with expectations of neutral data vectors for the emotion. This probably is the most emotion which is felt to have been distributed over the days.

Table 2. Confusion matrix with the average values for mouse features classification

Intended Emotions	**Detected Emotions**			
	Neutral	Fright	Sadness	Nervousness
Neutral	**0.930**	0.022	0.027	0.023
Fright	0.203	**0.787**	0.040	0.010
Sadness	0.084	0.012	**0.912**	0.015
Nervousness	0.175	0.015	0.065	**0.835**

If a PC user has the emotion of *annoyed*, he moves the mouse usually very fast and also fixed with short presses on the mouse button. The properties of mouse movement are *fast* and *brief*. The system can certainly capture and analyse these features. In the short pressing the mouse pointer doesn't move often. Thus generally no movement is detected during the mouse click.

But a question still remains that why the precision of the detection is still low. Here is the answer by analysing the volunteers. At the time of working in different situations with the computer, they are not sure about their own emotions. When they are asked to input their emotions, they are rather unsure what kind of emotion they have at the moment.

Thus an insecure person presses a little longer and deliberates on the mouse, where they will lead, and the person did not intend slight movement of the cursor. Data analysis of the features is shown for the recognition of emotion. It is not very meaningful, and this is probably one of the reasons for the lower values in the confusion matrix.

Finally, it can be concluded that although the recognition of emotion with a reliable performance works, unfortunately the lack of standard hardware with significant qualities cause a lower accuracy. It would be very important to have several data collection periods to increase the strength of the data. It also brings more clarity about the emotions; and it enables better detection.

Touch Screen

Table 3 shows the final average results over all the test subjects. The overall confusion matrix with an overall average of 0.76 (76%) for the correctly classified emotions values shown in bold font is achieved. After the evaluation of the existing system for the detection of the four emotions, it can be concluded that this system can be used for emotion recognition with a reliable accuracy.

Table 3. Confusion matrix with the average values for touch screen

Selected Emotion	**Detected Emotion**			
	Neutral	Fright	Sadness	Nervousness
Neutral	**0.71**	0.321	0.090	0.022
Fright	0.000	**0.900**	0.073	0.000
Sadness	0.008	0.113	**0.893**	0.000
Nervousness	0.071	0.354	0.122	**0.553**

4.3 Hybrid Analysis

By the combined results of the keyboard, mouse and touchscreen, the accuracy in the *Fright (Afraid)* emotion is the best among the others. Neutral and Nervousness have the lowest result, and these two emotions have the greatest rate of confusion with each other. These results are tabulated in Table 4.

As it can be seen in Table 4, all of the four emotions have been detected more accurately by using all three input devices (Keyboard, Mouse and Touch Screen) analysis methods. Also in some cases, the error has been increased a little bit, but the increase of performance is much higher than the error rates.

Table 4. Confusion matrix with the average values of Keyboard, Mouse and Touchscreen

Selected Emotion	**Detected Emotion**			
	Neutral	Fright	Sadness	Nervousness
Neutral	**0.851**	0.121	0.076	0.022
Fright	0.001	**0.932**	0.082	0.010
Sadness	0.008	0.118	**0.921**	0.004
Nervousness	0.091	0.254	0.122	**0.650**

5 Conclusion

Previously, researchers tried to gain more accurate results on human emotions recognition. This research could gain higher accuracy in comparison with the other researchers who worked with keyboard, mouse and touch screen. Especially by

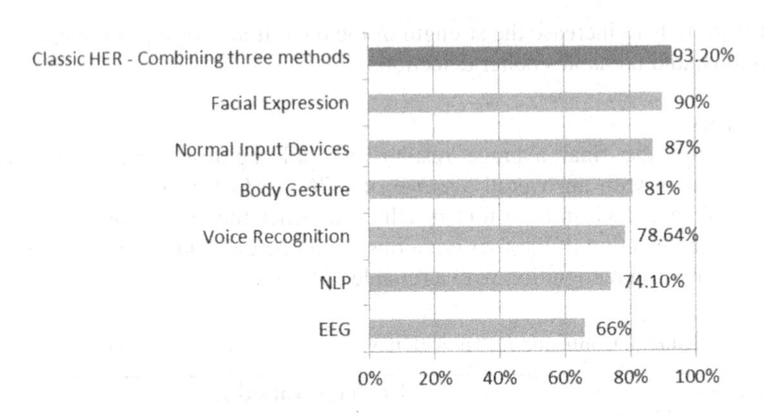

Fig. 2. Comparison of the best accuracies of different methods for HER

combining these three methods together, much better result of maximum 93.20% has been achieved, which is competitive with all other previous methodologies. Figure 2 compares the achieved results with the other methods and accuracies as discussed earlier in research literature.

The final best result of 93.20% among the achieved results of this research has been compared with the best results of the superior research on HER with different methods. In comparison only the best accuracies of the methods are considered. We cannot compare them based on the average accuracy of the emotions because of few problems. The problems are related to existing gap of the acquired accuracies, differences in number of emotions recognized, and different datasets for each research.

HER based on EEG has gained the accuracy of 66% in 2009 and 2010 [11, 12]. In 2008, *Lei et al.* got the result of 74.1% in the emotion of *anger* [13]. Voice processing in HER achieved the 78.64% of accuracy [14, 15].

HER systems based on body gesture recognition has been resulted in 81% by *Gunes & Piccardi* [16]. The most similar methods for this research has been done by *Milanova & Sirakov* and they gained 87% of emotions recognition [17]. The best competitive method is facial expression recognition which has improved a lot recently in 2012.

Among many researchers in facial expression recognition, *Konar et al.* and *Kao & Fahn* got 88.68% and 90% of accuracy respectively [3, 18]. Figure 2 shows that the method in this research has worked nearly from 5 to 6% better than the similar methods and more than 3% better than the superior method in facial expression recognition in 2012.

The evaluation criteria are used to consider for performing the evaluation. There were three main criteria for evaluation.

- The first criterion was recognition accuracy which is the most important item in the evaluation. The proposed methods of this research have been evaluated in terms of classification/recognition accuracy. Then at the end in Figure 2, they are compared with the similar research areas in measuring human emotions recognition accuracy.

- The second criterion was false positive rate which has been shown in every confusion table. However, the lack of enough information in the previous research papers, comparing the results of this study with the similar works was not possible.
- And the third criterion was computational/processing time. This is only related to the classification methods and the number of extracted features.

References

1. Cowie, R., Douglas-Cowie, E., Tsapatsoulis, N., Votsis, G., Kollias, S., Fellenz, W., Taylor, J.G.: Emotion recognition in human-computer interaction. IEEE Signal Processing Magazine 18(1), 32–80 (2001)
2. Ekman, P., Friesen, W.V.: Unmasking the face: A guide to recognizing emotions from facial clues. Malor Books (2003)
3. Konar, A., Chakraborty, A., Halder, A., Mandal, R., Janarthanan, R.: Interval Type-2 Fuzzy Model for Emotion Recognition from Facial Expression. In: Kundu, M.K., Mitra, S., Mazumdar, D., Pal, S.K. (eds.) PerMIn 2012. LNCS, vol. 7143, pp. 114–121. Springer, Heidelberg (2012)
4. Huang, T.: Audio-visual human computer interface. In: IEEE International Symposium on Consumer Electronics. IEEE, University of Illinois (2008)
5. Monrose, F., Rubin, A.D.: Keystroke dynamics as a biometric for authentication. Future Generation Computer Systems 16(4), 351–359 (2000)
6. Chang, M.: Iowa State engineers use keyboard, mouse and mobile device 'fingerprints' to protect data (November 18, 2013), `http://www.news.iastate.edu/news/2013/11/18/fingerprints`
7. Schuller, B., Rigoll, G., Lang, M.: Emotion recognition in the manual interaction with graphical user interfaces. In: IEEE International Conference on Multimedia and Expo. IEEE (2004)
8. Epp, C., Lippold, M., Mandryk, R.L.: Identifying emotional states using keystroke dynamics. In: Proceedings of the 2011 Annual Conference on Human Factors in Computing Systems, pp. 715–724. ACM, Vancouver (2011)
9. Kemp, F.: Applied Multiple Regression/Correlation Analysis for the Behavioral Sciences. Journal of the Royal Statistical Society: Series D (The Statistician) 52(4), 691–691 (2003)
10. Cohen, J.: Applied multiple regression/correlation analysis for the behavioral sciences. Lawrence Erlbaum Associates, Hillsdale (2003)
11. Liu, Y., Sourina, O., Nguyen, M.K.: Real-time EEG-based Human Emotion Recognition and Visualization. In: International Conference on Cyberworlds. IEEE, Alberta (2010)
12. Schaaff, K., Schultz, T.: Towards emotion recognition from electroencephalographic signals. In: Affective Computing and Intelligent Interaction and Workshops (ACII). IEEE, Memphis (2009)
13. Li, H., Pang, N., Guo, S., Wang, H.: Research on textual emotion recognition incorporating personality factor. In: IEEE International Conference on Robotics and Biomimetics. IEEE, Sanya (2008)
14. Amarakeerthi, S., Ranaweera, R., Cohen, M.: Speech-Based Emotion Characterization Using Postures and Gestures in CVEs. In: International Conference on Cyberworlds. IEEE, Alberta (2010)

15. Xiao, Z., Dellandrea, E., Dou, W., Chen, L.: Automatic hierarchical classification of emotional speech. In: Ninth IEEE International Symposium on Multimedia Workshops (2007)
16. Gunes, H., Piccardi, M.: Fusing face and body gesture for machine recognition of emotions. In: IEEE International Workshop on Robots and Human Interactive Communication. IEEE (2005)
17. Milanova, M., Sirakov, N.: Recognition of Emotional states in Natural Human-Computer Interaction. In: IEEE International Symposium on Signal Processing and Information Technology (ISSPIT). IEEE, Ho Chi Minh City (2008)
18. Kao, C.Y., Fahn, C.S.: A Design of Face Detection and Facial Expression Recognition Techniques Based on Boosting Schema. Applied Mechanics and Materials 121, 617–621 (2012)

An Item Bank Calibration Method for a Computer Adaptive Test*

Adrianna Kozierkiewicz-Hetmańska and Rafał Poniatowski

Institute of Informatics, Wroclaw University of Technology, Poland
adrianna.kozierkiewicz@pwr.wroc.pl, ravaelles@gmail.com

Abstract. Computer adaptive testing is a form of educational measurement that is adaptable to examinee's proficiency. The usage of a computer adaptive testing brings many benefits but requires creation of a big and a calibrated item bank. The calibration of an item bank made by statistical methods is expensive and time consuming. Therefore, in this paper we worked out an easy item bank calibration method based on experts' opinions. The proposed algorithm used the Consensus Theory. The researches pointed out that the proposed calibration procedure is efficient. As little as three experts' opinions were enough to obtain the calibrated item bank where values of items' parameters estimated by an expert-based method were not statistically different from values of items' parameter estimated by a statistical calibration method. The statistical calibration method required engaging over 50 persons.

1 Introduction

The growth of popularity of computers increases the interest of an adaptive testing in tutoring systems where the evaluation process is adapted to the actual student's proficiency level. The typical procedure of a computer adaptive testing (CAT) consists of four steps. Initially, the first item is selected. Next, the student's proficiency level is estimated based on all prior answers. In the third step the optimal item based on the current estimate of the examinee's ability is chosen. The procedure is finished when a termination criterion is satisfied. The computer adaptive testing very often applies the Item Response Theory (IRT). IRT is a theory for the design, analysis and scoring of tests, questionnaires and similar instruments measuring abilities, attitudes or other variables. IRT allows study examinees' test scores based on assumptions concerning the mathematical relationship between abilities and item responses. In IRT we can distinguish 3 types of models: one, two and three parameter logistic model [1],[15].

The automated evaluation process brings many benefits. The main advantages of the computer adaptive testing [5], [11] is to reduce opportunities to cheat by the possibility of choosing items from a big item bank. Additionally, items are selected to the user's actual level of proficiency, so they are not too hard nor too easy. It influences the student's level of stress and the motivation. In our research [10] we pointed out

* This research was financially supported by the Polish Ministry of Science and Higher Education.

N.T. Nguyen et al. (Eds.): ACIIDS 2014, Part I, LNAI 8397, pp. 375–383, 2014.

that examinees preferred the online version of the test rather than the traditional one and over 40% of the participants considered the online test to be less stressful. The CATs are shorter by 50% and give a higher level of precision than static tests. Furthermore, adaptive tests are able to show results immediately after testing.

However, the CATs are not free from defects. The biggest problem is connected with a calibration of an item bank. Depending on the selected model, test items are characterized by several parameters like: a difficulty, a discrimination and a guessing parameter. The first parameter describes how difficult an item is. The discrimination parameter is a degree which denotes how well the question is able to discriminate between examinees with slightly different abilities. The guess parameter is used in attempt to account for the effects of guessing on the probability of a correct response. After creating an item bank those parameters are unknown and should be appointed and analyzed. The most popular method for calibration of an item bank depends on pretesting all items with a sample large enough to obtain stable item statistics. This sample may be required to be as large as ~1000 examinees [15]. That solution is associated with the high cost of involving people for a pilot testing.

The above mentioned problem will be omitted in this work. Our objective is to work out item bank calibration method which reduces time and cost for preparing the well calibrated item bank. This method could be implemented in e-learning platforms where computer adaptive tests are used. In this paper a method for calibration of the item bank based on experts' opinions is proposed. For the final estimation of value of items' parameters the Consensus Theory will be applied. The worked out method will be compared with the statistical calibration method described by Sztejnberg and Hurek [11]. In this work, authors empirically demonstrate how many experts are required to obtain a well calibrated item bank. We check whether values of items' parameters estimated by the expert-based method are not statistically different from values of items' parameters estimated by the statistical calibration method.

The rest of this paper is organized as follows: Section 2 contains a short overview of different calibration methods together with systems where those have been applied. In section 3 an introduction to the Consensus Theory and a calibration method based on experts' opinions is described. Section 4 presents results of the experiment and their statistical analyses. Section 5 contains conclusions and further works.

2 Related Works

The fundamental concept of the Item Response Theory used as the basis of a CAT is that each test item is characterized by at least one parameter. The calibration of an item bank consists of determining the value of the item parameters. The calibration methods might be divided into three groups: a statistical calibration, an expert-based calibration and a hybrid method.

The statistical calibration is the most popular and gives a quite well calibrated item bank. Unfortunately, this method requires considerable financial outlays connected with students engaged in making the pre-test. In [2] author proposed method conducted in 5 steps. First, students' scores in the exam are sorted. Next, by P_H and P_L are denoted as the higher and the lower 25% of total students, respectively. In the third step students with correct answers and his/her percentage in the higher and the lower group of each question is calculated. The item's difficulty for each problem is calculated as $(P_H + P_L)/2$ and discrimination index is calculated as $P_H - P_L$. The statistics-based method was also used in [3].

An expert-based calibration method is very simple and cheap. The designer of the CAT system left the calibration of the item bank to teachers and test constructors. However, the opinion given by one expert/teacher is subjective, therefore it is possible to get an incorrectly calibrated item bank. The expert-based calibration method was applied in SIETTE [4]. SIETTE is a complete tool for creating adaptive tests where questions are selected as suitable for student's level of knowledge. SIETTE allows creating adaptive tests using different strategies of item selection and different termination criteria. Despite, that SIETTE is such a powerful tool, items' parameters are still estimated by teachers.

GenTAI is a module of ELSA system [8] which allows conducting the computerized adaptive testing. In GenTAI the item bank calibration task was assigned to tests' authors. Authors of GenTAI were aware of this solution's imperfection, therefore they proposed CALLIE [9] – the tool for the item bank calibration where both: the expert-based and the statistical calibration methods were applied.

The hybrid method was proposed in [7]. The initial difficulty level is a value from a range of 0 to 1 and is assigned by a test designer. Next, the difficulty parameter is re-evaluated as a combination of a designated initial value and historical information - the number of correct and incorrect answers. The same calibration method was applied in system INSPIRE [8]. In [12] authors proposed three strategies for automated calibration of an item bank when there is no available information about item parameters at the beginning of the testing process. Each of them is divided into two phases. In the random phase, tests are scored with the assumption that all items have a difficulty parameter equal to 0 and then, once sufficient data become available, an optimal item selection can be carried out with the Fisher's information function.

In our work we want to create a simple procedure for calibration of an item bank based on expert's opinions. We proposed a method which used the Consensus Theory, therefore our solution have a mathematical background and justification. The advantages of the proposed algorithm is its simplicity, relatively low cost of the calibration process and non-subjective results.

3 Item Bank Calibration Method

As it was mentioned, the problem of calibration of an item bank is very expensive because it requires engaging a big group of students to fill-in the pre-test. In this paper we propose another solution based on expert's opinions.

In IRT we can distinguish 3 types of models: one, two and three parameter logistic model. Depending on the model selected it is required to estimate one, two or three values of the parameters for each item. The guessing parameter does not require any estimation because it is possible to calculate it i.e. a four-option multiple choice item where all options are equally plausible, is equal to 0.25. The proposed method is dedicated to estimating the parameters: *b*- a difficulty parameter and *a*- a discrimination parameter.

For estimation of items' values of parameters the group of experts is needed. The experts have to demonstrate domain knowledge in which items from the bank will be evaluated. As experts the teacher or other certificated specialist, in the considered knowledge domain, should be engaged. Each expert is asked to assess the values of parameters of each item. This values are basis for final bank item calibration. For each item and each parameter the values given by experts are put in an ascending order. The middle value of this order is the sought value of the parameter.

Let us assume that we ask *n* experts about a value of a given parameter (*a* or *b*) of an item *i* coming from an item bank. The value of the given parameter a_i* or b_i* is appointed as the consensus of values of that parameter given by experts. The problem of determination of values of item's parameters is defined in the following way:

Definition 1. *For values of parameters* $p^{(1)}, p^{(2)}, ..., p^{(n)}$ *given by experts, one should determine a value p* such that the following condition is satisfied:*

$$\sum_{i=1}^{n} d(p^*, p^{(i)}) = \min_{p} \sum_{j=1}^{n} d(p, p^{(j)})$$

where: d- the defined distance function between two values
p- is interpreted as a or b.

The theoretical range of the values of *a* and *b* is $-\infty \leq a, b \leq \infty$ but typically values are in the range of $-3 \leq a \leq 3$ and $-2.8 \leq b \leq 2.8$ [15]. We assumed that expert gives a value of a parameter *a* or *b* as the number with the two decimal places from the range $U \in [-3,3]$. The distance between two values is calculated in the following way:

Definition 2. *By* $d: U \times U \rightarrow [0,1]$ *we mean a distance function between two values* $p^{(1)}$ *and* $p^{(2)}$. *The distance function is calculated as:*

$$d(p^{(1)}, p^{(2)}) = \frac{1}{6} | p^{(1)} - p^{(2)} |$$

where: p- is interpreted as a or b.

Such defined problem requires the solution (estimated value of a parameter) to be as near as possible to values of the parameter given by all experts and it can be understood as the best representative of all experts' opinions. For a defined problem the estimation of the value of a parameter *a* or *b* could be brought down to the problem of choosing a median from value of parameters given by experts which was shown in [13]. In the consensus theory there exists other postulate which requires the sum of the squared distances between the solution and the experts' opinions to be minimal and then the problem of estimating the value of a parameter *a* or *b* could be brought down to the problem of a choosing an average value of parameters given by experts. However, if the values of parameters are estimated by a small group of experts, median is a better choice than the average because median is more robust in the presence of outlier values.

The expert-based calibration method is conducted in three steps:

Item Bank Calibration Method

Input: $X = \{p^{(1)}, p^{(2)}, ..., p^{(n)}\}$ given by experts where $p^{(i)} \in U$ for $i \in \{1,2,3,...,n\}$

Output: p^*

BEGIN

1. Put the value from X in the ascending order
2. $k := \left\lceil \frac{n+1}{2} \right\rceil$; $k' := \left\lceil \frac{n+2}{2} \right\rceil$;
3. Choose p* that: $p^{(k)} \le p^* \le p^{(k')}$

END

Example 1. Let us ask 7 experts about the item's difficulty level. We obtain the following values: $X = \{-2.50, -1.85, 0, -1.24, -1.85, -0.3, -2.05\}$. In the first step we put the value from X in the ascending order. We get the following sequence: $-2.50, -2.05 - 1.85, -1.85, -1.24, -0.3, 0$. Then, $k := \left\lceil \frac{7+1}{2} \right\rceil = 4$ and $k' := \left\lceil \frac{7+2}{2} \right\rceil = 4$ therefore p^*=-1,85.

4 Experimental Results

For conducting an experiment a system called Hog-Art [10] was prepared. This is an easy web-based system for an adaptive testing. The system allows the calibration of item bank by using the statistical method described in [11]. The system was created by using technologies such as PHP, MySQL, HTML, AJAX and Java Script. The item bank consists of 92 multiple-choice questions concerning knowledge of English grammar and vocabulary. Initially, we do not know questions' difficulty levels. Our experiment was divided into two steps: the calibration of item bank based on the statistical method [11] and the calibration based on the method described in this work. In the first phase of our experiment over 50 students were asked to solve the pre-test. Each test consists of 20 questions chosen from the item bank in a way which promotes selection of questions with the smallest number of answers. The average number of answers for each question was equal to 11, therefore we collected over 550 answers. The results were processed according to the procedure described in [11].

In the second phase we asked a few experts to assess the difficulty level of each question from the item bank by using a scale ranging from 0 to 1000. The value near 0 meant the very easy questions, values around 500 meant the middle level of item difficulty and increased value in a direction of 1000 meant more and more difficult questions. The obtained results were metricated into a scale typical for difficulty parameter and treated as the input of the algorithm described in this work. People engaged in a role of experts were English teachers, possessed a Certificate in Advanced English (CAE) or a Certificate of Proficiency in English (CPE).

In our analysis we want to know if the results of the calibration obtained by the statistical [11] and the experts-based methods are statistically equal. Additionally, we try to answer the following question: how many experts are needed to obtain a well calibrated item bank.

In our research we consider the number of experts n=3, 5 and 7. This assumption is an implication that the method proposed in Section 3 is based on the Consensus Theory, therefore the result of the bank items calibration method can be called a consensus. In [13] the author pointed out that for the distance space (U,d) a set X is susceptible to a consensus if the cardinality of a set X is an even number. If a set is susceptible to a consensus it means that for this set it is possible to determine a reasonable result. For a better understanding of this problem let us consider a situation where in X we have only two opinions: -1 and 1. It is obvious, that neither -1 nor 1 is a reliable result and cannot be accepted as a common opinion of experts.

All analysis are made for a significance level $\alpha = 0.05$. First, we check the normality of the distribution of analyzed features using the Lilliefors test. For results obtained by the method described in [11] we achieve p-value equal to 0,015651 (the value of the statistic: 0,103995), therefore we reject the null hypothesis and decide that the sample does not come from the normal distribution. For the further analysis we use a non-parametric Wilcoxon test for a depend group.

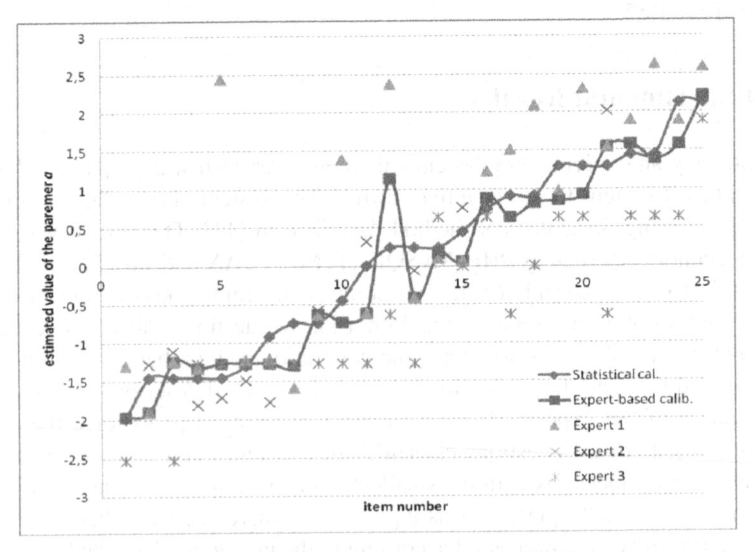

Fig. 1. The obtained results for n =3 and 25 items

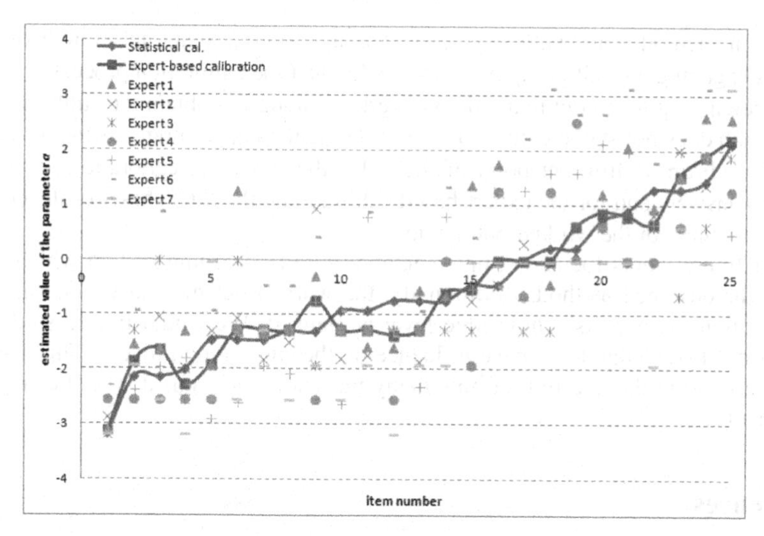

Fig. 2. The obtained results for n =7 and 25 items

In the next step we check that the calibration made by 3 experts gives the same results as the calibration made by the statistical method. Figure 1 presents difference between calibration obtained by a statistical and an expert-based method for 25 selected items and only 3 experts.

The value of statistic equal to 1807,5 and p-value equal 0,198 suggests that the median of a difference between the pairs of measurements is equal to 0, which means that values of the parameter estimated by both methods are not statistically different.

The same conclusions can be drawn from the analysis made for n=5 and n=7 (Wilcoxon test), where p-value is equal to 0,483 and 0,146 and value of statistic equal to 1957,5 and 1725 respectively. Figure 2 presents difference between calibration obtained by the statistical and the expert-based method for 25 selected items and 7 experts.

The results of the analysis demonstrate that values of items' parameters estimated by the expert-based method are not statistically different from values of items' parameters estimated by the statistical calibration method (the median of a difference between the pairs of measurements is equal to 0). Only three experts opinions are enough to obtain a well calibrated item bank. For this task we need only 3 persons and 276 answers. In comparison with the statistical calibration method (about 50 persons, 550 answers) it is a huge saving of not only cost but also time.

5 Conclusions and Further Work

The recent researches show that a collective is more intelligent than a single person [14]. We considered this fact and proposed the expert-based method for the item bank calibration. The algorithm for the item bank calibration used the Consensus Theory. The worked out procedure eliminates problems which occur when using the statistical calibration method. Our method is easy, has a low computational complexity and requires engaging a small group of people to conduct the calibration process.

Our research pointed out that only 3 experts is enough to obtain the calibrated item bank which does not statistically differ from the item bank calibrated using a statistical method. From a different point of view, for the statistical calibration we needed over 550 answers and for the expert-based calibration only 276. These results confirm the effectiveness of the worked out method.

Our future works focus on implementation of an e-learning system which incorporates the designed method. Additionally, the worked out method which takes into consideration the experts' knowledge level is planned. We would like to study how the experts' knowledge level have influence on the calibration process. This research allows answering the question of how many inaccurate opinions does not change the final results.

References

1. Baker, F.: The Basic of Item Response Theory, Eric Clearinghouse on Assessment and Evaluation, USA (2001)
2. Chang, W.-C., Yang, H.-C.: Applying IRT to Estimate Learning Ability and K-means Clustering in Web based Learning. Journal of Software 4(2) (2009)
3. Chen, C.-M., Duh, L.-J., Chao-Yu, L.: Personalized Courseware Recommendation System Based on Fuzzy Item Response Theory. In: Proceedings of the 2004 IEEE International Conference on e-Technology, e-Commerce and e-Service (EEE 2004), pp. 305–308 (2004)
4. Conejo, R., et al.: SIETTE: A Web–Based Tool for Adaptive Testing. International Journal of Artificial Intelligence in Education 14, 1–33 (2004)
5. Fetzer, M.: The Next Evolution of Computer Adaptive Testing. In: Talent Managment, vol. 5 (2009)
6. Gouli, E., Kornilakis, H., Papanikolaou, K., Grigoriadou, M.: Adaptive Assessment Improving Interaction in an Educational Hypermedia System. In: Proceedings of the Pan-Hellenic Conference with International Participation in Human-Computer Interaction, pp. 217–222 (2001)
7. Huang, S.X.: Content-Balanced Adaptive Testing Algorithm for Computer-Based Training Systems. In: Lesgold, A., Frasson, C., Gauthier, G. (eds.) ITS 1996. LNCS, vol. 1086, pp. 306–314. Springer, Heidelberg (1996)
8. López-Cuadrado, J., Armendariz, A., Pérez, T.A.: Adaptive evaluation in an E-Learning System Architecture. In: Mendez-Vilas, A., et al. (eds.) Current Developments in Technology-Assisted Education, Formatex, Badajoz, Spain, pp. 1507–1511 (2006)
9. López-Cuadrado, J., Armendariz, A., Pérez, T.A., Arruabarrena, R.: Helping Tools For Item Bank Calibration And Development Of Computrized Adaptive Tests. In: Chova, L.G., Belenguer, D.M., Torres, I.C. (eds.) Proc. of International Technology, Education and Development Conference (INTED 2008), Valencia (España). International Association of Technology, Education And Development, IATED (2008)
10. Poniatowski, R.: Project and implementation of Computerized Adaptive Testing system, Master of Thesis, Wrocław (2013)
11. Sztejnberg, A., Hurek, J.: Improving computer tests. Opole University Press (2010) (in polish)

12. Makransky, G.: An automatic online calibration design in adaptive testing. In: Weiss, D.J. (ed.) Proceedings of the 2009 GMAC Conference on Computerized Adaptive Testing (2009)
13. Nguyen, N.T.: Consensus Choice Methods and their Application to Solving Conflicts in Distributed Systems. Wroclaw University of Technology Press (2002) (in Polish)
14. Nguyen, N.T.: Inconsistency of knowledge and collective intelligence. Cybernetics and Systems: An International Journal 39(6), 542–562 (2008)
15. Wainer, H., Mislevy, R.J.: Item response theory, calibration, and estimation. In: Wainer, H. (ed.) Computerized Adaptive Testing: A Primer. Lawrence Erlbaum Associates, Mahwah (2000)
16. Conejo, R., et al.: SIETTE: A Web–Based Tool for Adaptive Testing. International Journal of Artificial Intelligence in Education 14, 1–33 (2004)
17. Chang, W.-C., Yang, H.-C.: Applying IRT to Estimate Learning Ability and K-means Clustering in Web based Learning. Journal of Software 4(2) (2009)
18. Chen, C.-M., Duh, L.-J., Chao-Yu, L.: Personalized Courseware Recommendation System Based on Fuzzy Item Response Theory. In: Proceedings of the 2004 IEEE International Conference on e-Technology, e-Commerce and e-Service (EEE 2004), pp. 305–308 (2004)
19. López-Cuadrado, J., Armendariz, A., Pérez, T.A.: Adaptive evaluation in an E-Learning System Architecture. In: Mendez-Vilas, A., et al. (eds.) Current Developments in Technology-Assisted Education, Formatex, Badajoz, Spain, pp. 1507–1511 (2006)
20. López-Cuadrado, J., Armendariz, A., Pérez, T.A., Arruabarrena, R.: Helping Tools For Item Bank Calibration And Development Of Computrized Adaptive Tests. In: Chova, L.G., Belenguer, D.M., Torres, I.C. (eds.) Proc. of International Technology, Education and Development Conference (INTED 2008), Valencia (España). International Association of Technology, Education And Development, IATED (2008)
21. Nguyen, N.T.: Inconsistency of knowledge and collective intelligence. Cybernetics and Systems: An International Journal 39(6), 542–562 (2008)

Hybrid Approach to Web Based Systems Usability Evaluation

Piotr Chynał

Institute of Informatics, Wroclaw University of Technology
Wyb.Wyspianskiego 27, 50-370 Wroclaw, Poland
piotr.chynal@pwr.wroc.pl

Abstract. This paper presents a concept for a design of a new, hybrid method for web systems usability evaluation. This method will combine various elements of other, well-known usability methods, and will be enhanced with a mechanism to evaluate the use of the system, by the users, against the model of its desired use. This way it will be possible to determine the usability of a tested system, using various metrics and techniques, during a single test.

Keywords: Usability, Human-Computer Interaction, Eye Tracking.

1 Introduction

Usability testing is one of the methods of software testing [10]. It is a dynamically growing field that is used to evaluate the quality of various types of interactive systems – from basic desktop applications, through websites to mobile applications. The ISO9241-11 norm defines usability as "extent to which a product can be used by specified users to achieve specified goals with effectiveness, efficiency and satisfaction in a specified context of use" [7]. Web systems usability testing dates back to the middle of the 1990's when people started to analyze their website designs. Since then many evaluation methods were created and many of them are constantly being improved and adopted to meet current website design trends and implementation technologies.

Conducting a complex usability audit of a given system, using various techniques, allows to increase its effectiveness, eliminate errors and increase user satisfaction. The main goals for usability testing of web systems:

- Enlarge the number of active clients
- Build clear and reliable message content for your website
- Enlarge satisfaction of your clients
- Provide qualitative feedback and helps improve your interactive experience
- Focus on the features that really matter
- Reduce costs by anticipating and eliminating potential user roadblocks
- Decrease user acclimation time and errors

N.T. Nguyen et al. (Eds.): ACIIDS 2014, Part I, LNAI 8397, pp. 384–391, 2014.

Conducting usability tests at every stage of the design process of the system, and also after its implementation, provides the possibility to perform rapid changes and eliminate usability issues.

In this paper, the following section presents an overview of various usability testing methods. In section 3 an idea of hybrid method and research methodology for its development is described. Section 4 presents completed research and section 5 - summary.

2 Usability Testing Methods

Currently there are many methods and tools that are used to perform web systems usability evaluation. However each method is constantly being improved and new methods are being introduced to match the evolution of user interfaces.

One of the most interesting usability testing techniques, which require user's participation, is eye tracking [5]. This method enables to track the movement of user gaze on the screen, using a special device called eyetracker. In the result of such test we receive graphical reports of where users were looking during performing tasks in the application. This provides data for effectiveness and efficiency analysis. It may have however some disadvantages, such as head immobility during eye tracking, using a variety of invasive devices, a relatively high price of commercially available eye trackers and a difficult calibration [4]. However, it provides very valuable information for usability studies. All of them are based on the eye-mind hypothesis that what a person is looking at, is assumed to indicate the thought on top of the stack of cognitive processes. This technique is also dynamically evolving. New eye tracking solutions are created, using web cams or cheap infrared cameras [2].

In usability testing we also use other techniques like questionnaires, where users complete a survey regarding their experience with the system and also clicktracking - tracking and recording user's action while they browse the website (for example their clicks) using some dedicated tools and applications. It allows to gather information about how users worked with the systems, where did they click, which parts they did not notice and from where they have entered the system.

In modern usability testing, remote tests are becoming more and more popular. Because of lack of time and money, companies are looking for alternatives for standard tests with users. Those tests require a place to test, gathered users, moderator and equipment. This situation leads to idea of remote usability testing. Its main goal is to test users in their natural working place, without any sophisticated equipment. Users in their own environment are behaving more naturally, like they would normally use given website [9]. Moreover we do not need to gather all the users at one time, we can work with them when they want to. Also we can have participants from different cities or even countries that would normally not visit our laboratory. Furthermore remote tests can be performed automatically. There are various tools and applications that allow to gather the data from users while they are visiting our website. Most of those tools allow to perform clicktracking [10] – they record where users have clicked while browsing the website. This way we can see which elements of the website were not visible to users, which attracted the most attention etc. We can also track other actions, like for example the mouse movement or the user's path, while working with the system, between various sites in our system.

Using those mentioned methods we can perform a full usability audit of a given system. However applying all those methods can be time and effort consuming. The solution to this problem is to create one method, which would combine particular elements of other methods – hybrid method.

3 Research Scope

In this doctoral research a new, hybrid, usability testing method is being developed. It will combine different elements of other methods. Moreover the method will be enhanced by a method for comparing the desired usability model for evaluated system against the data obtained from the usability test. The effectiveness of the proposed method will be verified by comparing it with standard usability testing methods.

3.1 Hybrid Method

The scope of this research is to create a hybrid usability testing method. It would allow to perform complex usability test much quicker and would allow to gather all sort of data regarding user's work with the system. Other advantages of such method are: low cost, it does not require moderation, it would allow to test a large group of users at once, and also it could be used to perform fast, iterative tests at each step of software development. Furthermore this method could be used in agile software development, where the emphasis is put on quickness of development, correct work of the system, the possibility to introduce dynamic changes and to react quickly to problems and errors. In addition this method will be extended by a mechanism that will allow to evaluate the system against its desired model. In this method, at the beginning we will determine the desired result parameters for tasks performed in the systems.

The final version of the proposed hybrid method will gather data from the users, such as:

- Time of task completion
- Task success rate
- Events during performing a task
- Users' path (visited pages)
- Eye tracking data

3.2 Research Methodology

After the implementation of the system that will allow to perform tests using the hybrid method, this method will be evaluated by performing a tests with users on real and benchmark web systems. They will be the point of reference to check the accuracy, completeness, cost and time required to complete the test with this method. In those systems, users and user interaction is going to be modeled, and tasks for users will be created. The number of users will be selected in such way to ensure the statistical significance of the results.

The effectiveness of the created method will be verified by performing tests with this method and standard usability testing methods on the same systems and comparison of the results. For the purpose of this evaluation we can assume that properties of each method can be written as:

$$W = \langle D, K, c, t \rangle,$$

where D is the accuracy of the technique (whether only big usability problems have been found, or has the evaluation using this technique showed also smaller usability problems), K is the completeness of the technique (number of usability problems found), c is the cost of such method and t is the time that we needed to perform a test with the given technique. We say that a usability problem has occurred when the usability rules provided in the definition are broken in the evaluated system, for example where the system or its part is not effective, or there are some errors in it.

This will allow to compare usability problems found during the tests, cost of such tests and time that they took. While verifying the created method we should obtain the following dependency:

$$D_s \approx D_h,\ K_s \approx K_h,\ c_s >> c_h,\ t_s >> t_h,$$

where index s stands for standard usability testing methods and h index stands for hybrid method.

For further verification of the proposed method statistical verification of found usability problems can be performed, using for example Fisher's exact test [11].

3.3 Validation Mechanism

First of all the desired parameters for tasks performed in the system will be determined. Each task consists of:

- Purpose, for example send a message to specified user
- Starting point, for example main page of the website
- End point, for example displaying on the screen "message was send"

To determine the desired parameters we can use:

- Distance in the graph of transitions between pages
- Fitt's law, which enables to predict the time, that user will spend on choosing a particular object with the pointing device [8]
- Optimal list of events for completing the task
- Number of steps for optimal task completion
- Design patterns for particular types of systems

After that a comparison between the desired parameters with the results obtained from the usability tests will be performed. For that purpose for example the "Lostness" measure [13] can be used which is defined as: N – number of unique visited pages,

S – total number of visited sites, R – desired (optimal) number of pages that user should visit, and then we can calculate it:

$$L=\sqrt{\left[\left(\frac{N}{S}-1\right)^2+\left(\frac{R}{N}-1\right)^2\right]} \quad (1)$$

A perfect Lostness score would be 0. Smith [12] found that participants with a Lostness score less than 0.4 did not exhibit any observable characteristics of being lost. On the other hand, participants with a Lostness score greater than 0.5 definitely did appear to be lost. For example, if user is given a task to find some contact information on a website, the optimal number of visited pages (R parameter) should be 1. If during the test user has visited for example 4 pages (S parameter) and 3 of them were unique (N parameter), before managing to find the contact information, we can calculate that the Lostness measure will be equal 0.71 (L>0.5), so there are some usability problems for this particular task and user's might get lost.

3.4 Research Problems

The first encountered problem was to formally define what a usability problem is. It was described as violation of rules presented in the usability definition [7].

Another problem was how to evaluate the effectiveness of the hybrid method. It was solved by formal definition of properties of each usability method and by introduction of calculating its efficiency for the given system [1]. Moreover, another way of verifying the hybrid method will be an implementation of systems that will have specific usability problems. They will be used during tests with users and they will work as a benchmark for evaluation the efficiency of each method used in those tests.

4 Completed Research

So far a simple application that enables to perform remote usability tests using eye tracking was created and presented at the Interact 2011 conference [3]. It uses simple web camera and modified Opengazer software to track the users' gaze movement. Using remote desktop software and an application to communicate with user, for example Skype, we can perform a remote usability test, knowing where user is looking at the screen all the time (it is shown by a blue square on the screen that is following the users' gaze). The data (coordinates of the points on the screen that user has looked on) is recorded during the test and can be used to generate a heat map, that shows the areas on which users looked, and with what intensity.

This system will be used as part of an application that performs a hybrid method usability testing.

Another research milestone was a method that was created and published on ISAT 2012 conference [1], which allows to compare the efficiency of various methods of usability testing. Proposed model of presenting each usability method properties was enhanced with fuzzy logic. To all of those parameters we will assign a value from 0 to 1.

Fig. 1. Application after the calibration process. Blue square shows where the participant is looking and in the background we have a window that shows the coordinates of the gaze.

The 1 value will be assigned to the parameter of the technique that had the better results in the usability test, for example if using a technique we would found more usability problems than with using the other one, the K parameter of the first one will be 1, and the K parameter of the second one will be calculated in proportion, base on the number of usability problems found. We can also attach weights to those parameters, depending on which parameter is the most important for us. The weights should sum up to 1. If the cost is the most important we can give it the highest weight, so even though another method turns out to be more accurate and complete the final effectiveness for the cheaper technique will be higher. At the end the effectiveness of a technique can be counted as:

$$E=D*weight_1+K*weight_2+c*weight_3+t*weight_4 \quad (2)$$

This method will be used to evaluate the efficiency of the hybrid method and compare it with standard usability testing techniques.

Furthermore, a paper describing application of the intelligent data analysis to the eye tracking research was published on HCI 2013 conference [2]. It presents a method for evaluating how the standard data analysis, which is usually made manually by experts, may be enhanced by application of intelligent data analysis. We applied well known expert system, which is using fuzzy reasoning. To build such a system we should first define a model of "desired" eye tracking record for a given poster, or more general web page or the whole application. We have presented how we can enhance the process of gaze tracking data analysis by application of fuzzy reasoning. The most important advantage of this method is that it may be used to analyze the data gathered for single user, several users or even hundreds of users, which will give the statistical significance. However one of the drawbacks of this method is necessity of definition or redefinition of set of rules for the different experiments. This problem may be solved by application of determining fuzzy rules out of experimental data [6]. The presented method will be used in the final version of the hybrid method as well, for analysis of the eye tracking data obtained from the tests with users.

5 Summary and Future Work

To sum up, this doctoral research is focused on creating a new, more effective method for web systems usability testing. There are some steps that need to be taken to achieve that. First of them is the introduction of additional usability measures for hybrid method. There are many usability measures in the literature and it is possible to design some new once. The next thing to do is the creation and selection of existing web-based systems that will serve as a benchmark to evaluate the effectiveness of the method created, modeling of users and interaction in these systems, the development of tasks for these systems. Moreover determination of desired parameters for those tasks is needed. This will allow to fully prepare the testing environment for the hybrid method. After that there is the most important step - design and implementation of a system for usability testing with the hybrid method that will enable to perform tests with users and gather all the necessary data from them. When this system will be ready the next step will be to conduct tests with users. It will have to be an adequate amount of users to achieve the statistical significance for such tests. On the basis of the data received a mechanism that will allow to identify the problems with the usability of the evaluated systems will be developed. The last thing to do will be to verification the designed hybrid method with the existing methods. It will be achieved by performing tests on the same systems as with the hybrid method using, using standard usability methods, and comparing the results. This will allow to determine the effectives of the proposed method.

Different types of systems will also be evaluated using this method, like for example touch screen applications. In the future it is also possible to use this method for mobile applications usability testing.

Acknowledgements. This paper is co-financed by the European Union as part of the European Social Fund.

References

1. Chynał, P.: A method for comparing efficiency of the different usability evaluation techniques. In: Information systems architecture and technology: Web Engineering and High-Performance Computing on Complex Environments Oficyna Wydawnicza Politechniki Wrocławskiej, pp. 51–57 (2012)
2. Chynał, P., Sobecki, J., Szymański, J.M.: Remote Usability Evaluation Using Eye Tracking Enhanced with Intelligent Data Analysis. In: Marcus, A. (ed.) DUXU/HCII 2013, Part I. LNCS, vol. 8012, pp. 212–221. Springer, Heidelberg (2013)
3. Chynał, P., Szymański, J.M.: Remote Usability Testing Using Eye tracking. In: Campos, P., Graham, N., Jorge, J., Nunes, N., Palanque, P., Winckler, M. (eds.) INTERACT 2011, Part I. LNCS, vol. 6946, pp. 356–361. Springer, Heidelberg (2011)
4. Chynał, P., Sobecki, J.: Comparison and analysis of the eye pointing methods and applications. In: Pan, J.-S., Chen, S.-M., Nguyen, N.T. (eds.) ICCCI 2010, Part I. LNCS (LNAI), vol. 6421, pp. 30–38. Springer, Heidelberg (2010)

5. Duchowski, A.T.: Eye tracking methodology: Theory and practice, pp. 205–300. Springer-Verlag Ltd., London (2003)
6. Guillaume, S.: Designing Fuzzy Inference Systems from Data: An Interpretability-Oriented Review. IEEE Transactions on Fuzzy Systems 9(3), 426–443 (2001)
7. International Standard ISO 9241-11. Ergonomic requirements for office work with visual display terminals (VDTs) – Part 11: Guidance on Usability. ISO (1997)
8. MacKenzie, S.: Fitts' law as a research and design tool in human-computer interaction. Hum.-Comput. Interact. 7(1), 91–139 (1992)
9. Moha, N., Li, Q., Seffah, A., Michel, G.: Towards a Platform for Usability Remote Tests via Internet, http://www.ptidej.net/Members/mohanaou/paperOZCHI2004/OZCHI2004_Moha.pdf (accessed September 20, 2013)
10. Pearrow, M.: Web Site Usability Handbook. Charles River Media (2007)
11. Routledge, R.: Fisher's Exact Test. Encyclopedia of Biostatistics (2005)
12. Smith, P.A.: Towards a practical measure of hypertext usability. Interacting with Computers 8(4), 365–381 (1996)
13. Tullis, T., Albert, B.: Measuring the user experience. Morgan Kaufmann (2008)

Application of Network Analysis in Website Usability Verification

Piotr Chynał, Janusz Sobecki, and Jerzy M. Szymański

Engine Centre, Wrocław University of Technology
Wybrzeże Wyspiańskiego 27, 50-370 Wrocław, Poland
{piotr.chynal,janusz.sobecki,jerzy.szymanski}@pwr.wroc.pl

Abstract. This paper describes the application of network analysis in website usability verification. In particular we propose a new automatic usability testing method by means of application of network motifs analysis. In our method motifs are constructed according to the patterns of visited web-pages by particular users, which define the relationships between the pairs of users.

Keywords: Website usability, network analysis, network motifs.

1 Introduction

Web along with mobile services are nowadays predominant types of interactive systems. One of the most important features of web-based systems that influence their efficiency is their usability. Usability may be defined in quite many different ways. Taking only into account the ISO norms we may find several definitions. For example in the norm ISO 9126, it is defined as a collection of attributes that the product should bear and are needed for the assessment of the use of the product, they are the following: understandability, learnability, operability, explicitness, customizability, attractivity, clarity, helpfulness and user-friendliness. In the following norm ISO 9241, usability is defined as an extent to which a product can be used by specified users to achieve specified goals with effectiveness, efficiency and satisfaction in a specified context of use. The usability of the websites has a very important influence on their efficiency, including different economic and marketing features.

Websites are used by many users, which could visit several web pages in each session. Taking into account records of visited pages we can create some patterns, which in turn may be used to define directed relationships among pairs of users. Secondly, the networks created by means of communication technologies are highly dynamic and huge structures. When investigating the topological properties and structure of complex networks we must face a number of complexity-related problems. In large social networks, tasks like evaluating the centrality measures, finding cliques, etc. require significant computing overhead and are inherently connected with a level of uncertainty – it is not easy to gather information about the overall network structure. In a result the methods which proved to be useful for a small and medium sized networks fail when applied to the huge structures [9].

N.T. Nguyen et al. (Eds.): ACIIDS 2014, Part I, LNAI 8397, pp. 392–401, 2014.

In this case data-driven techniques may be proposed for social networks analysis [10]. They enable us to infer not only static but also dynamic structure from local characteristics, because it has been observed that local topologies of social networks are significantly different from those of standard network models. One of the ways of the social network topology representation is network motifs (subgraphs). A triad is considered the smallest non-trival network subgraph, which is a directed network with exactly three nodes.

Having extracted triads we may determine the frequency of each motif or is statistical significance [9]. Then they may be used for further analysis of web site usability according to the scores of each motif or their groups. In general higher scores of motifs with isolated nodes mean that the number of users, whose visit patterns are different from other users is significant, what may be interpreted that they more or less randomly view the website pages.

The structure of the paper is as follows. In the second section the problem of website usability verification is presented. The third section gives some information on the sample website which was considered as a basis for analysis. The fourth section presents the proposal of the method and summary is given in the fifth section.

2 Web Systems Usability Evaluation

Usability evaluation is one of the methods of software testing. Such tests are becoming more and more popular in recent years [2]. Usability testing allows us to increase learnability, satisfaction and efficiency for the given system and also thanks to such test we can find and eliminate errors.

With such tests we can evaluate usability of web systems, desktop and mobile applications. Web systems usability testing dates back to the middle of the 1990's when people started to analyze their website designs. Since then many evaluation methods were created and many of them are constantly being improved and adopted to meet current website design trends and implementation technologies.

There are many various usability evaluation methods, and still new methods are being introduced to improve the process of validating particular website. These methods may be divided into several groups according to some criteria. The most popular methods are those that require user's participation and those which do not. In methods without users' participation the most popular are expert evaluation, heuristic evaluation and cognitive walkthrough [1]. In those methods an expert or group of experts evaluate website usability base on their experience and knowledge of design principals and standards. As for methods that require users' participation, there are a lot of them and we can further divide them into two groups – moderated and unmoderated. Moderated methods require at least one moderator who is working with the users and gathers information about the evaluated website from them. Unmoderated methods allow to perform usability tests automatically, using applications that gather the data from users. The most commonly used methods are presented in the table below:

Table 1. Usability evaluation methods with users' participation

<table>
<tr><th>Moderated</th><th>Unmoderated</th></tr>
<tr><td>Focus Group Research - Test in which a group of users works with a system and gives feedback to the moderator</td><td rowspan="4">Action tracking- Tracking and recording user's action while they browse the website (for example their clicks) using some dedicated tools and applications.</td></tr>
<tr><td>Survey Research - Users complete a survey regarding their experience with a system</td></tr>
<tr><td>Individual Test - Moderator is watching user work with a system and discusses his experience with it</td></tr>
<tr><td>Eyetracking Test - Similar to the individual test, but the participants are connected to the eyetracking equipment, that enables recording of the participants gaze on the screen during the test</td></tr>
<tr><td>Remote Test - Individual test in a remote environment</td><td>Remote Test – User performs tasks in the system using an online form that is shown on the screen while working with the website.</td></tr>
</table>

In modern usability testing, remote tests are becoming more and more popular. Because of the lack of time and money, companies are looking for alternatives for standard tests with users, which require a place to test, gathered users, moderator and equipment. This situation leads to the idea of remote usability testing. Its main goal is to test users in their natural working place, without any sophisticated equipment. Users in their own environment are behaving more naturally, like they would normally use given website [4]. Moreover we do not need to gather all the users at one time, we can work with them when they want to. Also we can have participants from different cities or even countries that would normally not visit our laboratory. Furthermore comparison of the results for standard laboratory and remote testing have shown, that participants find same usability issues on tested pages with both methods as shown in [5], [6] and [7].

Furthermore remote tests can be performed automatically. There are various tools and applications that allow to gather the data from users while they are visiting our website. Most of those tools allow to perform clicktracking [8] – they record where users have clicked while browsing the website. This way we can see which elements of the website were not visible to users, which attracted the most attention etc. We can also track other actions, like for example the mouse movement or the user's path of visiting various webpages.

The most popular tool for obtaining such data is Google Analytics. This tool enables to generate statistics about website's traffic and traffic sources. It also measures conversions and sales. Google Analytics can track visitors from all referrers, including search engines and social networks, direct visits and referring sites. This tool can be installed on a website by inserting a JavaScript code into its webpages. Google Analytics main features include generating around 80 types of reports related to the collection of data on Internet traffic, conversion rates and ROI. Google Analytics has also the ability to create user segments according to traffic source or user behavior. Conversion rates are goals that we set up on our website, for users to achieve. Goals might include sales, lead generation, viewing a specific page, or downloading a particular file. The term ROI stands for the Return of Investment and describes how much profit our website generates in comparison to the costs of investments in it, for example in advertisements that promote it in search engines. Thanks to these features we can optimize our website, select the most important referral sources to our website and control its profitability.

Google Analytics is the most widely used website statistics service[1], currently in use on around 55% of the 10,000 most popular websites. However Google Analytics has some drawbacks. The most important is that it does not allow viewing the data for

Fig. 1. Mixpanel streams - separate visits and the sequence of visited pages during users' visits

[1] http://techcrunch.com/2012/04/12/google-analytics-officially-at-10m/

a particular user. It only shows transitions between the pages for groups of users. That is why we looked for other tools for web analysis and we found an application called Mixpanel.

Mixpanel is also a web and mobile analytics tool, which uses JavaScript code pasted into website to gather data from users. It offers the following features:

1. Streams – this option enables to see how individual users are using your website, what actions do they take. It is done by tracking events in JavaScript code.
2. Trends - analyze trends in actions your users are taking on your website.
3. Segmentation – this feature enables segmentation of the data to learn how certain groups of users are using the website.
4. Funnels – allows to analyze recorded data and tracking different events to see where the biggest drop off rates are.

3 Web Service Instancje.pl

Our research has been done with website called *instancje.pl.* We selected this website because it has been designed and implemented by one of the authors of this paper. This website provides classified ads of flats for rent, especially aimed for students and other young people from big cities in Poland. It has been developed mainly in PHP Zend Framework, HTML, Java Script and AJAX technologies.

3.1 General Description

The whole website *instancje.pl* consists of three main views for standard user and system for submitting offers, for property owners. Figure 2 shows the main view. We can see there the following elements: website logo, main menu, breadcrumbs, image slider, and buttons for adding offers, search panel, map showing location of flats and a list of most recent offers.

Main menu enables users to navigate through the whole service. It contains links to pages such as: main page, offer list, add flat, add room, add place in room, tips and news. It is visible in every view in the system. Users can also log in to their account at every page.

Search panel helps users to find satisfactory offer. In the basic mode users can choose the type of offer (flat, room, place in room) and specify the number of rooms, residents and price range.

List of offers includes the latest offers, which are shown chronologically. User can view the most important information such as: offer type, street, price, date and newness factor (taking into account the publication time, interest, number of interested people and other information). It is important that users can quickly scan through newest offers without opening individual pages. Thus users can rapidly scan only newest offers and close the website without leaving any special activity information, to be collected in log files. Map located on the left of the screen can only boost that effect. It shows location of offered flat or room on the city map.. Thanks to this, users who don't know topography of the city can quickly locate offers and decide whether to click for details.

In the full list of offers we can see a big map of the city with offers marked at corresponding locations.

Fig. 2. Main page of web service instnajce.pl

Map navigation uses standard Google Maps mechanism, which is enhanced with offer grouping. Thanks to this map users can search for offers in their interest area and by clicking at a mark, they can get a short overview of them, without reloading the page. Below the map there is a list of offers, that presents some enriched information. There is a photo, type of offer, street, price, pictograms indicating available space, area, date and newness factor. Individual offer view contains all the information about flat or room provided by owner. Presented details are grouped into several categories: localization, price, contact details, general information, residents, rooms, equipment, photos and description.

3.2 Users and Common Behaviors

The main users are students who are looking for an apartment or room for rent. Service operates in three big academic cities in Poland. There is a very popular trend among polish students to rent an apartment with a group of friends. Thus most wanted are flats with few bedrooms, in the center, and of course, at a low price. Due to the cycle of the academic year, the traffic in the service is varied.

Peak of the interest occurs during the three months of summer holidays. At that time a large number of students are looking for accommodation for the next academic year. The main tools for this are various Internet services. There are several leading sites, which post up to few hundred offers a day. Typically, users browse all sites simultaneously. They define their price range and size of the apartment, and then periodically refresh all services.

Considering only one of the parts of the website *instancje.pl* in Wrocław in high season 2013 it had more than 6 thousand visitors and nearly 50 thousand page views monthly. Average user has spent on this website over 6 minutes and opened up about 7 pages.

Based on the log files and user observation, we concluded, that typical actions performed by users are as follows: Entrance to the main page, a quick overview of new offers, opening of selected offers in new web browser tabs, closing tabs, use the search form, opening of selected offers in new tabs. However these are not the only possible scenarios. Others include viewing a single offer page or viewing several offer pages, which may be modified by using search somewhere in between.

3.3 Mixpanel Integration

Mixpanel differs for example from Google Analytics, it helps to analyze the actions that users take in a web application. Mixpanel allows tracking users' path during browsing web service and logging their actions. To start making analysis prior integration with the web site is required. It is quick and comes down to just pasting the Java Script code into all of the website pages. After placing the code, users are tracked automatically. We can get access to single users' path. We can also track events done be users and links clicked by them.

The sample screen of Mixpanel service showing several users of *instancje.pl* (denoted as Guest #number) is presented in Fig. 1. For all the users we can view their individual visits and the sequence of visited. The structure of the website instancje.pl consists of several distinct pages and a lot of pages with separate offers. To simplify the usability analysis we will treat each offer page as the same page.

4 Network Motifs in Website Usability Verification

Networks that originated from different fields have very different local topological structure, despite all structural and statistical similarities [9]. It was shown recently that concentration of network motifs may help to distinguish and classify complex biological, technological and social networks. The definition of network motifs has been presented in the introduction. Smaller motifs (those which have fewer nodes) need less computational power during analysis. The smallest non-trivial subgraph has three nodes and is called a triad. In our analysis we are taking into account four states of the relation between pairs of nodes: no relation, a relation in one direction, a relation in the other direction and a relation in both directions. In consequence we may distinguish 64 ($=4^3$) distinct triad configurations [10]. In the case when we make no distinctions among the triad nodes, the number of distinct triads reduces to 16, they are presented in Fig. 3, which is based on the position [10].

The first step in motifs analysis is the relation determination. In our case -website user traffic analysis, we have at first a set of web pages P, the set of users U, and streams of each users visited pages, which is a tuple belonging to the following set $S=U\times P^k$, where k depends on the length of the stream. The number k may vary from 1 to w and w is a natural number. Then we should filter out the streams which are shorter than two pages and also those, which do not contain any of the defined goal pages.

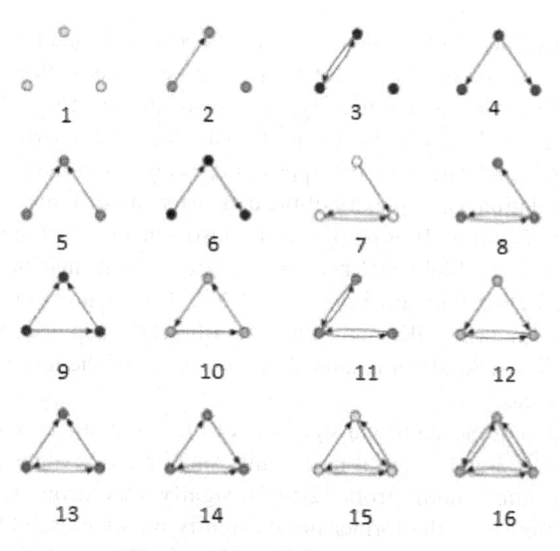

Fig. 3. Triad configurations [10]

The relation $R=U\times U$ is defined as follows, $(u_i,u_j)\in R$, if $i\neq j$ and stream of pages $s(u_i, p_{i,1},\ldots, p_{i,k})$ is a sub-stream of $s(u_i, p_{i,1},\ldots, p_{i,n})$, where $k\leq n$ and $p_{i,1}= p_{j,t},\ldots, p_{i,k}= p_{j,t+k-1}$ for t is natural number such as $t\leq n$.

The interpretation of this relation in the usability domain may be as follows. The lack of the relation means that these two users have different page visit patterns, so they use it in a different way. The existence of one direction relation means that one users' visit pattern was followed by the other user, no matter, in which part of his or her visit path. The symmetric relation signifies that both users have exactly the same page visit patterns or they each have at least two registered page visit patterns and one pattern is a sub-stream of the page visit pattern of the other user.

Having defined the relation R we can determine the specific significance profiles (SPs). This is achieved by measuring the concentration of individual motifs and comparing to their concentration in a number of random networks. The statistical significance of each motif $m\in M$ is defined by its Z-score that is denoted by Z_m, which is defined in the following way [11]:

$$Z_m = \frac{n_m - n_m^{rand}}{\sigma_m^{rand}},$$

where n_m is the frequency of the motif m, n_m^{rand} and σ_m^{rand} are the mean and standard deviations of m occurrences in the set of random networks, respectively. Along with the Z-score we can also use other coefficients, such as betweeness or clustering coefficients [9].

There are quite many different algorithms for detecting network motifs and most of them are based on an exhaustive search of all sub-graphs with specified number of nodes [9], which in consequence require more computational overload when the number of nodes increases. However it was shown that it is possible to estimate network motifs by random sampling of the given network. For example we may use the algorithm asymptotically independent of network size presented in the work [12], which is fast and may be used even for very huge networks with hundreds of thousands of nodes. This is of course usually the case of the most popular websites that have millions of users. However also in our website *instancje.pl*, which is having only thousands of users, application of the approximate algorithm may be necessary because of the computational complexity of the exhaustive algorithms.

The motifs signify not only topological but also functional properties of the network, for example in the biological networks motifs exhibit their informational role, where motif called Feed-Forward Loop (motif No. 9 in Figure 4) has the following meaning: to perform tasks like sign-sensitive filtering, response acceleration and pulse-generation. Those kinds of results show potentials of the motif analysis also in other application areas.

The interpretation of the statistical significance of each motif in usability domain is as follows. Motifs 1, 2 and 3 reveal no or quite small relation among user page visit patterns. In the contrary motifs from 12 to 16 signify very strong relations between user page visit patterns. So the former motifs signify possible usability problems and the further signify very good website usability. The motifs 4 to 11 may signify at least good quality of the website.

5 Summary and Future Works

In this paper we have presented the conception of the application of network motifs analysis in website usability verification. The ever increasing importance of web technology requires also the increase of the quality of these services. One of the website quality factors is its usability. The presented method enables to verify the usability using unmoderated method. What is also important that the proposed usability method enables to be launched pretty often, or be used even permanent. The permanent verification may be useful in the cases when users or their characteristics change over time. In the future we plan to verify the method on other websites. We will use this method on a given website and then compare the results with the data obtained from usability tests performed with standard usability testing methods. This way we can check if the interpretation of triads 1-3, 4-11 and 11-16 corresponds to the levels of usability of the tested website.

Acknowledgement. The research was partially supported the European Commission under the 7th Framework Programme, Coordination and Support Action, Grant Agreement Number 316097, ENGINE - European research centre of Network intelliGence for INnovation Enhancement (http://engine.pwr.wroc.pl/).

References

1. Barnum, C.M.: Usability testing and research. Longman (2002)
2. Barnum, C.M.: Usability Testing Essentials: Ready, Set...Test! Elsevier (2010)

3. International Standard ISO 9241-11. Ergonomic requirements for office work with visual display terminals (VDTs) – Part 11: Guidance on Usability. ISO (1997)
4. Moha, N., Li, Q., Seffah, A., Michel, G.: Towards a Platform for Usability Remote Tests via Internet, `http://www.ptidej.net/Members/mohanaou/paperOZCHI2004/OZCHI2004_Moha.pdf` (accessed March 20, 2013)
5. Brush, B., Ames, M., Davis, J.: A Comparison of Synchronous Remote and Local Usability Studies for an Expert Interface, `http://delivery.acm.org/10.1145/990000/986018/p1179-brush.pdf` (accessed March 21, 2013)
6. Oztoprak, A., Erbug, C.: Field versus Laboratory Usability Testing: a First Comparison, `http://www.aydinoztoprak.com/images/HFES_Oztoprak_.pdf` (accessed March 21, 2013)
7. Tullis, T., Fleischman, S., McNulty, M., Cianchette, C., Bergel, M.: An Empirical Comparison of Lab and Remote Usability Testing of Web Sites, `http://home.comcast.net/~tomtullis/publications/RemoteVsLab.pdf` (accessed March 21, 2013)
8. Pearrow, M.: Web Site Usability Handbook. Charles River Media (2007)
9. Juszczyszyn, K., Musiał, K., Kazienko, P., Gabryś, B.: Temporal Changes in Local Topology of an Email-Based Social Network. Computing and Informatics Computing and Informatics 28(6), 763–779 (2009)
10. Musiał, K., Juszczyszyn, K., Budka, M.: Triad transition probabilities characterize complex networks. Awareness Magazine Awareness Magazine (2012), doi:10.2417/3201209.004369
11. Barabasi, A.L., Albert, R.: Emergence of Scaling in Random Networks. Sciences 286, 509–512 (1999)
12. Kashtan, N., Itzkovitz, S., Milo, R., Alon, U.: Efficient Sampling Algorithm for Estimating Subgraph Concentrations and Detecting Network Motifs. Bioinformatics 20(11), 1746–1758 (2004)

Travel Password: A Secure and Memorable Password Scheme

Nattawut Phetmak[1], Wason Liwlompaisan[2], and Pruet Boonma[3,*]

[1] Faculty of Engineering, Kasetsart University, Bangkok, Thailand
neizod@gmail.com
[2] Blognone Co. Ltd.
wason@blognone.com
[3] Faculty of Engineering, Chiang Mai University, Chiang Mai, Thailand
pruet@eng.cmu.ac.th

Abstract. There is a trade-off between password security and usability; longer password provides higher security but can reduce usability, as it is harder to remember. To address this challenge, this paper proposed a novel password scheme, called "Travel Password", which is memorable and also secure. The proposed scheme is designed to aid human memory by using mnemonic device, e.g., pictures and symbols, and story telling. Mnemonic device aids memory because human can remember pictures better than text. Story telling, on the other hand, allows users to make connection between each part of the password. The experiment with eighty users shows that the proposed scheme allows users to have better password recall. Compared with traditional textual password which has about 0.8 recall rate for strong passwords, users with the proposed scheme can achieve 1.0 recall rate. Moreover, the proposed scheme is more memorable than the traditional textual one. 90% of users can promptly remember strong passwords in the proposed scheme, compared with 58% of the textual one.

1 Introduction

Security is a foundation of computer systems. Losing personal data or identity through security breach can lead to serious damage. Generally speaking, there are two types of security attack: attack on authentication system and attack by reverse engineering a security system. In particular, because human is the weakest link in the security system; attacking on the authentication systems is easier and more widespread than reverse engineering security systems.

Textual password is a widely accepted mean of authentication; however, a human can hold roughly seven objects, e.g. characters in a password, in his/her memory [1]. As a consequence, users tend to set their password shorter or equal to seven/eight characters, which is too short and insecure. Besides, if the users are forced to use longer passwords, the password tends to be easy to remember [2]. These passwords are insecure, for example, they are prone to dictionary attack. Therefore, designing a password scheme which is secure and memorable is a challenge task.

* Corresponding Author.

N.T. Nguyen et al. (Eds.): ACIIDS 2014, Part I, LNAI 8397, pp. 402–411, 2014.

There are many approaches to mitigate the attacks on weak passwords. Smartphones generally introduce a delay to accept next input after users enter wrong passwords a few times. Websites may utilize verification mechanism to confirm whether the password is input by human: Captcha is one the most popular method. However, for an offline computer system that requires, for example, complete harddisk encryption or secure private key files, the data is protected by the strength of the password only.

Achieving 128 bit entropy is one of the desirable goal for password schemes. Many strong encryptions still use 128 bit key, e.g., according to NSA's Suite B [3], US government requires to use AES128 for all secret documents. Also in RFC3766 [4], the life time of 128 bit symmetric encryption will be survive to 2053, which is long enough for many applications.

This paper proposes a new password scheme called "Travel Password" which, when has comparable strength with a textual password, is easier to remember than the textual one. In contrast with the current imaged based password schemes, as discussed in Section 2, this password scheme employs a mnemonic device, i.e., pairs of *city* and *activity*, in order to help users to retain password in their memory [5]. Section 3 discusses the proposed scheme in detail. The experiment, in Section 4, with eighty users shows that the proposed scheme allows users to have better password recall. Compared with traditional textual password which has about 0.8 recall rate for strong passwords, users with the proposed scheme can achieve 1.0 recall rate. Moreover, the proposed scheme is more memorable than the traditional textual one. 90% of users can promptly remember strong passwords in the proposed scheme, compared with 58% of the textual one.

2 Related Works

Password's entropy is a widely accepted password's strength indication. When the entropy is too small, the password becomes prone to brute-force attacks. In particular, strength of a password comes from the number of combination it can produce [6]. For example, an eight characters alphanumeric password will have entropy of $\log_2 62^8 = 47.6$ bits.

McDonald proposed the *dice word*, which use words instead of characters to compose a password [7]. For instance, a user can picks words, to compose a password, from the list of the most commonly used words in English. So, if the list of the words is large, the password's entropy will increase very fast for each additional word in the password. In case of the 2048 most commonly used word list, the user will need just twelve words to get 128 bits entropy. An alphanumeric password will require about 21 characters to achieve the same security.

In addition to textual password, image-based password was introduced in 1996 [8]. Based on the observation that human can remember and recognize images better than words [5], image-based password schemes have generally higher memorability, compared with a textual password. PassMap [9] was proposed to utilize users' traveling experience, for user could easily remember where they have been or where they want to visit. The vulnerable of PassMap is brute-force attack, as attackers may know travel history or the desired places of the victim, the attacker could easily guessed the password. Gani proposed Graphical User Authentication (GUA) [10]. This system shows images to users in multi-line grid and let the users choose images from different grid arrangement during the login process. While images from the grid can be easily to remember, the entropy of the password could be too low in a small grid.

Generally speaking, the image-based password schemes can be categorized into two kind as recognition based and recall based.

1. **Recognition Based**
   In this approach, a password is a set of user-selected images. In particular, users select a set of images from a collection as his/her password. Thus, to authentication, users are required to decide whether a set of presenting images is correct or not. An example of this kind of scheme is Déjà Vu [11]. This scheme has relatively low entropy because in order to increase the entropy, the image collection has to be large and that will take time to set up and authenticate.
2. **Recall Based**
   This approach, on the other hand, focus on the details of single image. For example, in DAS [12], users have to draw a picture to be used as a password. In PassPoint [2], users choose areas of an image as their password. The challenge of this approach is to handle the deviation of the drawing or the location of the areas, because human cannot specify the location on the screen in the pixel level. This deviation can greatly reduce the entropy of the password on this scheme.

In Windows 8, Microsoft uses a recall based picture password [13] as a user-friendly password scheme. While picture password seems to be user friendly, the password has limitation that it cannot be stored in hashed form. To check an input, the stored password will need to be compared to the input whether the differed among them is within a threshold. This checking scheme limit the usage of picture password to online systems only and the password must be stored in the recoverable way.

[14] suggests that PassPoints is the most efficient method because it is easy to learn, fast to use, has high entropy, and can manage data in hash form. Moreover, [15] suggests that unordered password is easier to use but the entropy is lower than the ordered one.

Recall based password scheme, e.g., PassPoints, needs to handle the deviation of the location on picture, which generally reduce the entropy of the password. Particularly, instead of using individual pixel, the pixels might be grouped into areas. Thus, the entropy of the password will reduce from the number of pixels to the number of areas.

3 Travel Password Scheme

In Travel Password scheme, passwords are randomly generated for users to avoid birthday attack. To enter a password, users have to choose a city and an associated activity, which is a part of their given passwords, from a map. In particular, they will choose a location closed to the desired city from the map. Then, a list of the nearby cities will be provided; therefore, users can choose the target city from the list. By using the combination of location and city list, deviation in the recall based approach can be handled. Next, users will have to choose an activity associated with the location. This can be seen as recognition based approach because a collection of activities will be provided. By using a sequence of city/activity pairs to represent a password, this Travel Password will be easier for users to remember. Detail of this scheme will be discussed in the following sections.

3.1 Map and City

The proposed approach uses geographical map, e.g., world map or country maps, because they are well known and recognizable. For the case of the world map, even though, there are more than 200 countries in this world, most of them are unknown by general public. Thus, a list of cities which are travel destinations or have large population is used instead. The similar approach will be applied when a country map is used.

3.2 Activity

The list of cities might not be enough to archive high entropy with a short password. To address this issue, a list of activities is introduced in combination with the list of cities. Each selected city must be accompany with an activity.

The list of 28 activities that are generally performed when people are traveling is selected. They are divided into groups with unique color for each group. These activities will be pinned into the map, according to the location of the associated cities.

3.3 Password Strength

As mentioned before, the password strength is a measure of its entropy. For Travel Password X, the entropy $H(X)$ in bit, when choose s cities from the list of c cities with the list of a activities, can be measured using following equation,

$$H(X) = \log_2 (P(c,s) \times a^s) \tag{1}$$

For example, if four cities are chosen from 247 cities, then,

$$H(X) = \log_2 \left(P(247,4) \times 28^4\right) = 50.98 \tag{2}$$

On the other hand, if eleven cities are chosen, then the entropy will be increased to more than 128 bit, i.e.,

$$H(X) = \log_2 \left(P(247,11) \times 28^{11}\right) = 139.98 \tag{3}$$

3.4 User Interface Design

To evaluate the proposed password scheme, a web-based authentication system was implemented. In this system, Google Maps is used as the main interface, together with some custom JavaScript code. The location of the cities come from Google Reverse Geocoding API.

Figure 1 shows the user interface of the proposed system. In this figure, six locations are already selected. Each represent as information card, which shows its mini-map,

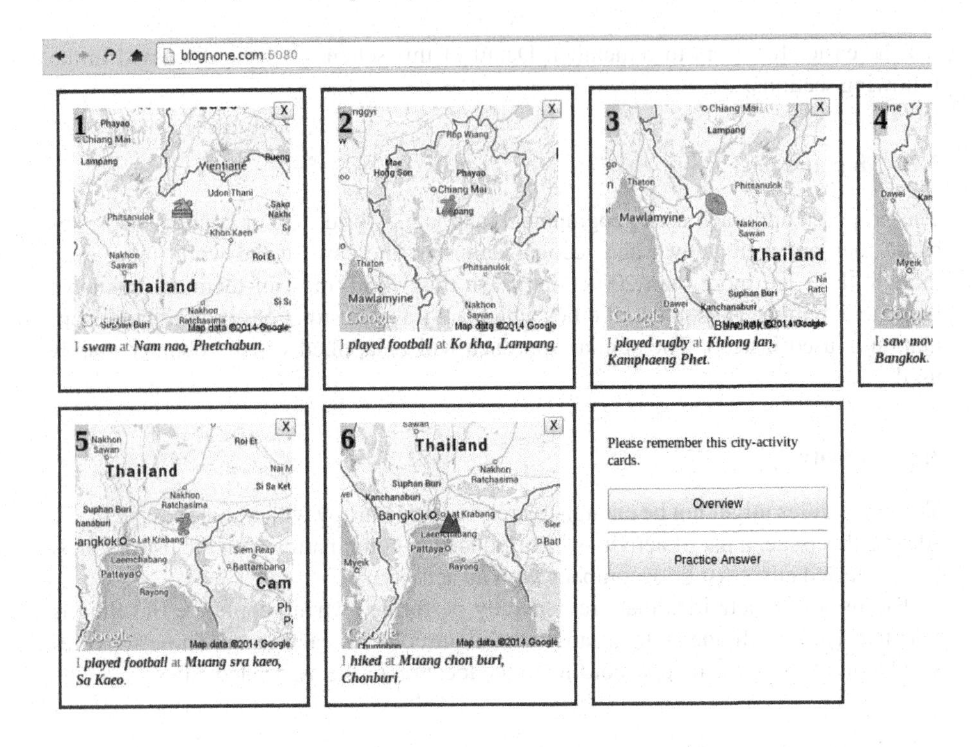

Fig. 1. Travel Password User Interface

order of occurrence, and a description of activity and city name where the activity happens. Also, when users click overview button (the above button in the third card of the second row), an overview map with pairs of city/activity will be visible as shown in Figure 2. In this figure, the swimming icon shows the information of the first card in the previous figure, where *swimming* is selected as the activity performed at *Nam Nao, Petchabun*.

To create a new pair of city and activity, users click add location button, a map will be shown on screen. Then, the users click a location on map, list of nearby cities, ordered by the distance, is generated using Haversine's equation [16]. Figure 3 shows such map and list, also, the list of activities is shown under the list of cities. The users then choose desired city and activity to create a new pair of city/activity for the password.

4 Experiment

A preliminary experiment was carried online. Users of Blognone[1], a well-known technology news website in Thailand, who has registered on the site more than a month were asked to participate in this experiment. These volunteers are active members of

[1] `http://www.blognone.com`

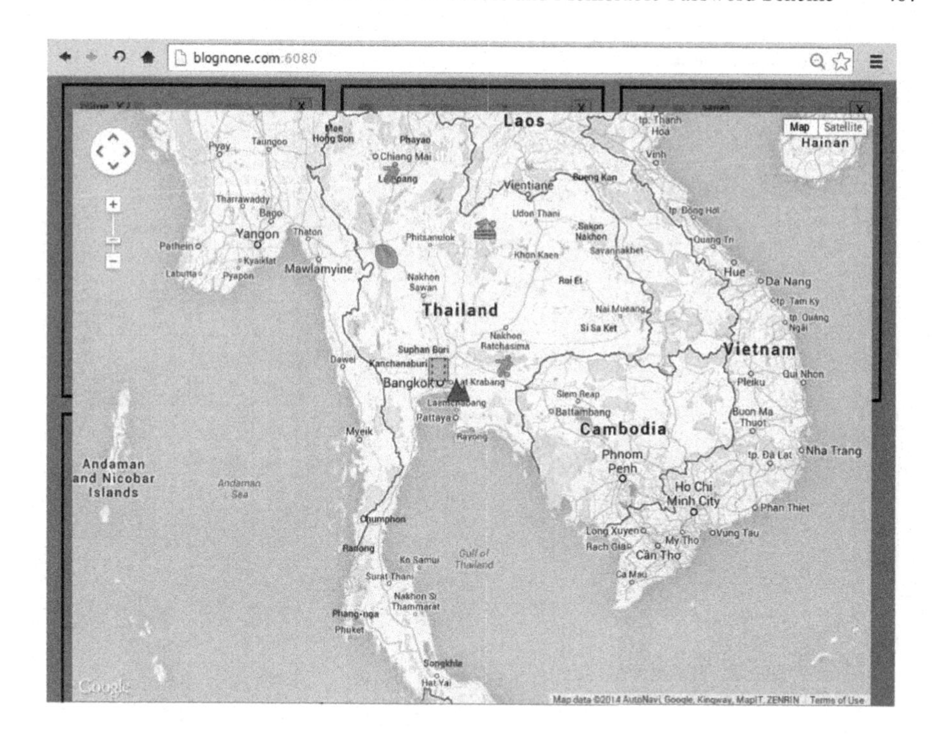

Fig. 2. Overview map

the website and should have higher than standard IT skill. About eighty users participated in this experiment in a period of two weeks. The map in used is Thai map together with a list of well known Thai cities. Following is the steps of the experiment:

1. For each user, a traditional textual password or the proposed password scheme is chosen for him/her randomly.
2. Then, the user has to remember a randomly generated password and perform a *confirmation test* that asks the user to enter the correct password. This step confirms that the user cans remember the passwords.
3. After the confirmation, the user has to watch a music video for about three minutes. This part is used to diverse the focus of the user from the password they just remembered.
4. Next, the user has to recall the password. Three incorrect password attempts is consider as fail. If the user can pass this level, he/she will go to the next level by repeating the whole process from the step 2 with a longer password.

In the experiment, as it was conducted with Thais, the list of large or well known cities in Thailand was used. And the overview map shows to users is the map of Thailand. Familiarity to cities in the map could help users to memorize the password. There are 247 cities in the list selected from the major city from each provinces in Thailand and also touristic cities.

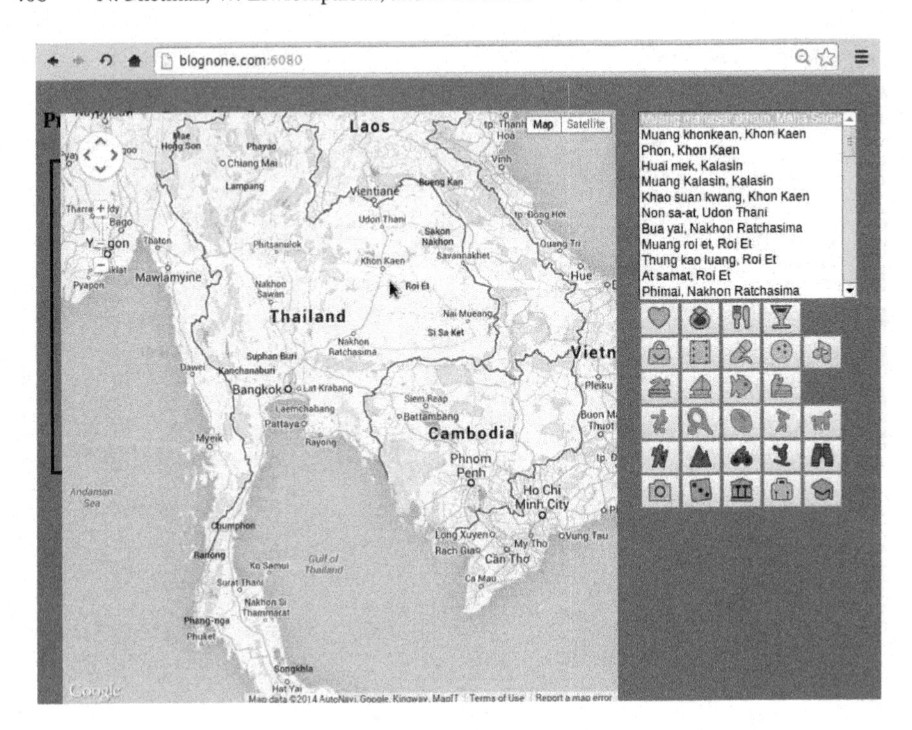

Fig. 3. List of nearby cities and list of activities

Table 1 shows the length and entropy of the password used in this experiment. For example, if a user is challenged with textual password scheme, at first, the user has to remember a password with five characters length (level 1). If the user can pass this level, he/she will go to the second level with a password with eleven characters (level 2). Likewise to Travel Password, but the length of a password is measured in the number of city/activity pairs instead.

4.1 Recall Rate

This section evaluates Travel Password scheme in term of recall rate. The recall rate is measured from the ratio of the number of users who can pass the test in each level (see step 4 in Section 4), i.e., able to recall the password, over the total number of attempted users.

Table 2 compares the recall rate of the traditional textual password and Travel Password scheme. The table presents the number of attempted users who can pass the confirmation test (see step 2 in Section 4) in that level (*Attempt*), the number of user who can recall and pass the level (*Recall*) and the ratio between the two numbers (*Recall Rate*). In particular, because some users might fail the confirmation test, the number of attempted users in a level might be smaller than the number of successful users from the previous level. From the table, the Travel Password scheme performs better in almost every level, especially, the level 3 and 4 where the entropy are high, 96 and 128 bits,

Table 1. Comparison of Length and Entropy

Level	Desired Entropy (bit)	Textual Password Length (Characters)	Textual Password Entropy (bit)	Travel Password Length (Pairs)	Travel Password Entropy (bit)
1	32	5	32.4	3	38.5
2	64	11	71.4	6	76.4
3	96	16	103.9	8	101.9
4	128	21	136.3	11	139.9

Table 2. Number of Attempt, Recall and Recall Rate

Level	Textual Password			Travel Password		
	Attempt	Recall	Recall Rate	Attempt	Recall	Recall Rate
1	40	33	0.8250	27	27	1.0000
2	30	26	0.8667	20	16	0.8000
3	21	17	0.8095	12	12	1.0000
4	12	10	0.8333	10	10	1.0000

respectively. In particular, at level 3 and 4, all users who use Travel Password scheme can remember and recall their password. The result shows that Travel Password scheme allows users to recall with greater success when compared with the traditional textual password scheme.

4.2 Memorability

Besides recallable, a password should also be memorable, i.e., easy to remember. In order to measure the memorability, the number of attempt to pass the confirmation test (see step 2 in Section 4) is considered. If a password is easier to remember, the users will take less attempt to pass the confirmation test. However, if the password is hard to remember, the users might need to perform confirmation test many times before they can pass it.

Table 3 shows the number of users who can pass the confirmation test according to the number of attempts, i.e., in the columns *1*, *2* and *3*, and the number of users who cannot pass the confirmation test in the column *x*. This table compares the results of the traditional textual password and Travel Password scheme in each level. The table clearly shows that the number of attempt of Travel Password scheme is lower than that of textual password. In Travel Password, most of the users can pass the confirmation test in the first attempt. In particular, 90% of users can pass the confirmation test in the first attempt in the Travel Password, compared with 58% of the textual one. The result confirms that Travel Password scheme allows users to use less attempt in order to remember the password, compared with traditional textual password scheme.

From Table 2, at the second level, the recall rate of Travel Password is lower than the other levels. Table 3 gives an insight into this issue; there are four persons who did not pass the confirmation test in that level. The experiment records show that their randomly generated password contains a popular city name, i.e., Muang, which is in many provinces. This replicated name confuses those users so they answer incorrect Muang. This indicates the city names have to be carefully chosen such that there are no replication among them.

Table 3. Number of Users According to the Number of Attempt

Level	Textual Password				Travel Password			
	1	2	3	x	1	2	3	x
1	25	7	1	7	26	1	-	-
2	19	6	1	4	14	2	-	4
3	14	2	1	4	12	-	-	-
4	7	2	1	2	9	1	-	-

5 Conclusion

In this work, a new password scheme, called Travel Password, is purposed. This scheme allows a strong password, i.e., password with high entropy, to be easier to remember and recall. By augmenting mnemonic device with a password, users can remember it better than just a sequence of characters, i.e., as in traditional textual password. The mnemonic device in the password is presented as pairs of city and activity. The experimental results confirm that the proposed scheme allow users to better remember and recall the password. In the future works, firstly, the experiment will be performed offline with more tightly control on the experimental process and samples. Moreover, the impact of Travel Password scheme on the long-term memory will be evaluated.

References

1. Miller, G.A.: The magical number seven, plus or minus two: Some limits on our capacity for processing information. The Psychological Review 63, 81–97 (1956)
2. Wiedenbeck, S., Waters, J., Birget, J.C., Brodskiy, A., Memon, N.D.: Passpoints: Design and longitudinal evaluation of a graphical password system. International Journal of Man-Machine Studies 63(1-2), 102–127 (2005)
3. NSA (2009), `http://www.nsa.gov/ia/programs/suiteb_cryptography/index.shtml`
4. Orman, H., Hoffman, P.: Determining Strengths For Public Keys Used For Exchanging Symmetric Keys. RFC 3766 (Best Current Practice) (2004)
5. Gruneberg, M.M.: The role of memorization techniques in finals examination preparation–A study of psychology students. Educational Research 15(2), 134–139 (1973)
6. Shannon, C.E.: A mathematical theory of communication. Bell System Technical 27(3), 379–423 (1948)
7. McDonald, D.: A Convention for Human-Readable 128-bit Keys. RFC 1751 (Informational) (December 1994)

8. Blonder, G.E.: Graphical passwords. United State Patent 5559961 (1996)
9. Yampolskiy, R.: User authentication via behavior based passwords. In: IEEE Long Island Systems, Applications and Technology Conference (2007)
10. Gani, A.: A new algorithm on graphical user authentication (gua) based on multi-line grids. Scientific Research and Essays 5(4), 3865–3875 (2010)
11. Dhamija, R., Perrig, A.: Déjà vu: A user study using images for authentication. In: USENIX Security Symposium (2000)
12. Jermyn, I., Mayer, A., Monrose, F., Reiter, M.K., Rubin, A.D.: The design and analysis of graphical passwords. In: USENIX Security Symposium (1999)
13. Microsoft Corporation: Personalize Your PC (2013), http://windows.microsoft.com/en-us/windows-8/personalize-pc-tutorial/
14. Hafiz, M.D., Abdullah, A.H., Ithnin, N., Mammi, H.K.: Towards identifying usability and security features of graphical password in knowledge based authentication technique. In: Second Asia International Conference Modeling Simulation (2008)
15. Komanduri, S., Hutchings, D.R.: Order and entropy in picture passwords. In: Graphics Interface Conference. Oxford University Press (2008)
16. Robusto, C.C.: The cosine-haversine formula. The American Mathematical Monthly 64(1), 38–40 (1957)

Performance Measurement of Higher Education Information System Using IT Balanced Scorecard

Nunik Afriliana[1] and Ford Lumban Gaol[2]

[1] Universitas Multimedia Nusantara, Jakarta, Indonesia
[2] Bina Nusantara University, Jakarta, Indonesia
nunik@umn.ac.id, fgaol@binus.edu

Abstract. Extensive research was conducted at a private university in Indonesia into the performance of the higher education information system, called SIPERTI. The IT Balanced Scorecard framework consisting of its four perspectives: Corporate Contribution, User Orientation, Operational Excellence, and Future Orientation was employed to assess the system. The study was accomplished in the form of a questionnaire composed of five-point Likert scale statements. The questionnaire was addressed to the members of the faculties and staff of the university who used SIPERTI in their everyday work. The data obtained was statistically analysed including the tests of reliability and validity. A structured interview which followed the questionnaire allowed for the formulation of recommendations on SIPERTI performance improvement.

Keywords: performance measurement, higher education, information system, IT balanced scorecard, SIPERTI.

1 Introduction

The Balanced Scorecard (BSC) is the most widely adopted performance management framework used in business and industry, government, and nonprofit organizations. BSC was introduced by Kaplan and Norton [1], [2], [3], [4] and based on the assumption that the evaluation of a company should not be confined to a traditional financial measures but should supplemented with other crucial operational measures which determined financial success. They stated that the financial evaluation should be extended to include the measures concerning additional perspectives such as customer satisfaction, internal processes and the ability to innovate. Keeping all four perspectives in balance should result in assuring future financial outcome and leading the company toward its strategic goals. According to the Balanced Scorecard Institute (BSI) the Balanced Scorecard has evolved over time from a simple performance measurement framework to a full strategic planning and management system [5].

Van Grembergen adopted BSC for use in the department of information technology in organizations [6], [7], [8], [9]. The department of information technology is the internal service provider so that perspective used to be changed and adjusted. By seeing that their users are internal, they proposed to transform the traditional BSC perspectives: Financial, Customer, Internal Business Process, and

N.T. Nguyen et al. (Eds.): ACIIDS 2014, Part I, LNAI 8397, pp. 412–421, 2014.

Learning & Growth into Corporate Contribution, User Orientation, Operational Excellence, and Future Orientation perspectives, respectively. The resulting IT Balanced Scorecard (IT BSC) is shown in Table 1 The User Orientation perspective represents the user assessment of IT. The Operational Excellence perspective delineates the IT processes utilized in developing and supporting applications. The Future Orientation perspective indicates the human and technology resources needed by IT to deliver its services over time. The Business Contribution perspective characterizes the business value acquired from the IT investments. These perspectives should be translated into corresponding metrics and measures that allow for the assessment of the current situation. The evaluations should be carried out periodically and aligned with pre-established goals and benchmarks.

Table 1. Standard IT Balanced Scorecard

USER ORIENTATION	**BUSINESS CONTRIBUTION**
How do users view the IT department? **Mission** To be the preferred supplier of information systems. **Objectives** • Preferred supplier of applications • Preferred supplier of operations • Partnership with users • User satisfaction	How does management view the IT department? **Mission** To obtain a reasonable business contribution from IT investments. **Objectives** • Control of IT expenses • Business value of IT projects • Provision of new business capabilities
OPERATIONAL EXCELLENCE	**FUTURE ORIENTATION**
How effective and efficient are the IT processes? **Mission** To deliver effective and efficient IT systems and services. **Objectives** • Efficient and effective developments • Efficient and effective operations	How well is IT positioned to meet future needs? **Mission** To develop opportunities to answer future challenges. **Objectives** • Training and education of IT staff • Expertise of IT staff • Research into emerging technologies • Age of application portfolio

IT BSC is considered as a measurement and management system that supports the IT governance, i.e. the processes that ensure the effective and efficient utilization of IT to attain the mission and strategic goals by an organization [10], [11], [12]. IT BSC is also used as a tool to measure the success of the IT department in the implementation of an information system, its performance level and its contribution to the organization [13], [14], [15]. Many works are devoted to the extension of IT BSC to measure and control the implementation and usage of Enterprise Resource Planning (ERP) systems [16], [17], [18], [19], [20]. The Balanced Scorecard was also employed in higher education to measure IT performance [21] and to support a management information system [22].

Along with the rapid growth of information technology, an information system to support the administrative processes in a college, university or higher education institution is absolutely needed nowadays. Higher Education Information System called SIPERTI is an online computer-based information system built with the aim to organize the academic data in the colleges in Indonesia. Its main functions include: management of student registration system, financial administration, faculty and employee data administration, library administration, admissions, tuition system, enrolment system, lectures monitoring, exam and grading system, and issuance of graduate diplomas. The study reported in this paper was motivated by problems with the usage of the SIPERTI information system and the need to streamline it. The goal of the research is to measure and evaluate the performance of the SIPERTI information system and formulate recommendations on its performance improvement.

2 Research Methodology

The research focused on the measurement and evaluation of the performance of the SIPERTI higher education information system using IT BSC with its four perspectives, i.e. Corporate Contribution, User Orientation, Operational Excellence, Future Orientation. It was a case study conducted at a private university located in Jakarta, Indonesia. The IT BSC framework was chosen for the following reasons:

- IT BSC is commonly used performance evaluation methodology and is regarded as the most flexible one.
- IT BSC was proven to increase the competitive advantage of the organization.
- IT BSC allows for determining appropriate measures and metrics for evaluating the effectiveness of the system.
- IT BSC is simpler than any other framework.
- IT BSC can be easily tailored to organization needs.

IT BSC measurement

The IT BSC measurement determined for the purposes of the study is presented in Table 2. It was based of the literature review and implements the best practices developed by other researchers working on similar problems.

Likert Scale

A Likert scale is an ordered scale from which respondents choose one option that best corresponds their view. This scale was developed by Rensis Likert in 1932 [23]. It is often used to measure respondents' attitudes or feelings by asking the extent to which they agree or disagree with a particular question or statement. A typical scale is "*Strongly disagree, Disagree, Neutral, Agree, Strongly agree*". At the moment the Likert scale is the most widely used method of scaling in behavioural and social sciences due to its ease to construct and high reliability. The scale is also commonly utilized in study on the performance and effectiveness of information systems [24], [25], [26], [27].

Table 2. IT Balanced Scorecard measurement

Perspective	Objective	Indicator	Measurement
Corporate Contribution	Cost control	System development in the budget and implementation of the system lowering operational costs	- Decrease operational costs after implementing SIPERTI - Reduction in the number of staff after implementation of SIPERTI - Development of SIPERTI is within the budget
	Provision of new business capabilities	- System suitability for the business strategy - Application strategic planning	- Application in a line with the business strategy - Application supports the business strategy - Design of applications tailored to the business strategy
	Business value of IT projects	Payback period	Payback period at the time determined
User Orientation	Cooperation with users	The level of user involvement in the development and implementation of application	- The number of meetings with the users - User involvement in determining the needs of the application
	User satisfaction	The level of user satisfaction with the application	- Application ease of use - Data format generated by the application - Application timelines
	The best solution providers	- Intensity of use of the application - Other applications besides SIPERTI - Dependence on SIPERTI	- Duration of the application use - The number of menus / modules used - Other applications besides SIPERTI - The number of transactions completed according to schedule
Operational Excellence	Efficiency of application development	- Development of applications according to business and user needs - The application development (new module) on schedule	- Development of applications according to business and user needs - The application development (new module) on schedule
	Efficiency of computer operational	- Application availability level - Application downtime level - Application response time - Scope of automation - Facilitate the work - Save the work time	- Application availability level - Application downtime level - Application response time - Scope of automation - Facilitate the work - Save the work time
	Efficiency of help desk functions	- Speed of troubleshooting - Precision of problem handling	- Problem of applications resolved quickly and on time - IT staff understands application problems
Future Orientation	User training	- User training index - Training effectiveness	- Number of user training in one period - Training of a new module - Benefits of training

Population

The questionnaire was addressed to the members of the faculties and staff of the university who used SIPERTI in their everyday work. As for the faculty, the survey audience was composed of 14 lecturers including 7 heads and 7 deputy heads of study programs. In turn, the survey audience consisted of 33 SIPERTI users from HR, Finance, Marketing, Admissions, and Academic departments. Moreover, 5 persons from the top-level management were invited to complete the questionnaire. In total 52 respondents constituted the population of the survey.

Data Processing

Data analysis included both descriptive and inferential statistical methods. Percentage and frequency tables, charts, and graphs were used to collate, analyse and present data. Reliability and validity test were also made.

Each measurement including a questionnaire should be tested in respect of its reliability and validity [28], [29]. The former refers to how consistent the measurement is, i.e. to what extent a given survey can provide similar results if used again in similar conditions. A questionnaire is reliable if the results are consistent or stable over time. In turn, the latter refers to whether the investigation examines what it intends to examine. A questionnaire is valid if its questions and/or statements are able to reveal something that is expected to be measured. The validity of a survey relies primarily on reliability. The most recognized estimation of reliability is the Cronbach's alpha [30], [31]. The Cronbach's alpha coefficient ranges in value from 0 to 1 and the score of 0.7 or higher is considered acceptable reliability. The application of Cronbach's alpha to assess the reliability of surveys based on the Likert scale is presented in [32], [33].

The results of questionnaires were processed by calculating the average of each statement filled by the respondents. The results of the measurement were then projected onto the grading scale used in the organization in order to make them easily understood by the management. The IT Balanced Scorecard grading scale is presented in Table 3.

Table 3. IT Balanced Scorecard grading

Category	Index
Outstanding	4.41 – 5.00
Exceed Expectation	3.51 – 4.40
Meet Expectation	2.51 – 3.50
Need Improvement	1.61 – 2.50
Unacceptable	1.00 – 1.60

The weights of individual perspectives were calculated based on the number of statements in each perspective and the results of interviews with university management. The weights used to calculate the final score are shown in Table 4.

Table 4. IT Balanced Scorecard weighting

Perspective	No. of statements	Percentage	Weights
Corporate Contribution	7	23%	0.23
User Orientation	11	35%	0.35
Operational Excellence	10	32%	0.32
Future Orientation	3	10%	0.10

3 Analysis of IT Balanced Scorecard Results

Respondent Profile

Table 5 shows the profile of questionnaire respondents. The number of respondents in User Orientation, Operational Excellence, and Future Orientation perspectives was 52, while the respondents in the Corporate Contribution perspective were 26.

Table 5. Profile of respondents

Feature	Groups	Participants	Percentage
Gender	a. Man	21	40.38 %
	b. Woman	31	59.62 %
Age	a. <=25 years old	10	19.23 %
	b. 26-30 years old	16	30.77 %
	c. 31-35 years old	9	17.31 %
	d. 36-40 years old	10	19.23 %
	e. >40 years old	7	13.46 %
Position	a. Lecturer	14	27.45 %
	b. Staff	33	62.75 %
	c. Top Management	5	9.80 %
Education	a. Doctoral	5	9.62 %
	b. Graduate	16	30.77 %
	c. Undergraduate	31	59.62 %
Work Experience	a. <2 years	15	28.85 %
	b. 2-3 years	15	28.85 %
	c. >3 years	22	42.31 %

Table 6. Results of reliability test

Perspective	Cronbach's Alpha	Conclusion
Corporate Contribution	0.846	Reliable
User Orientation	0.833	Reliable
Operational Excellence	0.792	Reliable
Future Orientation	0.746	Reliable

Validity and Reliability Tests
In this research reliability was estimated using Cronbach's alpha. The values of the coefficient are shown in Table 6 and they are greater than 0.7 for all perspectives. The results revealed that all questions formulated in the questionnaire were reliable.

The validity test was conducted by analysing the correlation value of each statement in every IT BSC perspective to the total correlation value. The measurement results showed that all formulated questions were valid.

IT BSC Measurement Results
Results of IT BSC measurements of SIPERTI are depicted in Fig. 1. The individual scores for the Corporate Contribution, User Orientation, Operational Excellence, and Future Orientation perspectives amounted to 3:01, 2.65, 2.69, and 2.97, respectively. The aggregate result computed as the weighted average of the scores for individual perspectives was equal to (0.23 x 3.01) + (0.35 x 2.65) + (0.32 x 2.69) + (0.10 x 2.97) = 2.78. Taking into account the IT Balanced Scorecard grading scale used at the university (see Table 3) it falls into the "Meet Expectation" category.

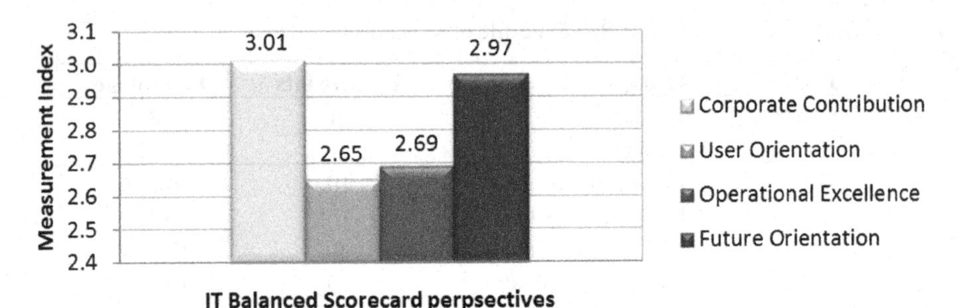

Fig. 1. Results of IT Balanced Scorecard measurement

4 Conclusions and Future Work

Extensive research was conducted at a private university in Indonesia into the performance of the higher education information system, called SIPERTI. The IT Balanced Scorecard framework consisting of its four perspectives: Corporate Contribution, User Orientation, Operational Excellence, and Future Orientation was employed to assess the system. The study was accomplished in the form of a questionnaire composed of five-point Likert scale statements. The members of the faculties and staff of the university who used SIPERTI in their everyday work completed the questionnaire. The data obtained was statistically analysed including the tests of reliability and validity. The weighted average of IT balanced Scoreboard measurement was equal to 2.78. It means that the SIPERTI information system was graded "Meet Expectation". It was concluded that the performance of information systems was not at the level required by the university management.

Therefore, based on a structured interview which followed the questionnaire a complex of recommendations on SIPERTI performance improvement was formulated. They are presented in Table 7.

It is planned to repeat the evaluation of the SIPERTI system with an extended set of metrics adjusted to the needs of the university after reviewing by the stakeholders. Moreover, research with a larger number of respondents and more varied in terms of demographic data might allow for considering demographic data as one of the factors that influence the assessment of the information system performance.

Table 7. Recommendations for SIPERTI performance improvement

Perspectives	Recommendations
Corporate Contribution	- Use waterfall model for SIPERTI development - Build an agreement with stakeholders about the scope of SIPERTI development. - Alignment of SIPERTI development with the organization business strategy
User Orientation	- More user involvement in determining the need of SIPERTI - Giving sufficient alert (error message) for SIPERTI users - Designing the SIPERTI menu in a coherent way from one process to another processes - Providing manual book for SIPERTI that has not existed until now - Providing help facilities for the SIPERTI user - Increased data accuracy generated by SIPERTI so there will be no error data interpretation and analysis - Provide tailored data format a line with the user requirements. - Speed up the improvements on the SIPERTI menus cannot be used by the user yet. - SIPERTI error handling to be done more quickly and accurately
Operational Excellence	- Specific training for IT staff responsible for SIPERTI - Increasing the number of the IT staff responsible for development and maintenance of applications - Decreasing the down time of SIPERTI
Future Orientation	- User training periodically - Training for the new modules - Training for the top level management a line with their need - More user involvement in determining the need of SIPERTI

References

1. Kaplan, R., Norton, D.: The balanced scorecard – measures that drive performance. Harvard Business Review, 71–79 (January-February 1992)
2. Kaplan, R., Norton, D.: Putting the balanced scorecard to work. Harvard Business Review, 134–142 (September-October 1993)
3. Kaplan, R., Norton, D.: Using the balanced scorecard as a strategic management system. Harvard Business Review, 75–85 (January-February 1996)

4. Kaplan, R., Norton, D.: The balanced scorecard: Translating vision into action. Harvard Business School Press, Boston (1996)
5. Balanced Scorecard Institute, `http://balancedscorecard.org/Resources/AbouttheBalancedScorecard/tabid/55/` (accessed December 30, 2013)
6. Van Grembergen, W., Van Bruggen, R.: Measuring and improving corporate information technology through the balanced scorecard technique. In: Proceedings of the Fourth European Conference on the Evaluation of Information Technology, Deflt, pp. 163–171 (October 1997)
7. Van Grembergen, W., De Haes, S.: The Balanced Scorecard and IT Governance. Information Systems Control Journal 2 (2000)
8. Van Grembergen, W., Saull, R., De Haes, S.: Linking the IT balanced scorecard to the business objectives at a major Canadian financial group. Journal of Information Technology Cases and Applications 5(1TY-JOUR) (2003)
9. Van Grembergen, W., De Haes, S.: Measuring and Improving IT Governance Through the Balanced Scorecard. Information Systems Control Journal 2 (2005)
10. Saull, R.: The IT Balanced Scorecard: A Roadmap to Effective Governance of Shared IT Organization. Information Systems Control Journal 2 (2000)
11. Keyes, J.: Implementing the IT Balanced Scorecard: Aligning IT with Corporate Strategy. Auerbach Publications, Boca Raton (2005)
12. Cram, A.: The IT Balanced Scorecard Revisited. Information Systems Control Journal 3 (2007)
13. Martinsons, M., Davison, R., Tse, D.: The balanced scorecard: A foundation for the strategic management of information systems. Decision Support Systems 25, 71–88 (1999)
14. Kim, D.J., Yue, K.-B., Al-Mubaid, H., Hall, S.P., Abeysekera, K.: Assessing Information Systems and Computer Information Systems Programs from a Balanced Scorecard Perspective. Journal of Information Systems Education 23(2), 177 (2012)
15. Brenner, W., Uebernickel, F., Győry, A., Cleven, A.: Finding Balanced Scorecards for Business Driven IT Service Portfolio Management: A Literature Review. International Journal of IT/Business Alignment and Governance 3(1), 63–78 (2012)
16. Rosemann, M., Wiese, J.: Measuring the Performance of ERP Software – A Balanced Scorecard Approach. In: Proceedings of the 10th Australasian Conference on Information Systems, pp. 773–784 (1999)
17. Chand, D., Hachey, G., Hunton, J., Owhoso, V., Vasudevan, S.: A balanced scorecard based framework for assessing the strategic impacts of ERP systems. Computers in Industry 56(6), 558–572 (2005)
18. Lin, H.-Y., Hsu, P.-Y., Ting, P.-H.: ERP Systems Success: An Integration of IS Success Model and Balanced Scorecard. Journal of Research and Practice in Information Technology 38(3), 215–228 (2006)
19. Kronbichler, S.A., Ostermann, H., Staudinger, R.: A Comparison of ERP-Success Measurement Approaches. Journal of Information Systems and Technology Management 7(2), 281–310 (2010)
20. Batada, I., Rahman, A.: Measuring System Performance & User Satisfaction after Implementation of ERP. In: Proceedings of Informing Science & IT Education Conference (InSITE), pp. 603–611 (2012)
21. Maria, E., Fibriani, C., Sinatra, L.: The Measurement of Information Technology Performance in Indonesian Higher Education Institutions in the Context of Achieving Institution Business Goals Using Cobit Framework Version 4.1 (Case Study: Satya Wacana Christian University, Salatiga). Researchers World – Journal of Arts Science & Commerce 3:3(3), 9–19 (2012)

22. Kettunen, J., Kantola, I.: Management information system based on the balanced scorecard. Campus-Wide Information Systems 22(5), 263–274 (2005)
23. Likert, R.: A Technique for the Measurement of Attitudes. Archives of Psychology (140) (1932)
24. Ray, M.N., Houston, T.K., Yu, F.B., Menachemi, N., Maisiak, R.S., Allison, J.J., Berner, E.S.: Development and Testing of a Scale to Assess Physician Attitudes about Handheld Computers with Decision Support. Journal of the American Medical Informatics Association 13(5), 567–572 (2006)
25. Wong, M.S., Nishimoto, H., Philip, G.: The Use of Importance-Performance Analysis (IPA) in Evaluating Japan's E-government Services. Journal of Theoretical and Applied Electronic Commerce Research 6(2), 17–30 (2011)
26. Gupta, D.: Impact of Computer Based Information Systems on Organisation Performance in Videocon Ltd. Research Cell: An International Journal of Engineering Sciences 8, 36–43 (2013)
27. Suzar, J.A.S.: End-user satisfaction with the Integrated System of the Federal Government Financial Administration (SIAFI): A case study. JJISTEM - Journal of Information Systems and Technology Management 10(1), 145–160 (2013)
28. Boulianne, E.: Empirical Analysis of the Reliability and Validity of Balanced Scorecard Measures and Dimensions. In: Epstein, M.J., Lee, J.Y. (eds.) Advances in Management Accounting, vol. 15, pp. 127–142. Emerald Group Publishing Limited (2006)
29. Sekaran, U., Bougie, R.: Research Methods for Business, 5th edn. John Willey & Sons Ltd., West Sussex (2009)
30. Cronbach, L.J.: Coefficient alpha and the internal structure of tests. Psychometrika 16, 297–334 (1951)
31. Nunnaly, J.: Psychometric theory. McGraw-Hill, New York (1978)
32. Gliem, J.A., Gliem, R.R.: Calculating, Interpreting, and Reporting Cronbach's Alpha Reliability Coefficient for Likert-Type Scales. In: Proceedings of the 2003 Midwest Research to Practice Conference in Adult, Continuing, and Community Education, pp. 82–88 (2003)
33. Santos, J.R.A.: Cronbach's Alpha: A Tool for Assessing the Reliability of Scales. Journal of Extension 37(2) (1999)

Decisional DNA Based Framework for Representing Virtual Engineering Objects

Syed Imran Shafiq[1], Cesar Sanin[1], Edward Szczerbicki[2], and Carlos Toro[3]

[1] The University of Newcastle, University Drive, Callaghan, 2308, NSW, Australia
{syedimran.shafiq,Cesar.Maldonadosanin}@uon.edu.au
[2] Gdansk University of Technology, Gdansk, Poland
Edward.Szczerbicki@newcastle.edu.au
[3] Vicomtech-IK4, San Sebastian, Spain
ctoro@vicomtech.org

Abstract. In this paper, we propose a frame-work to represent the Virtual Engineering Objects (VEO) utilizing Set of Knowledge Experience Structure (SOEKS) and Decisional DNA. A VEO will enable the discovery of new knowledge in a manufacturing unit and the generation of new rules that drive reasoning. The proposed VEO framework will not only be knowledge based representation but it will also have its associated experience embedded within it. This concept will evolve and discover implicit knowledge in industrial plant, which can be beneficial for the engineers and practitioners. A VEO will be a living representation of an object; capable of adding, storing, improving and sharing knowledge through experience, similar to an expert of that area.

Keywords: Decisional DNA (DDNA), Set of Experience Knowledge Structure (SOEKS), Virtual Engineering Objects (VEO).

1 Introduction

The manufacturing companies are facing intense pressure from the market, demanding customers and technological advancement. So, need of the hour is to take effective decisions, to develop products within time and cost constraints without compromising on the quality. In order to achieve these objectives, companies are taking support of various technologies, Knowledge based engineering is one such technology. Organizations generate new knowledge by solving problems, however due to lack of appropriate knowledge management techniques; knowledge has to be reprocessed in order to solve new problems with similar conditions. Knowledge representation reuses this information to make future decisions in a more efficient way, without wasting time and resources [1].

The use of knowledge based manufacturing and industrial design is arguably unexplored. Researches have shown that a large percentage of time during industrial design is spent on routine tasks. It is also observed that around 20% of the designer's time is spent searching for and absorbing information, and '40% of all design information requirements are met by personal stores [2]. It is clear that industrial design

N.T. Nguyen et al. (Eds.): ACIIDS 2014, Part I, LNAI 8397, pp. 422–431, 2014.

and manufacturing information is not represented in a shared and easily accessible knowledge base. Knowledge based industrial design techniques have been used in the past with fair bit of success. Nevertheless they have their share of limitations like they may be time consuming, costly, domain specific and at times not very intelligent.

This work proposes a novel concept of Virtual Engineering Objects (VEO). A VEO will have all the knowledge of the engineering artifact along with the associated experience embedded in it. This will help the practitioners in effective decision making based on the past experience. A novel technique of knowledge representation called Set of Experience knowledge structure (SOEKS) and Decisional DNA [3] is used for developing VEO.

The structure of this paper is as follows-section 2 describes the concept of SOEKS and DDNA. In section 3, we introduce the idea of VEO. In section 4, implementation and formulation of the VEO is discussed. In section 5, we conclude this paper and section 6presents ideas for further research.

2 Set of Knowledge Experience Structure (SOEKS) and Decisional DNA

As discussed in section 1, a large amount of previous knowledge is needed to design and manufacture a new component; the information may not be exactly the same but may be from the family of the related object. However, it has been observed that not much effort is made in the past to retain the knowledge. Knowledge and experience are lost indicating that there is a clear deficiency on its collection and reuse [4]. Some of the reasons for lack of knowledge base are:

- the non-existence of a common knowledge-experience structure which is able to collect multi-domain formal decision events, and
- the non-existence of a technology able to capture, store, improve, retrieve and reuse such collected experience [5].

Sanin and Szczerbicki proposed a new smart knowledge based decision support tool called Set of Knowledge Experience Structure (SOEKS) and Decisional DNA, having three important elements:

- a knowledge structure able to store and maintain experiential knowledge.
- a solution for collecting experience that can be applied to multiple applications from different domains.
- a way to automate decision making by using such experience, that is, retrieve collected experience by answering a query presented [3, 6, 7].

The SOEKS is a compound of variables (V), functions (F), constraints (C) and rules (R), which is uniquely combined to represent a formal decision event. Functions define relations between a dependent variable and a set of input variables; therefore, SOEKS uses functions as a way to establish links among variables and to construct multi-objective goals (i.e., multiple functions). Similarly, constraints are functions that act as a way to limit possibilities, restrict the set of possible solutions, and control the performance of the system with respect to its goals. Finally, rules are used to represent inferences and correlate actions with the conditions under which they should be executed. Rules are relationships that operate in the universe of variables and express the connection between a condition and a consequence in the form 'if then else' [3, 8].

Chromosomes are groups of (Set of Experience) SOE that can accumulate decisional strategies for a specific area of an organization. Multiple SOE can be collected, classified, and organized according to their efficiency, grouping them into decisional chromosomes. Finally, sets of chromosomes comprise what is called the Decisional DNA of the organization as shown in Figure 1.

DDNA is a metaphor of human DNA, because of its ability to capture and carry information (experience and knowledge).It has four elements provided by the SOEKS (variables, functions, constants and rules) which can be related in the same manner to the elements in DNA (adenine, thymine, guanine and cytosine).

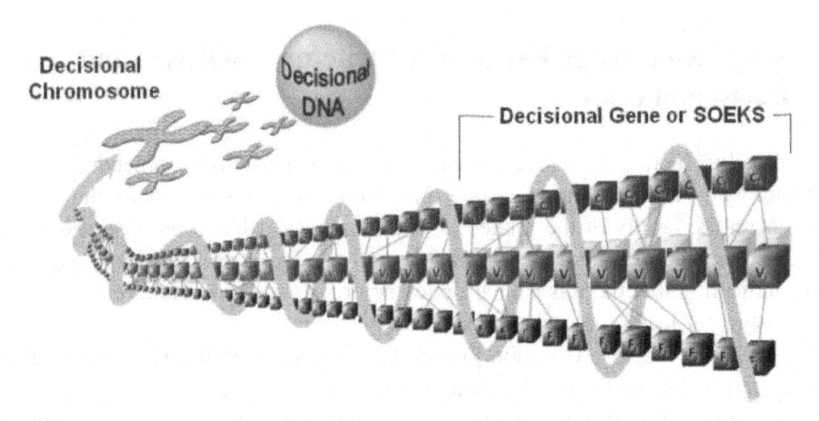

Fig. 1. SOEKS and Decisional DNA [5, 6]

3 Virtual Engineering Objects (VEO)

A Virtual Engineering Object (VEO) is a living representation of an artefact having knowledge and experience embedded within it and can behave like an expert of that area [9].

3.1 Is VEO knowledge Representation?

In this section we examine that whether a VEO can be termed as knowledge representation. According to Davis [10], a knowledge representation notion can be understood in terms of five fundamental roles that it play:

- A Knowledge Representation is a surrogate.

Viewing knowledge representations as surrogates leads naturally to two important questions. The first question about any surrogate is its intended identity: what is it a surrogate for? There must be some form of correspondence specified between the surrogate and its intended referent in the world; the correspondence is the semantics for the representation. In VEO case, this surrogate will be mainly oriented to decision making process for industrial plants.

The second question is fidelity: how close is the surrogate to the real thing? Which attributes of the original does it captures and make explicit, and what does it omit? Perfect fidelity is generally impossible, both in practice and in principle. It is impossible in principle because the only completely accurate representation of an object is the object itself.

VEO model intends to be the most complete possible model for a specified domain. Probably, it won't be necessary to have a perfect geometric model of each element, but many physical characteristics are fundamental, as they are the set of requirements and characteristics specified for each artifact. What these characteristics and requirements are that will be discussed in section 3.2. It is also important to highlight the experience that can be obtained from the use of that artifact.

- A Knowledge Representation is a set of ontological commitments.

Selecting a representation means making a set of ontological commitments. The commitments are in effect a strong pair of glasses that determine what we can see, bringing some parts of the world into sharp focus, at the expense of blurring others.

In VEO case, our domain is the industrial plant, being the focused part the industrial processes, and the blur part, all the administrative work done in a real plant.

- A Knowledge Representation is a fragmentary theory of intelligent reasoning.

In VEO case, we will use technologies like Reflexive Ontologies, SOEKS and Semantic Reasoners. This set of tools will allow us to infer new knowledge from both explicit knowledge and experience [11].

- A Knowledge Representation is a medium for efficient computation.

In VEO case, as we said before, we will use semantic tools like Reflexive Ontologies. To model these ontologies, we will use a methodology based on engineering standards. Using this methodology, we can assure that information is organized in a standard way, being the easiest way to facilitate the inference. In addition, Reflexive Ontologies are more efficient when queried than standard ontologies.

- A Knowledge Representation is a medium of human expression.

Using a methodology based on engineering standards assures that a VEO is a way of communication among people related to industry.

From the above deliberation it is established that VEO fulfills the requirements to be qualified as knowledge representation.

3.2 VEO Structure

As discussed previously, VEO is a knowledge representation for an engineering artifact. We must take into account that when we say 'an engineering artifact', we can be talking about something simple like a valve, or we can be talking about something complex like a painting cell. For such reason, the VEO specification must have a complexity level according to its functionality. In our very first approximation, we identified four different levels of VEO, as can be seen in Figure 2.

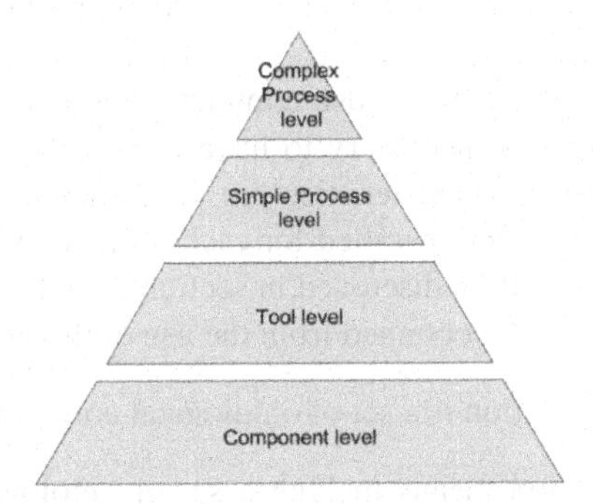

Fig. 2. VEO complexity pyramid

At Component level, VEO represent just a component (usually from any kind of machinery). By itself, this component has not any functionality that can be considered "useful" in a production process. Of course, it has its functionality in the machinery where it is part of. Examples of VEO at this level can be valves, printed circuit boards, etc.

Above the Component level is the Tool level. VEOs placed here represent those artifacts that have a basic functionality, being considered as useful unities in an industrial process. Nevertheless, they do not constitute an industrial process by itself. An example of VEO at this level can be a robot that pick an object and move it to another position.

Next level is Simple Process level. In this level, we consider that VEO represent artifacts which accomplish a full simple process. We consider a simple process those processes that made a simple change in the 'product' that is involved in it. An example could be a painter cell (where the simple process is painting; the product enters in one color and exits in another one).

Finally, at the top of the complexity pyramid is the Complex Process level. The complex process level VEO is a combination of various simple process level VEOs. An example could be car door manufacturing (where many simple processes take place, like welding, painting etc.).

After classifying VEO on the basis of complexity level, in the next step, we propose a structure for a VEO. A VEO comprises of knowledge of 5 different aspects/parts of an object to be represented, as depicted in Figure 3. Characteristics, Functionality, Requirements, Connections and Present State are designed in such a way that they consists various variables, which can represent maximum facets of the object. Besides having knowledge these different modules of a VEO have their past experience embedded in them. The main features of these modules are as follows:

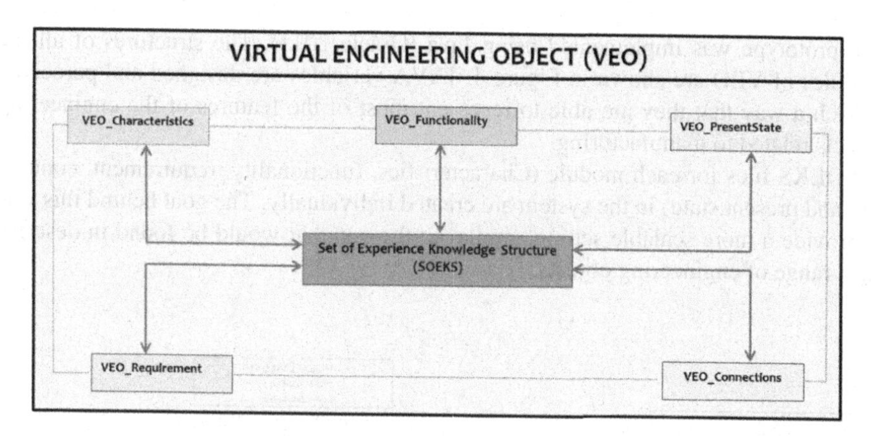

Fig. 3. Proposed VEO Structure

- *Characteristics* describe the set of expected benefits offered by the artefact represented by the VEO. Such characteristics will depend on what kind of artefact are we considering. For example, if our VEO is a SCARA robot, some characteristics will be the action range, functionalities, max speed, etc., but if our VEO is a software application, characteristics focused will be input/output formats etc.
- *Functionality* describes the basic working of the object and principle on which it works and accomplishes its operation.
- *Requirements* describe the set of necessities of the VEO for its correct work. In a similar case like characteristics, the set of requirements depend on what kind of VEO we are considering. For the SCARA Robot we can mention power supply, required space, etc. for the software application the requirements will be associated to operative system, memory etc.
- *Connections* describe how the VEO is related with other VEOs. These connections can be of different types. Some of them can be a *need* relationship, e.g. a robot that needs a computer application to control it. Other kind of relation can be, of course, *part of* relationship, e.g. a gear is part of an engine.
- The present state of the VEO indicates parameters of the VEO in the current moment. For example, information like for how much time has been this machine powered on? Or the machine is busy or idle at present? [9]

SOEKS is connected with the rest of the parts i.e. characteristics, functionality, requirements, connections and the present state and is able to capture and store new information. SOEKS contains the knowledge and experience acquired about a particular VEO over a span of time.

4 Formulation and Implementation

In the previous sections, we discussed the complexity levels and the structure of the VEO. This section presents the proposed structure of SOEKS conceived for the VEO.

The prototype was implemented using Java (Oracle 2011). The structures of all the modules of VEO are shown in Figure 4. JAVA variables are designed and perceived in such a way that they are able to represent most of the features of the engineering objects related to manufacturing.

SOEKS files for each module (Characteristics, functionality, requirement, connection and present state) in the system are created individually. The goal behind this was to provide a more scalable setting, similar to the one that would be found in describing a range of engineering objects.

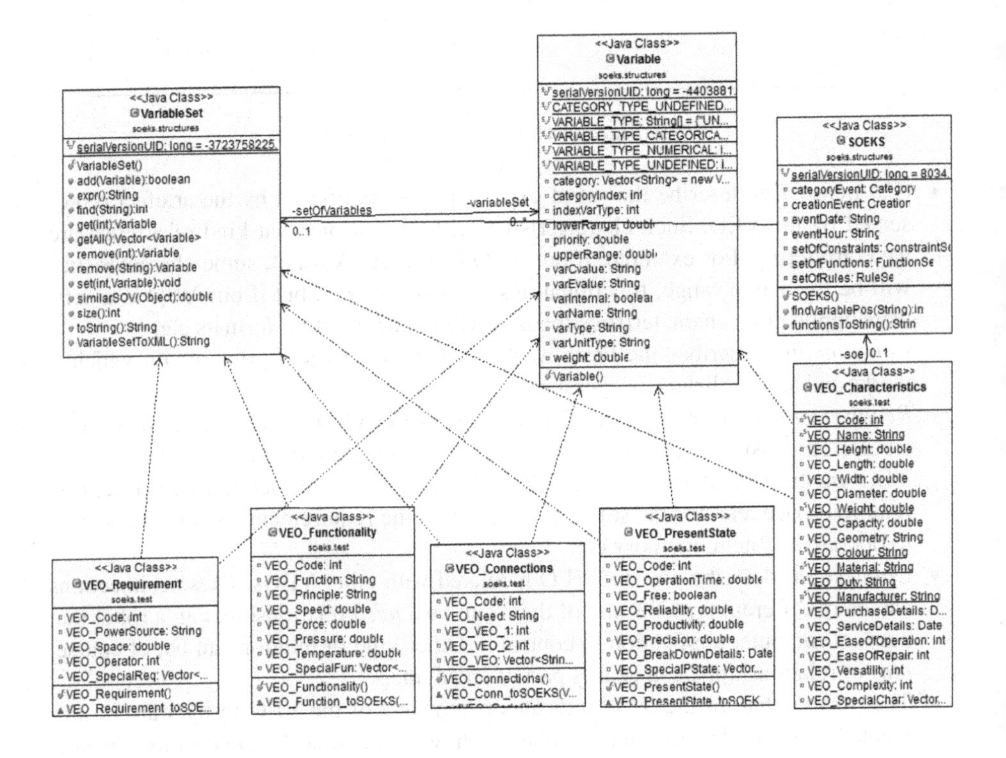

Fig. 4. Class diagram for VEO

Weights are assigned to the attributes of the variables of the above mentioned modules of an artefact, and then the five sets of SOEKS are generated. These individual SOEKS are combined under an umbrella (VEO), representing experience and prediction.

The output given below shows the XML representation of the attributes of one of the variable (Height of the VEO). In the similar fashion all the variables of the five modules of a VEO are assigned.

```
<set_of_variables>
<!-- Variables included in the model -->
<variable>
<var_name>VEO_Height</var_name>
<var_type>Numerical</var_type>
<var_cvalue>20.0</var_cvalue>
<var_evalue>20.0</var_evalue>
<unit>Metres</unit>
<internal>true</internal>
<weight>1.0</weight>
<l_range>1.0</l_range>
<u_range>1.0</u_range>
<categories>
<category>CATEGORY UNDEFINED</category></categories>
<priority>1.0</priority>
</variable>
```

Once VEO of all the engineering artifacts of a manufacturing plant is developed, then a network of the interconnected VEOs, shown in Figure 5 is established. These relations/connections will be based on the connection made in the VEO-Connections in section 3.2.

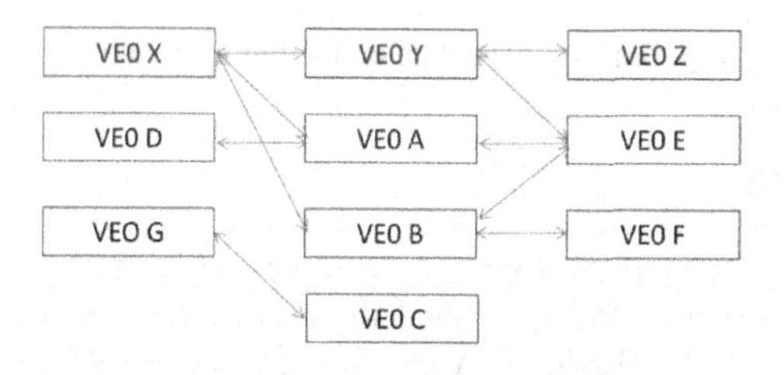

Fig. 5. Network of Interconnected VEO's

The next step in our plan, after getting the knowledge representation of all the engineering artifacts of a manufacturing unit is to extract the right knowledge out of it. Figure 6 shows the flowchart of achieving it. When a query is loaded in the DDNA, it will search its SOEKS and according to it will propose the VEO's that can perform the desired task or process. VEOs will be having their knowledge and past experience implanted in them, which in turn will help in deciding specific parameters for the queried task. So DDNA will give its complete and explicit recommendation to perform that specific task. If the recommendation is accepted and executed by the user then SOEKS will update itself by including the information of this particular task in its repository. If it is not, then DDNA will execute the query again and will present the next best solution according to its experience.

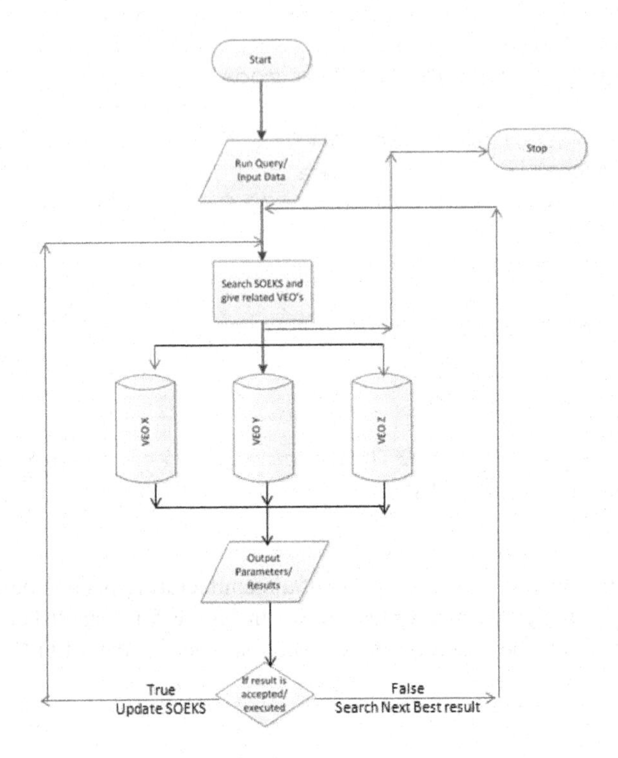

Fig. 6. Flowchart of an extracting information

5 Conclusion

In this article, we presented an approach to represent engineering artifact based on knowledge and experience. We described the architecture of our approach and implementation that uses SOEKS/DDNA to represent VEO. We demonstrated this approach through some initial tests. As the illustrative result shows, we can model and represent engineering artifact virtually, which can capture, store and reuse the associated knowledge and experience of the object. The Decisional DNA, as a novel knowledge representation structure, not only can be easily applied to the concept of VEO but indeed improves the decision making process in manufacturing units using experience.

6 Future Work

To continue with this idea, further research and refinements are required, and our efforts are currently directed towards:

- Refinement of variables that can represent VEO in a more general way.
- Further development of the VEO rule base database.
- Further development of user management based on gathered experience.

References

1. Verhagen, W.J.C., Garcia, P.B., Van Dijk, R.E.C., Curran, R.: A critical review of Knowledge-Based Engineering: An identification of research challenges. Advanced Engineering Informatics 26, 5–15 (2012)
2. Baxter, D., Gao, J., Case, K., Harding, J., Young, R., Cochrane, S., Dani, S.: An engineering design knowledge reuse methodology using process modelling. International Journal of Research in Engineering Design 18, 37–48 (2007)
3. Sanin, C., Szczerbicki, E.: Set of Experience: A Knowledge Structure for Formal Decision Events. Foundations of Control and Management Sciences 3, 95–113 (2005)
4. Sung, R.C.W., Ritchie, J.M., Lim, T., Kosmadoudi, Z.: Automated generation of engineering rationale, knowledge and intent representations during the product life cycle. Virtual Reality 16, 69–85 (2012)
5. Sanin, C., Szczerbicki, E.: Towards the Construction of Decisional DNA: A Set of Experience Knowledge Structure Java Class within an Ontology System. Cybernetics and Systems 38, 859–878 (2007)
6. Sanin, C., Toro, C., Haoxi, Z., Sanchez, E., Szczerbicki, E., Carrasco, E., Peng, W., Mancilla-Amaya, L.: Decisional DNA: A multi-technology shareable knowledge structure for decisional experience. Neurocomputing 88, 42–53 (2012)
7. Sanin, C., Szczerbicki, E.: Extending Set of Experience Knowledge Structure into a Transportable Language extensible Markup Language. Cybernetics and Systems 37, 97–117 (2006)
8. Sanín, C.: Smart Knowledge Management System, Thesis of Doctor of Philosophy Degree From The University of Newcastle Department Of Mechanical Engineering, Newcastle, Australia (2007)
9. Shafiq, S.I., Sanin, C., Szczerbicki, E., Toro, C.: Using Decisional DNA to Enhance Industrial and Manufacturing Design: Conceptual Approach. In: 34th International Conference on Information Systems Arhitecture and Technology, Szklarska Poreba, Poland, pp. 23–32. Wroclaw University of Technology (2013)
10. Davis, R., Shrobe, H., Szolovits, P.: What is a Knowledge Representation? AI Magazine 14, 17–33 (1993)
11. Toro, C., Sanín, C., Szczerbicki, E., Posada, J.: Reflexive Ontologies: Enhancing Ontologies With Self-Contained Queries. Cybernetics and Systems 39, 171–189 (2008)

A Data Quality Index with Respect to Case Bases within Case-Based Reasoning

Jürgen Hönigl and Josef Küng

Institute for Application-Oriented Knowledge Processing
Johannes Kepler University
Linz, Austria
{juergen.hoenigl,josef.kueng}@jku.at

Abstract. Within Case-Based Reasoning (CBR), terms concerning quality of a case base are mentioned in publications, but partially without clarifications of criteria. When developing a CBR system from scratch, an index for case base quality supports an assessment of the actual cases. In this approach, both theory and an application are demonstrated. An index was defined and subsequent applied within a current CBR project, which is under development. In addition, various approaches concerning case base quality are demonstrated. Big data occurs within a combination of high velocity, great volume and variety of incoming data. Defining an index to measure the case base quality copes with that.

1 Introduction

Within this section, the introduction was divided into several parts to demonstrate the motivation, a few statements about CBR and an outline.

1.1 Motivation

When reading literature about case-based reasoning, it was written about the quality of a case base and avoiding too redundant cases within case base. Various approaches are existing but partially with fuzzy definitions and primarily without clear results. Especially when researching towards an eventual re-use of an index. Therefore, the authors were defining an index to describe case base quality. This was applied within the first author's doctoral thesis as part within the proof of concept. Closing the gap between big data and CBR can be seen as a drive towards an easy to apply index for new relevant cases with respect to the size of a case base. The significance of a data quality index can be seen within the next annotations.

1.2 Significance towards a Case Base

A case base contains knowledge, which will be used for the reasoning process of a case-based reasoning system. An index, which states the quality of a case

N.T. Nguyen et al. (Eds.): ACIIDS 2014, Part I, LNAI 8397, pp. 432–442, 2014.

base, can be used within different steps of the CBR model given by Aamodt and Plaza.[1] A deletion strategy for too similar cases has to applied to a CBR system to keep the quality of a case base. A deletion strategy is one possible point to deal with the size of case base concerning the maintenance. Another point of view, establish rules for pre-processing to avoid not suitable reasoning efforts and impaired cases. For instance, a typo could cause an impaired case when not using pre-processing assertion rules. A customer with an age of 92 years (instead of 29 years) could be reasoned within a CBR system, but it would be an outlier within the case base. Subsequent, this case would be removed according to a deletion strategy, which uses the not recently used paradigm for instance. Within CBR, applying an index can combined with committing a database state. When receiving many new cases within a CBR approach, the advantage of an index can be seen to init a rollback of the database state, which reflects the case base, according to a modified index value with a percentage of minus 20 for instance. Big data occurs if a great volume, a high velocity and variety (structured and unstructured data) will be received. Even two of them can decrease the quality of a case base. A great volume of data with a high velocity can contain too many redundant and obsolete cases. Within a CBR system, pre-processing and similarity measures can avoid many inadequate data, but an assessment of the case base has to be applied in addition. When working on case mining, a complete case base without missing values should be seen as a pre-condition. For instance, gaining association models requires complete cases.[10] When considering an evolution such as IBM's (Industrial Business Machines) research projects Watson and DeepBlue within a decade, it is obvious that these projects can cope with missing values within their knowledge bases.[6], [14] In contrast, a CBR approach requires data within the case base because a CBR system is not intended to implement various application programming interfaces to download information on the fly.[12] In addition, the knowledge base of IBM's Watson contained a huge amount of text volumes, databases and journals.[7], [9]

1.3 Outline

To briefly present a red line regarding this paper, firstly, related work is demonstrated. Then, three sub-indices are demonstrated, which are required, to build the main index of this approach. Subsequent, the index will be calculated on a top level. Afterwards the application of the index will be explained within a case-based reasoning prototype. Subsequent, a discussion is presented regarding various sights when using thresholds for instance. At the end, a conclusion and eventual future work are enumerated.

2 Related Work

This section demonstrates chronological various possibilities concerning the term case base quality within literature. In 1997, an approach was stated to combine decision theory and CBR. This idea could be used if many missing values would occur to use CBR together with decision theory within an area like unfinished alternatives. Therefore, considering of quality weakness within a case base could be compensated. On the other side, their approach was an experiment and explained difficulties when combing two kind of decision support technologies. For instance, they have detected obstacles when using normative models due to the application of probability and utility for preference and judgement in combination with CBR.[19]

A historic approach given in 1998 refers to non-functional requirements regarding CBR systems. Their approach was applied within the medical domain. The efforts made were primarily focused on a CBR system instead of the managed data. An intersection between their system-related approach and a data-related approach can be seen within their work on confidentiality and integrity of data.[11]

The quality improvement paradigm (QIP) refers to steps to consider when developing a CBR system. Basili presents a cycle to gain a good combination of technical and managerial solution to achieve a professional CBR application development. The experience factory refers within various steps to different issues, which seems like a waterfall structure at first sight. However, these steps can be partially used in an iterative way, which avoids that. To give a brief explanation concerning this paradigm, two quality-related steps are stated. Within *characterize* (QIP1), the scope of the project will be defined, which results into a context for a goal definition. In addition, experience from the experience base can be selected. The experience base is a knowledge base of past projects related to achieved experience. *Set goals* (QIP2) consider different viewpoints such as customer, project manager and user. The defined goals must be measurable.[4]

Within an old approach presented in 2000, quality measures were defined to assess the case base quality with criteria such as correctness, consistency, uniqueness, minimality, and incoherence. They implemented their approach within a framework, but there is a lack concerning eventual other projects when considering application of their approach. In addition, they clearly stated that similarity measures would improve the performance of their assessment. On the other hand, clustering was defined as an issue to perform if their assessment would not be able to process too many cases *in a reasonable amount of time.*[17]

Within an approach concerning the maintenance, existing CBR approaches were applied to summarize them into a new approach. On the basis of the Aamodt and Plaza approach [1] and various INRECA research activities [5], terms were reused and combined. They divided their theoretic generic approach into three stages named retain, review and restore. For instance, retain refers to complete a case. Review points to an assessment of a case and restore implies modifying a case.[18]

Within INRECA (Induction and Reasoning from Cases), case base quality was mentioned, but not concrete stated within a definition of eventual solutions. For instance, a term like *define clear objectives* sounds too unclear to consider it within a concrete index towards case base quality from the authors point of view.[5]

Another approach tried to solve and improve maintenance issues with CBR classifiers. They used clustering and logistic regression to build their classifiers. Their approach was not applied within a generic way. Apart of that, the adaptation feature was neglected. Assigning a string label was their *simple adaptation.* When having the focus on maintenance, then adaptation must be carefully integrated into a CBR system from the authors point of view.[2]

An approach namely *Assessing Case Base Quality* states interesting notes, but some critical points towards their approach could be seen such as a missing portability and too much effort to integrate their approach. Their main goals were to assess and measure inherent problem-solution irregularity within a case base to improve using cases especially with respect to the accuracy concerning solutions. The Mantel Test (or Mantel's Randomisation Test) was applied together with different ratios to assess the quality of their case base. Therefore, their approach was not implemented in a generic way.[16]

Within [15], they stated an approach towards a case-mining algorithm. This generates a *competent* case base when processing existing cases. The stated two issues within their approach. On the one hand, processing nearest cases, which are not containing correct solutions. And another point of view, an uneven case distribution was named as potential obstacle. In addition, they proposed an algorithm to mine within cases, which includes avoiding the previous mentioned problems. Concerning their case-mining approach, they stated two points, which are worth to mention. With respect to the approach in this paper and their approach, their points are overlapping concerning an idea behind when searching for an intellectual intersection between different approaches.

- *Each case should cover as much of the problem space as possible to reduce the potential bias, and*
- *The cases should be as diverse as possible to reduce co-variance in producing errors.*

[15] When reading these items, a brief comparison to the quality index can be made. The first item above can be seen as avoiding missing values within this approach (third sub-index) concerning an index. The second item above can be seen within similar retained queries in this approach (second sub-index). In addition, the second item above can be partially seen within the first sub-index when assessing average solutions per case.

3 Building Sub-indices

Three indices are used to build an index for the quality of case base. Each of these sub-indices uses an interval from 0 to 1.

3.1 Index I: Average Solutions per Case

When using a revision graph for solutions, then an entire revision graph will be defined as 1 solution concerning this index. Null adaptation implies only one solution for a problem, but using a revision graph implies more than one solution for a query. At the end, only one solution is defined as an actual solution for a problem when using a revision graph. Therefore, using revision graphs must not aggravate this index. Multiple solutions are considered as an additional processing effort. In addition, maintenance of a case base can be more difficult with increasing similar solutions. A threshold concerning the maximum number of solutions per case has to be defined within a theoretical interval [1,count of solutions]. A practical interval would be from 3 to 9. For each case, the count of bad solved cases (argument *cc*), concerning too many solutions, will be incremented if the given threshold was reached. Subsequent, the sub-indices can be calculated with respect to all cases (argument *c*).

$$Idx_I = 1 - \frac{cc}{\sum c} \tag{1}$$

3.2 Index II: Count of Similar Retained Queries

To define similar retained queries, a similarity measure has to be applied with a certain threshold. A problem to problem similarity measure must exist with a known interval to define a threshold for a case base. If a threshold was reached, then the count has to be incremented. Subsequent, an index can be calculated with following formulae:

$$Idx_{II} = 1 - \frac{csrq}{\sum qc} \tag{2}$$

The count of similar retained queries is given by argument *csrq* and the query comparisons are denoted as *qc*.

3.3 Index III: Missing Values

The count of missing values (*cmv*) within cases, with respect to the count of occurrence, has to be calculated. The actual sum of fields (*f*) can be achieved within the persistence of a case base when counting all table fields.

$$Idx_{III} = 1 - \frac{cmv}{\sum f} \tag{3}$$

4 Calculating the Main Index

To clearly state the formulae, this section presents the integration of the three sub-indices stated above.
The case base quality index (CBQ) uses an interval from 0 to 100. 100 per cent states the best possible value for a case base and 0 per cent refers to a impaired value of a case base. The previous mentioned indices are subsequently weighted.

$$CBQ = 100 \cdot \frac{Idx_I \cdot Weight_I + Idx_{II} \cdot Weight_{II} + Idx_{III} \cdot Weight_{III}}{\sum_{i=1}^{i=3} Weight_i} \tag{4}$$

The weight factors can be applied concerning a concrete case base within a given domain. For instance, if avoiding of missing values is more important than the case redundancies, then $weight_{III}$ will receive another argument in comparison to $\frac{1}{3}$.

5 Application of the Index within Loaner

This section covers the practical aspects of the implementation regarding the index described above. Within code name Loaner, an application written in C# and LINQ (Language Integrated Querying), the approach of this paper was implemented. The visualization was made when using Windows Presentation Foundation (WPF). The training set of the data was analyzed due to the actual implementation state.[8] It is complete and without multiple solutions, which refers to a good value concerning the case base quality.

5.1 I - Solutions per Case

The used threshold for solutions per case was 7. Zero cases are reached this threshold. This generates a value of 1.

5.2 II - Similar Retained Queries

The chosen threshold was defined as 80 per cent. This was detected within prior experiments based on development of similarity measures. When using a high value such as 95 per cent, zero similar queries would occur. Within the screenshot of Loaner, a page depicts the counting process of sub-index II. 28 similar retained queries were achieved within 498501 query comparison iterations. This implies a temporal value of $\frac{28}{498501}$, which will be subsequently subtracted from 1. Therefore, the value within this sub-index results into $\frac{498473}{498501}$.

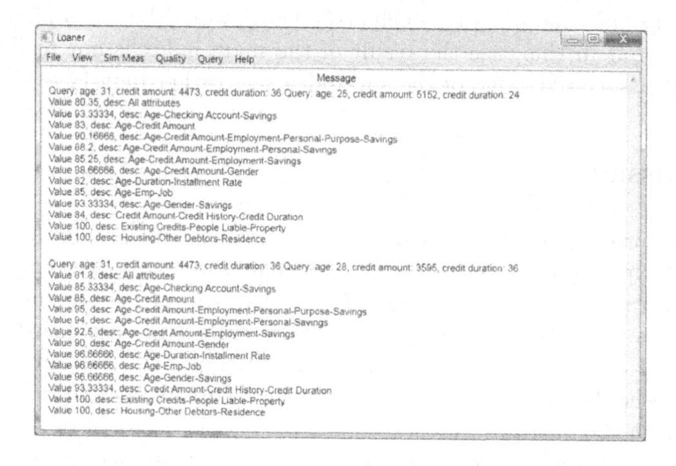

Fig. 1. Loaner 0.4 α - Measurement Index II

5.3 III - Missing Values

In fact, the train set of the actual approach is complete concerning the values. Each tuple contains a value for each column. Zero missing values occurred within the data. This generates an excellent sub-index III, value 1.

5.4 Using the Main Index

To avoid to fall into oblivion, the train set is complete without identical cases. This refers to a high quality concerning the case base in prior to an assessment of the quality.

$$CBQ = 100 \cdot \left(1 \cdot \frac{1}{3} + \frac{498473}{498501} \cdot \frac{1}{3} + 1 \cdot \frac{1}{3}\right) \tag{5}$$

In this application, the case base quality index refers to 99.9981277202.

5.5 Experiments with Weights

Weights were considered for similarity measures and the formulae above.

In experiments concerning similarity measures, it was observed that only the attribute gender should be weighted with $\frac{1}{3}$. Otherwise, a simple similarity measure, which uses only a few attributes could increase or decrease the value of the result too much. Therefore, all attributes (except gender) are using the weight 1.

All sub-indices were associated with a weight of $\frac{1}{3}$ within the main index. In this case, increasing the weight for sub-index II would decrease the index value. Another point of view when consider additional data with missing values, this would wrongly increase the index value. Therefore, a cautious weighting was applied. When using another weight for sub-index II such as $\frac{5}{6}$, the value of main index is marginally modified to 99.9953193006. $\frac{5}{6}$ would be a too high value for a sub-index, but in this case the result of the main index is not really affected because the associated value of the sub-index was rather high $\left(1 - \frac{28}{498501}\right)$.

6 Discussion

This section provides a few notes about circumstances concerning the prototype Loaner and explanations with respect to the quality index. Concerning sub-index III, the natural assumption for this index is that an application code prevents to store cases with primarily null values. Otherwise bad case-based reasoning results would occur beside of low values in sub-index III. Within an interval [0,100], thresholds were tested against the case base to see various similarity values. Within the diagram, thresholds and an associated count of similar query comparisons are presented. The ordinate presents the count of query comparisons from 0 to 498501. The abscissa presents thresholds from 0 to 100.

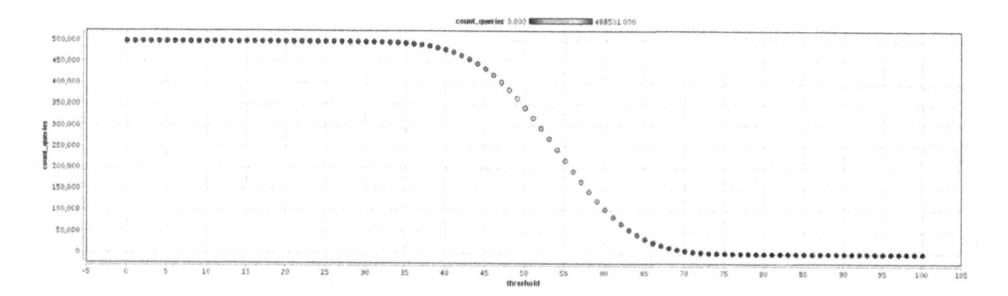

Fig. 2. Plot Thresholds 0 to 100

Within the threshold interval [0,100], the plot above presents that 57 per cent are a point to distinguish between the nearest queries and not related queries. Concerning sub-index II, 80 per cent was used because a threshold lower than 60 per cent would deliver many queries related to the concrete example within Loaner. For instance, the threshold 57 per cent refers to a count of 168570 queries. To use an adequate threshold for sub-index II, the concrete data such as a comma separated value file has to be analyzed. To give an excerpt within the higher threshold values regarding the second sub-index, a few relations are stated as follows.

- Threshold $\rightsquigarrow$ Count queries
- 75 $\rightsquigarrow$ 710
- 76 $\rightsquigarrow$ 407
- 77 $\rightsquigarrow$ 220
- 78 $\rightsquigarrow$ 117
- 79 $\rightsquigarrow$ 65
- 80 $\rightsquigarrow$ 28
- 81 $\rightsquigarrow$ 11
- 82 $\rightsquigarrow$ 6
- 83 $\rightsquigarrow$ 3
- 84 $\rightsquigarrow$ 2
- 85 $\rightsquigarrow$ 0

In addition, it is clearly presented that a percentage of 100 refers to zero similar retained queries. Therefore, 100 per cent is not suitable as threshold when using a similarity measure. Another point of view, a similarity with 100 per cent would be identical tuples, which has to be avoided when inserting data into a schema. In the second scatter plot, thresholds within the range [50,85] are depicted, which states an excerpt of the first scatter plot. The count of similar query comparisons starts with 0 and ends with 343038. When comparing this range to the full query range within the first scatter plot, it is clearly stated that within the range [50,85] a higher variability occurs concerning the similar query comparisons.

The second scatter plot presents that similarity values are reduced with various different steps in a range 50 to 85. Within Loaner, different similarity measures are using various attributes. For gaining the similarity value concerning

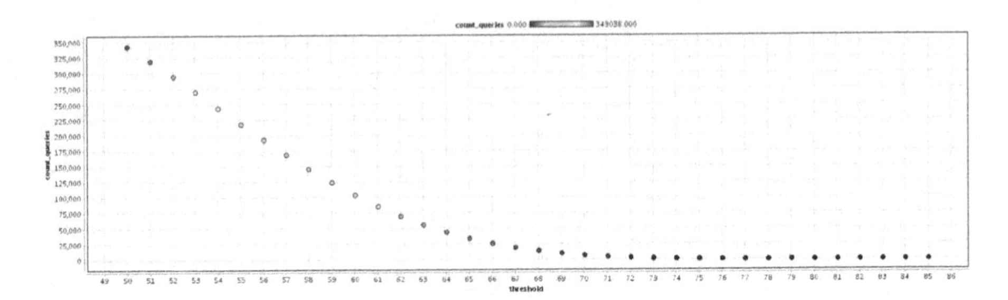

Fig. 3. Scatter Plot Thresholds 50 to 85

sub-index II, a similarity measure was applied, which uses all attributes. Those are age, credit amount, credit duration, number of people liable, other installment plans, gender, personal state, purpose of the loan, credit history, employment duration, job level, other credits, duration of the current residence, installment rate concerning disposable monthly income to give an excerpt. When using all attributes, no aspect such as personal-related issues (age, gender) or credit-related considerations (credit history, credit amount) will be neglected. Sub-index II calculated 28 similar retained queries within 498501 unique comparisons between different queries. Identical tuples are not persisted. Reflexive comparisons are avoided. Double comparisons are avoided in addition. For instance, the similarity between query id 100 and id 770 is calculated, but not vice versa.

A second data set was integrated into Loaner and analyzed. The additional data set was retrieved by the author of [3]. It was more numeric-based and contained more tuples in comparison to the first one. At first sight, the data set was evaluated with the same similarity measures which are applied towards the German data set. It contained no redundant solutions per case, only marginal similar retained queries and no missing values.

$$CBQ = 100 \cdot \left(1 \cdot \frac{1}{3} + \left(1 - \frac{193931}{4871881}\right) \cdot \frac{1}{3} + 1 \cdot \frac{1}{3}\right) \tag{6}$$

Hence, the result was stated as 98.6731271419 per cent. For determining the similar retained queries, 80 per cent was applied again as similarity value towards all query comparisons within the case base — except reflexive and redundant (id 31⇝94 but without 94⇝31 for instance) comparison steps.

7 Conclusion

Within Loaner, the application regarding sub-indices I and III was fatly achieved due to a complete training set. Sub-index II required an implementation, which refers to similarity measures. To avoid overlooking about similarities within queries, all attributes are applied to consider different aspects within a loan application. Concerning the theory, the three sub-indices are easy to use. When using weighting with the index formulae described above, agility can be attached to fit specific requirements of a given domain. In this approach, the weighting of the sub-indices within the formulae above was stated with $\frac{1}{3}$. For sub-index II, a generic threshold cannot be inferred due to many different domains, which are suitable for case-based reasoning. These are car mechanic, structural health monitoring, employee support, call center tools and text retrieval software for instance to refer to this diversity. To infer this approach within three steps namely The Good, the Bad and the Ugly.[13]

- The Good - it clearly presents an index within a defined interval [0,100]
- the Bad - even a generic index needs implementation effort
- the Ugly - using wrong weights to hide weakness of a case base would be possible

Big data can be applied to CBR, but not using an index concerning the case base quality could lead to obstacles. Especially if a deletion strategy was not applied within a CBR approach. A case base with redundant and unused cases impairs the performance in reasoning processes. The proposed index can be applied to prevent these performance obstacles.

8 Future Work

When a loan application simulator will be finished, the case base quality index can be applied to new cases for further testing with weights. Apart of that: An automatic evaluation feature can be implemented to avoid outlier values for an index. For instance, detection of bad used weighting when using a weight such as $\frac{7}{10}$ for excellent managed similar retained cases when applying a weight like $\frac{1}{10}$ for too many missing values.

Acknowledgement. Appreciation goes to the reviewers for their comprehensive hints and feedback. Thanks to Prof. Baesens (who was not a reviewer) for providing an additional data set to enhance the research for both thesis and this paper.

References

1. Aamodt, A., Plaza, E.: Case-based reasoning: Foundational issues, methodological variations, and system approaches. AI Commun. 7(1), 39–59 (1994)
2. Arshadi, N., Jurisica, I.: Maintaining case-based reasoning systems: A machine learning approach. In: Funk, P., González Calero, P.A. (eds.) ECCBR 2004. LNCS (LNAI), vol. 3155, pp. 17–31. Springer, Heidelberg (2004)
3. Baesens, B., Setiono, R., Mues, C., Vanthienen, J.: Using neural network rule extraction and decision tables for credit-risk evaluation. Management Science 49(3) (2003)
4. Basili, V.R.: The experience factory: Packaging software experience (1999)

5. Bergmann, R., Althoff, K.-D., Breen, S., Göker, M.H., Manago, M., Traphöner, R., Wess, S.: Developing Industrial Case-Based Reasoning Applications, 2nd edn. LNCS (LNAI), vol. 1612. Springer, Heidelberg (2003)
6. DeCoste, D.: The future of chess-playing technologies and the significance of kasparov versus deep blue. Papers from the 1997 AAAI Workshop (1997)
7. Ferrucci, D.A.: Ibm's watson/deepqa. SIGARCH Computer Architecture News 39(3) (2011)
8. Frank, A., Asuncion, A.: UCI machine learning repository (2010), `http://archive.ics.uci.edu/ml`
9. Hönigl, J., Kosorus, H., Küng, J.: On reasoning within different domains in the past, present and future. In: 23rd Database and Expert Systems Applications (DEXA), 2nd International Workshop on Information Systems for Situation Awareness and Situation Management - ISSASiM 2012 (September 2012)
10. Hönigl, J., Nebylovych, Y.: Building a financial case-based reasoning prototype from scratch with respect to credit lending and association models driven by knowledge discovery. In: Central & Eastern European Software Engineering Conference in Russia (November 2012)
11. Jurisica, I., Nixon, B.A.: Building quality into case-based reasoning systems. In: Pernici, B., Thanos, C. (eds.) CAiSE 1998. LNCS, vol. 1413, pp. 363–380. Springer, Heidelberg (1998)
12. Leake, D.B.: Cbr in context: The present and future. In: Reasoning From Remindings, pp. 3–30. MIT Press (1996)
13. Leone, S.: The good, the bad and the ugly. il buono, il brutto, il cattivo (original title) (1966)
14. Newborn, M., Newborn, M.: Deep blue establishes historic landmark. In: Beyond Deep Blue, pp. 1–26. Springer, London (2011)
15. Pan, R., Yang, Q., Pan, S.J.: Mining competent case bases for case-based reasoning. Artificial Intelligence 171(16-17), 1039–1068 (2007)
16. Rahul Premraj, M.S.: Assessing case base quality. Bournemouth University and Brunel University (2005)
17. Reinartz, T., Iglezakis, I., Roth-Berghofer, T.: On quality measures for case base maintenance. In: Blanzieri, E., Portinale, L. (eds.) EWCBR 2000. LNCS (LNAI), vol. 1898, pp. 247–260. Springer, Heidelberg (2000)
18. Roth-Berghofer, T., Reinartz, T.: Mama: A maintenance manual for case-based reasoning systems. In: Aha, D.W., Watson, I. (eds.) ICCBR 2001. LNCS (LNAI), vol. 2080, pp. 452–466. Springer, Heidelberg (2001)
19. Tsatsoulis, C., Cheng, Q., Wei, H.Y.: Integrating case-based reasoning and decision theory. IEEE Expert 12(4), 46–55 (1997)

Agent's Autonomy Adjustment via Situation Awareness

Salama A. Mostafa[1], Mohd Sharifuddin Ahmad[1], Alicia Y.C. Tang[1], Azhana Ahmad[1], Muthukkaruppan Annamalai[2], and Aida Mustapha[3]

[1] College of Information Technology, Universiti Tenaga Nasional, Putrajaya Campus 43000, Selangor Darul Ehsan, Malaysia

[2] Faculty of Computer and Mathematical Sciences, Universiti Teknologi MARA, Shah Alam, Selangor Darul Ehsan, Malaysia

[3] Faculty of Computer Sciences and Information Technology, Universiti Putra Malaysia, Selangor Darul Ehsan, Malaysia

semnah@yahoo.com, {sharif,aliciat,azhana}@uniten.edu.my, mk@tmsk.uitm.edu.my, aida_m@upm.edu.my

Abstract. Interactions between autonomous agents (humans and software) are necessary to increase the system's awareness, support agents' decision-making abilities and subsequently reduce the risks of failure. The challenge, however, is to formulate a mechanism that specifies when an agent or a human should take the initiative to interact. In this paper, we propose a Situation Awareness Assessment (SAA) model of autonomy adjustment with situation awareness capabilities in a decentralized environment. The SAA model systematizes the interactions to assist agents' decision-making and improve the systems' performance. The model inspects if an agent has the required awareness of a situation to be satisfactorily autonomous, otherwise, it intervenes the agent with feedback about the situation. An example scenario demonstrates the SAA ability to facilitate improved autonomous behavior in agents.

Keywords: Autonomous agent, Active Perception (AP), Situation Awareness (SA), Context Awareness (CA), autonomy assessment, decision-making.

1 Introduction

In complex and dynamic environments, autonomous systems' awareness of the prevailing conditions would greatly improve their ability to manifest critical decisions. However, it is crucial to model interactions between humans and software agents in autonomous systems in order to exchange some primary knowledge about the system [1], [2]. This fact instigates the need of adjustable autonomy to implement such interactions [3], [4], [5], [6]. The interaction, however, is a costly process and prone to disturbances especially in dynamic systems such as unmanned systems [7]. Hence, the challenge is to formulate an efficient adjustable autonomy model that specifies when an agent or a human should take the initiative to interact [2], [8].

Situation Awareness (SA) is one solution that provides reliable assessment of system performance at any point in time [9]. We adopt the formal SA definition for dynamic environment by Endsley [10] as, "*the perception of the elements in the environment within a volume of time and space, the comprehension of their meaning, and the projection of their status in the near future*."

N.T. Nguyen et al. (Eds.): ACIIDS 2014, Part I, LNAI 8397, pp. 443–453, 2014.

The phases of *perception*, *comprehension* and *projection* assist in improving decision-making and action-performing [11]. SA phases entail the analysis of events, surrounding environment, corresponding tasks and time-constrained actions. It encompasses perceiving information of situations' elements, the elements' dynamics in a specific period of time, understanding the elements, and projecting the understood elements into the near future [9].

In many existing SA methodologies, the aim is to enhance humans' awareness of situations by assessing the surrounding environment [9], [12]. Subsequently, SA and assessment privileges are frequently used in computer science, especially in computer automation aspects (e.g. [13] and [14]). However, Endsley and Connors [11] argued that tangible SA is an emergence of the human mind. As SA mainly involves humans' interactions with systems (e.g. [7]), the real challenge is to design a system that performs these interactions and manifests significant SA at human understanding level [2], [15].

In this paper, we focus on the SA from a multi-agent system (MAS) perspective and assume an environment that is equipped with effective interactive systems. Consequently, we investigate the compatibility of existing awareness techniques' core phases with an agent-based architecture. We then show the correlations between these techniques and their applicability in multi-agent systems (MAS). Meanwhile, we emphasize the role of the SA techniques in MAS performance assessment via proposing a human-centered Situation Awareness Assessment (SAA) model. The SAA model provides an independent mechanism that exploits SA phases to manifest SA capabilities in agents' actions. Interaction rules are formulated to deliver autonomy awareness and supervision in order to enhance agents' autonomy and decision-making capabilities.

2 Literature Review

2.1 Agent Implicit Awareness of Situations

An agent's practical reasoning methodology encompasses implicit SA capabilities to some limited depth. Since, one of the agent architecture core components is observation, then, the perception of the surrounding is embodied in its design [4]. An agent's understanding of an event is built upon its knowledge of the event's situational elements in an environment and its interpretation (or beliefs) of the situation [5], [15], [16]. The agent's desires dictate its existing or newly generated intentions. The agent's intention processing forms its projection of the situation's future scenarios [6]. However, understanding the context of the perceived aggregated knowledge of an event by the agent is a challenging process [13]. The agent's decision in a particular event is formed based on its interpretation of the situation's context of the event [16]. Therefore, explicit SA activities can be modeled and exploited for the agent to further improve its autonomic features especially when dealing with uncertainties in the event. An SA model is needed to enable the agent to derive precise conclusions from an observed situation [6].

2.2 The Awareness Techniques

In this section, we discuss the boundaries of the main awareness techniques proposed in the literature and studied in this paper, which are: Active Perception (AP), Situation Awareness (SA) and Context Awareness (CA). We define each as: *an independent component that intersects with agents' functioning cycle to form the critical inputs to the agents' decision-making process in order to improve their actions.* It is a result of a comprehensive literature review on the topics to reveal their contributions to agent-based systems. It is found that, there is a close relationship between AP, SA and CA, which serve the agents during their reasoning processes. In addition, there are obvious overlaps between the phases of the three techniques which make them tightly coupled. Figure 1 is a preliminary representation of the correlations between AP, SA and CA techniques in agent-based systems.

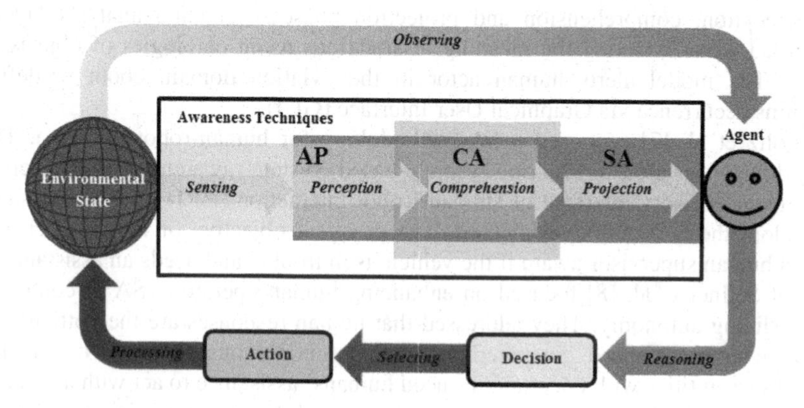

Fig. 1. The representation of the awareness techniques in an agent's action cycle

The AP technique is more concerned about the effectiveness of observation and quality of perceptions. It begins prior to SA and CA. According to Weyns et al. [17], AP is a process of directing an agent to the most relevant interpretation of its observations in an environment to assist the agent in completing the task (deriving accurate beliefs, e.g. [15]). It concerns with sensing the surrounding and observing the situations, representing the observed situations, understanding the states of the situations through their context descriptions, and finally, filtering the understood states to improve the perceptions and the decisions.

SA projects the understood states of the situation to reason and form future actions [9]. For instance, Hoogendoorn et al. [15] deployed an SA model on an agent's belief optimization in which the agent's degree of awareness on a situation is signified by an activation value of belief. Their aim is to generate complex beliefs from the observed beliefs that enable the agent to perform projection to future situations.

CA is a term used to describe the process of extracting the semantics of the context information to stimulate the behavior of an agent such as triggering an action [16]. The context of an event is a set of facts that describes the elements of a situation for a particular event at a time point [12]. In both AP and SA, CA is used as a tool to enhance the understanding of the context. For a specific domain, context ontology is one of the best proven approaches to formalize, reason, and share the semantics of data. Context ontology is used in many agent-based awareness approaches to give the context its meaning [13], [18].

2.3 Related Work

Many different attempts to exploit SA in systems have been recorded in the literature. To the best of our knowledge, there is no close related work that uses Endsley [9] model of SA as independent component for agents' autonomy assessment and integrated ontology-based semantic approach in the SA processes. Lili et al. [14] use different Artificial Intelligence (AI) techniques including Fuzzy Logic (FL), Bayesian Network and Case-Based Reasoning (CBR) for situation reasoning and autonomy adjustment. Baader et al. [13] adopted Endsley's SA model and propose data aggregation, semantic analysis and alert generation layers to accordingly correspond the perception, comprehension and projection phases. Semantic analysis layer is concerned with extracting the meaning of situations using ontologies of objects and events. The model alerts human actor in the aviation domain about predefined situations occurrence via Graphical User Interface (GUI).

Scholtz et al. [7] proposed a SA methodology for human-robot interface. Their methodology is used in robotic vehicle-based system awareness assessment to evaluate supervisory interface of Human-Robot Interaction (HRI). In SA assessment, they adopt the SAGAT style method. The practical objective of the research is to make a human supervisor aware if the vehicle is in trouble and needs an assistant. The work of Sellner et al. [8] focused on enhancing human operators' SA in controlling robots sliding autonomy. They addressed that human responses are the bottleneck of system responses' speed compared with other autonomous entities. Nevertheless, autonomous entities with uncertainties need humans' assistance to act with awareness.

Wardziński [19] emphasized the importance of SA assessment in improving an agent's knowledge and minimizing its action risk especially in dynamic and uncertain environments. Patrón et al. [1] used context ontology to represent and share the knowledge of situations among software and human agents which demonstrated system awareness improvement when dealing with unexpected situations. McAree and Chen [20] went further by claiming that agent's SA capability enhancement leads to its autonomy improvement.

3 A Proposed Situation Awareness Assessment Model

To illustrate our proposed Situation Awareness Assessment (SAA) model, we specify two autonomic entities of a system; software agent and human, each works on particular tasks or parts of tasks. The agent is responsible for performing different types of actions with different degrees of autonomy based on the given task and the corresponding actions types. The agent's action implementation life cycle proceeds through five steps which are observing, reasoning, action selection, autonomy setting and action processing [3], [4].

The human acts as a supervisor who monitors the agent's and system's performance. He/She interacts with the agent via a GUI to provide assistance to the agent and the system whenever is needed. Whenever necessary, the agent interacts to seek assistance from the human while the human intervenes whenever a risk is observed or anticipated. The human's intervention contributions are captured in the system's storage and progressively analyzed by the agent to be used in future circumstances [15].

Such setting provides the advantages of studying the agent's behavior in different situations at different levels of autonomy, determining the autonomy level of each action of the agent, feeding the assessment's information to the agents to be processed and reasoned and ultimately imposing the agent to a level of SA that maximizes its autonomy [20].

3.1 The Architecture of the SAA Model

The SAA model architecture comprises of three entities which are the human supervisor, system software, and software agent. The system software consists of pattern tuning, inference rules, semantic synthesis and GUI as depicted in Figure 2. The software agent has two supported functions which are *autonomy update function* (Φ) and *situation awareness function* (♀) that interact with the SAA model.

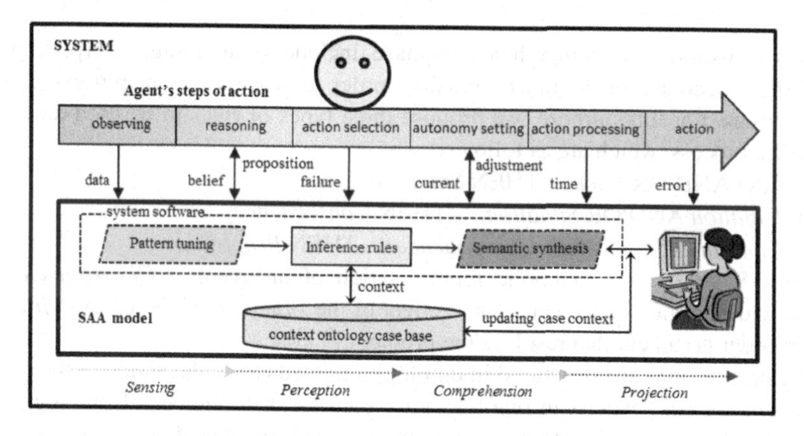

Fig. 2. The architecture of the SAA model

Sensing. The sensing phase encompasses the hardware and software components for an agent and responsible of data gathering [6]. It includes the operational tools and functions that support the agent's observation quality in an environment. The outputs of this phase, which are the collected data regarding an event, the errors, and the failures of previous attempts along with the agent's beliefs, form the inputs to the perception phase.

Perception. The data from the observation of an event forms the perceptions of the event's situation elements [17]. The perceptions are meaningful representations of the gathered data, which are refined in the agent's reasoning step to update its beliefs (feed-forward process). The agent's beliefs represent its current state in the environment to make a decision which might imply selecting an action (see Figure 2).

The SAA is built upon recognizing an error or a failure in a system [11], [12]. The errors and the failures are collectively determined based on system performance and represent system previous state (feedback process) as both errors and failures can only be detected after their occurrence. We define an error as a flawed action that results from incorrect action selection or incorrect action execution. The source of errors is mainly due to incorrect or ambiguous information [9].

Meanwhile, a failure is the inability to decide on actions to perform a task due to constraints. Failure may results from the agent's lack of knowledge and authority to perform an action [3], [5].

The pattern tuning classifies the collected beliefs into specific representations in order to perceive its semantic information during the comprehension phase. The odds in the data are removed through context inference process and the filtered data is treated as facts. The errors of the previous runs are fed-back to the SAA to be verified by the supervisor and the supervisor adjusts the agent autonomy parameters to meet with the agent performing abilities. The supervisor assistance is applicable if there is a regression in the agent's performance due to failure.

Comprehension. The comprehension phase includes combining, interpreting and retaining procedures of the tuned beliefs in order to process new or retrieve existing propositions. For this purpose, we propose three types of rules to be used during the inference process, which are as follows:

- IF *Fact* AND/OR *Fact* … THEN *Situation*
- IF *Situation* AND/OR *Situation* … THEN *Consequence*
- IF *Consequence* AND/OR *Consequence* … THEN *Proposition*

where a *Situation* is a symbolic representation of an event, a *Consequence*, is a semantic description of the causes and effect of the situation, while a *Proposition* is a higher-order argument that resolves the problem.

The propositions are structured in ontology-oriented cases and reside in the context ontology case base where an ontology case consists of a number of ontologies and their related classes and instances. Ontology is a well-organized conceptual structure that consists of contexts in specific hierarchy including superclass (e.g. hole) classes (e.g. manhole), relations (e.g. access), attributes (e.g. shape), properties (e.g. round) and axioms (e.g. a manhole might has a round shape and is used to access a sewer). The symbolic representation of the ontology concept is understood by both humans and software agents [1]. Figure 3 shows the structure of the context ontology case base, the case, the ontology and the related classes and instances of the ontology.

During the comprehension phase, a semantic synthesis procedure works with the inference rules of the ontology to produce the semantics of contexts as higher order solution arguments to provide awareness [13], [18]. First, the contexts are extracted from the matching of the perceived data with the context ontology by applying the inference rules (see Figure 2). Then, the semantics of the contexts are produced via the semantic synthesis procedure from the corresponding triggered rules. Some other aspects of the semantic synthesis procedure are detailed in the projection phase.

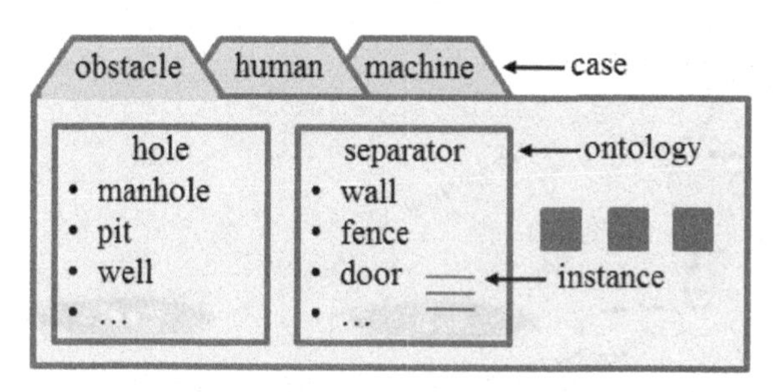

Fig. 3. Context ontology case base

Projection. The role of a human is to provide assessment of an agent's performance based on the observed risk of errors and failures by extrapolating the performed actions [7], [19]. The assessment modifies some aspects of the agent such as adjusting the autonomy parameters to intervene or block some actions [2], [6], [8]. The intervention is calculated based on the corresponding agent performance progress by:

$$\lambda \Leftarrow \Phi_i(z_i, (\text{♀}_{e_x}(f(s, t)))) \quad (1)$$

where λ is the intervention bias whose value is either 0 or 1, Φ is the function of the autonomy adjustment (*autonomy update function*) on track i of the agent ag_i and its autonomy level z_i. Φ adjusts z_i based on the agent's *situation awareness function*, ♀, which returns truth value by checking ag_i performance progress when handling event e_x. f is the satisfactory fluent that returns a truth value based on the provided satisfaction state, s at time, t.

The human also works to prevent or to minimize the automation errors or failures besides performing the assessment tasks [13], [18], [19]. He/She might remove any ambiguities from the contexts by updating a case if the case does not provide satisfactory context or inserting new cases in the context ontology case base if a context is not covered by the existing cases. The modifications are made by the semantic synthesis process to formulate the meaning of a context to be understood by the agent. The semantic synthesis provides template frames that allow the human user to insert new cases or an instance of a case. These modifications aim to increase the agent's awareness of situations through covering more events' contexts. Hence, the successful sequence of *Situation*, *Consequence* and *Proposition* that is adopted by the agent and retained in its knowledge base is to be applied in similar situations.

3.2 Example Scenario

A robot is moving forward towards its goal and carrying out a task to be achieved. Unexpectedly, the robot encounters a *hole* in its path (i.e. sensing event). The robot must know what a hole is (i.e. perception event) to be able to comprehend it. The awareness of the event represents the ability of the robot to detect that moving forward will cause it to fall into the *hole* (i.e. projection event) as shown in Figure 4.

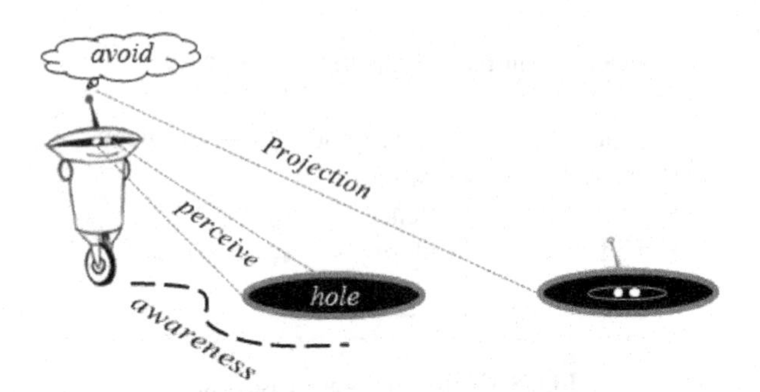

Fig. 4. Robot Situation Awareness example

The robot needs to change its direction to avoid the *hole*. This projection to the event motivates the robot to change its desire and adopt new intentions. Subsequently, understanding the situation triggers the robot to decide in performing an *avoid_obstacle* task. The *avoid_obstacle* task is a representation of a goal state of obstacle avoidance that is yet to be achieved by a set of discrete and/or repetitive movement of pre-compiled actions. The actions change the robot's path to a direction that enables the robot to avoid the *hole*, thus, manifesting high-level of situation awareness. The following axioms give an overview of the robot's required actions for the *hole* event and the support of the SAA model during the robot's actions.

```
Robot(L1, do(F, G))↔(Robot(F, G) ∧ move → L2))
Robot(L2, do(S, X))↔(Robot(S, X) ∧ see → true))          ┐
true ⇐ (Ϙ_ex(f (s, t))) [if false, human intervene]      │
Robot(L2, do(S, X))↔(Robot(S, X) ∧ locate → L3))         │
Robot(L2, do(P, X))↔(Robot(P, X) ∧ search → X) ∧         │ Extracting beliefs
                    (match(X, wall) → false) ∨           │
                    (match(X, column) → false) ∨         │
                              ⋮                          │
                    (match(X, hole) → true) ∨⋯)          ┘
Robot(L2, do(U, X))↔(Robot(U, X) ∧ context (X):→hole)
% Argument: "avoid_obstacle"
inference(X, obstacle)                                    ┐
#Case{obstacle}(                                          │
@hole & <Scope>(f1: thing)                                │
@hole & <Context>(f2: visual_object)                      │ Ontology process
@hole & <Shape>(f3: round)                                │
@hole & <Position>(f4: ground)                            │
@hole & <Size>(f5: Diameter > 20cm → big)                 │
@hole & <Type>(f6: uncovered_manhole))                    ┘
% end ontology process
```

```
IF Context(visual_object) ∧ Shape(round) ∧
Position(ground) ∧ Size(big) THEN Type
(uncovered_manhole)
IF Type(uncovered_manhole) THEN State(damage_robot)
IF State(damage_robot) THEN Avoid(uncovered_manhole)
Proposition(L3, avoid(X))
% end inference process
true ⇐ (♀ex(f(s, t))) [if false, human intervene]
Precondition: Robot(X = uncovered_manhole, on(L3, G),
avoid(X))
Robot(L2, do(D, X)) ↔ (Robot(D, X) ∧ act(X))
Robot(L2, do(<B ∨ L ∨ R>, G))↔(Robot(<B ∨ L ∨ R>, G) ∧
select → R))
Robot(L2, do(R, G))↔(Robot(R, G) ∧ move → L4))
true ⇐ (♀ex(f(s, t))) [if false, human intervene]
Robot(L4, do (S, X))↔(Robot(S, X) ∧ see → false))
Robot(L4, do (F, G))↔(Robot(F, G) ∧ move → L5))
true ⇐ (♀ex(f(s, t))) [if false, human intervene]
Robot(L5, do(L, G))↔(Robot(F, G) ∧ move → L6))
Robot(L6, do (S, X))↔(Robot(S, X) ∧ see → false))
Robot(L6, do (F, G))↔(Robot(F, G) ∧ move → G))
```

SAA proposition

where L is the location variable; G is the goal; X is the observed input; F, B, L and R are directional movement for *Forward*, *Backward*, *Left* and *Right* actions; S is the observation sensor; P and U are perceiving and understanding the context of X; and D is the deliberation on X. The satisfactory fluent *f* tunes the interaction between human and agent by indicating satisfaction for both human and agent in the system. The default of *f* is true and both interact if and only if there is no satisfaction.

In summary, autonomy awareness requires a system to be aware of its own capability (internal state), the surrounding capabilities (external states) and how to use these capabilities to participate in its surrounding. In the example, we have shown that the agent can behave with high-level of autonomy when the intervention is directed to enhance agent's awareness via third party situation awareness assessor.

4 Conclusion and Future Work

The principle of involving humans and software agents to carry out some system's initiative manifests the notion of adjustable autonomy. However, building a mechanism that controls the behavior of agents toward an optimized adjustment is the main challenge in modeling adjustable autonomy, especially, in dynamic systems. One promising solution is to establish an autonomy assessment mechanism that assists the system in the autonomy distribution process. A situation awareness methodology is found to be a very effective approach in enhancing human's decisions-making as knowing a situation is the first mile of knowing the solution.

In this paper, we employ the Endsley's [10] situation awareness model in agent's autonomy assessment via proposing the Situation Awareness Assessment (SAA) model. The SAA model is an independent assessor that is used for clarifying situations, assisting agents via providing situation ontology, and assessment of agent performance. We also show the correlation between different agent-based awareness techniques which are Active Perception (AP), Situation Awareness (SA) and Context Awareness (CA). Active perception in our approach is embodied in pattern tuning and the inference rules processes. The approach further demonstrates active perception by feeding-forward the collected data and back-propagating the errors and the failures as perceptions to be understood and processed. In addition, context awareness is visible in the comprehension phase where a context ontology case base approach is proposed. Eventually, the paper demonstrates a mechanism of agents' autonomy adjustment that works based on SA assessment. We found that the SA phases of perception, comprehension and projection are compatible with agents' automation improvement cycle.

In our future work, we will validate the proposed concepts of autonomy adjustment based on situation awareness assessment in an agentized unmanned system. In this work, we shall apply a CBR technique to enhance the SAA capabilities.

Acknowledgments. This project is sponsored by the Malaysian Ministry of Higher Education (MoHE) under the Exploratory Research Grant Scheme (ERGS) No. ERGS/1/2012/STG07 /UNITEN/02/5.

References

1. Patrón, P., Miguelanez, E., Cartwright, J., Petillot, Y.R.: Semantic Knowledge-based Representation for Improving Situation Awareness in Service Oriented Agents of Autonomous Underwater Vehicles. In: OCEANS 2008, pp. 1–9. IEEE Press (2008)
2. Fleming, M., Robin, C.: A Decision Procedure for Autonomous Agents to Reason about Interaction with Humans. In: AAAI 2004, pp. 81–86 (2004)
3. Mostafa, S.A., Ahmad, M.S., Annamalai, M., Ahmad, A., Gunasekaran, S.S.: A Conceptual Model of Layered Adjustable Autonomy. In: Rocha, Á., Correia, A.M., Wilson, T., Stroetmann, K.A. (eds.) Advances in Information Systems and Technologies. AISC, vol. 206, pp. 619–630. Springer, Heidelberg (2013)
4. Mostafa, S.A., Ahmad, M.S., Annamalai, M., Ahmad, A., Basheer, G.S.: A Layered Adjustable Autonomy Approach for Dynamic Autonomy Distribution. In: The 7th KES-AMSTA Conference, vol. 252, pp. 335–345. IOS Press, Hue City (2013)
5. Mostafa, S.A., Ahmad, M.S., Ahmad, A., Annamalai, M., Mustapha, A.: A Dynamic Measurement of Agent Autonomy in the Layered Adjustable Autonomy Model. In: Badica, A., Trawinski, B., Nguyen, N.T. (eds.) Recent Developments in Computational Collective Intelligence. SCI, vol. 513, pp. 25–35. Springer, Heidelberg (2014)
6. Mostafa, S.A., Ahmad, M.S., Annamalai, M., Ahmad, A., Gunasekaran, S.S.: A Dynamically Adjustable Autonomic Agent Framework. In: Rocha, Á., Correia, A.M., Wilson, T., Stroetmann, K.A. (eds.) Advances in Information Systems and Technologies. AISC, vol. 206, pp. 631–642. Springer, Heidelberg (2013)

7. Scholtz, J., Antonishek, B., Young, J.: Evaluation of a Human-robot Interface: Development of a Situational Awareness Methodology. In: The Proceedings of the 37th HICSS Conference, vol. 5, p. 9. IEEE Press (2004)
8. Sellner, B.P., Hiatt, L.M., Simmons, R., Singh, S.: Attaining Situational Awareness for Sliding Autonomy. In: Proceedings of the 1st ACM SIGCHI/SIGART Conference on Human-robot Interaction, pp. 80–87. ACM (2006)
9. Endsley, M.R., Bolté, B., Jones, D.G.: Designing for Situation Awareness: An Approach to User-Centered Design. Taylor and Francis, New York (2003)
10. Endsley, M.R.: Situation Awareness Global Assessment Technique (SAGAT). In: Aerospace and Electronics Conference, NAECON 1988, pp. 789–795. IEEE Press (1988)
11. Endsley, M.R., Connors, E.S.: Situation awareness: State of the Art. In: 2008 IEEE Power and Energy Society General Meeting-Conversion and Delivery of Electrical Energy in the 21st Century, pp. 1–4. IEEE Press, Pittsburgh (2008)
12. Naderpour, M., Lu, J.: A Human Situation Awareness Support System to Avoid Technological Disasters. In: Decision Aid Models for Disaster Management and Emergencies Atlantis CIS, vol. 7, pp. 307–325. Atlantis Press (2013)
13. Baader, F., et al.: A Novel Architecture for Situation Awareness Systems. In: Giese, M., Waaler, A. (eds.) TABLEAUX 2009. LNCS (LNAI), vol. 5607, pp. 77–92. Springer, Heidelberg (2009)
14. Lili, Y., Rubo, Z., Hengwen, G.: Situation Reasoning for an Adjustable Autonomy System. International Journal of Intelligent Computing and Cybernetics 5(2), 226–238 (2012)
15. Hoogendoorn, M., Van Lambalgen, R.M., Treur, J.: Modeling Situation Awareness in Human-like Agents Using Mental Models. In: Proceedings of the 21st International Joint Conference on Artificial Intelligence, vol. 2, pp. 1697–1704. AAAI Press (2011)
16. Ferrando, S.P., Onaindia, E.: Context-Aware Multi-agent Planning in Intelligent Environments. Information Sciences 227, 22–42 (2013)
17. Weyns, D., Steegmans, E., Holvoet, T.: Towards Active Perception in Situated Multi-agent Systems. Applied Artificial Intelligence 18, 867–883 (2004)
18. Gehrke, J.D.: Evaluating Situation Awareness of Autonomous Systems. In: Performance Evaluation and Benchmarking of Intelligent Systems, pp. 93–111. Springer (2009)
19. Wardziński, A.: The Role of Situation Awareness in Assuring Safety of Autonomous Vehicles. In: Górski, J. (ed.) SAFECOMP 2006. LNCS, vol. 4166, pp. 205–218. Springer, Heidelberg (2006)
20. McAree, O., Chen, W.H.: Artificial Situation Awareness for Increased Autonomy of UAS in the Terminal Area. Journal of Intelligent & Robotic Systems 70(1-4), 545–555 (2013)

The Development of a Decision Support Model for the Problem of Berths Allocation in Containers Terminal Using a Hybrid of Genetic Algorithm and Simulated Annealing

Zeinebou Zoubeir[1] and Abdellatif Benabdelhafid[2]

[1] PhD student at laboratory of Applied Mathematics in University of Le Havre-France

[2] Director of Research "Information System Integrated Logistics", University of Le Havre- France

Abstract. The berths allocation problem (BAP), aims to allocate the space along the waterfront, for incoming ships in a container terminal, to minimize an objective function. In this paper, we propose a multi- objective model for decision support, for the assignment problem the incoming ships on the quays in a containers terminal. The model we propose,seeks an assignment that simultaneously minimizes the time spent by vessels in the port and the distances traveled by containers imports/exports. We propose a mathematical model to achieve our goals by respecting imposed constraints. This model is solved by using, a hybrid of genetic algorithm and simulated annealing. Calculation results are presented in this article.

Keywords: Container terminal, BAP, genetic algorithm, simulated annealing, waiting time, flow containers.

1 Introduction

The BAP is to assign incoming vessels of berths along the dock. There several constraints to be taken into consideration, such as water depth for berthing of vessels, priorities assigned to ships, preferred docking areas, etc.

The BAP can be modeled depending on the state of the dock, in cas discrete or continuous, and, the schedule plan static or dynamic.
In the discrete BAP, the platform is considered a finite set of places (this mode is designed for example,by, Imai and al. (2001)[3], (2003)[2];. Monaco and Samara, (2007)[4]; Hansen and al. (2008) [10];. Golias and al. (2009c)[9]). In the continuous BAP, ships can dock anywhere on the quay (eg, Kim and Moon, (2003) [5]; Guan and Cheung, (2004) [6]; Imai and al. (2005) [8];. Moorthy and Teo, (2006) [7];, Lee and al. (2010)[11]). The majority of published research, consider the discrete case. For static BAP, all ships to be served must be at the port at the start of schedule plan. This method of planning is studied by: Imai et al. (2001)[3], Hansen and Oguz.(2003) [12], XU and al. (2012) [14] . While in

N.T. Nguyen et al. (Eds.): ACIIDS 2014, Part I, LNAI 8397, pp. 454–463, 2014.

the dynamic case, vessels may come after the start of planning plan. It is most studied case, for example,by,Imai and all. (2001)[3], (2003)[2];, Nishimura and all. (2001) [13].

In this paper, we consider a discrete distribution platform, with the static ships arrived (DSBAP). Our objective function minimizes the one hand, the residence time of the ships in the port, and on the other hand, the distances traveled by container imports / exports in the port area.

We proposes a mathematical model and solved by a hybrid of genetic algorithm and simulated annealing.

Next, we present an outline of the paper, we highlight its contributions. The description of the problem and the formulation of the model are presented in Section 3. Section 4, describes the proposed solutions and the numerical results. The last section conclude the paper.

2 The Berth Allocation Problem Static and Discrete (BAPSD)

In the berth allocation problem static (BAPSD) , all ships are already in port when the mooring plan is determined. That guarantees the assignment of each vessel on a berth in a given order.

The multi-objective model that we propose, on the one hand reduces the total transportation cost of loaded / unloaded containers ,and, on the other hand, the residence time of ships in the port.

The cost of transportation of containers equals to the number of containers (loaded and unloaded) by the distances between vessels and storage areas or vice versa.

After a long search in the literature dealing with the BAP, we are aware that the constraint of the distances traveled by the containers in the port area has not been taken in consideration. That's why we are proposing this multi-objective model, inspired by the models proposed proposed by Hansen et all. (2003) [12] and Imai et all. (2001) [3] , for the part which minimizes the residence time of ships in port and Zeinebou, Z. and Benabdelhafid, A. (2013) [1], for the second part.

2.1 The Notations Used in This Study Are Summarized Below

Indices

$i(= 1, ..., I) \in B$ set of berths

$j(= 1, ..., T) \in V$ set of the ships

$n(= 1, ..., N) \in P$ set of storage areas

$K(=1, ..., T) \in O$ set of service orders

Note: P the sources and destinations of all imports and exports containers (either for storage or for loading in another transport mode).

Parameters

S_i: time or berth i is free for the berth allocation planning ,

A_j: the arrival time of the vessel j,

Cd_{jn}: the number of containers associated with ship j, which will be discharged in n,

Cc_{jn}: the number of containers associated with n, which will be loaded into the ship j,

d_{in}: the distance between i and n,

Cc_j: the number of containers loaded in the ship j,

Cd_j: the number of containers unloaded from ship j,

C_{ij}: processing time of the ship j on the berth i.

Note: the processing time of each ship, is calculated based on the number of containers loadeds and unloadeds and the number of quay cranes available on the berth and their productivity

Variables Decisions

$$X_{ijk} = \begin{cases} 1, \text{if the ship } j \text{ is assigned to the berth } i \text{ in the order } k \\ 0, \text{otherwise} \end{cases}$$

2.2 The Assumptions Considered for the Formulation of the Problem

1-The planning process is considered static (BAPS), $\max A_j \leq \min S_i$,

2-Each berth can accommodate one ship at a time,

3-A ship can not be assigned to more than one berth,

4-The handling time of vessel depends on the berth assigned,

5- Once a ship is moored on a berth, he will remain in office until the end of his stay in port.

2.3 The Proposed Model

We present our model BAPSD, followed by step by step explanation:

$$F = \min \sum_{i \in B} \sum_{j \in V} \sum_{k \in O} (K.C_{ij} + S_i - A_j + \sum_{n \in P} (Cc_{jn} + Cd_{jn})d_{in})X_{ijk} \quad (1)$$

Subject to:

$$\sum_{i \in B} \sum_{k \in O} X_{ijk} = 1 \quad \forall j \in V, \quad (2)$$

$$\sum_{j \in V} X_{ijk} \leq 1 \quad \forall i \in B, k \in O, \quad (3)$$

$$\sum_{n \in P} Cc_{jn} = Cc_j \quad \forall j \in V, \quad (4)$$

$$Cd_j = \sum_{n \in P} Cd_{jn} \quad \forall j \in V, \quad (5)$$

$$X_{ijk} \in \{0, 1\} \quad \forall i \in B, j \in V, k \in O, \quad (6)$$

$$S_i \geq A_j \quad \forall i \in B, j \in V, \quad (7)$$

The objective function 1, minimizes the total cost of transport containers (imports / exports), in the port area and the times waiting and handling of incoming ships.

The constraint 2, ensures that each vessel will be served on a berth in given service order .

The constraint 3, ensures that each berth can only accommodate one ship at a time.

The constraint 4, ensures that the number of containers loaded in the vessel j , equals the sum of the containers intended for this vessel.

The constraint 5, ensures that the sum of the containers assigned to different terminals from a ship j, equals the number of containers unloaded from the ship.

The constraint 6,gives values ??that takes the decision variable.

The constraint 7, ensures that the vessels must arrive, before the availability of berths.

3 Experimental Results

Numerical experiments are conducted, to test the performance of the proposed approach.

The approach is coded in Matlab R2012b on a DELL PRECISION T3500 intel machine 5 GHz processor.

The model proposed in Section 2, is solved, using hybridization GA/SA presented in Section 3.1 . The data used in the experiments are generated randomly. We conducted two scenarios, depending on the charges ships. The first scenario, for small vessels with a load less than 160 TEU and the second for containerships with a load that goes to $18 * 10^3$ TEU . The succession plan, was considered over a period of 48 hours.

Several instances were used, to test the effectiveness of our approach. These instances vary according to the numbers of entering ships and available berths. The results for each instance are compared with those obtained by CPLEX 12.5, for the same instance.

3.1 The Solutions Process by Hybridization of GA / SA

We used a hybridization of the genetic algorithm and simulated annealing, for solve our model with realistic data and a reasonable time.

3.1.1. The Genetic Algorithm Used
The general procedures that we follow for our GA are:

Representation of Chromosomes
For the chromosome representation (coding), we choose a real coding to fully exploit the problem characteristics.. Figure 1, shows the chromosomal representation, of an example of five ships and two berths.

For this problem, each chromosome has ten cells chromosome length = number of berths * number of vessels .

The first five cells represent the five levels of service possible, in the number one berth.The last five cells , represent the five possible levels of service in the number two berth.

In this assignment the ships 2,4,5 are served at berth 1 as the first, second and third respectively vessels , and the vessels 1 and 3 are served in the berth 2 as the first and the second vessel, respectively. No vessel will be served after the vessels 5 and 3 (value 0 of the cell).

The Initial Population
After several tests, we set a number of individuals $Nd = 50$.

Fitness
For a minimization problem the fitness function is presented as follows:

$F_{fitness} = 1/f_{ob}$

f_{ob} : the objective function.

The berths	1	1	1	1	1	2	2	2	2	2
The ships	2	4	5	0	0	1	3	0	0	0
The orders	1	2	3	4	5	1	2	3	4	5

Fig. 1. Representation of chromosome

Selection
We use a method of roulette wheel selection.

Crossover
We used a 2-point crossover, with a probability of crossover $Pc = 0.8$.

Mutation
We use a point mutation, with probability $Pm = 0.1$.

The Stopping Criteria
The algorithm stops after a number of generations, fixed by $Ng = 45$. After his stopping , she outing the best found in every individual and his objective function.

3.1.2. The Simulated Annealing

The simulated annealing algorithm is applied to the final solution obtained by our GA (ie, the initial state of SA is the solution generated by the GA). The different steps of our algorithm SA are:

Initial State: the initial state of our algorithm SA is the output generated by GA.

Initial Temperature: we set the initial temperature $Ti = 400$,and,$\alpha = 0.001$.

Stopping Criteria: we set a threshold temperature as a stopping criterion $S = 0.0025$.

3.2 The Used Data

For each scenario the data are generated as follows:

1-The time of arrival of the ships A_j, and availability of berths S_i,are randomly generated in a planning horizon of 48h,

2-The numbers of ships and berths available are modified for each instance,
3-The number of storage areas is generated as follows:$P = (T + I)/2$,
4-The distances between the areas of storage and berths available d_{in}, are randomly generated between 200 and 1200m,
5-Processing times of ships C_{ij}, are randomly generated between 2 and 24.

3.3 The Obtained Results

For the two scenarios the results are presented as follows: in tables (1,2), we summarize the results of the two scenarios. The first columns, representing the instances used (Berths* ships). The second, third and fourth columns shown, the objective functions determined by GA, the hybridization of GA and SA, and CPLEX, respectively. The fifth and sixth columns, represent the deviation between the objective functions. The seventh and eighth columns, represent the computation time by hybridization (GA / SA) and CPLEX.

ET=Execution time,

CP=CPLEX,

D=Deviation,

Deviation=(The results obtained by our algorithm -The results obtained by CPLEX)/(The results obtained by CPLEX)(%).

AD: the average deviation.

Table 1. Summary of results for the scenario 1

I*T	GA	GA/SA	CP	D (GA/CP)	D ((GA/SA)/CP)	ET (GA/SA)	ET (CP)
2*5	29880120	29880110	29880110	3	0	89	102
2*10	69885008	69141677	69141677	1.05	0	175	198
5*10	106882837	104821772	104821772	1.9	0	213	286
5*15	61539962	57741359	56618834	6.33	1.76	489	345
5*25	156648737	118514748	116910133	33, 99	1.37	579	417
8*30	2052459	1700188	1683261	21.9	1.0056	802	618
10*30	1714739	1702404	1654747	3.62	2.88	913	810
10*50	3749143	3463680	3402977	10.17	1.78	1113	989
AD				10.24	1.09		

From the results shown in tables (1,2), we can conclude that the proposed algorithms give satisfactory and quasi-optimal results. We note that the deviations of the results obtained by GA are greater than those obtained by hybridization of GA / SA. But these deviations are relatively low.

Table 2. Summary of results for the scenario 2

I*T	GA	GA/SA	CP	D (GA/CP)	D ((GA/SA)/CP)	ET (GA/SA)	ET(CP)
2*5	168067314	168067304	168067304	$\simeq 0$	0	92	79
2*10	246888134	236304965	236304965	4.47	0	141	115
2*15	363662000	353859210	353859210	2.77	0	302	241
5*10	310308870	298486589	2880151169	7.74	3.635	263	212
5*15	470573127	434050051	394959475	19.14	9.89	496	345
5*20	902882223	834122734	827857441	9.06	0.75	526	418
8*30	999176241	976650564	953419632	4.79	2.43	816	612
10*30	$1.0139 * 10^9$	777685507	749789693	35.22	3.72	919	803
10*50	$1.89 * 10^9$	$1.44 * 10^9$	$1.31 * 10^9$	44.27	9.54	1098	989
AD				14.16	3.32		

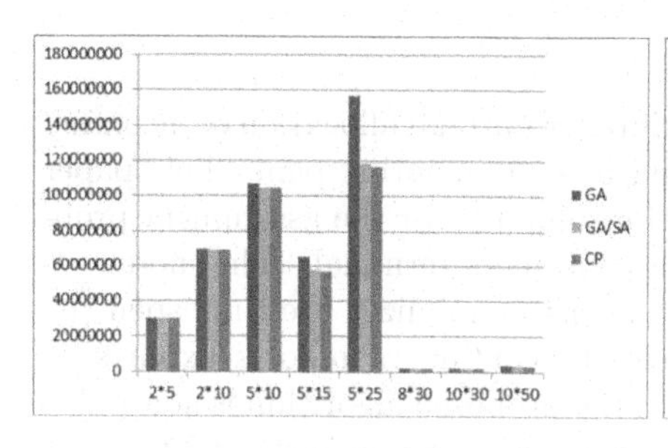

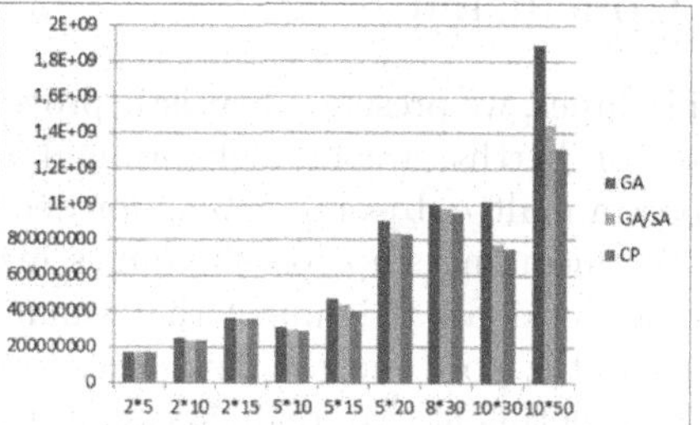

Fig. 2. Functions objectives calculated by the scenarios 1,2

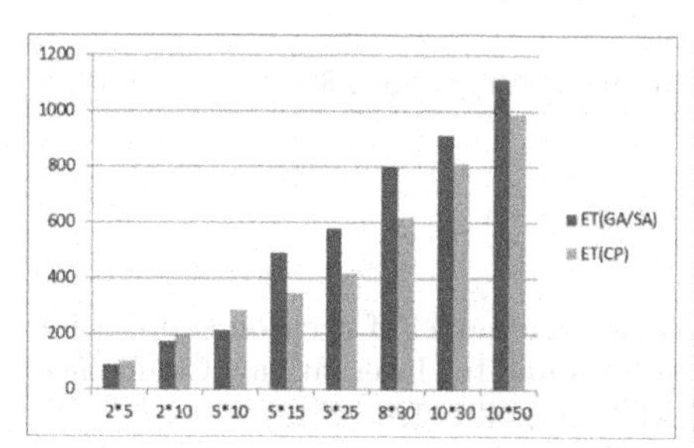

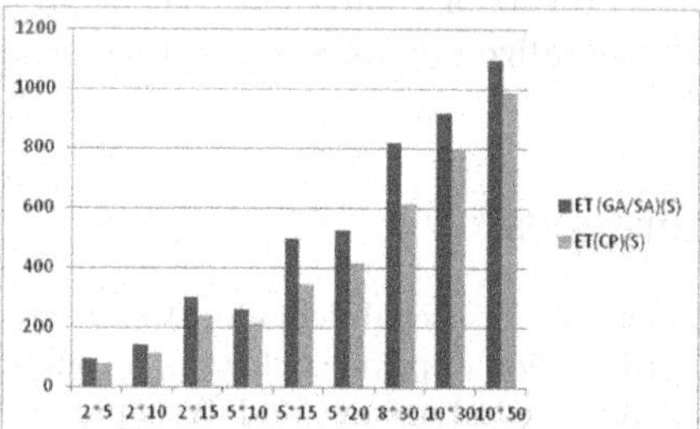

Fig. 3. Execution time for the scenarios 1,2

Figure (2), represented the objective functions calculated for the scenarios (1et2). The results obtained by GA are satisfactory, while those obtained by hybridization of GA/SA are near optimal.

Figure (3), represents changes in execution time for the hybridization GA/SA and by CPLEX , for the two scenarios respectively. We observe that the execution time, varies proportionally with the number of ships and berths. The execution times are still reasonable.

Figure (4), represent the deviations for the objective functions computed by our algorithms , and those calculated by CPLEX for scenarios 1 and 2 respectively. We can see that the deviations becomes larger with increasing number of ships and berths.

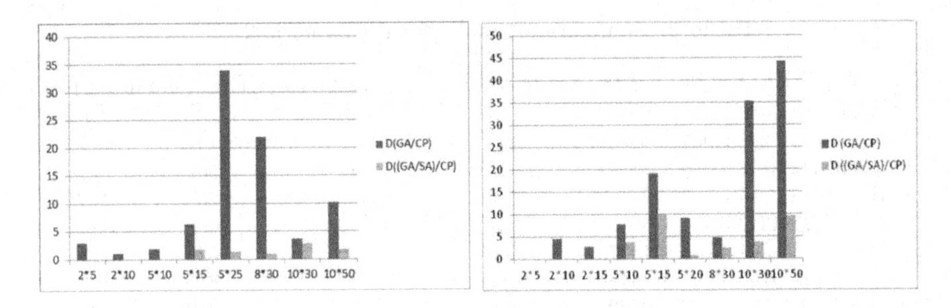

Fig. 4. The deviation for the scenarios 1,2

4 Conclusion

In this paper, we present a heuristic procedure for optimal allocation of incoming vessels on berths, a container terminal and a static mooring plan. This paper propose a multi- objective model for decision support, for the assignment problem the incoming ships on the quays in a containers terminal. The model we propose, seeks an assignment that simultaneously minimizes the time spent by vessels in the port and the distances traveled by containers imports/exports.

The experiments we conducted show that the proposed algorithm is adaptable to different realistic situations for the problem. The comparison with CPLEX and the deviations calculated for each instance, show the effectiveness of our model in terms of results and execution time.

To generalize our work,we will propose in the future a case study of dynamic BAP.

References

1. Zeinebou, Z., Benabdelhafid, A.: Development of a model of decision support for optimization of physical flows in a container terminal. In: International Conference on Advanced Logistics and Transport (ICALT), pp. 421–426 (2013)
2. Imai, A., Nishimura, E., Papadimitriou, S.: Berth allocation with service priority. Transportation Research Part B 37, 437–457 (2003)
3. Imai, A., Nishimura, E., Papadimitriou, S.: The dynamic berth allocation for a container terminal. Transportation Research Part B 35, 401–417 (2001)
4. Monaco, F.M., Sammarra, M.: The Berth Allocation Problem: A Strong Formulation Solved by a Lagrangean Approach. Transportation Science 41(2), 265–280 (2007)
5. Kim, K.H., Moon, K.C.: Berth scheduling by simulated annealing. Transportation Research Part B 37, 541–560 (2003)
6. Guan, Y., Cheung, K.R.: The Berth Allocation Problem: Models and Solution Methods. Operations Research Spectrum 26, 75–92 (2004)
7. Moorthy, R., Teo, C.-P.: Berth management in container terminal: The template design problem. OR Spectrum (2006), doi:10.1007/s00291-006-0036-5
8. Imai, A., Sun, X., Nishimura, E., Papadimitriou, S.: Berth Allocation in a Container Port: Using Continuous Location Space Approach. Transportation Research Part B 39, 199–221 (2005)

9. Golias, M.M., Boile, M., Theofanis, S.: Berth scheduling by customer service differentiation: A multi-objective approach. Transportation Research Part E 45, 878–892 (2009)
10. Hansen, P., Oguz, C., Mladenovic, N.: Variable neighborhood search for minimum cost berth allocation. European Journal of Operational Research 131(3), 636–649 (2008)
11. Lee, D.-H., Chen, J.H., Cao, J.X.: The continuous Berth Allocation Problem: A Greedy Randomized Adaptive Search Solution. Transportation Research Part E46, 1017–1029 (2010)
12. Hansen, P., Oguz, C.: A Note on Formulations of Static and Dynamic Berth Allocation Problems (2003) ISSN: 0711-2440 Les Cahiers du GERAD
13. Imai, A., Nishimura, E., Papadimitriou, S.: The dynamic berth allocation problem for a container port. Transportation Research Part B 35, 401–417 (2001)
14. Xu, D., Li, C.-L., Leung, J.Y.-T.: Berth allocation with time-dependent physical limitations on vessels. European Journal of Operational Research 216, 47–56 (2012)

Sensitivity Analysis of a Priori Power Indices

František Turnovec[1] and Jacek Mercik[2]

[1] Charles University in Prague, Czech Republic
frantisek.turnovec@tiscali.cz
[2] Wroclaw University of Technology, Poland
jacek.mercik@pwr.wroc.pl

Abstract. Power index analysis is very important in all group decision making processes. There is no better way how to evaluate the power to act of a decision maker than a priori power index. The formal analysis of measuring sensitivity of the index are presented in the paper. Constant power partition of weight allocation space with respect to a given quota is defined. Responses of the measures of power to changes of allocations of weights, quota and/or number of voters are considered. Proposed methods are base for algorithms of sensitivity analysis of a priori power indices.

Keywords: a priori power index, sensitivity, quota, weight, number of players.

1 Introduction

Since 1954 when a priori power index of Shapley and Shubik (1954) was introduced there is a consistent belief among the researchers that such index is a proper way to evaluate the role and significance of a player in a group decision making process. Confrontation with the real world applications very soon show that changes in preliminary assumptions lead to different a priori power indices: Banzhaf (1965), Coleman's indices (1971), Brams-Lake (1977), Johnston (1978), Deegan-Packel (1979) and Holler (1982) when presenting the most popular ones only. All a priori indices are based upon the concept of winning and losing coalitions within the frame of weighted simple games. Their nature is non-linear due to changes in parameters such as dimension, quota or/and weights.

Power index analysis is very important in all group decision making processes (see last vivid discussion on representation of countries in enlarging European Union) but introducing automated procedures into the process of decision making changes the significance of such evaluation of a player's (a decision maker's) role in this process: the role and position of the decision maker may change the final outcome not only in politics but in, for example, trading or mobile telephones networks. As there is no better way how to evaluate the power to act of a decision maker than a priori power index, the formal analysis and algorithms of measuring sensitivity of the index are very essential. This was done previously by many researchers starting from different points, for example: paradoxes of voting (Brams, 1995), epsilon - stability of power distribution (Mercik et al. 2004) or winning coalition analysis (Felsenthal and Machover M., 1998).

N.T. Nguyen et al. (Eds.): ACIIDS 2014, Part I, LNAI 8397, pp. 464–473, 2014.

What we present in this paper is formal attempt to the sensitivity analysis of a priori power index with emphasis given to practical results leading to potential algorithms of the analysis. Such algorithms could be applied also in automated decision making systems.

The paper is set up as follows. The next section outlines the preliminaries of simple games and power indices. The second section is devoted to constant power partition by introducing so called allocation space. This concept is in use in the next sections devoted to problems of stability of power indices. Finally, there are some conclusions and suggestions for future research.

2 Preliminaries

Let N be a finite set of committee's members, q be a quota, w_j be a voting weight of member $j \in N$. A game on N is given by a map $v : 2^N \rightarrow R$ with $v(\emptyset) = 0$. The space of all games on N is denoted by G. A coalition $T \in 2^N$ is called a carrier of v if $v(S) = v(S \cap T)$ for any $S \in 2^N$. The domain $SG \subset G$ of simple games on N consists of all $v \in G$ such that

(i) $v(S) \in \{0,1\}$ for all $S \in 2^N$,

(ii) $v(N) = 1$,

(iii) v is monotonic, i.e. if $S \subset T$ $then$ $v(S) \leq v(T)$.

A coalition S is said to be winning in $v \in SG$ if $v(S) = 1$ and losing otherwise. Therefore, the voting upon a bill is equivalent to formation of a winning coalition consisting of voters. A simple game (N,v) is said to be proper, if and only if it is satisfied that for all $T \subset N$, if $v(T) = 1$ $then$ $v(N \backslash T) = 0$.

We analyse only simple and proper games where players may vote yes or no.

By $(N,q,w) = (N, q, w_1, w_2, \ldots, w_n)$ we shall denote a committee (weighted voting body) with member set N, quota q and weights w_j, $j \in N$. We shall assume that w_j are nonnegative integers. Let $t = \sum_{j=1}^{n} w_j$ be the total weight of the committee.

A power index is a mapping $\varphi: SG \rightarrow R^n$. For each $i \in N$ and $v \in SG$, the ith coordinate of $\varphi(v) \in R^n, \varphi(v)(i)$, is interpreted as the voting power of player i in the game v. In the literature there are two dominating power indices: the Shapley-Shubik power index and the Banzhaf power index. Both indices are based on the Shapley value concept (Shapley, 1953) and belong to so called family of a priori power indices.

The Shapley-Shubik (Shapley, Shubik (1954)) power index for simple game is the value $\varphi: SG \rightarrow R^n, v \rightarrow (\varphi_1(v),\varphi_2(v),\ldots,\varphi_n(v))$ where for all $i \in N, card\{N\} = n; card\{S\} = s$

$$\varphi_i^{SS}(v) = \sum_{S \subset N, i \notin S} \frac{s!(n-s-1)!}{n!}. \quad (1)$$

The Banzhaf (1965) power index for simple game is the value: $\varphi: SG \rightarrow R^n$, $v \rightarrow (\varphi_1(v),\varphi_2(v),\ldots,\varphi_n(v))$ where for all $i \in N; card\ \{N\} = n; card\{S\} = s$

$$\varphi_i^B(v) = \frac{1}{2^{n-1}} \sum_{S \subseteq N\{i\}} [v(S \cup \{i\}) - v(S)] \tag{2}$$

The above definitions of power indices are directly obtained from characteristic function games where a marginal value of power excess introduced into winning coalition is calculated[1].

3 Constant Power Partition

By weight allocation space we shall call a set $\Omega_N^t = [\mathbf{w} : \mathbf{w} \in R_n, \sum_{j \in N} w_j = t, \ w_j \geq 0]$ of all weight allocations in committees with the same member set N and total weight t.

Let us introduce a quota into this picture. By $W(N,q,\mathbf{w}) = \left\{ S \subseteq N : \sum_{j \in S} w_j \geq q \right\}$ we shall denote the set of all winning configurations for a committee [N, q, **w**]. We shall say that two committees [N, q, **u**] and [N, q, **v**] are strategically equivalent if W(N, q, **u**) = W(N, q, **v**). A partition $P_1, P_2, \ldots, P_m$ of weight allocation space Ω^t_N such that for any two allocations **u**, **v** $\in P_i$ it holds that $W(N,q,\mathbf{u}) = W(N,q,\mathbf{v})$ and for any $\boldsymbol{u} \in P_r$ and $\boldsymbol{v} \in P_k$ $(r \neq k)$ $W(N,q,\mathbf{u}) \neq W(N,q,\mathbf{v})$ we shall call a constant power partition. All committees from the same set P_i of constant power partition are strategically equivalent. It is easy to show that any well-defined measure of power of the same member in all of these committees is the same.[2]We receive a partition of the weight triangle (allocation space) into a finite number of subsets, each of them having the same "power status" from the point of view of ability of the committee members to form winning coalitions. In all committees corresponding to the same subset each particular member of the committee is the member of the same winning coalitions (and the same losing coalitions). Moving from one subset to another the structure of winning and losing coalitions is changing. So we can use this partition as a starting point for some general considerations about distribution of power in a committee system.

4 Sensitivity Analysis: Changing the Quota

Having a power index π of a committee $[N, \gamma, \boldsymbol{\omega}]$ with a total weight t, quota γ_0 and an allocation $\boldsymbol{\omega}^0$, let us consider a response of the measure of power to changes of quota. We

[1] In Turnovec et al. (2008) one can find introduction of power indices without theory of games but based on concept of permutations and of their probability.

[2] It follows from the fact that in characteristic function representation of the weighted voting body (simple game) the characteristic function for all committees is the same, so power indices defined in terms of characteristic function (Shapley-Shubik, Banzhaf, Holler-Packel, Johnston, Deegan-Packel) will generated the same values.

shall assume integer weights and quotas. By $y_S(\gamma,\boldsymbol{\omega}) = \sum_{i\in S} \omega_i - \gamma$ we shall denote the surplus of the total weight of a coalition $S \in N$ over the quota γ with respect to the allocation $\boldsymbol{\omega}$.

Let us suppose that the quota is changing: $\gamma(\delta) = \gamma_0 + \delta$. We shall say that an allocation $\boldsymbol{\omega}^0$ is stable with respect to a change δ of quota if the structure of winning coalitions remains the same. In this case any a priori power index does not change.

Let us denote $\delta^+ = min_{y_S \geq 0} y_S(\gamma_0, \omega^0)$, $\delta^- = min_{y_S < 0} y_S(\gamma_0, \omega^0)$.

Lemma 1: an allocation $\boldsymbol{\omega}^0$ is stable with respect to changes of quota $\gamma(\delta) = \gamma_0 + \delta$ in an interval $\gamma(\delta) \in [1 + \gamma_0 + \delta^-, \gamma_0 + \delta^+]$.

For proof it is enough to show that for any $S \subset N$ and $\delta \in [1+\delta^-, \delta^+]$,

$\sum_{i\in S} \omega_i^0 \geq \gamma_0 \Rightarrow \sum_{i\in S} \omega_i^0 \geq \gamma_0 + \delta$, $\sum_{i\in S} \omega_i^0 < \gamma_0 \Rightarrow \sum_{i\in S} \omega_i^0 < \gamma_0 + \delta$ what follows immediately from the definition of δ^- and δ^+.

Example 1. Let $[\gamma; \boldsymbol{\omega}] = [60; 25, 25, 50]$. In our case $N = \{1, 2, 3\}$, $\gamma_0 = 60$, $t = 100$. To analyse stability of this distribution of power we shall list all configurations and corresponding surpluses y_S:

S	y_S	
{1}	-35	
{2}	-35	$\delta^- = -10$
{3}	-10	
{1,2}	-10	
{1,3}	15	
{2,3}	15	$\delta^+ = 15$
{1,2,3}	40	

Hence, the power distribution in our committee is stable with respect to the interval of quotas $[1 + \gamma_0 + \delta^-, \gamma_0 + \delta^+] = [51, 75]$

For all quotas from this interval the power indices of all members will not change.

Let us define for $k = 1, 2, \ldots$: $\delta_0 = int\left(\frac{n}{2}\right)$, $S(\delta_k) = \{S \subseteq N: \sum_{i\in S} w_i^0 \geq 1 + \delta_{k-1}\}$, $\delta_k = min_{S \subseteq S(\delta_k)} \sum_{i\in S} \omega_i^0$.

Lemma 2: for any power index, there exist intervals of stability for an allocation $\boldsymbol{\omega}^0$ with respect to changes of quota $\delta_0 + 1 \leq \delta \leq \delta_1, \ldots, \delta_{k-1} + 1 \leq \delta \leq \delta_k, \ldots, \delta_{r-1} + 1 \leq \delta_r = t$ such that power index stays constant for all quotas from each of those intervals.

The proof follows immediately form the above definitions.

Example 2. Let us consider a committee $[\gamma, \boldsymbol{\omega}] = [\gamma; 0.2, 0.3, 0.15, 0.35]$ where $\delta_0 = 0.5$ We shall evaluate all intervals of stability.

a) For $\delta_0 = 50$ we have $S(\delta_0) = \left\{ \{1,4\}, \{2,4\}, \{1,2,3\}, \{1,3,4\}, \{2,3,4\}, \{1,2,4\}, \{1,2,3,4\} \right\}$

Then $\delta_1 = \min [55, 65, 65, 70, 80, 85, 100] = 55$ and we have the first interval of stability $[1 + \delta_0, \delta_1] = [51, 55]$.

b) For $\delta_1 = 55$ $S(\delta_1) = \{\{2,4\},\{1,2,3\},\{1,2,4\},\{1,3,4\},\{2,3,4\},\{1,2,3,4\}\}$. Then $\delta_2 = \min[65,65,70,75,100] = 65$ and we have the second interval of stability $[1+\delta_1, \delta_2] = [56,65]$.

c) For $\delta_2 = 65$ $S(\delta_2) = \{\{1,2,4\},\{1,3,4\},\{2,3,4\},\{1,2,3,4\}\}$. Then $\delta_3 = \min[70,80,85,100] = 0.70$ and we have the third interval of stability $[1+\delta_2, \delta_3] = [66,70]$.

d) For $\delta_3 = 70$ $S(\delta_3) = \{\{2,3,4\},\{1,2,4\},\{1,2,3,4\}\}$. Then $\delta_4 = \min[80,85,100] = 80$

e) For $\delta_4 = 80$ $S(\delta_4) = \{\{1,2,4\},\{1,2,3,4\}\}$. Then $\delta_5 = \min[85,100] = 85$

f) For $\delta_5 = 85$ $S(\delta_5) = \{\{1,2,3,4\}\}$ and $\delta_6 = 100$. We obtained the last interval of stability $[1+\delta_5, \delta_6] = [86,100]$.

5 Sensitivity Analysis: Changing the Weights

Now we shall assume that in a committee $[N, \gamma_0, \omega^0]$ quota γ_0 is fixed but allocation of weights ω^0 is changing in the following way: for any two selected members $k, r \in N$ ($k \neq r$) we shall consider allocation $\omega(\delta)$ such that

$$\omega_i(\delta) = \begin{cases} \omega_i^0 & for\ i \neq k,r \\ \omega_i^0 + \delta & for\ i = k \\ \omega_i^0 - \delta & for\ i = r \end{cases} \tag{3}$$

(transfers of weights between k and r). Clearly interval of feasible changes of δ (keeping weights $\omega_k(\delta)$ and $\omega_r(\delta)$ non-negative) is $\langle -\omega_k^0, \omega_r^0 \rangle$.

We want to answer two questions:

a) Given an allocation ω^0 for what interval of δ the voting power of committee members remains the same?
b) Given an allocation ω^0 for what δ is the voting power of committee members changing on the interval of feasible changes $\delta \in \langle -\omega_k^0, \omega_r^0 \rangle$?

To answer the first question we have to find an interval $\langle \Delta_{kr}^-, \Delta_{kr}^+ \rangle$ of δ not changing the voting power of all committee members.

Let us denote by C the set of all configurations $S \subseteq N$ (all subsets of N), and by $C(i,j)$ the set of all configurations $S \subseteq N$ such that $i \in S$ and $j \notin S$. For given k, r the set $C(k,r)$ consists of all S with k and no S with r. Respectively, $C(r,k)$ consists of all S with r and no S with k. Then the set $C \setminus \{C(k,r) \cup C(r,k)\}$ consists of all configurations with both k and r and configurations without both k and r.

For a δ the committees $[N, \gamma_0, \omega(\delta)]$ and $[N, \gamma_0, \omega^0]$ are strategically equivalent (providing all members with the same voting power) if the sets of all winning configurations are the same. Clearly, any feasible δ will not change the winning and losing configurations in the set $C \setminus \{C(k,r) \cup C(r,k)\}$. Henceforth, we have to check sets $C(k,r)$ and $C(r,k)$.

Let $\delta_S(\gamma_0,\omega^0) = \gamma_0 - \sum_{s\in S} \omega_s^0$. Clearly, if $\delta_S(\gamma_0,\omega^0) = \gamma_0 - \sum_{s\in S} \omega_s^0 \le 0$ then S is a winning configuration, and if $\delta_S(\gamma_0,\omega^0) = \gamma_0 - \sum_{s\in S} \omega_s^0 > 0$ then S is a losing configuration.

Considering $C(k,r)$, winning and losing configurations for ω^0 and $\omega(\delta)$ will remain the same if

$$\delta_S(\gamma_0,\omega^0) \le 0 \Rightarrow \delta_S(\gamma_0,\omega(\delta)) \le 0\text{, and} \tag{4}$$

$$\delta_S(\gamma_0,\omega^0) > 0 \Rightarrow \delta_S(\gamma_0,\omega(\delta)) > 0 \tag{5}$$

for all $S \in C(k,r)$. From (4), for all $S \in C(k,r)$ such that $\delta_S(\gamma_0,\omega^0) = \gamma_0 - \sum_{s\in S} \omega_s^0 \le 0$ we get for δ: $\delta_S(\gamma_0,\omega(\delta)) = \gamma_0 - \sum_{s\in S} \omega_s^0 - \delta = \delta_S(\gamma_0,\omega_s^0] - \delta \le 0$. From feasibility of resulting allocations we get $\delta \ge -\omega_k^0$. Define

$$\delta_{kr}^- = \max[\max_{S\in C(k,r),\delta_S\le 0} \delta_S(\gamma_0,\omega^0), -\omega_k^0] \tag{6}$$

From (4), for all $S \in C(k,r)$ such that $\delta_S(\gamma_0,\omega^0) = \gamma_0 - \sum_{s\in S} \omega_s^0 > 0$ we get for δ $\delta_S(\gamma_0,\omega(\delta)) = \gamma_0 - \sum_{s\in S} \omega_s^0 - \delta = \delta_S(\gamma_0,\omega) - \delta) > 0$. From feasibility of resulting allocations we get $\delta \le \omega_r^0$. Define $\delta_{kr}^+ = \min\left[\min_{S\in C(k,r),\delta_S>0} \delta_S(\gamma_0,\omega^0) - 1, \omega_r^0\right]$. For any δ from the interval $\delta_{kr}^- \le \delta \le \delta_{kr}^+$ the changed allocation $\omega(\delta)$ remains feasible and structure of winning in losing configurations in $C(k,r)$ remains the same.

By symmetric considerations for $C(r,k)$ we obtain $\delta_{rk}^+ = \min\left[\min_{S\in C(r,k),\delta_S\le 0} -\delta_S(\gamma_0,\omega^0), \omega_r^0\right]$; $\delta_{rk}^- = \max\left[\max_{S\in C(r,k),\delta_S>0} -\delta_S(\gamma_0,\omega^0) + 1, -\omega_k^0\right]$.

For any δ from the interval $\delta_{rk}^- \le \delta \le \delta_{rk}^+$ the changed allocation $\omega(\delta)$ remains feasible and structure of winning and losing configurations in $C(r,k)$ remains the same.

To find an interval $\langle \Delta_{kr}^-, \Delta_{kr}^+ \rangle$ of δ not changing the voting power of all committee members we have to take intersection of the both intervals $\langle \delta_{kr}^-, \delta_{kr}^+ \rangle$ and $\langle \delta_{rk}^-, \delta_{rk}^+ \rangle$: $\Delta_{kr}^- = \max[\delta_{kr}^-, \delta_{rk}^-]$, $\Delta_{kr}^+ = \min[\delta_{kr}^+, \delta_{rk}^+]$

Lemma 3: for any δ from the interval $\Delta_{kr}^- \le \delta \le \Delta_{kr}^+$ allocation $\omega(\delta)$ remains feasible and structure of winning and losing configurations in committee $[N, \gamma_0, \omega(\delta)]$ remains the same as in committee $[N, \gamma_0, \omega^0]$, what implies that power indices of all committee members are the same in both committees.

We shall call $\langle \Delta_{kr}^{-}, \Delta_{kr}^{+} \rangle$ an interval of stable power with respect to transfers of votes between k and r.

Example 3. Let us consider a committee $[\gamma_0, \omega^0] = [60; 20, 30, 15, 35]$. We shall evaluate interval of stability with respect to transfers between 1 and 2. In our case S(1, 2) = {{1}, {1,3}, {1,4}, {1,3,4}} and S(2,1) = {{2}, {2,3}, {2,4}, {2,3,4}}.

We have to find $\langle \Delta_{12}^{-}, \Delta_{12}^{+} \rangle$. For S(1,2) we have

	$\delta_{S(\gamma 0, \omega 0)}$	δ^{-}_{12}	δ^{+}_{12}	
{1}	40			
{1,3}	25			
{1,4}	5		4	min [min [40, 25, 5] – 1, 30] = 4
{1,3,4}	-10		-10	max [max [-10], -20] = -10

For S(2,1) we have

	$\delta_{S(\gamma 0, \omega 0)}$	δ^{-}_{21}	δ^{+}_{21}	
{2}	30			
{2,3}	15		-14	max [max [-20, -15] + 1, -20] = -14
{2,4}	-5	5		min [min [5, 20], 30] = 5
{2,3,4}	-20			

Then $\Delta^{-}_{12} = \max[-10, -14] = -10$ and $\Delta^{+}_{12} = \min[4, 5] = 4$.

For transfers of weights between 1 and 2 we have the interval of stability ⟨-10, 4⟩. Ceteris paribus, “selling” at most 10 votes to member 2 member 1 will have the same voting power as in original allocation, “selling” at most 4 votes to member 1 member 2 will keep its voting power. Also power of other members is not changing.

Table 1. Transferring weights between *1* and *2*: interval of stability $\delta \in \langle -10, 4 \rangle$

Member	*Weight*	*Power index in %*		
		SS	PB	HP
1	20+δ	16.67	16.67	25
2	30-δ	33.33	33.33	25
3	15	16.67	16.67	25
4	35	33.33	33.33	25
Σ	100	100	100	100

6 Sensitivity Analysis: Changing the Number of Players

Before we start with sensitivity analysis of power index due to changes in the number of players it is worth to describe a real life example. i.e. changes of a structure of a parliament (at least the Polish but not only that one) during a term. What we may observe during a term: there are a few dissidents from parties, acting as independent members of parliament or forming a parliamentary group(s). Therefore, total weight t is fixed and equals to the total number of parliamentary members, the quota is unchanged (in most cases it is 50% +1 vote), but the number of players (parties/groups) is changed from time to time and consequently the distribution of weights is changed too. Potentially, analysis of power index sensitivity in this case may reflect not only the actual situation but also it may allow manipulation within the parliamentary structure.

Having a power index π of a committee $[N, \gamma, \boldsymbol{\omega}]$ with a total weight t, quota γ_0 and an allocation $\boldsymbol{\omega}^0$, let us consider a response of the measure of power to changes of number of players.

As we assumed in preliminaries, w_j are nonnegative integers. Hence, the number of players may vary within the interval $\langle 1 \leq N \leq t \rangle$. Let N_0 denotes number of players before changes. The following situations may occur with the current number of players N:

1) $For\ N_0 \geq 2,\ N = N_0 - 1$. This means that two players (parties, groups, etc.) have united. This new situation is completely different as players are different. Any sensitivity analysis should be done from the very beginning.
2) $For\ 2 \leq N_0 \leq t - 1,\ N = N_0 + 1$. This means that a new player has appeared and at least one of the remaining players has different weight.

In the following we analyse only the second situation, i.e. the increasing number of players.

Similarly to sensitivity analysis with changing weights we accept allocation of weights ω^0 being changing in the following way:

Case I: there is one and only one more player (transfers of weights between player k and player $N+1$). The following distribution of weights may occur for $k = 1, \dots, N_0$: $(\omega_1, \dots, \omega_k - \delta, \dots, \omega_{N_0}, \omega_{N_0+1} = \delta)$. Following axioms of power index (for example Holler, Li, 1995) any dummy player should not change the value of the coalition. So, expanding initial (before transfer) committee $[N_0, \gamma, \boldsymbol{\omega}]$ to committee $[N_0 + 1, \gamma, \boldsymbol{\omega}]$ by adding $(N_0 + 1)$-th player with weight $\omega_{N_0+1} = 0$ we realize enlargement of the number of players. Now, it's time for transfer of weights between k-th and $(N_0 + 1)$-th players. This may be analysed the same way as it was made during the analysis of sensitivity due to changes of weights.

Definition: *A priori* power index for player k from committee $[N_0 + 1, \gamma, \boldsymbol{\omega}]$ with allocation $\boldsymbol{\omega} = (\omega_1, \dots, \omega_k - \delta, \dots, \omega_{N_0}, \omega_{N_0+1} = \delta)$ is stable if it is equal to the k-th value of the same index for committee $[N_0, \gamma, \boldsymbol{\omega}]$ and $\pi_{N_0+1} = 0$.

From the above definition one may find the interval of stability (at least one) of the k-th player, $[0, \delta]$, against changes in the number of players as long as $\pi_{N_0+1} = 0$ for committee $[N_0 + 1, \gamma, \boldsymbol{\omega}]$ with allocation $\boldsymbol{\omega} = (\omega_1, \dots, \omega_k - \delta, \dots, \omega_{N_0}, \omega_{N_0+1} = \delta)$. This can be solved via iteration for $\delta = 1, 2, \dots$ till the moment when power index of the new player becomes positive, i.e. $\pi_{N_0+1} > 0$, fixing on the way intervals with the same value of the given power index.

Case II: there are more than one extra player (transfers of weights between k and new players). This is similar to the case I but significantly more complicated as we have to consider all possible permutations of potential new players and their allocation of weights respectively. For k-th player's transfers there are $p = \sum_{j=1}^{j=\omega_k - 1} \binom{\omega_k - 1}{j}$ such permutations. Each permutation gives interval of stability when value of the a priori power index equals zero: $[0, \delta_i]$, $i = 1, \dots, p$. This leads to the interval of absolute stability for *k-th* player as equal to $[0, min_i \delta_i]$.

Both cases (I and II) may produce an empty interval of stabilization for k-th player as every $\delta \geq 1$ could give $\pi_{N_0+1} > 0$ in committee $[N, \gamma, \boldsymbol{\omega}]$ for $N > N_0$.

Example 4. Let us consider a committee (enlarged but before transfer starts) $[\gamma_0, \omega^0] =$ [60; 20, 30, 15, 35, 0]. We shall evaluate interval of stability with respect to transfers between player 1 and player 5. We start iteration with $\delta = 1$.

Table 2. Sensitivity to changes of number of players with transfer of weight from the first player for $\delta = 1, 2$

Member	*Initial weights* $\delta = 0$	*SS*	$\delta = 1$	*SS*	$\delta = 2$	*SS*
1	20-δ	25.00	19	20.00	18	20.00
2	30	25.00	30	20.00	30	20.00
3	15	8.33	15	11.67	15	11.67
4	35	41.67	35	45.00	35	45.00
5	0+δ	0	1	3.33	2	3.33
Σ	100	100	100	100	100	100

Looking at Shapley-Shubik value for 3-th and 4-th players we may observe so called a paradox of size (Brams, 1975) when increasing number of voters increases (against intuition) S-S value for one of the previous players. One should rather expect proportional decreasing as standardized value should be distributed among greater number of beneficiaries. Such paradox may also disturb our analysis.

7 Conclusions and Future Work

Presented in the paper sensitivity analysis of a priori power indices is formalized according to the three dimensions of a group decision, namely: quota, weights and number of players. Obtained results are good base for practical algorithms and may be used in automated decision making systems.

The future research should be done towards two potential areas:

- unification of those three dimensions in one, leading finally into a concept of "index of stability",
- application of the proposed methods to other power indices, not only a priori ones.

References

1. Banzhaf III, J.F.: Weighted voting doesn't work: a mathematical analysis. Rutgers Law Review 19, 317–343 (1965)
2. Brams, S.J.: Game Theory and Politics. Free Press, New York (1975)
3. Coleman, J.S.: Control of collectivities and the power of a collectivity to act. In: Lieberman, B. (ed.) Social Choice, New York, pp. 277–287 (1971)
4. Deegan Jr., J., Packel, E.W.: A new index of power for simple n-person games. International Journal of Game Theory 7, 113–123 (1979)
5. Felsenthal, D., Machover, M.: The measurement of voting power, Theory and Practice, Problems and Paradoxes. Edward Edgar, Cheltenham (1998)
6. Holler, M.J.: Forming coalitions and measuring voting power. Political Studies XXX(2), 266–271 (1982)
7. Holler, M.J., Li, X.: From public good index to public value: an axiomatic approach and generalization. Control and Cybernetics 24, 257–270 (1995)
8. Johnson, R.J.: On the measurement of power: Some reactions to Lawer. Environment and Planning A 10, 907–914 (1978)
9. Mercik, J., Turnovec, F., Mazurkiewicz, M.: Epsilon - stability of power distribution, Operation and System Research. In: Kulikowski, R., Kacprzyk, J., Słowiński, R. (eds.), Warszawa, pp. 129–141. Exit (2004)
10. Shapley, L.S.: A Value for n-person Games. In: Kuhn, H.W., Tucker, A.W. (eds.) Contributions to the Theory of Games, vol. II (1953); Annals of Mathematical Studies 28, 307–317
11. Shapley, L.S., Shubik, M.: A method of evaluating the distribution of power in a committee system. American Political Science Review 48(3), 787–792 (1954)
12. Turnovec, F., Mercik, J., Mazurkiewicz, M.: Power Indices Methodology: Decisiveness, Pivots, and Swings. In: Braham, M., Steffen, F. (eds.) Power, Freedom, and Voting, pp. 23–37. Springer, Heidelberg (2008)

Artificial Neural Network Based Prediction Model of the Sliding Mode Control in Coordinating Two Robot Manipulators

Parvaneh Esmaili[1] and Habibollah Haron[2]

[1] Department of Computer Science, Faculty of Computing, Universiti Teknologi Malaysia, 81310 UTM Skudai, Johor, Malaysia
p.esmaili1984@gmail.com
[2] Department of Computer Science, Faculty of Computing, Universiti Teknologi Malaysia, 81310 UTM Skudai, Johor, Malaysia
habib@utm.my

Abstract. The design of a decentralized controlling law in the coordinated transportation area of an object by multiple robot manipulators employing implicit communication between them is a specific alternative in synchronization problems. A decentralized controller is presented in this work which is combination of the sliding mode control and artificial neural network which guarantees robustness in the system. Implicit communication among robot manipulators considers the light weight beam angle in this controller. A multi layer feed forward neural network based prediction model is presented not only to improve trajectory tracking of multiple robots but also to solve the chattering phenomena in the sliding mode control. The simulation results show the effectiveness of the proposed controller on two cooperative PUMA 560 robot manipulators.

Keywords: Decentralized control, neural network, cooperative robot manipulators, handles an object, synchronization.

1 Introduction

Recently, in cooperative robot manipulator systems which can accurately control the system in the presence of structured and unstructured uncertainties have become a challenging issue. Several types of controllers have been categorized into different groups. Cooperative robot controlling methods are categorized into one of two methods: Centralized and Decentralized. Centralized controlling methods are approaches which consider all robot arms and the handling object as one closed chain system. Conversely decentralized methods almost entirely use implicit behavior to communicate instead of explicit information which helps robots to act individually.

In addition, there are several kinds of controlling methods within both decentralized and centralized classification methods such as kinematics, dynamics, master slave, sliding mode, adaptive robust methods and adaptive neural network based sliding mode control. Several researchers have considered kinematics and

N.T. Nguyen et al. (Eds.): ACIIDS 2014, Part I, LNAI 8397, pp. 474–483, 2014.

dynamics method for control, but in cooperative robot system, due to complexity problems, these methods were not sufficient [1] [2] [3]. Some studies used various combinations with force control and position/force control [4-5] [6] [7]. Component based methods are considered as [8-9]. Impedance control was proposed [10] to achieve accurate system control and there are several works based on impedance controller in a multiple manipulator structure [11]. However, impedance control method was not sufficient to handle the dynamic properties of the real world environment such as stiffness, damping and inertia.

A variable structure system with sliding mode control was first suggested in the 1950's, and after 1970 the sliding mode control became commonly used because it consistently showed good performance for nonlinear systems in multi input multi output systems and has been usefully implemented in discrete time systems as well. It is robust controller and is also insensitive in the presence of parameter changes and external disturbances. However, the sliding mode control has two disadvantages. The first is chattering phenomena which is high output frequency of the controller. The second is that the calculation of known sections of nonlinear system as equivalent control is difficult [12] when solving these kinds of problems. Some researchers have been investigated on some soft computing methods which have been adapted from artificial intelligence such as fuzzy logic, neural network, evolutionary computing and so on [12][13] [14][15]. A hybrid sliding mode control with neural network has been investigated by several researchers to develop one robot manipulator which was initially proposed [13] and investigated by some researchers [12] [14] [16].

In a similar manner, some works have focused only on the dual arm robot as a cooperative system based on the sliding mode with adaptive neural network controlling method. Also, in these kinds of systems, the system is considered a closed chain kinematic or dynamics equation. However, the ability of act separately faces problems as they must use a centralized controlling method which is not sufficient and accurate enough for multiple robotic systems in the presence of structured and unstructured uncertainties in the real world.

The aim of this work is to design a decentralized control law by using implicit communication, the robot manipulators and the object transfer simultaneously with the same constant yet limited velocity. First, the sliding mode control schema for PUMA 560 is then implemented. The artificial neural network based prediction model for the dynamic section of PUMA 560 is presented. Finally the proposed controller in this work is extended for two decentralized robot manipulators and the results of the controller are shown as follows.

2 Modeling of the System

The decentralized control law uses the implicit information between robots. As implicit information for each robot manipulator, the edge angle of the object, reveals the objects position and the fixed end- effectors on the object by using set points which are mentioned in Fig.1. By using the dynamic equation, the motion of the robot manipulator will be derived. As implicit information, by changing the angle of the object and related end-effector, the other robot can understand information about another robot. For this reason, robot manipulators work by implicit communication with respect to each other. The purpose of this work is to design a decentralized artificial neural network based prediction model of sliding mode control law by using implicit communication, the robot manipulators and the object transfer simultaneously with the same constant limited velocity.

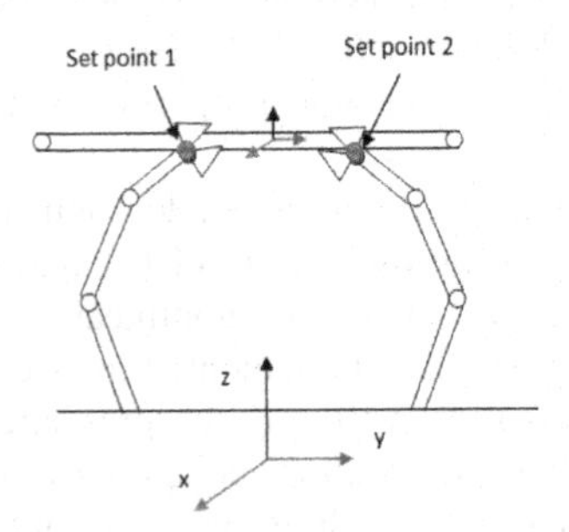

Fig. 1. The construction of the two cooperative robot manipulators to handle object

Dynamic equation of PUMA 560 robot based on Lagrange – Euler [17] formula is as follows in Eq. (1).

$$M(\theta)\begin{bmatrix}\ddot{\theta}_1\\ \ddot{\theta}_2\\ \ddot{\theta}_3\\ \ddot{\theta}_4\\ \ddot{\theta}_5\\ \ddot{\theta}_6\end{bmatrix}+B(\theta)\begin{bmatrix}\dot{\theta}_1\dot{\theta}_2\\ \dot{\theta}_1\dot{\theta}_3\\ \vdots\\ \dot{\theta}_4\dot{\theta}_6\\ \dot{\theta}_5\dot{\theta}_6\end{bmatrix}+C(\theta)\begin{bmatrix}\dot{\theta}_1\\ \dot{\theta}_2\\ \dot{\theta}_3\\ \dot{\theta}_4\\ \dot{\theta}_5\\ \dot{\theta}_6\end{bmatrix}+G(\theta)=\tau \quad (1)$$

Where M is a [6×6] mass matrix, B is the Carioles- Coefficient which is [6×15], C is Centrifugal- Coefficient which is [6×6] matrix, G is Gravity term [6 ×1] matrix and τ is Torque [6 ×1] matrix. The measured parameters of PUMA 560 are driven using the Denavitt- Hartenberg [18]. Therefore, $\theta = [\theta_1, \theta_2, \ldots, \theta_6]$ is the vector of the angle of the 6-DOF robot manipulator and $\dot{\theta} = [\dot{\theta}_1, \dot{\theta}_2, \ldots, \dot{\theta}_6]^T$, $\ddot{\theta} = [\ddot{\theta}_1, \ddot{\theta}_2, \ldots, \ddot{\theta}_6]^T$ is the velocity and acceleration of the robot manipulator. In general, Eq. (2) is used for robot manipulator dynamics.

$$M(\theta)\ddot{\theta} + B(\theta,\dot{\theta}) + C(\theta,\dot{\theta}) + G(\theta) = \tau \quad (2)$$

This work attempts to develop a decentralized robust sliding mode control with neural network for two cooperative manipulators in a cooperative manner.

3 The Controller Design

3.1 The Sliding Mode Control

The sliding mode controller will be designed based on the dynamic equation of the cooperative system. There are two steps needed for the sliding mode controller [12]. In the first step, a sliding surface for the cooperative system is chosen. The second step involves the equivalent controller. In this work, to omit the chattering phenomenon, the sliding mode controller uses the neural network based prediction model. Some researchers used soft computing methods [13] [19]. In general, the dynamic second order nonlinear model of multi input multi output system can be explained in Eq. (3) [19].

$$x_2 = \dot{x}_1 \quad, \dot{x}_2 = \ddot{x}_1 = f(x) + B(x)u(t) \tag{3}$$

Where x= $[x_1 , x_2, \dots, x_n]^T$ is the vector of the generalized coordinates such as position of the n DOF mechanical system and $\dot{x} = [\dot{x}_1, \dot{x}_2, \dots, \dot{x}_m]^T$, $\ddot{x} = [\ddot{x}_1, \ddot{x}_2, \dots, \ddot{x}_m]^T$ is the velocity and acceleration of the robot manipulators. In addition, $f(x)$a bounded nonlinear vector is a function of the state vector to present the nonlinear term of the system and $B(x)$ is a bounded positive definite nonlinear function over the entire state space known as the control gain and $u(t)$ is the control input of the system. Therefore, the proposed dynamic equation for the robot manipulator to handle an object as mentioned above is mapped into Eq. (4) as follows.

$$M(\theta)\ddot{\theta} + B(\theta,\dot{\theta}) + C(\theta,\dot{\theta}) + G(\theta) = \tau \Rightarrow$$

$$\ddot{\theta} = M(\theta)^{-1}\tau - \left(M(\theta)^{-1}\left(B(\theta,\dot{\theta}) + C(\theta,\dot{\theta}) + G(\theta)\right)\right)$$

$$x_1 \equiv \theta, \qquad x_2 = \dot{x}_1 \equiv \dot{\theta}, \qquad \dot{x}_2 \equiv \ddot{\theta}$$

$$B(x) \equiv M(\theta)^{-1}, f(x) \equiv \left(M(\theta)^{-1}\left(B(\theta,\dot{\theta}) + C(\theta,\dot{\theta}) + G(\theta)\right)\right)$$

$$u(t) \equiv \tau \tag{4}$$

The control inputs are as follows:

$$\ddot{\theta} = f(\theta) + B(\theta)\,u \tag{5}$$

In the SMC design, tracking control problems can be investigated by holding the system trajectory on the sliding surface. To switch the function design in the first step, the PD type sliding surface is chosen.

$$s(t) = \dot{e} + \lambda e \tag{6}$$

Where $s(t)$ is a 6× 1 vector, λ is a constant. So, $e = \theta - \theta_d$, $\dot{e} = \dot{\theta} - \dot{\theta}_d$ are tracking error vector and the rate of tracking error vector. The second step of the controller design is to define the control law with variable parameters. The following control law is considered as:

$$u(t) = u_c + u_{eq} \tag{7}$$

Where u_c and u_{eq} terms are defined as follows. The u_{eq} term is the equivalent control term which is proposed for the approximately known section of the system in the presence of perturbations. This section makes the derivative of the sliding surface equal to zero in order to remain on the sliding surface that has been previously mentioned as the low frequency control law. The u_c term is the corrective control term which is proposed to compensate the derivatives from the sliding surface known as the high frequency control law. The u_{eq} term helps the system to set on the sliding surface.

$$s = \dot{e} + \lambda e, \dot{s} = \ddot{e} + \lambda \dot{e} \tag{8}$$

So, by using $e = \theta - \theta_d$, $\dot{e} = \dot{\theta} - \dot{\theta}_d$, there is:

$$\dot{s} = \ddot{\theta} - \ddot{\theta}_d + \lambda \dot{e} \tag{9}$$

To keep the system on the sliding surface requires that $s(t) = 0$ and $\dot{s}(t) = 0$. So, u_{eq} is as follows as expressed in Eq. 10.

$$u_{eq} = B^{-1}(\ddot{\theta}_d - \lambda \dot{e} - f) \tag{10}$$

To overcome the dynamic uncertainty in the controlling system caused by system uncertainty, external disturbances, friction and parameter variation, the approximation of f is termed $\hat{f}$[19]. It should be mentioned that an approximate of an artificial neural network based prediction model has been designed. So, the equivalent control is as follows:

$$f = \hat{f} + \Delta f < F , u_{eq} = B^{-1}(\ddot{\theta}_d - \lambda \dot{e} - \hat{f}) \tag{11}$$

Where, Δf is the difference between real and approximated f function. Thus, the F is a positive function which is bounded the values of f and $\hat{f}$. The approximation in this work uses a predictive neural network model to find accurate values for dynamic uncertainty in the control input to achieve good trajectory tracking in the presence of uncertainties. Discontinues control unit (corrector) is used to achieve good trajectory tracking performance in very fast switching.

The u_c term: here is considered a u_c function to compensate the derivations from the sliding surface and reach to sliding surface.

$$u_c = -B^{-1} K \, sgn(s), sgn(s_i) = \begin{cases} 1 & s_i > 0 \\ -1 & s_i < 0 \end{cases} \tag{12}$$

To stabilize such systems, the differential equation solutions of describing dynamic systems can be achieved by Lyapunov stability theory [19] which is concerned with the stability of solutions near the equilibrium point. It also lends qualitative results to the stability equations which may be useful in designing stabilizing controllers of nonlinear dynamical systems. By using the Lyapunov stability method, the parameter boundaries of sliding mode controllers to stabilize the proposed system can be determined. Positive definite Lyapunov function is as follows:

$$V(x) = \frac{1}{2} s(t)^T s(t) \tag{13}$$

Thus, for stability the Lyapunov function V(x) should be positive and the derivative of the Lyapunov function should be negative.

$$\dot{V} = S.\dot{S} > 0$$

$$\dot{s} = \ddot{e} + \lambda\dot{e} \Rightarrow \dot{s} = \ddot{\theta} - \ddot{\theta}_d + \lambda\dot{e}$$

$$\dot{V} = s\left(f + \left(\ddot{\theta}_d - \lambda\dot{e} - \hat{f} - K\,sat(s)\right) - \ddot{\theta}_d + \lambda\dot{e}\right) \tag{14}$$

As mentioned before$f - \hat{f} \leq F$:

$$\dot{V} = s\left(f - \hat{f} - K\,sat(s)\right) < F - K|s| \tag{15}$$

By choosing $k = F + \eta$ as follows:

$$\dot{V} \leq -\eta\,|s| \tag{16}$$

The interest of using artificial neural network arises from non linearity in the system which caused disorder in the performance of the system. One of the common and most applicable neural networks is multi layer neural network. The important key of using multi layer neural network is that this type of neural network can solve continuous non linearity function (mapping between input and output) of the system [13] [16].

3.2 The Artificial Neural Network Based Prediction Model

The predictive model based neural network is implemented by Matlab 2013 software which uses a neural network model of the nonlinear plant to predict plant performance. The controller also computes the control input to increase plant performance. The Multi layer feed forward neural network is used in this model. Because of the 6 inputs and outputs in the robot manipulator, 6 separate but same prediction models for each robot manipulator angle are considered here. The 10 hidden layers are used in all 6 neural network layers. Therefore, it uses both Tangsig and Pureline activation functions as seen in Fig 2. The controller can act in two steps which are Identification and Predictive control. The first one determines the neural network plant model to identify of the system. The second step is to use the controller to predict future system performance in terms of predictive control. In system Identification, the prediction model is used to train the plants neural network which uses the prediction error between the plant and neural network outputs as the neural networks training signals. Based on the previous inputs and outputs, future outputs of the plant will be predicted by the neural network. For training, multilayer back propagation neural network algorithm is used. A multi layer network is an artificial neural network model that includes multiple layers of fully connected nodes to map the input data into the desired outputs. To learn multilayer perception, the model applies a learning algorithm such as back propagation algorithm. Back propagation or backward error propagation is a training algorithm that consists of two steps: propagation and weight update in the propagation step activities, feed forward propagation in terms of training input patterns are used. The backward propagation of output activities propagation is then used to create the rule of gradient descent learning for weight updating of hidden and output layers (neurons). In the second step, to achieve the gradient of the weights, the outputs data set and input activation are multiplied. Thus, the weights are set in the opposite direction. The second step in designing the prediction model based on neural network is predictive control. The prediction section includes the neural network and the optimization part which corrects the data coming from the neural network for the plant.

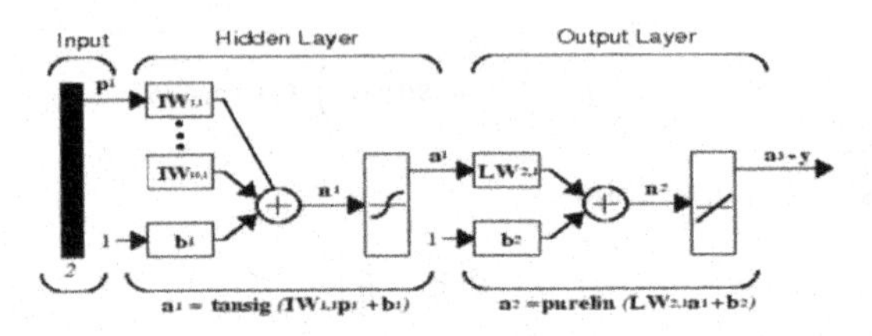

Fig. 2. The NN model of prediction model

So, the proposed controller for two robot manipulators is extended as follows as in Fig.2. The linkage between two robots as implicit communication is object's angle.

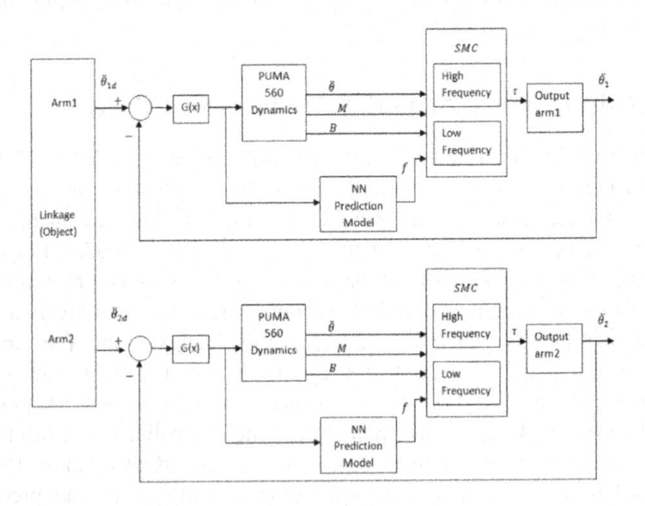

Fig. 3. The Decentralized proposed controller

4 Simulation Results

Multi layer feed forward neural network with 10 hidden layers (which is based on the Baum-Haussler rule [20]) is used for f function approximation. As there are six robot manipulator angles in PUMA 560, neural networks also use six inputs and outputs. To all six neural network prediction boxes is applied 0.05 control weighting (ρ), 0.001 search parameter (α), 2 iteration per sample time, 10 cost horizon (N2), 2 cost Horizon (Nu), 0.2 (Sec) sampling interval, 2 delayed plant input, 2 delayed plant output and 1000 training epochs.

Table 1. Results of regression, training state and Neural network with validation check=6

Angle No.	Epoch No.	Gradient	Mu
1	17	483.6533	0.1
2	54	155.4274	0.001
3	47	1.2332	0.1
4	40	0.096128	0.0001
5	21	0.016548	0.001
6	497	9.9687e-11	1e-13

Table 2. The regression results for all six angles of PUMA in neural network with training performance

Angle No.	Training	validation	Test	All regression	Training Performance
1	0.954662	0.88318	0.71784	0.83915	55.4822
2	0.99968	0.98219	0.97627	0.98763	22.5304
3	0.99982	0.96905	0.96917	0.95569	9.306
4	0.99992	0.88538	0.7119	0.88382	0.47042
5	0.99442	0.94824	0.74094	0.9355	0.32236

The performance of the proposed controller just for one robot is compared with classic sliding mode controller at fig. 4. The x axis of the each plot shows the angle of each joint of robot and the y axis of the plots shows time(second). The dashed line is classic sliding mode controller and the regular line is proposed controller.

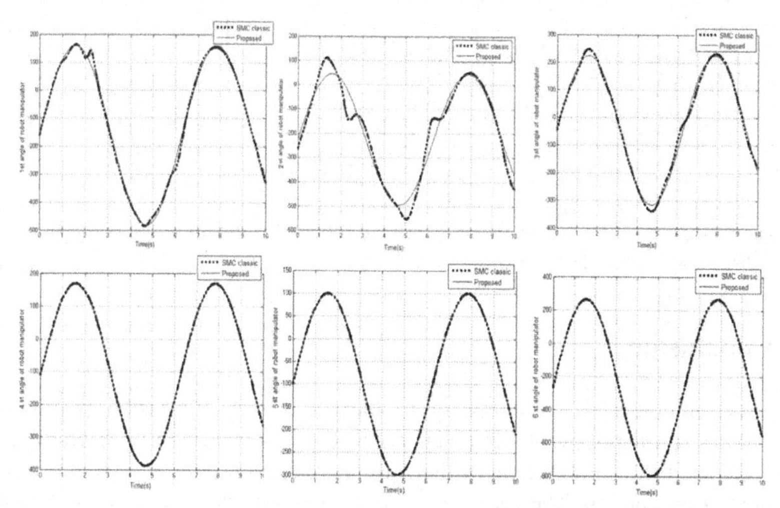

Fig. 4. The results of the decentralized controller for two cooperative robots

As can see in fig.4 the proposed controller has regular motion in the sine input for angle of robot joint's. The purpose of the controller is to achieve minimum tracking error which obtains as follows in Fig. 4. Then, the output of the proposed scheme for two robot manipulators which is mentioned at fig.3 is presented at fig. 5(a). for evaluation the proposed controller at fig.5(b) the classic sliding mode controller for two robot manipulators is extended which is revealed the proposed scheme has as regular and reliable motion in the environment to handle an object in the desired trajectory.

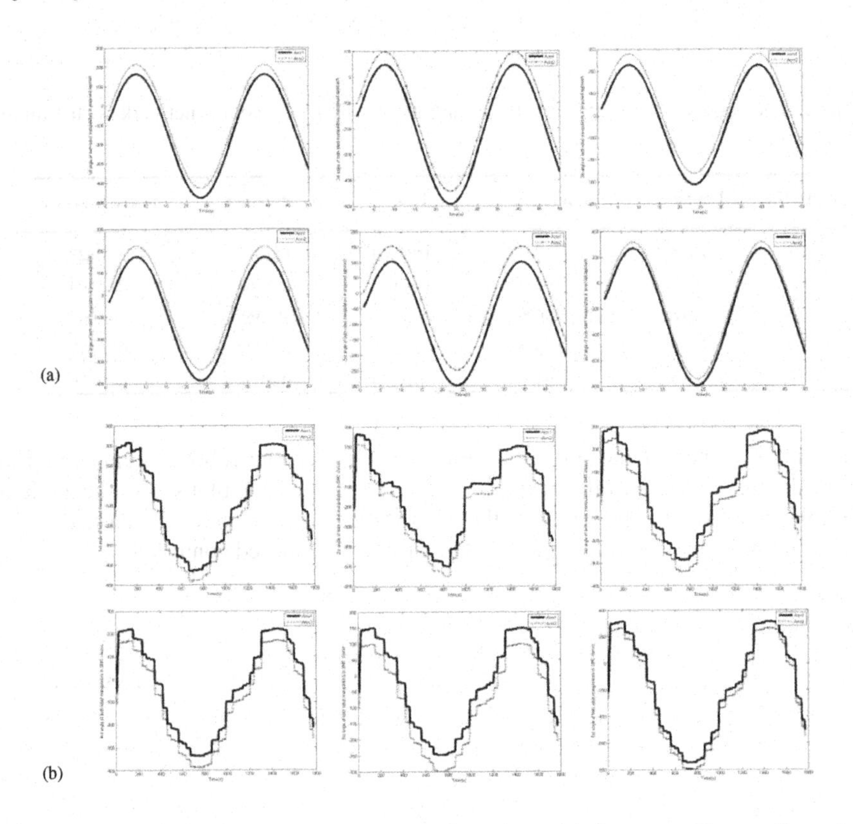

Fig. 5. The results of controller for two cooperative robots (a) the proposed controller and (b) classic sliding mode controller

5 Conclusion

A decentralized control law is proposed to allow two PUMA560 robot manipulators to cooperate when handling objects. This controller is robust with no chattering controller because of the combination of the sliding mode controller with neural network based prediction model. The angle of the lightweight beam is used as implicit communication in a cooperative manner to solve the synchronization problem. The simulation results guarantee the validity of the proposed controller.

References

1. Callegari, M., Suardi, A.: Hybrid kinematic machines for cooperative assembly tasks. In: Proceedings of the International Workshop: Multiagent Robotic Systems: Trends and Industrial Applications, Padova, Italy, pp. 223–228 (2003)
2. Tarn, T.J., Bejczy, A.K., Yun, X.: Design Of Dynamic Control Of two Cooperative Robot Arms: Closed Chain Formulation. In: Proceedings of the IEEE International Conference on Robotics and Automation, ICRA 2004, pp. 7–13 (1987)
3. Zhao, Y.S., Ren, J.Y., Huang, Z.: Dynamic loads coordination for multiple cooperating robot manipulators. Mech. Mach. Theory 35, 985–995 (2000)
4. Chen, G.: Study on Dynamic Hybrid Position/Force Control of Two Coordinated Manipulators. Adv. Mater. Res. 468, 1224–1230 (2012)
5. Chiaverini, S., Sciavicco, L.: The Parallel Approach to ForcePosition Control of Robotic Manipulators. IEEE Trans. Robot. Autom. 9, 361–373 (1993)
6. Ishida, T.: Force control in coordination of two arms. In: Proc. 5th Int. Conf. on Artificial Intelligence, pp. 717–722 (1977)
7. Tarn, T.J., Bejczy, A.K., Yun, X.: Co-ordinated control of two robot arms. In: Proc. IEEE Int. Conf. on Robotics and Automation, pp. 1193–1202 (1986)
8. Jawawi, D., Deris, S., Mamat, R.: Software Reuse for Mobile Robot Applications Through Analysis Patterns. Int. Arab J. Inf. Technol. 4, 220–228 (2007)
9. Jawawi, D.N.A., Deris, S., Mamat, R.: Early-Life Cycle Reuse Approach for Component-Based Software of Autonomous Mobile Robot System. In: 2008 Ninth ACIS International Conference on Software Engineering, Artificial Intelligence, Networking, and Parallel/Distributed Computing, pp. 263–268 (2008)
10. Hogan, N.: Impedance control: An Approach to Manipulation, Parts I-Theory, II Implementation, III-Applications. ASME J. Dyn. Syst. Meas. Control 107, 1–24 (1985)
11. Esakki, B.: Modeling and Robust Control of Two Collaborative Robot Manipulators (2011)
12. Yagiz, N., Hacioglu, Y., Arslan, Y.Z.: Load transportation by dual arm robot using sliding mode control. J. Mech. Sci. Technol. 24, 1177–1184 (2010)
13. Panwar, V., Kumar, N., Sukavanam, N., Borm, J.-H.: Adaptive neural controller for cooperative multiple robot manipulator system manipulating a single rigid object. Applied Soft Computing 12, 216–227 (2012)
14. Hacioglu, Y., Arslan, Y.Z., Yagiz, N.: MIMO fuzzy sliding mode controlled dual arm robot in load transportation. J. Franklin. Inst. 348, 1886–1902 (2011)
15. Moosavian, S.A.A., Papadopoulos, E.: Cooperative Object Manipulation with Contact Impact Using Multiple Impedance Control. 8, 314–327 (2010)
16. Tsai, C.-C., Cheng, M.-B., Lin, S.-C.: Robust Tracking Control for A Wheeled Mobile Manipulator With Dual Arms Using Hybrid Sliding-Mode Neural Network. Asian Journal of Control 9(4), 377–389 (2007)
17. Lewis, F.L., Abdallah, C.T., Dawson, D.M.: Control of robot manipulators. Macmillan Pub. Co. (1993)
18. Armstrong, B., Khatib, O., Burdick, J.: The Explicit Dynamic Model and Inertial Parameters of the PUMA 566 Am, pp. 510–518. IEEE (1986)
19. Zeinali, M., Notash, L.: Adaptive sliding mode control with uncertainty estimator for robot manipulators. Mech. Mach. Theory 45, 80–90 (2010)
20. Baum, E.B., Haussler, D.: What size net gives valid generalization? Neural Computation 1, 151–160 (1989)

Comparison of Reproduction Schemes in Spatial Evolutionary Game Theoretic Model of Bystander Effect

Andrzej Świerniak and Michał Krześlak

Silesian University of Technology, Department of Automatic Control, Gliwice, Poland
andrzej.swierniak@polsl.pl, krzeslak.michal@gmail.com

Abstract. We compare results of different reproduction schemes used in modelling of radiation induced bystander effect in normal cells. The model is based on the theory of spatial evolutionary games on a lattice and pay-offs are defined by changes in fitness measure resulting from intersections between cells representing different phenotypes. We discuss also qualitative differences between results of simulations of spatial evolutionary games and steady state solutions of replicator dynamics equations for relevant mean field games.

Keywords: spatial evolutionary games · bystander effect · cellular automata · reproduction · systems biology.

1 Introduction

Recent studies have shown that cells exposed to ionizing radiation can trigger reactions affecting non-targeted neighbouring cells. Phenomenon known as radiation based bystander effect has been widely reviewed in literature see e.g.[1]. Irradiated cells release signals which lead to damage in nearby, non-irradiated cells and reduction in survival of adjacent cells. Moreover the signalling is mutual, so the irradiated cells can also receive signals from non-irradiated neighbours. The induced bystander effect in non-irradiated cells can be exposed in couple ways: reduction of survival, delay of cell's death, oxidative damage in DNA and the genomic instability, micronuclei induction, lipid peroxidation and apoptosis. One of the well-known environmental conditions for bystander effect are biological systems exposed to low doses of alpha, gamma and X radiation. Although such processes are documented, the mechanisms responsible for bystander phenomena are still unknown because their complexity and dependence on many circumstances. Intercellular signalling, so important for bystander effect, is correlated with reactive oxygen species, nitric oxide, cytokines such as interleukin 8 or TGF-β. Also significant role is played by gap junction communication and presence of soluble mediators [2].

The bystander effect can be induced also by chemotherapy, ultraviolet radiation and photo-chemotherapy. The UV radiation may lead to skin cancer, basal and squamous-cell carcinoma and to the emergence of malignant melanoma. Oxidative stress, which can also be a result of UVB radiation, is an important mediator of bystander effect induced by ionizing radiation [3].

The radiation induced bystander may have positive and negative effects, since the factors issued by irradiated cells may lead to mutation and second neoplasia.

N.T. Nguyen et al. (Eds.): ACIIDS 2014, Part I, LNAI 8397, pp. 484–492, 2014.

If healthy cells are damaged then the effect is harmful and could increase the adverse effect of radiotherapy in the form of actinic complications.

There are a number of different reports showing that 1% of radiated cells with the increase in dose results in death of 30% of the cells, and then above a certain dose threshold effect disappeared [1], [4]. On the other hand, other sources indicate that this effect is visible at both low and high doses [2].

Genetic instability, the delayed effect of the changes and the death of cells in distant generations previously irradiated is also observable.. Furthermore, radiation-induced cell clones emit cytotoxic agents and resistance to radiation can be acquired at low doses, but above a prespecified threshold.

Apart from described phenomena some others could be listed too:

- irradiated cells harm surrounding cells.
- it is possible to increase a count of non- irradiated neighbours.
- a growth of cells that have received a high dose of radiation through signalling from cells irradiated by low-dose may appear.
- paracrine fashion of intercellular interaction.

From the modelling point of view some of the conditions and phenomena detected within bystander effect could be covered by Tomlinson's models of growth factors [5] and production of cytotoxic substances [6]. The former is a very basic model showing the bystander effect in a way of paracrine signalling between cells, incurring losses due to factors productions and some passive group of individuals that receive benefits from other actions. The latter shows the phenomenon of cytotoxic substances production, which perfectly illustrates production of free radicals and immunity induction (also observed in bystander effect studies). In a similar way of basic and overall modelling fashion we present game theoretic model strictly assigned to radiation induced bystander effect. Adding to this different ways of solving game theoretical models, it allows for observations of complicated and various responses and results of intercellular signaling and communication.

The primary goal of this paper is to extend our model of radiation induced bystander effect based on the theory of spatial evolutionary games presented in [7] and check how different reproduction schemes used to generate spatial distribution of phenotypes affect results of simulations.

2 Evolutionary versus Spatial Evolutionary Games

Evolutionary game theory (EGT) initiated by John Maynard Smith and George Price [8] linked the mathematical tools of game theory with Darwinian fitting and species evolution. The individuals, without any rationality, compete or cooperate with each other to obtain better position in the population (access to food supplies, life space or females). The players and their strategies should be treated as phenotypes of mentioned individuals. Payoffs measure a change in the degree of fitness resulting from interactions of the individuals. To define the results of such a game, the concept of Evolutionary Stable Strategy (ESS) was introduced in [9]. A phenotype, which is ESS, is resistant to the emergency and impact of the other including new phenotypes (as a result of environmental migration or mutation) and cannot be repressed by them.

On the other hand, the opposite situation is possible, so that whenever ESS arises within a population then it can coexist stably with other phenotypes or dominate the population. In addition ESS is always in Nash equilibrium, but reverse implication is generally not true.

Application of evolutionary game theory to mathematical modelling of processes during carcinogenesis is based on the following assertions:

- in an organism, cells compete for space and nutrients, while cells with different phenotypes are players in the game.
- cooperation and even evolutionary altruism can occur.
- mutation (appearing in tumour cells) occurs in cell division due to various reasons.
- an advantage of tumour cells over healthy ones is a signature of cancer.
- environmental factors can affect different cells to varying degrees.

To our knowledge the first tumour's phenomena modelled by evolutionary game theory are avoidance of apoptosis and production of growth factors presented by Tomlinson and Bodmer [5]. The authors proposed three kinds of phenotypes: two of them produce a factor to prevent apoptosis either in paracrine or autocrine fashion and the third one is neutral gaining benefits from paracrine factors. This paper has triggered a series of other studies, where evolutionary game theory has been applied to present different tumour phenomena (see [11, 12] for survey). On the other hand, game theory models show only single phenomena occurring in a very complicated process of cancer evolution (results represent quantitative, but not qualitative description). To track the evolution of different phenotypes in the population it is feasible to simulate replicator dynamics equations [13]. They show how frequencies of different strategies change in time, thereby effecting the composition of studied population. Results from replicator dynamics and models used for the simulation are commonly called mean-field models. Another way to track the phenotypes' evolution is created by methodology of spatial evolutionary games which enables study of players' allocation. A crucial difference between non-spatial and spatial models is lack of perfect mixing. Due to this feature the various local structures of cells could have an impact on the course of the game and the results. Comparison of the results between the mean-field and the spatial model was presented by Bach et al [14].

We followed the line of reasoning presented in[14] adding new kinds of cells reproduction and applying spatial games to various, carcinogenesis models (see [12]) including a model of radiation induced bystander [7]. In this paper we extend the model from [7] and discuss effects of different reproduction schemes in comparison with results of mean field model.

Spatial games follow one global algorithm, which is used every single interaction among players placed on the lattice forming torus:

- payoff updating – determination of local fitness for players in their neighbourhoods.
- cell mortality – choosing 10% of the players from the lattice.
- reproduction by competition – defining a new player on the empty place.

In this paper we are using semi-synchronous updating. This technique allows for biologically realistic situation. The authors [14] have suggested two kinds of reproduction, which in their general understanding could be applied for both spatial games, presented by us:

- deterministic reproduction – in competition of empty place the winner is the strongest player with highest local adaptation.
- probabilistic reproduction – each player's local adaptation is divided by the sum of the local scores among its neighbourhood. According to the authors this reproduction shall give a chance to strategies with lower fitness, but in better location and locally superior in numbers to predominate in population.

Additionally we have introduced [7] two other ways of reproduction: quantitative (assumes cooperation between players with the same strategies) and switching (depending on the size of diversity between scores, quantitative or deterministic reproduction is chosen). Every competition results giving tie are settled randomly. Simulations can be extended by studies of effects of the size and type of neighborhood (e.g. von Neumann or Moore neighborhood).

In our paper for spatial simulations two initial lattices (30 x 30 cells) were generated randomly, providing substitute of perfect mixing. For the purpose of this paper we have not studied effects of the choice of initial lattices and their different sizes.

3 Model of Radiation Induced Bystander Effect and Simulation Results

We consider three different strategies/phenotypes of cells:

- escape to apoptosis - frequency of appearance: X
- production of growth and mutation factors - frequency of appearance: Y
- neutrality - frequency of appearance: Z

Pay offs are defined by the following pay-off matrix:

Strategies	X	Y	Z
X	1-k	1-i+j-p	1-p
Y	1-k+j	1-i+j	1+j
Z	1-k	1-i+j	1

Parameters of the model:

k – cost of apoptosis/profit from bystander effect
j – profit of cell contact with growth factors
i – cost of producing the growth factors
p – cost/advantage from resistance to bystander effect

Results from the replicator dynamics equations (and the same, from the mean-field, non-spatial analysis) show that population can reach different states [15]. The resulting trimorphic, dimorphic or monomorphic populations may be both independent from initial frequencies (in case when the equilibrium point is an attractor), or dependent on them(equilibrium point is a repiler).

Within this paper main aim focuses on comparison between different replicator schemes, however some references to replicator dynamics results should be also considered.

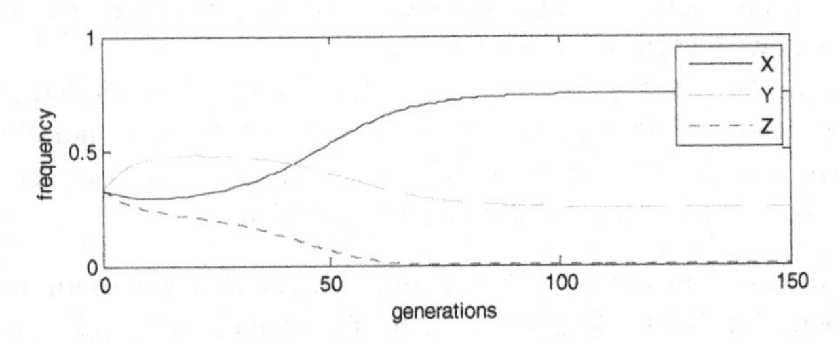

Fig. 1. Replicator dynamics results for parameters i=0.4, j=0.8, k=0.1, p=0.4

First set of parameters leads to dimorphic population with X and Y cells (Z cells have been repressed) for non-spatial game (Fig. 1). This is a case when the equilibrium point (X=0.25, Y=0.5, Z=0.25) is a repiler, which means that final result depends on initial frequencies. In case of spatial simulation, starting frequencies are equal for all three phenotypes. Changing initial frequencies (for instance X, Y to 0.3 and Z to 0.4) may result in dimorphic population consisting of Z and Y cells (Fig. 2). Such change does not have any simple impact on results of spatial games. First of all, the change is not so significant for spatial consideration. Such alteration could affect the results while changing predefined clusters of the phenotypes within the initial lattice (but not in the case of lattices generated randomly).

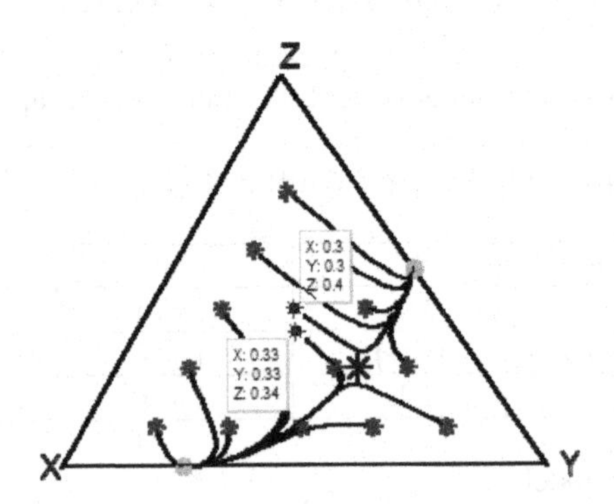

Fig. 2. Different initial frequencies leading to different final distribution of phenotypes. Parameters i=0.4, j=0.8, k=0.1, p=0.4.

Coming back to the spatial results, the outcomes are different than in the mean-field game for the same initial frequencies (Fig. 3). Except of probabilistic reproduction the dominant are X and Y cells, however qualitative reproduction in population games allows to set up some clusters of pure phenotypes and for the switching one we have also small cluster of phenotype Z..

On the other hand probabilistic reproduction leads to monomorphism of phenotype Z. The trimorphic population (see Fig. 4) in non-spatial game is feasible for the second set of parameters.

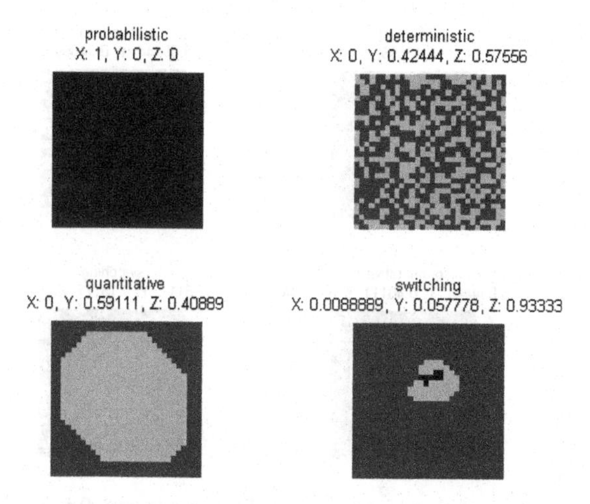

Fig. 3. Results of simulation of spatial game for parameters i=0.4, j=0.8, k=0.1, p=0.4

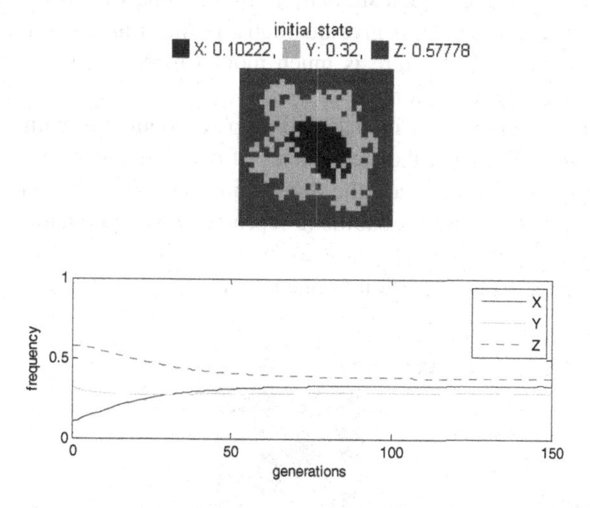

Fig. 4. Initial distribution and replicator dynamics for parameters i=0.5, j=0.7, k=-0.1, p=-0.3

Probabilistic reproduction gives similar results., however the leadership within the population is swamp (Fig. 5). Interesting phenomenon is observed in the case of deterministic reproduction. Z-cells are repressed from the population in case of the spatial game. First thought could be that this difference is a result of semi-synchronous player choosing, which in fact introduces some randomization into the algorithm. However vast amount of simulations have shown that the results are the same even for different initial lattices. It indicates that despite of mentioned randomness and even various configurations for some set of pay-off parameters the behavior

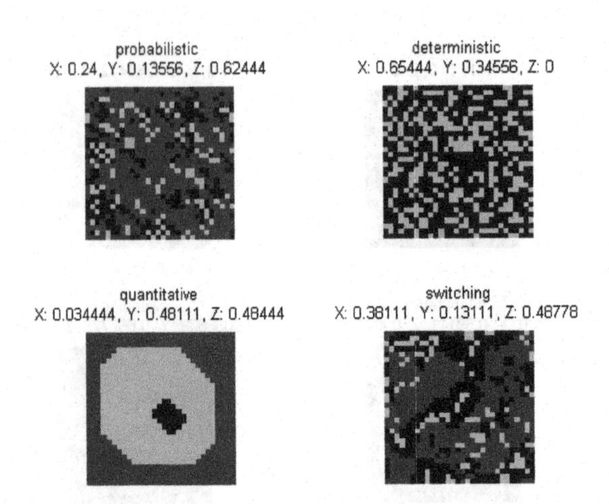

Fig. 5. Results of simulation of spatial game for parameters i=0.5, j=0.7, k=-0.1, p=-0.3

and results are the same every each simulation. In the same way as for previous set of parameters the other reproductions give qualitatively the similar results as probabilistic one (again the players' structure is much more variable). Once more qualitative reproduction leads to clusterization.

Last set of parameters results in stable dimorphic population with X and Z in all cases. In the mean-field model the frequencies of occurrences for X and Z are equal (Fig.6), in spatial games Z-cells are the dominating ones (Fig. 7), but the advantage over X is not so great. Within probabilistic reproduction in population game the difference between these two phenotypes is even smaller. Once more qualitative reproduction scheme leads to clusterization of phenotypes.

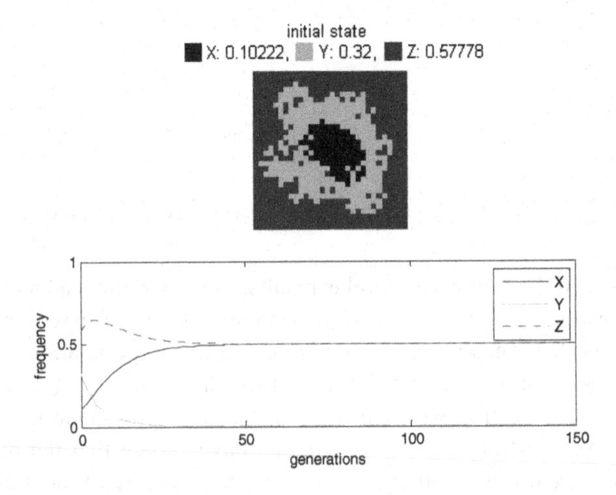

Fig. 6. Initial distribution and replicator dynamics for parameters i=0.6, j=0.5, k=-0.2, p=-0.4

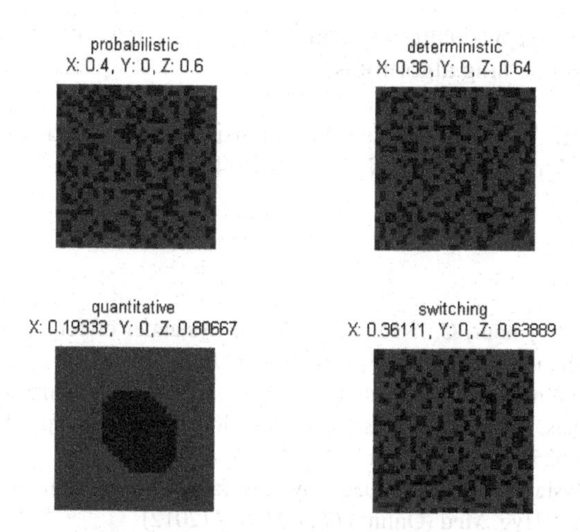

Fig. 7. Results of simulation of spatial game for parameters i=0.6, j=0.5, k=-0.2, p=-0.4

4 Remarks

We have proposed an extension of the evolutionary spatial games introduced by Bach et all [14] and studied by us in the context of bystander effect modeling in [7]. The role and use of different reproduction schemes has been studied. Besides two main kinds of reproduction some new methods have been proposed. So called qualitative reproduction allows for formation of stable clusters. In some cases spatial evolutionary games give qualitatively the same results as obtained in the mean field approach. However in many cases the results are both qualitatively and quantitatively different. It seems that the stronger evolutionary adjustment the more stable results of the simulations. Within this paper we have neglected some chosen sets of parameters like initial lattice, the size of the neighbourhood or the way of choosing players. Despite numerous parameters and possible configurations of spatial games other possibilities for different results are provided by various values within payoff table. We showed three sets of model parameters which lead to in some sense similar results, in terms of phenotypes quantities, for spatial games (still a bit different than in mean-field games). The reason of similarities could be the evolutionary game factor, which in fact should have the biggest and the most basic impact on the results. Such analysis of quantitative results is appropriate for making a comparison between different spatial games and their mean-field counterparts. New possibilities of spatial evolutionary games have been presented by using the game theoretic model of radiation induced bystander effect. The effect mechanisms could be better understood by spatial factor, the more that some results may be matched with biological phenomena. The spatial evolutionary games seem to be the next stage (introducing players allocation) in carcinogenesis phenomena modeling, especially if we remember that various scenarios could be studied according to different configurations. Recently [16] evolutionary game theory has been applied for image processing and pattern recognition. We hope that spatial evolutionary games theory could be used for the same purposes.

The proposed new reproduction schemes may give new possibilities of different classes of image processing algorithms.

Acknowledgment. This study has been partially supported by NCN (National Science Centre,Poland) Grant no. N N519 6478 40 in year 2013.

References

1. Mothersill, C., Seymour, C.: Radiation-induced bystander effects: past history and future directions. Radiation Research 155, 759–767 (2001)
2. Rzeszowska-Wolny, J., Przybyszewski, W.M., Widel, M.: Ionizing radiation-induced bystander effects, potential targets for modulation of radiotherapy. European Journal of Pharmacology 625, 98–107 (2009)
3. Wideł, M.: Bystander effect induced by UV radiation; why should we be interested? Advances Exper. Hyg. Med (Online) 66, 828–837 (2012)
4. Iyer, R., Lehnert, B.E.: Low dose, low-LET ionizing radiation-induced radioadaptation and associated early responses in unirradiated cells. Mutation Research 503, 1–9 (2002)
5. Tomlinson, I.P.M., Bodmer, W.F.: Modelling the consequences of interactions between tumour cells. British J. Cancer 75, 157–160 (1997)
6. Tomlinson, I.P.M.: Game-theory models of interactions between tumour cells. European Journal of Cancer 33, 1495–1500 (1997)
7. Krześlak, M., Świerniak, A.: Spatial evolutionary games and radiation induced bystander effect. Archives of Control Sciences 21, 135–150 (2011)
8. Maynard Smith, J., Price, G.R.: The logic of animal conflict. Nature 246, 15–18 (1973)
9. Maynard Smith, J.: Evolution and the theory of games. Cambridge University Press, Cambridge (1982)
10. Mesterton-Gibbons, M.: An Introduction to Game Theoretic Modelling. AMS, 2nd edn., Ann Arbor (2001)
11. Basanta, D., Deutsch, A.: A game theoretical perspective on the somatic evolution of cancer. In: Bellomo, N., Chaplain, M., Angelis, E. (eds.) Selected Topics in Cancer Modeling: Genesis, Evolution, Immune Competition, and Therapy, pp. 1–16. Birkhauser, Boston (2008)
12. Świerniak, A., Krześlak, M.: Application of evolutionary games to modeling carcinogenesis. Mathematical Biosciences and Engineering 10, 873–911 (2013)
13. Hofbauer, J., Shuster, P., Sigmund, K.: Replicator dynamics. J. Theor. Biol. 100, 553–588 (1979)
14. Bach, L.A., Sumpter, D.J.T., Alsner, J., Loeschcke, V.: Spatial evolutionary games of interactions among generic cancer cells. Journal of Theoretical Medicine 5, 47–58 (2003)
15. Świerniak, A., Krześlak, M.: Game theoretic approach to mathematical modeling of radiation induced bystander effect. In: Proc. of the 16 Nat. Conf. on Appl. Mathematics in Biology and Medicine, Krynica, Poland, pp. 99–104 (2010)
16. Bullo, S.R.: A game-theoretic framework for similarity-based data clustering. PhD. Thesis. Dipartimento di Informatica, Universita Ca' Foscari di Venezia. Italia (2009)

Ant Colony Optimization Algorithm for Solving the Provider - Modified Traveling Salesman Problem

Krzysztof Baranowski, Leszek Koszałka,
Iwona Poźniak-Koszałka, and Andrzej Kasprzak

Department of Systems and Computer Networks,
Wroclaw University of Technology,
Wroclaw, Poland
krzysztofbaranowski2@gmail.com, {leszek.koszalka,
iwona.pozniak-koszalka,andrzej.kasprzak}@pwr.wroc.pl

Abstract. The paper concerns the introduced and defined problem which was called the Provider. This problem coming from practice and can be treated as a modified version of Travelling Salesman Problem. For solving the problem an algorithm (called ACO) based on ant colony optimization ideas has been created. The properties of the algorithm were tested using the designed and implemented experimentation system. The effectiveness of the algorithm was evaluated and compared to reference results given by another implemented Random Optimization algorithm (called RO) on the basis of simulation experiments. The reported investigations have shown that the ACO algorithm seems to be very effective for solving the considered problem. Moreover, the ACO algorithm can be recommended for solving other transportation problems.

Keywords: modified TSP, ant colony optimization, random optimization, tuning algorithm, experimentation system, simulation.

1 Introduction

The considered problem is a modification of well-known Traveling Salesman Problem [1]. The basic TSP required visiting all given places while the chosen route should be as shortest as it is possible, the main constraint is to visit each place only once (finding Hamiltonian cycles in a graph [1] and [2]). For solving the classic TSP, meta-heuristic and evolutionary approaches are applied [3-7]. In literature, the case when time of delivering goods to the customer has impact on the value of particular product and as consequence on the obtained profit by the transportation company - is considered very rarely. Solving the modified TSP, in which the additional factors from the real world are present, might cause savings for entities which find opportunity of using the proposed solution in their own companies. In the paper, the introduced problem assumes more complex constraints which require significantly modifying not only the cost function, but also an approach for solving.

The objective of this paper is to introduce the Provider problem, to create the algorithms for solving the problem, and to show algorithms' properties. Two algorithms

N.T. Nguyen et al. (Eds.): ACIIDS 2014, Part I, LNAI 8397, pp. 493–502, 2014.

have been implemented: the ACO algorithm based on ant colony optimization idea [2-8], and the RO algorithm finding a random solution. Moreover, the way of tuning ACO algorithm is proposed. The results of the obtained solutions to various instances of the problem (found with the tuned ACO and RO) are shown, compared and discussed. The investigations are based on two-stage simulation experiments made using an experimentation system with properly designed computer programs - implemented in C++ programing language.

The rest of the paper is organized as follows: Section 2 contains the problem formulation. In Section 3, the two proposed algorithms are described. Section 4 presents the experimentation system. In Section 5, the results of some investigations concerning the quality of solutions are presented and discussed. In this section the tuning process of ACO is also shown. Finally, in Section 6, the conclusions appear.

2 Problem Statement

2.1 Practical Problem

The formulation of the problem is based on the practical situation. Let us consider the case where the ordered food products are delivered from a restaurant to the customers. The customers are located in different but known places. In order to improve the competiveness, the manager of the restaurant decides to relate the price of the delivered food (paid by the customer) to the satisfaction of the customer. It is assumed that this satisfaction is related to the temperature of the product (e.g., pizza, meal dish) in the time-moment of receiving by the customer. The decrease of the temperature is related to some factors, e.g., the distance between restaurant and customer. At the same time several orders should be completed.

The question is – how the manager should plan the routes for courier (who delivers the products to customers) – to get the maximum profit ? The considered problem may be presented as follows:

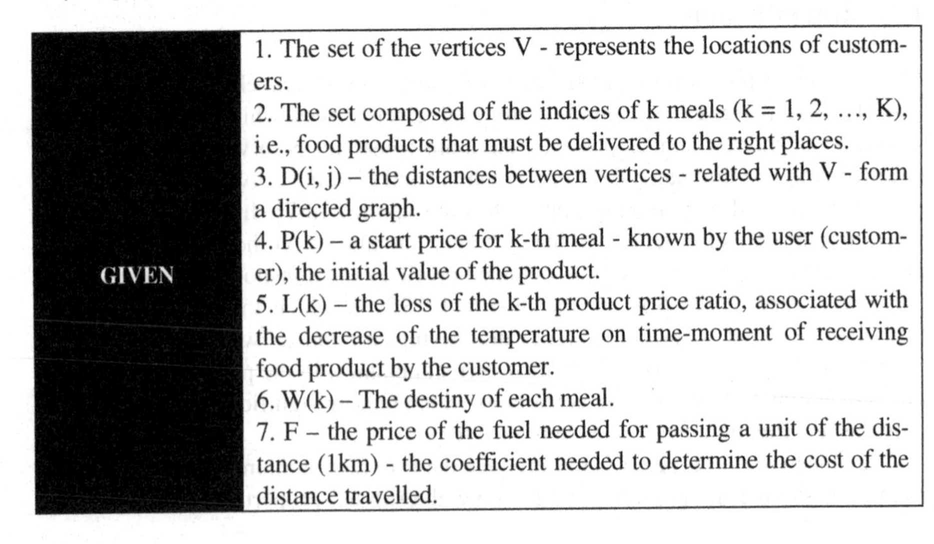

GIVEN	1. The set of the vertices V - represents the locations of customers. 2. The set composed of the indices of k meals (k = 1, 2, …, K), i.e., food products that must be delivered to the right places. 3. D(i, j) – the distances between vertices - related with V - form a directed graph. 4. P(k) – a start price for k-th meal - known by the user (customer), the initial value of the product. 5. L(k) – the loss of the k-th product price ratio, associated with the decrease of the temperature on time-moment of receiving food product by the customer. 6. W(k) – The destiny of each meal. 7. F – the price of the fuel needed for passing a unit of the distance (1km) - the coefficient needed to determine the cost of the distance travelled.

TO FIND	Sn - a permutation of the vertices which represents a sequence of the all visited places (the route for a courier).
SUCH THAT	To maximize the profit P which is obtained by sold meals (decreased by the cost of the fuel used during travel and the loss of the customers' satisfaction caused by lower temperature of the meal).
SUBJECT TO CONSTRAINTS	1. Meals must be provided even when it brings no profit. 2. Time of travel (per kilometer) is always the same. 3. Dissatisfaction of customer is caused by the lower temperature of food product than expected.

The three introduced constraints simplify model to avoid too much complexity which effect can be averaged. The other ordinary factors like, e.g., traffic (with an impact on the time of travel) are not considered in this paper. The necessity of delivering all goods even if it is unprofitable refers directly to the constraints given by the original TSP.

2.2 Objective

For the given scenario and subject to the constraints (described in Section 2.1) we need to find the permutation of vertices Sn such that the function, called profit P defined by the equation (1), reaches the maximal value.

$$P = P(Sn) = \{[\sum (Pi (k) - Li (k))] - \sum Fij \} \quad (1)$$

where

Fij is the cost of the fuel to travel from point i to j;
Pi (k) is the start price of the meal which has to be delivered to the k-th customer located in k-th place;
Li (k) is the financial penalty due to cool down of a given food product.

3 Algorithms

In order to solve the stated problem two algorithms have been implemented. The simple random optimization [9] is a kind of reference to check the quality of the proposed algorithm which is based on ant colony optimization (ACO) approach. The ACO was chosen from the methods based on artificial intelligence and inspired by the nature - as the most flexible tool (see [6] and [8]). Such approach can also allow for the development of the statement for Provider problem, e.g., by taking into account (in the future) more factors and constructing the expanded expression (more complicated than that given by (1)) for calculating profit. Finally, it could give a look into various factors anticipated improving cost function and for scenarios closer to the real complicated world.

3.1 Random Optimization

The proposed simple algorithm is working in such a way: (i) to designate randomly first available permutation; (ii) to calculate the profit along with (1); (iii) to select randomly two places (vertices) which are swapped (Fig. 1); (iv) to calculate the profit for new permutation; (v) to compare the old profit and the new one;(vi) to memorize the solution (permutation) with better profit.; (vii) to repeat (iii) to (vi) as the stop condition is reached. The main advantage of this algorithm is speed allowing checking a huge number of the possible permutations.

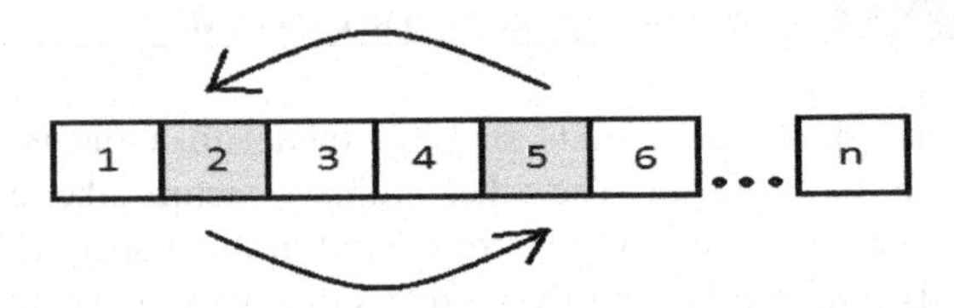

Fig. 1. Principle of the RO algorithm

3.2 Ant Colony Optimization

Ant colony optimization based on two main phenomena. The first is the stigmergy [6], which describes a way, how single organisms communicate each other using environment in which they live in – the ants communicate by pheromone left by them. The second is a positive feedback telling about strengthening importance of particular path which relates to the more frequently used paths. In the real world, ants initially wander randomly, and upon finding food they return to their colony while laying down pheromone trails. If other ants find such a path, they follow the trail causing that the pheromone trail evaporates. However, for a short path it can be observed less evaporation. Thus, this process can aid the search for the shortest path. From mathematical side the main part of the algorithm is expressed in (2). The formula (2) defines the probability of choosing path i-th being in position j-th in the neighborhood which was not visited yet [6].

$$p_{ij}^{k}(t) = \frac{[f_{ij}(t)]^{a} * [\frac{1}{d_{ij}(t)}]^{b}}{\sum_{l \in N_i^k}([f_{il}(t)]^{a} * [\frac{1}{d_{il}(t)}]^{b})} \tag{2}$$

The equation (2) contains the value of pheromone left by ants, in previous iterations, and heuristic knowledge like inversion of distance between the two considered places. It is worth to mention of two parameters ('a' and 'b') which can have a remarkable impact on the efficiency of the algorithm based on this formula (see Section 5.1). The detailed description of ACO approach can be found in [6] and [8].

The algorithms for finding the shortest path, based on ACO approach, are also described by the authors of this paper in [10].

4 Experimentation System

4.1 Input – Output System

The block-scheme of the experimentation system is shown in Fig.1. The system can be treated as input-output system. There are distinct three categories of inputs, including problem parameters : the set V, and the set W(k); the cost parameters : D(i,j), P(k), L(k), F – all described in Section 2, and algorithms RO and ACO for finding the solution. The ACO algorithm is represented by the four tuned parameters: a, b, An, Er. The outputs are: Sn – the permutation composed of the indices corresponded to the vertices, and P - the value of the profit computed with the expression (1).

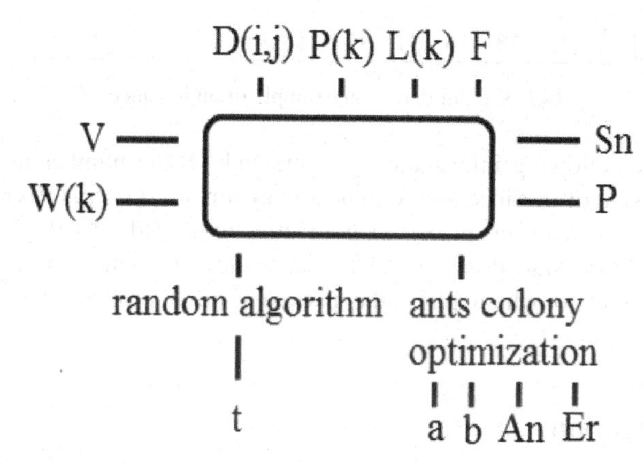

Fig. 2. Experimentation system as input-output system

The two algorithms were implemented in C++ programing language as window applications. All main parameters can be modified in the application code. The RO algorithm has only one parameter which can be set, namely time of working. The ACO algorithm has four main parameters which can be modified also in source code - the chosen values of these parameters can be found in the course of subsequent research (see Section 5.1). It is possible to measure the operating time of ACO performance. In order to conduct experiments for testing properties of the algorithms solving Provider problem, it was introduced a special format of input data which allows generating data at random (see an example in Section 4.2).

4.2 Instances of the Problem

In order to conduct experiments for testing properties of the algorithms solving Provider problem, it was introduced format of input data for generating instances of data at random. For the research presented in this paper fifteen instances were generated, starting from the set V composed of ten places up to one hundred and fifty places. An example of an instance composed of input data (problem parameters and cost parameters) is shown in Fig. 3.

```
10
 0  3  7  8  10  1  6  1  5  7
 9  0  8  6  1  8  7  3  3  5
 2  6  0  4  9  2  8  8  6  10
 4  10  8  0  5  9  6  1  4  3
 4  8  8  2  0  3  2  9  1  10
 3  6  8  3  6  0  10  4  2  5
 3  10  10  3  6  3  0  4  6  3
 7  8  6  6  6  7  3  0  3  6
 3  5  4  10  1  8  5  10  0  3
 2  10  7  3  6  8  4  3  3  0

7  13  18  16  31  33  11  7  30  27
```

Fig. 3. Input data – an example of an instance

In Fig. 3, the following information is contained: (i) the number in the first row determines the size of instance (set V); (ii) non-symmetric matrix (10x10) gives data about the distances between places in both directions, and (iii) the bottom vector shows the data of the start prices of each meal while the position in the vector gives also information in to which place the particular food product (k=10) has to be delivered.

5 Investigation

The investigation was made using the two stage approach proposed by the authors of this paper in [11] and [12]. At the first stage, the simulation experiments were conducted in order to find the best parameters of the ACO algorithm. At the second stage, the RO algorithm and the ACO algorithm with the best parameters were tested with the same instances of input data in order to compare the algorithms' quality.

5.1 The First Stage. Tuning the ACO Algorithm

It was shown, e.g., in [13] and [14], that tuning the parameters of algorithms based on ACO idea is very effective. Therefore, at the first stage of the investigations the proposed ACO algorithm has been tuned. The objective was to choose the values of:

An – the number of all ants (treated as agents), Er – the evaporation ratio; and two important parameters denoted as 'a' and 'b' which decide about the probability of choosing a given admissible path along with the equation (2). From theoretical point of view, the experiments should be made for all instances available and for each instance the best parameters can be found. However, we took into account that following reasoning can be justified: "if data has familiar behavior, then we can average the results obtained in all instances and recommend as the best the averaged values of parameters without loss of the quality". Thus, such a methodology was undertaken: for each instance were checked different values of the particular factors - starting from 'a' parameter, through 'b', next An - parameter, and finally the parameter Er. All the results were averaged and the best values were chosen to be used at the second stage of research. In Table 1, Table 2, Table 3, and Table 4 the sample results from two instances for 'a' parameter, for 'b' parameter, for An parameter, and for Er parameter are shown, respectively.

Table 1. Search of a parameter

	b = 1		**An = 20**		**Er = 0.1**	
a	0.5	1	2	3	4	5
Size (20)	376.6	377.094	377.094	376.158	376.6	376.6
Size (100)	1964.66	1965.8	1963.31	1964.66	1966.1	1963.49

Table 2. Search of b parameter

	a = 4		**An = 20**		**Er = 0.1**	
b	2	3	4	5	6	7
Size (20)	376.681	376.648	377.094	377.094	377.094	377.094
Size (100)	1963.39	1964.75	1964.44	1962.8	1967.38	1963.52

Table 3. Search of An parameter

	a = 4		**b = 6**		**Er = 0.1**
An	10	15	20	25	30
Size (20)	377.094	377.094	377.094	377.094	377.094
Size (100)	1962.28	1964.48	1964.66	1964.26	1963.11

Table 4. Search of Er parameter

	a = 4		**b = 6**		**An = 20**
Er	0	0.1	0.2	0.3	0.4
Size (20)	375.152	377.094	377.094	377.094	377.094
Size (100)	1965.32	1964.66	1968.71	1964.32	1964.66

After establishing the value of the given parameter, this value was used in searching the next one. Finally, the following values of the parameters were found:

- a = 4.
- b = 6.
- An = 20.
- Er = 0.2.

It was observed, that the number of ants has strong impact on how time-consuming algorithm is, and the evaporation rate prevents to stop in a local minimum.

5.2 The Second Stage. Comparison of the Algorithms

Checking the quality of the algorithms was made using two criteria. The first criterion was the profit **P** expressed by (1). The second criterion was the number of the checked solutions, denoted by **N**. The experiment was designed in the following way: At first, the ACO algorithm was conducted and the time of working (**T**) was measured. Next, the RO algorithm was used for the same time. This sequence of operations was made for all fifteen instances. The obtained results are given in Table 5, in the second column P (RO) and in the third column P (ACO); both are shown in Fig. 4.

Table 5. Comparison of two algorithms

Size of instance	P (RO)	P (ACO)	T [sec]	N (RO)	N (ACO)
10	180.62	181.91	0.023	16568	400
20	333.61	358.56	0.084	86983	400
30	583.04	634.82	0.252	217129	400
40	601.59	691.31	0.585	367722	400
50	816.35	946.68	1.128	644354	400
60	1041.11	1234.49	1.925	1000475	400
70	944.83	1194.37	3.039	1395725	400
80	1214.05	1534.41	4.528	1768348	400
90	1349.00	1745.01	6.424	2329186	400
100	1432.53	1964.77	8.780	2848565	400
110	1413.19	1961.75	11.621	3511567	400
120	1538.60	2227.35	15.434	4367894	400
130	1535.03	2333.57	19.220	5032234	400
140	1701.37	2653.69	23.969	5797987	400
150	1533.78	2602.32	29.252	6875673	400

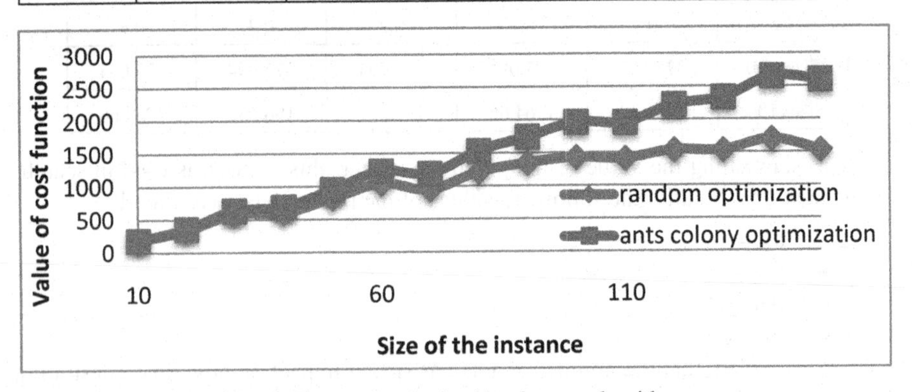

Fig. 3. The results obtained by the two algorithms

Moreover, the total number of permutations is also presented in Table 5. It may be observed in the fifth column, that the total numbers of permutations N checked by the RO algorithm increase significantly. According to the fixed stop criterion for ACO algorithm – the number N for this algorithm was the same for all instances (see the sixth column of Table 5).

6 Conclusion

It was observed, that for all tested instances the ACO algorithm gives better results than RO algorithm. For the small sizes of the problem the differences between results produced by the algorithms are almost not remarkable. For the big sizes of the problem (for $k \geq 70$) the advantage of using ACO is significant. However, it can be anticipated that increasing of the algorithms' working time would probably cause reaching better results by RO for very small instances because it would be possible for this algorithm to check all admissible solutions and to find the optimal solution. From the other side, the exact solutions for big problems are not known so it is hard to recognize whether ACO is able to reach very high accuracy, but undoubtedly it was shown that this algorithm performs pretty well for the considered problem. The obtained number of checked permutations also shows the superiority of ACO – this number is dependent only on the size of the problem and the number of ants. Random optimization is able to overview a large number of solutions but it becomes relatively smaller according to all possible permutations in a particular instance. It can be also concluded that for similar input data there is possibility to generalize solutions to ACO tuning process. We can justify that for real cases (e.g., a given restaurant and customers located rather in the vicinity) the changes of the environment, in the sense of distances, are not frequent.

The studies presented in this paper show that the own implementation of ACO algorithm can improve solving the real problems. We recommend such a way to improve profits when the problems similar to the introduced Provider problem are under consideration.

In the further research in this area, we plan an implementation of the additional algorithms to solve the introduced Provider problem. It is planned to follow the approaches based on tabu search [15], on genetic ideas [16], and on cuckoo search [17]. We would like to create the environment allowing the comparison of more algorithms and in consequence, for answering the question which of meta-heuristic and evolutionary approaches is the most convenient for solving the considered problem. Moreover, we are working on the improved version of the experimentation system with a module structure [18] which would give opportunities for taking into account the ideas presented in [19] and conducting complex experiments in automatic manner.

Acknowledgement. This work was supported by the statutory funds of the Department of Systems and Computer Networks, Faculty of Electronics, Wroclaw University of Technology, Wroclaw, Poland.

References

1. Zuhori, S.T.: Traveling Salesman Problem. Lambert Academic Publishing (2012) ISBN:3846583057
2. Applegate, D.L., Bixby, R.B., Chvatal, V., Cook, W.J.: The travelling salesman problem: A computational study. Princeton Series in Applied Mathematics. Princeton University Press (2007)

3. Wong, K.-C., Wu, C.-H., Mok, R.K.P., Peng, C., Zhang, Z.: Evolutionary multimodal optimization using the principle of locality. Information Sciences 194(1), 138–170 (2012)
4. Bonabeau, E., Dorigo, M., Theraulaz, G.: Swarm Intelligence: From Natural to Artificial Systems. Oxford University (1999)
5. Yang, X.S., Cui, Z.H., Xiao, R.B., Gandomi, A.H., Karamanoglu, M.: Swarm Intelligence and Bio-Inspired Computation: Theory and Applications. Elsevier (2013)
6. Dorigo, M., Stützle, T.: Ant Colony Optimization. MIT Press (2004)
7. Fister, I., Fister Jr., I., Yang, X.S., Brest, J.: A comprehensive review of firefly algorithms. Swarm and Evolutionary Computation (2013), `http://dx.doi.org/10.1016/j.swevo.2013.06.001`
8. Lizárraga, E., Castillo, O., Soria, J.: A method to solve the traveling salesman problem using ant colony optimization variants with ant set partitioning. In: Castillo, O., Melin, P., Kacprzyk, J. (eds.) Recent Advances on Hybrid Intelligent Systems. SCI, vol. 451, pp. 237–246. Springer, Heidelberg (2013)
9. Cormen, T.C., Leiserson, C., Rivest, R.L.: Introduction to algorithms. McGraw Hill (2001)
10. Kubacki, J., Koszalka, L., Pozniak-Koszalka, I., Kasprzak, A.: Comparison of heuristic algorithms to solving mesh network path finding problem. In: Proceedings to 4th International Conference on Frontier of Computer Science and Technology, Shanghai. IEEE (2009)
11. Regula, P., Pozniak-Koszalka, I., Koszalka, L., Kasprzak, A.: Evolutionary algorithms for base stations placement in mobile networks. In: Nguyen, N.T., Kim, C.-G., Janiak, A. (eds.) ACIIDS 2011, Part II. LNCS (LNAI), vol. 6592, pp. 1–10. Springer, Heidelberg (2011)
12. Kakol, A., Pozniak-Koszalka, I., Koszalka, L., Kasprzak, A., Burnham, K.J.: An experimentation system for testing bee behavior based algorithm to solving a transportation problem. In: Nguyen, N.T., Kim, C.-G., Janiak, A. (eds.) ACIIDS 2011, Part II. LNCS (LNAI), vol. 6592, pp. 11–20. Springer, Heidelberg (2011)
13. Neyoy, H., Castillo, O., Soria, J.: Dynamic fuzzy logic parameter tuning for ACO and its application in TSP problems. In: Castillo, O., Melin, P., Kacprzyk, J. (eds.) Recent Advances on Hybrid Intelligent Systems. SCI, vol. 451, pp. 259–271. Springer, Heidelberg (2013)
14. Castillo, O.: ACO-tuning of a fuzzy controller for the ball and beam problem. In: Castillo, O. (ed.) Type-2 Fuzzy Logic in Intelligent Control Applications. STUDFUZZ, vol. 272, pp. 151–159. Springer, Heidelberg (2012)
15. Basu, S.: Tabu search implementation on traveling salesman problem and its variations: a literature survey. American Journal of Operations Research 2(2), 163–173 (2012)
16. Bhattacharyya, M., Bandyopadhyay, A.K.: Comparative study of some solution methods for traveling salesman problem using genetic algorithms. Cybernetics and Systems 40(1), 1–24 (2008)
17. Ouaarab, A., Ahiod, B., Yang, X.S.: Discrete cuckoo search algorithm for the traveling salesman problem. Neural Computing and Applications (April 2013), `http://link.springer.com/article/10.1007%00521-013-1402-2`
18. Ohia, D., Koszalka, L., Kasprzak, A.: Evolutionary algorithm for solving congestion problem in computer network. In: Velásquez, J.D., Ríos, S.A., Howlett, R.J., Jain, L.C. (eds.) KES 2009, Part I. LNCS (LNAI), vol. 5711, pp. 112–121. Springer, Heidelberg (2009)
19. Martinez, A.C., Castillo, O., Montiel, O.: Comparison between ant colony and genetic algorithms for fuzzy system optimization. In: Castillo, O., Melin, P., Kacprzyk, J., Pedrycz, W. (eds.) Soft Computing for Hybrid Intelligent Systems. SCI, vol. 154, pp. 71–86. Springer, Heidelberg (2008)

Controlling Quality of Water-Level Data in Thailand

Pattarasai Markpeng[1], Piraya Wongnimmarn[1], Nattarat Champreeda[1], Peerapon Vateekul[1], and Kanoksri Sarinnapakorn[2]

[1]Department of Computer Engineering, Faculty of Engineering, Chulalongkorn University, Phayathai Road, Pathumwan, Bangkok, Thailand 10330
{pattarasai.m,piraya.w,nattarat.c}@student.chula.ac.th, peerapon.v@chula.ac.th
[2]Hydro and Agro Informatics Institute, Ministry of Science and Technology, Thailand, 10400
kanoksri@haii.or.th

Abstract. Climate change has increased the number of occurrences of extreme events around the world. Warning and monitoring system is very important for reducing the damage of disasters. The performance of the warning system relies heavily on the quality of data from automated telemetry system (ATS) and the accuracy of the predicting system. Traditional quality management systems cannot discover complicated cases, such as outliers, missing patterns, and inhomogeneity. This paper proposes novel procedures to handle these complex issues in hydrological data focusing on water level. In the proposed system, DBSCAN, which is a clustering algorithm, is applied to discover outliers and missing patterns. The experimental results show that the system outperforms a statistical criterion, mean $\pm$ $n\times$SD, where n is a constant. Also, all missing patterns can perfectly be discovered by our approach. For the inhomogeneity problem, several statistical approaches are compared. The comparison results suggest that the best homogenization tool is *changepoint*, a method based on F-test.

Keywords: Data quality control, outlier detection, homogeneity check, missing pattern detection, data improvement.

1 Introduction

Due to the global warming issue, extreme weather and climate disastrous events have occurred more frequently and severely in Thailand. One of the recent examples is the great flooding in 2011, which is the most critical natural disaster in Thai history. To prevent such damages, a warning system must be implemented. The system is usually built around many forecast models, where input data is assumed to be clean. Unfortunately, this assumption is not true in reality. Most of collected data usually have lots of errors, which leads to unsatisfactory prediction performance. Therefore there is a need for a data quality management system [1, 2].

Hydro and Agro Informatics Institute (HAII) is a public organization under the Ministry of Science and Technology. It has implemented Hydro-Information Systems

N.T. Nguyen et al. (Eds.): ACIIDS 2014, Part I, LNAI 8397, pp. 503–512, 2014.

(HIS) that collect hydro-meteorological data, such as rainfall, water level, temperature, humidity, light intensity, etc., which then are used as input to several forecast models.

To collect water level data, there are 200 telemetry stations installed across the whole country that automatically collect data and continuously send them to a central server every ten minutes, so the amount of data is quite large. However, no matter how good and how advanced the data acquisition system is, there are always some unexpected and unavoidable problems that could affect the quality of data, e.g., communication failures. From our preliminary analysis on water level data from HAII since 2010, there are many types of errors found as shown in Fig. 1-3. On the one hand, some anomalies can be easily detected by a simple thresholding strategy, e.g., missing values (Fig. 1 (a)) and flat-line data (Fig. 1 (b)). On the other hand, some are complex and hard to discover: outliers, homogeneity, and missing patterns. Fig. 2 (a) shows an example of data with outliers. Many tools, e.g., *RClimDex* [3], provide a feature to detect outliers in daily temperature data using just a simple statistics criterion, mean $\pm$ $n\times$SD, where n is a constant. Fig. 2 (b) demonstrates an example of time series data whose trend is not steady (a sudden change). This issue mostly happens because an instrument is replaced or there is a change in the location, so the data before the installation must be adjusted to match the level of new one. Although there are many tools with homogeneity check function, such as *RHtest* [4], *cpm* [5], and *changepoint* [6], there are no comparative studies suggesting what the best one is.

Furthermore, another complex problem called "missing patterns" has been discovered in this work. The problem happens because the machine cannot transmit the data smoothly which results in missing patterns in data. Fig. 3 illustrates this issue at one of the telemetry station. In the dashed rectangle, the data are always missing from noon to midnight, and that lasts for two months! It would have been better if the system could detect the problem and reported it to admin to repair a machine.

This paper presents a novel procedure to assure a quality of hydrological data. We focus on a collection of "water level" data, which is collected by HAII since 2010. There are three contributed parts designed to tackle different complex problems in the data: (*i*) outliers, (*ii*) inhomogeneity, and (*iii*) missing patterns. First, outliers are discriminated by employing a clustering technique; our clustering choice is *"Density-Based Spatial Clustering of Applications with Noise"* (*DBSCAN*). The experiment showed that the clustering approach was superior to the simple statistics criterion. Second, many homogeneity check tools, i.e., *RHtest*, *cpm*, *changepoint*, were compared, and *changepoint* is recommended. Finally, we propose a new algorithm to detect missing patterns. The method transforms time series data into a missing frequency graph and, then, applies *DBSCAN* to find all groups of missing patterns. The output is a set of start and end points with missing patterns. The experimental result showed that our approach is efficient in terms of accuracy and time.

The paper is organized as follows. Section 2 explains related works and relevant performance criteria. Section 3 describes our proposed methods. The experiments are reported in Section 4. Conclusion and future work are the topics for Section 5.

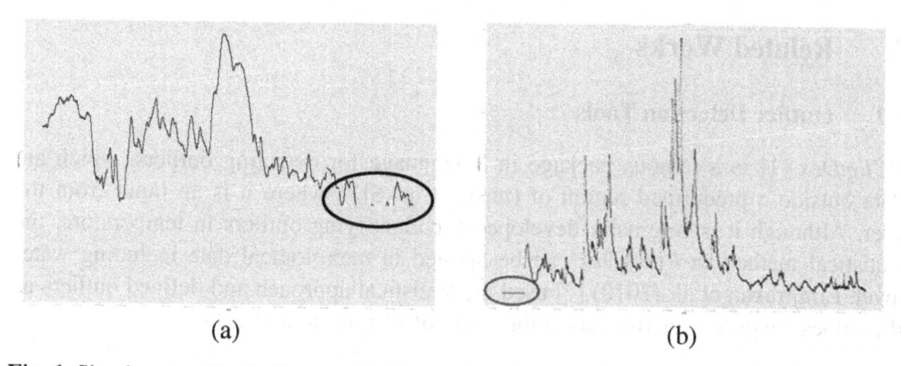

Fig. 1. Simple anomalies in time series of water level data. X-axis represents time in hours, and Y-axis shows a water level in meters. Anomalies are emphasized in a circle: (a) data with missing values and (b) flat-line type anomaly.

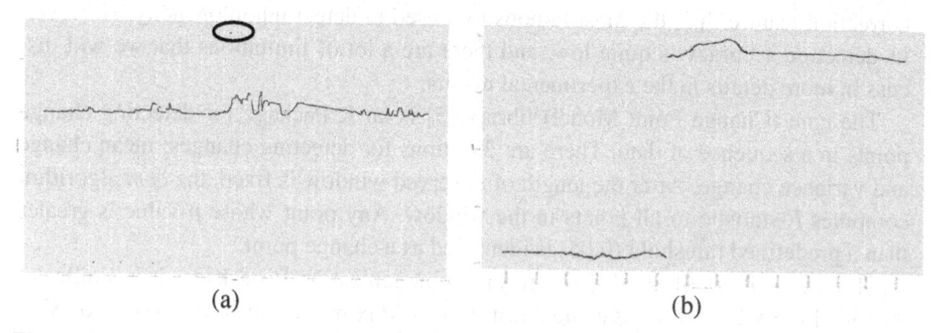

Fig. 2. Complex anomalies in time series of water level data. X-axis represents time in hours, and Y-axis shows a water level in meters. Anomalies are emphasized in a circle: (a) data with outliers and (b) inhomogeneity data.

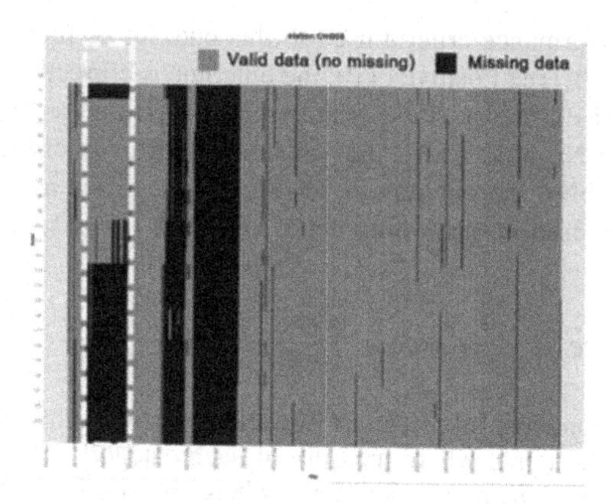

Fig. 3. Missing pattern of CHI005 station. X-axis represents date, and Y-axis represents time (hour) in a day.

2 Related Works

2.1 Outlier Detection Tools

RClimDex [1] is a famous package in R-language for detecting outliers, which are data outside a predefined region of (mean ± n×SD), where n is an input from the user. Although it is originally developed for identifying outliers in temperature, the statistical method in *RclimDex* can be applied in metrological data including water level. Limjirakan et al. (2010) [7] used the statistical approach and defined outliers as the values which are encroaching a threshold of 15σ in rainfall data.

2.2 Homogeneity Detection Tools

RHtest [4] is a tool for homogeneity checking based on *t*-statistic, along with a data correction feature. It is the most famous tool used to detect inhomogeneity. However, its detection accuracy is quite low, and there are a lot of limitations that we will discuss in more details in the experimental results.

The *cpm* (Change Point Model) library [5] is an R Package for detecting change points in a sequence of data. There are 2 options for detecting changes: mean change and variance change. After the length of observed window is fixed, the *cpm* algorithm computes F-statistic to all points in the window. Any point whose p-value is greater than a predefined threshold (0.05) is identified as a change point.

The *changepoint* library [6] is a recent homogenization R package. It is similar to the *cpm* library except the statistical test. It uses Maximum Log-likelihood ratio (ML) test for comparing the fit of two sequences of data and detecting change points.

2.3 Performance Evaluation

To establish the performance criteria used to this end, let us first introduce four elementary quantities: (*i*) True Positive (TP): the outcome is correctly classified as "yes", (*ii*) False Positive (FP): the outcome is incorrectly classified as "yes", when it is in fact "no", (*iii*) True Negative (TN): the outcome is correctly classified as "no", and (*iv*) False Positive (FN): the outcome is incorrectly classified as "no", when it is in fact "yes". They are used in the definitions of *precision (Pre)*, *recall (Rec)*, and F_1 as follows:

$$Pre = \frac{TP}{TP+FP}, Rec = \frac{TP}{TP+FN}, F_1 = \frac{2\times Pre\times Rec}{(Pre+Rec)} \quad (1)$$

For multi-class domains, [8] proposed two alternative ways to generalize the above criteria: (i) macro-averaging, where measure of each category is computed separately and, then, averaged, and (*ii*) micro-averaging, where each basic element, {TP, TN, FP, FN}, for each category is first combined and, then, the desired measures are computed using the original equations.

Table 1. The macro and micro averaging of *preicsion*, *recall*, and F_1 (c is the number of classes)

	Precision	*Recall*	*F1*
Macro-averaging	$Pre^M = \frac{\sum_1^c Pre_i}{c}$	$Rec^M = \frac{\sum_1^c Rec_i}{c}$	$F_1^M = \frac{\sum_1^c F_{1,i}}{c}$
Micro-averaging	$Pre^\mu = \frac{\sum_1^c TP_i}{\sum_1^c (TP_i + FP_i)}$	$Rec^\mu = \frac{\sum_1^c TP_i}{\sum_1^c (TP_i + FN_i)}$	$F_1^\mu = \frac{2 \times Pre^\mu \times Rec^\mu}{Pre^\mu + Rec^\mu}$

Another performance measure is the overall accuracy rate (Accuracy) shown in Equation 2. It is a popular measure due to its simplicity. Its range is between 0 and 1.

$$Accuracy = \frac{TP}{TP+FP+TN+FN} \tag{2}$$

3 The Proposed Data Quality Management

In the hydrological data collected at HAII, there are three problems that we target at: outliers, inhomogeneity, and missing patterns. In this section, details of the proposed algorithms to handle outliers and missing patterns problems are described. For the homogeneity issue, we compare homogenization tools that are available and suggest the best one.

3.1 Outlier Detection Algorithm

In time series of water stage, we propose to use a density-based clustering algorithm, *DBSCAN*, for outlier removal. However, no one has applied it to precipitation data.

DBSCAN, which was developed by Ester et al. (1996) [9], requires two parameters specified by user: distance epsilon (*eps*) and minimum number of points (*minpts*). A data point in time series is referred to as a neighbor if its distance to another data point is less than *eps*. A group of data point is a cluster if the number of neighboring points is more than *minpts*. Fig. 4 shows an example of how to detect outliers from time series data. Let *minpts* be 2 and *eps* be 1, Group 1 is determined as a cluster, while Group 2 is assigned as outlier because Group 2 does not contain adequate number of points to be a cluster.

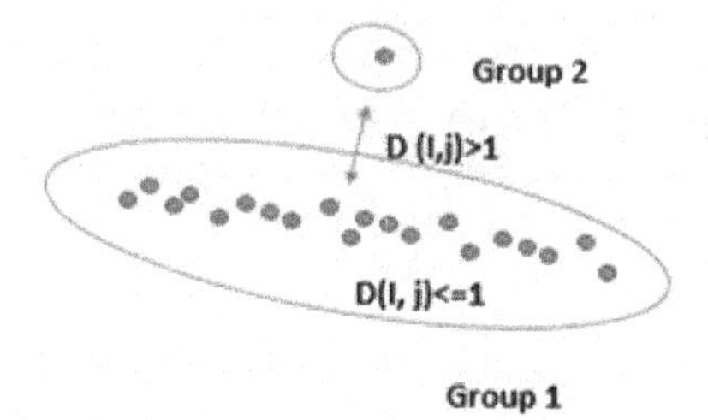

Fig. 4. A sample data set for detection outliers by DBSCAN

3.2 Missing Pattern Detection Algorithm

An issue of missing patterns is a new problem discovered in this paper. It is an event which data is failed to transmit to a central database server at some periods of time in a day, and this occurs in the same pattern every day for a while. A graph in Fig. 3 shows this circumstance found in the CHI005 station. The missing pattern appears in the dashed box showing a form of twelve-hours missing data every day during January to February, 2012.

In this work, a new algorithm is proposed to detect missing patterns resulting in a set of start-end points for each missing form. Fig. 5 demonstrates the pseudo code which composes of three main steps.

First, the program aims to detect missing patterns in "the whole data". Observing the pattern in Fig. 3, it can easily be figured by human since it is obvious and contains shape cutting edges. Therefore, the proposed algorithm focuses on those shape cutting edges. However, it is hard to find those edges in the original time series data, so we need to transform it into a non-missing frequency space. As shown in Fig. 6, three cutting edges can obviously noticed. Then, DBSCAN is applied to find a set of missing patterns (clusters) in the frequency space. If more than one cluster is generated, there are missing patterns in data.

Second, the program targets to find a smaller missing pattern in "a subset of data". To solve this problem, a divide-and-conquer strategy is presented by dividing one long period into two short periods. Then, we combine the result from those short periods to represent an overall detection result. In the `combineResult` function, there are 8 possible cases as shown in Table 2.

Finally, all of results are refined by merging two adjacent missing patterns if their gap is smaller than a predefined threshold gap (*merge_gap*).

```
Program PatternDetection(data){
  // Step1: initialization
  if(checkDataCondition(data)) return (-1, -1)
  tranformedData = dataTransformation(data)
  cluster = DBSCAN(transformedData, eps, minPts=1)
  isFullPeriodPatternDetected = moreThanOne(cluster)
  // Step2: divide data and detect pattern
  (leftSideData, rightSideData) = divideIntoHalf(data)
  leftPeriod = PatternDetection(leftSideData)
  rightPeriod = PatternDetection(rightSideData)
  resultList = combineResult(leftPeriod,rightPeriod)
  // Step3: merge any missing patterns with a small gap
  result = refineTooCloseGap(resultList, merge_gap)
  return result
}
```

Fig. 5. A pseudo code of the proposed missing pattern detection

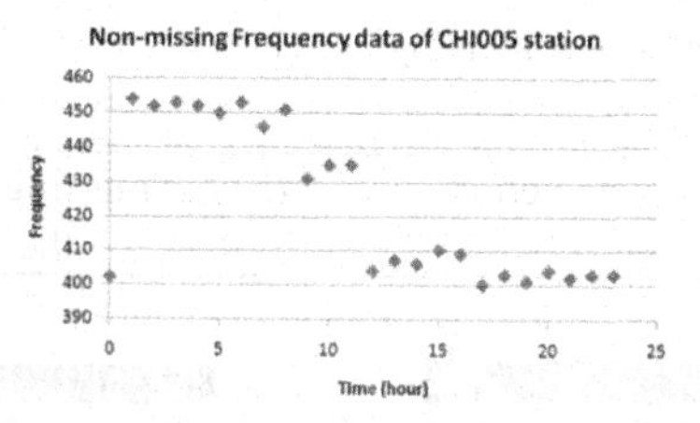

Fig. 6. Non-missing frequency data of CHI005 station. X-axis shows time in hours, and Y-axis represents the number of non-missing records.

4 Experiments

Although there are 200 telemetry stations collecting water level data, there are 29 stations chosen for the experiments. Many stations do not have enough data since they have just installed less than 2 years, while others' data include missing values more than 5%. After the selection, we manually cleaned the data in order to create "a complete (answer) data set" for the sake of comparison. Then, "a testing data set" for each experiment can be simulated by imputing errors to the complete data.

4.1 Experiment on Outlier Detection

The testing data for outlier detection is simulated by randomly add noises to the complete data. The range of noises is between a river bottom minus 2 and the highest river bank plus 4. The reasons are that dredging may decrease water level by 1-2 meters, and flood situation increase the upper range, normally less than 4 meters. Note that the number of imputed outliers is random and varied by station.

In this experiment, our approach (*DBSCAN*) in Section 3.1 is compared to the statistical method. The parameters in DBSCAN are assigning by setting *eps* to 1.05 and *minpts* to 3[1]. The comparison result in Table 3 shows that our method is superior to the statistical method in any measures. The reason is that the statistical method created a lot of false predicted outliers as an example in Fig. 7.

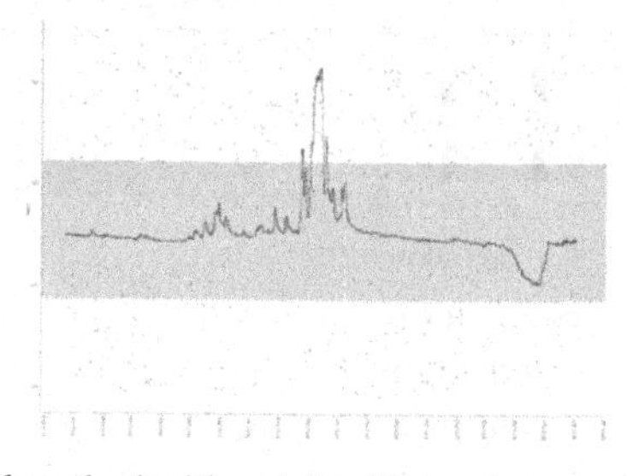

Fig. 7. A result of statistical method with $\mu \pm 2\sigma$. The points outside gray bar are predicted as outliers.

[1] The *minpts* parameter is suggested to be greater than 2 to speed up the clustering process as shown in `http://en.wikipedia.org/wiki/DBSCAN#Parameter_estimation`.

Table 2. All possible cases of combining two sides in the `combineResult` function

(a) Missing pattern is detected in the full period	(b) *No* missing pattern is detected in the full period
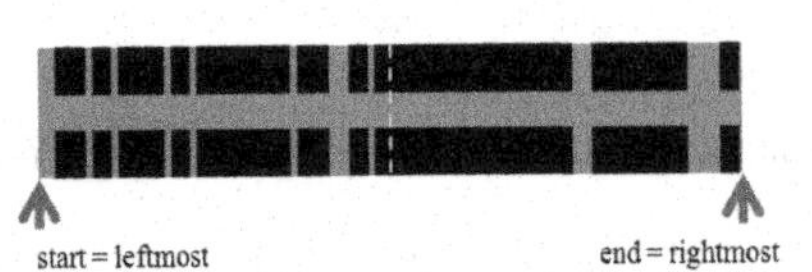 (i) Case 1: If the missing pattern is detected in the left and right period, there is a missing pattern from the left to the right period.	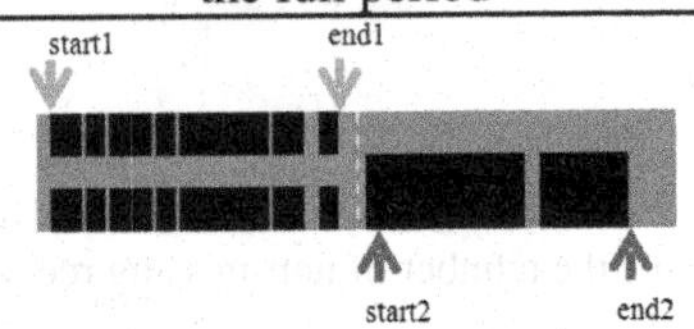 (v) Case 5: If the missing pattern is detected in the left and right period, there are more than one missing pattern period.
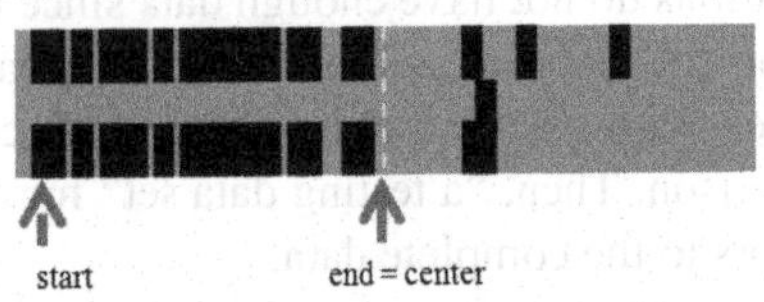 (ii) Case 2: If the missing pattern is detected in the left period but not in the right period, there is a missing pattern in the left period.	 (vi) Case 6: If the missing pattern is detected in the left period but not in the right period, there is a missing pattern in the left period.
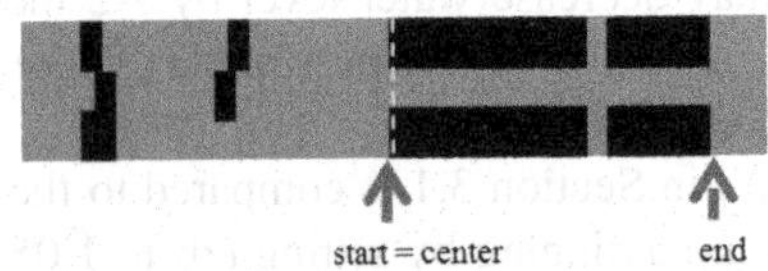 (iii) Case 3: If the missing pattern is detected in the right period but not in the left period, there is a missing pattern in the right period.	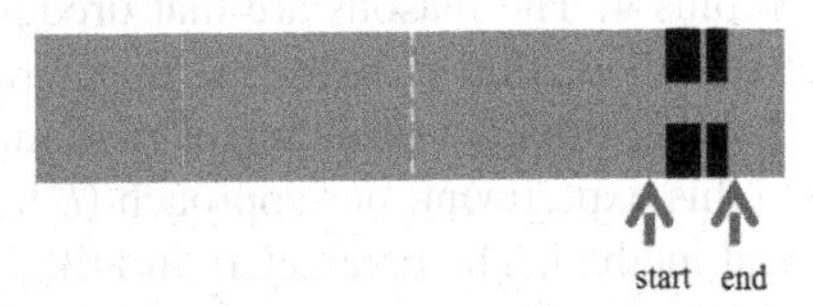 (vii) Case 7: If the missing pattern is detected in the right period but not in the left period, there is a missing pattern in the right period.
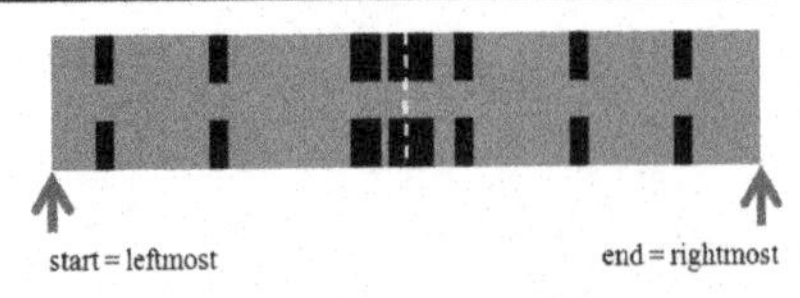 (iv) Case 4: If no missing patterns are detected in the left and right period, there is a missing pattern in the full period, but it is not obvious.	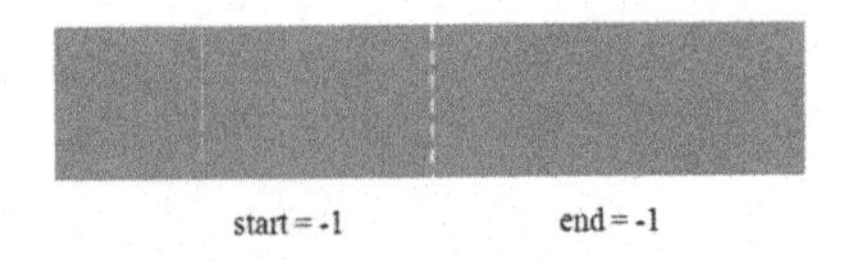 (viii) Case 8: If no missing patterns are detected in the left and right period, there are no missing patterns in the full period.

Table 3. The result of outlier detection comparing between our approach (*DBSCAN*) and statistical method with many choices of standard deviation, $\{2\sigma, 3\sigma, 4\sigma\}$

Method	Macro-averaging				Micro-averaging			
	Acc	Pre	Rec	F_1	Acc	Pre	Rec	F_1
DBSCAN	**0.9972**	**0.9253**	**1.0000**	**0.9608**	**0.9972**	**0.9065**	**1.0000**	**0.9509**
$\mu \pm 2\sigma$	0.9705	0.5659	0.7160	0.5641	0.9714	0.4863	0.6929	0.5715
$\mu \pm 3\sigma$	0.9818	0.8204	0.5106	0.6037	0.9815	0.7480	0.4949	0.5957
$\mu \pm 4\sigma$	0.9809	0.9222	0.3692	0.5487	0.9805	0.8833	0.3333	0.4840

4.2 Experiment on Homogeneity Check

The homogeneity check is tested based on data of 29 stations, where 16 stations are simulated to have the homogeneity issue, while the remaining 13 stations do not have the problem. Inhomogeneity data is created by first randomly selecting change points to split data into many parts and, then, adding a random shift value to a part of data. Two most common homogeneity scenarios are simulated including a single change point and two change points.

Table 4 illustrates the experimental results comparing among 3 homogenization tools: *cpm*, *changepoint*, and *RHtest*. From the results, the *changepoint* library is the best algorithm. It is surprising that *RHtest*, which is the most common tool, cannot be the winner in any scenarios. Fig. 8 demonstrates the detailed result of the YOM009 station with two change points. Fig. 8 (a) proves that the *changepoint* library can accurately detect all the change points, whereas Fig. 8 (b) shows that *RHtest* cannot perform well since it tends to give many incorrect change points.

Table 4. Performance evaluation on inhomogeneity data with single change point and two change points

Tools	Single Change Point				Two Change Points			
	Acc	Pre	Rec	F_1	Acc	Pre	Rec	F_1
cpm	**1.0000**	**1.0000**	**1.0000**	**1.0000**	0.9164	0.9615	0.0009	0.0018
changepoint	0.9999	**1.0000**	**1.0000**	**1.0000**	**0.9999**	**1.0000**	**0.9630**	**0.9811**
RHtest	0.9995	**1.0000**	0.0703	0.1313	0.9994	0.9231	0.1148	0.2043

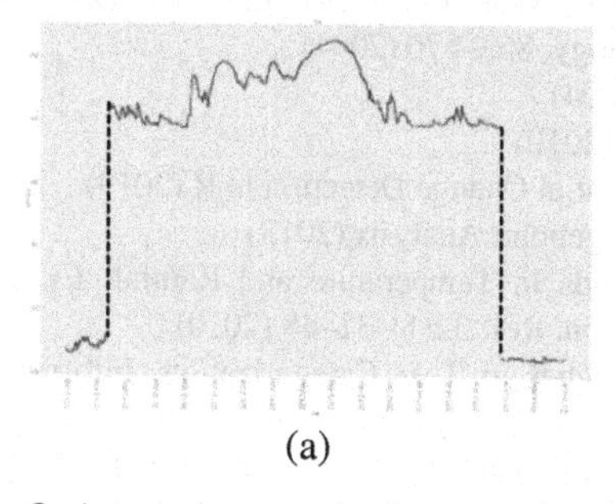

(a)

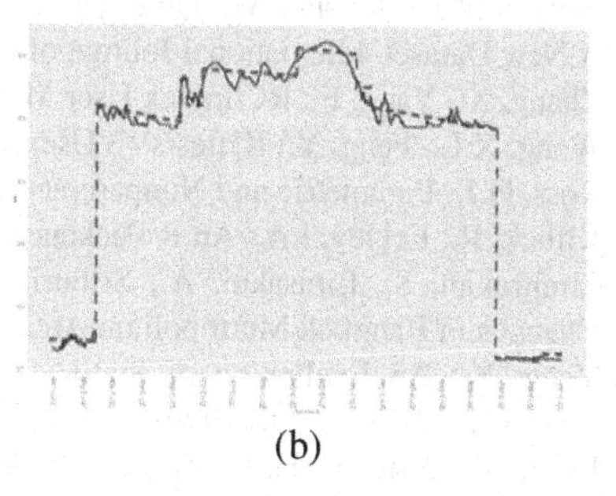

(b)

Fig. 8. A two-change-point homogeneity detection in the YOM009 station: (a) the result of *changepoint* and (b) the result of *RHtest*

4.3 Experiment on Missing Pattern Detection

There are 20 stations included in this experiment, where 7 stations are founded to have the missing pattern issue, while the remaining 13 stations do not have this problem. In the proposed algorithm, there are 3 input parameters: distance epsilon (*eps*), minimum points (*minpts*), and thresholding gap (*merge_gap*). We set *minpts* is to 1, and varied the values of *eps* and *merge_gap* are. Table 5 shows that the perfect result is possible when *eps* is 50% of the total data and *merge_gap* is within 15 days.

Table 5. The result of missing pattern detection on varied sets of (*eps*, *merge_gap*)

(*eps*, *merge_gap*)	Acc	Pre	Rec	F_1
(70%, 10)	0.8000	1.0000	0.4286	0.6000
(50%, 15)	**1.0000**	**1.0000**	**1.0000**	**1.0000**
(30%, 15)	0.8000	0.6364	1.0000	0.7778

5 Conclusion

This paper presents a novel data quality management system in hydrological data particularly in water level. There are three targeted problems: outliers, inhomogeneity, and missing patterns. The experiments were conducted on the data collected by a public organization called Hydro and Agro Informatics Institute (HAII). First, the result of the outlier detection shows that our clustering based approach outperforms a statistical method of mean and standard deviation, which can be applied from *RClimDex*. Second, our comparison study illustrates that the *changepoint* library is the best homogeneity detection tool. Finally, all missing patterns can be detected by employing our proposed divide-and-conquer solution.

References

1. Branisavljević, N., Prodanović, D., Arsić, M., Simić, Z., Borota, J.: Hydro-Meteorological Data Quality Assurance and Improvement. Journal of the Serbian Society for Computational Mechanics 3(1), 228–249 (2009)
2. Feng, S., Hu, Q., Qian, W.: Quality Control of Daily Metrological Data In China, 1951-2000: A New Dataset. International Journal of Climatology, 853–870 (2004)
3. Zhang, X., Yang, F.: RClimdex User Manual (2004)
4. Wang, X.L., Feng, Y.: RHtestsV3 User Manual (2010)
5. Ross, G.J.: Parametric and Nonparametric Sequential Change Detection in R (2013)
6. Killick, R., Eckley, I.A.: An R Package for Changepoint Analysis (2013)
7. Limjirakan, S., Limsakul, A., Sriburi, T.: Trends in Temperature and Rainfall Extreme Changes in Bangkok Metropolitan Area. J. Environ. Res. 32(1), 31–48 (2010)
8. Yang, Y.: An Evaluation of Statistical Approaches to Text Categorization. Information Retrieval 1(1/2), 69–90 (1999)
9. Ester, M., Kriegel, H.-P., Sander, J., Xu, X.: A density-based algorithm for discovering clusters in large spatial databases with noise. In: KDD 1996 Proceedings, pp. 226–231 (1996)

Application of Nonlinear State Estimation Methods for Sport Training Support

Krzysztof Brzostowski, Jarosław Drapała, and Jerzy Świątek

Wrocław University of Technology, Wyb. Wyspiańskiego 27, 50-370 Wrocław, Poland
{brzostowski.krzysztof,jaroslaw.drapala,jerzy.swiatek}@pwr.wroc.pl

Abstract. Typical understanding of healthcare concerns treatment, diagnosis and monitoring of diseases. But healthcare also includes well-being, healthy lifestyle, and maintaining good body condition. One of the most important factor in this respect is physical activity. Modern techniques of data acquisition and data processing enable development of advanced systems for physical activity support with use of measurement data. The need for reliable estimation routines stems from the fact, that many widely available (for bulk customers) measurements devices are not reliable and measured signals are contaminated by the noise. One of the most important variables for physical activity monitoring is the velocity of a moving object (e.g. velocity of selected parts of a body such as elbows). Apart from intensive use of system identification, optimization and control techniques for physical training support, we applied Kalman filtering technique in order to estimate speed of moving part of a body.

1 Introduction

Typical understanding of healthcare concerns treatment, diagnosis and monitoring of diseases. But healthcare also includes well-being, healthy lifestyle, and maintaining good body condition. One of the most important factor in this respect is physical activity. Prevention and treatment of chronic diseases, such as diabetes, cardiovascular and respiratory diseases involves leading an active lifestyle. Modern techniques of data acquisition and processing, together with wireless communication technologies, enable development of advanced systems for physical activity support. The role of data processing is to support decision making process. Decisions concern training intensity adjustment and correctness of exercises execution.

Wide availability of mobile devices that may measure such variables as heart rate, EMG, ECG, acceleration, [23], allows to develop solutions available for bulk consumers. Many companies offer cheap and small wearable sensors to to acquire human physiological and kinematic data. Both commercial and research teams develop solutions to support professional and recreational sports. Polar belongs to the most important companies offering wearable devices together with data processing tools [14]. Wrist worn watch and chest worn heart rate sensor are frequently use to acquire measurement data and present results of computations to the user. Polar provides equipment for fitness improvement and for

N.T. Nguyen et al. (Eds.): ACIIDS 2014, Part I, LNAI 8397, pp. 513–521, 2014.

maximization of performance. These devices can be used for example in motivational feedback i.e. to generate beeps every time when certain amount of calories is burnt. Based on the same data and specific signal processing techniques it is possible to prevent injuries and overtraining. Moreover the Polar's software provides tools to optimize training intensity. Another solution offered by Suunto [17], provides devices that generate personalized training plan. Based on results of user's training monitoring the system is capable of making recommendations for training volume i.e. the frequency, duration and intensity of exercises. Moreover, proposed training plan may be adjusted to the user's current capabilities i.e. when the user's activity level decreases. Adidas miCoach product is advanced training tool [13,15] that can be used to optimize training plan for endurance, strength and flexibility. Data are measured from stride and heart rate monitor. The website allows the user and a trainer to manage the training process. Important feature of the system is digital coaching, which serves to motivate user – through feedback – by giving voice notifications such as „speed up" or „slow down", tracking and informing about the workout progress. MOPET is an example of advanced academic project which is still under development. It uses measurement data to work out the user's mathematical model. The model has ability to adapt to the user data and is used to predict user's performance. On the basis of the model analysis, advices concerning the user's health state and safety issues are generated. Such a model-based prediction plays important role in applying such functionalities as personalization and context-awareness in systems to support sport training [21].

We have developed a system providing communication and computational services for healthcare applications. The system acquires in a real time measurement data from wireless sensors, sends data to a computer center, performs advanced processing (filtering, system identification, pattern recognition, system control), and uses results to support the training process.

We decided to use devices originating from the wireless sensors system called Shimmer Research. The user equipped with Shimmer measurement devices can measure the following physiological signals: heart rate, breath rate, temperature, ECG and others such as acceleration. Shimmer Research offers the following sensors: ECG (Electrocardiography), EMG (Electromyography), GSR (Galvanic skin response), Acceleration, Gyroscope, GPS etc. The sensors are still under development but many successful applications has been reported by both researchers and engineers [22].

Most of widely available (for bulk customers) wireless measurement devices are not reliable and measurements are contaminated by a noise. One of the most important variables for physical activity monitoring purposes is the velocity of a moving object (e.g. velocities of selected parts of the body such as trunk). The problem is that even when velocity is zero, a noise causes non-zero values returned by the sensor. This phenomena results in bias of velocity estimation procedure. Therefore, advanced filtering algorithms and the Kalman filters are important part of physical activity support system.

2 Applications – Performance Analysis in Sports

Performance analysis in sports relies on vital signs measurements and makes use of such equipment as video systems, radar, treadmill etc. There are two approaches to perform such analysis: qualitative and quantitative. Qualitative analysis relies on subjective interpretation of registered data concerning human movement. The advantage of qualitative analysis is that it can be performed even by a trainer that is not familiar with with signal processing techniques. The disadvantage of such analysis is that it is subjective and strongly depends on experience of a trainer. Results may be unreliable because human eyes are not capable of capturing quick events such as tennis strokes or golf swings. Unlike qualitative methods, quantitative analysis is transparent because it is based on precisely defined objective function and algorithms. The most common motion analysis tools for quantitative analysis are opto-reflective systems. One of the main disadvantages of quantitative analysis using opto-reflective systems is lack of a real-time feedback to the coach and athletes. Moreover, advanced and time consuming post-processing is required to extract and analyse the collected data. Therefore, such systems have limited area of application: they can be applied only in laboratory. They cannot be used in sport hall, tennis court, golf course etc.

Another example of technique widely used in quantitative analysis of sport training is radar measurements. This method allows to precisely measure the speed of moving object (e.g. athlete) but it is limited to a motion along a straight line, e.g. the speed (not velocity!) of athlete running on the stretch track.

Aforementioned methods are widely applied in sport analysis field, but are limited to laboratories and closed areas. Only some aspects of athlete's movement during sport performance may be measured and analysed.

Those limitations maybe overcame by employing modern technologies such as microelectronic and microelectromechanical systems (MEMS). This technology allows to build relatively inexpensive sensors with wireless interface to transfer data. It means that we can measure various aspects of athlete movements without wiring systems. It is very important because it means that athlete during tests can perform naturally and do not watch out wires. It is important feature that helps to design and build systems to analyze athlete performance from many points of view.

In this paper we propose and investigate a method to estimate speed of moving athlete. The speed values estimated during a test are used to generate the so called *speed curve*. This curve is used, for example, to make assessment of anaerobic performance [12]. Analysis of the speed curve allows to extract some features of a movement that can be useful to assess athlete's physical state. In our system we applied wireless sensor (called IMU) which consist of three-axis accelerometers, gyroscopes and magnetometers.

Attitude Estimation. To determine current speed of athlete we make use of attitude estimation method. Measurements are gathered by the wireless three-axis accelerometers, gyroscopes and magnetometers attached to the athlete's chest. It is worth mentioning that accelerometers in conjunction with gyroscopes are used to estimate attitude variations. These sensors contain sufficient amount of information for estimation purposes but obtained results suffer from the so called sensor drift. Sensor drift means that sensor output slowly varies, independently of measured variable dynamics. In order to overcome this problem and to improve attitude estimation magnetometers are usually applied. Magnetometers measure local earth magnetic field. Combining these data with data from accelerometers and gyroscopes allows to estimate attitude with high precision and cancel out the sensor drift. Non-linear state estimation and filtering have to be applied at this stage of data processing.

At the first step, attitude of sensor attached to athlete's chest is determined. Measurements from accelerometer (denoted as a_x, a_y, a_z for each axis), gyroscope (denoted as ω_x, ω_y, ω_z) and magnetometer (denoted as m_x, m_y, m_z) are collected. These measurements udergo filtering process with the Extended Kalman Filter (EKF). As a result, attitude of sensor attached to the athlete's chest is estimated. This step allows to cancel out the gravity part of acceleration measurements. At the second step we can estimate the speed of moving athlete with high precision.

Let us define a discrete-time state equation of the process that describe evolution of the state denoted as x_k. The system dynamics is characterized by the matrix A_k. We have inputs u_k acting on the state according to the matrix B_k. The state equation is:

$$\hat{x}_k^- = A_k \hat{x}_{k-1} + B_k u_k, \tag{1}$$

where hat means that the real state of the system is unknown and we estimate it with use of Kalman filter. Superscript *minus* stands for *a priori* state estimation which is determined according to equation (1).

The next step is to establish the process model representing body motion dynamics. This model is a particular case of equation (1) and is described below:

$$\dot{q}_k^B = \frac{1}{2} \Omega_{B,k}^N q_N^B \tag{2}$$

where q_N^B is the quaternion representing the rotation of the *body frame* with respect to the *navigation frame* and $\Omega_{B,k}^N$ is the rotation matrix. Quaternion q is composed of a real number q_0 and a vector: $[q_1 \quad q_2 \quad q_3]^T$ which can be rewritten in the following compact form:

$$q = [q_0 \quad q_1 \quad q_2 \quad q_3]^T \tag{3}$$

The rotation matrix $\Omega_{B,k}^N$ is defined as follows:

$$\Omega_{B,k}^N = \begin{bmatrix} 0 & -\omega_x & -\omega_y & -\omega_z \\ \omega_x & 0 & \omega_z & -\omega_y \\ \omega_y & -\omega_z & 0 & \omega_x \\ \omega_z & \omega_y & -\omega_x & 0 \end{bmatrix} \tag{4}$$

where ω_x, ω_y, ω_z are measurement collected from gyroscope. It is worth mentioning that there is no external input u_k, which means that the second element of equation (2) is neglected.

In Kalman Filter, after determination of *a priori* value of state vector (prediction) the second stage relies on calculating its *a posteriori* value (update). To this end it is necessary to know residual between calculated and measured signals. Here, it is applied to gravity and magnetic vector. For the sake of simplicity this phase is decomposed into two phases: the first for gravity vector and the second for magnetic vector.

The estimated gravity vector $\hat{g}$ is calculated using direction cosine matrix R_N^B, assuming that the g-force is constant:

$$\hat{g} = R_N^B \begin{bmatrix} 0 \\ 0 \\ g \end{bmatrix} = g \begin{bmatrix} 2q_1q_3 - 2q_0q_2 \\ 2q_2q_3 + 2q_0q_1 \\ q_0^2 - q_1^2 - q_2^2 + q_3^2 \end{bmatrix} \tag{5}$$

where R_N^B has the following form:

$$R_N^B = \begin{bmatrix} q_0^2 + q_1^2 - q_2^2 - q_3^2 & 2q_1q_2 + 2q_0q_3 & 2q_1q_3 - 2q_0q_2 \\ 2q_1q_2 - 2q_0q_3 & q_0^2 - q_1^2 + q_2^2 - q_3^2 & 2q_2q_3 + 2q_0q_1 \\ 2q_1q_3 + 2q_0q_2 & 2q_2q_3 - 2q_0q_1 & q_0^2 - q_1^2 - q_2^2 + q_3^2 \end{bmatrix} \tag{6}$$

Let us denote $\hat{g}$ as $h_1(q) = \hat{g}$.

Similar operations must be performed for the second measurement vector i.e. for the Earth magnetic field. This vector is normalized to unity and only x-axis and z-axis are considered. To estimate magnetic field vector we apply the following equation:

$$\hat{m} = R_N^B \begin{bmatrix} 1 \\ 0 \\ 1 \end{bmatrix} = \begin{bmatrix} q_0^2 + q_1^2 - q_2^2 - q_3^2 & 2q_1q_3 - 2q_0q_2 \\ 2q_1q_2 - 2q_0q_3 & 2q_2q_3 + 2q_0q_1 \\ 2q_1q_3 + 2q_0q_2 & q_0^2 - q_1^2 - q_2^2 + q_3^2 \end{bmatrix} \tag{7}$$

By analogy, we denote $\hat{m}$ as $h_2(q) = \hat{m}$.

In order to implement Extended Kalman Filter, the Jacobian of equation equations (5) and (7) must be determined. For the first equation, the Jacobian has the form:

$$H_{1,k} = \frac{\partial h_1(q)}{\partial q_i} = 2 \begin{bmatrix} -q_2 & q_3 & -q_0 & q_1 \\ q_1 & q_0 & q_3 & q_2 \\ q_0 & -q_1 & -q_2 & q_3 \end{bmatrix} \tag{8}$$

for $i = 1 \dots 4$. In the second case (7) the Jacobian is:

$$H_{2,k} = \frac{\partial h_2(q)}{\partial q_i} = 2 \begin{bmatrix} q_0 - q_2 & q_1 + q_3 & -q_2 - q_0 & -q_3 + q_1 \\ -q_3 + q_1 & q_2 + q_0 & q_1 + q_3 & -q_0 + q_2 \\ q_2 + q_0 & q_3 - q_1 & q_0 - q_2 & q_1 + q_3 \end{bmatrix} \tag{9}$$

for $i = 1 \dots 4$.

Having defined the process, measurement models and the Jacobians we are ready to estimate values of quaternions that are used to describe attitude of sensor attached to the athlete's chest. Further on, we cancel gravity part of acceleration measurements. This is important step since it allows to estimate the speed of running athlete. In the next subsection details of this procedure are given.

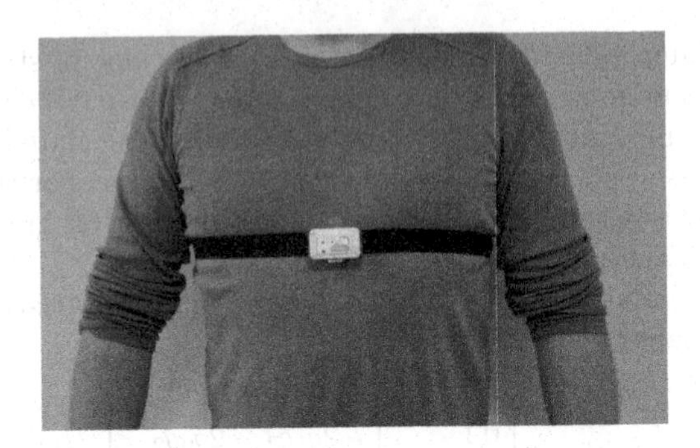

Fig. 1. Experimental set-up: sensor attached to the athlete's chest

Estimation of Runner's Velocity. In the previous section a method of attitude estimation of sensor attached to athlete's chest was presented. Based on obtained results we can determine the speed of moving athlete's on the basis of previously obtained results and acceleration measurements. In order to estimate the speed of moving object (e.g. running athlete) it is necessary to cancel gravity element of acceleration measurements. To this end we apply the following formula [9]:

$$\tilde{\mathbf{a}} = C(q)\mathbf{a} - \mathbf{g}; \tag{10}$$

where $\mathbf{a}$ is acceleration measurements vector composed of $\mathbf{a} = [a_x \quad a_y \quad a_z]^T$, $\mathbf{g}$ stands for gravity vector ($\mathbf{g} = [0 \quad 0 \quad -9.81]^T$) and $C(q)$ is the direction cosine matrix representation of the estimated quaternion q (see the previous section). This matrix is defined as:

$$C(q) = (q_4^2 - \mathbf{e}^T\mathbf{e})\mathbf{I_3} + 2\mathbf{e}\mathbf{e}^T - 2q_4[\mathbf{e}\times] \tag{11}$$

where $q = [q_1 \quad q_2 \quad q_3 \quad q_4]^T$. Substituting $\mathbf{e} = [q_1 \quad q_2 \quad q_3]^T$ we obtain: $q = [\mathbf{e}^T \quad q_4]^T$. Element of equation (11): $[\mathbf{e}\times]$ is a skew-symmetric matrix.

Having acceleration measurements with gravity element cancelled, it is possible to estimate the speed of a moving object.

Experimental Verification. In order to verify the proposed method of running athlete speed estimation, the experiment was designed and performed. In Fig. 1 location of sensor attached to the body is illustrated. A subject was asked to run along a straight line, which was necessary because for verification purposes we used simultaneously radar to estimate the speed of a subject. We have used the Shimmer 9DoF sensor [18] and radar Stalker [19].

Measurements from Shimmer 9DoF sensor are collected in personal computer and processed with use of software developed in MATLAB environment. In Fig. 2 results of estimation of speed curve for running athlete is presented. In experimental procedure two different tests are performed. During the first one a subject was asked to run slowly at first, then faster and eventually slowly again. In the second test the Wingate procedure was performed [12].

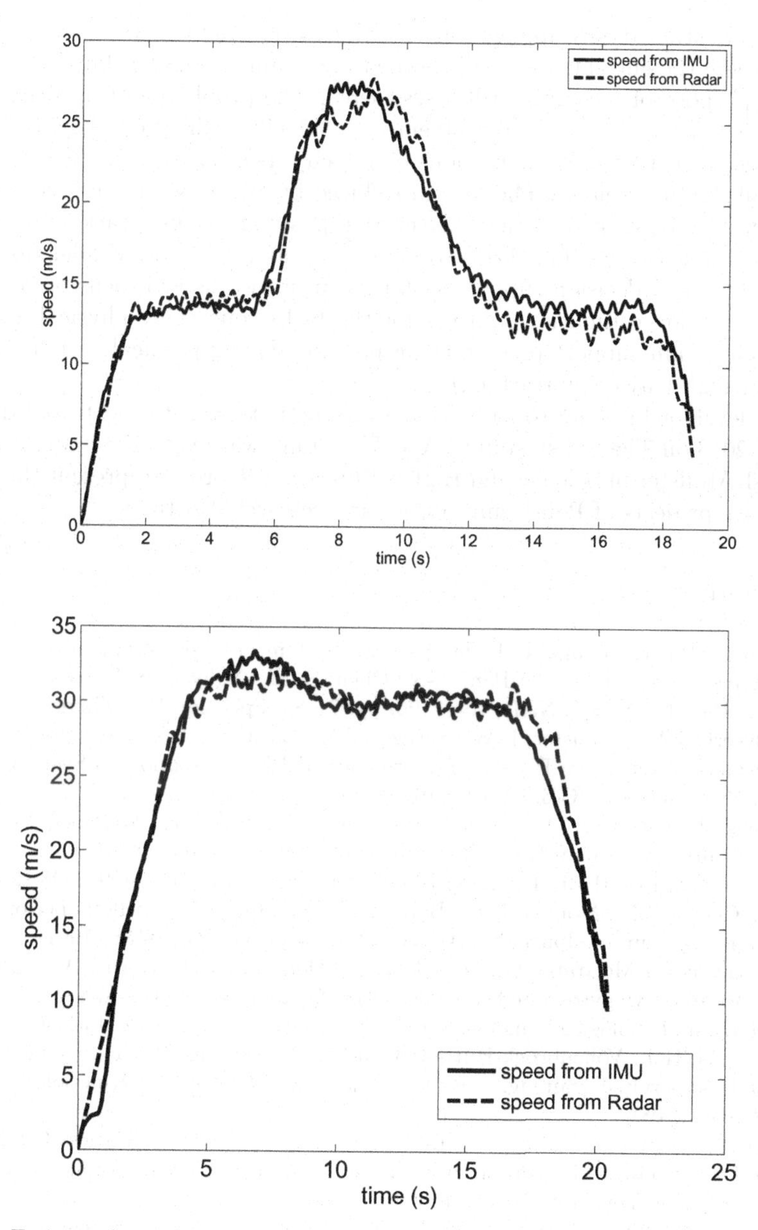

Fig. 2. Estimated speed curve for two tests: a) slow, fast, slow run (above), b) Wingate test (below)

3 Summary

In the work, applications of data filtering and state estimation routines for eHealth system to support planning training protocol for exerciser are presented.

We have mentioned only applications utilizing velocity, but there are some situations, where the current position is needed (e.g. human limb tracking trajectory for the purpose of technical skill assessment). The problem with position estimation is twice as hard as the problem with velocity estimation. This is due to the problems concerning calculations of velocity as integral of the acceleration signal affected by a noise. The bias introduced by the noise may be reduced by the filter, but in order to estimate the current position, the integration procedure must be applied twice. This leads to higher bias and it is more challenging task for developers of decision support systems. Future works will focus on applications of position estimation routines for the tasks concerning physical activity or technical skill support (e.g. tracking position during physical activity, lower or upper limb trajectory tracking).

The developed system to support physical activity was demonstrated during CEBiT 2013 on The Polish stand LAB. This stand was organized under auspice of Polish Minister of Science and Higher Education in order to present the most innovative projects of Polish universities and research institutes.

References

1. Brzostowski, K., Drapała, J., Świątek, J.: System analysis techniques in ehealth systems: a case study. In: Pan, J.-S., Chen, S.-M., Nguyen, N.T. (eds.) ACIIDS 2012, Part I. LNCS (LNAI), vol. 7196, pp. 74–85. Springer, Heidelberg (2012)
2. Calveert, T.W., Banister, E.W., Savage, M.V., Bach, T.: A Systems model on the Effects on Training on Physical Performance. IEEE Transaction on Systems, Man and Cybernetics SMC-6, 94–102 (1976)
3. Cheng, T.M., Savkin, A.V., Celler, B.G., Su, S.W., Wang, L.: Nonlinear Modeling and Control of Human Heart Rate Response During Exercise With Various Work Load Intensities. IEEE Trans. on Biomedical Engineering 55, 2499–2508 (2005)
4. Fitz-Clarke, J.R., Morton, R.H., Banister, E.W.: Optimizing athletic performance by influence curves. Journal of Applied Physiology 71, 1151–1158 (1991)
5. Greene, B.R., McGrath, D., O'Neill, R., O'Donovan, K.J., Burns, A., Caulfield, B.: An adaptive gyroscope-based algorithm for temporal gait analysis. Journal of Medical and Biological Engineering and Computing 48, 1251–1260 (2010)
6. Hermens, H.J., Vollenbroek-Hutten, M.M.R.: Towards remote monitoring and remotely supervised training. Journal of Electromyography and Kinesiology 18(6), 908–919 (2008)
7. Lim, J.E., Choi, O.H., Na, H.S., Baik, D.K.: A Context-Aware Fitness Guide System for Exercise Optimization in U-health. IEEE Trans. On Information Technology in Biomedicine 13, 370–379 (2009)
8. Lorincz, K., Chen, B.R., Challen, G.W.: Mercury: A Wearable Sensor Network Platform for High-Fidelity Motion Analysis. In: Proc. of the 7th ACM Conference on Embedded Networked Sensor Systems, Berkeley, California, pp. 183–196 (2009)
9. Meng, X., Zhang, Z.Q., Wu, J.K., Wong, W.C.: Hierarchical Information Fusion for Global Displacement Estimation in Microsensor Motion Capture. IEEE Transaction on Biomedical Engineering 60, 2052–2063 (2013)
10. Moxnes, J.F., Hausken, K.: The dynamics of athletic performance, fitness and fatigue. Mathematical and Computer Modelling of Dynamical Systems 14, 515–533 (2008)

11. Pantaelopoulos, A., Bourbakis, N.G.: A Survey on Wearable Sensor-Based Systems for Health Monitoring and Prognosis. IEEE Transactions on Systems, Man, and Cybernetics, Part C: Applications and Reviews, 1–12 (2010)
12. Smołka, ł., Ochmann, B.: A Novel Method of Anaerobic Performance Assessment in Swimming. Journal of Strength and Conditioning Research 27, 533–539 (2013)
13. adidas miCoach: The Interactive Personal Coaching and Training System (2011), `http://www.micoach.com` (accessed: September 14, 2013)
14. Polar, `http://www.polar.fi` (accessed: September 15, 2014)
15. Porta, J.P., Acosta, D.J., Lehker, A.N., Miller, S.T., Tomaka, J., King, G.A.: Validating the Adidas miCoach for estimating pace, distance, and energy expenditure during outdoor over-ground exercise accelerometer. International Journal of Exercise Science 2 (2012)
16. Simon, D.: Optimal State Estimation: Kalman, H_{inf} and Nonlinear Approaches. John Wiley & Sons (2006)
17. Suunto, `http://www.suunto.com` (accessed: September 14, 2014)
18. Shimmer Sensing, `http://www.shimmersensing.com` (accessed: September 14, 2014)
19. Stalker, `http://www.stalkerradar.com` (accessed: September 14, 2014)
20. Świątek, J.: Some problems of complex static systems identification. Wrocław University of Technology Publishing House, Poland (2009)
21. Świątek, P., Klukowski, P., Brzostowski, K., Drapała, J.: Application of Wearable Smart System to Support Physical Activity. In: Frontiers in Artificial Intelligence and Applications. Advances in Knowledge-Based and Intelligent Information and Engineering Systems, vol. 243, pp. 1418–1427 (2012)
22. Twomey, N., Faul, S., Marnane, W.P.: Comparison of accelerometer-based energy expenditure estimation algorithms. In: Proc. of the 4th International Conference on Pervasive Computing Technologies for Healthcare (PervasiveHealth), pp. 1–8 (2010)
23. Vales-Alonso, J., et al.: Ambient Intelligence Systems for Personalized Sport Training. Sensors 10(3), 2359–2385 (2010)

Multiple Object Tracking Based on a Hierarchical Clustering of Features Approach

Supannee Tanathong[1] and Anan Banharnsakun[2]

[1]Laboratory for Sensor and Modeling, Department of Geoinformatics, University of Seoul, Seoul 130-743, South Korea
[2]Laboratory for Computational Intelligence, Faculty of Engineering at Si Racha, Kasetsart University Siracha Campus, Chonburi 20230, Thailand
stanathong@yahoo.co.uk, anan@eng.src.ku.ac.th

Abstract. One challenge in object tracking is to develop algorithms for automated detection and tracking of multiple objects in real time video sequences. In this paper, we have proposed a new method for multiple object tracking based on the hierarchical clustering of features. First, the Shi-Tomasi corner detection method is employed to extract the feature points from objects of interest and the hierarchical clustering approach is then applied to cluster and form them into feature blocks. These feature blocks will be used to track the objects frame by frame. Experimental results show evidence that the proposed method is highly effective in detecting and tracking multiple objects in real time video sequences.

Keywords: Multiple Object Tracking, Feature Extraction, Shi-Tomasi Corner Detection, Hierarchical Clustering.

1 Introduction

Object tracking [1] has played an essential role in the field of computer vision over the decades. It is very useful in a variety of important application areas including surveillance [2], vehicle tracking [3], robotics [4], medical imaging [5], and manufacturing [6]. The objective of object tracking is to faithfully locate the targets throughout a series of consecutive video frames. The increasing need for automated video analysis has generated a great deal of interest in object tracking algorithms.

Basically, the task of estimating the motion path of an object in successive frames of a video includes two steps: detection of objects and tracking. The commonly used detection methods in the context of object tracking include point detectors [7], background subtraction [8], segmentation [9], and supervised learning [10], while the popular approaches for tracking an object include point tracking [11], kernel tracking [12], and silhouette tracking [13].

In detection methods, point detectors are used to find interest points in images which have an expressive texture in their respective localities. Background subtraction is a method typically used to segment moving regions in image sequences taken from a static camera by comparing each new frame to a model of the scene background, after

N.T. Nguyen et al. (Eds.): ACIIDS 2014, Part I, LNAI 8397, pp. 522–529, 2014.

which the image's foreground is then extracted for further processing. Segmentation is typically used to locate objects and boundaries in images by partitioning them into multiple segments. Supervised learning is the machine learning task of inferring a function from labeled training data. It can be used for detecting objects by learning different object views automatically from a set of examples.

In tracking approaches, point tracking can be formulated as the correspondence of detected objects represented by points across frames. While kernel tracking is typically performed by computing the motion of the object represented by a primitive object region from one frame to the next, silhouette tracking aims to find the object region in each frame by means of an object model generated using the previous frames.

Over the past decade, numerous methods of object tracking have been proposed as suitable for specific target applications. Sbalzarini and Koumoutsakos [14] proposed a method for two-dimensional feature point tracking and trajectory analysis for video imaging in cell biology. A model-based tracking algorithm which can extract the trajectory information of a target object by detecting and tracking a moving object from a sequence of images was presented by Jang and Choi [15]. To improve the robustness of tracking mechanisms, a visual tracking algorithm based on a multiple feature fusion method under a semi-supervised learning framework was introduced by Zhou et al. [16]. Since the scale of the target objects often varies irregularly, a multi-scale information measurement for images was introduced by Qian et al. [17] to probe the changes in size of tracked objects. The use of proximate distribution densities of the local regions was presented by Huang and Jiang [18] for tracking an object from a video sequence of a moving background. However, when a problem occurs while tracking more than one object, known as multiple object tracking, these aforementioned methods are not applicable due to the increased complexity of the problem.

Currently, there are several methods available for multiple object tracking. The most well-known of these methods include the Kalman filter [19], the particle filter [20], and the mean-shift [21]. The Kalman filter is used to predict the possible location of the selected object. However, it requires linear models for state dynamics and does not perform well in high dimensional problems. The Particle filter is used to track multiple objects by generating hypotheses on a top-view plan for position estimation of a target object, but this method does not yield good results for real time video sequences. The Mean-shift algorithm is an approach to tracking objects whose appearance is defined by histograms. Although the Mean-shift method performs very quickly, it sometimes gets stuck at the local minimum and is also difficult to use for handling abrupt motion.

In this paper, we aim to propose a new method for tracking multiple objects by using the feature points from objects of interest based on the Shi-Tomasi corner detection method to detect the objects first and the hierarchical clustering approach is then employed to cluster and form them into feature blocks. These feature blocks will be used to track the objects in succeeding frames.

The remainder of this paper is organized as follows: Section 2 describes the background and knowledge used in this work. Section 3 proposes our tracking method.

Section 4 presents the experiments and results. Section 5 summarizes the conclusions of the work.

2 Background and Knowledge

2.1 Feature Extraction

A feature extraction method [22] is a process that attempts to discover the dominant characteristics or dimensions in unstructured data. The method uses several mathematical functions to transform the high-dimensional input data into lower dimensional data which can be represented as a set of features. In the image processing domain, there are several methods used to extract the features from the image such as edge detection [23], corner detection [24], blob detection [25], and ridge detection [26].

Corner detection is a popular approach used within computer vision systems to extract certain kinds of features and infer the contents of an image, because corner points are interesting features as they are formed from two or more edges and edges usually define the boundary between two different objects or parts of the same object. We thus employ the method for feature extraction based on corner detection proposed by Shi and Tomasi [27] in this work.

The Shi-Tomasi corner detection method aims to find corners with large eigen-values in the image. The function first calculates the corner quality measure at every source image pixel. Then, it performs non-maxima suppression (only the local maxima in 3x3 neighborhoods are retained). The next step rejects the corners with a minimal eigen-value less than a predefined quality level. Finally, the function ensures that the distance between any two corners is not less than a predefined minimal distance. The weaker corners (with a smaller minimal eigen-value) that are too close to the stronger corners are rejected.

2.2 Hierarchical Clustering

Hierarchical clustering [28] is a method of cluster analysis which seeks to build a hierarchy of clusters without knowing the predefined number of clusters. Generally, it can be categorized into two types: agglomerative and divisive. In the agglomerative method, each individual element starts in its own cluster, and pairs of clusters are merged as one moves up the hierarchy; whereas all individual elements in divisive method start in one cluster, and splits are performed recursively as one moves down the hierarchy. In this work we focus on the agglomerative hierarchical clustering.

Given a set of N items to be clustered, the algorithmic steps for agglomerative hierarchical clustering can be described as below:

1. Start by assigning each element to its own cluster, so that if we have N elements, we now have N clusters, each containing just one element. Let the distances (similarities) between the clusters equal the distances (similarities) between the elements they contain.

2. Find the closest (most similar) pair of clusters and merge them into a single cluster, so that now we have one less cluster.
3. Compute distances (similarities) between the new cluster and each of the old clusters.
Repeat steps 2 and 3 until the distances of all of the closest pairs of clusters are less than the predefined threshold value.

3 Proposed Work

There are two main parts in our proposed work. The first is the detection of objects of interest by using their feature points and the second is clustering and tracking them by using the hierarchical clustering approach. These two steps are used to detect and track these objects from frame to frame. The work flow of our proposed method is illustrated in Fig. 1.

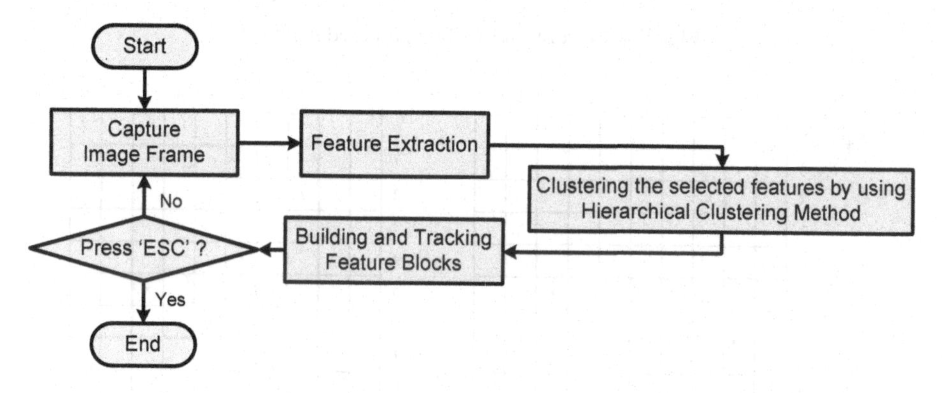

Fig. 1. Object tracking based on the Hierarchical Clustering of Features Approach

First, we capture an image frame from a video sequence which is then sent to the feature extraction process. The Shi-Tomasi corner detection method is used in this step to select interesting feature points from this image frame. Next, we build the feature block for each object based on the feature points selected from the previous step by using the hierarchical clustering approach. The number of detected objects is also obtained by the hierarchical clustering method at this step.

More clearly, Fig. 2 illustrates the artifact result when the objects in the image frame are detected and processed by our proposed method. Fig. 2(a) shows the initial image frame. Fig. 2(b) presents the interesting feature points of objects detected by the Shi-Tomasi corner detection method. Fig. 3(b) shows the feature block on each detected object built by the hierarchical clustering approach.

The details of building feature blocks can be described as follows: Let (FB_{xj},FB_{yj}) be the planar coordinates of the top left corner of the feature block for object j with a size of M x N. We can calculate FB_{xj}, FB_{yj}, M_j, and N_j as below:

$$
\begin{aligned}
FB_{xj} &= \min(x_i \mid x_i \in C_{xj}) \\
FB_{yj} &= \min(y_i \mid y_i \in C_{yj}) \\
M_j &= \max(x_i \mid x_i \in C_{xj}) - \min(x_i \mid x_i \in C_{xj}) + 1 \\
N_j &= \max(y_i \mid y_i \in C_{xj}) - \min(y_i \mid y_i \in C_{xj}) + 1
\end{aligned}
$$

where C_{xj} and C_{yj} are the set of features in dimension x and y of object j respectively.

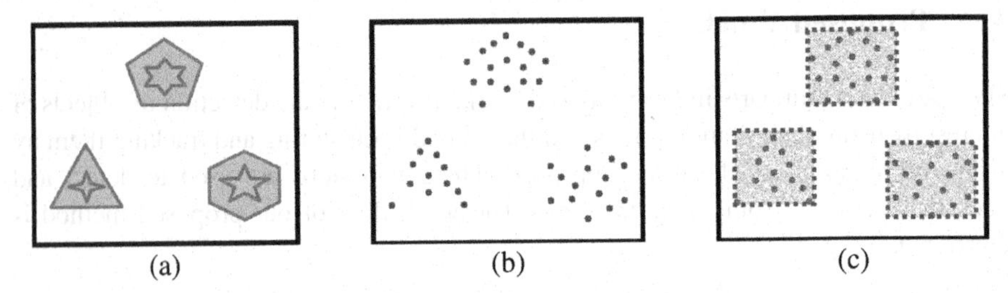

Fig. 2. Artifact result by using proposed method

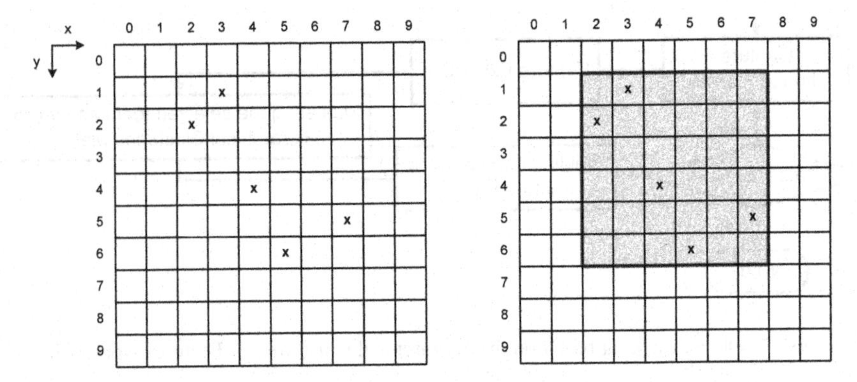

Fig. 3. Feature block building

For example in Fig. 3, if we have 5 feature points in a cluster consisting of (2,2) , (3,1) , (4,4) , (5,6) , and (7,5), the set of C_{xj} will be {2,3,4,5,7} and C_{yj} will be {1,2,4,5,6}. The FB_{xj}, FB_{yj}, M_j, and N_j can thus be calculated as below:

$$
\begin{aligned}
FB_{xj} &= \min(x_i \mid x_i \in \{2,3,4,5,7\}) && = 2 \\
FB_{yj} &= \min(y_i \mid y_i \in \{1,2,4,5,6\}) && = 1 \\
M_j &= \max(x_i \mid x_i \in C_{xj}) - \min(x_i \mid x_i \in C_{xj}) + 1 && = 7 - 2 + 1 = 6 \\
N_j &= \max(y_i \mid y_i \in C_{xj}) - \min(y_i \mid y_i \in C_{xj}) + 1 && = 6 - 1 + 1 = 6
\end{aligned}
$$

Now that we know the number of objects and the positions of the feature blocks representing the positions of the objects in the current image frame, we can thus use this information to track the objects in the next image frame.

4 Experiment Setting and Results

In this section, we will show the multiple object tracking results obtained by using the hierarchical clustering of features approach. To validate the effectiveness of the proposed tracking method, multiple objects with different sizes and colors undergo tracking by using a webcam with a resolution of 640 × 480 pixels. This experiment was programmed in C++ and all experiments were run on a PC with an Intel Core i7 CPU, 2.8 GHz and 16 GB memory. The subsequent frames are used to track the objects as shown in Fig. 4. The recorded frames in Fig. 4 have been taken at different points in time, while the tracking of objects is performed in real time. The red box illustrates the feature block on each object of interest.

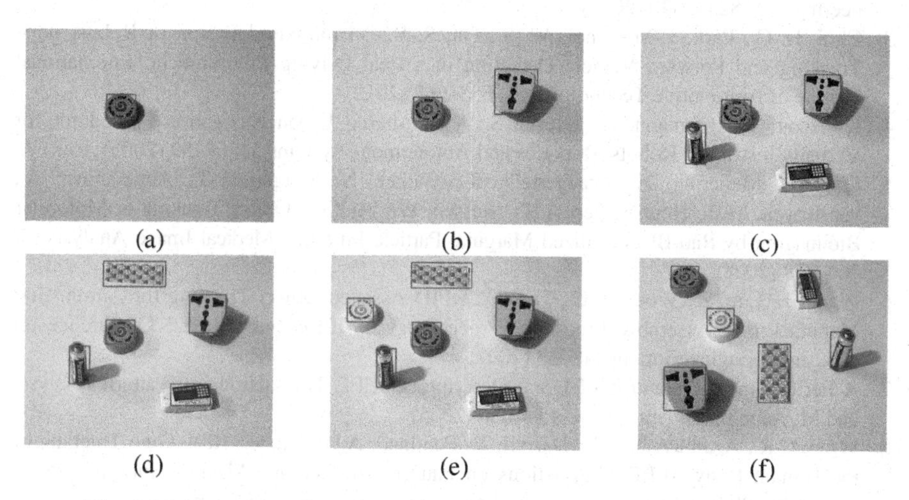

Fig. 4. Multiple object detection and tracking results by using proposed method

First, in Fig. 4(a) we start the evaluation of our proposed method with a single object. We then add more objects into the scene as shown in Fig. 4(b) to Fig. 4(e). We can see that every time we add an object to the scene, the object will be detected and a feature block will be built to cover each detected object. Finally, we move each object from its previous position to a different position as illustrated in Fig. 4(f). The results show that the objects can still be detected and tracked by our proposed method.

5 Conclusions

In this work, we have proposed an algorithm to detect multiple objects against a static and plain background by using their feature points based on the Shi-Tomasi corner detection method. After the detection of the object, we employed the hierarchical clustering approach to group these feature points and represent them as feature blocks. These feature blocks are then used to track the objects in the succeeding frames.

The preliminary experiments demonstrate the effectiveness of the algorithm. Multiple objects can be detected and tracked by our proposed method. In the next work, we will aim to improve the ability of this proposed method for tracking objects in complex background environments. Training and learning strategies for tracking desirable and specific objects will also be addressed.

References

1. Yilmaz, A., Javed, O., Shah, M.: Object Tracking: A Survey. ACM Computing Surveys 38, 1–45 (2006)
2. Li, C., Hua, T.: Human Action Recognition Based on Template Matching. Procedia Engineering 15, 2824–2830 (2011)
3. Choi, H.-C., Park, J.-M., Choi, W.-S., Oh, S.-Y.: Vision-based Fusion of Robust Lane Tracking and Forward Vehicle Detection in a Real Driving Environment. International Journal of Automotive Technology 13, 653–669 (2012)
4. Kyriacou, T., Bugmann, G., Lauria, S.: Vision-based Urban Navigation Procedures for Verbally Instructed Robots. Robotics and Autonomous Systems 51, 69–80 (2005)
5. Smal, I., Meijering, E., Draegestein, K., Galjart, N., Grigoriev, I., Akhmanova, A., van Royen, M.E., Houtsmuller, A.B., Niessen, W.: Multiple Object Tracking in Molecular Bioimaging by Rao-Blackwellized Marginal Particle Filtering. Medical Image Analysis 12, 764–777 (2008)
6. Wang, J., Luo, Z., Wong, E.C., Tan, C.: RFID Assisted Object Tracking for Automating Manufacturing Assembly Lines. In: Proceedings of IEEE International Conference on e-Business Engineering, pp. 48–53 (2007)
7. A Performance Evaluation of Local Descriptors. IEEE Transactions on Pattern Analysis and Machine Intelligence 27, 1615–1630 (2005)
8. Wren, C.R., Azarbayejani, A., Darrell, T., Pentland, A.P.: Pfinder: Real-Time Tracking of the Human Body. IEEE Transactions on Pattern Analysis and Machine Intelligence 19, 780–785 (1997)
9. Shi, J., Malik, J.: Normalized Cuts and Image Segmentation. IEEE Transactions on Pattern Analysis And Machine Intelligence 22, 888–905 (2000)
10. Viola, P., Jones, M., Snow, D.: Detecting pedestrians using patterns of motion and appearance. In: Proceedings of the Ninth IEEE International Conference on Computer Vision, pp. 734–741 (2003)
11. Veenman, C.J., Reinders, M.J.T., Backer, E.: Resolving motion correspondence for densely moving points. IEEE Transactions on Pattern Analysis And Machine Intelligence 23, 54–72 (2001)
12. Comaniciu, D., Ramesh, V., Meer, P.: Kernel-Based Object Tracking. IEEE Transactions on Pattern Analysis and Machine Intelligence 25, 564–577 (2003)
13. Blake, A., Isard, M.: Active Contours: The Application of Techniques From Graphics, Vision, Control Theory and Statistics to Visual Tracking of Shapes in Motion. Springer, London (1998)
14. Sbalzarini, I.F., Koumoutsakos, P.: Feature point tracking and trajectory analysis for video imaging in cell biology. Journal of Structural Biology 151, 182–195 (2005)
15. Jang, D.-S., Choi, H.-I.: Active models for tracking moving objects. Pattern Recognition 33, 1135–1146 (2000)

16. Zhou, Y., Rao, C., Lu, Q., Bai, X., Liu, W.: Multiple Feature Fusion for Object Tracking. In: Zhang, Y., Zhou, Z.-H., Zhang, C., Li, Y. (eds.) IScIDE 2011. LNCS, vol. 7202, pp. 145–152. Springer, Heidelberg (2012)
17. Qian, H., Mao, Y., Geng, J., Wang, Z.: Object Tracking with Self-Updating Tracking Window. In: Yang, C.C., et al. (eds.) PAISI 2007. LNCS, vol. 4430, pp. 82–93. Springer, Heidelberg (2007)
18. Huang, Z.Q., Jiang, Z.: An Object Tracking Scheme Based on Local Density. In: Cham, T.-J., Cai, J., Dorai, C., Rajan, D., Chua, T.-S., Chia, L.-T. (eds.) MMM 2007. LNCS, vol. 4351, Part I, pp. 166–175. Springer, Heidelberg (2006)
19. Welch, G., Bishop, G.: An introduction to the Kalman filter. Technical Report-TR 95-041, University of North Carolina, Department of Computer Science, USA (1995)
20. Ristic, B., Arulampalam, S., Gordon, N.J.: Beyond the Kalman Filter: Particle Filters for Tracking Applications. Artech House (2004)
21. Comaniciu, D., Meer, P.: Mean Shift: A Robust Approach Toward Feature Space Analysis. IEEE Transactions on Pattern Analysis and Machine Intelligence 24, 603–619 (2002)
22. Guyon, I., Gunn, S., Nikravesh, M., Zadeh, L.A.: Feature Extraction, Foundations and Applications. STUDFUZZ, vol. 207. Springer, Heidelberg (2006)
23. Canny, J.: A computational approach to edge detection. IEEE Transactions on Pattern Analysis and Machine Intelligence 8, 679–698 (1986)
24. Mokhtarian, F., Suomela, R.: Robust image corner detection through curvature scale space. IEEE Transactions on Pattern Analysis and Machine Intelligence 20, 1376–1381 (1998)
25. Danker, A., Rosenfeld, A.: Blob detection by relaxation. IEEE Transactions on Pattern Analysis and Machine Intelligence 3, 79–92 (1981)
26. Subirana-Vilanova, J.B., Sung, K.K.: Ridge Detection for the Perceptual Organization Without Edges. In: Proceedings of the Fourth International Conference on Computer Vision, pp. 57–64 (1993)
27. Shi, J., Tomasi, C.: Good features to track. In: Proceedings of the IEEE Conference on Computer Vision and Pattern Recognition, pp. 593–600 (1994)
28. Hastie, T., Tibshirani, R., Friedman, J.: The Elements of Statistical Learning, 2nd edn., pp. 520–528. Springer, New York (2009)

A Copy Move Forgery Detection to Overcome Sustained Attacks Using Dyadic Wavelet Transform and SIFT Methods

Vijay Anand, Mohammad Farukh Hashmi, and Avinash G. Keskar

Department of Electronics Engineering,
Visvesvaraya National Institute of Technology, Nagpur, 440010, India
{vijjanand117,farooq78699}@gmail.com,
agkeskar@ece.vnit.ac.in

Abstract. In the present digital world integrity and trustworthiness of the digital images is an important issue. And most probably copy- move forgery is used to tamper the digital images. Thus as a solution to this problem, through this paper we proposes a unique and blind method for detecting copy-move forgery using dyadic wavelet transform (DyWT) in combination with scale invariant feature transform (SIFT). First we applied DyWT on a given test image to decompose it into four sub-bands LL, LH, HL, HH. Out of these four sub-bands LL band contains most of the information we intended to apply SIFT on LL part only to extract the key features and using these key features we obtained descriptor vector and then went on finding similarities between various descriptors vector to come to a decision that there has been some copy-move tampering done to the given image. In this paper, we have done a comparative study based on the methods like (a).DyWT (b).DWT and SIFT (c). DyWT and SIFT. Since DyWT is invariant to shift whereas discrete wavelet transform (DWT) is not, thus DyWT is more accurate in analysis of data. And it is shown that by using DyWT with SIFT we are able to extract more numbers of key points that are matched and thus able to detect copy-move forgery more efficiently.

Keywords: Copy-move forgery, Dyadic Wavelet Transform (DyWT), DWT, SIFT.

1 Introduction

In present scenario due to current developments in the digital technologies, traditional concept that "seeing is believing" is no more valid. Since most of the information is preserved in digital form, and digital images can be manipulated very easily thus image forgery has become topic of serious concern. The image editing software such as Adobe Photoshop is readily available thus increasing the amount of doctored images. This phenomenon leads to serious consequences, reducing trustworthiness and creating false belief in many real-world applications. Therefore truthfulness of the images

N.T. Nguyen et al. (Eds.): ACIIDS 2014, Part I, LNAI 8397, pp. 530–542, 2014.

has to be taken seriously and thus verification of reliability and wholeness of digital images is a major issue.

Most of the forgery detection techniques are categorized into two major domains: intrusive/non-blind and non-intrusive/blind [1]. Intrusive method also known as non-blind method requires some digital information to be embedded in the original image when it is generated and thus it has a limited scope. Some of the examples of these methods are watermarking and using digital signature of the camera and not all the digital devices can provide this feature. Whereas, non-intrusive method also known as blind method does not requires any embedded information. A digital image is said to be forged when its original version is tampered by applying various transformation like that of rotation, scaling, resizing etc. It may also happen that an image is tampered by adding noise or by removing or adding some objects to hide the real information [1]. Most commonly used image forgery method is copy-move forgery in this a part of original image is copied and pasted on other part once or may be multiple times to hide some information. The ease and effectiveness of copy-move forgery makes it the most common forgery that is used to alter the content of an image [2]. Figure 1 represents a general copy- move forgery example.

Fig. 1. An example of Copy-Move Forgery

Currently we are focusing on copy-move forgery detection because of the complexity involved in forgery detection in this case. This is due to the fact that when we copy a region from an image and paste it on that very image they have similar characteristics of that of the original image. We decomposed the given test image using dyadic wavelet transform (DyWT). DyWT is shift invariant and is better than discrete wavelet transform (DWT) for data analysis [3]. After the decomposition we apply SIFT algorithm to extract the features, it obtains descriptor for color as well as boundary images as wavelet produces both. We now perform a searching to search for occurrence of same features at different part of images. Image blocks that returns similar SIFT features from all four images are marked as forged regions.

The remaining paper is presented in following manner. Next section deals with all previous work related to image forgery detection. Section 3 completely explains the proposed method. In section 4 we have done a comparative study of presented method with some of the previous methods and section 5 deals with results and simulation and in the end we have conclusion and references.

2 Related Work

The main aim of copy-move forgery is to detect the different copied regions and their pasted one. This is not an easy job because the copied part has mostly the same characteristic as that of original image and there are chances that these copied part are processed by operation like noise addition, filtering and geometrical distortion. Thus in order to correctly detect the copied region forgery detection techniques should detect the copied region even if they are slightly different.

In this section we will deal with blind methods of copy-move forgery detection. Fridrich et al. [4] introduced a block matching forgery detection method based on discrete cosine transform (DCT). It used quantized DCT coefficients as feature of image overlapping block, and then these DCT coefficients are arranged lexicographically to find matched block. It uses DCT because most of the signal energy will be contained in first few DCT coefficients and most of other coefficients are nearly zero. Thus changes made in high frequency region due to operations like noise addition and compression will not affect the low frequency coefficients. Popescu et al. [5] proposed a similar method as that of Fridrich's. It utilized Principal Component Analysis (PCA) instead of DCT so that dimension representations are reduced. Numbers of features generated are half the number of that generated by Fridrich's, thus proving to be more effective method. But it has some drawback that this method fails when the copied region is re-sampled through scaling or rotation. Li et al. [6] discussed methods which reduce the overall computation load. It first applied Discrete Wavelet Transform (DWT) to decompose the given image into four different sub-bands LL, LH, HL, and HH. Since most of the information is present in the low frequency band thus low frequency sub-band i.e. LL band is divided into overlapping blocks. By doing this number of blocks have been reduced and the overall process has speed up. On these blocks they applied SVD (singular value decomposition). Fourier Mellin Transform (FMT) was introduced by Bayram et al [7] which is applied on the image blocks. First the Fourier transform representation of each image block is obtained, and then magnitude value is re-sampled to result in log-polar coordinates. Vector representation is obtained by projecting these log-polar values onto 1-D. X. Kang et al.[8] proposed a simple method based on singular value decomposition (SVD)which is used to obtain singular values feature vectors as blocks representation of different blocks in which the given image is portioned. These feature vectors are then sorted lexicographically. This algorithm works well even in the images with uniform regions. H. Huang et al. [9] used SIFT algorithm (Scale Invariant Feature Transform) to represent the features. SIFT algorithm is invariant to changes in illumination, rotation, scaling, etc. Amerini et al. [10], deals with detecting whether an image has been forged or not specially using copy-move forgery. In this a novel method for detecting image forgery using Scale Invariant Features Transform (SIFT) is proposed. SIFT allow to understand that copy-move forgery has occurred and it also recover from geometric transformation used for cloning. Using this method we can also deal with multiple copy-move forgery. Muhammad et al. [3] proposed a robust and blind copy-move forgery detection technique using dyadic (undecimated) wavelet transform (DyWT). Al-Qershi et al. [15] provided a state-of-art of passive detection of copy-move forgery in digital images. The key currents issues are discussed for developing robust passive copy-move forgery detection. Leida Li et al. [16] proposed an efficient method for detecting copy-move forged images using local binary patterns, in this image is first image is filtered and then divided into overlapping circular blocks, features of these blocks are calculated using local binary patterns to detect the forged regions. Birajdar et al [17] summarized a complete survey on digital image forgery detection using passive techniques, which discussed currently available forgery detection techniques and also provide further recommendation for future research. Mahalakshmi et al [18] provided digital image forgery detection by exploring basic image manipulations. Here they have presented techniques to detect image is manipulated using basic method like copy-move, region duplication, splicing etc.

3 Proposed Algorithm

Further we analyze copy-move forgery by introducing a new technique for forgery detection. On the contrary to SIFT features that are used for feature matching, we introduce Wavelet features for the same. Image is transformed to Wavelet domain and SIFT features are extracted from decomposed images. As wavelet produces multispectral images, features are more predominant [11]. Obtaining Similarity of features within same image is extracted as basis of copy paste forgery detection. Our works confirm that SIFT features are an optimal solution because of their high computational efficiency and robust performance. An ideal detection algorithm should detect image forgery after some geometrical transformation like rotation or scaling and also other manipulations like addition of Gaussian noise, JPEG compression etc. Since our method is Shift invariant, forgery can be detected even if the copied part is scaled or rotated or both.

3.1 Dyadic Wavelet Transform (DyWT)

In section 2 we have seen that many previous techniques use DWT for detecting copy-move forgery detection. But DWT has its own drawback like it is shift invariant and thus less optimal for data analysis. In order to overcome this drawback of DWT, Mallat and Zhong [12] introduced the DyWT. DyWT is shift invariant and is different from DWT because in DyWT there is no down-sampling like that of DWT. Next we are going to explain in detail about DyWT. Let us begin with taking I as the given image for decomposition, and h[k] be scaling (low pass) and g[k] be wavelet (high pass) filters. The DyWT of an image can be computed using the following algorithm [1].

Start at scale j = 0, and take $I^{o} = I$, and compute the scaling and wavelet coefficients at scales j = 1, 2, …, J using Equations. (1) and (2):

$$c^{j+1}[n] = \sum_{k} h[k] c^{j}[n + 2^{j} k] \tag{1}$$

$$d^{j+1}[n] = \sum_{k} g[k] c^{j}[n + 2^{j} k] \tag{2}$$

Let $h^j[k]$ and $g^j[k]$ be the filters obtained by inserting $2^j - 1$ zeros between the terms of h[k] and g[k]. Then we can perform DyWT using filtering as follows:

- Start with *I*, which is assumed to be at scale zero, i.e. $I^o = I$
- To obtain the scaling and wavelet coefficients I^j and D^j at scales *j* = 1, 2, …, *J*
- Filter I^{j-1} with $h^{j-1}[k]$
- Filter I^{j-1} with $g^{j-1}[k]$

Single level decomposition of DyWT is shown in figure 2 and figure 3 represents one level DyWT decomposition of a 2-D image.

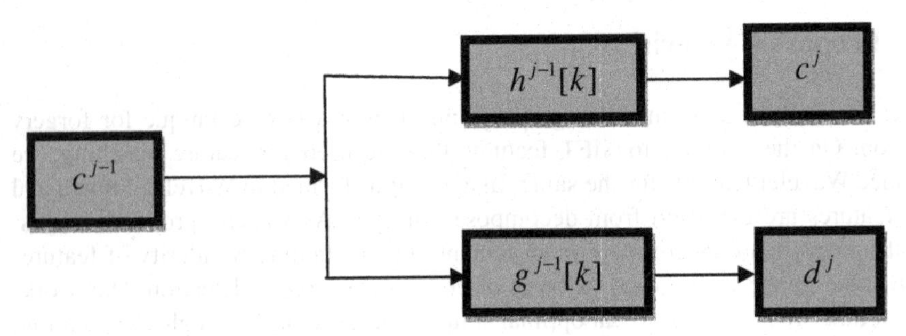

Fig. 2. DyWT single level decomposition

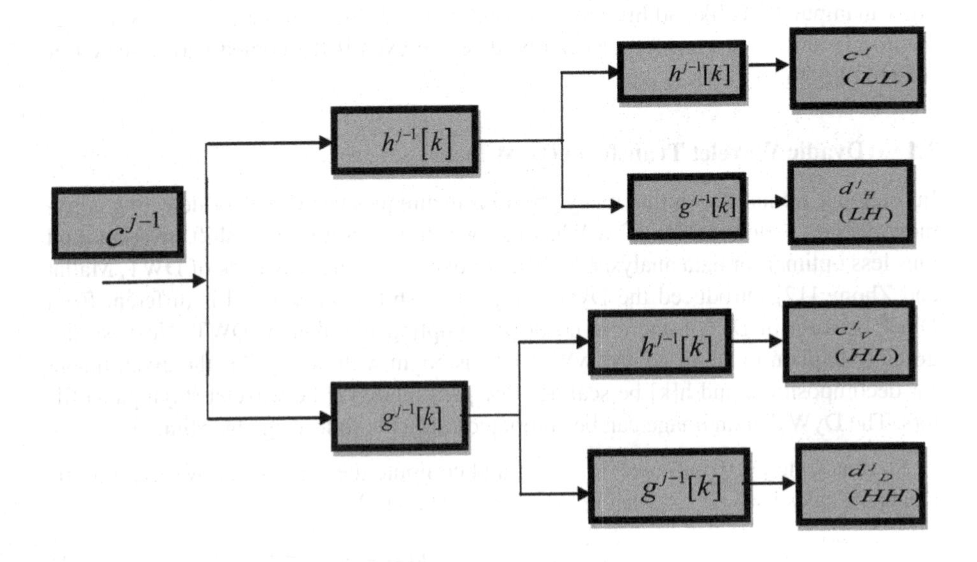

Fig. 3. Single level DyWT decomposition of a 2D image

3.2 Algorithm Combining DyWT and SIFT

For copy-move forgery detection, we first apply dyadic wavelet transform to test image. It produces four different components i.e. HH, HL, LH and LL. We perform SIFT descriptor extraction on LL part only. It obtains descriptor for color as well as boundary images as wavelet produces both. We now perform a searching to search for occurrence of same features at different part of images. Image blocks that returns similar SIFT features are marked as forged regions. The process flow chart is as shown in figure 4.

Step I.
The image inputted by the user is read. Proposed algorithm uses RGB image but can also be applied on gray-scale images.

Step II. *Dyadic Wavelet Transform (DyWT):*
As shown in figure 2 and 3, which depicts how DyWT is different than DWT. Size of the image is reduced at every level by the DWT transform. But using DyWT the size of image remains same at different level since there is no down sampling as it is in DWT. The image is reduced in to 4 sub images at each level which are labeled as LL, HL, LH, and HH. Most of the data is concentrated in the LL sub-image and it is considered as the approximation of the image. It corresponds to the coarse level coefficients of the original image. It is the LL sub-image which is decomposed in to four sub-images at the next level.

Step III. *Scale Invariant Feature Transform (SIFT):*
SIFT is used to extract distinctive features from an image. A very important advantage of this transform is that it provides a set of features which are invariant to scaling and rotation [14]. These features are extracted by applying cascade filtering approach which consists of four stages as explained below.

- Scale-space Extrema Detection
  The scale space of an image is defined as a function L(x, y, σ), that is produced by convolving G(x, y, σ) which is a variable-scale Gaussian with an input image I(x, y):

$$L(x,y,\sigma) = G(x,y,\sigma) * I(x,y) \tag{3}$$

Lowe et al. [14] proposed to use scale-space extrema in the difference-of-Gaussian function D(x, y, σ) to detect stable key point locations in scale-space, which can be calculated by taking difference of two nearby scales separated by a constant multiplication factor k as shown by equation (4).

$$D(x,y,\sigma) = L(x,y,k\sigma) - L(x,y,\sigma) \tag{4}$$

After getting difference-of-Gaussian we need to find the local extrema which is nothing but the key point or interest point, for this each pixel is compared to its eight neighboring pixel on the same scale and with nine neighboring pixels on the scale above as well as scale below it. And the pixel is selected only if it is either maximum or minimum of all its 26 neighbors.

- Key-point Localization

This step is required to remove key points which have low contrast or poor localized pixel along the edge. For low contrast points we use Taylor's series expansion of DoG (difference-of-Gaussian) , maxima or minima of the expansion up to two term is calculated and only those points are retained which are greater than some threshold. And for edge response elimination we use Hessian detector. We first calculate Hessian of DoG, from the Hessian matrix we obtain trace as well as determinant of Hessian matrix which are sum and multiplication of eigenvalues. We can remove the point by taking ratio of trace square and determinant and setting a particular threshold.

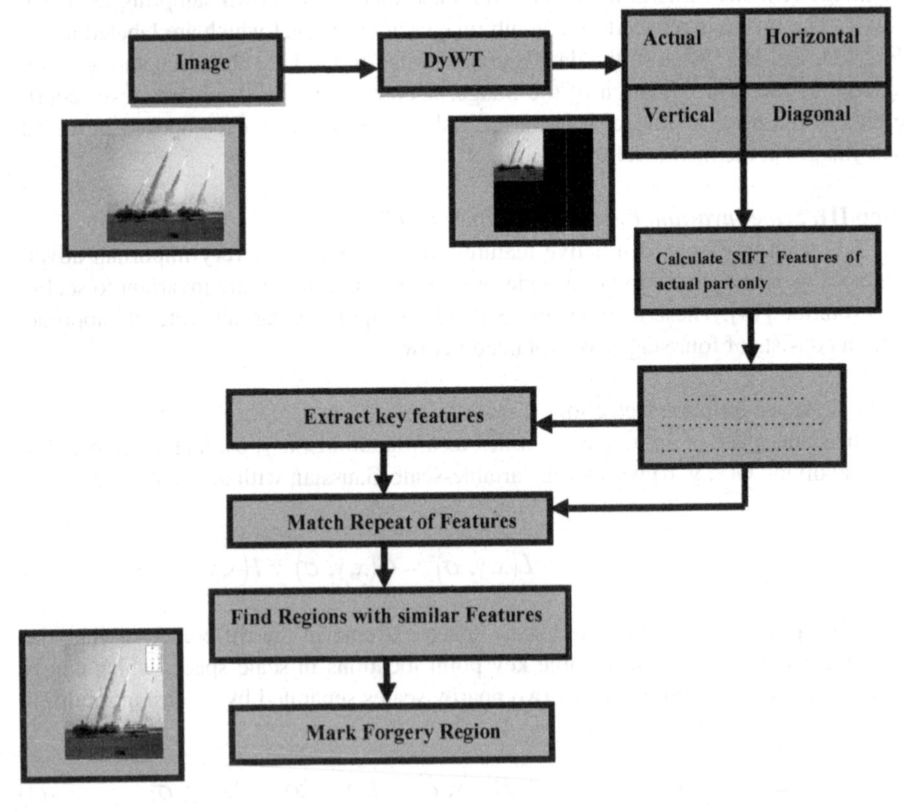

Fig. 4. Block diagram of proposed algorithm

- Orientation Assignment

Orientation is assigned to each key point based on local image properties, thus by assigning orientation to key point we are making it invariant to rotation. For each image sample L(x, y) at a scale the gradient magnitude m(x, y) and orientation θ(x, y) is by the equation given below [14]:

$$m(x,y)=\sqrt{(L(x+1,y)-L(x-1,y))^2+(L(x,y+1)-L(x,y-1))^2} \quad (5)$$

$$\theta(x,y)=tan^{-1}\left(L(x,y+1)-L(x,y-1)/L(x+1,y)-L(x-1,y)\right) \quad (6)$$

- Key-point Descriptor

For Key-point descriptors gradient orientation histogram is used, this will provide robust representation. First compute relative orientation and magnitude in 16*16 neighbourhoods at key points, divide the 16*16 region into 4*4 blocks so we have a6 such blocks. Then find histogram of each block consisting of 8 bins. Finally concatenate these 16 histograms in one single long vector of dimension 128.

4 Comparative Analysis

We have compared our proposed algorithm with DyWT, DWT combined with SIFT Methods for detecting copy-move forgery

4.1 Dyadic Based Algorithm

Ghulam Muhammad in [13] proposed an algorithm based on DyWT, which first decomposes the given image by DyWT and then LL region is divided into overlapping blocks and then similarities is checked based on Euclidean distance between the blocks where as in HH part that is high frequency region dissimilarities is checked based on Euclidean distance. Since this method uses block matching techniques it is quite computationally complex and time consuming above all we have fix the threshold in this method . This method is not robust to various attacks like scaling and rotation.

4.2 DWT and SIFT

In previous literature we have seen that SIFT algorithm is used to extract key features and this algorithm can withstand geometrical attacks. Next we tested the given database with a technique which combines DWT and SIFT. DWT is a multi-resolution technique which extract feature like edges and corner perfectly. First DWT is applied on the given image which decomposes the given image into four sub-bands low frequency band containing most of the information and high frequency band containing details. Then we apply SIFT on the low frequency band part to first extract the features and then key-points matching is done to determine the regions where the forgery has occurred [19].

4.3 DyWT and SIFT

This algorithm is based on dyadic wavelet transform [DyWT]. DyWT is invariant to shift and captures the structural information in a better way than discrete wavelet transform. First DyWT is applied on the image and then for feature extracting feature we use SIFT algorithm. It may happen that two regions of the given image may have same features so we go for feature clustering and forgery detection as discussed in [10]. We conclude that key feature extracted using DyWT and SIFT combined with RANSAC algorithm are more than what we got using DWT and SIFT. And the proposed method can sustain various attacks like scaling, rotation, noise addition and any combination of these attacks.

4.4 Testing Robustness against Various Attacks

In order to verify our proposed method we applied our method on MICC-F220 database images which contains non-tampered images as well as tampered one having scaled version and some have rotation and some have combination of two or more attacks. Figure 5 shown below describe the forgery detection results under various attacks.

Thus from above results we are going to compare our proposed method with previous methods in Table 1.

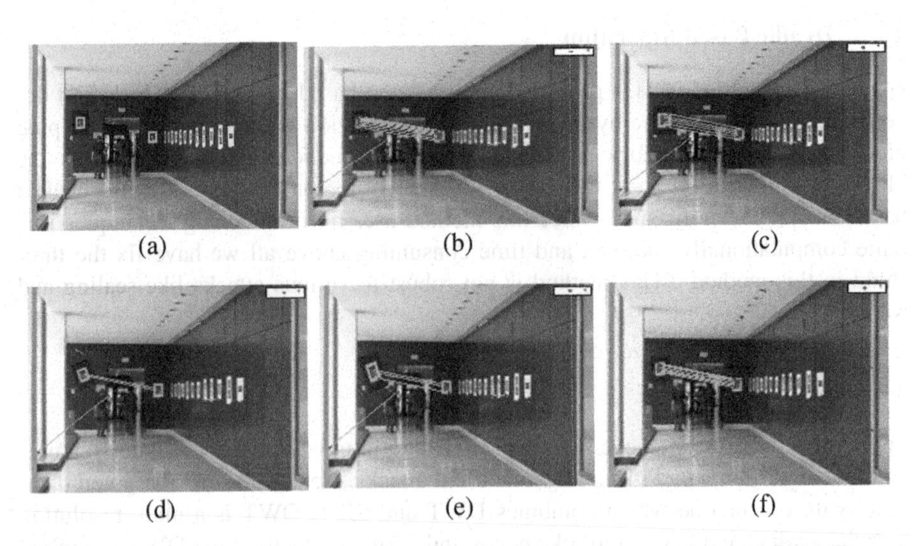

Fig. 5. (a) Tampered test image, **5.** (b) Forgery detection of tampered image, **5.** (c) Forgery detection of scaled version, **5.** (d) Forgery detection of rotated version, **5.** (e) Forgery detection of combination of attacks, **5.** (f) Forgery detection of tampered image with addition of noise

Table 1. Comparison of proposed method with previous existing methods

Method	Technique used	Feature extracted	Rotation	Scaling	Noise	Combination of attacks	Computational complexity
Muhammad et.al [1]	DyWT and block matching	Intermediate	Less robust	No	No	No	More
DWT and SIFT [19]	DWT and then SIFT on LL part	High	Intermediate	average	Low	Low	Less
DyWT and SIFT	DyWT and then SIFT on LL part	Highest	More robust	More robust	High	High	Intermediate

5 Simulation Results and Discussions

In result analysis, we tested over a 100 test images collected from the web sources and database of different formats in MATLAB. This simulation has been performed on MATLAB 2010a software with 4GB Ram and core i5 processor. Test images are taken and then tampered using Adobe Photoshop software. This algorithm is also tested and verified with database of MICC-F220 [10]. DyWT of the test image is first calculated and on the low frequency component of DyWT we apply SIFT to extract the features which are nothing but descriptor vectors of the object of interest in the test image and the final step is to go matching these features to detect copy- move forgery.

Images shown in figure 6 describe various process involved in proposed method, since the test image as shown in figure 6(a) is a forged image because original image as shown in figure 1 has only three missile, so we applied the proposed algorithm on the test image shown in figure 6(a) we got 748 key-points and total match of 46 in the elapsed time of 0.962806 seconds.

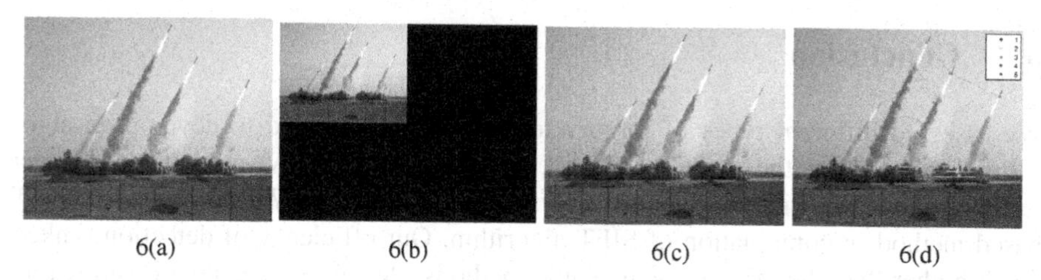

6(a) 6(b) 6(c) 6(d)

Fig. 6. (a) Tampered test image, **6.**(b) DyWT of test image, **6.**(c) LL or approximation part of DyWT, **6.**(d) Matching of similar feature using DyWT and SIFT

Images shown in figure 7 describe various process involved in DWT and SIFT method, and the test image remain as shown in figure 7(a) is a forged image, so we applied DWT and SIFT algorithm on the test image shown in figure 7(a) we got 567 key-points and total match of 34 in the elapsed time of 0.832993seconds.

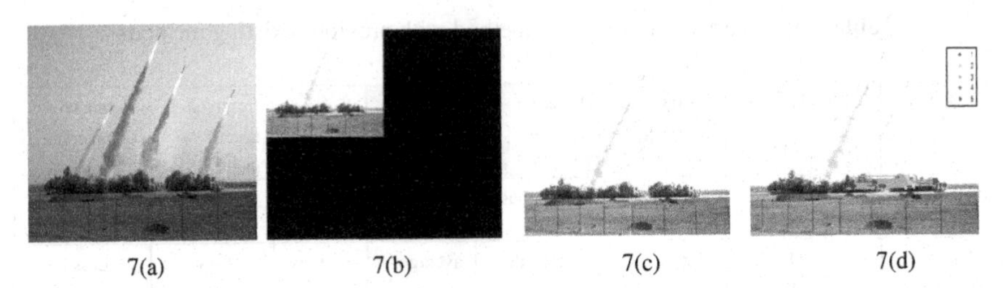

7(a) 7(b) 7(c) 7(d)

Fig. 7. (a) Test image, **7.**(b) DWT of test image, **7.**(c) LL or approximation part of DWT, **7.**(d) Matching of similar features using DWT and SIFT

Thus after performing experiment on database of MICC-F220 which consists of 220 images of which half of the images are tampered. We obtained Figure-of-merit such as precision rate, recall rate and false positive rate (FPR) and tabulated the result obtained in Table 2.

Table 2. Figure-of-merit for copy-move forgery detection algorithm

Method	PRECISION RATE (p)	RECALL RATE (r) or TPR(TRUE POSITIVE RATE)	FPR(FALSE POSITIVE RATE)
SIFT	95%	74%	4%
DWT + SIFT	97%	66%	2%
DyWT + SIFT	88%	80%	10%

6 Conclusion

We assessed different types of forgery techniques and decided to develop an algorithm for detecting the most common copy-move forgery. We reviewed different techniques and algorithm developed previously for the same. We proposed a DyWT based method in combination of SIFT algorithm. Our efficiency of detection ranked much higher than the previously available methods. Recall rate i.e. True Positive rate of our proposed algorithm is far better than previous results. Since DyWT does not perform down sampling so the image size is intact and low frequency components are having most of the information thus we applied SIFT on it to extract features and then match between the descriptor vector of different key-points will determine that copy-move forgery has been done to the given image. Our algorithm has higher matching rate and it is robust to most of the attack and pre-processing techniques we can conclude that it's a feasible one.

References

1. Muhammad, N., Hussain, M., Muhammad, G., Bebis, G.: Copy-move forgery detection using dyadic wavelet transform. In: Proceedings of IEEE Eighth International Conference on Computer Graphics, Imaging and Visualization (CGIV 2011), pp. 103–108 (2011)
2. Jing, L., Shao, C.: Image Copy-Move Forgery Detecting Based on Local Invariant Feature. Journal of Multimedia 7(1), 90–97 (2012)
3. Muhammad, G., Hussain, M., Khawaji, K., Bebis, G.: Blind copy move image forgery detection using dyadic undecimated wavelet transform. In: Proceedings of IEEE 17th International Conference on Digital Signal Processing (DSP 2011), pp. 1–6 (2011)
4. Fridrich, A.J., Soukal, B.D., Lukáš, A.J.: Detection of copy-move forgery in digital images. In: Proceedings of Digital Forensic Research Workshop, Cleveland, OH, USA, pp. 55–61 (August 2003)
5. Popescu, A.C., Farid, H.: Exposing digital forgeries by detecting duplicated image regions, Department of Computer Science, Dartmouth College. Technical Report. TR2004-515 (August 2004)
6. Li, G., Wu, Q., Tu, D., Sun, S.: A sorted neighbourhood approach for detecting duplicated regions in image forgeries based on DWT and SVD. In: Proceedings of IEEE International Conference on Multimedia and Expo, pp. 1750–1753 (2007)
7. Bayram, S., Sencar, H.T., Memon, N.: An efficient and robust method for detecting copy-move forgery. In: Proceedings of the IEEE International Conference on Acoustics, Speech and Signal Processing (ICASSP 2009), pp. 1053–1056 (2009)
8. Kang, X., Wei, S.: Identifying tampered regions using singular value decomposition in digital image forensics. Proceedings of the IEEE International Conference on Computer Science and Software Engineering 3, 926–930 (2008)
9. Huang, H., Guo, W., Zhang, Y.: Detection of copy-move forgery in digital images using SIFT algorithm. In: Proceedings of Pacific-Asia Workshop on Computational Intelligence and Industrial Application (PACIIA 2008), vol. 2, pp. 272–276 (2008)
10. Amerini, I., Ballan, L., Caldelli, R., Bimbo, A.D., Serra, G.: A sift-based forensic method for copy–move attack detection and transformation recovery. IEEE Transactions on Information Forensics and Security 6(3), 1099–1110 (2011)
11. Amerini, I., Ballan, L., Caldelli, R., Bimbo, A.D., Tongo, L.D., Serra, G.: Copy-move forgery detection and localization by means of robust clustering with J-linkage. Signal Processing: Image Communication 28, 659–669 (2013)
12. Mallat, S., Zhong, S.: Characterization of signals from multiscale edges. IEEE Transactions on Pattern Analysis and Machine Intelligence 14(7), 710–732 (1992)
13. Muhammad, G., Hussain, M., Bebis, G.: Passive copy move image forgery detection using undecimated dyadic wavelet transform. Digital Investigation 9(1), 49–57 (2012)
14. Lowe, D.G.: Distinctive image features from scale-invariant keypoints. International Journal of Computer Vision 60(2), 91–110 (2004)
15. Al-Qershi, O.M., Khoo, B.E.: Passive detection of copy-move forgery in digital images: State-of-the-art. Forensic Science International 231(1), 284–295 (2013)
16. Li, L., Li, S., Zhu, H., Chu, S.-C., Roddick, J.F., Pan, J.-S.: An Efficient Scheme for Detecting Copy-move Forged Images by Local Binary Patterns. Journal of Information Hiding and Multimedia Signal Processing 4(1), 46–56 (2013)
17. Birajdar, G.K., Mankar, V.H.: Digital image forgery detection using passive techniques: A survey. Digital Investigation 10(3), 226–245 (2013)

18. Devi Mahalakshmi, S., Vijayalakshmi, K., Priyadharsini, S.: Digital image forgery detection and estimation by exploring basic image manipulations. Digital Investigation 8(3), 215–225 (2012)
19. Hashmi, M.F., Hambarde, A.R., Keskar, A.G.: Copy Move Forgery Detection using DWT and SIFT Features. In: Proceeding of 13th IEEE International Conference on Intelligent Systems Design and Applications (ISDA 2013), pp. 188–193 (December 2013)

Methods for Vanishing Point Estimation by Intersection of Curves from Omnidirectional Image

Danilo Cáceres Hernández, Van-Dung Hoang, and Kang-Hyun Jo

Intelligent Systems Laboratory, Graduate School of Electrical Engineering,
University of Ulsan, Ulsan 680-749, South Korea
{danilo,hvzung}@islab.ulsan.ac.kr, acejo@ulsan.ac.kr
islab.ulsan.ac.kr

Abstract. In this paper, the authors propose solutions for finding the vanishing point in real time based on the Random Sample Consensus (RANSAC) curve fitting and density-based spatial clustering of applications with noise (DBSCAN). First, it was proposed to extract the longest segments of lines from the edge frame. Second, a RANSAC curve fitting method was implemented for detecting the best curve fitting given the data set of points for each line segment. Third, the set of intersection points for each pair of curves are extracted. Finally, the DBSCAN method was used in estimating the VP. Preliminary results were gathered and tested on a group of consecutive frames undertaken at Nam-gu, Ulsan, in South Korea. These specific methods of measurement were chosen to prove their effectiveness.

Keywords: Autonomous ground navigation, RANSAC, DBSCAN, Vanishing point, Omnidirectional image.

1 Introduction

Autonomous ground navigation is still facing important challenges in the field of robotics and automation due to the uncertain nature of the environments, moving obstacles, and sensor fusion accuracy. Therefore, for the purpose of ensuring autonomous navigation and positioning along the environments aforementioned, a visual based navigation process is implemented by using an omnidirectional camera. Based on the perspective drawing theory, one VP is a point in which a set of parallel lines converge and disappear into the horizon. Ultimately, by using VP for indoor and outdoor 2D images and roads, as well as corridors can be described as a set of orthogonal lines. These particular sets of parallel lines typically exhibit the structures in the 3D scenes shown in figure 1. In the case of omnidirectional scenes, parallel lines are projected as curves that converge and disappear into the horizon. Therefore, in order to address the challenges of VP detection in omnidirectional scenes, the authors decided to present an iterative VP detection based on RANSAC [1] and DBSCAN [2] approaches. To this end, the main contributions of the presented method are:

N.T. Nguyen et al. (Eds.): ACIIDS 2014, Part I, LNAI 8397, pp. 543–552, 2014.

- Implementation of real time RANSAC curve fitting algorithm for VP detection.
- Implementation of DBSCAN, due that the number of cluster in the image are not specified.

In fact, there are several approaches for estimating VP in the field of Intelligent Systems. However, those estimation approaches are based on the constraint that the estimate vanishing point is localized in the front of the designed system. In order to decrease these limitations, omnidirectional camera systems were used. The authors are proposing to extract the cluster which contains the large amount of points located in either the front most or rear most part of the mobile robot systems. The rest of this paper is structured as follows: (2) Related Works (3) Proposed Method, (4) Experimental Result, (5) Conclusions and Future Works.

2 Related Works

In the field of autonomous navigation systems, efficient navigation, guidance and control design are critical in averting current challenges. When referring to the need for estimation of rotation angle for automatic control, one VP plays an important role. By detecting VP in the 2D image, autonomous unmanned systems are able to navigate towards the detected VP. Several approaches for VP detection can be used for detection, for example, Hough transform (HT), RANSAC algorithm, dominant orientation, and lastly the equivalent sphere projection were all during experimentation. In the case of HT, for estimating lines into the 2D images [3] propose the randomized HT to estimate the lane model based VP voting, [4] propose a VP detection method based HT and K-mean algorithm mainly based on the straight lines orientation given by the edges of corridors. In [5] authors use the RANSAC approach to describe a parametric model of the lane marking into the image. The clear examples of dominant orientation approaches are [6][7] in which authors detected the VP by using a bank of 2D Gabor wavelet filters with a voting process. In the case of spherical representation, [8] [9] authors used the 3-D line RANSAC approach in real time.

3 Proposed Method

Essentially, the proposed method consists mainly of extracting the information surrounding the ground plane. Frames are extracted in short time intervals that started just before the earliest detection from the video-capture sequence. In this section, the proposed algorithm has three steps: (1) extracting line segments, (2) RANSAC curve fitting, (3) cluster extraction by DBSCAN.

3.1 Extracting Line Segments

Road scenes can be described as structures that contain lane marking, soft shoulders, gutters, or a barrier curb. Therefore, in 2D images these features are represented as a set of connected points. The main idea of this section consisted of extracting the longest line segments around the ground by applying the canny edge detector [10] from a group of consecutive frames captured during research in Nam-gu, Ulsan, South Korea. After the edge detection step, the subsequent task was to remove the smallest line segments by extracting the longest line segments after applying canny edge detection. As a result a set of contours are extracted from the road scene. In Fig. 1(b) clearly we can see that most of these contours are not closed curves. In that sense, for each possible curve candidate the extracted edges are coded by using a 3-connectivity scheme. In this step the algorithm scan the binary image from bottom to top until it reaches an edge pixels. Then, each pixel value which is located in front of the current position is classified as a member or not of the line segment. This process is repeated until the values in front of the pixel no remaining an edge pixels. , the code were able to extract the basic information such as: length, number of point per line segments, as well as the pixel position location of the points of each line segments into the image plane. As a result, extraction from an image sequence the set of longest line segments was completed, see Fig.1(c).

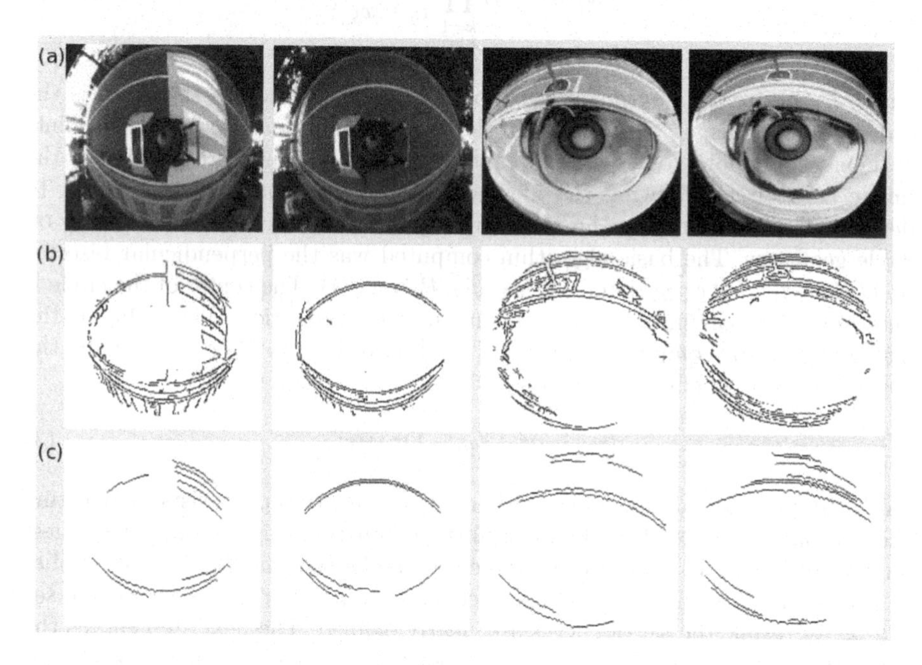

Fig. 1. Determining longest line segments.(a) Selected frames sequences taken at different time steps. (b) Edge detection results. (c) Extracting longest line segments after applying edge detection step.

3.2 RANSAC Curve Fitting

At this point the set of line segments in the image plane are known. Hence, the new task consists of defining the function for each line segment. The process starts at every iteration by selecting a set of three random points for each line segment. This will help to describe the function that determines the curvature of the lines. As far as we know, in numerical analysis there are various approaches for solving a polynomial system. For example, in [9] authors use polygonal approximation to extract that lines segments. On the other hand, in [11] authors use an Auzinger and Stetter method. In that sense, the main contribution of the authors work is to estimate VP in real time processing, required by autonomous navigation. To this end, we compare two different methods for curve fitting. First, Lagrange polynomials are used to develop the fitting polynomial. For example: given the data set of points $(x_1, y_1 = f(x_1)), (x_2, y_2 = f(x_2)), ..., x_n, y_n = f(x_n))$ the interpolation polynomial $(f_n(x)$is defined as follows:

$$f_n(x) = \sum_{j=1}^{n} L_j(x) f(x_j) \tag{1}$$

$$L_j(x) = y_j \prod_{\substack{k=1 \\ k \neq j}}^{n} \frac{x - x_k}{x_j - x_k} \tag{2}$$

where $L_j(x)$ are the Lagrange basis polynomials associated to the x_k, n is the order polynomials, y_j is a value at the corresponding x_j for all data points j. The second point relies on the concept of parallel lines projected into the image as a circle following the idea of sphere projection model. In order to define the curve model by the three given points, researchers relied mainly on circle geometry. The basic algorithm computed was the perpendicular bisector of: $(P_1(x_1, y_1), P_2(x_2, y_2))$ and $(P_1(x_1, y_1), P_3(x_3, y_3))$, The center of the circle is determined by solving the perpendicular bisectors equation. The radius is the distance from the computed center of a candidate circle to $P_1(x_1, y_1)$, finally the circle equation is defined as follows:

$$(x_i - a)^2 + (y_i - b)^2 = r^2 \tag{3}$$

Once the polynomials have been disclosed, the next step consists of selecting the appropriate curve fit model by finding the function that contains the largest number of inliers. In other words, the curve model that has the largest number of points based on the given data set becomes the points that are extracted, see Fig. 2. As result, the set of curves are clearly defined. The next step entails the extraction of the data set of intersection point by finding the same for each pair of defined curves at the point where they intersect, see figure 3.

3.3 Cluster Extraction by DBSCAN

Given the data set of intersection point from the previous step, the algorithm should be able to extract clusters. From figure 3, it is clear that the projected data into the image plane give us vital information about the data set points, as follows:

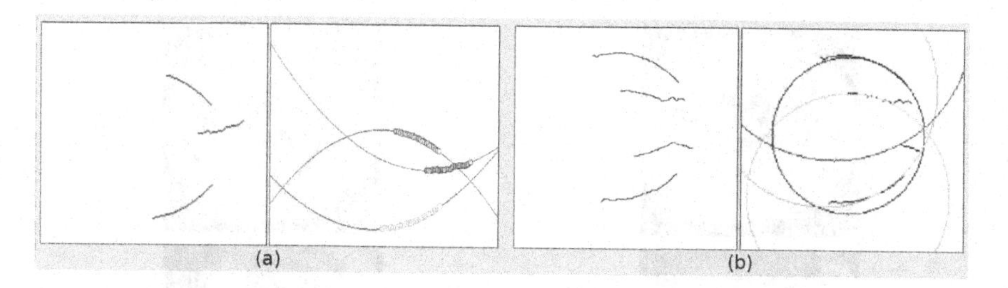

Fig. 2. Polynomial fitting curve results. (a) Lagrangian polynomial interpolation. (b) Circle model.

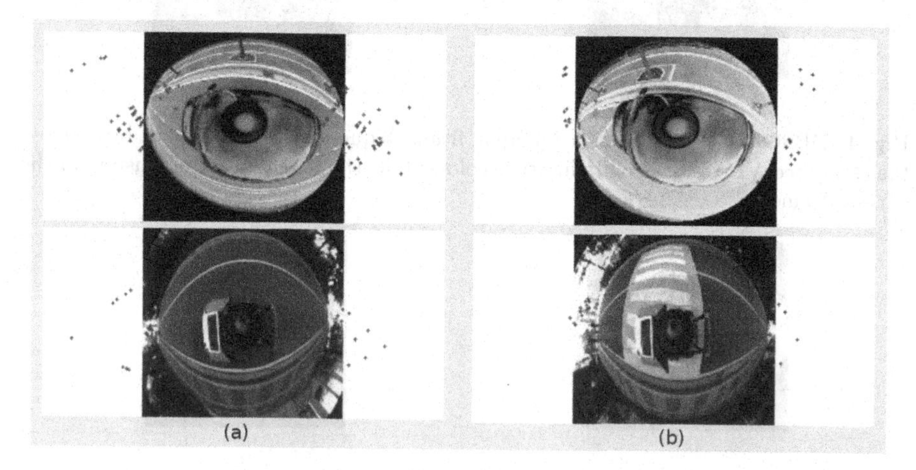

Fig. 3. Intersection point results. (a) Lagrangian polynomial interpolation. (b) Circle model.

- The data does not depict a well-defined shape.
- Presence of noise in the data due to the lack of a pre-processing model for road analysis. To compensate, all line segments are considered as a part of the road.
- The number of cluster could not be described in advance.

Considering the above mentioned occurrences, among the various clustering algorithms it was proposed to use the DBSCAN algorithm Shown in [9] ; an unsupervised method was used, due to the algorithm having achieved a good performance with respect to some other algorithms. In essence, the main idea of DBSCAN is that for each point of a cluster, the neighborhood of a given radius has to contain a minimum number of points. The DBSCAN algorithm depends mostly on two parameters:

- Eps: number of points within a specified radius
- MinPts: minimum number of points belonging to the same cluster.

After the candidate clusters are formed, the idea is to define the cluster that contains the largest amount of points. Then, the centroids of the cluster are

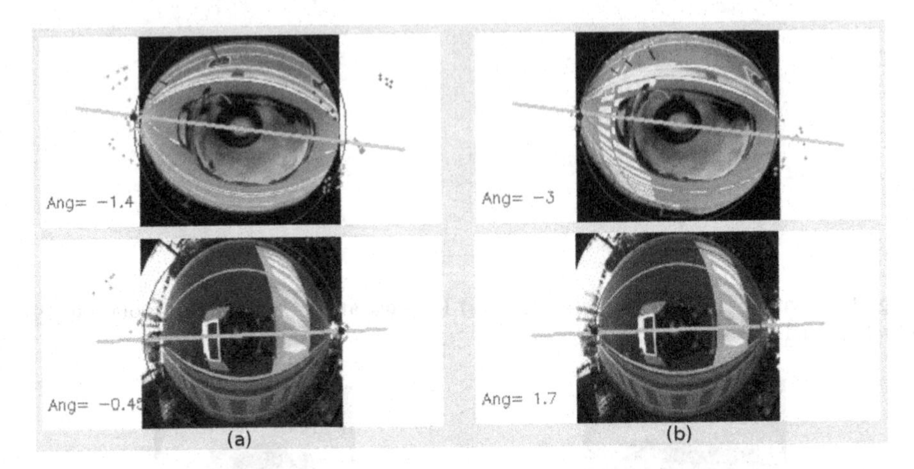

Fig. 4. DBSCAN Result from a different frame sequences. The point in blue shows the estimated VP result. Points in other colors show the detected set of clusters in the current frame.

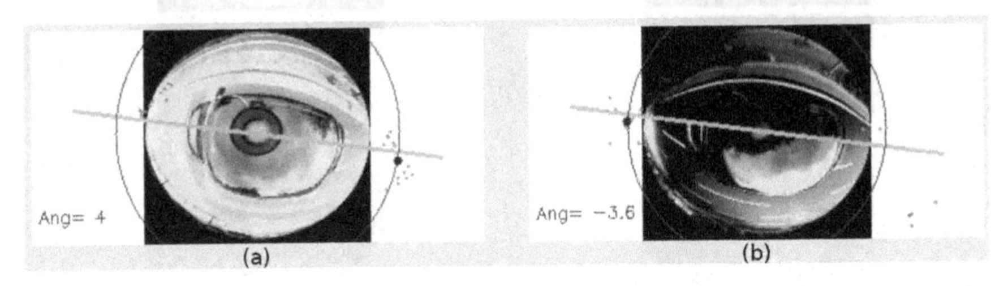

Fig. 5. Experimental results for a frame sequences with problems given by the illumination distribution. (a)frame sequences with lens flare problem. (b)frame sequences with strong sunlight and dark shadows areas.

calculated, see figure 4. It is important to remark that one of the main advantages of using omnidirectional cameras lies in the fact that the possibility to extract clusters which are located either in the front most or rear most part of the mobile robot systems, thereby helping the estimation of the VP. Thus, it is of the utmost importance in determining the VP in frames which are affected by an illumination distribution, such as strong sunlight or dark shadows, see figure 5. On the other hand, once the VP has been extracted, the angle between the line formed by the VP and the center point of the omnidirectional camera, in respect to the x-axis of the coordinate system (which passes through the center point of the omnidirectional camera) is now able to be accurately computed, see figure 4 and figure 5.

4 Experimental Result

In this section, the ending results of the experiment will be introduced. All of the experiments were done on Pentium Intel Core 2 Duo Processor E4600, 2.40

GHz, 2 GB RAM. Implementation was done in C++ under Ubuntu 12.04. The algorithm used a group of 4,355 consecutive frames taken at Namgu (Munsu) as well as 1,933 frames taken at the University of Ulsan (UOU), South Korea with a frame resize to 160x146 pixels. The consecutive frame depicts urban road scene images. The chosen path contains a set of good examples of different road geometric designs, such as crossroad, turning lanes, sag and crest curves, fork, straight roads, and both underpass and overpass sections. On the other hand, the scene also contains different urban objects like cars, trees, buildings, as well as frames with problems given by the illumination distribution, strong sunlight and dark shadows. The depicted results are to the implementation of the RANSAC curve fitting and DBSCAN. The experimental result of the proposed method is shown in the closing set of images, see table 1 as well as figure 6 and figure 7. Table 1 shows the computing time of processing in msec. Figure 6 shows the result of time processing for a set of 100 consecutive frames.

Table 1. Computing time For main phase of processing

Process (msec.)	UoU[a]	UoU[b]	Munsu[a]	Munsu[b]
Capture	3.30	2.62	3.05	3.09
Preprocessing	5.01	4.56	3.47	4.18
Line extracting	24.23	23.04	24.87	23.29
RANSAC	9.15	9.50	11.27	6.75
DBSCAN	2.98	3.27	6.23	4.97
Total	44.67	42.99	48.91	42.28

[a] Polynomial case.
[b] Circle case.

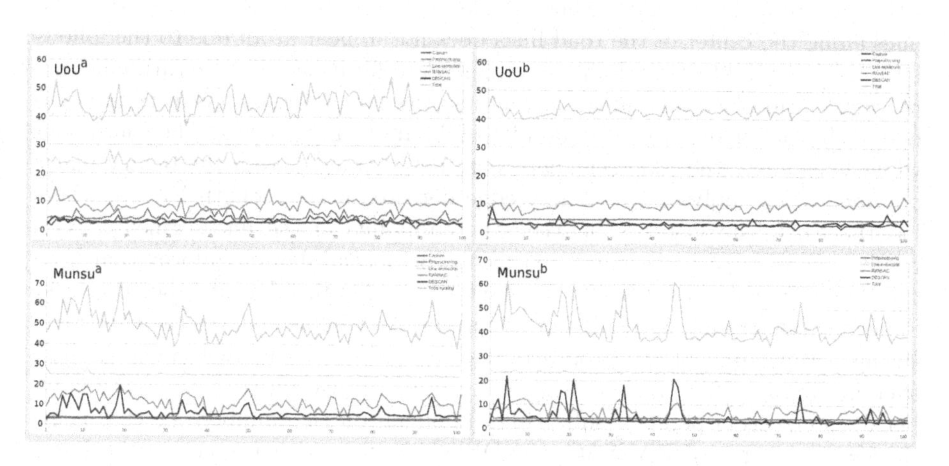

Fig. 6. Frame-rate image sequences; [a] shows the polynomial case, while [b] shows the circle case. Data in orange represent the preprocessing time. Data in yellow represent the line extraction processing time. Data in green represent the RANSAC processing time. Data in red represent the DBSCAN processing time. Data in blue sky represent the total running time.

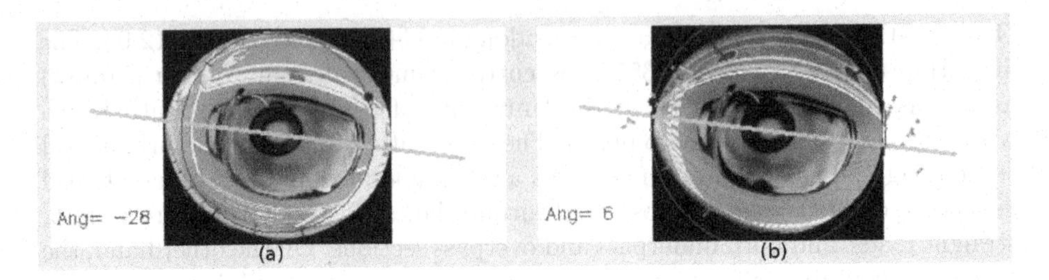

Fig. 7. Experimental results for a frame sequences with false VP estimation. Points in blue represent the VP candidate.

5 Conclusions and Future Works

5.1 Conclusions

The preliminary results of the proposed method presented relevant information for finding the VP estimation in road scenes. Specified imaging results show the performance of our proposed work over a set of possible scenarios. For example, in Figure 5 the proposed algorithm works well despite the problems given by the illumination distribution, strong sunlight, and dark shadows. Initially, it could be concluded that these problems did not appear to affect the image. Thus, the algorithm was able to estimate the VP efficiently. In comparison to, Figure 8 shows the result of a false VP estimation. The images show the VP extracting from the largest amount of polynomial segments located on the top of the camera system in contrast to the bottom side. These results indicate that the vanishing point has an error in places where single or double continuous white or yellow lines along the center of the road lines does not appear at all due to the following: weather conditions, wildlife, pavement deterioration or where turn lanes are not marked in places like junctions, intersection, etc. Part of this discrepancy is due to the fact that picture looses a lot of detail during resizing. The continuous white lines along the road segments were not extracted from the binary image obtained after applying the Canny edge detector with a length $l \geq 25$, where l is the Euclidean distance between the endpoints of the line segments. As a result, the VP had to be extracted using the spatial density information from the side strip areas (bicycle and/or planter) or sidewalks. Finally, the circle case indicates a better fitting model for estimating VP. Table 1 shows the time measuring performance of our proposed algorithm; for two different cases the best average response time had been given by the circle case model. Since the number of intersection points results (Fig.3) depend on the number detected edge, the response time of RANSAC and DBSCAN vary with respect to the intersection points number involved in the process. On the other hand, the detected horizon region show better response in the circle model than the polynomial model. Finally, VP estimation should be consider crucial for autonomous navigation. To this end, having analysed and evaluated the VP estimation by using Lagrange interpolation polynomial and and circle approaches, and taking into consideration both the processing time (Table 1) and the detected horizon region, the circle case shows good performance for real-time vanishing point estimation.

5.2 Future Works

As a result, the proposed algorithm is able to determine the VP in the 3D space based on RANSAC curve fitting and DBSCAN. However, in order to improve the VP robustness, there will be improvements to the performance and the experiment by using road model features, surface as well path planning approaches. Regarding autonomous navigation, guidance and control design this result reinforces the point of usage of a set of sensor (GPS, IMU, LRF, and online interactive maps) for dealing with this problem in real time.

Acknowledgments. This research was supported by the MOTIE (The Ministry of Trade, Industry and Energy), Korea, under the Human Resources Development Program for Convergence Robot Specialists support program supervised by the NIPA(National IT Industry Promotion Agency) (H1502-13-1001).

References

1. Fischier, M.A., Bolles, R.C.: Random Sample Consensus: A Paradigm for Model Fitting with Applications to Image Analysis and Automated Cartography. In: Communications of the he Association for Computing Machinery, vol. 24, pp. 381–395. ACM, New York (1981)
2. Ester, M., Kriegel, H.-P., Sander, J., Xu, X.: A Density-Based Algorithm for Discovering Clusters in Large Spatial Databases with Noise. In: Proceeding of 2nd International Conference of Knowledge Discovery and Data Mining, pp. 226–231. AAAI Press, Oregon (1996)
3. Samadzadegan, F., Sarafraz, A., Tabibi, M.: Automatic Lane Detection in Image Sequences for Vision-Based Navigation Purposes. In: Proceeding of the International Society for Photogrammetry and Remote Sensing Commission V Symposium Image Engineering and Vision Metrology, vol. XXXVI Pt. 5. Copernicus Publications, Dresden (2006)
4. Ebrahimpour, R., Rassolinezhad, R., Hajiabolhasani, Z., Ebrahimi, M.: Vanishing point detection in corridors: using Hough transform and K-means clustering. IET Computer Vision 6, 40–51 (2012)
5. López, A., Cañero, C., Serra, J., Saludes, J., Lumbreras, F., Graf, T.: Detection of Lane Markings based on Ridgeness and RANSAC. In: Proceeding of the 8th International IEEE Conference on Intelligent Transportation Systems, pp. 254–259. IEEE Press, Vienna (2005)
6. Miksik, O., Petyovsky, P., Zalud, L., Jura, P.: Detection of Shady and Highlighted Roads for Monocular Camera Based Navigation of UGV. In: IEEE International Conference of Robotics and Automation, pp. 64–71. IEEE Press, Shanghai (2011)
7. Kong, H., Audibert, J.-Y., Ponce, J.: Vanishing point detection for road detection. In: IEEE Computer Society Conference on Computer Vision and Patter Recognition, pp. 96–103. IEEE Press, Miami Beach (2009)

8. Bosse, M., Rikoski, R., Leonard, J., Teller, S.: Vanishing Point and 3D Lines from Omnidirectional Video. In: Proceeding of the International Conference on Image Processing, vol. 3, pp. 513–516. IEEE Press, New York (2002)
9. Bazin, J.-C., Pollefeys, M.: 3-line RANSAC for Orthogonal Vanishing Point Detection. In: IEEE/RSJ International Conference on Intelligent Robots and Systems, pp. 4282–4287. IEEE Press, Algarve (2012)
10. Canny, J.: A Computational Approach to Edge Detection. IEEE Transactions on Pattern Analysis and Machine Intelligence 8, 679–698 (1986)
11. Faraz, M., Mirzaei, F.M., Roumeliotis, S.I.: Optimal Estimation of Vanishing Points in a Manhattan World. In: IEEE International Conference on Computer Vision, pp. 2454–2461. IEEE Press, Barcelona (2011)

Human Detection from Mobile Omnidirectional Camera Using Ego-Motion Compensated

Joko Hariyono, Van-Dung Hoang, and Kang-Hyun Jo

Graduate School of Electrical Engineering, University of Ulsan, Ulsan 680–749 Korea
{joko,hvzung}@islab.ulsan.ac.kr, acejo@ulsan.ac.kr

Abstract. This paper presents a human detection method using optical flows in an image obtained from an omnidirectional camera mounted in a mobile robot. To detect human region from a mobile omnidirectional camera achieved through several steps. First, a method for detection moving objects using frame difference. Then ego-motion compensated is applied in order to deal with noise caused by moving camera. In this step an image divides as grid windows then compute each affine transform for each window. Human shape as moving object is detected from the background transformation-compensated using every local affine transformation for each local window. Second, in order to localize the region as a human or not, histogram vertical projection is applied with specific threshold. The experimental results show the proposed method achieved comparable result comparing with similar methods, with 87.4% in detection rate and less than 10% in false positive detection.

Keywords: Human detection, Omnidirectional camera, Mobile robot, Ego-motion compensated.

1 Introduction

Nowadays, one of the critical challenges in video surveillance systems is how to reliably detect moving objects to perform high level tasks. In general, cameras are getting adopted because it is relatively economic than other sensors so that it is more appropriately used for indoor and outdoor surveillance. Omnidirectional cameras by which provides a 360° horizontal field of view in a single image attract wider attention. It is an important advantage in many application areas such as surveillances and tracking. In real-time application, a common approach to identify the moving objects is background subtraction, where each video frame is compared against a reference or background model. A block of pixels in the current frame that deviates significantly from the background is considered a moving object. As regarded that background subtraction is the first step in many computer vision applications, it is important to extract foreground pixels accurately correspond to the moving objects of interest. But it is not easy to segment out only moving object areas when the camera also moving caused by camera ego-motion. The frame difference represents all motions caused by camera ego-motion and moving object in the scene. It needs to compensate this effect from frame difference to segment out only moving object motion.

N.T. Nguyen et al. (Eds.): ACIIDS 2014, Part I, LNAI 8397, pp. 553–560, 2014.

Several researchers [1], [2] tried to measure camera ego-motion itself using omnidirectional vision. They used Lucas Kanade optical flow tracker and obtained corresponding features of background in the consecutive two omnidirectional images. Use analyzing the motion of feature points, camera ego-motion was calculated, but it was not used for moving object detection. They set up an omnidirectional camera on a mobile robot and obtained panoramic image transformed from omnidirectional image. They obtained camera ego-motion compensated frame difference based on an affine transformation of two consecutive frames where corner features were tracked by Kanade-Lucas-Tomasi (KLT) optical flow tracker [3]. But detecting moving objects resulted in a problem that only one affine transformation model could not represent the whole background changes since the panoramic image has many local changes of scaling, translation and rotation of pixel groups. For this problem, each affine transformation of local pixel groups should be tracked by KLT tracker. The local pixel groups are not a type of image features such as corner or edge. We use grid windows-based KL T tracker by tracking each local sector of panoramic image while other methods use sparse features-based KLT tracker. Therefore we can segment moving objects in panoramic image by overcoming the nonlinear background transformation of panoramic image.

In this paper, a method for detection moving human in omnidirectional image is proposed based on the ego-motion compensated. Though there have been proposed several human detection methods, those overall detection rate are not higher than 60%. Unless combining multiple methods, the rate has not shown significantly high and after all it achieves less than 90% [4]. In this proposal, the result shows a small improvement from background subtraction and ego-motion compensated.

Fig. 1. An omnidirectional camera mounted on the mobile robot

2 Mobile Robot with Omnidirectional Camera

This section presents the omnidirectional camera system which used in this work. An omnidirectional camera mounted on the mobile robot is calibrated to perform high accuracy task. The mobile robot is shown in Fig. 1. The omnidirectional camera consists of perspective camera and hyperboloid mirror as shown in Fig. 2. It captures an image reflecting from the mirror so that the image obtains reflective scene and not perspective. It is easier to recognize whether image contains moving object or not, so it is necessary to transform the obtained image into panoramic image [5].

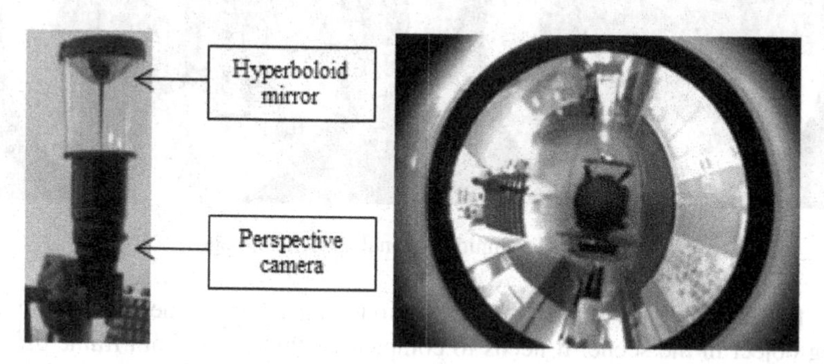

Fig. 2. The structure of omnidirectional vision and its image

In order to use the omnidirectional camera in mobile robot, it needs to calibrate and investigate accuracy when it is applied in structure from motion [7]. In this experiment the omnidirectional camera has been calibrated using chessboard as a pattern with control points [6]. We used a flexible calibration method for omnidirectional single viewpoint sensors from planar grids. However this method is based on an exact theoretical projection function and some parameters as distortion were added to consider real-world errors [10].

The sphere model was used by [6] and didn't consider the image flip. This approach adds to this model distortion parameters to consider real world errors. This method is multi views that mean it requires several images of the same pattern containing as many points as possible. This method needs the user to provide prior information to initialize the principal point and the focal length of the catadioptric system. The principal point is computed from the mirror center and the mirror inner border. The focal length is computed from three or more collinear non-radial points. The intrinsic and extrinsic parameters are initialized a non-linear process is performed. From this step we obtain the intrinsic and extrinsic camera parameters that are useful to apply this omnidirectional camera system in real application for mobile robot system [7].

3 Ego-Motion Compensated

In order to obtain moving object from omnidirectional image in mobile robot, it is not easy to segment out only moving object areas when the camera also moving caused by camera ego-motion. KLT Optical Flow Tracker [3] is applied in order to deal with several conditions. Brightness constancy which projection of the same point looks the same in every frame, small motion that points do not move very far and spatial coherence that points move like their neighbors.

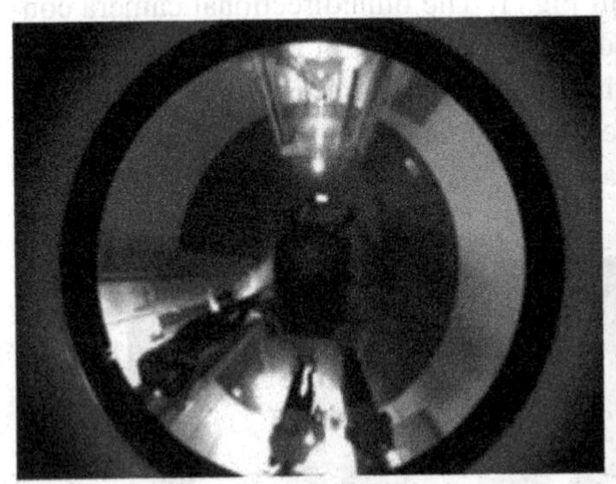 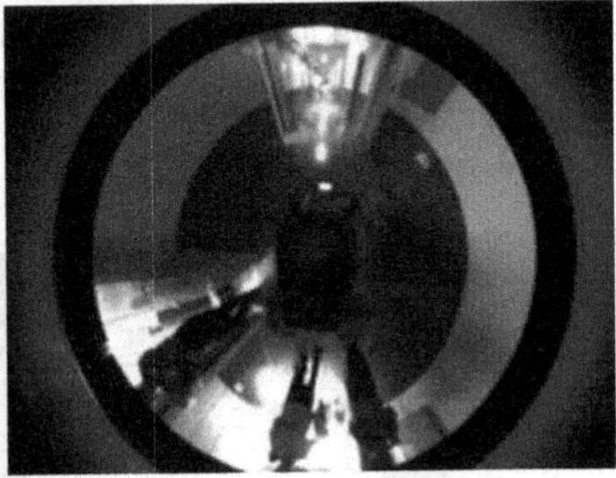

Fig. 3. Two omnidirectional sequence images

The frame difference represents all motions caused by camera ego-motion and moving object in the scene. It needs to compensate this effect from frame difference to segment out only moving object motion, so how much the background image has been transformed in two sequence images. Affine transformation represents the pixel movement between two sequence images as in (1),

$$P' = AP + t \tag{1}$$

where P′ is pixel location in second image and P is pixel location in first image. A is transformation matrix and t is translation vector. Affine parameters can be calculated by least square method using at least three corresponding features in two images.

From the image, we decide grid windows and then compare and track each windows in the next image. Using method from [3], find the motion $d(i,j)$ of each group $g_{t-1}(i,j)$ by finding most similar group $g_t(i,j)$ in the next image.

$$g_{t-1}(i,j) = g_t(i + d_x(i,j), j + d_y(i,j)) \tag{2}$$

It represented as affine transformation of each group as (3)

$$g_t(i,j) = \mathrm{I}g_{t-1}(i,j) + d_x(i,j) \tag{3}$$

The camera motion compensated frame difference I_d is calculated based on the tracked corresponding pixel groups using (4)

$$I_d(i,j) = |g_{t-1}(i,j) - g_t(i,j)| \tag{4}$$

where $I_d(i,j)$ is a pixel group located at (i,j) in the grid.

Suppose two sequence images shown in Fig. 3. It is not easy to segment out moving object using frame difference, as shown in Fig. 4(a), then apply frame difference with ego-motion compensate could obtain moving objects area shown in Fig. 4(b).

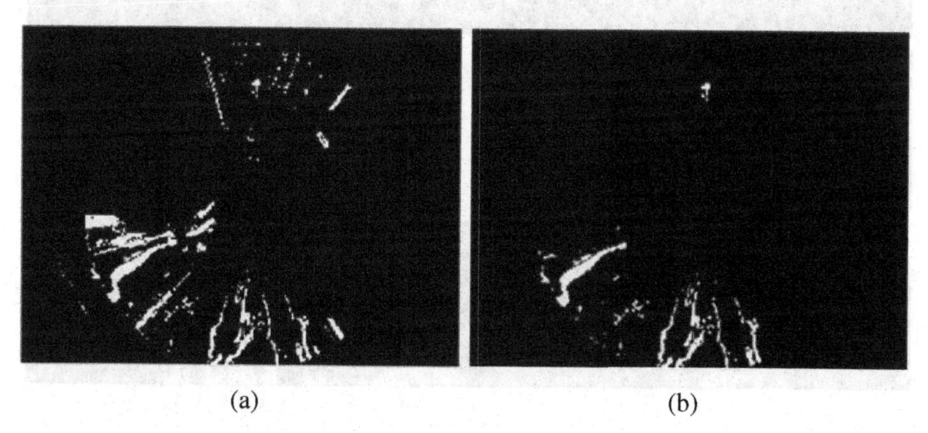

(a) (b)

Fig. 4. (a) frame difference, (b) frame difference with ego-motion compensated

3.1 Object Localization

Each pixel output from frame difference with ego-motion compensated could not show clearly silhouette. It just gives information of motion area from moving object. In order to get detected object, it's important to localize moving object area from the image. The omnidirectional camera captures an image reflecting from the mirror so that the image obtains reflective scene and not perspective. It is easier to recognize whether image contains human shape as moving object or not, so it is necessary to transform the obtained image into panoramic image.

In this work, we define detected moving objects are represented by the position and width in x-axis. Using projection histogram h_x by vertically project image intensities onto x-coordinate.

$$h_x = P_x I_d = [I, \dots, I] I_d \tag{5}$$

Where P_x is a projection vector which size is same as the height of panoramic image. An obtained h_x, is shown in Fig. 5.

We detect moving object based on the constraint of moving object existence that the bins of histogram in moving object area must be higher than a threshold and the width of these bins should be higher than a threshold as below,

$$h_x(i \pm 10) > A\, max(h_x) \tag{6}$$

Where A is a control constant and the threshold of bin value is dependent on the maximum bin's value. In order to get threshold of bin value, in this work using 10

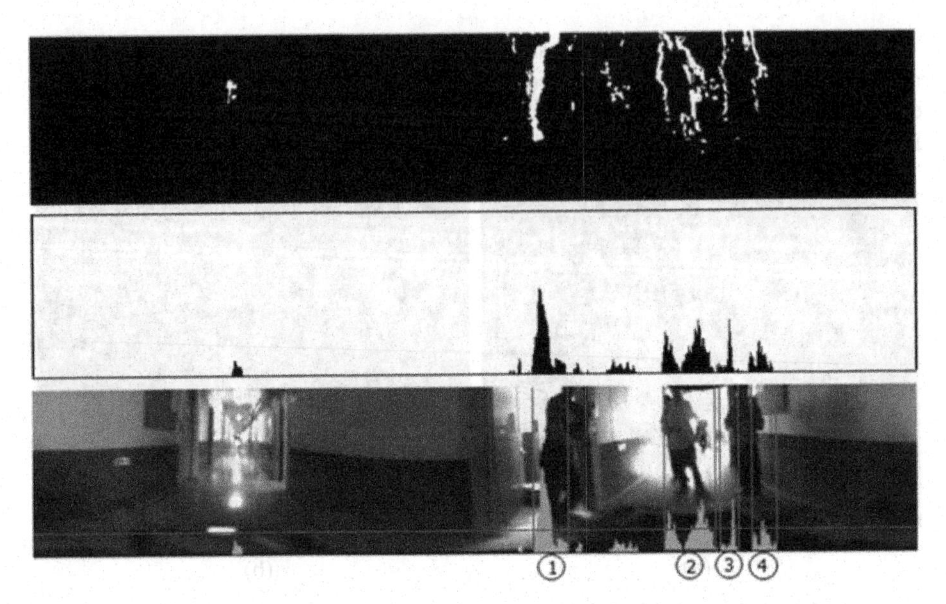

Fig. 5. Detection result

omnidirectional images for training. Each image contains one or more human shape with different shape and high, it's related to the distance from camera to object.

From Fig. 5 shows localization results. Top image show image result from frame difference with ego-motion compensated has been transform to panoramic image. Middle image shows histogram vertical projection from above image, and bottom image shows there are four detected objects as moving object, obtain from the region where have the number of bin above of the horizontal line as threshold value.

4 Experiment Results

In this work, our robot system is run in corridor with constant speed and detected object moving surround its path. Then evaluate our method from those image sequences, and compare with other methods proposed by [2], [8] and [9]. Proposed algorithm was programmed in MATLAB and executed on a Pentium – IV 2.80 GHz with 2 GB Random Access Memory.

From omnidirectional camera we took image sequences contain 500 frames with average number of person in each frame is three. The accuracy of human detection result shown in table 1. The number of human shape in each image frames has been computed, then calculate exactly number of detection rate. False positive detection rate is the ratio of non-human examples that have been wrongly classified as human. The proposed method compared with several methods [2], [8] and [9] shown that the proposed method higher detection rate.

Table 1. Detection comparison

Method	The number of human	True positives	False positive	Detection rate
H. Liu [2]	260	219	31	84.23%
B. Jung [8]	172	114	12	66.28%
C. M. Oh [9]	1371	1169	194	85.26%
Proposed method	1500	1297	172	86.46%

In Fig. 6 shows detection results several images taken from omnidirectional camera.

Fig. 6. Successful detection results

5 Conclusion

This paper presents a human detection has been apply omnidirectional camera mounted in mobile robot. In order to obtain moving objects from omnidirectional image in mobile robot using frame difference background subtraction. To deal with noise caused by motion camera, improved frame difference has been proposed and ego-motion compensation applied. In this step the image divides as grid windows then compute each affine transform for each window. Human shape as moving objects can be detected from the background transformation-compensated using every local affine transformation for each local window. The proposed method achieved comparable results with similar methods, with 87.4% in detection rate and less than 10% in false positive detection.

In order to improve detection rate of the system, in the future work it need consider to combines object detection based on moving object detection with geometrical approach to calculate object position or kinematic model of robot movement relative to static objects environment.

Acknowledgement. This research was supported by the MOTIE (The Ministry of Trade, Industry and Energy), Korea, under the Human Resources Development Program for Convergence Robot Specialists support program supervised by the NIPA (National IT Industry Promotion Agency) (H1502-13-1001).

References

1. Vassallo, R.F., Santos-Victor, J., Schneebeli, H.: A General Approach for Egomotion Estimation with Omnidirectional Images. In: Proceedings of the Third Workshop on Omnidirectional Vision (2002)
2. Liu, H., Dong, N., Zha, H.: Omni-directional Vision based Human Motion Detection for Autonomous Mobile Robots. Systems Man and Cybernetics 3, 2236–2241 (2005)
3. Tomasi, C., Kanade, T.: Detection and Tracking of Point Features. In: Proceedings of Fourteenth International Conference on Pattern Recognition, vol. 2, p. 1433 (1998)
4. Beleznai, C., Bischof, H.: Fast Human Detection in Crowded Scenes by Contour Integration and Local Shape Estimation. In: IEEE Conference on Computer Vision and Pattern Recognition, Miami, pp. 2246–2253 (2009)
5. Hoang, V.D., Vavilin, A., Jo, K.H.: Fast Human Detection Based on Parallelogram Haar-Like Feature. In: The 38th Annual Conference of the IEEE Industrial Electronics Society, Montreal, pp. 4220–4225 (2012)
6. Mei, C., Rives, P.: Single view point omnidirectional camera calibration from planar grids. In: International Conference on Robotics and Automation, pp. 3945–3950 (2007)
7. Hariyono, J., Wahyono, Jo, K.H.: Accuracy Enhancement of Omnidirectional Camera Calibration for Structure from Motion. In: International Conference on Control, Automation and Systems, Korea (2013)
8. Jung, B., Sukhatme, S.: Detecting Moving Objects using a Single Camera on a Mobile Robot in an Outdoor Environment. In: International Conference on Intelligent Autonomous Systems, pp. 980–987 (March 2004)
9. Oh, C.M., Lee, Y.C., Kim, D.Y., Lee, C.W.: Moving Object Detection in Omnidirectional Vision based Mobile Robot. In: The 38th Annual Conference of the IEEE Industrial Electronics Society, Montreal, pp. 4220–4225 (2012)
10. Hong, T.-P., Liou, Y.-L., Wang, S.-L., Vo, B.: Feature selection and replacement by clustering attributes. Vietnam Journal of Computer Science, doi:10.1007/s40595-013-0004-3

Simple and Efficient Method for Calibration of a Camera and 2D Laser Rangefinder

Van-Dung Hoang, Danilo Cáceres Hernández, and Kang-Hyun Jo

Graduated School of Electrical Engineering, University of Ulsan Ulsan, Korea
{hvzung,danilo}@islab.ulsan.ac.kr, acejo@ulsan.ac.kr

Abstract. In the last few years, the integration of cameras and laser rangefinders has been applied to a lot of researches on robotics, namely autonomous navigation vehicles, and intelligent transportation systems. The system based on multiple devices usually requires the relative pose of devices for processing. Therefore, the requirement of calibration of a camera and a laser device is very important task. This paper presents a calibration method for determining the relative position and direction of a camera with respect to a laser rangefinder. The calibration method makes use of depth discontinuities of the calibration pattern, which emphasizes the beams of laser to automatically estimate the occurred position of laser scans on the calibration pattern. Laser range scans are also used for estimating corresponding 3D image points in the camera coordinates. Finally, the relative parameters between camera and laser device are discovered by using corresponding 3D points of them.

Keywords: Camera- laser rangefinder calibration, extrinsic parameters, sensor fusion, perspective n points.

1 Introduction

Nowadays, systems using camera and laser rangefinder (LRF) have been widely applied to various intelligent vehicles, robot applications, intelligent transport systems namely automotive navigation, motion estimation, path planning and mapping, and quality control systems[19]. In order to use multiple devices simultaneously[1][2], the relative pose between devices is required for systems. In recent years, there are many proposed methods for discovering the rigid-body transformation between a camera and LRF. Those methods are separated into several categories of research. The first group of methods focuses on extracting the relative parameters between a camera and 3D laser scanner. The second one tries to deal with the problem between a camera and 2D laser scanner. The difficulty of this task is how to extract corresponding information between devices, due to the laser beams, which are not visualized on the calibration pattern. Therefore, the discontiguous calibration pattern is used to emphasize and visualize the laser beams, which is helpful for extracting corresponding points of a camera and LRF. Corresponding points are used to estimate the absolute rig-body transformation of a camera and LRF.

N.T. Nguyen et al. (Eds.): ACIIDS 2014, Part I, LNAI 8397, pp. 561–570, 2014.

This paper proposes a method for automatic discovering the relative position and direction of a camera with respect to a LRF by using the special calibration pattern. The calibration pattern consists of discontiguous depth of right-angled triangles, which emphasizes feature points of laser ranges to extract corresponding feature points between images and laser range scans. The 3D points in the LRF coordinates are used for estimating the 3D points of image in the camera coordinates by using the well-known perspective-n-point (PnP) method. Then, the corresponding 3D points are used for estimating the relative rotation and translation between a camera and LRF.

2 Related Work

Related works are separated into several categories. In the first group of methods, researches proposed methods for extracting rigid-body transformation between a camera and 3D LRF. The full calibration parameters of a camera- 3D LRF system was autonomously recovered by using special calibration pattern [3]. The authors used the center marks of the circle's physical as a calibration pattern. That method does not rely on the corner extraction, but rather on the simple target, which extracted from the center marks of a circle. Experimental results demonstrated that the effectiveness of the method. Other work estimated rigid-body transformation between a camera and 3D LRF by discovering the statistical dependence of two sensors [4]. They proposed the method, which uses maximal mutual information of vision-LRF to estimate the extrinsic parameters. Mutual information was used as registered criterion. The method is able to work in position without any specific calibration targets. On the contrary, a new calibration method based on nature scenes was proposed in [5]. It does not need any special object required for calibration. In their method, the corresponding points are manually selected on common scenes, which are viewed by both sensors. Their paper also presented a novel method to visualize range information, which obtained from a 3D LRF. Discontiguous information of laser ranges is emphasized and superimposed on a color image, which bring the simplicity and efficiency to selected corresponding points between color image and laser range image. Practical experiment had been processed on a camera and rotating laser scanner. The results showed that the method is suitable for real applications. Recently, a real-time calibration of multiple cameras and a 3D laser scanner was proposed in [6]. In this task, the authors described the approach using special checkerboard, which captured by both sensors. The corresponding points were manually selected. The paper did not present how to extract corresponding features between cameras and a LRF. Most recently, [7] presented a web interface toolbox for calibration multiple visions and laser rangefinders, especially for the Kinect 3D and Velodyne HDL-64 device. In that work, authors used multiple checkerboard patterns, which located at difference positions, to deal with solution of information, outliers matching for improving accuracy of results. Usually, most of calibration methods require manual selection of corresponding points. The advantage of a vision and 3D LRF calibration is easy to emphasize a depth image of laser sensor with respect to color image of vision sensor. This task supports to extract corresponding points, which occur on calibration pattern. Authors in [8] presented an algorithm for extracting the relative pose between color cameras and depth cameras, which requires only a planar surface to be imaged from various poses. The authors did not use depth discontinuities of depth image for making flexible and robust method. Similarly with previous methods, new method was proposed to efficiently extract the extrinsic parameters of a 3D LRF and stereo cameras. That system is an application for autonomous navigation of mobile robots in [9]. However, the approach required other additional devices, which support to calibrate the system. Difference with previous works, natural line matching method [10] was proposed to discover the extrinsic parameters of camera-range devices. However, this method requires more than three distinct line pairs, which are expensive computational cost than that of points based method.

The second group of methods focuses on calibration of a camera and 2D LRF. In this field, researches proposed methods for extracting rigid-body transformation between a camera and 2D laser rangefinder based on special calibration pattern. The early work in [11] improved a camera calibration method for extrinsic calibration of a camera and 2D LRF. In this paper, the method is based on observing a planar checkerboard and solving constraints between views of a planar checkerboard from a camera and 2D LRF. The authors assumed that the normal of calibration plane is aligned to parallel with optical flow in camera coordinates. This assumption simplifies the geometry constraints for estimating of vision-LRF transformation. A simple and effective method was proposed to extract the extrinsic parameters of a camera and two LRFs in [12]. This method solved the problem by using a linear solution method. The special configuration of two LRF devices obtains particular constraint information, which suitable for simple and accurate estimating of the relation between a camera and laser devices. In addition, authors in [13] presented a special method for calibrating of a camera and 2D LRF. In order to obtain plane poses in camera coordinates and depth readings in a LRF reference frame, this method was applied for moving pattern.

In general, almost of the former methods are not easily for automatically extracted corresponding features between a camera and LRF to estimate relative transformation between them. Herein, we will present a novel calibration method, which uses a calibration pattern based on right-angled triangles to automatically estimate corresponding features between a camera and laser sensor. Depth information of LRF is used to estimate world points in camera coordinates by using the perspective-three-point (P3P) method. Finally, the set of corresponding world points is used for discovering the position and direction of a camera with respect to a LRF.

3 Calibration Pattern Design for Automatic Calibration

This section describes a new calibration pattern, which is discontinuous and contains two adjacent right-angled triangles. One of them is a triangular hold, which makes facilitating confident extraction of the border between the triangles. Based on the proposed calibration pattern, a computationally efficient automatic calibration method is developed. The methodologies for automatic extracting the common features as well as determining the relation of a camera with respect to a LRF are also presented.

A method for automatic estimation the laser scan position on the pattern is described in this section. The main idea of proposed method is taking advantage of discontiguous calibration pattern and triangular properties. The discontiguous pattern is used to emphasize the laser beams, which supports for detecting segmental parts on the pattern. The triangular properties are used for estimating the position of laser scans on the pattern as well as detecting corresponding features between image and laser beam.

The rectangular calibration pattern consists of two right-angled triangles, ABC and ADC in Fig. 1(a). The grey triangular hole is used to make discontiguous laser range, which easily determines the common boundary of two triangles. Four black rectangles at four corners of the pattern are used to conveniently detect vertices A, B, C, D on image from camera.

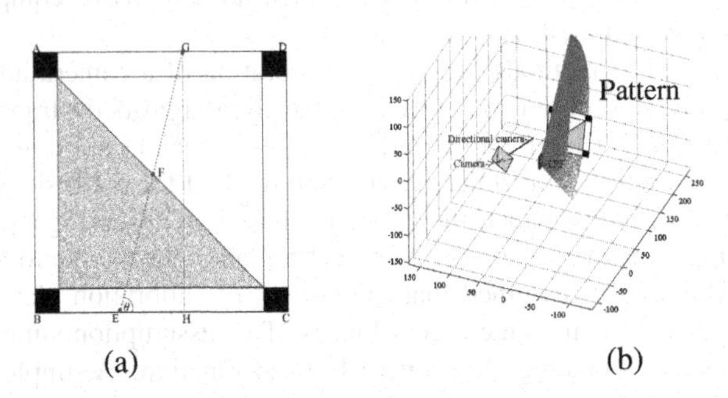

(a) (b)

Fig. 1. (a) Structure of calibration pattern, (b) visually the configuration calibration system

The calibration pattern places in front of both the camera and LRF, how to it belongs in FOV of camera and laser scan can be crossed two sides of the pattern as Fig. 1(b). The objective of this step is to be able to automatically estimate the points *E*, *F*, and *G* on calibration pattern.

First, the located point of *F* is directly estimated based on laser beams. Since laser range information, the distance from the center of LRF to the points *E*, *F*, and *G* are measured. Then the magnitudes of line segments *EF*, *FG*, and *EG* are computed. Considering two similar triangles *AFG* and *CFE* for estimating *AF*, there is a relationship:

$$\frac{AF}{FG} = \frac{AC - AF}{FE} \Rightarrow AF = \frac{AC * FG}{FE + FG} \tag{1}$$

Second, the located points of *G* and *E* on the pattern are discovered by using triangular properties. In order to discover the ratios of line segments *AG* and *CE* on the pattern, the gradient θ angle is required for the task. There are two solutions for θ angle in (2). This problem is described in late section by using RANSAC filter[14].

$$\theta = \begin{cases} a\sin(\dfrac{GH}{GE}) \\ 180^{o} - a\sin(\dfrac{GH}{GE}) \end{cases} \tag{2}$$

The location points of *E* and *G* are discovered by applying the law of cosines. Considering two triangles of *AFG* and *CFE*, since known of two edges and opposite angle, we have (3) and (4), respectively.

$$AG = \sqrt{AF^2 + FG^2 - 2AF \times FG \times \cos(\pi - \widehat{DAC} - \theta)} \quad (3)$$

$$CE = \sqrt{CF^2 + EF^2 - 2CF \times EF \times \cos(\pi - \widehat{ACB} - \theta)} \quad (4)$$

4 Estimation Camera-Point Distance

This section describes a method to estimate the world point in the camera coordinates based on PnP method. In general, the relative of camera-laser system can be estimated in two ways. First way directly estimates using corresponding homogeneous laser beams and camera points using the methods in [15],[16]. The second way solves the problem in two stages: compute world points in the camera coordinates based on world points in the laser coordinates, and then estimates the relative rotation and translation by using corresponding world points in the LRF and camera coordinates. The method, in this paper, follows the second one. Therefore, it is necessary to recover world-points in camera coordinates.

Summary of the method for estimating world points is described as following. Each pair of corresponding points $P_i \leftrightarrow p_i$ and $P_j \leftrightarrow p_j$ gives a constraint on the camera- world point distance. Notice that p_i and p_j are the norm vectors, which are rays from the camera center to points in the world scene. The points of p_i and p_j are estimated based on intensity image of camera by using image process method [17]. The points of P_i and P_j are world points in LRF coordinates (they are also the same in camera coordinates). All corresponding points in system (LRF and camera), they have to satisfy constraints of $P_i = d_i p_i$ and $P_j = d_j p_j$, where d_i, d_j are scalar values.

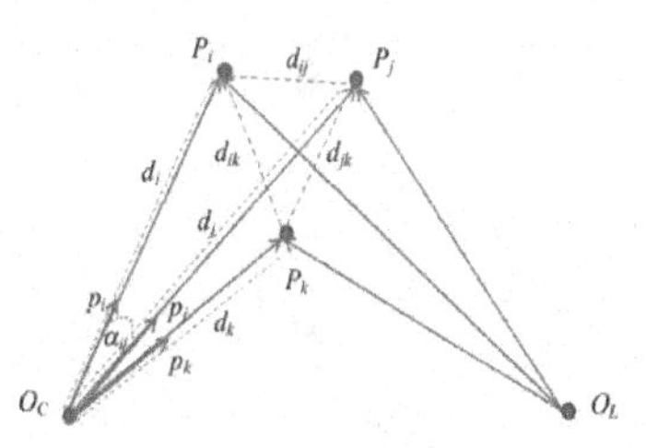

Fig. 2. Recovering real world points using perspective 3 points

$$d_i = \|P_i - O_c\| \quad (5)$$

$$d_{ij}^2 = d_i^2 + d_j^2 - 2d_i d_j \cos(\alpha_{ij}) \quad (6)$$

where $d_{ij}=\|P_i-P_j\|$ is the Euclidean distance between P_i and P_j in the camera coordinates as well as the RLF coordinates. That means d_{ij} will be directly computed from two laser range beams.

$$d_{ij} = \|P_i^L - P_j^L\| \quad (7)$$

Alternatively, the angle α_{ij} is also computed by using pair of reflection rays from the center of camera to the world point $p_i=[x_i, y_i, z_i]$ and $p_j=[x_j, y_j, z_j]$.

Equation (6) can be rewritten as:

$$f_{ij} = d_i^2 + d_j^2 - 2d_i d_j \cos(\alpha_{ij}) - d_{ij}^2 = 0 \quad (8)$$

There are two unknown variables of d_i and d_j, which are required to solve by using two known variables of α_{ij} and d_{ij}. In order to solve this equation, two additional points will be added to construct the equation system, which consists of three equations with three unknown variables. Camera- point distances can be discovered by solving the equation system (9).

$$\begin{cases} f_{ij} = d_i^2 + d_j^2 - 2d_i d_j \cos(\alpha_{ij}) - d_{ij}^2 = 0 \\ f_{ik} = d_i^2 + d_k^2 - 2d_i d_k \cos(\alpha_{ik}) - d_{ik}^2 = 0 \\ f_{kj} = d_k^2 + d_j^2 - 2d_k d_j \cos(\alpha_{kj}) - d_{kj}^2 = 0 \end{cases} \quad (9)$$

5 Extrinsic Calibration Extraction

5.1 Geometry Constraint

The objective of this section is to describe geometry constraints for estimating the rig-body transformation between a camera and laser scanner. In canonical problem, a landmark in world coordinates $P=[X, Y, Z]^T$ is projected into image is $I=[u,v]^T$ (image point). Image point is converted to homogenous in camera coordinates $p=[x,y,z]^T$ by using intrinsic parameter of camera [17]. The relative transformation is represented as

$$p \sim R_I P + t_I \quad (10)$$

where R_I and t_I are rotation matrix and translation vector, respectively.

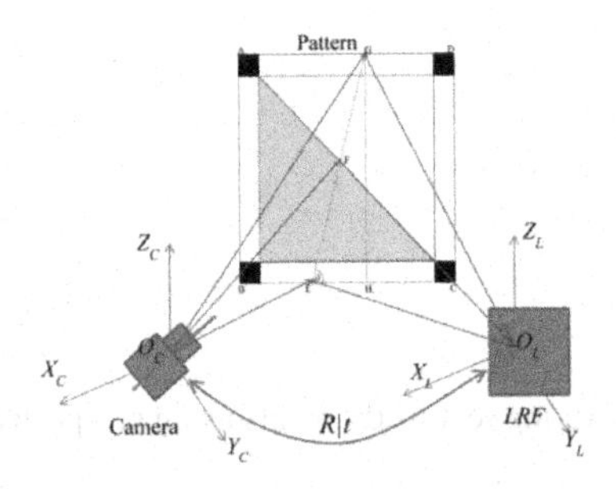

Fig. 3. Geometry constraints for estimating the relative pose between a camera and LRF

On the other hand, suppose that the world point P (denoted by 3D point) is represented in LRF coordinates as P^L. The nonhomogeneous point of P^L is described by $P^L = [X^L, Y^L, Z^L]^T$. Due to LRF is 2D scanner, without loss of generality, the non-homogeneous point P^L is represented with the constant Y (Y=0). For convenience, the camera coordinates is defined as being the system coordinates and the LRF coordinates as the reference coordinates. Therefore, the rigid transformation from RLF to camera coordinates is described as

$$P = RP^L + t \quad (11)$$

where R is a rotation matrix and t is a translation vector. Notice that the pair of points P^L and P is nonhomogeneous points. Euclidean displacement from corresponding nonhomogeneous points P and P^L *is* used to estimate the rotation and translation between the camera and LRF.

5.2 Rotation and Translation Estimation

The extrinsic parameters of transformation between a camera and LRF are determined from the known corresponding 3D points by using the method in [18]. The method is briefly described as following.

Step 1: The centroid of the set of 3D points P_i and P_i^L in camera and LRF coordinates are computed by following equations, respective.

$$P = \frac{1}{n}\sum_{i=1}^{n} P_i \text{ , and } P^L = \frac{1}{n}\sum_{i=1}^{n} P_i^L \tag{12}$$

where n is the number of corresponding 3D points.

Step 2: The vectors from the centroid to 3D points are computed

$$q_i = P - P_i \text{, and } q_i^L = P^L - P_i^L \text{, } \forall i = 1..n \tag{13}$$

The corresponding points p_i and p_i^L should satisfy the constraint

$$q_i^L = Rq_i \tag{14}$$

Due to noise of measurement and matching, the equation is not always fixed. Therefore, the objective of estimation rotation R, how to minimize the error:

$$e = \sum_{i=1}^{n} \| q_i^L - Rq_i \| \tag{15}$$

Step 3: Estimating rotation by using SVD method. First, computing the matrix H:

$$H = \sum_{i=1}^{n} q_i^L q_i^t \tag{16}$$

Then, the singular value decomposition of H is UAV', computing the matrix X

$$X = VU' \tag{17}$$

The rotation matrix R is X if the determinant of X is one, and otherwise for failure solution in degeneracy case. This is due to noise of measurement, the world points in coplanar, or collinear. To deal with the collinear case, in experiment, the set of world points is collected from at least two poses of calibration pattern. In the cased of coplanar, the method in [18] is used.

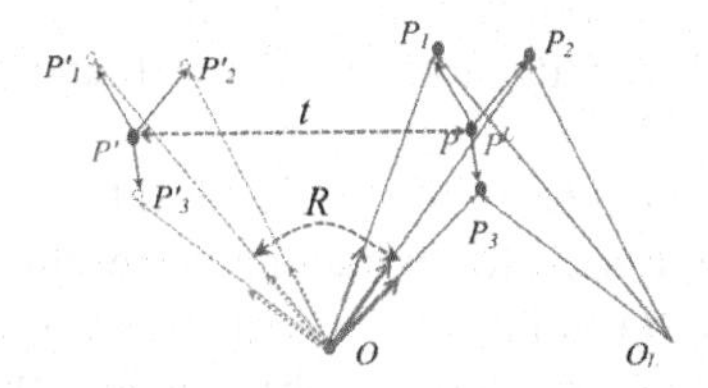

Fig. 4. Estimating rotation and translation by using corresponding 3D points, where P_i' is the rotated 3D point P_i in camera coordinates

Finally, the translation is recovered by using the displacement between two centroids P' and P^L on the rotated camera coordinates and LRF coordinates, respectively.

$$t = P^L - RP \tag{18}$$

The problem is how to estimate the rotation R and translation t with minimum error (15). To deal with noise of measurement and error of matched corresponding points, the well-known iterative RANSAC technique is used to filter outliers out of the corresponding points. The basic of the RANSAC technique is at reference [14].

6 Evaluation Results

This section evaluates the expected performance of our proposal on synthetic data with addition Gaussian noise with σ setting 1*pixel* and 1*cm* for a camera and LRF, respectively. The platform equipped the fix system consists of the LMS-200 laser sensor, camera and calibration pattern. The calibration pattern is 70*cm*×70*cm*. The size of legs of the right triangle hole in pattern is 50*cm*×50*cm*. LRF device works at configuration of resolution of 0.25^O, metric unit *mm*, and FOV $40^O \div 140^O$. The camera works at configurations of 778×1032-pixels resolution and FOV 70°. In this experiment, the LRF coordinate is used as reference coordinate. The camera is located at t= [-150.9305; 52.4871; 53.2558]*cm*, with direction [*yaw*; *pitch*; *roll*]= [20.8803; 8.2738; 8.4993]^O.

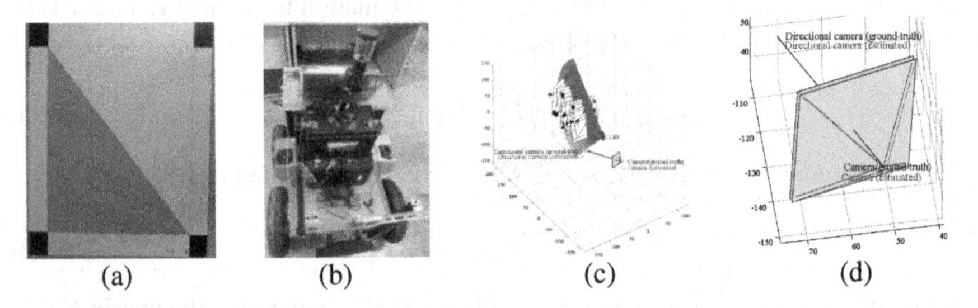

(a) (b) (c) (d)

Fig. 5. (a) Calibration pattern, (b) robot system, (c) visualization the calibration of LRF- camera system, (d) location and direction result of a camera with respect to LRF

The poses of calibration pattern are randomly generated (see also Fig. 5a). The calibration pattern is located in front of both the camera and LRF device. The distance from LRF to pattern is about 100*cm* ÷ 200*cm* with the rotation is ($-20^O \div 20^O$) for each of angles yaw, pitch, roll. The configuration of calibration system is shown in Fig. 5(c), which consists of 12 poses of pattern, LRF position, laser beams and position, direction of camera. Fig. 5(d) plots ground-truth and estimated position and direction of camera result.

This evaluation analyzed calibration errors with difference the number of calibration pattern poses, which are used for extracting corresponding features. The sum of error after transformation 3D points in camera with respect to 3D points in LRF is used as criterion for evaluation. Figs. 6(a, b) represent the affection of the number of calibration pattern poses on translation and rotation angle results. The number of poses versus with evaluated error is shown in Fig. 6(c). The experimental results show that the error of translation and rotation are not proportional due to criterion for evaluation based on the error of reprojection points. The best result is present in table 1, which uses 12 poses of calibration pattern.

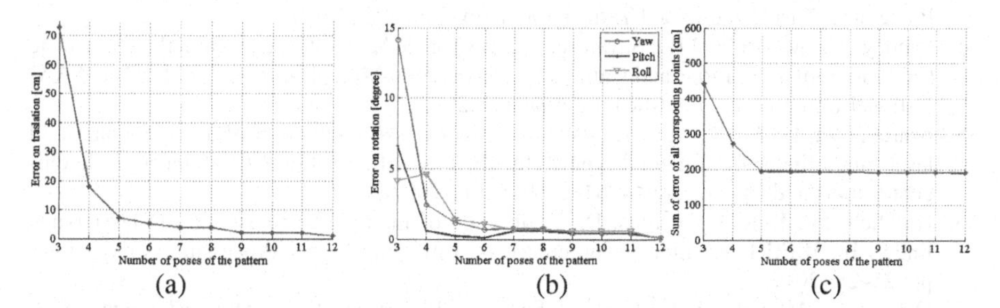

Fig.6. An error versus the number of calibration pattern poses using for estimation

Table 1. Extrinsic parameters estimation result

	Translations (*cm*)			Rotation angles (*degree*)		
	X	Y	Z	Yaw	Pitch	Roll
Ground truth	-150.9305	52.4871	53.2558	20.8803	8.2738	8.4993
Estimated value	-151.9214	52.0436	53.2052	20.9942	8.2056	8.5261
Error	0.9909	0.4435	0.0506	0.1139	0.0682	0.0268

7 Conclusion

This paper presents a new method for automatic calibration of a camera and laser rangefinder using the special calibration pattern. The new calibration pattern using depth discontinuities of right-angled triangle, which provide helpful for automatic estimation position of laser beam on the calibration pattern. The depth laser ranges are used to extract 3D points of an image with respect to LRF beams. Then corresponding 3D points are used for recovering extrinsic parameters of position and direction of a camera with respect to a LRF device. The evaluation result shows that the proposed method is simple and efficient.

Acknowledgment . This research was supported by the MOTIE (The Ministry of Trade, Industry and Energy), Korea, under the Human Resources Development Program for Convergence Robot Specialists support program supervised by the NIPA(National IT Industry Promotion Agency) (H1502-13-1001).

References

1. Hoang, V.-D., Hernández, D.C., Le, M.-H., Jo, K.-H.: 3D Motion Estimation Based on Pitch and Azimuth from Respective Camera and Laser Rangefinder Sensing. In: IEEE/RSJ International Conference on Intelligent Robots and Systems (IROS), pp. 735–740 (2013)

2. Hoang, V.-D., Le, M.-H., Jo, K.-H.: Planar Motion Estimation using Omnidirectional Camera and Laser Rangefinder. In: International Conference on Human System Interactions (HSI), pp. 632–636 (2013)
3. Alismail, H., Baker, L.D., Browning, B.: Automatic Calibration of a Range Sensor and Camera System. In: Second International Conference on 3D Imaging, Modeling, Processing, Visualization and Transmission, pp. 286–292 (2012)
4. Pandey, G., McBride, J.R., Savarese, S., Eustice, R.M.: Automatic Targetless Extrinsic Calibration of a 3D Lidar and Camera by Maximizing Mutual Information. In: The AAAI National Conference on Artifical Intelligence (2012)
5. Scaramuzza, D., Harati, A., Siegwart, R.: Extrinsic self calibration of a camera and a 3d laser range finder from natural scenes. In: IEEE/RSJ International Conference on Intelligent Robots and Systems, pp. 4164–4169. IEEE (2007)
6. Haselich, M., Bing, R., Paulus, D.: Calibration of multiple cameras to a 3D laser range finder. In: IEEE International Conference on Emerging Signal Processing Applications, pp. 25–28 (2012)
7. Geiger, A., Moosmann, F., Car, O., Schuster, B.: Automatic camera and range sensor calibration using a single shot. In: IEEE International Conference on Robotics and Automation, pp. 3936–3943 (2012)
8. Daniel Herrera, C., Kannala, J., Heikkil, J.: x00E, Janne: Joint Depth and Color Camera Calibration with Distortion Correction. IEEE Transactions on Pattern Analysis and Machine Intelligence 34, 2058–2064 (2012)
9. Aliakbarpour, H., Nunez, P., Prado, J., Khoshhal, K., Dias, J.: An efficient algorithm for extrinsic calibration between a 3D laser range finder and a stereo camera for surveillance. In: International Conference on Advanced Robotics, pp. 1–6 (2009)
10. Moghadam, P., Bosse, M., Zlot, R.: Line-based extrinsic calibration of range and image sensors. In: IEEE International Conference on Robotics and Automation, vol. 2, p. 4 (2013)
11. Zhang, Q., Pless, R.: Extrinsic calibration of a camera and laser range finder (improves camera calibration). In: IEEE/RSJ International Conference on Intelligent Robots and Systems, vol. 3, pp. 2301–2306 (2004)
12. Lixia, M., Fuchun, S., Ge, S.S.: Extrinsic calibration of a camera with dual 2D laser range sensors for a mobile robot. In: IEEE International Symposium on Intelligent Control, pp. 813–817 (2010)
13. Vasconcelos, F., Barreto, J.P., Nunes, U.: A Minimal Solution for the Extrinsic Calibration of a Camera and a Laser-Rangefinder. IEEE Transactions on Pattern Analysis and Machine Intelligence 34, 2097–2107 (2012)
14. Fischler, M.A., Bolles, R.C.: Random sample consensus: a paradigm for model fitting with applications to image analysis and automated cartography. Communications of the ACM 24, 381–395 (1981)
15. Kneip, L., Scaramuzza, D., Siegwart, R.: A novel parametrization of the perspective-three-point problem for a direct computation of absolute camera position and orientation. In: IEEE Conference on Computer Vision and Pattern Recognition, pp. 2969–2976 (2011)
16. Xiao-shan, G., Xiao-Rong, H., Jianliang, T., Hang-Fei, C.: Complete solution classification for the perspective-three-point problem. IEEE Transactions on Pattern Analysis and Machine Intelligence 25, 930–943 (2003)
17. Mei, C., Rives, P.: Single View Point Omnidirectional Camera Calibration from Planar Grids. In: IEEE International Conference on Robotics and Automation (ICRA), pp. 3945–3950 (2007)
18. Arun, K.S., Huang, T.S., Blostein, S.D.: Least-Squares Fitting of Two 3-D Point Sets. IEEE Transactions on Pattern Analysis and Machine Intelligence PAMI-9, 698–700 (1987)
19. Abaei, G., Selamat, A.: A survey on software fault detection based on different prediction approaches. Vietnam Journal of Computer Science, doi:10.1007/s40595-013-0008-z

Iris Image Quality Assessment Based on Quality Parameters

Sisanda Makinana[1,2,⋆], Tendani Malumedzha[1], and Fulufhelo V. Nelwamondo[1,2]

[1] CSIR Modeling and Digital Science, Pretoria, South Africa
[2] Faculty of Engineering and the Built Environment, University of Johannesburg, South Africa
{smakinana,tmalumedzha,fnelwamondo}@csir.co.za
http://www.csir.co.za

Abstract. Iris biometric for personal identification is based on capturing an eye image and obtaining features that will help in identifying a human being. However, captured images may not be of good quality due to variety of reasons e.g. occlusion, blurred images etc. Thus, it is important to assess image quality before applying feature extraction algorithm in order to avoid insufficient results. In this paper, iris quality assessment research is extended by analysing the effect of entropy, mean intensity, area ratio, occlusion, blur, dilation and sharpness of an iris image. Firstly, each parameter is estimated individually, and then fused to obtain a quality score. A fusion method based on principal component analysis (PCA) is proposed to determine whether an image is good or not. To test the proposed technique; Chinese Academy of Science Institute of Automation (CASIA), Internal Iris Database (IID) and University of Beira Interior (UBIRIS) databases are used.

Keywords: Image quality, Iris recognition, Principal Component Analysis, Quality measures.

1 Introduction

Iris recognition is an automated method of biometric identification that analyses patterns of the iris to identify an individual [7]. It is said to have high reliability in identification because each individual has unique iris patterns [11], [10]. However, due to the limited effectiveness of imaging, it is of principal importance that image of high-quality is selected/produced in order to improve iris recognition performance. Some advanced preprocessing algorithms can process poor quality iris images, which might produce inefficient results . For example, if it's known that the image may be slightly out of focus, then it can be enhanced before

⋆ S Makinana is with CSIR: Modeling and Digital Science,Pretoria, South Africa and Faculty of Engineering and Built Environment,University of Johannesburg, South Africa. phone number: +27 12 841 3342, email : smakinana@csir.co.za.

N.T. Nguyen et al. (Eds.): ACIIDS 2014, Part I, LNAI 8397, pp. 571–580, 2014.

matching. Hence, image quality assessment may improve the performance of iris recognition.

Various quality assessment methods have been adapted to enhance the quality of acquiring an iris image online [4] . These approaches are good for quick elimination of poor quality images, but even images which are capable of accurate segmentation may be assigned quality levels as some iris images are more discriminating than others. Daugman [4] evaluated the optical defocus model and proposed an 8x8 convolution kernel to extract the high frequency of an image. The convolution kernel acted as a bandpass filter. Chen *et al.* [2] divided the entire iris region into eight concentric bands and measured the frequency content using Mexican hat wavelet. The quality score of the entire iris image is computed by obtaining the weighted average of the quality score of individual bands. The bands closer to the pupil are given higher weight. This approach has good spatial adaptability on local image quality [2].

Generally, an iris sample is of good quality if it successfully provides enough information for reliable identification of an individual [1]. The ISO/IEC have developed three quality components which together defines biometric sample quality, these are; character, fidelity and utility [12]. In this paper, the focus is on character component of a biometric sample quality due to the fact that available algorithms utilises and focuses on fidelity and utility components [4], [6], [8] . There are various quality parameters that may affect iris images; these include defocus, motion blur, entropy, contrast, occlusion and dilation. Several authors have proposed algorithms that address individually the following parameters: defocus blur, motion blur and occlusion[8] or pair them with other quality parameters [4], [6]. A complete assessment of iris image quality is still a challenge because the overall iris image quality score is determined by multiple quality parameters and there is currently no well-defined standard for determining the weight of each individual quality parameter.

This paper proposes an algorithm that assesses the quality of an iris image based on multiple quality parameters. Firstly, image quality parameters are estimated, i.e. mean intensity, sharpness, blur, dilation, area ratio, entropy and occlusion. Thereafter, a fusion technique based on principal component analysis is used to weight each quality parameter and obtain a quality score for each image. The remainder of this paper is organised as follows. Section 2 describes the overview of the proposed method. Section 3 provides the estimations of individual quality parameters and discusses the implementation of the proposed fusion method. Last sections provides experimental results and a conclusion.

2 Iris Image Quality Overview

The approach taken in this paper is to estimate individual quality parameters and fuse them using principal component analysis to obtain a quality score. Fig. 1 illustrates the overview of the iris assessment plan.

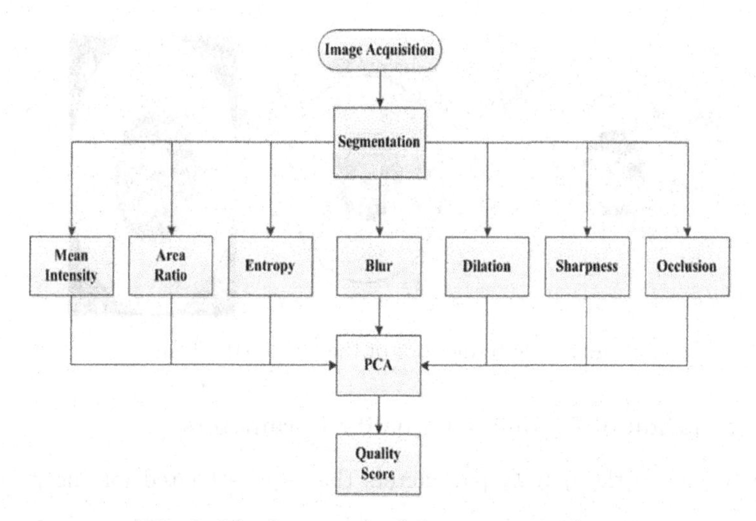

Fig. 1. The framework of the proposed scheme

3 Estimation of Quality Parameters

Implementation of assessment algorithm is carried out two steps: namely, localization and estimation of quality parameters. The subsections below details how this is done.

3.1 Localization

This stage involves segmenting the iris image, which is carried out by invoking Masek's [11] publicly available automatic segmentation method. It is based on the Hough transform, and is able to localise the circular iris and pupil region, occluding iris-lids, iris-lashes and reflections. Hough Transform is an extraction technique that uses the radius values to detect circles in an image. The circular Hough transform was used for detecting the iris and pupil boundaries. This involves first employing canny edge detection to generate an edge map. The Canny edge detector is an operator that uses a multi-stage algorithm to detect a wide range of edges in images. The range of radius values was set manually, depending on the database used. For CASIA and IID databases the values of the iris radius range from 80 to 150 pixels, while the pupil radius ranges from 28 to 75 pixels. For UBIRIS database the range for pupil radius is 5 to 11 pixels and for the iris radius it ranges from 39 to 49 pixels. The Hough transform determines the iris boundary is first, and then determines the pupil boundary within the iris region. After getting the iris and pupil boundaries, the iris and pupil image is displayed with everything beyond the iris radius masked out. Fig. 2 illustrates the segmented image of CASIA.

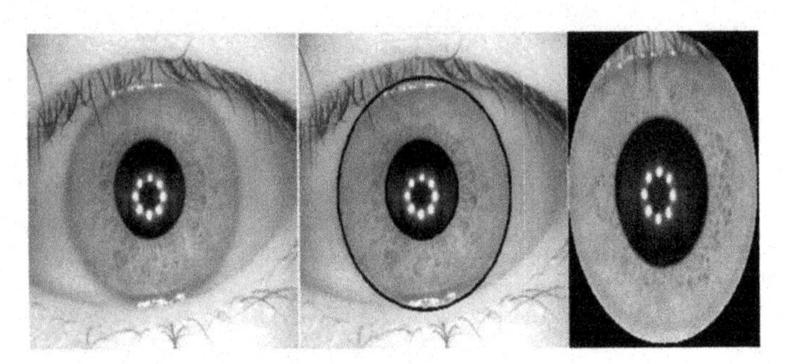

Fig. 2. The framework of the proposed scheme

3.2 Estimation of Predefined Quality Parameters

The following are the quality parameters that are estimated for the proposed algorithm:

Mean Intensity. The term intensity refers to the brightness of a pixel in an image. The intensity of a pixel is determined by several quantities including sensitivity of the light sensor in the imaging system [5]. The average intensity measures the directional change in colour of an image. This may have an effect on performance of the recognition system as it may be difficult to extract clear iris features.It measures the total intensity in an image by summing all of the pixel intensities. The function measures the scalar specifying the mean of all the intensity values in the iris region. Mean Pixel Intensity is denoted by:

$$M_{MPI} = \sum \frac{P_{INT_{i,j}}}{N} \tag{1}$$

Area Ratio. The representation of pattern recognition should be invariant to the change in size, position and orientation of the iris image. If the captured iris was far from the capturing device, the image might be small, which means fewer features will be resolvable for extraction. Also, if the subject is too close to the capturing device that may cause the captured image to be blurry. Thus, it is of utmost importance to assess the iris area ratio; which is defined as:

$$M_A = \frac{A_I}{A_E} \tag{2}$$

where A_I is the area of the iris and A_E is the area of the entire image.

Entropy. Entropy is a statistical measure of randomness that can be used to characterize the texture of the input image. It is the quantity of information contained in the image. It's also given as the uncertainty associated with random variables. Entropy is defined as:

$$M_E = -\sum_{i=0}^{255} p_i \log p_i \tag{3}$$

where

$$p_i = \frac{N_i}{N} \tag{4}$$

In 4 N_i is the number of pixels with grey level and N is the total number of pixels in the image. p_i is the probability of occurrence of grey level intensities.

Blur. Blur may result from many sources, but in general it occurs when the focal point is outside the depth of field of the captured object. The further the object is from the focal point of the capturing device, the higher degree of blur in the output image [8]. On the databases used there is no significant blur, so a low-pass Gaussian filter was convolved to the iris input image in order to estimate blur. First, the intensity variations between neighbouring pixels of the input image and blurred image are computed. Then, the intensity variations of these two images are compared; the results obtained are used as the estimation of blur. This blur estimation is based on Crete et al [3] approach. The blur estimation procedure is illustrated in Fig. 3.

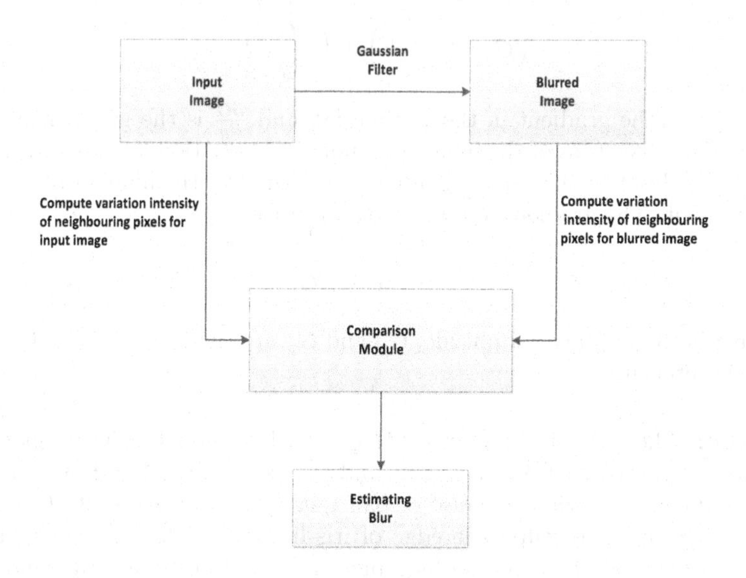

Fig. 3. The flowchart of the blur estimation [3]

Dilation. The variation in pupil dilation between the enrolment image and the image to be recognised or verified may affect the accuracy of iris recognition system. A degree of dilation was measured for each iris image. The segmentation results provided the radius of the pupil and of the iris. To measure dilation, a ratio of radius of pupil to radius of iris was calculated. Since the pupil radius is always less than the radius of iris, the ratio will fall between 0 and 1. The dilation measure M_D is calculated by:

$$M_D = \frac{P_R}{I_R} \tag{5}$$

Where P_R is the pupil radius and I_R is the iris radius.

Sharpness. Images are usually affected by distortions during acquisition and processing, which may result in loss of visual quality. Therefore, image focus assessment is useful in such applications. Sharpness is arguably the most important quality parameter because it determines the amount of readable information an image may hold. Sharpness generally attenuates high frequencies. Due to that factor, sharpness can be assessed by measuring high frequencies in the image. Daugman illustrated this in [4] by proposing a (8 X 8) convolution kernel. To assess the appearance of the blur effect, a sharpness metric is proposed in the literature. Sharpness is estimated based on the gradient of the image to determine whether the image is in focus or not, because the gradient of an image is the directional change in the intensity of an image. The gradient of the image is given by:

$$\nabla G = \left(\frac{\partial G}{\partial x}\hat{x}\right) + \left(\frac{\partial G}{\partial y}\hat{y}\right) \tag{6}$$

where $\frac{\partial G}{\partial x}$ is the gradient in the x direction and $\frac{\partial G}{\partial y}$ is the gradient in the y direction. The sharpness of the image is denoted by equation 6. The sharpness is calculated by dividing the sum of gradient amplitude by the number of elements of the gradient. The gradient amplitude is given by:

$$M_S = \sqrt{G_x^2 + G_y^2} \tag{7}$$

where S is the gradient amplitude, G_x and G_y are the horizontal and vertical changes in intensities.

Occlusion. The occlusion measure (M_{Occ}) is the amount of iris region that is invalid due to obstruction by eyelids and eyelashes. Eyelid and eyelashes occlusion problem is a primary cause of bad quality in iris image [9]. Compared with the edge of iris texture, the edge of iris-lid and iris-lash is much sharper and usually considered to contain high pixel values. To estimate the amount of occlusion at each level an occlusion is measured by calculating the total grey value of the image[14]. It is defined as:

$$M_{Occ} = \frac{1}{A_I \times A_E} \tag{8}$$

where A_I and A_E in are the area of the iris and Area of iris-lids. The higher the metric value the greater is the chance for occlusion by iris lid.

4 Implementation of Fusion Method

A unique quality score is of value to the prediction step of iris recognition system. To obtain this quality score, a fusion technique based on principal component analysis (PCA) [13] is proposed. Prior to applying the PCA, quality parameters need to be normalised. The quality parameters is normalised using the Z_s before obtaining the first PCA, which is:

$$Z_s = \frac{x - \mu}{\sigma} \tag{9}$$

Where μ is the mean and σ is the standard deviation of the estimated measures of the entire database.

Suppose n independent observation are given on $X_1, X_2, ..., X_k$, where the covariance X_i and X_j is

$$Cov(X_i, X_j) = \sum i,j \tag{10}$$

for $i, j = 1, 2, ..., k$ in equation 10. Then the eigenvalues and eigenvectors of the covariance matrix are calculated. Since the observations are normalised then

$$v_j^T v_j = 1 \tag{11}$$

where v is the eigenvectors of the covariance matrix. Then W is defined to be the first principal component. It is the linear combination of the $X's$ with the largest variance:

$$W = a_1^T X_i \tag{12}$$

where $i = 1, 2, ...k$ and a is the eigenvector of the covariance which also implies that $a_j^T a_j = 1$. Then the quality score is obtained by multiplying normalised measures of parameters with weights for each quality parameter of the image. The fusion quality index is given as:

$$Q_s = \sum_{i=1}^{N} Q_p W_p \tag{13}$$

Where Q_s is the quality score, Q_p is the estimated quality parameter and W_p is the amount of influence each parameter has on the quality score. The scores represent the global quality score of the iris segmented images.

5 Dataset Used for Analysis

In this paper, the CASIA and UBIRIS databases which are available free online and Internal Iris Database (IID) database were used, to estimate the quality parameters and their scores. For CASIA a subset of images called 'interval' was used. It contained 525 images which were captured at a resolution of 320 x 280 pixels. UBIRIS consists of 917 images captured at a resolution of 200 x 150 pixels. IID consists of 116 images captured at a resolution of 640 x 480 pixels.

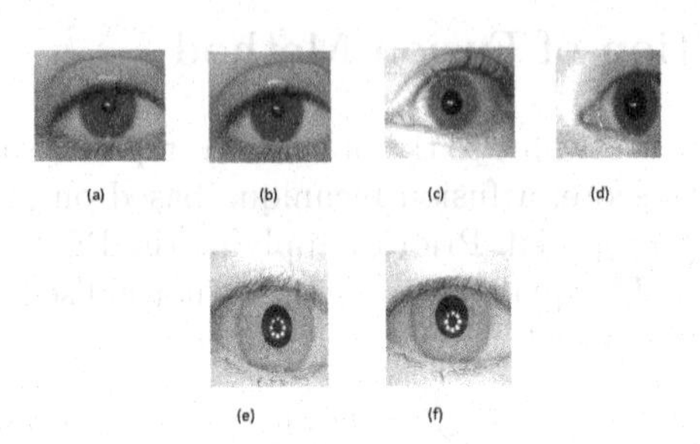

Fig. 4. Sample eye images from UBIRIS, IID and CASIA Interval database. (a) - (b) UBIRIS. (c) - (d) IID. (e) - (f) CASIA.

6 Experimental Results

In this paper, analysis was performed on three databases, namely, CASIA Interval, UBIRIS and IID. The sample iris images in Fig. 4 are from UBIRIS, IID and CASIA databases. Based on human visual assessment sample (a) represents good quality from UBIRIS database and (c) represent good quality from IID database. Sample (e) and (f) represent good and bad quality respectively, from CASIA Interval database. Image (b) and (f) represent degraded image quality that is affected by occlusion, blur and dilation. Sample image (b) is also affected by area ratio quality parameter. Table 6 illustrates the estimated quality parameters of the images in Fig. 4. The quality scores are normalized to the values between zero and one, with one implying good quality and 0 bad quality. The overall quality distribution for CASIA, UBIRIS and IID databases are illustrated in Fig. 5 respectively. CASIA has the highest quality distribution, followed by IID and the UBIRIS. IID suffers quality degrading with respect to entropy, occlusion, sharpness and mean intensity. These parameters have high weight on the quality score which accounts to low quality. For UBIRIS the quality score is influenced more by sharpness, entropy and mean intensity. When grading these data sets in terms of quality scores obtained on the plots, CASIA scores the highest, followed by IID and then UBIRIS.

Table 1. Estimated quality parameters of images in Fig. 4

Image	M_{MPI}	M_E	M_O	M_S	M_A	M_D	M_B	Quality score
a	0.9188	0.8897	0.4137	0.0271	0.0729	0.5182	0.3242	0.8089
b	0.6240	0.6013	0.0316	0.0502	0.0636	0.4352	0.4134	0.3095
c	0.5261	0.8273	0.0707	0.0052	0.6773	0.3714	0.3103	0.8930
d	0.7126	0.9810	0.3574	0.0299	0.5958	0.4222	0.285	0.3471
e	0.7725	0.8456	0.2927	0.0252	0.6689	0.3818	0.3212	0.8339
f	0.7981	0.9113	0.7552	0.0711	0.0229	0.4286	0.2924	0.3768

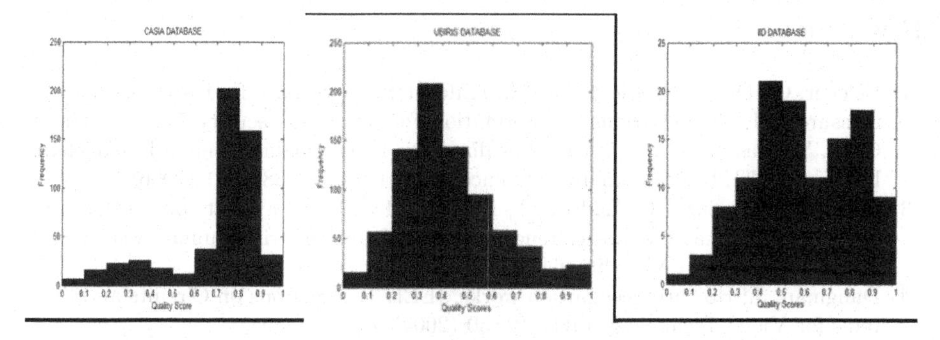

Fig. 5. Overall Quality Distribution of CASIA, UBIRIS and IID Databases

7 Classification of Quality Scores

The quality scores were classified using the K-Nearest Neighbour classifier. The scores were divided into two data-sets (60% training and 40% testing sample). There were two output classes (1 for low quality and 2 for high quality). Table 2 contains the error rate of the classified data of the three data-sets. The mean quality of the three databases is also represented, with CASIA containing highest quality scores, with the mean value of 0.7056. Following CASIA is IC with the mean value 0.5892 and then UBIRIS with the mean value of 0.5305.

Table 2. Summary performance statistics of the three databases

Databases	Error rate	Mean Quality	Count
CASIA	0.0047	0.7056	525
UBIRIS	0.0062	0.5305	802
IID	0	0.5892	116

8 Conclusion

In order to guide the selection of image of good quality, a quality model that evaluates the results of segmented iris image based on richness of the texture, shape and amount of information in the iris image has been developed. A fusion approach is presented for fusing all quality measures to a quality score. This is necessary because in order to successfully identify an individual on iris recognition systems an iris image must have sufficient features for extraction. We extend iris quality assessment research by analysing the effect of various quality parameters such as entropy, mean intensity, area ratio, occlusion, blur, dilation and sharpness of an iris image. The aim of this paper is to present a method that could be used to produce high quality images, which may improve iris recognition performance. The major benefit of this paper is that assessment is done before feature extraction, so only high quality images will be processed therefore saving time and resources.

References

1. Belcher, C., Du, Y.: A selective feature information approach for iris image-quality measure. IEEE Transactions on Information Forensics and Security, 572–577 (2008)
2. Chen, Y., Dass, S.C., Jain, A.K.: Localized Iris Inage Quality Using 2-D Wavelets. In: Proc. IEEE International Conference Biometrics, pp. 387–381 (2006)
3. Crete, F., Dolmiere, T., Ladret, P., Nicolas, M.: The blur effect: perception and estimation with a new no-reference perceptual blur metric. Human Vision and Electronic Image in XII 6492, 64920I (2007)
4. Daugman, J.: How iris recognition works. IEEE Transactions on Circuits and Systems for Video Technology 14(1), 21–30 (2004)
5. Fairchild, M.D.: Color Appearance Models. In: Slides from a tutorial at the IST/SID 12th Color Imaging Conference (2004)
6. Fatukasi, O., Kittler, J., Poh, N.: Quality controlled multi-modal fusion of biometric experts. In: Rueda, L., Mery, D., Kittler, J. (eds.) CIARP 2007. LNCS, vol. 4756, pp. 881–890. Springer, Heidelberg (2007)
7. Gulmire, K., Ganorkar, S.: Iris recognition using Gabor wavelet. International Journal of Engineering 1(5) (2012)
8. Kalka, N.D., Dorairaj, V., Shah, Y.N., Schmid, N.A., Cukic, B.: Image quality assessment for iris biometric. In: Proceedings of the 24th Annual Meeting of the Gesellscha it Klassikation, pp. 445–452. Springer (2002)
9. Li, Y.H., Savvides, M.: An automatic iris occlusion estimation method based on high-dimensional density estimation. IEEE Transactions on Pattern Analysis Machine Intelligence 35(4), 784–796 (2013)
10. Ma, L., Tan, T., Wang, Y., Zhang, D.: Personal identification based on iris texture analysis. IEEE Transactions on Pattern Analysis and Machine Intelligence 25(12), 1519–1533 (2003)
11. Masek, L.: Recognition of human iris patterns for biometric identification. PhD thesis
12. Tabassi, E.: Biometric Quality Standards, NIST, Biometric Consortium (2009)
13. Tipping, M.E., Bishop, C.M.: Mixtures of probabilistic principal component analysers. Neural Computation 11(2), 443–482 (1999)
14. Yalamanchili, R.K.: Occlussion Metrics. West virginia University (2011)

Contextual Labeling 3D Point Clouds with Conditional Random Fields

Anh Nguyen and Bac Le

University of Science, Vietnam
{nqanh,lhbac}@fit.hcmus.edu.vn

Abstract. In this paper we present a new approach for labeling 3D point clouds. We use Conditional Random Fields (CRFs) as an objective function, with unary energy term assessing the consistency of points with labels, and pairwise energy term between points and its neighbors. We propose a new method to learn this function from a collection of trained labels using JointBoost classifier formalism. By using CRFs with different geometric and contextual features, we show that our method enables the combination of semantic relations and achieves higher accuracy. We validate and demonstrate the efficiency of our method on complex urban laser scans and compare it with several alternative approaches.

Keywords: 3D point cloud labeling; conditional random fields; Joint-Boost.

1 Introduction

With the development of scanning technologies, 3D point clouds are now widely available. Professional scanners such as Light Detection and Ranging (LIDAR) and Microsoft Kinect provide enormous amount of data need to be analyzed. In 2011, Point Cloud Library (PCL) [10] which contains state of the art algorithms for 3D understanding was introduced. As scanning technologies advance and the development of PCL, processing point clouds gains more and more attraction in computer vision and robotics community. 3D perception is an important problem and has many applications in autonomous systems such as intelligent vehicles, autonomous mapping and navigation.

3D point cloud labeling is a process that assigns a label to all objects in an observed scene. This is an important task because the its results could be helpful in many autonomous robot navigation tasks such as locating and recognizing objects, obstacle avoidance, and and environment modeling. Point cloud data contain useful geometric information which is ambiguous and unreliable if we reconstruct from 2D images or stereo images. However, labeling objects from point clouds is complicated and challenging because the point cloud data are noisy, unorganized and sparse. Moreover, the sampling density of point clouds is typically uneven due to varying linear and angular rates of the scanner. In addition, the surface shape can be arbitrary with sharp features and there is no statistical distribution pattern in the point cloud data [9].

N.T. Nguyen et al. (Eds.): ACIIDS 2014, Part I, LNAI 8397, pp. 581–590, 2014.

A labeling algorithm should always consider the trade off between accuracy and speed. Because the over or under segmentation when labeling is inevitable, it is helpful for an algorithm to accept additional intuitive parameters describing assumptions or can include contextual information. In general, there are two basic approaches for segmenting and labeling 3D point clouds:

Geometric Reasoning: This approach uses purely mathematical model and geometric reasoning techniques such as region growing or model fitting methods; and a robust estimators to fit linear and nonlinear models to point cloud data. This approach achieves good results in simple scenario and not time consuming. However, its limitations are it cannot deal with incomplete or uneven distribution data, difficult to choose the size of model when fitting objects, and not working well in complex scenes.

Machine Learning: This approach uses feature descriptors to extract 3D features from point cloud data, then machine learning techniques are used to build a classifier to learn different classes of objects. Afterwards, different points are labeled by applying this classifier. Points can be treated independently or contextually with their neighbors.

The machine learning techniques usually outperform techniques purely based on geometric reasoning. The reason is due to uneven density, occlusions in point cloud data, it is very difficult to find and fit complicated geometric primitives to objects. Although machine learning techniques give better results, they are usually slow and rely on the result of feature extraction step [9].

We propose a graph based method to simultaneous segmentation and labeling 3D point clouds. Our method works by extracting geometric and contextual features from the point clouds. We use CRFs as the combination of a set of unary terms, defined for each point individually, and a set of pairwise terms, defined as a function of neighbor points. Instead of solving the model as a discrete energy minimization task over a graph containing nodes corresponding to individual points, we use JointBoost [15] classifier formalism to learn CRFs.

The remainder of the paper is organized as follows: In the next section, we summarize related work on similar initiatives. We review the formulation of CRFs used for labeling 3D point clouds in section 3. Section 4 describes the learning problems and techniques using JointBoost classifier algorithm. Section 4 reports the experimental results followed by a discussion and future work suggestions.

2 Previous Work

Several methods have been proposed to segment and label 3D point clouds. In [9], authors provided a good survey for segmenting and labeling point cloud data. In this section, we only summarize previous works that use machine learning techniques.

Graphical models such as Markov Random Field (MRF) and Conditional Random Field (CRF) are commonly used in computer vision tasks (e.g. image denoising, optical flow, segmentation, etc). Many work on labeling point clouds field use these models. Rusu et al. [1] proposed an approach based on surface segmentation for labeling points with different geometric surface primitives using CRF and feature descriptor called Fast Point Feature Histograms (FPFH) [6]. By defining classes of 3D geometric surfaces, and making use of contextual information using CRF, this method successfully segment and label 3D points based on their surfaces even with noisy data. Golovinskiy [8] used k-nearest neighbours (KNN) to build a 3D graph on the point clouds. This method introduces a penalty function to encourage smooth segmentation where the foreground is weakly connected to the background.

The MRF framework provides a natural way of incorporating contextual information. Schoenberg et al. [3] used MRF to segment 3D point clouds achieved by the combination of an optical camera and a laser scanner. Texture of point clouds are generated from interpolating from laser range and aligned optical image. The weights of MRF model are computed as a fusion of Euclidean distances, pixel intensity differences and angles between surface normals estimated at each point. This method showed good results in urban environment but requires a complex camera system.

Shapovalov [7] introduced a cutting-plane training of Non-associative Markov Network for 3D point cloud segmentation. This method applied kernel trick as the non-linear method to train non-associative Markov networks in a principled manner using the structured Support Vector Machine (SVM) formalism. The work of Munoz et al. [4] use Max-Margin Markov Networks and adapt a functional gradient approach for learning high dimensional parameters in order to perform discrete, multi-label classification. This method showed how the model can be incorporated with robust potentials to preserve less dominant labels.

Anguelov et al. [5] used Associative Markov Networks (AMN) for segmentating 3D point clouds. The Associative Markov Networks encourage neighboring points to have same class labels. This method uses maximum-margin framework to train the model from a set of labeled scans. The energy is a sum of potential functions of the features corresponding to individual nodes extracted from the scan. The inference is performed using graph cut and the parameters of the energy are learned by reducing the training problem to quadratic programming.

Munoz [11] introduced an efficient learning algorithm of random field with higher-order cliques by using subgradient optimization. A context approximation is proposed to make the model usable on a mobile vehicle for environment modeling. This work is not time consuming and can be run onboard of a mobile robot.

Koppula [13] proposed a new learning algorithm for scene understanding that exploits rich relational information derived from 3D point cloud for object labeling. In particular, this work used graphical model that naturally captures the geometric relationships of a 3D scene. Each 3D segment is associated with a node, and pairwise potentials model the relationships between segments. The model is trained using a maximum-margin approach that globally minimizes an upper bound on the training loss.

3 CRFs for Labeling 3D Point Clouds

Generative graphical models represent a joint probability distribution $p(x, y)$, where x expresses the observations and y expresses the label. These approaches require to model the observations which lead to erroneous independence assumptions among features. A solution is to use discriminative models such as CRFs, which represent a conditional probability distribution $p(y|x)$. The distribution is defined by the dependencies of the random variables represented in an undirected graph where each vertex represents a random variable and the edges represent a dependency between two variables. In general, we can formulate a CRF model as:

$$p(y|x) = \frac{1}{Z(x)} \exp \psi(x, y) \tag{1}$$

where $Z(x)$ is the normalizing partition function.

In this work, we consider CRF as a pairwise potentials. Our model includes the unary term E_1 measures consistency between the unary features x_i, which includes descriptors such as curvatures, FPFH, etc, of point i and its label c_i; the pairwise term E_2 measure the consistency between adjacent point labels c_i and c_j. To avoid under or over segmentation which leads to false labeling, for each adjacent pair of points we define pairwise features y_{ij} to provide cues whether adjacent points should have the same label. Given C as a predefined set of labels, our goal is to label each point i with a label in C. This task is equivalent to minimize the following objective function:

$$E(c, \theta) = \sum_i E_1(c_i; x_i, \theta_1) + \sum_{i,j} E_2(c_i, c_j; y_{ij}, \theta_2) \tag{2}$$

where $\theta = (\theta_1, \theta_2)$ are the parameters of the model.

Our model is in contrast to MRF model, which defines a joint probability over the labels, from which the conditional may then be derived. For segmenting and labeling 3D point clouds, MRF model may have worse labeling performance, while CRF learning algorithms is optimized for labeling performance. Moreover, the pairwise term in CRF model expresses connection between adjacent objects, which is not true in MRF model. Our CRF model is similar to Kalogerakis [12] and Huang [14]. However, we use different 3D features for both the unary and pairwise potentials which are more suitable for labeling 3D point clouds.

3.1 Unary Energy Term

The unary classifier is the most important component of our system because it evaluates a classifier. The classifier takes the feature vector x as input, and returns a probability distribution of labels for that point: $P(c|x, \theta_1)$. Same as [14], we use JointBoost classifier [15] to learn this term. Then, the unary energy of a label c is equal to its negative log-probability:

$$E_1(c; x, \theta_1) = -\log P(c|x, \theta_1) \tag{3}$$

Features: We use geometric features as described in [11] [20] and FPFH descriptor to produce adequate 3D features. The estimation of a FPFH descriptor includes two steps. The first step is to compute the histogram of the three angles between a point i and its k-nearest neighbors to produce the Simplified Point Feature Histogram (SPFH). Then, for each point i, the values of the SPFH of its k neighbors are weight by their distance w to produce the FPFH:

$$FPFH(i) = SPFH(i) + \frac{1}{k}\sum_{i=1}^{k}\frac{SPFH(i)}{w_i} \tag{4}$$

Next, we describe the pairwise energy term in our model.

3.2 Pairwise Energy Term

The main role of the pairwise energy term is to prevent incompatible segments from being adjacent and take contextual information into account. In general, the pairwise term penalizes neighboring parts of point being assigned different labels:

$$E_2(c_i, c_j; y, \theta_2) = L(c_i, c_j)G(y_{i,j}) \tag{5}$$

where the term L enables the label compatibility, and term G describe a geometry dependent.

The label compatibility L measures the consistency between two adjacent labels. This term is represented as a matrix of penalties for each possible pair of labels, which allows different pairs of labels to incur different penalties. The geometry dependent term G measures the likelihood of there being a difference in labels. Similar to the unary energy term, this term also evaluates a JointBoost classifier as follow:

$$G(y_{i,j}) = -\lambda \log P(c_i, c_j | y_{i,j}, \theta_2) \tag{6}$$

where λ controls the contribution of the pairwise term. In implementation, we use Point Pair Feature (PPF) [2] as a pairwise feature. Given two points i_1 and i_2 and their normals n_1 and n_2, the PPF is given by:

$$PPF(i_1, i_2) = (|d|_2, \angle(n_1, d), \angle(n_2, d), \angle(n_1, n_2)) \tag{7}$$

where $\angle(a, b) \in [0, \pi]$ represents the angle between a and b and $d = i_2 - i_1$. A point pair (i_1, i_2) is aligned to a scene pair (s_1, s_2) that has the same feature vector.

3.3 JointBoost Classifier

JointBoost is a boosting algorithm that automatically performs feature selection process with a large numbers of input features for multiclass classification. This algorithm jointly trains multiple classifiers so that they share as many features as possible. The result is a classifier that runs faster and requires less data to train than original classifiers. In particular, the number of features required to reach a fixed level of performance grows sub-linearly with the number of classes, as opposed to the linear growth observed with original classifiers [15]. JointBoost has a fast sequential learning algorithm, and produces output probabilities suitable for combination with other terms in the CRF model.

The JointBoost classifier takes as input a feature vector z, and outputs a probability $P(c = l|z)$ for each possible class label $l \in C$, where C is the set of possible labels. In the unary energy term, a classifier computes the likelihood of a part label c given unary features x. In the pairwise energy term, a second JointBoost classifier is used to determine the likelihood whose adjacent points have different classes given pairwise features y. This model reduces generalization error for multiclass recognition when classes overlap in feature space. So, we believe that JointBoost is the best available classifier for this task.

Similar to Torralba [15], we use a decision stumps in our classifier. A decision stump is a simple classifier that scores each possible class label l. Given a feature vector z and its threshold z_f of f-th entry, a JointBoost decision stump can be written as:

$$h(z, l; \phi) = \begin{cases} \alpha & \text{if } z_f > \tau \text{ and } l \in C' \\ \beta & \text{if } z_f \leq \tau \text{ and } l \in C' \\ k_l & \text{if } l \notin C' \end{cases} \quad (8)$$

Each decision stump stores a set of classes C'. If $l \in C'$, then the stump compares z_f against a threshold τ. It returns a constant α if $z_f > \tau$, and β otherwise. If $l \notin C'$, then the comparison is ignored, and a constant k_l is returned instead. There is only one k_l for each $l \notin C'$. The parameters ϕ of a single decision stump are f, α, β, τ, the set C', and k_l for each $l \notin C'$.

The probability of a given class l is then computed by summing the decision stumps and then performing a softmax transformation:

$$H(z, l) = \sum_j h(z, l; \phi_j) \quad (9)$$

$$P(c = l|z, \xi) = \frac{\exp(H(z, l))}{\sum_{l \in C} \exp(H(z, l))} \quad (10)$$

The parameters ξ consist of the parameters ϕ_j of all the individual decision stumps.

4 Learning CRFs

A well known approach for learning CRFs is maximize the log-likelihood of $p(y|x)$ as described in Lim [16]. This nonlinear optimization problem is solved by applying the Broyden-Fletcher-Goldfarb-Shannon (BFGS) method. Unfortunately, computing the normalization $Z(x)$ is intractable and the computational cost is expensive. Instead of using BFGS, we use the work of Shotton et al. [17] by to apply these following steps. First, we randomly split the training data into an exemplar set and a validation set in a proportion of approximately 3:1. The JointBoost classifiers for the unary term and the pairwise term is learned from the exemplar set. Finally, the remaining CRF parameters are learned by iteratively optimizing segmentation performance on the validation set. The performance of our CRF model relies on the classification accuracy of the classifiers used to define both the unary and pairwise terms.

4.1 Learning JointBoost Classifiers

We learn the JointBoost classifiers as described in Torralba [15]. The input to the algorithm is a collection of N training pairs (z_i, c_i), where z_i is a feature vector and c_i is the corresponding class label. For the unary terms, the training pairs are the feature vectors and their labels (x_i, c_i) for all points in the exemplar set. For the pairwise terms, the training pairs are the pairwise feature vectors. Each training pair is assigned a per-class weight $w_{i,c}$.

JointBoost classifier minimizes the weighted multiclass exponential loss over the exemplar set by solving the following equation:

$$J = \sum_{i=1}^{M} \sum_{l=1}^{C} w_{i,c} \exp(-I(c_i, l) H(z_i, l)) \tag{11}$$

where $I(\delta, \delta')$ is an indicator function:

$$I(\delta, \delta') = \begin{cases} 1 & \text{if } \delta = \delta' \\ -1 & \text{otherwise} \end{cases} \tag{12}$$

We stores a set of weights $\tilde{w}_{ic}$ that are initialized to the weights $w_{i,c}$. Then, at each iteration, one decision stump is added to the classifier. The parameters ϕ_j of the stump at iteration j are computed to optimize the following weighted least squares problem at each iteration:

$$J_{wse}(\phi_j) = \sum_{l=1}^{C} \sum_{i=1}^{N} w_{i,l} (I(c_i, l) - h(z_i, l; \phi_j))^2 \tag{13}$$

Once the parameters ϕ_j are determined, the weights are updated as:

$$\tilde{w}_{i,c} \leftarrow \tilde{w}_{i,c} \exp(-I(c_i, l) h(z_i, l; \phi_j)) \tag{14}$$

and the algorithm continues with the next decision stump.

4.2 Learning the Remaining Parameters

Once the JointBoost classifiers have been learned, our model learn the remaining parameters of the pairwise term by cross validation. Specifically, for any particular setting of these parameters, we can apply the CRF to all of the validation points, and evaluate the classification results.

The parameter λ controls the contribution of the pairwise term. We learn this parameter using the validation data set by trying different λ and picking the λ with the smallest testing error. Other parameters are optimized in two steps. Firstly, the segmentation error is minimized over a coarse grid in parameter space by brute-force search. Then, starting from the minimal point in the grid, optimization continues by using preconditioned conjugate gradient with numerically estimated gradients.

5 Experimental Results

We evaluate our model on the VMR-Oakland [18] point cloud dataset. This dataset contains labeled point cloud data collected from a moving platform around Carnegie Mellon University campus. The points were collected using laser scanner and are saved in text format, three real valued coordinates of each point are written in each line on airborne and terrestrial laser scans. Seven classes are used: *wire, pole, ground, vegetation, trunk, building,* and *car*. Examples of classified scenes from this dataset are shown in Fig. 1.

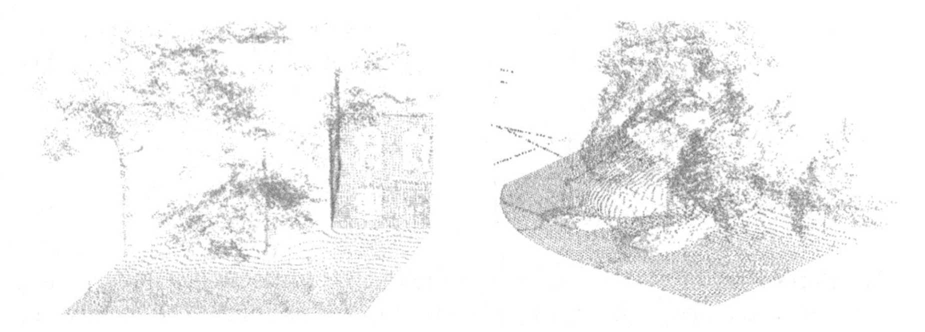

Fig. 1. Example classification results from VMR-Oakland dataset

In our experiments, the neighborhood of points is defined within a $1.0m$ radius. We use a fixed radius support region to compute following features: spectral and directional features to capture the local orientation, distribution of heights and related features [20], spin images [19] around z axis. We use spin images with 3×3 in the bottom level and 10×10 in the top level, with each cell being $0.3m \times 0.3m$. We run unexponentiated variant of the algorithm during 100 iterations and estimate the parameters on a validation set. All computations were performed on a Core 2 Duo @ 2 GHz CPU with 3 GB memory.

We compare our method with the Stacked 3D Parsing (S3DP) algorithm of Xiong at al. [18] and the linear, associative Max-Margin Markov Network (M3N) model of Munoz et al. [4]. Overall, methods that take contextual information into

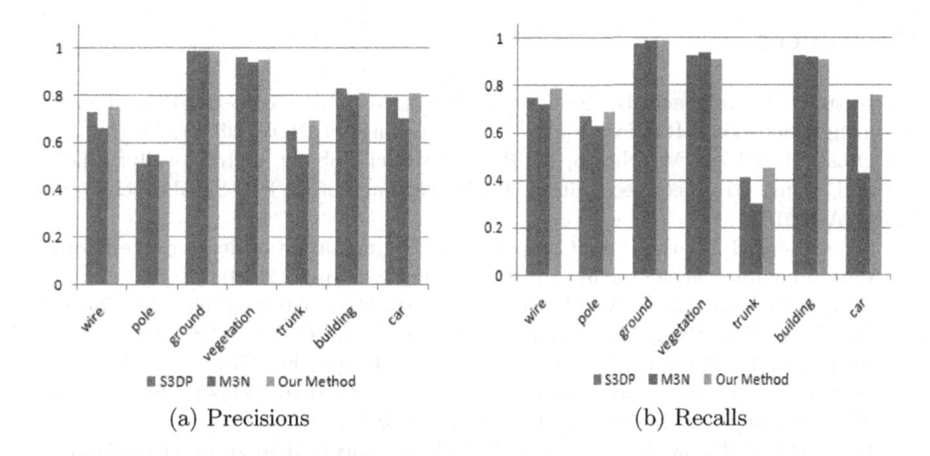

(a) Precisions

(b) Recalls

Fig. 2. Precisions and Recalls for seven classes in VMR-Oakland dataset

account such as our method and S3DP outperform methods that use only local features. A complete comparison is shown in Fig.2. Some objects are easier to label: ground, vegetation, building are often isolated and have a lot of neighbor points so our algorithm performs well. Other objects, such as pole and trunk, are only have weak neighbor connection or often close to background clutter, so the precision and recall remain low.

6 Conclusion

We propose a new method for semantic labeling 3D point clouds. We consider CRF as a pairwise potentials, with the unary term measures consistency between features and the pairwise term between neighbor points. The JointBoost classifier is used to learn unary and pairwise terms. This approach reduces the size of feature vectors while still maintains the accuracy and encodes neighboring information. We include various features and contextual relations in our model. The experiments show that our approach achieves good performance when compared with state-of-the-art methods in complex urban scenes.

The main limitation of our approach is the need for labeled training data. As future work, we plan investigate this problem. Another way to improve our method is to add new features. Additional features such as geometric features or symmetry based features should significantly improve the results.

Acknowledgments. This work was supported by The National Foundation for Science and Technology Development, and Computer Science Department, University of Science, Vietnam. Author wishes to thank the authors who kindly provided the copies of their papers as references.

References

1. Rusu, R.B., Holzbach, B., Blodow, N., Beetz, M.: Fast Geometric Point Labeling using Conditional Random Fields. In: Proceedings of IROS, USA (2009)
2. Drost, B., Ulrich, M., Navab, N., Ilic, S.: Model globally, match locally: Efficient and robust 3d object recognition. In: Proceedings of CVPR, San Francisco, CA, USA (2010)
3. Schoenberg, J., Nathan, A., Campbell, M.: Segmentation of dense range information in complex urban scenes. In: Proc. of IROS, Taipei (2010)
4. Munoz, D., Bagnell, J., Vandapel, N., Hebert, M.: Contextual classification with functional Max-Margin Markov Networks. In: Proc. of CVPR, USA (2009)
5. Anguelov, D., Taskar, B., Chatalbashev, V., Koller, D., Gupta, D., Heitz, G., Andrew, Y.: Discriminative Learning of Markov Random Fields for Segmentation of 3D Range Data. In: Proc. of CVPR, USA (2005)
6. Rusu, R.B., Blodow, N., Beetz, M.: Fast Point Feature Histograms (FPFH) for 3D Registration. In: Proceedings of ICRA (2009)
7. Shapovalov, R., Velizhev, A.: Cutting-Plane Training of Non-associative Markov Network for 3D Point Cloud Segmentation. In: International Conference on 3D Imaging, Modeling, Processing, Visualization and Transmission (2011)
8. Golovinskiy, A., Funkhouser, T.: Min-cut based segmentation of point clouds. In: IEEE Workshop on Search in 3D and Video at ICCV (2009)
9. Nguyen, A., Le, B.: 3D Point Cloud Segmentation: A survey. In: Proceedings IEEE International Conference on Robotics, Automation and Mechatronics (RAM), Philippines (2013)
10. Rusu, R.B., Cousins, S.: 3D is here: Point Cloud Library (PCL). In: Proceedings of ICRA, China (2011)
11. Munoz, D., Vandapel, N., Hebert, M.: Onboard contextual classification of 3D point clouds with learned high-order Markov Random Fields. In: Proceedings of ICRA (2009)
12. Kalogerakis, E., Hertzmann, A., Singh, K.: Learning 3D Mesh Segmentation and Labeling. In: Proceedings of SIGGRAPH (2010)
13. Koppula, H.S., Anand, A., Joachims, T., Saxena, A.: Semantic Labeling of 3D Point Clouds for Indoor Scenes. In: Proceedings of NIPS (2011)
14. Huang, Q., Han, M., Wu, B., Ioffe, S.: A hierarchical conditional random field model for labeling and segmenting images of street scenes. In: Proceedings of CVPR (2011)
15. Torralba, A., Murphy, K.P., Freeman, W.T.: Sharing Visual Features for Multiclass and Multiview Object Detection. IEEE Trans. Pattern Anal. Mach. Intell. (2007)
16. Lim, E.H., Suter, D.: Conditional Random Field for 3D Point Clouds with Adaptive Data Reduction. In: International Conference on Cyberworlds (2007)
17. Shotton, J., Winn, J., Rother, C., Criminisi, A.: TextonBoost for Image Understanding: Multi-Class Object Recognition and Segmentation by Jointly Modeling Texture, Layout, and Context. Int. J. Comput. Vision 81(1)
18. Xiong, X., Munoz, D., Bagnell, J.A., Hebert, M.: 3D scene analysis via sequenced predictions over points and regions. In: Proceedings of ICRA (2011)
19. Johnson, A.E., Hebert, M.: Surface matching for object recognition in complex three-dimensional scenes. Image Vision Comput. 16 (1998)
20. Munoz, D., Vandapel, N., Hebert, M.: Directional associative markov network for 3D point cloud classification. In: Proceedings of 3DPVT, Atlanta, GA (2008)

Categorization of Sports Video Shots and Scenes in TV Sports News Based on Ball Detection

Kazimierz Choroś

Institute of Informatics, Wrocław University of Technology,
Wybrzeże Wyspiańskiego 27, 50-370 Wrocław, Poland
kazimierz.choros@pwr.wroc.pl

Abstract. Content-based indexing of TV sports news is based on the automatic temporal segmentation, recognition, and then classification of player shots and scenes reporting the sports events in different disciplines. Automatic categorization of sports in TV sports news is a basic process in video indexing. Many strategies how to recognize a sports discipline have been proposed. It may be achieved by player scenes analyses leading to the detection of playing fields, of superimposed text like player or team names, identification of player faces, detection of lines typical for a given playing field and for a given sports discipline, recognition of player and audience emotions, and also detection of sports objects specific for a given sports category. The paper examines the usefulness of ball and ball colour detection for the categorization of sports video shots and scenes in TV sports news. This approach has been verified and its efficiency has been analyzed in the Automatic Video Indexer AVI.

Keywords: content-based video indexing, video indexing strategies, TV sports news analyses, sports video categorization, temporal aggregation, player scenes analysis, ball detection, AVI Indexer.

1 Introduction

An automatic processing of television broadcast is one of the most frequent application of content-based video indexing. TV news and especially TV sports news is one of the most viewed video content on the Web. For effective retrieval of video data not only standard text indexing and retrieval procedures should be used but also more and more sophisticated content-based video indexing and retrieval methods. The main goal of research and experiments with sports videos is to propose and develop automatic methods such as automatic detection or generation of highlights, video summarization and content annotation, player detection and tracking, action recognition, ball detection and tracking, kick detection such as penalty, free, and corner kick, replay detection, player number localization and recognition, text detection and recognition for player and game identification, detection of advertisement billboards and banners, authentic emotion detection of audience, and so on. Because of a huge commercial appeal sports videos became nowadays a dominant application area for video automatic indexing and retrieval.

N.T. Nguyen et al. (Eds.): ACIIDS 2014, Part I, LNAI 8397, pp. 591–600, 2014.

The main purpose of sports news processing is to categorize sports video shots and scenes in TV sports news. This processing should take into account that the shots in news are relatively short and they come from various sports. Furthermore, studio discussions, commentaries, interviews, charts, tables, announcements of future games, discussions of decisions of sports associations, etc., so non-sports parts are presented at the same broadcasted news. Due to the automatic categorization of sports events videos can be automatically indexed. The retrieval of individual sports news and sports highlights such as the best or actual games, tournaments, matches, contests, races, cups, etc., special player behaviours or actions like penalties, jumps, or race finishes, etc. in a desirable sports discipline becomes more effective.

A scene is defined as a group of consecutive shots sharing similar visual properties and having a semantic correlation – following the rule of unity of time, place, and action. A player scene is a scene presenting the sports game, i.e. a given scene was recorded on the sports fields such as playgrounds, tennis courts, sports hall, swimming polls, ski jumps, etc. All other non-player shots and scenes usually recorded in a TV studio such as commentaries, interviews, tables, announcements of future games, discussions of decision of sports associations, etc. are called studio shots or studio scenes. Studio shots are slightly useful for indexing and therefore can be rejected. It was observed that the studio scenes may be even two thirds of TV sports news [1]. This rejection of non-player scenes before starting content analyses can significantly reduce computing time and conduct these analyses more effective.

TV sports news should be indexed with the names of sports disciplines presented in this news. The time-consuming analysis process of the recognition of sports disciplines is usually performed for all frames of TV sports news although it is not necessary. The best highlights of sports events seem to be the most adequate for automatic categorization of sports events. Then similarly to text document indexing where the indexing is limited only to abstracts the sports news indexing can be limited to the highlights presented in the news headlines [2]. If in a single frame we recognize a sports discipline with a very high probability the whole video shot or the whole scene can be classified. Therefore, the methods of sports classifications should not be necessarily effective for every single frame. The unquestionable recognition of even only one frame of a scene is sufficient for the indexing purpose.

The paper is organized as follows. The next section describes some related works in the area of automatic sports video categorization. Strategies in sports categorization are discussed in the third section. The identification of sports objects, mainly ball detection in digital videos is outlined in the forth section. The fifth section presents the experimental results of the categorization of sports video shots based on ball and colour ball detection in TV sports news obtained in the AVI Indexer. The final conclusions and the future research work areas are discussed in the last sixth section.

2 Related Works

There has been much research carried out in the area of automatic recognition of video content and of visual information indexing and retrieval [3-4]. To retrieve efficiently videos stored in more and more huge multimedia databases new methods are being sought oriented to visual data, methods much more effective than traditional textual techniques applied frequently for videos. An automatic semantic categorization of TV sports news videos is one of the most popular content-based video indexing. Sports games are very popular in TV broadcasts, a huge amount of broadcast sports videos is generated every day, and the sports video materials are an important part in multimedia databases. Then, a great commercial appeal for sports video automatic indexing and retrieval systems is observed.

Sports equipment detection is one of the techniques proposed for sports shot categorization. Characteristic and specific objects for a given sports category can be used to detect the sports discipline. Of course in many cases the results are ambiguous because a ball is used in several sports disciplines, similarly bicycles, skis, or skates. Nevertheless, the detection of balls could significantly reduce the number of potential sports and then the balls used in basketball or tennis are different in sizes and colours.

Different solutions have been proposed for ball detection. For example a scheme to detect and locate the players and the ball on the grass playfield in soccer videos has been proposed in [5]. A shape analysis based approach was applied to identify the players and the ball from the roughly extracted foreground, which was obtained by a trained, colour histogram-based playfield. an approach for detecting ball in broadcast soccer videos. The results of a number of experiments obtained by using a modified version of the directional Circle Hough Transform with different lighting conditions have been presented in [6]. In [7] hybrid techniques for identifying ball in medium and long shots have been applied. Candidate ball positions were first extracted using features based on shape and size. Then for medium shots a ball was identified by filtering the candidates with the help of motion information, but in long shots first motion based filters were used and next a directed weighted graph was constructed for the remaining ball candidates.

A comparison of different feature extraction methods for automatically recognizing soccer ball patterns through a probabilistic analysis has been performed in [8]. The effectiveness of the different methodologies was demonstrated by a huge number of experiments on real ball examples under challenging conditions.

Video classification methods discussed in the related works as well as in this paper are the methods using visual features only. There are also audio-visual approaches analyzing not only visual information but also audio (see for example [9]).

3 Categorization of Sports Video Scenes

Many algorithms, methods, frameworks, and strategies have be proposed for content analyses of digital videos and for automatic sports shot categorization [3-4, 10-12]. They are based on the traditional comparison of single frames with images in pattern sets or on the comparison of their histograms as well as on the detection of different elements of digital videos typical for a given sports category. In the case of TV sports news such elements are: lines in playing fields, player faces, sports equipments, text imposed on the image, etc.

Histogram matching has become the most common technique for measuring the similarity between two images. Much research in content-based video indexing has examined the usefulness of histogram-based pattern representations. The comparison of histograms is effective in the case of such sports like tennis and ski jumping, but useless in the case of football and baseball, basketball and volleyball, or hockey and figure skating on ice. Therefore, one technique is not sufficient in the categorization of shots in TV sports news.

Text may be useful. Text is omnipresent in sports videos because in any sports broadcast we observe different words not only on playing fields but also on sports stadium grandstands, in the audience, and of course as publicity billboards or banners, etc. We find the names of players or teams on player sports wears. Game place names, stadium names, league tables, numeric results, time, etc. or names of sports commentators are usually superimposed on the images, or included as closed captions. Because these textual elements are characteristic for a specific sports discipline, so, they can serve as an additional criterion in shot categorization.

The next strategy for sports shot categorization can be the recognition of players. Faces of sportsmen are easily recognizable because of the great popularity of sports idols. Their pictures are common in printing materials and in the Web, not only on the website of sports clubs.

The detection and analysis of players and audience emotions is relatively a novel viewpoint and perhaps the most advanced and sophisticated approach. Such people reactions as "exciting", "happy", or "sad" while playing a game, cheering in the stadium grandstands, or observing a sports video broadcast are recognized. Emotions are produced by visual, vocal, and other physiological means. The strategy consists in creating an authentic expression database based on spontaneous emotions and then in comparing these patterns with video frames. Some audience behaviours are characteristic for a particular sports category.

Also the detection of a playing field of the game in a video shot, or of interesting area in a field, i.e. boundary lines, lines marking each end of a court, the penalty area, goal line – the end line between the goal posts in soccer, back boundary lines in tennis or basketball, etc. can be used for categorization. Playing field lines are painted on the playing surface and all lines are in the same colour, usually white, which clearly contrasts with the rest of the field. Generally colours of lines, balls, nets, and of other sports objects must be such to facilitate playing and watching sports.

4 Ball Detection

In many sports disciplines different objects are used such as ball, flying disc, cricket bat, javelin, tennis racket, hockey stick, net, soccer post, springboard, diving board, and many others. Players are using different sports equipments, protective equipments, wear, footwear, etc. The recognition of these objects in sports videos leads to the identification of content and to the categorization of sports video shots.

The main ball sports, i.e. sports where a ball is used are: soccer, basketball, tennis, volleyball, handball, and others. There are many factors with negative influence on the efficiency of ball detection. The colour of playfields in soccer is green, in tennis a playfield can be of different colours but it is a uniform background. In such sports like volleyball or handball a background usually provokes many false detections because the audience is close to the field and usually the publicity is very frequent.

The sizes of balls differ in these sports. For example the size of a soccer ball is roughly 22 cm (8.65 inches) in diameter. Rules state that a soccer ball must be 68 to 70 cm in circumference – standard size number 5. Basketballs for men's play have a circumference of 78 cm (30.7 inches) – size number 7, whereas in women's international play the maximum size is 74 cm (29 inches) – standard size number 6. And then the diameter of the tennis ball is between 6.35 cm (2.5 inches) and 6.67 cm (2.625 inches), so, the circumference is about 20.32 cm (8 inches). But the size of a ball in a given sports category does not determine the resolution of a ball in sports videos. The resolution of a ball depends more on the distance of a camera from a playing field. In general if view is wider a ball is smaller. Also the speed of a ball hit, kicked, or thrown influences significantly the resolution of a ball registered in a video.

Especially the tennis ball is difficult to be recognized, particularly in the case of a wide plan when the whole tennis court is presented. The tennis ball in such a case is only of several pixels, during the game the most often it is a blurred small dash or several blurred points. It means that on one hand the small balls should be also detected but on the other hand small objects looking like a ball lead to the great number of false detections.

The conclusion is that we cannot base our detection process only on the automatic shape classifier of a ball. To discriminate balls from the ball-like objects we should apply additional criteria. These additional filters should be rather restrictive because the sports video categorization does not require to successfully analyze all frames or even all shots of a given scene and to detect an extremely small ball, difficult to see. It is enough to detect even only several frames but with very height probability enabling us to recognize the sports category.

To significantly reduce the false detections three assumptions have been made. The detections are assumed to be false if:

- in one frame three or more objects have been detected as balls,
- a object detected as a ball is at the same position in more than ten consecutive frames – in sports news the player shots are very dynamic,
- it is a simple detection in 100 consecutive frames.

The next criterion will be a colour of a ball. The colours of balls in ball sports are formally defined by the sports rules. The tennis balls are yellow, the colour of basket balls is brown. Unfortunately, the colours of small balls in tennis videos are not pure yellow. The colour hue changes because of the changes in lighting and the camera distance. Therefore, it is desirable to determine the range of variation of a ball colour.

These ranges of colour variations have been determined by analyzing 100 balls automatically detected of a three sports disciplines. The following values have been calculated: minimal (min) and maximal (max) values, arithmetic means (x), and standard deviations (S). They are presented in Tables 1–6.

Table 1. Values of colour hue in main ball sports

	min	max	x	S
Basketball	5.4545	21.8181	12.7612	3.7262
Soccer	52.3404	131.5385	75.6312	17.6951
Tennis	61.4285	125.6604	87.1661	22.5293

Table 2. Range of colour hue in main ball sports

	x – S	x – 2*S	x – 3*S	x + S	x + 2*S	x + 3*S
Basketball	9.0349	5.3087	1.5824	16.4874	20.2136	23.9399
Soccer	57.9365	40.2414	22.5463	93.3268	111.0223	128.7171
Tennis	64.6368	42.1075	19.5781	109.6955	132.2248	154.7542

Table 3. Values of colour intensity in main ball sports

	min	max	x	S
Basketball	0.1255	0.5353	0.3276	0.0786
Soccer	0.3196	0.7843	0.5341	0.1470
Tennis	0.4196	0.5764	0.4784	0.0473

Table 4. Range of colour intensity in main ball sports

	x – S	x – 2*S	x – 3*S	x + S	x + 2*S	x + 3*S
Basketball	0.2489	0.1703	0.0916	0.4062	0.4849	0.5635
Soccer	0.3871	0.2401	0.0930	0.6811	0.8282	0.9752
Tennis	0.4310	0.3836	0.3362	0.5258	0.5732	0.6206

Table 5. Values of colour saturation in main ball sports

	min	max	x	S
Basketball	0.1459	0.4845	0.3178	0.0768
Soccer	0.0508	0.7787	0.3521	0.1654
Tennis	0.1982	0.4959	0.3030	0.1097

Table 6. Range of colour saturation in main ball sports

	x – S	x – 2*S	x – 3*S	x + S	x + 2*S	x + 3*S
Basketball	0.2410	0.1642	0.08745	0.3946	0.4714	0.5483
Soccer	0.1866	0.0212	-0.1442	0.5175	0.6830	0.8485
Tennis	0.1933	0.0836	-0.0260	0.4128	0.5225	0.6322

Assuming that the samples are described by the normal distribution up to 99.5% of all cases fall within two standard deviations of the mean (x±2*S). The confidence interval (x±S) is too small because the results are very poor whereas (x±3*S) leads to too many false detections. The interval (x±2*S) is the best compromise between the number of lost cases and the number of false cases.

5 Tests in AVI Indexer

The method of ball detection applied in the experiments uses the Haar feature-based cascade classifier [13, 14]. The classifier with additional ball colour detection process has been tested and its efficiency has been analyzed using four TV sports news broadcasted in the Polish First National Channel (TVP1) and recorded in January 2012: 05 Jan. (video 1), 12 Jan. (video 2), 14 Jan. (video 3), and 17 Jan. (video 4). The main purpose of the experiments carried out using the Automatic Video Indexer AVI [15] was to verify how the ball detection method with medium ability to detect a ball in single frames can be in spite of all applied for shots and scenes categorizations.

Table 7. Results of the detection of ball frames

	Video 1	Video 2	Video 3	Video 4
	FRAMES IN A VIDEO			
Total number of frames	12 850	13 875	11 000	12 700
Number of frames with ball	315	957	1226	755
Percentage of ball frames	2.45 %	6.90 %	11.15 %	5.94 %
Number of frames with ball < 20px	315	576	885	609
Number of frames with ball > 20px	0	381	341	146
	FRAMES WITH BALL DETECTED			
Number of correctly detected frames with ball	0	237	97	101
Number of frames with ball < 20px	0	0	0	0
Number of frames with ball > 20px	0	237	97	101
Number of frames without ball but detected as ball frames	14	215	66	49
EFFICIENCY OF DETECTION OF PLAYER FRAMES WITH BALL				
Recall	0 %	24.76 %	7.91 %	13.38 %
Recall of detection of frames with ball > 20px	–	62.20 %	28.45 %	69.18 %
Precision	0 %	52.43 %	59.51 %	67.33 %

Table 8. Results of the detection player shots with ball

	Video 1	Video 2	Video 3	Video 4
	PLAYER SHOTS IN A VIDEO			
Total number of player shots	82	104	76	109

Number of player shots with ball	8	26	21	26
Percentage of player shots with ball	9.76 %	25.00 %	27.63 %	23.85 %
Number of player shots with ball < 20px	8	21	15	21
Number of player shots with ball > 20px	0	11	11	8
	DETECTED PLAYER SHOTS WITH BALL			
Number of correctly detected shots with ball	0	11	6	8
Number of detected player shots with ball < 20px	0	0	0	0
Number of detected player shots with ball > 20px	0	11	6	8
Number of player shots without ball but detected as ball shots	1	12	15	1
EFFICIENCY OF DETECTION OF PLAYER SHOTS WITH BALL				
Recall	0	42.31 %	28.57 %	30.77 %
Recall of detection of shots with ball > 20px	–	100.00 %	54.55 %	100.00 %
Precision	0	47.83 %	28.57 %	88.89 %

Table 9. Results of the detection of player scenes with ball

	Video 1	Video 2	Video 3	Video 4
	SCENES IN A VIDEO			
Total number of player scenes	15	16	16	13
Number of player scenes with ball	4	10	5	9
Percentage of scenes ball	26.67 %	62.50 %	31.25 %	69.23 %
Number of player scenes with ball < 20px	4	10	5	9
Number of player scenes with ball > 20px	0	6	3	4
	SCENES WITH BALL DETECTED			
Number of correctly detected scenes with ball	0	6	2	4
Number of detected player scenes with ball < 20px	0	0	0	0

Number of detected player scenes with ball > 20px	0	6	2	4
Number of scenes without ball but detected as ball scenes	1	3	2	1
EFFICIENCY OF DETECTION OF PLAYER SCENES WITH BALL				
Recall	0 %	60.00 %	40.00 %	44.44 %
Recall of detection of scenes with ball > 20px	–	100.00 %	66.67 %	100.00 %
Precision	0 %	66.67 %	50.00 %	80.00 %

Table 10. Aggregate results of the detection of player frames, player shots, and scenes with ball (averages for three videos with non-zero numbers of frames with ball greater than 20 pixels).

Average numbers	Frames	Shots	Scenes
Total	12 525	96.33	15.00
With ball	979.33	24.33	8.33
Percentage	7.82 %	25.26 %	55.56
With ball < 20px	690.00	19.00	8.33
With ball > 20px	289.33	10.00	4.67
Correctly detected balls > 20px	145.00	8.33	4.00
False detections	110.00	9.33	2.00
Recall	14.81 %	34.24 %	48.02 %
Recall of detection of balls > 20px	50.12 %	83.30 %	85.65 %
Precision	56.86 %	47.17 %	66.67 %

The results obtained in the tests (Table 10) carried out in the AVI Indexer confirm that ball detection methods are useful for the categorization of sports videos. It has been also shown that there is no need to detect all frames with balls in content based video indexing. The recognition of a sports discipline basing on the detection of ball is efficient, almost half of scenes with balls (48.02%) greater than 20 pixels are detected with the precision on the level of two thirds (66.67%). Furthermore, not all balls must be detected. The detection of balls greater than 20 pixels is sufficient for the categorization of sports video scenes.

6 Final Conclusion and Further Studies

Ball detection methods with additional colour filter are useful for the categorization of sports video shots and scenes. Not all ball in all frames must be detected. The method which ensures only 14.81% of the recall of ball detection enables us to detect more than one third of ball shots (34.24%) and almost a half (48.02%) of ball scenes. The recall of the detection of balls greater than 20 pixels for frames is 50.12%, whereas for shots as well as for scenes is greater than 80% (83.30% and 85.65% respectively).

The results of tests performed in the AVI Indexer have confirmed that the detection of sports objects is useful for the categorization of sports shots and scenes.

In further research the tests on more reach video material will be performed. Then, the detection of other sports objects in sports news is planed. Finally, new computing techniques will be still developed leading to new functions implemented in the Automatic Video Indexer.

References

1. Choroś, K.: Temporal aggregation of video shots in TV sports news for detection and categorization of player scenes. In: Bădică, C., Nguyen, N.T., Brezovan, M. (eds.) ICCCI 2013. LNCS (LNAI), vol. 8083, pp. 487–497. Springer, Heidelberg (2013)
2. Choroś, K.: Headlines usefulness for content-based indexing of TV sports news. In: Zgrzywa, A., Choroś, K., Siemiński, A. (eds.) Multimedia and Internet Systems: Theory and Practice. AISC, vol. 183, pp. 65–76. Springer, Heidelberg (2013)
3. Hu, W., Xie, N., Li, L., Zeng, X.: Maybank S.: A survey on visual content-based video indexing and retrieval. IEEE Transactions on Systems, Man, and Cybernetics, Part C: Applications and Reviews 41(6), 797–819 (2011)
4. Money, A.G., Agius, H.: Video summarisation: A conceptual framework and survey of the state of the art. Journal of Visual Communication and Image Representation 19, 121–143 (2008)
5. Huang, Y., Llach, J., Bhagavathy, S.: Players and ball detection in soccer videos based on color segmentation and shape analysis. In: Sebe, N., Liu, Y., Zhuang, Y., Huang, T.S. (eds.) MCAM 2007. LNCS, vol. 4577, pp. 416–425. Springer, Heidelberg (2007)
6. D'Orazio, T., Guarangnella, C., Leo, M., Distante, A.: A new algorithm for ball recognition using circle Hough transform and neural classifier. Pattern Recognition 37, 393–408 (2004)
7. Pallavi, V., Mukherjee, J., Majumdar, A.K., Sural, S.: Ball detection from broadcast soccer videos using static and dynamic features. Journal of Visual Communication and Image Representation 19(7), 426–436 (2008)
8. Mazzeo, P.L., Leo, M., Spagnolo, P., Nitti, M.: Soccer ball detection by comparing different feature extraction methodologies. Advances in Artificial Intelligence 6, 12 (2012)
9. Ionescu, B., Seyerlehner, K., Rasche, C., Vertan, C., Lambert, P.: Content-based video description for automatic video genre categorization. In: Schoeffmann, K., Merialdo, B., Hauptmann, A.G., Ngo, C.-W., Andreopoulos, Y., Breiteneder, C. (eds.) MMM 2012. LNCS, vol. 7131, pp. 51–62. Springer, Heidelberg (2012)
10. Choroś, K., Pawlaczyk, P.: Content-based scene detection and analysis method for automatic classification of TV sports news. In: Szczuka, M., Kryszkiewicz, M., Ramanna, S., Jensen, R., Hu, Q. (eds.) RSCTC 2010. LNCS (LNAI), vol. 6086, pp. 120–129. Springer, Heidelberg (2010)
11. Kowdle, A., Chang, K.-W., Chen, T.: Video categorization using object of interest detection. In: Proc. of 17th IEEE Int. Conf. on Image Processing, pp. 4569–4572 (2010)
12. Choroś, K.: Video structure analysis for content-based indexing and categorisation of TV sports news. International Journal of Intelligent Information and Database Systems 6(5), 451–465 (2012)
13. Lienhart, R., Maydt, J.: An extended set of Haar-like features for rapid object detection. In: Proc. of the IEEE Int. Conf. on Image Processing (ICEP), vol. I, pp. 900–903 (2002)
14. Nishimura, J., Kuroda, T.: Versatile recognition using Haar-like feature and cascaded classifier. IEEE Sensors Journal 10(5), 942–951 (2010)
15. Choroś, K.: Video structure analysis and content-based indexing in the Automatic Video Indexer AVI. In: Nguyen, N.T., Zgrzywa, A., Czyżewski, A. (eds.) Advances in Multimedia and Network Information System Technologies. AISC, vol. 80, pp. 79–90. Springer, Heidelberg (2010)

A Coral Mapping and Health Assessment System Based on Texture Analysis

Prospero C. Naval, Jr.[1], Maricor Soriano[2], Bianca Camille Esmero[1], and Zorina Maika Abad[1]

[1] Department of Computer Science, University of the Philippines
[2] National Institute of Physics, University of the Philippines
pcnaval@dcs.upd.edu.ph

Abstract. Corals have long played an important role in the environment with reefs hosting over four thousand species of marine animals all over the world. However, coral are very delicate in that slight unfavorable changes in environmental conditions cause them harm. With extreme changes in the environment happening ever more frequently, there is a need to efficiently monitor these corals for environmental efforts to keep pace. Manual monitoring of corals is expensive, tedious, and time-consuming. In this paper, an information system called the Coral Mapping and Health Assessment System (CMHAS) is proposed. The system aims to provide a digital repository of information that includes videos and images of corals, which are analyzed using image processing algorithms based on texture features to assess the health status of corals reefs. Evaluation on more than a hundred coral images show promising results with recognition rates as high as 82%.

Keywords: Coral Recognition, Texture Analysis, Gabor Filters.

1 Introduction

Corals play an integral role in the ecological balance of many marine ecosystems. Coral presence benefits many marine species, serving as host to over four thousand species of organisms, aside from providing food and shelter for numerous marine fauna. Corals are very delicate creatures. It takes decades for corals to grow, but only needs an unfavorable situation for them to get damaged. A slight change in environmental conditions is sufficient to kill them. For example, a two-degree (Celsius) rise in water temperature in an area is enough for corals to die *en masse*. Even a slight mishandling can cause irreparable damage to their body structures with the self-repair process also taking several years.

For these reasons, the importance of monitoring corals health is imperative. Currently, the prevalent method in the Philippines requires manual monitoring and assessment. A mounted camera pulled by a boat is submerged into the water and videos of the coral reefs are taken at certain monitoring sites. This process is time consuming, requiring several hours just to map a portion of a single site. The video shots are then manually analyzed by a team of marine biologists. In order

N.T. Nguyen et al. (Eds.): ACIIDS 2014, Part I, LNAI 8397, pp. 601–609, 2014.

to verify the results, several experts must agree on these results. The analysis is not only tedious and expensive but very slow and constitutes a bottleneck in the coral health assessment process.

We introduce the Coral Mapping and Health Assessment System (CMHAS), an information system for mapping of coral sites in the Philippines and machine-assisted assessment of coral reef health. It also serves as an online repository of coral images and videos taken from monitored sites at different times. CMHAS users can compare different images in a site across time and determine the changes that happened in that specific location. The criteria for assessment are based on recognition of these three classes: *porites, acropora*, and sand. Having a relatively high percentage of *porites* and *acropora* in a site indicates that the particular coral reef is healthy.

Given the urgency posed by the deteriorating state of the environment, several groups have taken steps to monitor corals and take decisive actions to preserve the aquatic environment. We developed this system for use by local governments where corals are abundant for the purpose of providing objective justification and prioritization of environmental protection initiatives.

The implementation of our proposed system has two major components. The first involves providing functionalities for online storage of images and videos of corals obtained from different sites all over the country at different times and archiving them in such a way that they are retrieved with ease by authorized users. Our implementation uses the Google Map API that allows easy browsing of videos for sites indicated by pin markers. The second module provides coral health assessment functionality through comparisons through time of the relative abundance of *porites, acropora*, and sand. This requires accurate classification of these two coral species. The algorithm employed uses a Gabor filter which extracts texture features from the images and is known to be robust and efficient. For recognition of these classes, a k-nearest neighbor classification is performed on the extracted Gabor texture features. Due to space constraints, we only describe the details of the second module.

1.1 Previous Studies on Coral Classification

Soriano et. al. proposed a neural network-based coral image classifier using texture and color features. The authors classified corals into 5 benthic categories: dead coral, live coral, dead coral with algae, algae, and abiotics (sand/rock) [2]. For pre-processing, the coral RGB image is first transformed into normalized chromaticity coordinates. An RG chromaticity histogram with 32×32 bins is then formed. The authors observed that four color groups are dominant in coral reef images. The color features used for recognition consisted of a 4-component major color histogram and the average full 32×32 chromacity histogram. For texture features, the Local Binary Patterns (LBP) texture operator served as texture descriptors for the coral images. It was chosen for robustness against intensity and rotational variations which are inherent in coral images. The color and texture features were concatenated and submitted to a k-nearest neighbor classifier.

The same group developed another classifier using a feedforward backpropagation neural network which classifies images of coral reef components into three categories: live coral, dead coral, and sand [3]. The neural network produced a high recognition rate of 86.5% for test images. Color and texture features obtained from videos of the Great Barrier Reef were used as input to the neural network. Aside from the neural network, they also developed a rule-based decision tree classifier which achieved a lower recognition rate of 79.7% using the same data set.

1.2 Texture Analysis of Coral Image Data

Coral images obtained from a towed camera contain corals and their surroundings. Corals are "texture-rich" objects, and as such are not easily differentiated from other corals thus making recognition/classification difficult. The images are divided into fixed-size blocks of subimages. We need a window size that is large enough to contain a single texture but not so large as to contain more than one. In this work we use Gabor filters to allow representation of signals in both frequency and orientation domains.

Gabor filters are a group of wavelets with each wavelet tuned to a specific frequency and orientation. A two-dimensional Gabor filter is an oriented sinusoidal grating modulated by a two-dimensional Gaussian function. A Gabor-filtered image is obtained by convoluting the raw image with the Gabor function. Texture analysis often relies on banks of Gabor filters tuned at different frequencies and orientations to generate the texture features [4]. Different textures produce different features whose similarity may be computed using appropriate similarity metrics [1].

2 Coral Classification Using Textures

Marine scientists distinguish coral types through their shape, texture and color. In this study, texture was used as the main distinguishing feature of the coral reef components. Using an algorithm described in [6], patches from coral reef images are classified as *porite, acropora* or others. This is done in two steps: feature extraction and image classification. Gabor filters provide a robust and effective way for extraction of texture energy features. Using the texture features from the first step, k-nearest classification is done by measuring the Euclidean distance between the subimages and a set of pre-classified images of *porites, acropora* and others. In creating the system's graphical user interface, the Java programming language, along with the Google Web Tookit (GWT) were used.

2.1 Data Set

For the image data set, 143 coral images taken from San Diego, Batangas (Philippines) by marine scientists from the University of the Philippines Marine Science

Fig. 1. A sample 2560 x 1920 pixel image of a quadrat at San Diego, Batangas

Institute were used for the study. The distance between the camera and corals were kept at around 5 meters. A sample image is provided in Fig. 1.

Portions of the images were manually labeled by marine scientists as either *porite* or *acropora*. Unlabeled segments are labeled as sand. Image blocks from 2560 x 1920 pixel sized image frame were cut into 180 sub-images of different block sizes (60 per class) to comprise the training data set. Another 450 sub-images of different block sizes (150 per class) were used as the test set. The test and training sets are disjoint. Samples of sub-images are shown in Fig. 2.

2.2 Texture Feature Extraction

Gabor filters are particularly useful for discriminating among textured surfaces. However, a central problem in using Gabor filters is the determination of the appropriate parameters [5]. We want a set of filters diverse enough to cover all possible orientations of the textured surface.

Gabor filters are obtained by modulating a sinusoid by a Gaussian of the form [5]:

$$f(x,y) = e^{\left(\frac{-x^2+y^2}{2v}\right)} \cos(w(x\cos p + y\sin p) + \psi) \tag{1}$$

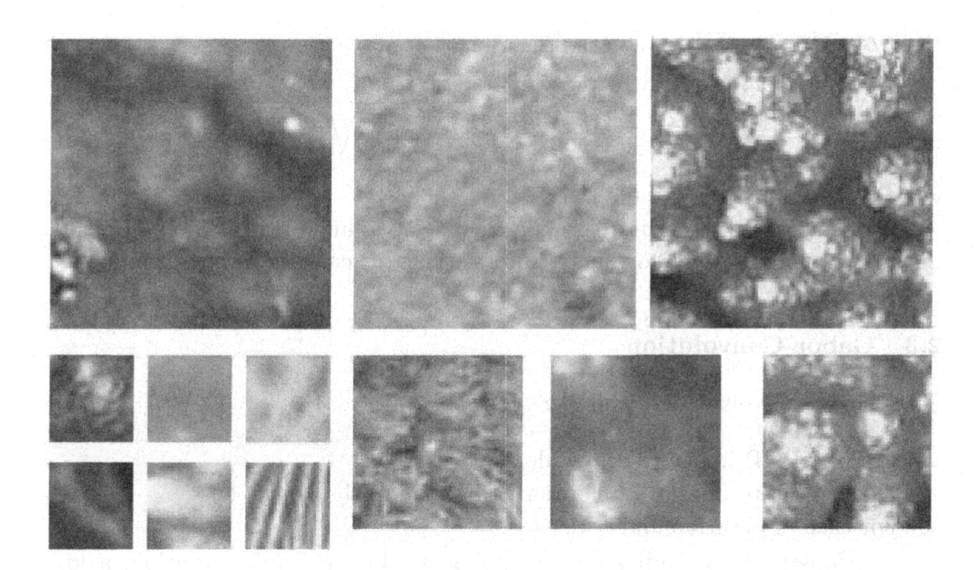

Fig. 2. Examples of cropped images of *porites*, *acropora*, and others. Different textures are observed from these three classes.

where v is the variance, w is the pulsation, p (for *phase*) is the orientation of the Gabor filter, and ψ is the phase offset. Convolving an image with a Gabor Filter produces a response for the image. This response is strong where an image's orientation coincides with that of the filter's, and weak otherwise.

Following [6], we define the discrete Gabor wavelet transform of a sub-image $I(x, y)$ of size $P \times Q$ as

$$G_{mn}(x, y) = \sum_s \sum_t I(x - s, y - t)\psi^*_{mn}(s, t) \quad (2)$$

where the filter mask variables s and t range from 0 to 60. The complex conjugate ψ^*_{mn} is a class of functions obtained from a mother wavelet:

$$\psi(x, y) = \frac{1}{2\pi\sigma_x\sigma_y} e^{\left(\frac{1}{2}\left[\frac{x^2}{\sigma_x^2} + \frac{y^2}{\sigma_y^2}\right]\right)} e^{j2\pi Wx} \quad (3)$$

where W is the modulation frequency. The resulting family of functions ϕ_{mn} from the dilation and rotation of this mother wavelet has the form:

$$\phi_{mn}(x, y) = a^{-m}\phi(\tilde{x}, \tilde{y}) \quad (4)$$

where m is the scale of the wavelet running from 0 to $M - 1$ and n is the orientation of the wavelet running from 0 to $N - 1$, all integers. The other variables are defined as follows:

$$\tilde{x} = a^{-m}(x\cos\theta + y\sin\theta) \qquad \tilde{y} = a^{-m}(-x\sin\theta + y\cos\theta) \quad \theta = n\pi/N$$

$$a = (U_h/U_l)^{\frac{1}{M-1}} \qquad W_{m,n} = a^m U_l$$

$$\sigma_{x,m,n} = \frac{(a+1)\sqrt{2\ln 2}}{2\pi a^m (a-1) U_l} \qquad \sigma_{y,m,n} = \frac{1}{2\pi \tan\frac{\pi}{2N}\sqrt{\frac{U_h^2}{2\ln 2} - (\frac{1}{2\pi\sigma_{x,m,n}})^2}}$$

where $U_l = 0.05$ and $U_h = 0.4$.

We define a grid of kernels based on their scale m and orientation n. For each image, we compute G_{mn} for many combinations of scale and orientation.

2.3 Gabor Convolution

The Gabor Convolve Algorithm is as follows:

1. Convert RGB image to grayscale
2. Apply FFT to obtain the Fourier Transform of the image
3. For each orientation and scale:
   - Construct the radial filter components. In our implementation, a low pass filter was applied to smoothen out the filter
   - Compute filter data specific to the current orientation
   - Normalize filters if L-norm > 0
   - Do the convolution and back transform using Inverse Fast Fourier transform
   - Save the values in a matrix

2.4 Texture Feature Representation

After performing Gabor convolution on the image, an array of magnitudes is obtained for different orientations and scales. The following equation is applied to get the energy matrix [6]:

$$E(m,n) = \sum_x \sum_y |G_{mn}(x,y)| \qquad m = 0,1,...,M-1; n = 0,1,...,N-1 \tag{5}$$

where $G(x,y)$ is the Gabor filter function. This image energy matrix contains the total energy of the image for the different orientations and different scales. The mean and deviation of the transformed coeffcients represent the homogeneous textures:

$$\mu_{mn} = \frac{E(m,n)}{PQ} \qquad \sigma_{mn} = \frac{\sqrt{\sum_x \sum_y (|G_{mn}(x,y)| - \mu_{mn})^2}}{PQ}$$

An energy signature vector is created with μ_{mn} and σ_{mn} as feature components. Figure 3 shows two energy signatures of two *acropora* subimages. Observe that their energy representations are very similar.

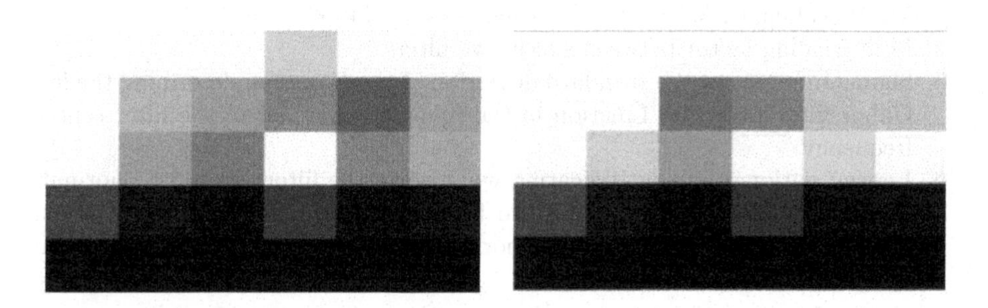

Fig. 3. Energy Feature Maps of two different *acropora* sub-images

2.5 Rotation Invariant Texture Similarity

The corals in the images are not assumed to have same orientation. A simple circular shift is applied on the feature map to make the images rotation-invariant. This is done by computing the total energy for each orientation and choosing the orientation with the highest total energy. The feature elements are then circularly shifted so that the dominant orientation occupies the first element of the feature vector [6].

After extracting the texture features, we measure the texture similarity of the query image and an image in the database. The distance between the query image Q and an image T is [6]:

$$D(Q,T) = \sum_{m} \sum_{n} d_{mn}(Q,T) \tag{6}$$

where

$$d_{mn}(Q,T) = \sqrt{(\mu_{mn}^{Q} - \mu_{mn}^{T})^2 + (\sigma_{mn}^{Q} - \sigma_{mn}^{T})^2} \tag{7}$$

Pre-classified subimages are stored in a database. Using the similarity equation (Eqn. (6)), we find the image R in the database with the highest texture similarity for the image query Q.

2.6 Gabor Filter Optimization

To obtain optimal parameters, many combinations for the Gabor filter parameters are tried out until the best recognition rate is obtained. The parameters and their respective descriptions are as follows:

1. Nscale: number of wavelet scales
2. Norient: number of filter orientations
3. MinWaveLength: wavelength of smallest scale filter
4. Mult: scaling factor between successive filters
5. SigmaOnf: ratio of the standard deviation of the Gaussian describing the log Gabor filter's transfer function in the frequency domain to the filter center frequency.
6. Lnorm: optional integer indicating what norm the filters should be normalized to. A value of 1 will produce filters with the same L1 norm, 2 will produce filters with matching L2 norm. The default value of 0 results in no normalization

For each set of parameters, the classification algorithm is run to get the recognition rate. Different sets of parameters were tried for each window size and the parameters with the highest recognition rates were obtained.

2.7 Results

For our experiments, we used the Gabor filter and Fourier Transform algorithms on a 150-image dataset made up of 50 images each of *porites, acropora*, and others, and used different parameters, as well as window sizes to determine which values for the parameters would result in the best recognition rate. Our best results, along with the values of their parameters, are shown in Table 1.

Table 1. Parameter Values vs Recognition Rates

Window Size	Nscale	Norient	MinWaveLength	Mult	SigmaOnf	Lnorm	Recognition Rate
30×30	5	6	3	2.1	0.6	0	64.67
60×60	5	6	4	2.1	0.6	0	73.33
100×100	**5**	**6**	**4**	**1.6**	**0.75**	**1**	**82.0**
150×150	5	6	4	1.8	0.6	1	71.739
250×250	5	6	4	1.6	0.75	1	69.799

2.8 Graphical User Interface

The Graphical User Interface of our system allows users to register and set up accounts. Administrators of the site may set permissions and access to other users' data. In the map page, the user is shown the map of the Philippines and a list of sites at the left-hand side of the page to where these images belong. Pins are placed on the map which correspond to the sites listed on the left-hand side. Once the user clicks on any of the pins, a window will appear which contains all of the images that are mapped to that particular location. Users are also allowed to view, analyze, and download images from their or other users' archived data.

3 Conclusion

We have successfully implemented and tested an information system for assessing the health of coral reefs based on the relative abundance of key coral reef components. Our system uses texture analysis to recognize three classes of coral reef components (*acropora*, *porites* and others) and their percentage cover in a coral image. An average recognition accuracy of 82% was achieved using frequency and orientation features generated by Gabor filters and a rotation-invariant histogram-based comparison method.

References

1. Manjunath, B.S., Wei-Ying, M.: Texture Features for Browsing and Retrieval of Image Data. IEEE Trans. Patt. Anal. and Mach. Intell. 18, 837–842 (1996)
2. Soriano, M., Marcos, S., Saloma, C., Quibilan, M., Alino, P.: Image Classification of Coral Reef Components from Underwater Color Video. In: MTS/IEEE Conf. Oceans, vol. 2. IEEE Press, New York (2001)
3. Marcos, S., Maricor, S., Saloma, C.: Classification of coral reef images from underwater video using neural networks. Opt. Exp. 13, 8766–8771 (2005)
4. Petrou, M., Sevilla, P.G.: Image Processing: Dealing with Texture, vol. 10. Wiley, Chichester (2006)
5. Weldon, T.P., William, E.H., Dennis, F.D.: Efficient Gabor Filter Design for Texture Segmentation. Patt. Recog. 29, 2005–2015 (1996)
6. Zhang, D., Wong, A., Indrawan, M., Lu, G.: Content-based Image Retrieval using Gabor Texture Features. In: IEEE Pacific-Rim Conf. Multimedia. University of Sydney, Australia (2000)

Navigation Management for Non-linear Interactive Video in Collaborative Video Annotation

Ivan Ariesthea Supandi[1], Kee-Sung Lee[1], Ahmad Nurzid Rosli[1], and Geun-Sik Jo[2]

[1] Department of Information Engineering, Inha University
{ivanaries,lks,nurzid}@eslab.inha.ac.kr
[2] School of Computer & Information Engineering, Inha University
gsjo@inha.ac.kr

Abstract. This paper proposes a method that enables the use of shared interactive videos to promote collaborative environments in authoring the nonlinear video process. The proposed method addresses a problem in applying the nested nonlinear flow in shared interactive videos. The system enables authors' collaboration in using existing interactive videos and allows them to have a full control in applying the nonlinear flow on top of each video. The security and policy issues regarding this full control is solved by displaying the nonlinear flow on an additional layer. We separate the video from the navigational elements to maintain the originality of the reused video source. The strength of the system is the collaborative approach to authoring nonlinear interactive videos. Hence, it helps authors to reduce and distribute the authoring workload by reusing existing interactive videos and enabling authors' collaboration, respectively.

Keywords: non-linear video, interactive video, authoring system, authoring tool, augmented objects, navigational element.

1 Introduction

Advances in video annotation technologies have led to the rapid growth of interactive videos. Many studies have enriched the user's viewing experience with more information and diverse services. For example, interactive videos are widely used through online video-sharing platforms such as YouTube and WireWAX [10].

In general, interactive elements in videos are represented by clickable objects through which users obtain or become linked to related information on particular objects; social networking services (SNSs); have discussions; purchase products or services; navigate to specific parts of videos; and engage in many other activities that enrich their video-watching experience. Therefore, this feature plays an important role in promoting commercial products by taking advantage of interactive video system features and creating business models for monetization [13, 14, 15].

One major concern in the growth of interactive videos is their ability to enable other video authors to reuse and reproduce new videos that retain original augmented objects. However, in nonlinear video environments, videos are restricted to particular structures, including (i) augmented objects (those objects linked to webpages, audio

N.T. Nguyen et al. (Eds.): ACIIDS 2014, Part I, LNAI 8397, pp. 610–618, 2014.

clips, and videos) and (ii) navigation elements (the representation of the nonlinear flow used to link the user to another scene or video) [6].

In general, existing methods allow navigation elements to be included only manually by the author of the shared video. Therefore, navigation links are always visible whenever the video is played. In fact, in some videos, navigation links still exist, which may mislead the user to the plot of a different video, making him or her lose the original or previous playback path.

To address the aforementioned problem, this paper proposes a method that enables the use of shared interactive videos in authoring the nonlinear video process. The proposed system allows the user to directly place navigation elements on top of interactive videos and ensures that the elements are displayed according to the context of the particular author of the nonlinear video. The method stores all data on navigation links as a representation of the nonlinear flow created by the user and then uses the data to create an additional layer to display the links on top of the video player. Therefore, the author of nonlinear videos does not need to obtain the original author's permission to insert a navigation element into his or her interactive video. In this way, the content of the shared interactive video is preserved.

The rest of this paper is organized as follows: Section 2 discusses current trends and related studies with respect to authoring interactive nonlinear videos and collaborative video annotation. Section 3 presents a detailed explanation of the problem. Section 4 explains how the proposed method addresses the problem, and Section 5 concludes with a short summary and some suggestions for future research.

2 Related Works

Many studies have considered nonlinear videos [1, 5, 6, 7, 8, 9] and collaborative video annotation [2, 3, 4]. Interactive videos basically allow the viewer to choose the order of alternative playback paths and scenes in videos [6]. That is, instead of the linear structure provided by conventional videos, the proposed method creates a nonlinear structure for the video playback. Nonlinear videos not only provide more information and services but also allow the user to select a plot or scene in the video. For example, the viewer may choose the type of skateboard trick he or she wants to see and watch different scenes for each option chosen (see Fig. 1(b)).

SIVA Producer [8] is an interactive video-authoring tool that allows users to create a nonlinear structure in video projects. The scene graph editor allows users to arrange composing scenes in a graph representing intersections and nonlinear structures. In addition, users can annotate scenes in a graph with images, videos, links, or HTML files/rich text. As an offline desktop application, SIVA Producer provides no collaborative work environment and allows other authors to reuse resources.

In terms of nonlinear structures, Barthel et al. [1] introduce a multi-path video concept referred to as Video Pathway, a collaborative knowledge-building system using a multi-path video representation. It allows users to collaboratively use YouTube videos to create multiple versions of a given video by providing alternative and interchangeable scenes forming different paths through video content. The final outcome of collaboration is a shared multi-path video representation consisting of several linear video paths. Here each linear video path represents a possible narrative about a particular topic.

There are various commercial products for interactive videos available on the Internet, including WireWAX and RAPT Media [11]. These products produce interactive videos for commercial purposes by providing interactive elements in videos. This is due to information richness and entertainment power. Interactive videos can increase the user's engagement and interest in the subject of the video. WireWAX provides an online video annotation tool that is user-friendly and has the ability to track faces automatically.

RAPT Media is an online nonlinear video-authoring tool. Similar to SIVA Producer, RAPT Media provides an intuitive scene graph editor and allows users to upload local videos as their video resources.

Figs. 1(a) and 1(b) show examples of presenting linear interactive videos on WireWAX and nonlinear interactive videos on RAPT-Media, respectively. Both provide clickable objects that enable interactivity. However, the clickable objects in both videos demonstrate different behaviors. Those objects in RAPT-Media are called navigation elements and are used to select scenes that users want to see next, whereas those in WireWAX are known as augmented objects that have no influence on the plot of the video.

In sum, SIVA Producer and RAPT Media are nonlinear video-authoring tools based on offline desktop and online web-based authoring tools, respectively. With respect to the proposed system, both do not allow users to employ existing interactive videos in their nonlinear video projects. WireWAX is a provider of commercially available interactive video services and does not provide any nonlinear structures in its interactive video products. Finally, Video Pathways [1] introduces a multi-path video concept that promotes a collaborative knowledge-building process to produce videos with several linear pathways.

(a) wireWAX [10] **(b)** RAPT-Media [11]

Fig. 1. The linear **(a)** and non-linear **(b)** interactive videos examples

3 Problem Description

In collaborative video annotation, multiple users can share or participate in the annotation process. One benefit of sharing the annotation task is to distribute the work load of annotating a video. The video annotation task may be too complicated to be handled by a single annotator, particularly for long videos with many points of interest to be annotated.

Another type of video collaboration is the sharing of videos between members. This enables members to reuse existing interactive videos by pointing to particular videos when they need to. However, this collaborative process has yet to be implemented in nonlinear interactive video creation tasks. As mentioned in Section 2, scholars and practitioners still exclude collaborative process in their nonlinear interactive video systems.

Fig. 2 shows a scenario in which a shared video is reused as a resource in nonlinear video projects. Three annotators create three different nonlinear interactive video projects, and each project has its own starting video (i.e., Video #4, Video #8, and Video #2). In the scenario, each author is assumed to arrange a nonlinear interactive video that may consist of local videos (e.g., Video #3, Video #5, and Video #6) and existing interactive videos (i.e., Video #1).

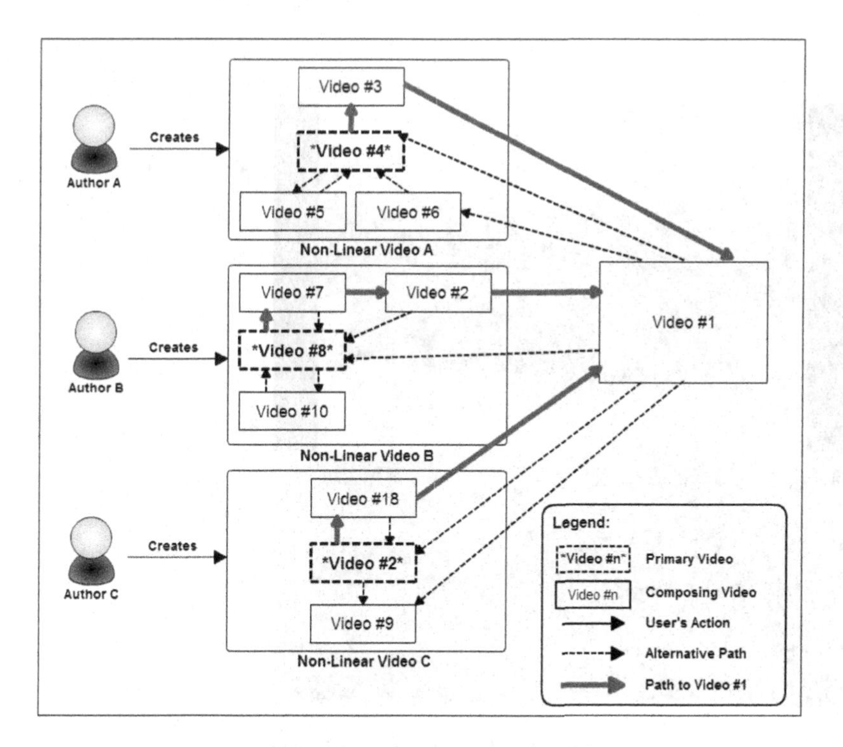

Fig. 2. Problem Description Scenario

Here, to implement the nonlinear flow in Video #1, each author must obtain permission from the author of Video #1 to add augmented objects that link his or her video to the author's video in his or her nonlinear video project. To demonstrate the scenario, the author of Video #1 is assumed to permit the request, and a navigation button is added to the interactive video (see Fig. 3). Placing additional augmented objects in interactive videos causes a problem that may affect the reused interactive video (in this case, Video #1). Because navigation buttons are added directly to the original video (i.e., Video #1), additional buttons are always visible in Video #1. However, each project (see Fig. 2) may have a different context. In addition, navigation links created by Author A for "Nonlinear Video A" are not likely to be meaningful to "Nonlinear Video B" and "Nonlinear Video C," which may have different contexts. Therefore, it may induce viewers to see incorrect paths or videos, and they may get lost in a nested nonlinear video structure.

Originally, those navigation buttons added by Author A in Video #1 are meant to follow the playback path arranged to create "Nonlinear Video A." However, if the viewer navigates to watch Video #1 from "Nonlinear Video B," then he or she can follow the playback path originally created for "Nonlinear Video A." This problem occurs by navigation buttons in "Nonlinear Videos A, B, and C," which are always visible in Video #1 (see Fig. 4).

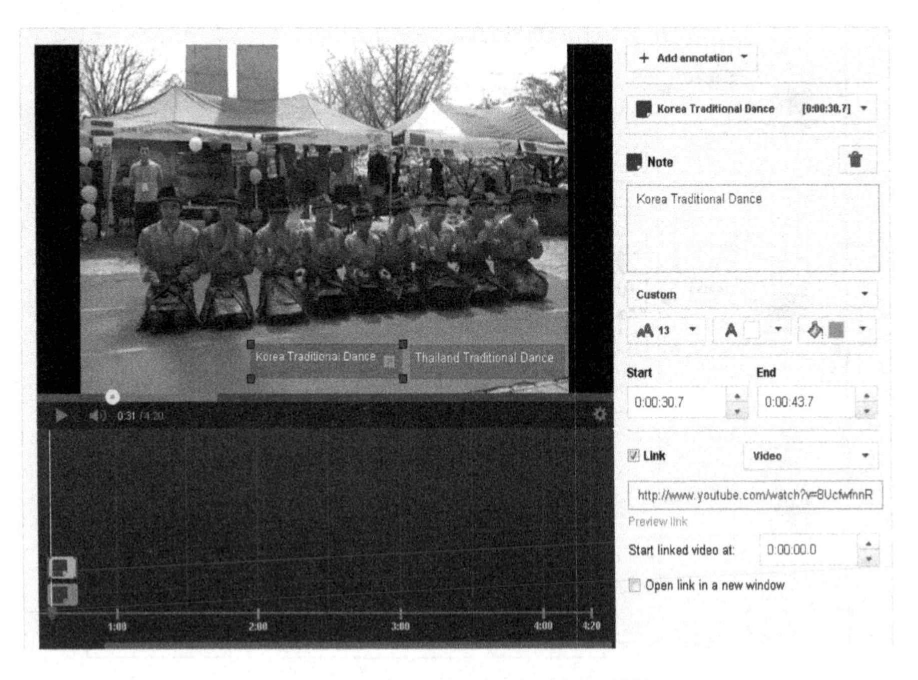

Fig. 3. YouTube Annotation Tools [12]

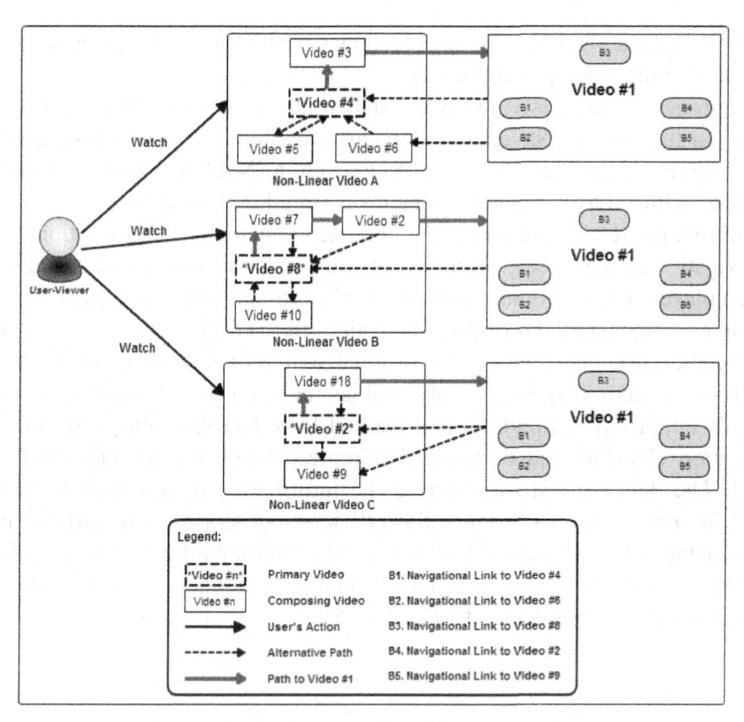

Fig. 4. Current approach way of displaying navigational buttons

4 Navigation Management for Non-linear Interactive Video

To address the problem mentioned in Section 3, this paper proposes a method that enables the content creator (annotator or author) to use available interactive videos in the nonlinear video-authoring process. The proposed system enables the author to include existing interactive videos to create new nonlinear interactive videos.

The method divides interactive elements in interactive videos (e.g., Video #1) and navigation elements representing the nonlinear flow of the "Nonlinear Video" project. The proposed approach stores all data on navigation elements in the system database, and video resources remain unchanged. Therefore, the method maintains the relevant nonlinear flow of Video #1. Because the main purpose is to enable the use of available interactive videos, it is important to note that Video #1 is made publicly available on the Internet. By assuming this, we skip the videos' privacy and policy issues in this paper.

Similar to other nonlinear interactive video systems, the proposed approach consists of two phases: (1) the authoring phase and (2) the viewing phase. Basically, a "Nonlinear Video" project is identified by a unique video's *project_id*, title, video category, keyword tags, and description. Each project has a collection of video resources called the video project library and a nonlinear flow represented by a collection of navigation elements. In the authoring phase, the method uses common procedures for authoring nonlinear videos [8, 11]. By contrast, the proposed method allows the author to upload video resources as well as to search for and include shared interactive videos in the video project library.

Video properties such as video titles, start times, end times, and original sources are stored in the system. Navigation elements are created in the authoring phase by defining source and destination videos. As a result, a "Nonlinear Video" consisting of several videos linked by navigation elements is stored in the database.

In addition, in the viewing phase, the proposed system adds one additional layer to show navigation elements (e.g., Buttons) of some given *project_id* on top of the original video player. The additional layer allows the original interactive element in Video #1 to remain unchanged. To display the video, the HTML <iframe> tag is used to load a specific URL that refers to the original source of Video #1 in the "Nonlinear Video" project. Here <iframe> simply enables the original video player to maintain the video's interactivity. In addition, a method for effectively retrieving information on navigation elements and displaying its representation on the additional layer is provided. The data structure for storing the information of navigational buttons is designed in such a way to enhance the efficiency of storing and retrieval process. Thus, we achieved a low payload for adding the additional layer. The proposed method retrieves only those navigation elements in "Nonlinear Video" *project_id* to display relevant buttons in Video #1 (see Fig. 5).

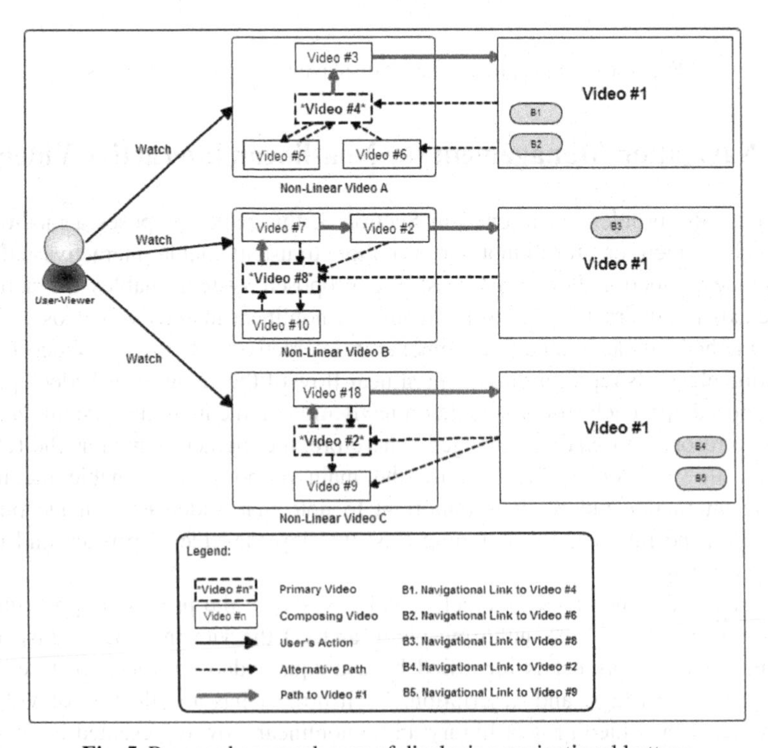

Fig. 5. Proposed approach way of displaying navigational buttons

5 Conclusion and Future Work

The proposed system is designed to enable a collaborative process in authoring nonlinear interactive videos. The highlighted collaboration method is the ability of multiple users to use shared interactive videos as part of their "Nonlinear Video" project. The proposed system allows authors to apply navigation elements as a representation of the nonlinear flow on top of shared interactive videos. In addition, the system ensures the display of links only when the video is played from a specific playback path according to the context of the "Nonlinear Video" created by the author.

Future research should apply the proposed method to web-based nonlinear interactive video-authoring platforms. In addition, an ontology-based video search feature should be considered for the system to help users find more suitable interactive videos for their nonlinear video projects.

Acknowledgement. This work was supported by the National Research Foundation of Korea (NRF) grant funded by the Korea government (MEST) (No.2011-0015484).

References

1. Barthel, R., Ainsworth, S., Sharples, M.: Collaborative knowledge building with shared video representations. International Journal of Human-Computer Studies 71(1), 59–75 (2013)
2. Zhai, G., Fox, G.C., Pierce, M., Wu, W., Bulut, H.: eSports: Collaborative and synchronous video annotation system in grid computing environment. In: Seventh IEEE International Symposium on Multimedia, pp. 1–9. IEEE (2005)
3. Hofmann, C., Hollender, N., Fellner, D.W.: Workflow-based architecture for collaborative video annotation. In: Ozok, A.A., Zaphiris, P. (eds.) Online Communities. LNCS, vol. 5621, pp. 33–42. Springer, Heidelberg (2009)
4. Mu, X.: Towards effective video annotation: An approach to automatically link notes with video content. Computers & Education 55(4), 1752–1763 (2010)
5. Meixner, B., Kosch, H.: Interactive non-linear video: Definition and xml structure. In: Proceedings of the 2012 ACM Symposium on Document Engineering, pp. 49–58. ACM (2012)
6. Meixner, B., Matusik, K., Grill, C., Kosch, H.: Towards an easy to use authoring tool for interactive non-linear video. Multimedia Tools and Applications, 1–26 (2012)
7. Meixner, B., Köstler, J., Kosch, H.: A mobile player for interactive non-linear video. In: Proceedings of the 19th ACM International Conference on Multimedia, pp. 779–780. ACM (2011)
8. Meixner, B., Hölbling, G., Stegmaier, F., Kosch, H., Lehner, F., Schmettow, M., Siegel, B.: SIVA Producer–A Modular Authoring System for Interactive Videos. In: Proceedings of I-KNOW and I-SEMANTICS, pp. 215–225 (2009)
9. Porteous, J., Benini, S., Canini, L., Charles, F., Cavazza, M., Leonardi, R.: Interactive storytelling via video content recombination. In: Proceedings of the International Conference on Multimedia, pp. 1715–1718. ACM (2010)
10. Wirewax – Interactive Video Tool, `https://www.wirewax.com`

11. Rapt Media – The creative platform for interactive enterprise video, http://www.raptmedia.com
12. YouTube Video Annotation, http://www.youtube.com/t/annotations_about
13. VideoClix – Turning Viewers into customers, http://videoclix.tv
14. Link To – Interactive Video made simple, http://www.linkto.tv
15. ClikThrough - shoppable video optimized for touch, http://www.clikthrough.com/

Author Index